Fundamentals of Medicine for Biomedical Engineering

Hamid Hosseinzadeh

Fundamentals of Medicine for Biomedical Engineering

Springer

Hamid Hosseinzadeh
Orthopedic Device Engineering
Rowan University
Glassboro, NJ, USA

ISBN 978-3-031-74084-8 ISBN 978-3-031-74085-5 (eBook)
https://doi.org/10.1007/978-3-031-74085-5

This Springer imprint is published by the registered company Springer Nature Switzerland AG
The registered company address is: Gewerbestrasse 11, 6330 Cham, Switzerland

To the future innovators, healers, and pioneers of Biomedical Engineering: This book is dedicated to you—the students whose curiosity and dedication drive the advancement of both engineering and medicine. Your passion for blending engineering principles with medical science's complexities inspires hope for a future where technology and human health converge seamlessly. May this text serve as a foundation for your studies, a catalyst for your discoveries, and a testament to your potential to transform lives. Your journey in Biomedical Engineering is a beacon of progress, and this book is a humble offering to support you on your path to excellence.

Preface

The interface between biomedical engineering (BME) and clinical medicine represents an interdisciplinary and profoundly impactful convergence of disciplines. It is within this confluence that the inspiration for this book was born. This text is crafted to bridge the gap between engineering principles and medical science, providing a comprehensive yet focused resource explicitly tailored for biomedical engineering students.

Over the past two decades, my journey as an orthopedic surgeon has been both challenging and rewarding. My experiences have been further enriched throughout these years by a nine-year close collaboration with the BME department. This dual perspective has afforded me a unique vantage point from which to observe the critical role that biomedical engineers play in advancing medical science and patient care. This unique clinical practice and engineering collaboration blend has fundamentally shaped this book's structure and content.

The primary aim of this book is to provide a foundational understanding of medical science that is both relevant and essential for biomedical engineering students. The content is meticulously curated to ensure that each topic covered is integral to the knowledge base required by BME professionals. In constructing this text, great care has been taken to avoid extraneous material that, while potentially interesting, does not serve the core educational objectives of BME students.

This book covers a broad spectrum of medical topics, each selected for its relevance to biomedical engineering. Each chapter is designed to offer insights that will enable BME students to appreciate the clinical context of their work.

One of the distinguishing features of this text is the use of illustrations drawn by the author. These illustrations are not mere artistic renditions but scientifically accurate depictions that have been rigorously reviewed for medical accuracy. These illustrations aim to facilitate a clearer understanding of complex medical concepts through visual representation. I believe that these visual aids will greatly enhance the learning experience, making intricate details more accessible and understandable.

I hope the text will be a valuable resource for students embarking on their journey in biomedical engineering. By providing a solid foundation in medical science, this book aims to equip BME students with the knowledge and skills necessary to contribute meaningfully to advancing medical technology and patient care.

In conclusion, I sincerely thank the students and professionals who will utilize this book. Your dedication to biomedical engineering is truly commendable, and I sincerely hope this text will support and inspire your efforts to bridge the gap between engineering and medicine. May it serve as a stepping stone in your educational journey, guiding you toward a future of innovation and excellence in the service of human health.

Glassboro, NJ, USA Hamid Hosseinzadeh

Contents

1 Medical Terminology

In this chapter, you will learn

- Roots, prefixes, and suffixes in medical terminology
- Combining forms
- Eponyms in medical terminology
- Acronyms in medical terminology
- Homonyms in medical terminology
- Importance of medical terminology

1.1 Introduction

Medical terminology is a crucial aspect of medical education.

Medical terminology is the foundation for clear communication in the healthcare field. Using terminology allows medical experts to precisely explain medical conditions, illnesses, and procedures and accurately communicate with others in the field. This uniformity guarantees the transfer of information while minimizing the chances of miscommunication and potential errors.

Learning the meaning of medical terms and how to use them requires knowledge of common medical roots and the prefixes and suffixes used to modify them. Most medical terms are a combination of two or more of these parts; hence, we will explore them in this chapter.

1.2 Roots

Roots are the basic building blocks of medical terms (Table 1.1). They are derived from Greek or Latin words and provide the core meaning behind a term.

H. Hosseinzadeh, *Fundamentals of Medicine for Biomedical Engineering*,
https://doi.org/10.1007/978-3-031-74085-5_1

Table 1.1 Common roots used in medical terminology

Root	Meaning	Root	Meaning
Abdomin/o	Abdomen	**Col/o**	Colon
Acou	Hearing	**Colp/o**	Vagina
Acr/o	Extremity, Peak	**Cor/e/o**	Pupil of eye
Aden/o	Gland	**Cost/o**	Rib
Adren/o	Adrenal gland	**Cry/o**	Cold
Adipo	Fat	**Crypt/o**	Hidden
Alveol/i/o	Air sac	**Cutane/o**	Skin
Andr/o	Male	**Cyan/o**	Blue
Angi/o	Vessel	**Cysti, cyst/o**	Bladder or cyst
Ankyl	Crooked, fusion	**Cyt/o**	Cell
Anter/o	Front	**Derm/a/o, dermat/o**	Skin
Arteri/o	Artery	**Dipl/o**	Double, twice
Arteriol/o	Arteriole	**Dors/i/o**	Back or posterior
Arthr/o	Joint	**Duoden/o**	Duodenum
Atri/o	Atrium	**Encephal/o**	Brain
Audi/o	Hearing	**Enter/o**	Intestine
Axill/o	Armpit, axilla	**Episi/o**	Vulva
Balan/o	Glans penis	**Erythr/o**	Red
Bili	Bile	**Esophag/o**	Esophagus
Bio	Life	**Faci/o**	Face
Blast	Embryonic state	**Fasci**	Bundle
Blephar/o	Eyelid or eyelash	**Fibr/o**	Fibers
Brachi/o	Arm	**Febri**	Fever
Brady	Slow	**Galact/o**	Milk
Bronch/i/o	Bronchus	**Gastr/o**	Stomach
Bronchiol/o	Bronchiole	**Ger/o, Geront/o**	Aging
Bucc/o	Cheek (on the face!)	**Gingiv/o**	Gums in mouth
		Gloss/o	Tongue
Calc	Heel	**Glyc/o**	Sugar
Canth/o	Angle of the eyelids	**Gnath/o**	Jaws
		Gynec/o	Female
Capit/o	Head	**Hem/a/o**	Blood
Carcin/o	Cancer	**Hemato**	Blood
Cardi/o	Heart	**Hepat/o, hepatic/o**	Liver
Carp/o	Wrist	**Heter/o**	Other, different
Caud/o	Tail/downward	**Hidr/o**	Sweat
Cellul/o	Cell	**Hist/o, histi/o**	Tissue
Cephal/o	Head	**Hom/o, home/o**	Same, alike
Cerebell/o	Cerebellum	**Hydr/o**	Water, wet
Cerebr/i/o	Cerebrum	**Hyster/o**	Uterus
Cervic/o	Neck or cervix (neck of uterus)	**Iatr/o**	Treatment
		Ile/o	Ileum (intestine)
Cheil/o, chil/o	Lip	**Ili/o**	Ilium (pelvic bone), flank
Cheir/o, chir/o	Hand		
Chol/e	Bile	**Inguin/o**	Groin
Cholecyst/o	Gallbladder	**Intestin/o**	Intestine
Choledoch/o	Common bile duct	**Irid/o**	Iris of eye
Chondr/i/o	Cartilage	**Ischi/o**	Hip, Ischium bone
Chrom/o	Color	**Jejun/o**	Jejunum
Cili/o	Eyelash or eyelid, or small hair-like processes	**Kerat/o**	Cornea of the eye, horny tissue
		Labi/o	Lips

(continued)

Table 1.1 (continued)

Root	Meaning	Root	Meaning
Lacrima	Tears	**Phren/o**	Diaphragm
Lact/o	Milk	**Pil/o**	Hair
Lapar/o	Abdomen, loin, or flank	**Plasm/o**	Liquid part of blood
Laryng/o	Larynx	**Pleur/o**	Pleura, rib (side)
Later/o	Side	**Pneum/a/o**	Lungs
Leuk/o	White	**Pneum/ato/ono**	Lungs
Lingu/o	Tongue	**Pod/o**	Foot
Lip/o	Fat	**Poli/o**	Gray matter of nervous system
Lith/o	Stone (in gallbladder or kidney)	**Proct/o**	Rectum, anus
		Prote/o	Protein
Lymph/o	Lymph vessels	**Psych/o**	Mind
Mamm/a/o	Breast	**Pulmon/o**	Lungs
Mast/o	Breast	**Py/o**	Pus
Melan/o	Black	**Pyel/o**	Pelvis of kidney
Men/o	Menses, menstruation	**Pyr/o**	Heat
		Rect/o	Rectum
Mening/o	Meninges	**Ren/i/o**	Kidney
Metr/a/o	Uterus	**Rhin/o**	Nose
My/o	Muscle	**Rub/r**	Red
Myc/o	Fungus	**Sangui**	Bood
Myel/o	Bone marrow or spinal cord	**Sacr/o**	Sacrum
		Salping/o	Fallopian tube
Myring/o	Eardrum	**Sarc/o**	Flesh
Myx	Mucus	**Scler/o**	Hard
Nas/o	Nose	**Sept/o**	Infection
Nat/o	Birth	**Somat/o**	Body
Necr/o	Death	**Splen/o**	Spleen
Nephr/o	Kidney	**Spondyl/o**	Vertebra
Neur/o	Nerve	**Stern/o**	Sternum
Occipit/o	Back of the head	**Steth/o**	Chest
Ocul/o	Eye	**Stomat/o**	Mouth
Odont/o	Teeth	**Tachy**	Rapid/swift
Omphal/o	Umbilicus	**Tal/o**	Ankle
Onych/o	Nails	**Tars/o**	Foot
Oophor/o	Ovary	**Tend/o, ten/o**	Tendon
Ophthalm/o	Eyes	**Testicul/o**	Testis
Optic/o, opt/o	Seeing, sight	**Therm/o**	Heat
Or/o	Mouth	**Thorac/o**	Chest
Orchid/o, orchi/o	Testis	**Thromb/o**	Clot
Oss/eo/i, ost/e/eo	Bone	**Thym/o**	Thymus
Ot/o	Ear	**Thyr/o**	Thyroid gland
Palat/o	Roof of mouth	**Tonsill/o**	Tonsils
Path/o	Disease	**Toxi/o**	Poison
Ped/o	Child	**Trache/o**	Trachea
Pelv/o	Pelvis	**Trich/o**	Hair or hairlike
Peritone/o	Peritoneum	**Tympan/o**	Eardrum
Pharmac/o	Drug	**Ur/e/o, urin/o**	Urine
Pharyng/o	Pharynx	**Ureter/o**	Ureter
Phleb/o	Vein	**Urethr/o**	Urethra

(continued)

Table 1.1 (continued)

Root	Meaning	Root	Meaning
Vas/o Vas	deferens	**Viscer/o**	Viscera (internal organs)
Vas/o, ven/o	Vein		
Ventr/i/o	Front of body	**Xanth/o**	Red, redness
Vesic/o	Bladder	**Xer/o**	Dry

- Cardi/o: related to the heart
- Gastro: referring to the stomach
- Osteo: relating to the bone

Roots cannot stand alone and need a modifier like a prefix, a suffix, or another root to complete their meaning. The root can appear at a word's beginning, middle, or end.

- *Arthro*centesis (*Arthro-* is a root that means joint)
- En*cephal*ic (*Cephal-* is a root that means head)
- Xero*derma* (*-derm* is a root that means skin)
- *Myocard*itis (*Myo* is a root that means muscle, *Card* is a root that means heart)

1.3 Prefixes

Prefixes are added to the beginning of a root word to modify its meaning (Table 1.2). They provide additional context or indicate a particular condition or procedure.

- *Anti-*: against or opposing (e.g., *anti*biotic)
- *Dys-*: difficult or painful (e.g., *dys*phagia)
- *Hyper-*: excessive or above normal (e.g., *hyper*tension)
- *Sub-*: below or under (e.g., *sub*cutaneous)

Table 1.2 Common prefixes used in medical terminology

Prefix	Meaning	Prefix	Meaning
A-, an-	Lack of, without, not	**Infra-**	Below or beneath
Ab-	Away from, take away	**Inter-**	Between
Ad-	Towards or to, near	**Intra-**	Within, inside
Ambi-, ambo-	Both	**Intro-**	Into or within
Ana-	Up, apart	**Leuko-**	White
Ante-	Before, in front of, or forward	**Macro-**	Large
Anterior-	Front	**Mal-**	Bad
Anti-	Opposing or against	**Melano-**	Black
Ap-, apo-	Separated or derived from	**Mes-,meso-**	Middle
Auto-	Self, by yourself	**Meta-**	Beyond or changing
Bi-	Double, two, twice, both	**Micro-,micr-**	Tiny, small
Bio-	Life	**Mono-,uni-**	One
Brachy-	Short	**Mult-,multi-**	Many, multiple
Brady-	Slow	**Necro-**	Death
Cata-	Lower, down, against	**Neo-**	New or recent
Caudo-	Tail (downward)	**Oligo-**	Scanty
Cephalo-	Head (upward)	**Pan-**	All
Circum-	Around	**Para-**	Beyond, beside, or after
Co-, con-, com-	Together or with	**Per-**	Through
Contra-	Against or opposed to	**Peri-**	Around
De-	Down or from	**Poly-**	Many, excessive
Di-	Twice or two	**Post-**	After, or following, behind
Dia-	Through, apart, across, between	**Posterior-**	Back
Dis-	Apart from or free from	**Pre-, pro-**	In front of, before, preceding
Distal-	Away from (think distance)	**Presby-**	Old age
Dys-	Difficult, unwell	**Proximal-**	Near (think proximity)
E-, ec-, ex-	From, away from, out of	**Pseudo-**	False
Ect-, exo-, ecto-	Outside, outer, on	**Quadri-**	Four
Em-, en-	In	**Re-**	Backward or again
End-,endo-,ent-,ento-	Within or inner	**Retro-**	Behind, backward
Epi-, ep-	Upon, over, or on	**Semi-**	Half
Eu-	Normal, well	**Sub-**	Under or beneath
Exo-	Outside	**Super-,supra-**	Above, excessive, superior
Extra-, extro-	Beyond, outside of, or outward	**Superior-**	Above
Hemi-, semi-	Half, half of	**Sym-,syn-**	With or together
Hyp-, hypo-	Below, beneath, deficient	**Tachy-**	Rapid or fast
Hyper-	Above, excessive, beyond	**Trans-**	Through or across
Im-, in-	Into, in, within	**Tri-**	Three
Inferior-	Below	**Ultra-**	Excessive, beyond

1.4 Suffixes

Suffixes are added to the end of a root word to modify its meaning or indicate a particular condition, disease, or procedure (Table 1.3).

Table 1.3 Common suffixes used in medical terminology

Suffix	Meaning	Suffix	Meaning
-ac, -ic, -al, -ous, - tic	Related to or pertaining to	**-ology**	Study of, process of study
-algia	Pain in fibrous tissue	**-oma**	Tumor or mass
-ate, -ize	Subject to, use	**-opsy**	To view
-cele	Protrusion (hernia)	**-ostomy, -stomy**	Artificial surgical opening
-centesis	Surgical puncture to withdraw or aspirate fluid	**-otomy**	To cut into
-cle, -cule, -ule, - ulus	Small	**-pathy**	Disease
-cyte	Cell	**-penia**	Lack of or deficiency
-desis	Surgical fixation	**-pexy, -pexis**	Surgical fixation
-dynia	Pain	**-phagia, -phagy**	Eating
-ectomy	Surgical removal or excision	**-phobia**	Morbid fear of or intolerance
-emesis	Vomit	**-plasia**	Formation, development
-emia	Pertaining to blood, a blood condition	**-plasty**	Surgical reconstruction, or shaping of
-ent, -er, -ist	Person, agent	**-plegia**	Paralysis
-esis, -ia, -iasis, -ity, -osis, -sis, - tion, -y	State of or condition of	**-pnea**	Breathing
-form, -oid	Looking like, resembling, or shaped like	**-poiesis**	Production or manufacture of
-genesis	A beginning process, origin of	**-ptosis**	Downward displacement, or drooping
-genic	Produced by	**-rrhage, -rrhagia**	Excessive flow or discharge
-gram	A record	**-rrhaphy**	Suturing in place, fixation
-graph	Instrument used to record	**-rrhea**	Flow or discharge
-graphy	The process of recording	**-rrhexis**	Rupture or breaking away
-ism	Condition or theory	**-sclerosis**	Hardening
-itis	Inflammation	**-scope**	Instrument used to examine visually
-lysis	Destruction, breakdown, or separation	**-scopy**	Process of visual examination
-malacia	Softening	**-spasm**	Sudden or involuntary
-megaly	Enlargement	**-tome**	Instrument
-meter	Instrument used to measure	**-tripsy**	Crushing
-metry	Process of measuring	**-trophic, -trophy**	Growth or development
-ologist	One who studies, a specialist		

- *-itis*: inflammation (e.g., arthr*itis*)
- *-ectomy*: surgical removal (e.g., append*ectomy*)
- *-logy*: study or science of (e.g., cardio*logy*)
- *-osis*: abnormal condition or disease (e.g., osteopor*osis*)

1.5 Combining Forms in Medical Terminology

Combining forms are created by combining a root word with a combining vowel, usually "o." This allows for more straightforward pronunciation and word formation.

- Cardi/o + logy = cardiology (study of the heart)
- Angi/o + edema = angioedema (swelling originating from vessels)

1.6 Eponyms

An eponym is a term named after a person, a place, or a thing. Previously, the eponyms were required to be used with a possessive apostrophe (Crohn's disease, Parkinson's disease); however, currently, it is acceptable to drop the possessive apostrophe (Crohn's disease, Parkinson's disease). If an eponym has an equivalent in medical terms, it is recommended to use the medical term instead of the eponym. Some common medical eponyms are provided in Tables 1.4.

Table 1.4 Common medical eponyms

Medical eponym	Description	Named after
Alzheimer	A Chronic neurodegenerative disease	Alois Alzheimer, 1864–1915
Apgar score	A scoring system that summarizes the health of the newborn (appearance, pulse, grimace, activity, respiration)	Virginia Apgar, 1901–1974
Asperger syndrome	A syndrome with difficulties in social interaction and nonverbal communication; autism spectrum disorder	Hans Asperger, 1906–1980
Baker cyst	An accumulation of joint fluid (synovial fluid) behind the knee	William Morrant Baker, 1839–1896
Barrett esophagus	An abnormal change in the lining of the lower third of the esophagus that has the potential for esophageal adenocarcinoma	Norman Rupert Barrett, 1903–1979
Bell palsy	Abnormal function of cranial nerve VII, causing facial paralysis	Charles Bell, 1774–1842
Crohn disease	Inflammatory bowel disease; cobblestone ulceration	Burrill Crohn, 1884–1983

(continued)

Table 1.4 (continued)

Medical eponym	Description	Named after
Cushing disease/ syndrome	A disease/syndrome presenting with symptoms caused by hyperactivity of the adrenal cortex	Harvey Williams Cushing, 1869–1939
Down syndrome	Also called trisomy 21, a chromosomal disorder with three 21 chromosomes	John Haydon Down, 1828–1896
Ebola virus	A virus causing a disease presenting with vomiting, musculoskeletal pain, weakness, diarrhea, hemorrhagic fever, and stomach pain	Ebola River, in Zaire
Gleason grade	A grading system for prostate cancer based on the degrees of primary and secondary growth	Donald F. Gleason, 1920–2008
Graves disease	Overproduction of thyroid hormones; exophthalmic goiter	Robert Graves, 1796–1853
Hodgkin disease	Cancer of the lymphatic tissue	Thomas Hodgkin, 1797–1865
Homan sign	Reproducing pain on dorsiflexion of the foot, which is a sign of thrombosis of deep veins of the calf	John Homans, 1877–1954
Legionnaire disease	A severe form of pneumonia caused by legionella bacterium	American Legion convention (outbreak) in 1976
Ligament of Treitz	A ligament located in the intestinal tract	Wenzel Treitz, 1819–1872
Lou Gehrig disease	Amyotrophic lateral sclerosis	Lou Gehrig, 1903–1941
Lyme disease	An infectious bacterial disease spread by ticks	Old Lyme, Connecticut
Parkinson disease	Degenerative disease affecting movement, including speech	James Parkinson, 1755–1824
Peyronie disease	An abnormal curvature of the penis caused by fibrous tissue within the tunica albuginea	Francois de la Peyronie, 1678–1747
Raynaud disease/ syndrome	Vasospasms (decreased blood supply to affected regions)	Maurice Raynaud, 1834–1881
Tourette syndrome	Neuropsychiatric disorder; childhood onset	Georges Albert Édouard Brutus Gilles de la Tourette, 1857–1904
West Nile virus	Virus transmitted by mosquitos: mild to severe symptoms	Uganda (eastern Africa), 1937
Wilms tumor	Malignant kidney tumor containing developing (germ) cells	Max Wilms, 1867–1918

1.7 Acronyms

A word (or abbreviation) created from the first letters or syllables of other words is known as an acronym. Most acronyms, but not all, are written in capital letters. Today, many medical acronyms are used as a word (Table 1.5). It is essential to learn the acronyms to become familiar with the context in which they are used because many are identical or sound similar but have entirely different meanings. In addition, each acronym may represent several different phrases.

Table 1.5 Common acronyms used in medical terminology

Acronyms	Meaning
A&O	Alert and oriented
A.A.R.O.M.	Active assistive range of motion
A.B.G	Arterial blood gas
a.c.	Before meals
A.D.L.	Activities of daily living
A.Fib.	Atrial fibrillation
A.R.O.M.	Active range of motion
A/C	Assist control
A/P	Anterior-posterior
ABG	Arterial blood gases
ACE	Angiotensin-converting enzyme
ACL	Anterior cruciate ligament
ADHD	Attention deficit hyperactivity disorder
AIDS	Acquired immunodeficiency syndrome
AKA	Above-knee amputation
ALP	Alkaline phosphatase
ALS	Amyotrophic lateral sclerosis.
ALT	Alanine aminotransferase
AMD	Age-related macular degeneration
AMI	Acute myocardial infarction
AODM	Adult-onset diabetes mellitus
ASAP	As soon as possible
ASD	Autism spectrum disorder
AST	Aspartate aminotransferase
AVM	Arteriovenous malformation
b.i.d.	Twice a day
B.L.BS	Bilateral breath sounds
BID	Twice a day
BKA	Below-knee amputation
BMI	Body mass index
BMR	Basal metabolism rate
BP	Blood pressure
BPH	Benign prostatic hypertrophy
BR	Bed rest
BRCA	Breast cancer gene
bs	Bowel sounds
BS	Breath sounds
BUN	Blood urea nitrogen
bx	Biopsy
C1, C2, etc.	First cervical vertebrae, second cervical vertebrae, etc.
CA	Cardiac arrest
CA, ca	Cancer, carcinoma
CA-125	Cancer antigen 125
CABG	Coronary artery bypass graft
CAD	Coronary artery disease
CAT	Computerized axial tomography
CBC	Complete blood count
cc	Cubic centimeter
CC	Chief complaint
CCU	Coronary care unit
CHD	Congenital heart disease
CHF	Congestive heart failure, chronic heart failure
CHI	Closed head injury
CMV	Cytomegalovirus
CN	Cranial nerve
CNS	Central nervous system
COPD	Chronic obstructive pulmonary disease
CP	Cerebral palsy
CPAP	Continuous positive airway pressure
CPK	Creatine phosphokinase
CPR	Cardiopulmonary resuscitation
CRF	Chronic renal failure
CRP	C-reactive protein
CSF	Cerebrospinal fluid
CT	Computerized tomography
CV	Cardiovascular
CVA	Cerebral vascular accident
CXR	Chest X-ray
D&C	Dilatation and curettage
d/c	Discontinue
DC	Discharge
DJD	Degenerative joint disease
DM	Diabetes mellitus
DNR	Do not resuscitate
DOA	Dead on arrival
DOB	Date of birth
DOE	Dyspnea on exertion
DTP	Diphtheria, tetanus, pertussis
DVT	Deep-vein thrombosis
DX	Diagnosis
ECG, EKG	Electrocardiogram
ED	Emergency department
EEG	Electroencephalogram
EENT	Eyes, ears, nose, throat
EMG	Electromyography

(continued)

Table 1.5 (continued)

Acronyms	Meaning	Acronyms	Meaning
ENT	Ears, nose, throat	**I&O**	Intake and output
ER	Emergency room	**IBD**	Inflammatory bowel disease
ERCP	Endoscopic retrograde cholangiopancreatography	**IBS**	Irritable bowel syndrome
		ICCU	Intensive coronary care unit
ESR	Erythrocyte sedimentation rate	**ICD**	Implantable cardioverter defibrillator
ESRD	End-stage renal (kidney) disease	**ICP**	Intracranial pressure
		ICU	Intensive care unit
f/u	Follow-up	**IDDM**	Insulin-dependent diabetes mellitus
FH	Family history		
FOB	Foot of bed	**IM**	Intramuscular
FSH	Follicle-stimulating hormone	**IPPB**	Intermittent positive pressure breathing
FWB	Full weight bearing	**IUD**	Intrauterine device
G.I.	Gastrointestinal	**IV**	Intravenous
G/E	Gastroenteritis	**IVP**	Intravenous pyelogram
GCS	Glasgow coma scale	**L.E.**	Lower extremities
GE	Gastroenterology	**L.O.C.**	Loss of consciousness, level of consciousness, laxative of choice
GERD	Gastroesophageal reflux disease		
GFR	Glomerular filtration rate		
GI	Gastrointestinal	**LBW**	Low birth rate
GP	General practitioner, general paralysis	**LDL**	Low-density lipoprotein
		LFT	Liver function tests
GSW	Gunshot wound	**LP**	Lumbar puncture
GTT	Glucose tolerance test	**LUE**	Left upper extremity
GU	Genitourinary	**Lx**	Larynx
H&P	History and physical	**MBC**	Maximum breathing capacity
h.d.	At bedtime		
H/A	Headache	**MBSS**	Modified barium swallow study
h/o	History of		
HAV	Hepatitis A virus	**MCA**	Middle cerebral artery
HB	Heart block	**MD**	Muscular dystrophy
Hb.	Hemoglobin	**MG**	Myasthenia gravis
HBP	High blood pressure	**MI**	Myocardial infarction
HBV	Hepatitis B virus	**MICU**	Medical intensive care unit
HCT	Hematocrit	**MMR**	Measles, mumps, and rubella
HCV	Hepatitis C virus		
HDL	High-density lipoprotein	**MRI**	Magnetic resonance imaging
HEENT	Head, eyes, ears, nose, throat	**MRSA**	Methicillin-resistant *Staphylococcus aureus*
HGB	Hemoglobin		
HIV	Human immunodeficiency virus	**MS**	Multiple sclerosis
HOB	Head of bed	**MVA**	Motor vehicle accident
HPV	Human papilloma virus	**N&V**	Nausea and vomiting
HR	Heart rate	**NAD**	No abnormality detected
HRT	Hormone replacement therapy	**NAD**	No apparent distress
		NG	Nasogastric
HTN	Hypertension	**NIC**	Neonatal intensive care
HVD	Hypertensive vascular disease	**NICU**	Neonatal intensive care unit
		NIDDM	Noninsulin-dependent diabetes mellitus
Hx	History		

(continued)

Table 1.5 (continued)

Acronyms	Meaning	Acronyms	Meaning
NKA	No known allergies	**PET**	Positron emission tomography
NKDA	No known drug allergies		
NOS	Not otherwise specified	**PFT**	Pulmonary function test
NPO	Nothing by mouth	**PH**	Past history
NSAID	Nonsteroidal anti-inflammatory drug	**PI**	Present illness, pulmonary insufficiency
NVD	Nausea, vomiting, diarrhea/ Normal vaginal delivery	**PICU**	Pulmonary intensive care unit
NWB	Non–weight bearing	**PID**	Pelvic inflammatory disease
O.M.	Otitis media	**PMH**	Past medical history
O.M.E.	Otitis media with effusion	**PMR**	Physical medicine and rehabilitation
O.T.	Occupational therapy, old tuberculin		
		PMS	Premenstrual syndrome
O/E	On examination	**PN**	Poorly nourished, practical nurse
O2 cap.	Oxygen capacity		
O2 sat.	Oxygen saturation	**PNA**	Pneumonia
OA	Osteoarthritis	**PNI**	Peripheral nerve injury
OB, OBG	Obstetrics	**PNX**	Pneumothorax
OB/GYN	Obstetrics and gynecology	**PPD**	Purified protein derivative
OBS	Organic brain syndrome	**pre-op**	Preoperative
OCD	Obsessive-compulsive disorder	**PRN**	As needed
		PROM	Passive range of motion
ODD	Oppositional defiant disorder	**PSA**	Prostate-specific antigen
		PSH	Past surgical history
OH	Occupational history	**PT**	Prothrombin time
OHD	Organic heart disease	**PT, P.T.**	Physical therapy
OOB, oob	Out of bed	**PTA**	Prior to admission
OR	Operating room	**PTA pulse**	Posterior tibial artery pulse
OTC	Over-the-counter (pharmaceuticals)	**PTH**	Parathyroid hormone
		PTSD	Post-traumatic stress syndrome
p&a	Percussion and auscultation		
P&N	Psychiatry and neurology	**PTT**	Partial thromboplastin time
P.M.	Afternoon, postmortem	**PUD**	Peptic ulcer disease
p.r.n., PRN	As often as necessary, as needed	**PVC**	Premature ventricular contraction
PA view	Posterior–anterior view on X- ray	**PVD**	Peripheral vascular disease
		PVT	Previous trouble
PACU	Post-anesthesia care unit	**PWB%**	Partial weight bearing with percent
PAD	Peripheral arterial disease		
PAF	Paroxysmal atrial fibrillation	**q.h.**	Every hour
		QID, q.i.d.	Four times a day
PAP	Papanicolau	**R to L&A**	React to light and accommodation
PAT	Paroxysmal atrial tachycardia		
		r.a.m.	Rapid alternating movements
PD	Parkinson disease		
PE	Physical exam, pulmonary embolism, pressure equalizer (tubes)	**R.A.S.**	Right arm sitting
		RA	Rheumatoid arthritis, right atrium
PEEP	Positive end-expiratory pressure	**RAtx**	Radiation therapy
		RBC	Red blood cell
PEG	Percutaneous endoscopic gastrostomy	**RCA**	Right coronary artery
		RCU	Respiratory care unit

(continued)

Table 1.5 (continued)

Acronyms	Meaning	Acronyms	Meaning
RD	Respiratory distress	**T3**	Triiodothyronine
RDS	Respiratory distress syndrome	**T4**	Thyroxine
RE	Reconditioning exercise	**TAH**	Total abdominal hysterectomy
RF	Rheumatic fever	**TB**	Tuberculosis
RLAS	Rancho los amigos scale	**TBI**	Traumatic brain injury
RLE	Right lower extremity	**THR**	Total hip replacement
RND	Radical neck dissection	**TIA**	Transient ischemic attack
RO, R/O	Rule out	**TIBC**	Total iron binding capacity
ROM	Range of motion, rupture of membranes, right otitis media	**TID**	Three times a day
		TKR	Total knee replacement
		TMJ	Temporomandibular joint
ROS	Review of symptoms	**TNM**	Tumor, nodes, and metastases
RSV	Respiratory syncytial virus		
RT	Radiation therapy, respiratory therapy	**TORCH**	A group of infections that may cause congenital disabilities if they happen during pregnancy (toxoplasmosis, rubella cytomegalovirus, herpes simplex, and HIV)
RUE	Right upper extremity		
RV	Residual volume		
Rx	Therapy, prescription		
s.c.	Subcutaneous(ly)		
s.gl.	Without correction (without glasses)	**TPN**	Total parenteral nutrition
s/s	Signs and symptoms	**TPR**	Temperature, pulse, respiration
SAD	Seasonal affective disorder		
Scc	Squamous cell carcinoma	**TSH**	Thyroid stimulating hormone
SCCA	Squamous cell carcinoma antigen	**TURP**	Transurethral resection of prostate gland
SCD	Sudden cardiac death		
SCI	Spinal cord injury	**Tx**	Treatment, traction
SCU	Special care unit	**U/A**	Urinalysis
SGA	Small for gestational age	**u/o**	Under observation
SH	Social history	**u/s, US**	Ultrasound
SI	Stroke index	**UG**	Upward gaze
SICU	Surgical intensive care unit	**URD**	Upper respiratory disease
SIDS	Sudden infant death syndrome	**URI**	Upper respiratory infection
		UTI	Urinary tract infection
SL	Under the tongue	**VA**	Visual acuity
SLE	Systemic lupus erythematosus	**VC, vit.cap.**	Vital capacity
		VD	Venereal disease
SLP	Speech-language pathologist	**VF**	Visual fields, ventricular fibrillation
SNF	Skilled nursing facility	**VFSS**	Videofluoroscopic swallowing study
SOAP	Subjective, objective, assessment, plan		
		VO	Verbal order
SOB	Shortness of Breath	**VS, V.S.**	Vital signs
sp&H	Speech and hearing	**W/C, wh.ch.**	Wheelchair
ST	Speech therapy	**w/u**	Workup
stat., STAT	Immediately	**WBC**	White blood cell
STD	Sexually transmitted disease	**WBT**	Weight-bearing tolerance
Sx	Symptoms	**WFL**	Within functional limits
Sz	Seizure	**WNL**	Within normal limits
T&A	Tonsils and adenoids, tonsillectomy, and adenoidectomy	**XRT**	Radiotherapy
		y.o.	Years old
		yrs.	Years

Table 1.6 Some common homonyms

Term	Meaning	Term	Meaning
Ail	To suffer poor health	**Mnemonic**	To assist in remembering
Ale	A beverage	**Pneumonic**	Relating to the lungs
Beat	Strike	**Mucus**	Secretion of the mucous membranes
Beet	A plant with red roots	**Mucous**	Resembling mucus
Cell	Cellule	**Pail**	Bucket
Sell	Vend	**Pale**	Not pink enough
Humerus	A long bone in the upper arm	**Plane**	Anatomic (imaginary) level
Humorous	Funny	**Plain**	Not fancy (plain x-rays)
Ileum	A portion of the colon	**Plural**	More than one
Ilium	A part of the pelvic bone	**Pleural**	Pertaining to the lung
Lice	A parasite	**Toe**	Foot appendage
Lyse	To break	**Tow**	Pull along
Loop	An oval or circular ring, by bending	**Weak**	Not strong
Loupe	Magnifying glass or lens	**Week**	Seven days

1.8 Homonym

A homonym is a word with the same pronunciation as another but a different meaning and a different spelling. There are many homonyms in the medical language, as it is in the English language, like ail and ale, bare and bear, air and heir, and bail and bale (Table 1.6).

Suggested Reading

Easy Medical Terminology for Everyday Use: Unlock the Language of health, Kunlektra S A, Kunlektra Publishing, 2024

Medical Terminology for Dummies, 3rd ed. Henderson, For Dummies Publication, 2019

Medical Terminology: A Quick & Easy Reference Book, G. Chen, Stephen Leesburg, PENN Medical Education Publishing, 2023

Medical Terminology: Learn to Pronounce, Understand and Memorize, Matt Clark, Independent Publisher, 2019

Medical Terminology: The Best and Most Effective Way to Memorize, Pronounce and Understand Medical Terms, 2nd Ed. David Andersson, et al. Independent Publisher, 2016

Human Body: General Characteristics

2

In this chapter, you will learn

- **Body organization:** understanding the hierarchical structure of the body
- **Anatomical position:** standard position is used as a reference when describing body parts and positions
- **Directional terms:** terms that describe the locations of structures in relation to other structures
- **Regional terms:** terms used to identify specific visible regions of the external body
- **Body planes:** imaginary planes that divide the body into sections for anatomical study
- **Body cavities:** major body cavities and their subdivisions that house organs
- **Homeostasis:** the body's ability to maintain a stable internal environment
- **Components of homeostasis:** elements like set points, sensors, control centers, and effectors
- **Feedback mechanisms:** processes that maintain homeostasis, including negative and positive feedback

Anatomy and physiology, two complementary science fields, provide most of the knowledge we need about the normal human body. Anatomy studies the body shape, from microscopic to gross (macroscopic) structures. Physiology is the science of functions. The physiology of any organ is mainly determined by its anatomy.

Throughout this textbook, when we discuss the normal values for structures and functional activities, we consider a healthy young man weighing about 155 lb. or a healthy woman weighing about 125 lb. Whatever we discuss as anatomy or physiology is humans' most common normal variation. It is obvious that human bodies are not totally similar, and there are many tiny differences in shape and function.

H. Hosseinzadeh, *Fundamentals of Medicine for Biomedical Engineering*,
https://doi.org/10.1007/978-3-031-74085-5_2

2.1 Part 1: Body Gross Structure and Orientation

Body organization starts with atoms, which form the molecules, organelles, cells, tissues, organs, organ systems, and the entire body. A cell is the smallest functional unit of all forms of life. Different functional parts of each cell are called organelles. The collection of several cells (not necessarily similar cells) that perform a similar function(s) is called a tissue, like epithelial or muscle tissues. An organ is a distinct body structure formed by at least two different tissues that perform a specific function, such as liver, kidney, or heart. A collection of several organs that carry out a common purpose is called an organ system, like the cardiovascular or gastrointestinal systems.

2.1.1 Anatomical Position

In all discussions in anatomy and physiology, unless mentioned otherwise, we consider a standard body position called anatomical position. In the anatomical position, the body is erect (standing) with feet slightly apart and pointing forward, the palms facing forward, and the thumbs pointing outward, laterally (Fig. 2.1).

In anatomy, the terms "right" and "left" refer to those sides of the model or the cadaver being viewed, not those of the observer.

2.1.2 Directional Terms

Directional terms (Fig. 2.2) describe the relative positions of various body parts and organs and their movements in relation to each other.

"Superior" or "cranial" means toward the upper part of the body or a structure (e.g., the head is superior to the chest). "Inferior" or "caudal" is the opposite of superior and means toward the lower part of the body or a structure (e.g., the abdomen is inferior to the chest).

"Anterior" or "ventral" denotes toward or at the front of the body or a structure (e.g., the nose is anterior to the ear). "Posterior" or "dorsal" is the opposite of anterior and means toward or at the back of the body or a structure (e.g., the ear is posterior to the mouth).

"Medial" means toward or at the middle of the body or a structure (e.g., the navel is medial to the shoulder). "Lateral" is the opposite of medial and denotes away from the middle of the body or a structure (e.g., the arm is lateral to the chest). "Intermediate" means between a more lateral and a more medial structure (e.g., the armpit is intermediate to the arm and the chest).

"Proximal" and "distal" are two terms used for extremities and any long structure with an origin and end. "Proximal" means closer to the attachment of an extremity or origin of a structure (e.g., the shoulder is proximal to the elbow). "Distal" is the opposite of proximal and shows a structure is farther from the attachment of an extremity or origin of a structure (e.g., the foot is distal to the knee).

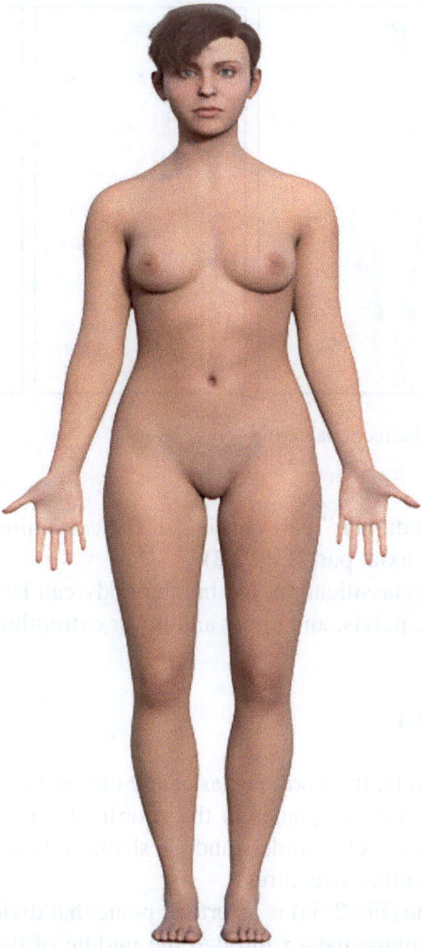

Fig. 2.1 Anatomical position

"Superficial" and "deep" show a structure's location relative to the body's surface or structure. "Superficial" means toward or the surface of the body or a structure (e.g., the skin is superficial to the subcutaneous fat). "Deep" is the opposite of superficial and means away from the body's surface or structure (e.g., the brain is deep to the skull bones).

2.1.3 Regional Terms

The regional terms describe the different regions of the body. In a broad classification, the human body is divided into two parts: the axial part and the appendicular part. The "axial" part constitutes the body's central axis and includes the head, neck,

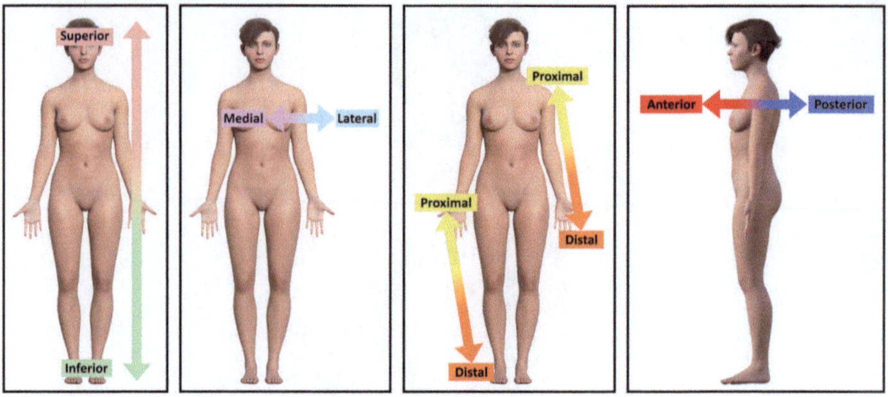

Fig. 2.2: Orientation and directional terms

and trunk. The "appendicular" part consists of the extremities or limbs (upper and lower) attached to the axial part (Fig. 2.3).

In a more detailed classification, the human body can be divided into the head, neck, chest, abdomen, pelvis, and upper and lower extremities (Fig. 2.4).

2.1.4 Body Planes

In human anatomy study, the body is classically cut by three perpendicular planes (similar to the xy, yz, and xz planes in the coordinate system). The cuts at these imaginary planes help us better understand the shape and location of a structure and its position relative to other structures.

The "sagittal" plane (Fig 2.5a) is a vertical plane that divides the body into right and left parts. If this plane passes through the middle of the body, it is called the "midsagittal" plane and divides it into two almost identical and symmetrical left and right halves. The other sagittal planes that do not pass through the midline are called "parasagittal" planes.

The "frontal" or "coronal" plane (Fig 2.5b) is a vertical plane that divides the body into anterior and posterior parts. The "transverse" or "axial" plane (Fig 2.5c) is a horizontal plane that divides the body into superior and inferior parts.

Since the body is not symmetric in superoinferior and anteroposterior directions, no midcoronal or midaxial planes exist.

Since the advent of cross-sectional imaging (like CT scan or MRI) and their ability to reconstruct images in any plane, oblique planes have been defined to delineate specific structures more accurately.

Fig. 2.3 Illustration depicting the major sections of the human body: the "axial part" and the "appendicular part"

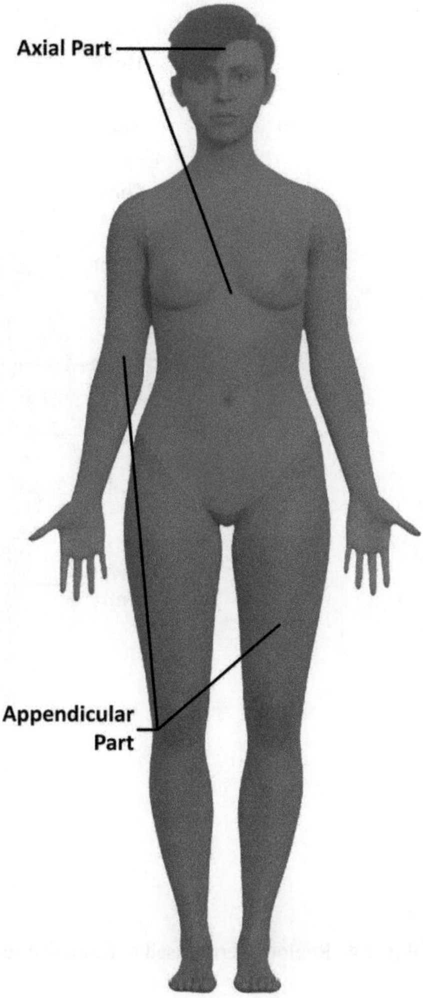

2.1.5 Body Cavities

The human body has two major closed cavities: the dorsal body cavity and the ventral body cavity. These cavities provide a protective space for vital organs and structures, allowing their movement and flexibility.

The "dorsal body cavity" (Fig. 2.6) is located within the skull and the spinal column and is formed by two connected cavities:

- The "cranial cavity" is surrounded by cranial bones and encases the brain.
- The "vertebral or spinal cavity" is located inside the vertebral column and contains the spinal cord and the origin of the peripheral nerves.

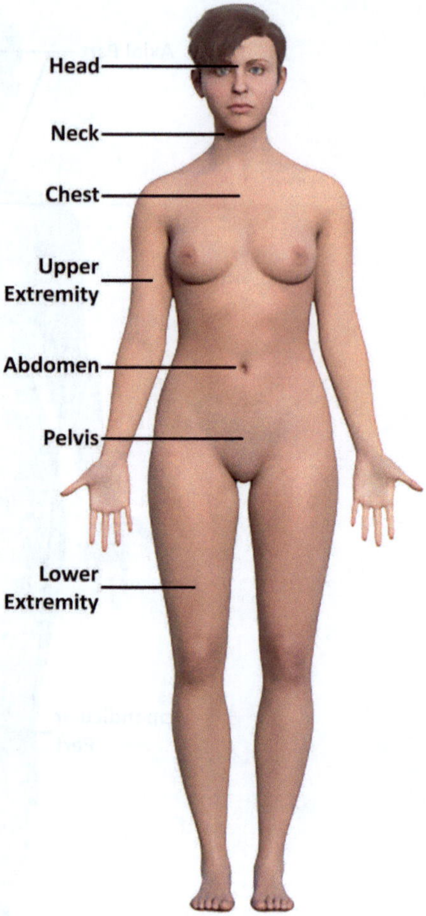

Fig. 2.4 Regional terms used to designate specific body areas

The "ventral body cavity" (Fig. 2.7) is more anterior and larger and houses internal organs called viscera. Two smaller cavities form the ventral body cavity:

- The superior one is called the "thoracic cavity."
- The inferior one is called the "abdominopelvic cavity."

These two cavities are separated by a dome-shaped muscle called the diaphragm. The chest wall surrounds the thoracic cavity (Fig. 2.7) and is further subdivided into three smaller cavities:

- The left pleural cavity, containing the left lung.
- The right pleural cavity, containing the right lung.
- The mediastinum, located in the center, between two lateral pleural cavities, containing the heart, the esophagus, the trachea, the thymus, and the large vessels.

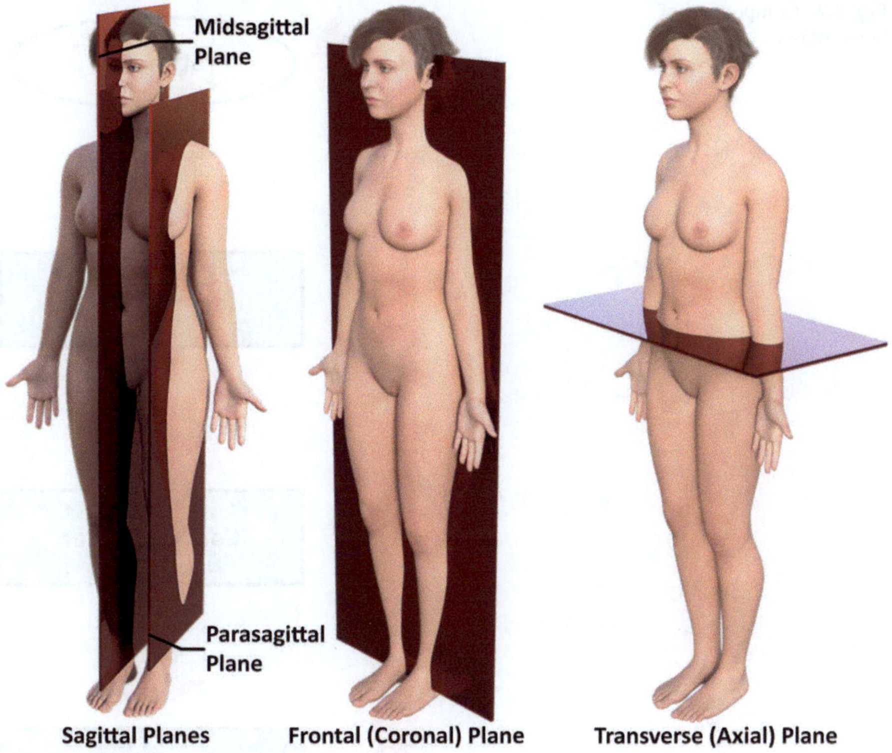

Fig. 2.5 Planes of the body

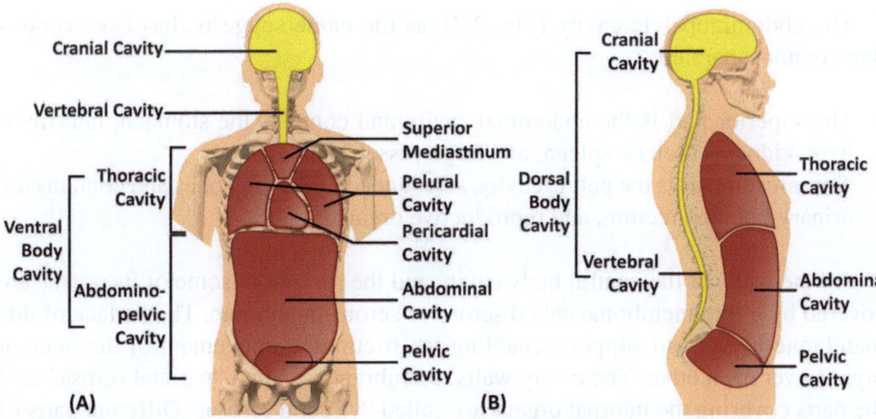

Fig. 2.6 Dorsal and ventral body cavities and their subdivisions: (**a**) anterior view; (**b**) lateral view

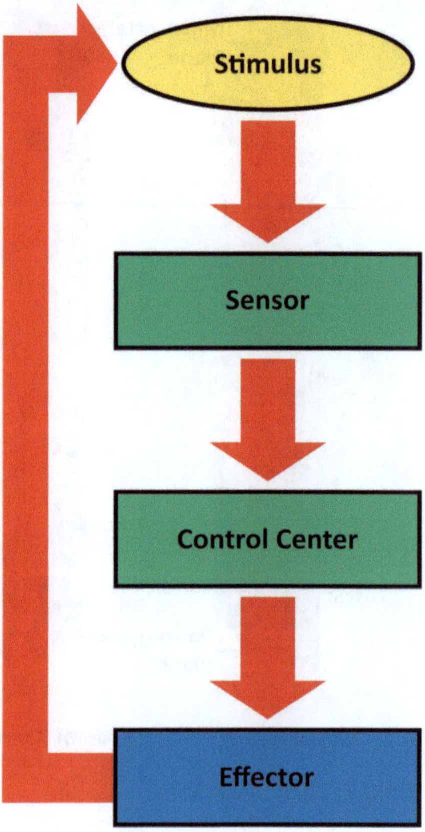

Fig. 2.7 Components of homeostasis

The abdominopelvic cavity (Fig. 2.7), as the name suggests, has two sections that are not separated:

- The superior part is the abdominal cavity and contains the stomach, intestines, liver, kidneys, ureters, spleen, and large vessels.
- The inferior part is the pelvic cavity, surrounded by pelvic bones and contains the urinary bladder, rectum, and reproductive organs.

All the walls of the ventral body cavity and the surface of some of its organs are covered by a thin membrane called serosa or serous membrane. The surface of this membrane is wet and slippery, enabling the frictionless movement of the internal organs over each other. The cavity walls' membrane is called "parietal serosa," and the parts covering the internal organs are called "visceral serosa." Different parts of these membranes have a specific name; the most important ones are the membranes around the lungs, in the lateral thoracic cavities, which are called the pleura (parietal and visceral pleura) (Section 15.1.3), and the one around the heart, called pericardium (Section 13.1.1.1).

2.2 Part 2: Homeostasis

Homeostasis is the dynamic equilibrium the human body maintains by continuously adjusting its physiological processes to ensure a stable internal environment. This equilibrium involves regulating variables such as temperature, pH, blood pressure, and glucose levels within narrow limits to support optimal cellular function. Without homeostasis, our cells would be exposed to constant environmental fluctuations, leading to dysfunction and organ failure.

2.2.1 Components of Homeostasis

The homeostasis components can be broadly categorized into key elements (Fig. 2.7):

Set points: Set points are the ideal or target values for physiological variables that the body strives to maintain. For instance, the body maintains a core temperature of approximately 98.6 °F (37 °C) as a set point.

Sensors: Sensors, also known as receptors, are specialized cells or tissues that detect changes in physiological variables and transmit this information to the control center. Sensors play a vital role in initiating feedback mechanisms.

Control center: The control center, typically located in the brain or specific endocrine glands, receives input from sensors and compares it to the set point. If a deviation is detected, the control center responds to correct it.

Effectors: Effectors are tissues or organs that carry out the corrective actions determined by the control center; for example, in temperature regulation, sweat glands and blood vessels in the skin act as effectors.

2.2.2 Types of Feedback Mechanisms

2.2.2.1 Negative Feedback

Negative feedback is an automated process in which a change in a physiological variable triggers a response that opposes or "negates" that change. The goal is maintaining the variable close to a set point, ensuring stability. Negative feedback mechanisms are essential for homeostasis. They ensure that variables such as temperature, blood pressure, and blood glucose levels remain within the optimal range for cellular function. Negative feedback systems are energy-efficient because they operate only when deviations from the set point occur. The corrective actions cease once the variable returns to the set point, conserving energy. Examples of negative feedback mechanisms include

Thermoregulation: Thermoregulation is a classic example of negative feedback in action. When the body's temperature deviates from the set point (usually around 98.6 °F or 37 °C), sensors in the hypothalamus detect the change. If the temperature

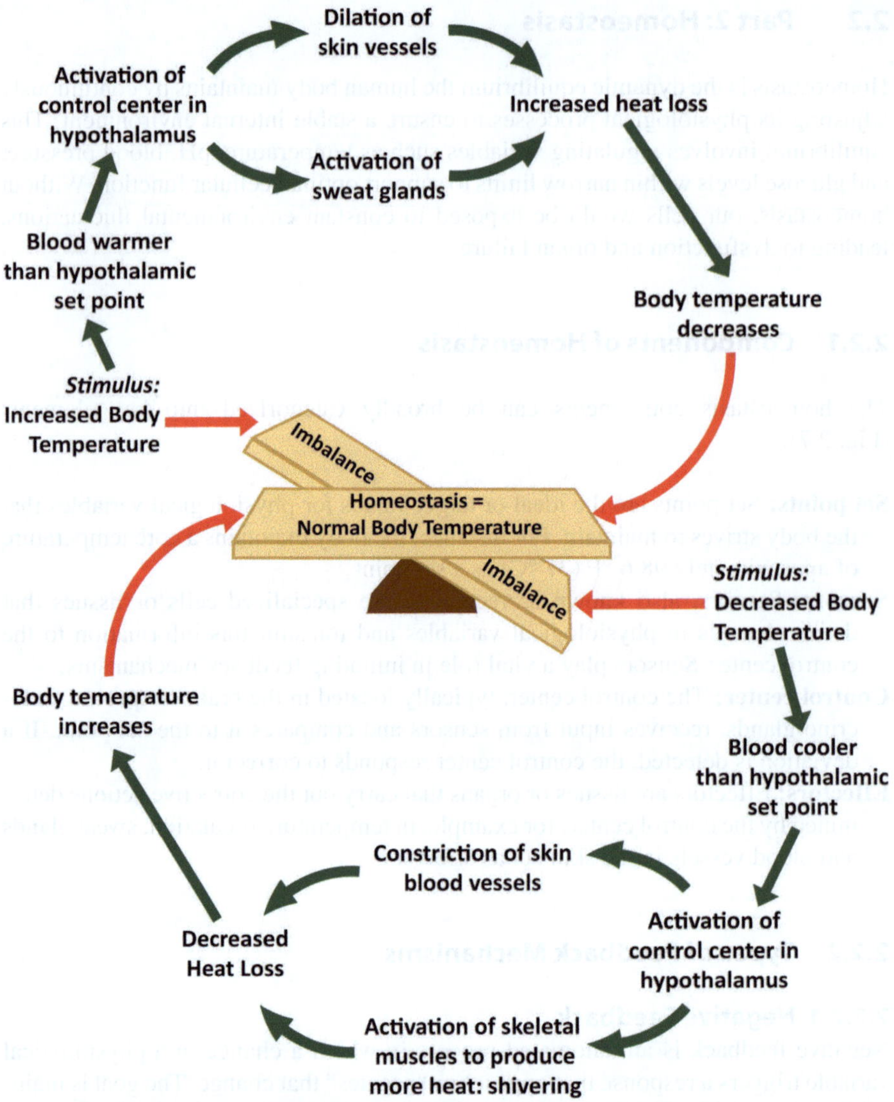

Fig. 2.8 Negative feedback mechanism in regulating the body temperature

rises, the body responds by initiating sweating and vasodilation to cool down. Conversely, if the temperature falls, shivering and vasoconstriction are triggered to generate heat (Fig. 2.8).

Blood pressure regulation: Blood pressure regulation is another essential function of negative feedback. Sensors in blood vessels monitor blood pressure, and if it rises above the set point, the control center in the medulla oblongata signals the heart to reduce its rate and the blood vessels to dilate, lowering the pressure. Conversely, if blood pressure falls below the set point, the heart rate increases, and vessels constrict to raise pressure.

Blood glucose control: Blood glucose levels are tightly controlled through negative feedback. When glucose levels rise after a meal, the pancreas releases insulin, which promotes glucose uptake into cells and reduces blood sugar levels.

When glucose levels drop, the pancreas releases glucagon, stimulating the release of glucose from the liver and raising blood sugar levels.

2.2.2.2 Positive Feedback

Positive feedback is a regulatory process in which a change in a physiological variable triggers a response that further amplifies the change, moving the variable away from its set point. Unlike negative feedback, which aims to maintain stability, positive feedback mechanisms drive a process to completion. Positive feedback is fundamental in specific processes like blood clotting and childbirth.

Blood clotting: One of the most prominent examples of positive feedback is blood clotting. When a blood vessel is damaged, platelets adhere to the site, releasing chemical signals. These signals stimulate further platelet aggregation and the release of clotting factors. This cascade amplifies until a stable clot forms, effectively stopping bleeding.

Childbirth: Positive feedback also plays a critical role in childbirth. As the baby's head presses against the cervix during labor, stretch receptors in the cervix send signals to the brain, stimulating uterine contractions. These contractions push the baby's head against the cervix, further activating stretch receptors and intensifying contractions. This cycle continues until the baby is delivered.

Suggested Reading

Anatomy & Physiology: The Unity of Form and Function, 10th Ed. Kenneth S. Saladin, McGraw Hill, 2023

Gray's Anatomy for Students, 5th Ed. Richard L. Drake, et al. Elsevier, 2020

Human Anatomy & Physiology, 11th Ed. Elaine N. Marieb and Katja Hoehn, Pearson, 2022

Moore's Clinically Oriented Anatomy, 9th Ed. Keith L. Moore, et al. Lippincott Williams & Wilkins, 2020

Biochemistry

3

In this chapter, you will learn

- Roles and properties of water in biological systems
- Significance and examples of salts in the human body
- How acids and bases are defined in a biochemical context
- Concept of organic chemistry and the formation of organic compounds
- Structure and functions of carbohydrates, including monosaccharides, disaccharides, and polysaccharides
- Importance of lipids and their classifications, such as triglycerides, phospholipids, and steroids
- Detailed structures and functions of proteins
- Roles of enzymes as biological catalysts
- Interaction of enzymes with coenzymes and cofactors
- Different types of enzyme inhibition, including competitive and noncompetitive inhibition
- Feedback mechanisms in metabolic pathways
- Structure of nucleotides and their role in forming nucleic acids
- Detailed structure of DNA and its function in genetic information storage
- Different types of RNA and their roles in protein synthesis
- Role of adenosine triphosphate (ATP) as a primary energy carrier in cells

Biochemistry is a branch of science focusing on the chemical reactions and pathways in living cells and the molecules that make up living organisms. It studies the structure, function, and interactions of molecules such as carbohydrates, lipids, proteins, and nucleic acids. Biochemistry has applications in many fields, including medicine, agriculture, biotechnology, and environmental science.

Chemicals in the body fall into one of two major classes: organic and inorganic compounds. Compounds containing carbon and hydrogen are organic and made by living organisms, while the other chemical compounds in organisms are inorganic (Fig. 3.1).

H. Hosseinzadeh, *Fundamentals of Medicine for Biomedical Engineering*,
https://doi.org/10.1007/978-3-031-74085-5_3

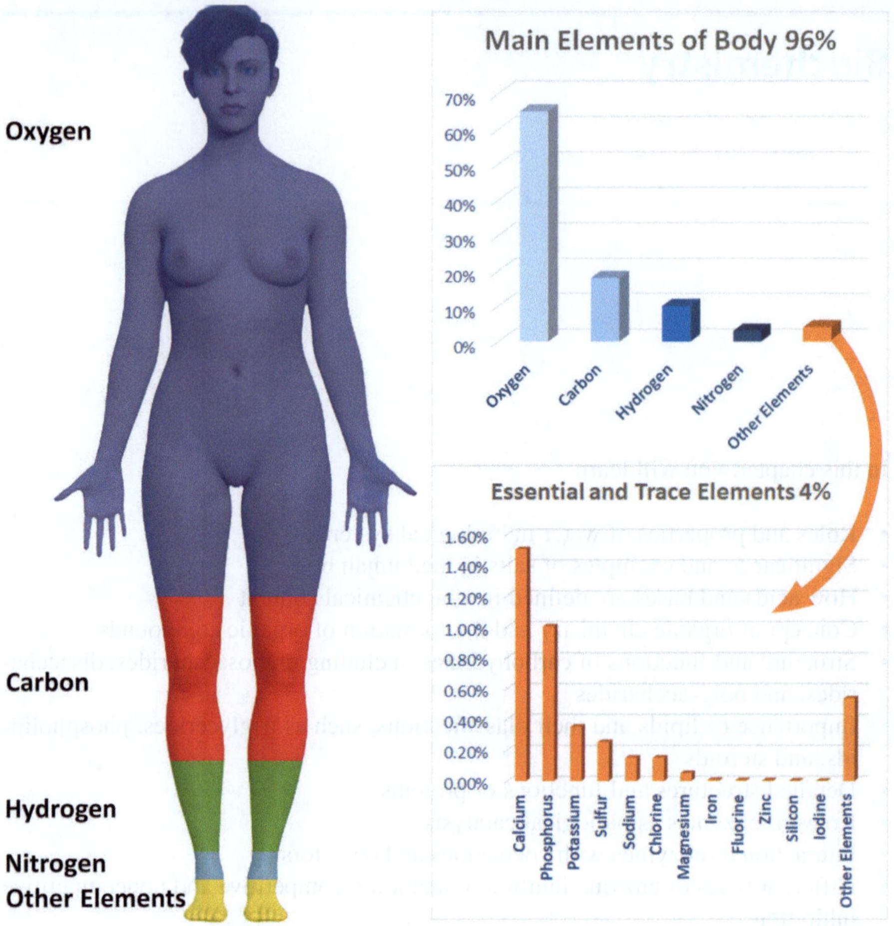

Fig. 3.1 Elemental components of the human body

3.1 Part 1: Inorganic Compounds

Inorganic compounds are essential for various biological processes, such as enzymatic reactions, energy production, and structural support. Inorganic compounds in the human body include water, salts, acids, and bases.

3.1.1 Water

Water is the most abundant and essential inorganic compound in living organisms, making up 60–80% of the volume of most living cells. The properties of water include

High heat capacity: Heat capacity refers to the amount of heat required to raise the temperature of a substance by a certain degree. Water's high heat capacity means it can absorb and store a large amount of heat energy without experiencing a significant temperature rise. This property results from the hydrogen bonding between water molecules, which needs to be broken first to allow the kinetic energy of the individual water molecules to increase. This high heat capacity of water will enable organisms to regulate their body temperatures, which is essential for maintaining metabolic processes.

High heat of vaporization: The heat of evaporation of water is approximately 40.7 kJ/mol at room temperature and atmospheric pressure. This value is much higher than the heat of evaporation for most other liquids, which typically range from 10 to 30 kJ/mol. The human body uses this property in the sweating mechanism to control its temperature when exposed to high heat (external or internal).

Polar solvent properties: As a polar solvent, water is an excellent solvent for many substances. The polar nature of water is due to the unequal sharing of electrons between the oxygen and hydrogen atoms in the molecule, resulting in a partial negative charge on the oxygen atom and partial positive charges on the hydrogen atoms (Fig. 3.2). In addition, water can solve some nonpolar molecules with polar functional groups. This ability to dissolve many different substances is critical for many biological processes, such as transporting nutrients and waste products in cells, and makes water the "universal solvent." Almost all chemical reactions in the body happen in a water environment and depend on water's solvent properties.

High reactivity: Water is an essential reactant in many chemical reactions. Many large organic molecules are broken down in the body by adding a water molecule to their bond (hydrolysis).

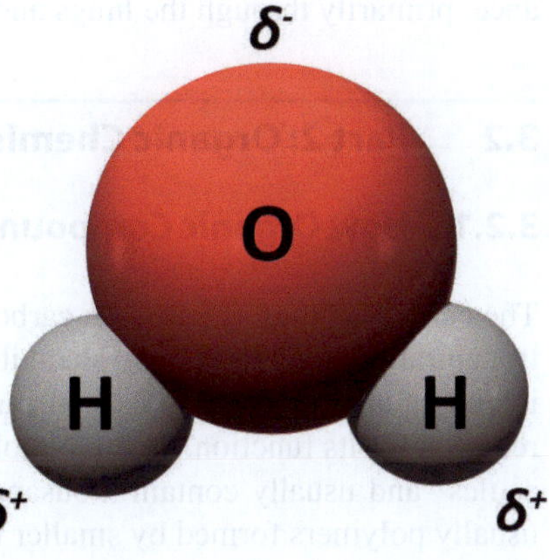

Fig. 3.2 Molecular structure of water

Cushioning: Like many fluids, water can flow easily but is not compressible. This property allows water to form elastic cushions around certain body organs, helping to protect them from physical trauma, like cerebrospinal fluid surrounding the central nervous system.

3.1.2 Salts

A salt is an ionic compound containing cations other than H^+ and anions other than the hydroxyl ion (OH^-). When salts are dissolved in water, they dissociate into their component ions. Salts commonly found in the body include sodium chloride (NaCl), calcium carbonate ($CaCO_3$), calcium phosphate [$Ca_3(PO_4)_2$], and potassium chloride (KCl). Calcium phosphate is the most abundant salt in the body, forming the major part of the bones and teeth. Some ions resulting from the dissolution of salts that play vital roles in the body include sodium, potassium, calcium, magnesium, phosphate, carbonate, and iron.

3.1.3 Acids and Bases

Acids release hydrogen ions (H^+) when dissolved in water, while bases release hydroxide ions (OH^-) when dissolved in water. The concentration of hydrogen ions determines the acidity of a solution, and the concentration of hydroxide ions determines the alkalinity of a solution. Essential acids in the body include hydrochloric acid (HCl), citric acid ($C_6H_8O_7$), lactic acid ($C_3H_6O_3$), and carbonic acid (H_2CO_3). Many other essential molecules, like several amino acids and fatty acids, are also acids. Some of the essential bases in the body are bicarbonate (HCO_3^-), and adenosines, which build the DNA and RNA.

Maintaining the proper acid–base balance is crucial for normal cellular function. Deviations from the normal pH range can have detrimental effects on proteins and other biomolecules. The body has intricate mechanisms to regulate acid–base balance, primarily through the lungs and kidneys (Section 18.4).

3.2 Part 2: Organic Chemistry

3.2.1 How Organic Compounds Are Made?

The earth's biology is based on carbon, a genuinely electroneutral atom. This electroneutrality allows carbon to share its four valence electrons with four other atoms, including another carbon. Carbon can form long molecules in almost any structure required for its function. Organic molecules are large molecules called "macromolecules" and usually contain thousands or more atoms. These macromolecules are usually polymers formed by smaller building blocks (monomers).

To form a polymer in the body, monomers are joined together by a chemical reaction called "dehydration," in which a molecule of water is formed by removing a hydrogen ion and a hydroxyl group from the monomers and then creating a covalent bond between the monomers. When the body needs to break down the macromolecules into smaller ones, it uses a chemical reaction opposite to dehydration, called "hydrolysis." In a hydrolysis reaction, a covalent bond between two monomers is broken, together with breaking a molecule of water into hydrogen and hydroxyl ions. Then, the hydrogen and hydroxyl ions are added to the monomers, and the smaller molecules are released (Fig. 3.3).

3.2.2 Carbohydrates

Carbohydrates are one of the three essential macronutrients for the human body, representing 1–2% of cell mass. Carbohydrates can be classified as a monosaccharide ("one sugar"), disaccharide ("two sugars"), or polysaccharide ("many sugars").

"Monosaccharides," or simple sugars (Fig. 3.4), are single-chain or single-ring structures containing three to seven carbon atoms. Usually, the carbon, hydrogen, and oxygen atoms occur in the ratio 1:2: 1, so a general formula for a monosaccharide is $(CH_2O)_n$, where n is the number of carbons in the sugar. They are the building blocks of larger carbohydrates. The most important monosaccharides in the body are the pentose (five-carbon) and hexose (six-carbon) sugars. Glucose has six carbon atoms; its molecular formula is $C_6H_{12}O_6$, and ribose has five carbons; the molecular formula is $C_5H_{10}O_5$.

"Disaccharides," or double sugars (Fig. 3.4), are formed when two monosaccharides are joined by dehydration synthesis. Important disaccharides in the diet are sucrose (glucose + fructose), which is cane or table sugar; lactose (glucose + galactose), milk sugar; and maltose (glucose + glucose), the malt sugar.

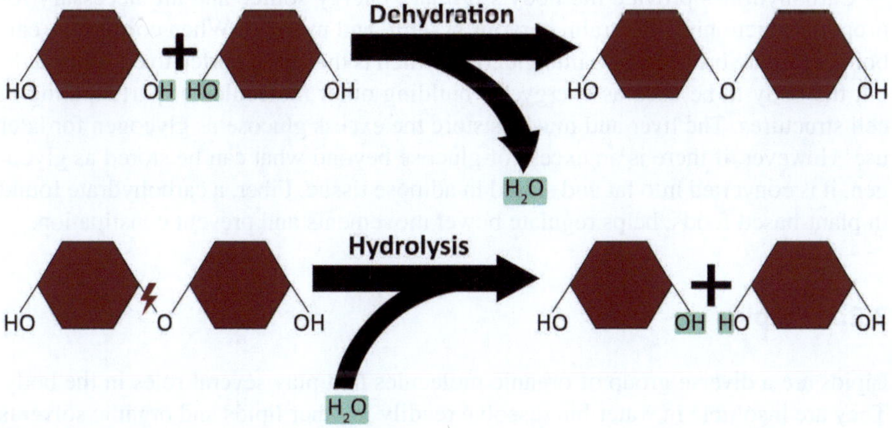

Fig. 3.3 Dehydration and hydrolysis chemical reactions

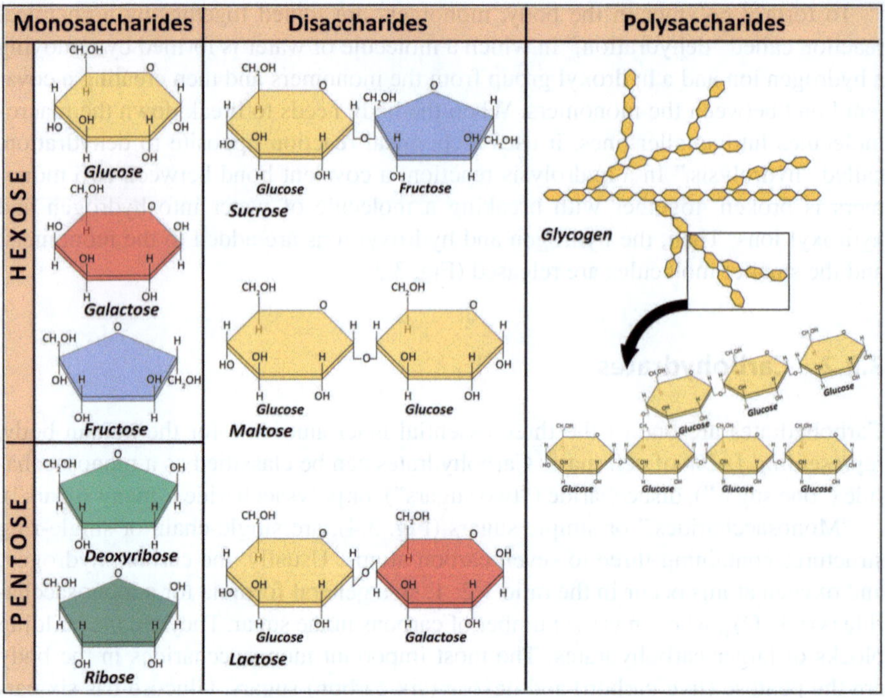

Fig. 3.4 Carbohydrate molecules

"Polysaccharides" (Fig. 3.4) are large polymers of monosaccharides linked together by dehydration synthesis. Their large size makes them fairly insoluble in water. Starch and glycogen are two polysaccharides of significant importance to the body, both glucose polymers. Plants form starch, and animals form glycogen, which function as storage molecules.

Carbohydrates provide the body's primary energy source and are necessary for properly functioning the brain, nervous system, and muscles. When consumed, carbohydrates are broken down into glucose, which is then transported to cells throughout the body to be used as energy for building other molecules or participating in cell structures. The liver and muscles store the excess glucose as glycogen for later use. However, if there is an excess of glucose beyond what can be stored as glycogen, it is converted into fat and stored in adipose tissue. Fiber, a carbohydrate found in plant-based foods, helps regulate bowel movements and prevent constipation.

3.2.3 Lipids

Lipids are a diverse group of organic molecules that play several roles in the body. They are insoluble in water but dissolve readily in other lipids and organic solvents such as alcohol and ether. All lipids contain carbon, hydrogen, and oxygen, but compared to carbohydrates, the proportion of oxygen in lipids is much lower. The lipids can be classified into triglyceride, phospholipid, steroid, and eicosanoid.

3.2.3.1 Triglycerides

Triglycerides (Fig. 3.5) consist of three fatty acid chains linked to a glycerol backbone. Fatty acids are linear chains of carbon and hydrogen atoms (hydrocarbon chains) with an organic acid group (-COOH) at one end. Glycerol is a sugar alcohol with the chemical formula of $C_3H_5(OH)_3$. Synthesis of a triglyceride molecule involves combining three fatty acid molecules with a single glycerol molecule by dehydration reactions. The glycerol molecule is the same in all triglycerides, but the fatty acid chains can vary in length and degree of saturation (number of double bonds between carbon molecules), giving rise to a diverse array of triglyceride molecules with unique physical and chemical properties. If a triglyceride has one double bond, it is called monounsaturated; if the number of double bonds is two or more, it is called polyunsaturated. The hydrocarbon chains of these three fatty acids make the triglycerides nonpolar molecules, insoluble in water.

When all bonds between carbon molecules in fatty acids of a triglyceride are single, the triglyceride is saturated and remains solid at room temperature (these

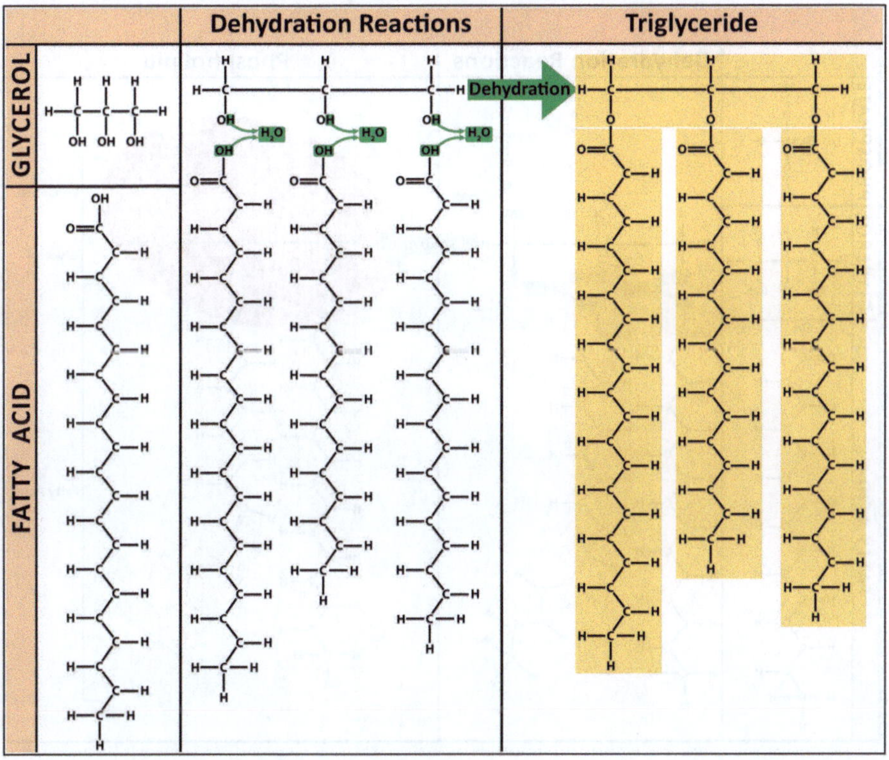

Fig. 3.5 Triglyceride molecular structure and formation

triglycerides are also called fats). Double bonds between carbon molecules in the fatty acids make the triglyceride unsaturated, which remains fluid at room temperature (these triglycerides are also called oils).

One of the primary roles of triglycerides is as an energy source. When the body needs energy, triglycerides stored in adipose tissue are broken down into fatty acids and glycerol, which can be used as fuel by cells throughout the body. Additionally, triglycerides are an essential component of cell membranes, where they help to maintain membrane fluidity and integrity. Triglycerides also play a role in the absorption and transport of dietary fats.

3.2.3.2 Phospholipids

Phospholipids (Fig. 3.6) comprise a glycerol backbone and two, rather than three, fatty acid chains. The third chain is replaced by a phosphate group (PO_4) with an attached nitrogen-containing group. Phospholipids have a hydrophilic head and a hydrophobic tail, giving them unique structure and function. The phosphate group and the glycerol molecule form the hydrophilic head of a phospholipid. This part of the molecule is polar and attracted to water, making it soluble in water-based solutions. The hydrophobic tail, on the other hand, is made up of two fatty acid chains

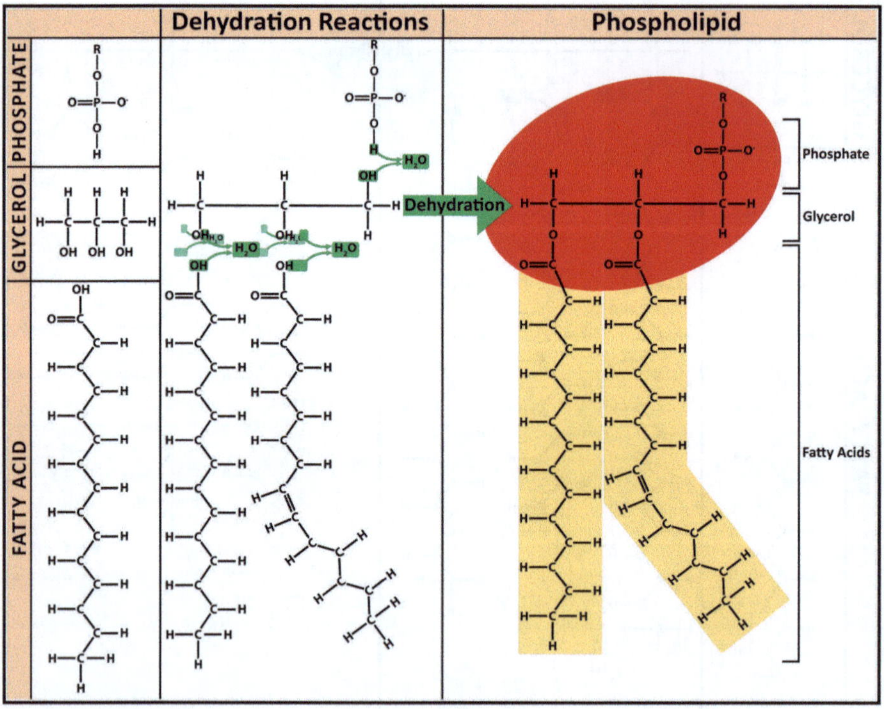

Fig. 3.6 Phospholipid molecular structure and formation

that are nonpolar and repelled by water. This part of the molecule is insoluble in water-based solutions.

Phospholipids are an essential component of cell membranes in the human body. The structure of phospholipids allows them to form a bilayer in cell membranes. The hydrophilic heads face outward, interacting with the aqueous environment inside and outside the cell, while the hydrophobic tails face inward, away from the water. This arrangement creates a barrier that separates the inside of the cell from the outside (plasma membrane), controlling what enters and exits the cell (Section 4.1.2.1). Phospholipids also play a crucial role in cell signaling. They act as signaling molecules, transmitting messages between cells and regulating cellular processes. Phospholipids are also involved in the metabolism of lipids and other nutrients, helping to break down and transport these molecules throughout the body.

3.2.3.3 Steroids

Steroids are a class of lipids with a flat molecule formed by four interlocking hydrocarbon rings (Fig. 3.7). The most common steroid in the human body is cholesterol ($C_{27}H_{46}O$). It is part of the cell membrane structure and is also the precursor for synthesizing steroid hormones (glucocorticoid, mineralocorticoid, and sex hormones), bile salts, and vitamin D. All these chemical compounds derived from cholesterol are also steroids.

3.2.3.4 Eicosanoids

Eicosanoids are formed by a twenty-carbon backbone, which is modified by adding functional groups such as hydroxyl (OH^-), carbonyl (-C=O), and carboxyl (-COOH). There are three main classes of eicosanoids: prostaglandins, thromboxanes, and leukotrienes. They play a crucial role in regulating various physiological processes in the human body, including inflammation, blood clotting, and immune response.

3.2.4 Proteins

Proteins compose 10–30% of the cell mass and are vital components of every living cell in the human body. They are complex molecules that perform a wide range of functions, from providing structural support to catalyzing chemical reactions. Proteins are made up of chains of building blocks called amino acids, which are linked together by peptide bonds.

3.2.4.1 Amino Acids

Amino acids are the building blocks of proteins. The human body has 20 different amino acids (Table 3.1). All amino acids contain carbon, oxygen, hydrogen, and nitrogen; two also contain sulfur. Amino acids are either essential or nonessential. The body cannot synthesize essential amino acids, which must be obtained through the diet. On the other hand, nonessential amino acids can be synthesized by the body and are not required in the diet.

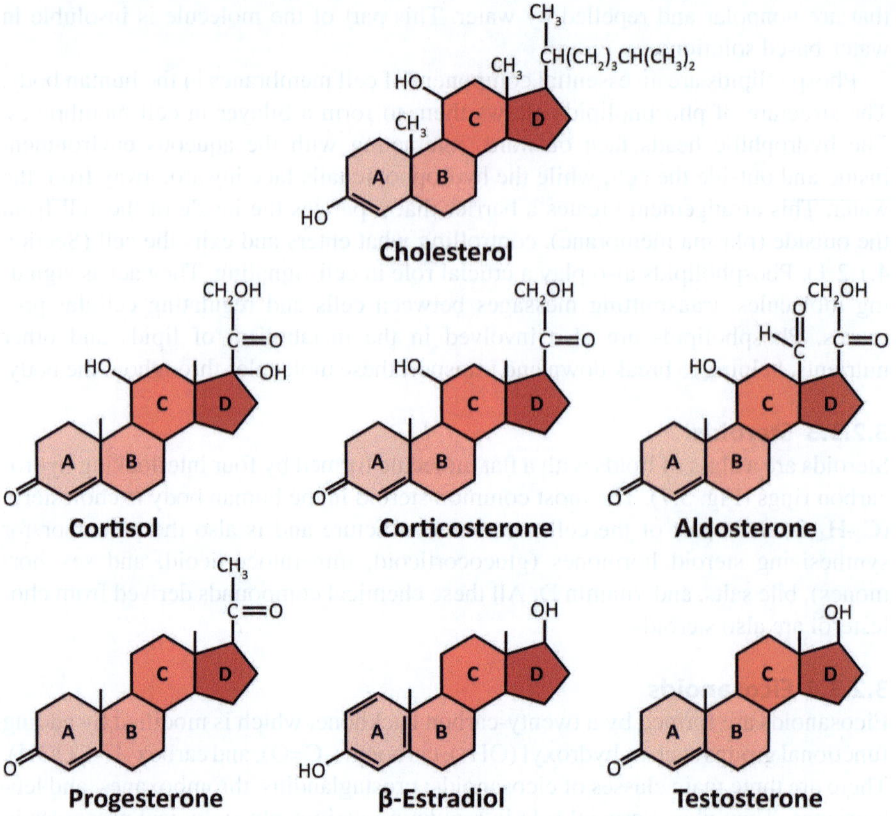

Fig. 3.7 Molecular structure of cholesterol and various steroid hormones

The structure of an amino acid consists of a central carbon atom, known as the alpha carbon, which is bonded to four different chemical groups. These groups include an amino group (-NH_2), a carboxyl group (-COOH), a hydrogen atom (–H), and a side chain or R-group. The side chain differentiates one amino acid from another, giving each amino acid its distinct properties (Fig. 3.8).

To build a protein, amino acids must join by a dehydration reaction, which connects one amino acid's amino group to another amino acid's carboxyl group, forming a peptide bond (Fig. 3.9). Two linked amino acids form a dipeptide, three linked amino acids form a tripeptide, and a peptide chain formed by 4–50 amino acids is called a polypeptide. The peptide chains containing more than 50 amino acids are called proteins.

The primary function of amino acids is to serve as the building blocks of proteins. Some amino acids may also serve as precursors for synthesizing neurotransmitters, which are chemical messengers that transmit signals between nerve cells (amino acid tryptophan is the precursor for serotonin). Amino acids also play a role in the regulation of gene expression. Furthermore, some amino acids have

Table 3.1 List of amino acids and their dietary sources

Amino acid name	Source	Essential/ nonessential
Alanine	Various proteins in meat, dairy, and plant sources	Nonessential
Arginine	Meat, dairy, nuts, seeds, legumes	Conditionally essential
Asparagine	Dairy, beef, poultry, eggs, fish, asparagus, potatoes, legumes, nuts, seeds	Nonessential
Aspartic acid	Plant proteins, avocado, asparagus, sprouts, oat flakes	Nonessential
Cysteine	Meat, garlic, onions, Brussels sprouts, oats, wheat germ, sprouted lentils	Conditionally essential
Glutamine	Beef, chicken, fish, dairy products, eggs, vegetables, wheat, papaya, Brussels sprouts	Conditionally essential
Glutamic acid	Cheese, meat, fish, eggs, corn, barley	Nonessential
Glycine	Protein-rich foods like meat, fish, dairy, legumes	Nonessential
Histidine	Meat, fish, poultry, nuts, seeds, grains	Essential
Isoleucine	Meat, fish, poultry, eggs, cheese, lentils, nuts	Essential
Leucine	Meat, fish, poultry, dairy, legumes, nuts, seeds	Essential
Lysine	Meat, eggs, soy, black beans, quinoa, pumpkin seeds	Essential
Methionine	Eggs, grains, nuts, seeds	Essential
Phenylalanine	Meat, fish, eggs, cheese, milk	Essential
Proline	Meat, dairy, gelatin	Nonessential
Serine	Meat, soybeans, nuts, eggs, chickpeas, lentils	Nonessential
Threonine	Meat, poultry, fish, dairy, lentils, black beans	Essential
Tryptophan	Turkey, chicken, milk, cheese, pumpkin and sesame seeds	Essential
Tyrosine	Cheese, almonds, avocados, bananas, dairy, beans, seeds	Nonessential
Valine	Meat, soy, cheese, peanuts, mushrooms, whole grains	Essential

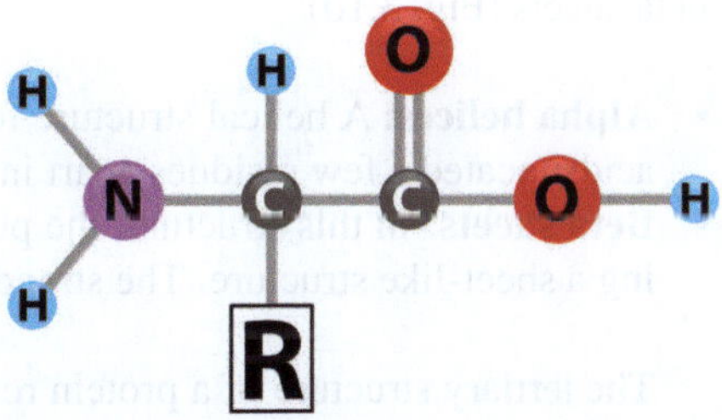

Fig. 3.8 General amino acid structure

antioxidant properties and help protect cells from oxidative damage. (cysteine and methionine contain sulfur atoms that can scavenge free radicals and prevent them from causing cellular damage).

3.2.4.2 Structure of Proteins

Proteins are large molecules with complex structure. Their structure can be described at four levels: primary, secondary, tertiary, and quaternary (Fig. 3.10). The different levels of protein structure are critical for their function, and any changes in protein

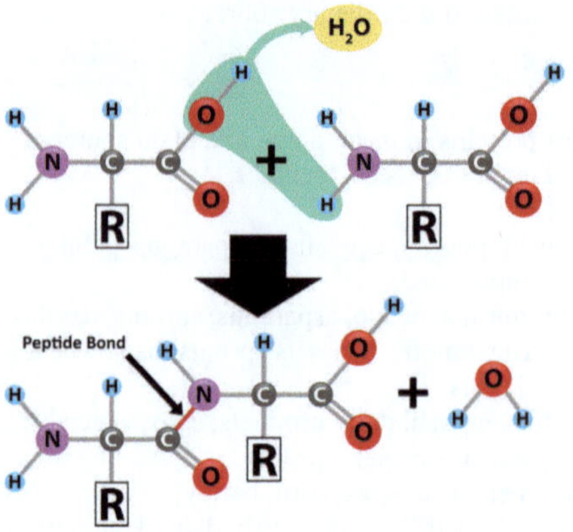

Fig. 3.9 Peptide bond formation through a dehydration synthesis reaction: the acid group of one amino acid is bonded to the amine group of the next, with the loss of a water molecule

structure can lead to alterations in protein function and contribute to various diseases.

The primary structure of a protein refers to the linear sequence of amino acids linked together by peptide bonds that make up the polypeptide chain. This sequence is determined by the genetic code specifying the order of amino acids in the protein.

The secondary structure of a protein refers to the local folding patterns within the polypeptide chain. These structures are stabilized by hydrogen bonds between one amino acid's amino group and another amino acid's carboxyl group in the same polypeptide chain. Two common types of secondary structure are alpha helices and beta sheets (Fig. 3.10).

- **Alpha helices:** A helical structure formed by hydrogen bonding between amino acids located a few residues apart in the sequence.
- **Beta sheets:** In this structure, the polypeptide chain folds back and forth, forming a sheet-like structure. The strands can be either parallel or antiparallel.

The tertiary structure of a protein refers to the overall three-dimensional folding pattern of a polypeptide molecule. This shape is determined by interactions between the protein chain's amino acid side chains (R groups). These interactions can include hydrogen bonds, ionic bonds, covalent bonds, and hydrophobic interactions. The quaternary structure of a protein refers to the arrangement of multiple polypeptide chains into a larger, functional multi-subunit protein complex. The subunits are held together by the same interactions that stabilize the tertiary structure.

3.2.4.3 Types of Proteins

Proteins can be classified based on their function or shape.

There are five types of proteins based on their function: enzymes, structural proteins, transport proteins, hormones, antibodies, receptor proteins, storage proteins, contractile proteins, defensive proteins, and regulatory proteins.

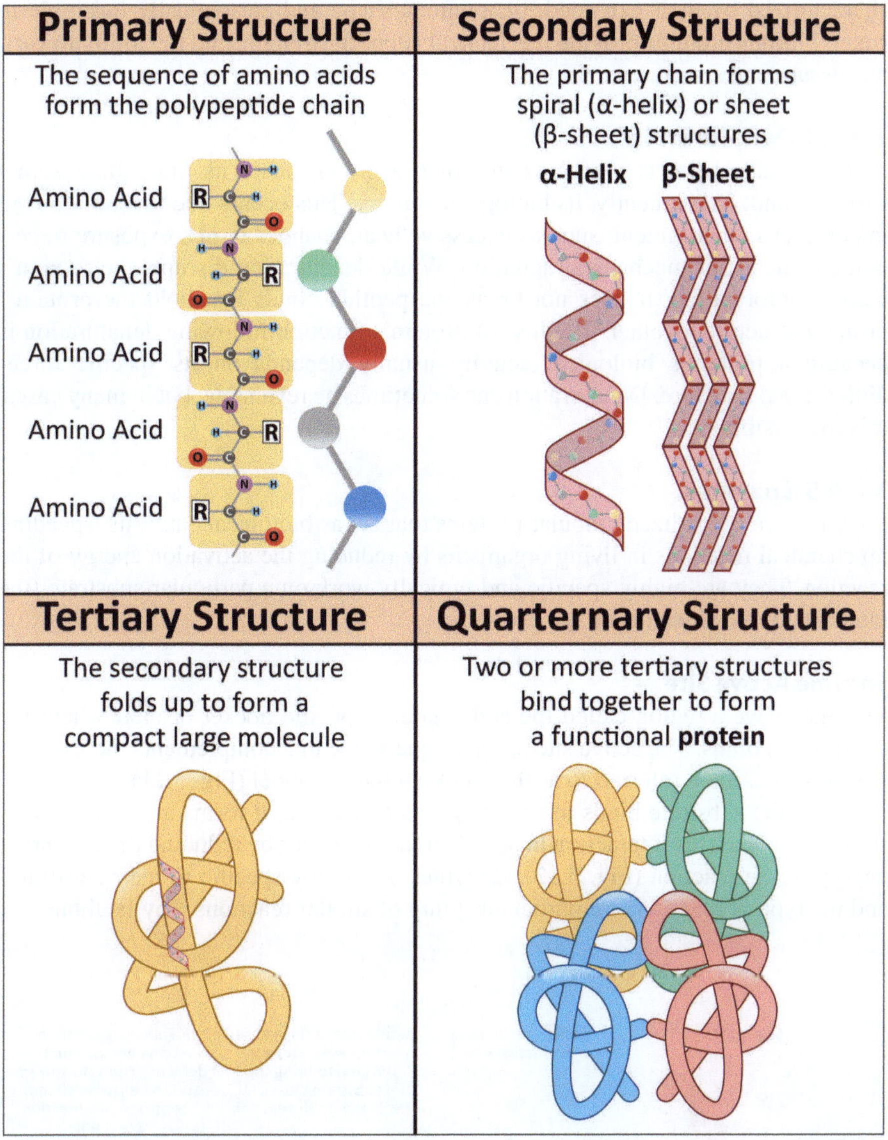

Fig. 3.10 Levels of protein structure

Based on the shape, proteins are classified into globular and fibrous proteins. Globular proteins have a spherical or globular shape, are water-soluble, and typically have complex, irregular three-dimensional structures. This structural arrangement allows them to carry out various biological functions, like enzymes, transport proteins, hormones, antibodies, and regulatory proteins. Fibrous proteins are characterized by their extended, filamentous shape and are typically insoluble in water. Fibrous proteins provide mechanical support, strength, and flexibility to various tissues and structures.

3.2.4.4 Denaturation

Protein denaturation is a process in which a protein loses its three-dimensional structure and, consequently, its biological activity. This occurs due to alterations in the protein's environment, such as excessive heat, changes in pH, exposure to certain chemicals, or mechanical agitation. While denaturation disrupts the protein's native conformation, it does not break the peptide bonds that hold the protein's primary structure together. The loss of protein's function following denaturation is because a protein's biological activity usually depends on its specific three-dimensional structure. Denaturation can sometimes be reversible, but in many cases, it is irreversible.

3.2.4.5 Enzymes

Enzymes are specialized globular proteins that act as biological catalysts, speeding up chemical reactions in living organisms by reducing the activation energy of the reaction. They are highly specific and typically work on a particular substrate (the molecule upon which the enzyme acts).

Enzyme Active Site

Enzymes have a region called the active site, a specific pocket or cleft where the substrate(s) binds. The active site has a unique shape that complements the shape of the substrate, often referred to as the "lock-and-key" model (Fig. 3.11).

When the substrate binds to the enzyme's active site, it forms an enzyme–substrate complex. This interaction makes the reaction occur by reducing the activation energy of the reaction (Fig. 3.12). Enzymes are highly specific to their substrates and the type of reaction or a particular group of similar reactions they facilitate.

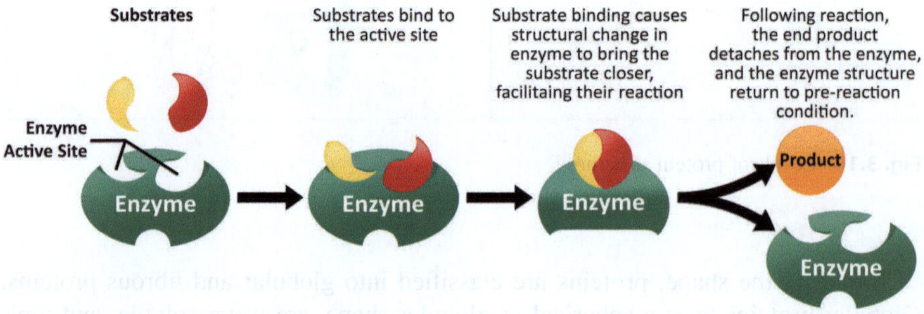

Fig. 3.11 Mechanism of enzyme action

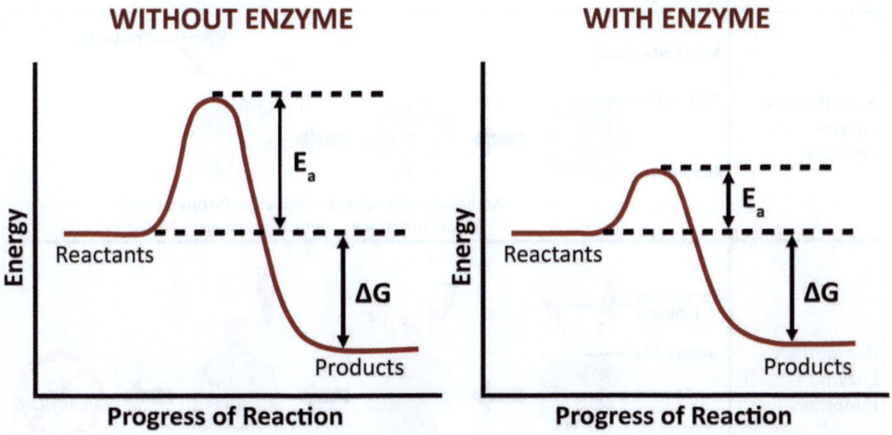

Fig. 3.12 Enzymes lower the activation energy required for a reaction. The graph on the left shows a reaction without an enzyme, where the activation energy (Ea) is high, representing the energy barrier that must be overcome for the reaction to proceed. The graph on the right depicts the same reaction in the presence of an enzyme, which significantly lowers the activation energy, thereby increasing the reaction rate. Both charts show that the reaction's free energy change (ΔG) remains unchanged with or without the enzyme

Coenzymes and Cofactors

Some enzymes are pure protein. The others consist of two parts: an apoenzyme (the protein portion) and a cofactor, an ion of a metal element such as copper or iron or an organic molecule needed to assist the reaction somehow. When the cofactor is an organic molecule, it is called a coenzyme.

Enzyme Inhibition

Enzyme activity can be modulated by molecules that interact with the enzyme. There are two main types of enzyme inhibition: competitive and noncompetitive inhibition.

Competitive inhibitors (Fig. 3.13) are molecules closely resembling the substrate and binding to the enzyme's active site. By doing so, they compete with the substrate for the active site. If a competitive inhibitor occupies the active site, the substrate cannot bind, and the reaction is inhibited. Increasing the substrate concentration can overcome the effects of a competitive inhibitor.

Noncompetitive inhibitors (Fig. 3.13) bind to a different site on the enzyme (allosteric site) than the active site. When a noncompetitive inhibitor binds to the enzyme, it induces a conformational change that makes the active site less effective at catalysis, even if the substrate is present. Increasing substrate concentration cannot fully overcome the effects of a noncompetitive inhibitor.

Feedback inhibition (Fig. 3.14) is a form of regulation in which a metabolic pathway's end product acts as an enzyme inhibitor earlier in the pathway. This helps prevent an over-accumulation of the final product.

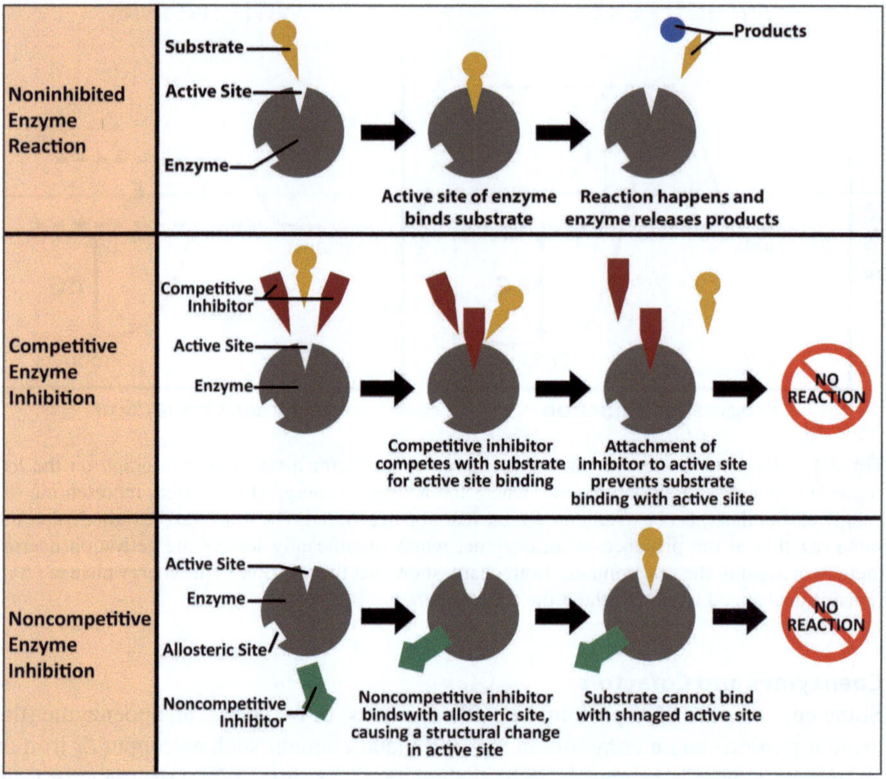

Fig. 3.13 Enzyme inhibition mechanisms. Comparison of noninhibited enzyme reactions, competitive inhibition, and noncompetitive inhibition, illustrating how inhibitors affect enzyme activity by blocking the active site or altering the enzyme's conformation

3.2.5 Nucleic Acids (DNA and RNA)

Nucleic acids are the largest molecules in organisms and are composed of carbon, oxygen, hydrogen, nitrogen, and phosphorus. They include two major classes: deoxyribonucleic acid (DNA) and ribonucleic acid (RNA).

The building blocks of nucleic acids are "nucleotides." Each nucleotide comprises a pentose sugar molecule (ribose or deoxyribose), a phosphate group, and a nitrogen-containing base. Five major varieties of nitrogen-containing bases contribute to nucleotide structure: adenine (A), guanine (G), cytosine (C), thymine (T), and uracil (U) (Fig. 3.15). Adenine and guanine are called purines; cytosine, thymine, and uracil are called pyrimidines. These nucleotides attach in two ways: one by linking the phosphate group of one nucleotide to the pentose sugar of another nucleotide, forming a long chain of nucleotides. The second way is connecting the nitrogen-containing group of one nucleotide to the other, which joins the two nucleotide chains together, happening only in DNA.

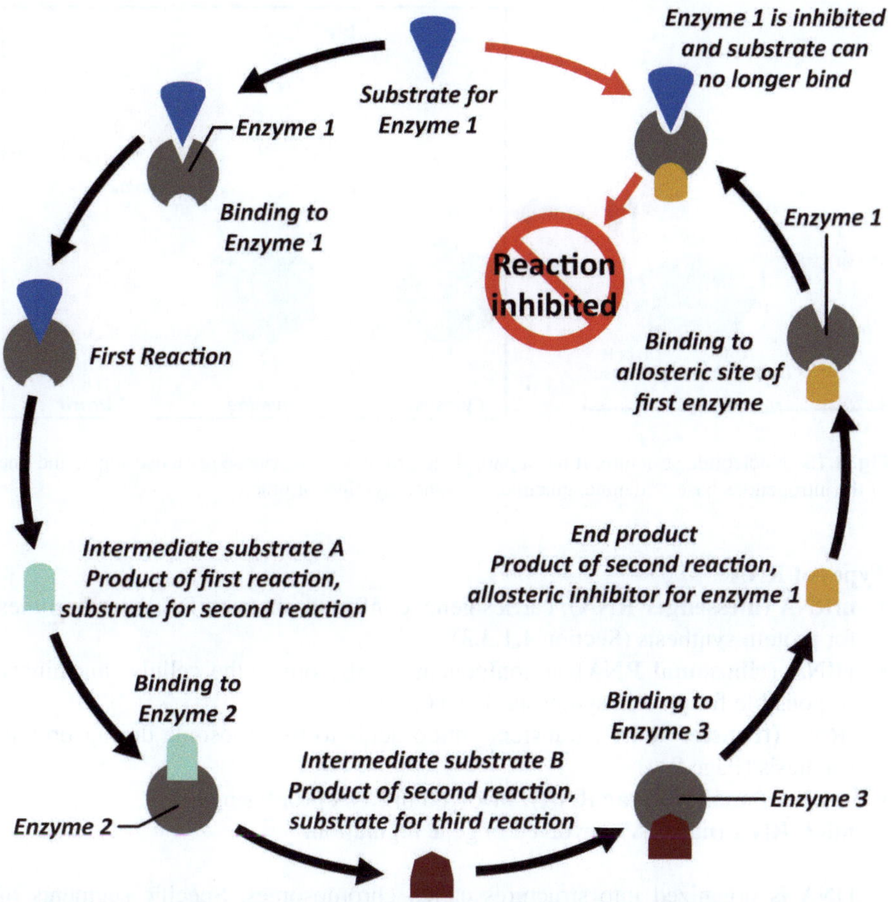

Fig. 3.14 Feedback inhibition. The end product of a metabolic pathway inhibits the first enzyme in the pathway by binding to its allosteric site, preventing further substrate binding and subsequent reactions

DNA is a double-stranded helical molecule (Fig. 3.16). It resembles a twisted ladder, with two long chains (polynucleotide strands) that run in opposite directions. The sugar in DNA is deoxyribose. The four nitrogen-containing bases in DNA are adenine, thymine, cytosine, and guanine. In DNA, adenine always pairs with thymine and cytosine pairs with guanine. This is known as complementary base pairing.

Unlike DNA, RNA is typically single-stranded (Fig. 3.17), although it can form secondary structures through intramolecular base pairing. RNA contains ribose instead of deoxyribose as its sugar component. RNA is formed by four nitrogen-containing bases: adenine, uracil, cytosine, and guanine. Uracil replaces thymine in RNA and pairs with adenine.

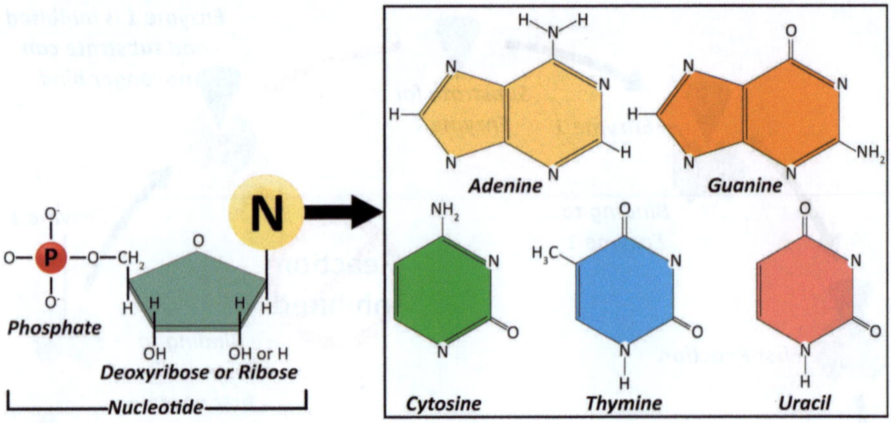

Fig. 3.15 Nucleotide structure. It has a phosphate group, a deoxyribose or ribose sugar, and one of the nitrogenous bases: adenine, guanine, cytosine, thymine, or uracil

Types of RNA

- **mRNA (messenger RNA):** carries genetic information from DNA to ribosomes for protein synthesis (Section 4.1.3.3).
- **rRNA (ribosomal RNA):** a component of ribosomes, the cellular machinery responsible for protein synthesis (Page 62).
- **tRNA (transfer RNA):** transfers amino acids to the ribosome during protein synthesis (Page 81).
- **snRNA (small nuclear RNA):** involved in RNA processing.
- **microRNA (miRNA):** involved in gene regulation.

DNA is organized into structures called chromosomes. Specific segments of DNA, known as genes, contain the information necessary to code for particular proteins. Genes determine an organism's traits.

3.2.6 Adenosine Triphosphate (ATP)

The energy released during glucose catabolism is coupled to adenosine triphosphate (ATP) synthesis. ATP is the primary energy-storing and energy-transferring molecule in cells, providing energy immediately usable by all body cells. ATP is easily regenerated through a cellular respiration reaction called oxidative phosphorylation in eukaryotic cells' mitochondria.

ATP is a nucleotide consisting of three main components: a nitrogen-containing base (adenine), a ribose sugar, and three phosphate groups (Fig. 3.18). The nitrogen-containing base in ATP is adenine. The ribose sugar is a five-carbon sugar that forms the backbone of the ATP molecule. ATP has three phosphate groups linked together. They are labeled α (alpha), β (beta), and γ (gamma) based on their proximity to the

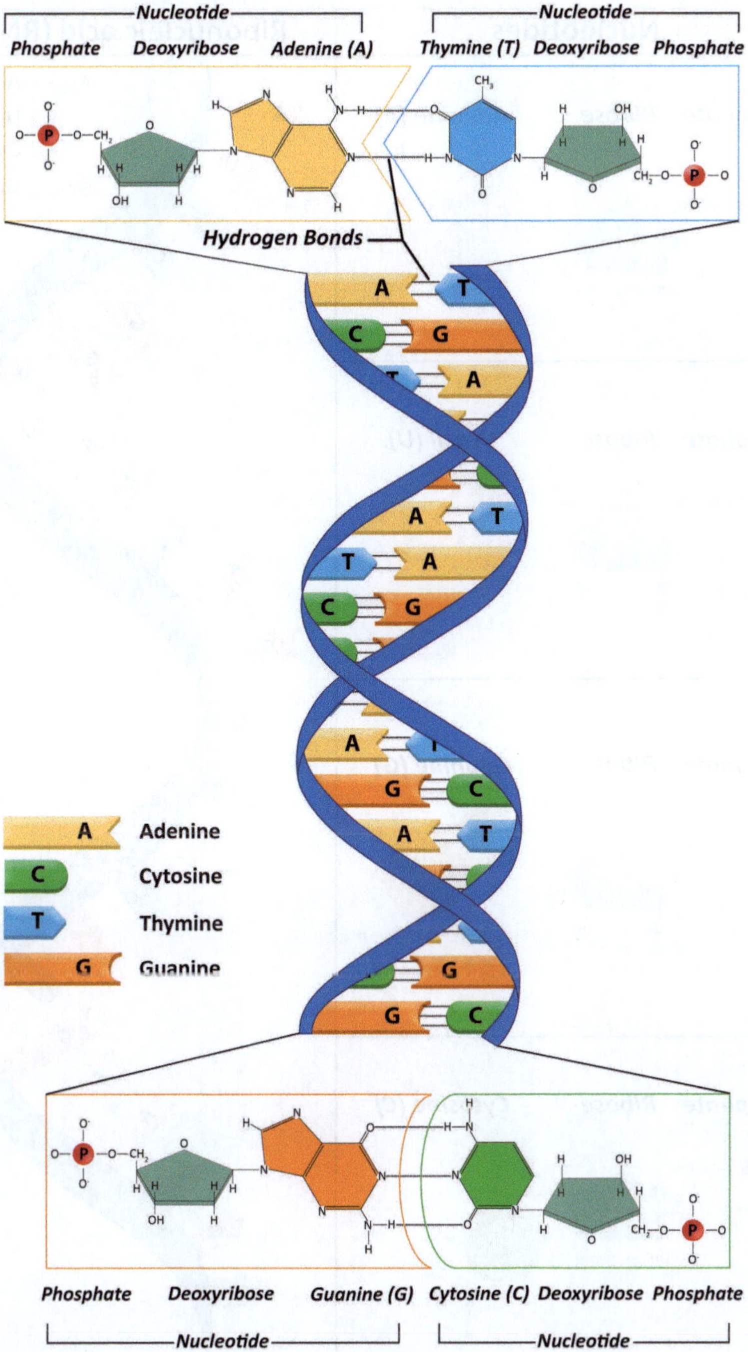

Fig. 3.16 Structure of DNA. Nucleotides, the structural units of DNA, are linked by hydrogen bonds (shown by dotted lines) between their complementary bases [adenine with thymine (upper section) and cytosine with guanine (lower section)]. DNA is a double helix: a coiled double polymer of nucleotides (middle section)

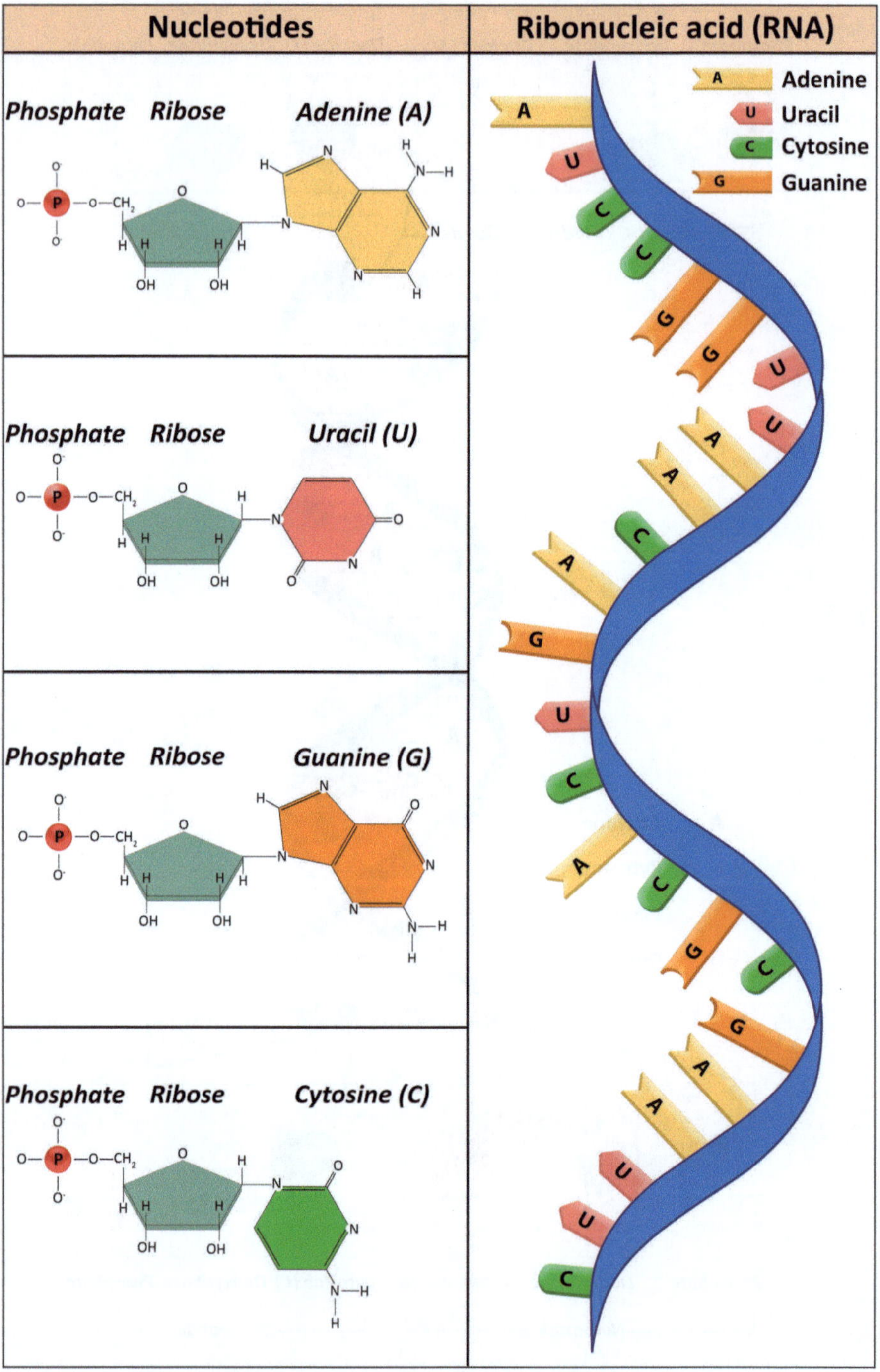

Fig. 3.17 Structure of RNA. The structural units of RNA (adenine, cytosine, guanine, and uracil) are single, forming a single-stranded polymer of nucleotides

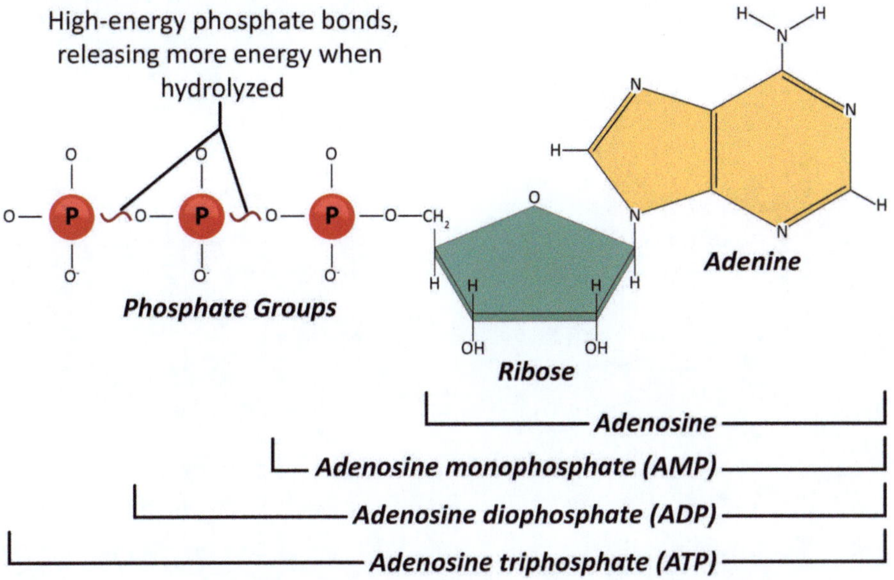

Fig. 3.18 Structure of ATP

ribose sugar. The bonds between the phosphate groups, particularly the β-γ phosphate bond, are high-energy bonds. When these bonds are broken, a significant amount of energy is released.

ATP is crucial for the survival and functioning of all living organisms. It serves as a universal energy currency, allowing cells to perform a wide range of tasks necessary for life. The constant recycling of ATP and ADP is fundamental to cellular energy metabolism.

Suggested Reading

An Introduction to Medicinal Chemistry, 7th Ed. Graham L. Patrick, Oxford University Press, 2023
Biochemistry, 9th Ed. Lubert Stryer, W.H. Freeman, 2019
Biochemistry, 8th, North American Ed. Emine E. Abali, at al. Lippincott Williams & Wilkins, 2021
Harper's Illustrated Biochemistry, 31st Ed. Victor W. Rodwell, et al. McGraw Hill, 2018
Lehninger Principles of Biochemistry, 8th Ed. David L. Nelson, W.H. Freeman, 2021

Cells

4

In this chapter, you will learn

- Definition of cells as the fundamental units of life
- Comparison of prokaryotic and eukaryotic cells
- Structure and function of the plasma membrane
- Role of the extracellular matrix and fluids
- Functions and types of cell junctions
- Mechanisms of membrane transport
- Composition and function of the cytoplasm
- Structure and function of mitochondria
- Role of ribosomes in protein synthesis
- Structure and function of the endoplasmic reticulum
- Golgi apparatus and its role in processing and sorting molecules
- Lysosomal function and its importance in waste management
- Overview of the cytoskeleton and its components
- Structure and function of centrosomes and centrioles
- Role of cellular appendages such as cilia and flagella
- Detailed structure and function of the nucleus
- Chromatin structure and its role in gene expression
- Cell cycle, including phases and regulatory mechanisms
- Details of the DNA replication process
- Process of translation and its importance in protein synthesis
- Mechanisms and stages of transcription in protein synthesis
- Post-translational modifications of proteins
- Bacterial cell structure and function
- Bacterial cell wall and its composition
- Differences between Gram-positive and Gram-negative bacteria
- Mechanisms of genetic exchange in bacteria
- Concept and role of biofilms in bacterial colonies

H. Hosseinzadeh, *Fundamentals of Medicine for Biomedical Engineering*,
https://doi.org/10.1007/978-3-031-74085-5_4

A cell is the smallest independently functioning unit of a living organism. It is enclosed within a membrane and contains genetic material (DNA or RNA) that encodes the information necessary for growth, reproduction, and function. Cells are capable of carrying out the essential processes of life, including metabolism, growth, response to stimuli, and reproduction.

Cells can exist in various forms, from single-celled organisms like bacteria and protists to highly specialized and differentiated cells in complex organisms like plants, animals, and humans. Each type of cell is uniquely adapted to its specific function, contributing to the overall functioning of the organism.

The human body has trillions of cells, with over 250 different types varying in shape, size, and function. Cells vary in length, ranging from 1–10 micrometers (1/25,000–1/2500 of an inch) in bacteria to over a meter in the nerve cells that provide sensation to the toes.

All living cells can be classified as either prokaryotic [pro (before) and karyon (nucleus)] or eukaryotic [eu (true) and karyon (nucleus)]. Prokaryotic cells lack a nucleus and other membrane-enclosed structures, whereas eukaryotic cells have such structures. In eukaryotic cells, DNA is in a nucleus surrounded by a membranous nuclear envelope, but in prokaryotic cells, DNA is in a nuclear region not surrounded by a membrane. All prokaryotes are single-celled organisms. Eukaryotes include all plants, animals, fungi, and protists.

4.1 Part 1: Eukaryotic Cells

All eukaryotic cells have three main parts:

- **Plasma membrane:** the outer boundary of the cell, separating the inside and outside environments of the cell.
- **Cytoplasm:** the intracellular fluid packed with organelles, tiny structures that perform specific cell functions.
- **Nucleus:** an organelle that contains genetic information and controls cellular activities.

4.1.1 Extracellular Materials

Extracellular materials refer to substances located outside the plasma membrane of cells. These materials support, communicate, and protect cells and tissues. The key components of extracellular materials are:

Extracellular fluid: Extracellular fluid, also called interstitial fluid, is the tissue fluid that bathes all cells, facilitating the substance exchange between cells. Like a rich, nutritious "soup," extracellular fluid contains thousands of ingredients, including amino acids, sugars, fatty acids, regulatory substances, and wastes.

Extracellular matrix (ECM): The ECM is a complex jellylike network of proteins, carbohydrates, and other molecules that form a supportive framework surrounding cells in tissues. These molecules provide structural support and elasticity.

4.1.2 General Cell Structure

Eukaryotic cells are incredibly diverse in form and function, but they share many structures and characteristics.

4.1.2.1 Plasma Membrane

The plasma membrane, also known as the cell membrane, forms the outer boundary of the cell, separating the internal environment from the external surroundings. The primary structural feature of the plasma membrane is a phospholipid bilayer, forming the basic fabric of the membrane. Phospholipids are molecules with a hydrophilic (water-attracting) "head" and two hydrophobic (water-repelling) "tails." In the bilayer, the hydrophilic heads face outward, toward the aqueous environment, while the hydrophobic tails are oriented inward, away from water, shaping two parallel sheets of phospholipid molecules (Fig. 4.1). The plasma membrane is often described using the "fluid mosaic model." This model suggests that the membrane is dynamic, with lipids, proteins, and other molecules able to move laterally within the bilayer. Cholesterol molecules are interspersed within the phospholipid bilayer. They help stabilize the membrane, maintain fluidity, and prevent fatty acid chains from packing too closely together.

A cell's plasma membrane contains several proteins scattered inside and over the membrane. Proteins in the plasma membrane have diverse functions, including receptor proteins for signaling, transport proteins for moving substances across the membrane, and enzymes for catalyzing specific reactions. There are two distinct populations of membrane proteins: integral and peripheral (Fig. 4.1).

Integral proteins are embedded within the lipid bilayer. These proteins often have hydrophobic regions that interact with the lipid tails. Peripheral proteins are associated with the membrane surface, usually attached loosely to integral proteins or interacting with the polar heads of phospholipids.

Carbohydrates are often bound to proteins (glycoproteins) or lipids (glycolipids) on the extracellular surface of the membrane (Fig. 4.1). They play roles in cell recognition, signaling, and adhesion.

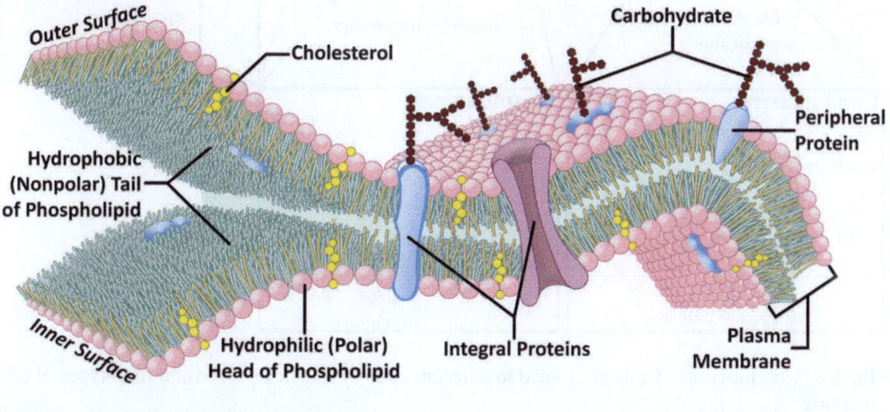

Fig. 4.1 Plasma membrane structure

4.1.2.1.1 Cell Junctions

Cell junctions (Fig. 4.2) are specialized structures facilitating communication, adhesion, and coordination among cells in multicellular organisms. There are several types of cell junctions, each with specific functions.

Tight junctions: Tight junctions form a continuous, impermeable barrier between adjacent cells. They consist of integral membrane proteins in the plasma membranes of adjacent cells, which interact to seal the intercellular space. Tight junctions prevent the leakage of substances between cells and restrict the movement of membrane proteins.

Desmosomes (macula adherens): Desmosomes form disk-shaped structures that link intermediate filaments of adjacent cells. They provide strong mechanical attachments between cells, especially in tissues subjected to mechanical stress, like skin and cardiac muscle. They contribute to the structural integrity of tissues. Hemidesmosomes are specialized adhesion structures found in epithelial cells. They anchor epithelial cells to the underlying basement membrane through integrin proteins and provide strong adhesion between the epithelial cells and the underlying tissue.

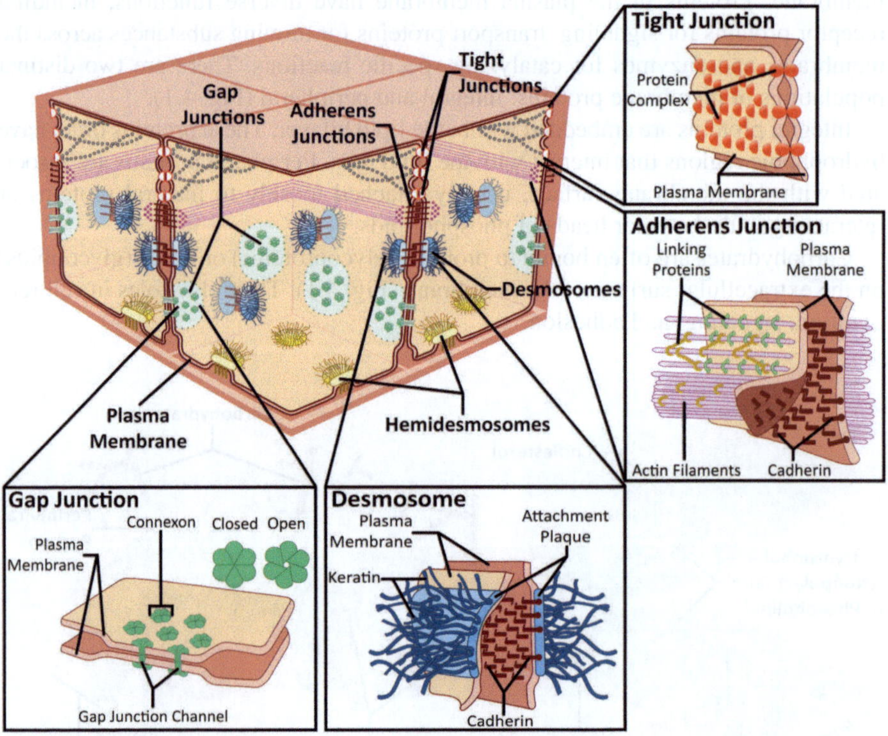

Fig. 4.2 Cell junctions. Cells are joined to adjacent cells by one or all three common types of cell junctions

Gap junctions (nexus): Gap junctions are specialized channels formed by connexin proteins that create narrow passages between adjacent cells. These channels allow for direct cytoplasmic communication, facilitating the rapid exchange of ions, small molecules, and signaling molecules between cells.

4.1.2.1.2 Membrane Transport

Membrane transport encompasses the processes by which various molecules cross the plasma membrane, including ions, nutrients, and signaling molecules. The plasma membrane's selective permeability is essential for maintaining cellular homeostasis and allowing cells to interact with their environment. Substances move through the plasma membrane in two ways: passive or active.

4.1.2.1.2.1 Passive Transport

Passive transport allows molecules to move across the plasma membrane without the cell consuming energy. It occurs down the concentration gradient, meaning substances move from higher concentration areas to lower concentration regions. There are three types of passive transport across the plasma membrane: simple diffusion, facilitated diffusion, and osmosis, all of which are types of diffusion (Fig. 4.3).

Simple diffusion: Simple diffusion (Fig. 4.3a) is the unassisted movement of molecules across the plasma membrane, down their concentration gradients. It occurs when molecules are small enough to pass directly through the lipid bilayer. The transport of small, nonpolar molecules like oxygen and carbon dioxide diffusing directly through the lipid bilayer is an example of passive diffusion.

Facilitated diffusion: Facilitated diffusion is the movement of molecules across the plasma membrane with the help of specific membrane proteins, such as channel proteins and carrier proteins (Fig. 4.3b, c). Channel proteins create hydrophilic pores through which specific ions and small molecules can pass. Carrier proteins undergo conformational changes to transport specific molecules across the membrane. There is no energy expenditure for this transport. Glucose transporters (GLUTs) are carrier proteins that facilitate the diffusion of glucose across the plasma membrane.

Osmosis: Osmosis (Fig. 4.3d) is the natural, passive diffusion of water across a selectively permeable membrane in response to differences in solute concentration. In their inherent nature, water molecules move from regions of lower solute concentration (higher water concentration) to areas of higher solute concentration (lower water concentration).

4.1.2.1.2.2 Active Transport

Active transport is a vital cellular process that enables the movement of substances across a cell membrane against their concentration gradient. Unlike passive transport, active transport necessitates cellular energy expenditure, usually adenosine

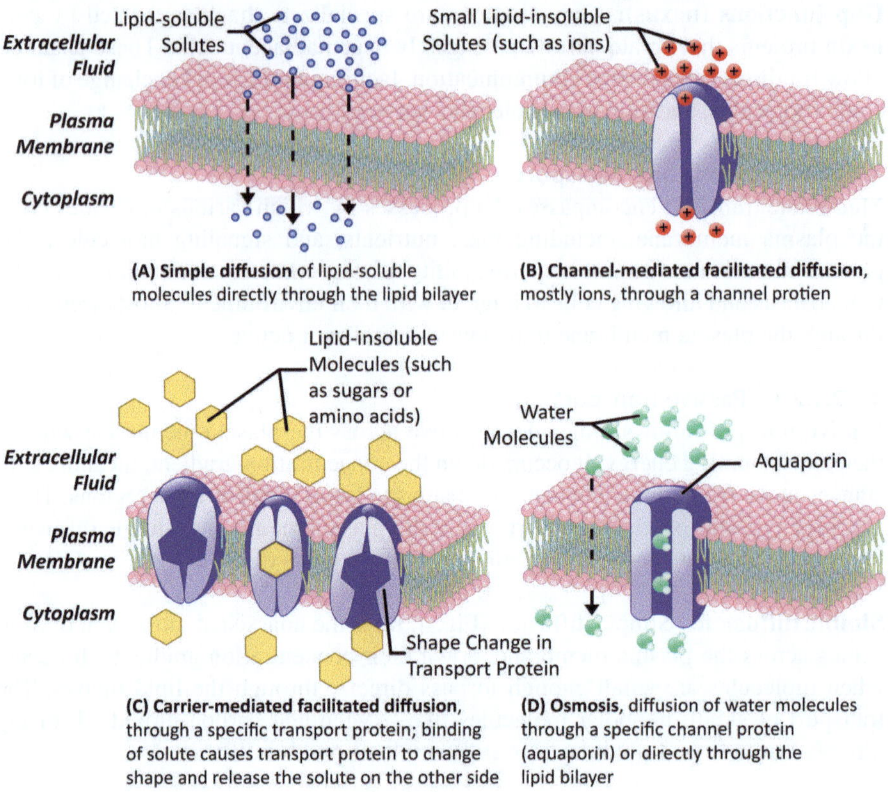

Fig. 4.3 Passive transport across the plasma membrane

triphosphate (ATP). This mechanism maintains cellular homeostasis, allowing cells to accumulate essential nutrients, expel waste products, and regulate ion concentrations.

The primary driving force behind active transport is specialized proteins embedded in the cell membrane known as transporters or pumps. These proteins act as molecular machines that facilitate the movement of specific substances across the membrane, typically from an area of lower concentration to an area of higher concentration.

Primary active transport: In primary active transport, the energy to do work comes directly from the hydrolysis of ATP by transport proteins called pumps (ATPase) (Fig. 4.4). The process begins with the binding of ATP to the ATPase protein. This binding triggers ATP hydrolysis, resulting in the release of a phosphate

Fig. 4.4 (continued) fluid. (5) Two extracellular K^+ ions bind to pump proteins at specific sites. (6) K^+ binding triggers the release of the phosphate, resulting in the dephosphorylation of pump protein. (7) Dephosphorylation causes the pump protein to resume its original shape, exposing K^+ ions to the cytoplasm. (8) A new ATP binds the pump protein, releases K^+ into the cytoplasm, and makes the Na^+ binding sites ready to bind Na^+ again to repeat the cycle

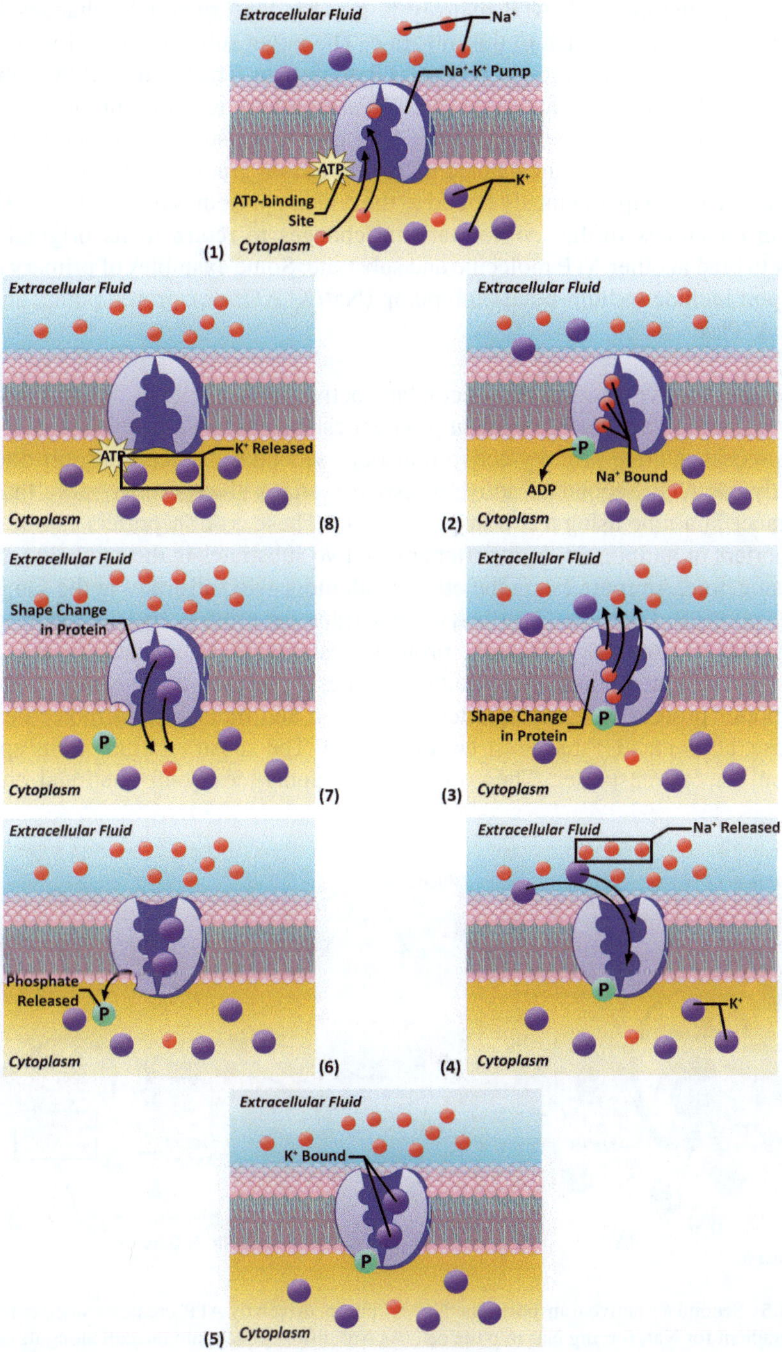

Fig. 4.4 Primary active transport: the Na^+-K^+ pump. (1) Three cytoplasmic Na^+ ions bind to pump protein at specific sites. (2) Na^+ binding promotes hydrolysis of ATP. The energy released from ATP phosphorylates the pump protein. (3) Phosphorylation causes the pump protein to change shape, exposing Na^+ to the extracellular fluid. (4) Three Na^+ ions are released into the extracellular

group. The energy from ATP hydrolysis powers conformational changes in the pump, allowing it to attach to the substrate. The target substrate (e.g., ions or molecules) binds to the specific site on the ATPase protein. The binding of the substrate induces conformational changes in the ATPase protein, resulting in the movement of the substrate across the membrane. As a result, the substrate is actively transported across the membrane against its concentration gradient. After translocation, the phosphate group previously bound to the ATPase protein is released. The ATPase protein undergoes further conformational changes to return to its original state, ready to bind another ATP molecule and substrate. Some examples of primary active transport include sodium-potassium pump (Na^+/K^+ ATPase), proton pump, and calcium ATPase.

Secondary active transport: Secondary active transport (Fig. 4.5) utilizes the energy stored in an electrochemical gradient established by a primary active transport process. Unlike primary active transport, secondary active transport does not directly use ATP. Secondary active transport systems always move more than one substance at a time using a cotransport protein. These co-transporters can be either symporters or antiporters. Symporters move two substrates in the same direction at the same time. Antiporters, on the other hand, move two substrates at the same time in two opposite directions. A primary active transport process, usually involving the movement of ions (e.g., sodium, protons), establishes a concentration gradient across the cell membrane. This gradient creates a potential energy source. The co-transporter protein has binding sites for the ion and the substrate molecule (e.g., glucose, amino acids) that must be transported. The substrate binds to its specific site on the co-transporter. The co-transporter undergoes conformational changes

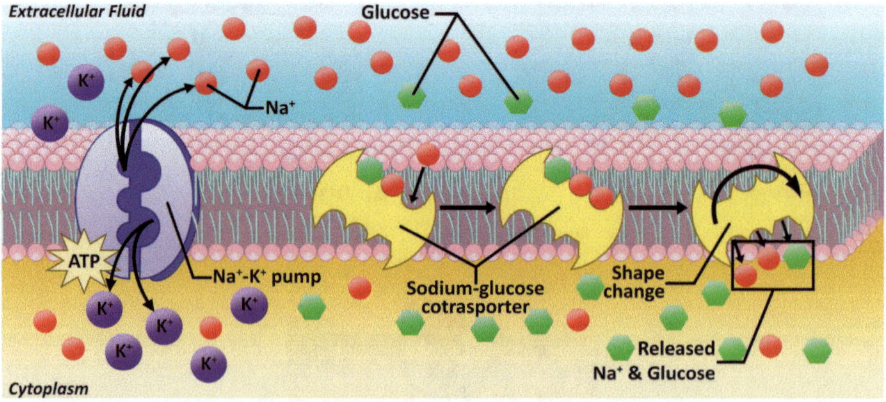

Fig. 4.5 Secondary active transport. The Na^+-K^+ pump, driven by ATP, creates a steep concentration gradient for Na^+, forcing Na^+ into the cell. As Na^+ diffuses back into the cell along its concentration gradient using a cotransport protein, it drives glucose against its concentration gradient into the cell. Once Na^+ ions and glucose molecules attach to their binding sites on the cotransport protein, it changes shape and moves the glucose and Na^+ into the cytoplasm

triggered by the binding of the ion. These changes allow the ion and the substrate to move across the membrane simultaneously, in the same or opposite directions. The movement of the ion down its electrochemical gradient provides the energy necessary to drive the uphill movement of the substrate against its concentration gradient.

4.1.2.1.2.3 Vesicular Transport

Vesicular transport moves large molecules and whole organelles across the cell membrane. Unlike active and passive transport, which occur directly through the phospholipid bilayer, vesicular transport involves the formation of vesicles— tiny, membrane-bound sacs—that enclose the cargo and facilitate its transport. ATP energizes vesicular transport processes. The vesicular transport moves substances into the cell (endocytosis) and out of the cell (exocytosis).

Endocytosis (Fig. 4.6): The process begins with a coated pit and continues to form a vesicle at the cell membrane that encloses extracellular material. There are three types of endocytosis:

Phagocytosis (Fig 4.7a): In phagocytosis, the cell engulfs large particles such as bacteria or cellular debris. Pseudopodia (cytoplasmic extensions) surround and enclose the particle, forming a phagosome. The phagosome usually fuses with a lysosome, and its contents are digested. In the human body, only cells called macrophages, commonly referred to as phagocytes, are experts at phagocytosis.

Pinocytosis (Fig 4.7b): Pinocytosis, also called fluid-phase endocytosis, involves the nonselective uptake of fluids and dissolved solutes from the extracellular environment.

Receptor-mediated endocytosis (Fig 4.7c): This is a highly specific form of endocytosis where specific molecules bound to cell surface receptors are taken up in clathrin-coated vesicles. This mechanism is responsible for the endocytosis of most macromolecules, like enzymes, insulin (and some other hormones), low-density lipoproteins (such as cholesterol attached to a transport protein), and iron.

Exocytosis (Fig. 4.6): Exocytosis involves the fusion of a vesicle with the cell membrane, releasing its contents into the extracellular space. This process is essential for processes like neurotransmitter release and hormone secretion.

4.1.2.2 Cytoplasm

The cytoplasm (Fig. 4.8) is a complex, gel-like substance that fills the interior of eukaryotic cells. It is a semi-fluid matrix enclosed by the cell membrane and surrounds the nucleus, where various cellular organelles are suspended. This dynamic region of the cell is vital for many cellular processes.

The cytoplasm contains cytosol, inclusions, and organelles. The cytoplasm comprises a viscous, clear, semi-fluid substance known as cytosol, in which the other cytoplasmic elements are suspended. It contains water, ions, small molecules, and macromolecules like proteins, nucleic acids, and carbohydrates. The cytosol

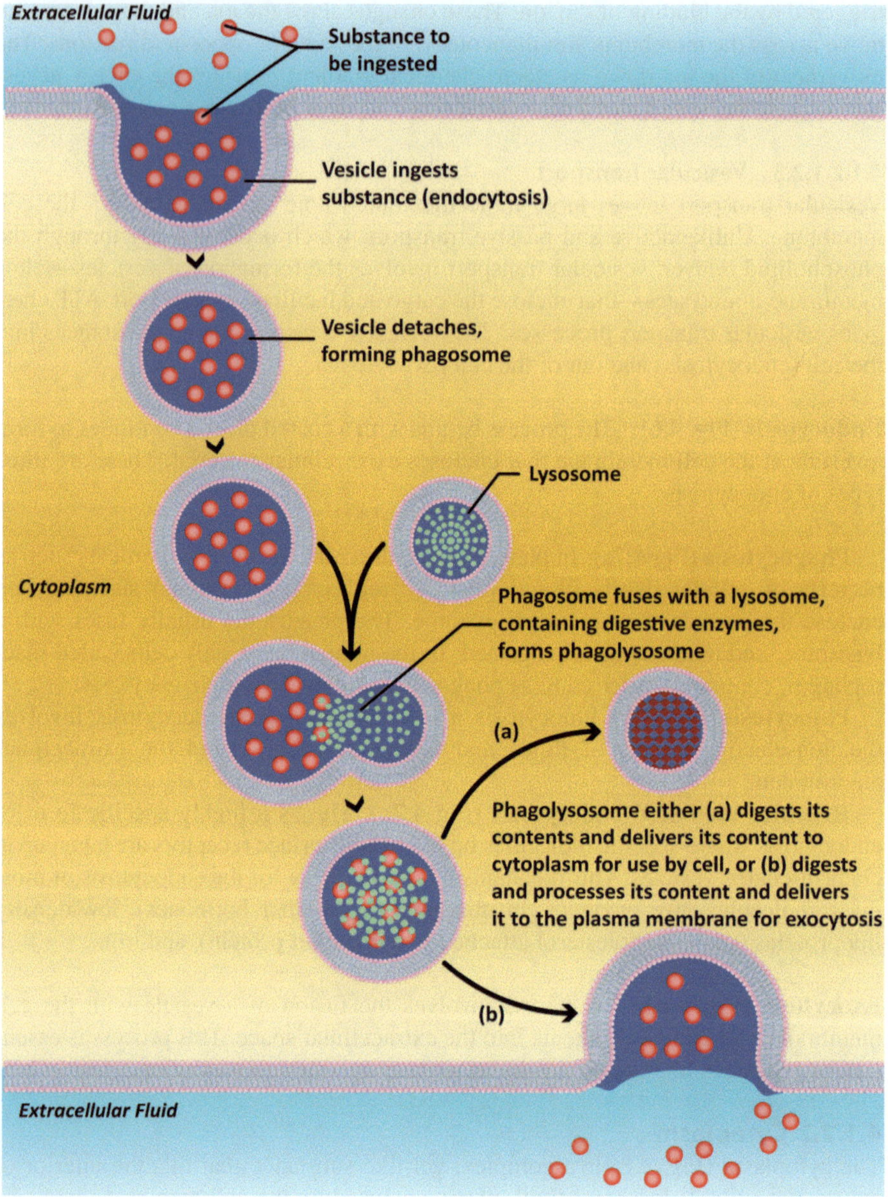

Fig. 4.6 Events of endocytosis

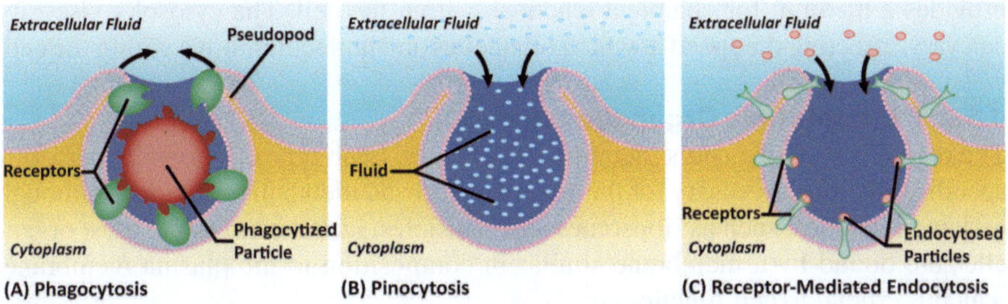

Fig. 4.7 Types of endocytosis. (**a**) **Phagocytosis:** Phagocytosis is a type of endocytosis that ingests and digests large particles and microorganisms. The membrane has receptors capable of binding to phagocytized particles. The cell engulfs a large particle by extending a pseudopod, or "false foot," around it and subsequently enclosing it within a membrane- bound vesicle called a phagosome. After its formation, the phagosome fuses with a lysosome, leading to the digestion of the phagosome's contents. (**b**) **Pinocytosis:** The cell ingests a droplet of extracellular fluid, which includes various solutes, into small vesicles. This process does not involve receptors, rendering it nonselective. (**c**) **Receptor-mediated endocytosis**: In this process, external substances attach to specialized receptor proteins on the cell surface, allowing the cell to internalize and concentrate particular substances within protein-coated vesicles. These substances can be released within the cell's interior or broken down by a lysosome

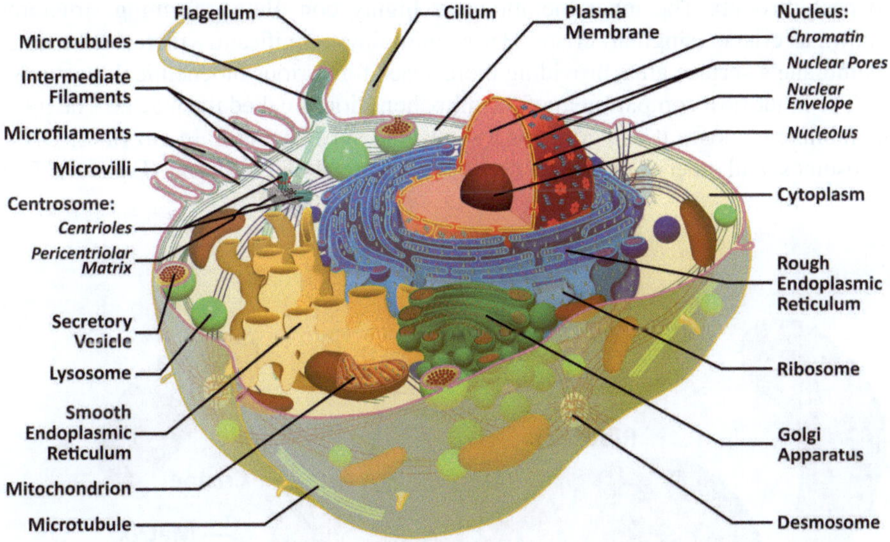

Fig. 4.8 The structure of a eukaryotic cell. Not all these structures are present in every cell

provides a medium for chemical reactions within the cell. The cytosol's viscosity can vary, affecting the speed at which organelles and molecules move within the cell.

4.1.2.3 Cytoplasmic Organelles

The organelles are the cell's metabolic machinery. Each type of organelle performs specific functions necessary for the cell's survival and metabolic activities. Although certain organelles, such as ribosomes and centrioles, lack a membrane, most organelles are bound by a membrane similar in composition to the plasma membrane, which is crucial to cell functioning.

Mitochondria

Mitochondria (Fig. 4.8) are membrane-bound organelles found in the cytoplasm of eukaryotic cells, often referred to as the "powerhouses" of the cell. The density of mitochondria in a particular cell reflects that cell's energy requirements. Busy cells like kidney and liver cells have hundreds of mitochondria, whereas relatively inactive cells (such as certain lymphocytes) have just a few, and the only cell in the human body that lacks mitochondria is sperm.

Mitochondria have two distinct lipid bilayer membranes: outer and inner membranes. (Fig. 4.9) The outer membrane is smooth and covers the organelle's surface. It acts as a protective barrier and contains porin proteins that allow small molecules to pass through. The inner membrane is highly convoluted, forming structures known as cristae (singular: crista). These infoldings significantly increase the inner membrane's surface area, providing more space for various biochemical reactions.

The innermost compartment of the mitochondrion is called the matrix. The inner membrane encloses it and contains a semi-fluid substance rich in enzymes, DNA, ribosomes, and other components necessary for metabolic reactions (Fig. 4.9). This

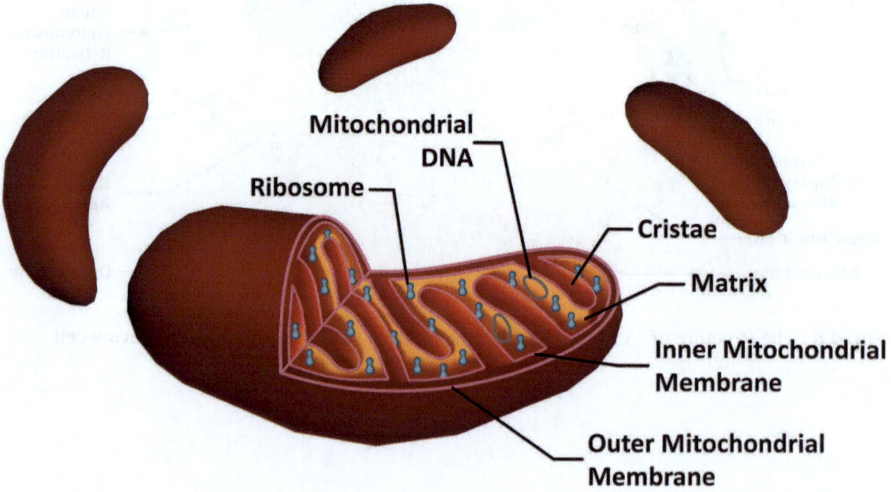

Fig. 4.9 Mitochondria: longitudinally sectioned mitochondrion showing its internal structure

genetic material encodes some of the proteins essential for mitochondrial function. However, most of the proteins required are encoded by nuclear DNA and then transported into the mitochondria. Mitochondria also have their own ribosomes, distinct from those found in the cytoplasm and similar to prokaryotic ribosomes (Section 4.2.2.4).

Ribosomes

Ribosomes (Fig. 4.8) are small, dark-staining granules composed of proteins and a variety of RNAs called ribosomal RNAs, responsible for the synthesis of proteins. The ribosome is the primary site for translating the genetic information provided by mRNA into the sequence of amino acids, hence building a polypeptide chain. Besides producing proteins, they play a role in quality control by ensuring that only correctly synthesized and folded proteins are released into the cell. Each cell has two ribosomal populations:

- Free ribosomes float freely in the cytosol. These ribosomes produce soluble proteins that function in the cytosol, and the proteins imported into mitochondria and some other organelles.
- Membrane-bound ribosomes are attached to membranes, forming a rough endoplasmic reticulum complex. They synthesize proteins destined for incorporation into cell membranes, lysosomes, or export from the cell.

The ribosomes consist of two subunits (Fig. 4.10): the small subunit [referred to as the 30S (Svedberg) subunit in prokaryotes and 40S in eukaryotes] and the large subunit (50S in prokaryotes and 60S in eukaryotes). These subunits come together during protein synthesis and dissociate after the process is complete. The whole eukaryotic ribosomes are 80S, and the prokaryotic ribosomes are 70S. (The sedimentation rate, measured in Svedberg units, represents the rate at which particle sediments when subjected to high centrifugal force. It is a measure of the particle's

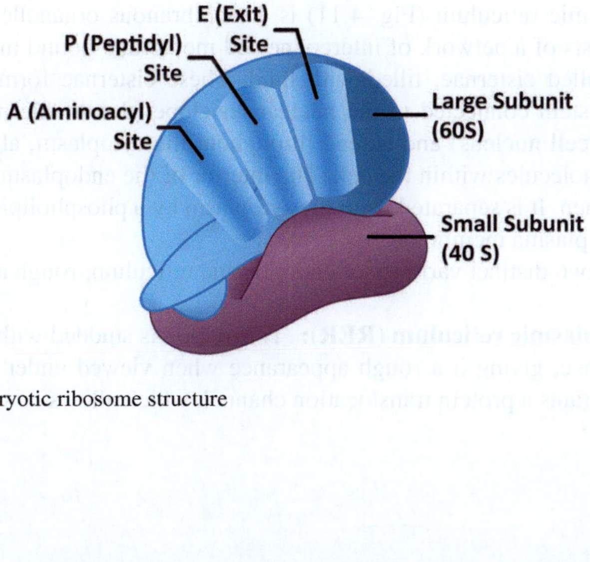

Fig. 4.10 Eukaryotic ribosome structure

size, shape, and density relative to the surrounding medium. The sedimentation rate is calculated using the formula:

$$S = \frac{\text{Sedimentation Velocity}\left(v\right)}{\text{Acceleration due to Centrifugal Force}\left(\omega^2 r\right)}$$

Here, v is the sedimentation velocity, ω is the angular velocity, and r is the radial distance from the axis of rotation. One Svedberg unit corresponds to 10^{-13} seconds. Therefore, a particle that sediments at a rate of 1 Svedberg will take 10^{13} seconds to travel a unit distance under a unit centrifugal field. It is important to note that the Svedberg unit is not additive. This means that the sedimentation rate of a complex composed of several macromolecules is not necessarily the sum of the Svedberg values of the individual components. This nonadditivity is due to changes in shape and hydration that can occur when individual subunits form a complex.)

Ribosomal RNA (rRNA) molecules form a significant portion of the ribosome's structure. The small subunit contains one rRNA molecule, while the large subunit contains three. Ribosomes have three key binding sites for RNA and amino acids (Fig. 4.10):

- **A site (aminoacyl-tRNA site):** this site holds the incoming aminoacyl-tRNA (transfer RNA), which carries the following amino acid to be added to the growing protein chain.
- **P site (peptidyl-tRNA site):** this site holds the tRNA attached to the growing polypeptide chain.
- **E site (exit site):** this site temporarily holds the tRNA that no longer carries an amino acid before it leaves the ribosome.

The large subunit of the ribosome contains a peptidyl transferase center, called a "catalytic center," where the formation of peptide bonds between amino acids occurs.

Endoplasmic Reticulum

The endoplasmic reticulum (Fig. 4.11) is a membranous organelle in eukaryotic cells. It consists of a network of interconnected membrane-bound tubules and flattened sacs called cisternae, filled with fluid. These cisternae form a continuous membrane system connected to the nuclear envelope (the double membrane surrounding the cell nucleus) and extend throughout the cytoplasm, allowing for the transport of molecules within the cell. The interior of the endoplasmic reticulum is called the lumen. It is separated from the cytoplasm by a phospholipid bilayer similar to the cell plasma membrane.

There are two distinct varieties of endoplasmic reticulum: rough and smooth.

Rough endoplasmic reticulum (RER): This region is studded with ribosomes on its outer surface, giving it a rough appearance when viewed under a microscope. The RER contains a protein translocation channel, which allows newly synthesized

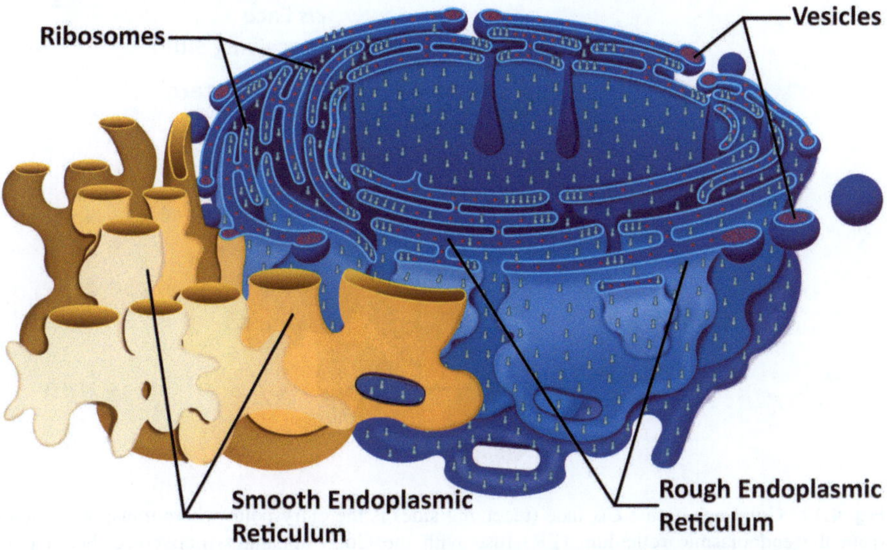

Fig. 4.11 Endoplasmic reticulum: The rough endoplasmic reticulum is characterized by the presence of ribosomes on its surface. The vesicles budding off from the ER transport synthesized proteins and lipids to various destinations within the cell

proteins to enter the lumen. The ribosomes on the surface of the RER are responsible for synthesizing proteins destined for secretion, membrane insertion, or lysosomal localization. As the proteins are synthesized, they enter the lumen of the RER for further processing, including folding, glycosylation, and disulfide bond formation. Misfolded or improperly synthesized proteins are identified and targeted for degradation. The RER is crucial for the biogenesis of membranes, including those of the plasma membrane and organelles.

Smooth endoplasmic reticulum (SER): This region lacks ribosomes and appears smooth. It is primarily involved in lipid metabolism and detoxification processes. The SER plays a role in calcium homeostasis. It stores and regulates the release of calcium ions, which are essential for cellular signaling and muscle contraction.

Golgi Apparatus

The Golgi apparatus (Fig. 4.12) is a membranous organelle consisting of stacked and flattened membranous sacs shaped like hollow plates, called cisternae, associated with swarms of tiny membranous vesicles. These cisternae are interconnected. It receives vesicles containing newly synthesized proteins from the ER. Then, various modifications are applied to proteins and lipids. Following this processing, the Golgi apparatus packages and sorts proteins and lipids into transport vesicles for

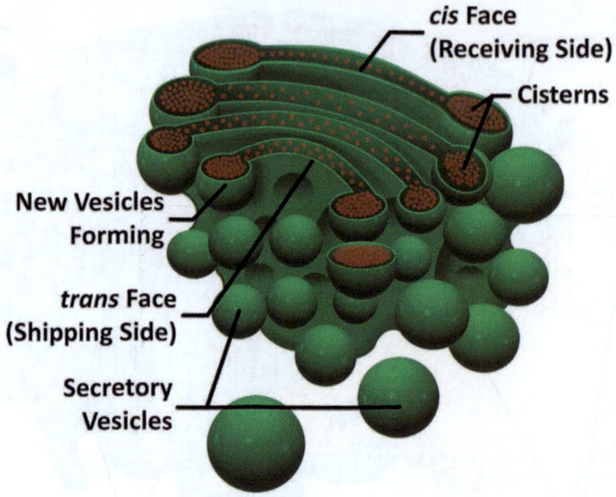

Fig. 4.12 Golgi apparatus: *Cis* face (receiving side) is the entry point where transport vesicles from the endoplasmic reticulum (ER) fuse with the Golgi apparatus, delivering their cargo. Cisterns are flattened, membrane-bound sacs where proteins and lipids undergo further modifications. These cisterns are organized in a stacked formation. *Trans* face (shipping side) is the exit point where processed proteins and lipids are packaged into new vesicles for transport to their final destinations within or outside the cell. Vesicles budding off from the cisterns carry modified molecules

delivery to their final destinations. The Golgi apparatus is the principal "traffic director" for cellular proteins. The Golgi apparatus forms lysosomes.

Lysosomes

Lysosomes (Fig. 4.13) are membrane-bound organelles found in the cytoplasm of eukaryotic cells. A single lipid bilayer membrane surrounds them. Lysosomes have an acidic interior, with a pH of about 4.5–5.0. Proton pumps in the lysosomal membrane maintain this low pH. The interior of lysosomes contains a variety of hydrolytic enzymes, including proteases (for protein digestion), nucleases (for nucleic acid digestion), lipases (for lipid digestion), and glycosidases (for carbohydrate digestion). Lysosomal enzymes can digest almost all kinds of biological molecules. Lysosomes are responsible for intracellular digestion, autophagy, allowing the cell to recycle damaged organelles and proteins, extracellular digestion (phagocytosis), cellular remodeling and differentiation, and programmed cell death (apoptosis) (Section 4.3.2).

Cytoskeleton

The cytoskeleton (Fig. 4.8), literally "cell skeleton," is a complex and dynamic network of protein filaments and tubules extending throughout eukaryotic cells' cytoplasm. This intricate system comprises three main types of protein filaments: microfilaments (actin filaments), intermediate filaments, and microtubules.

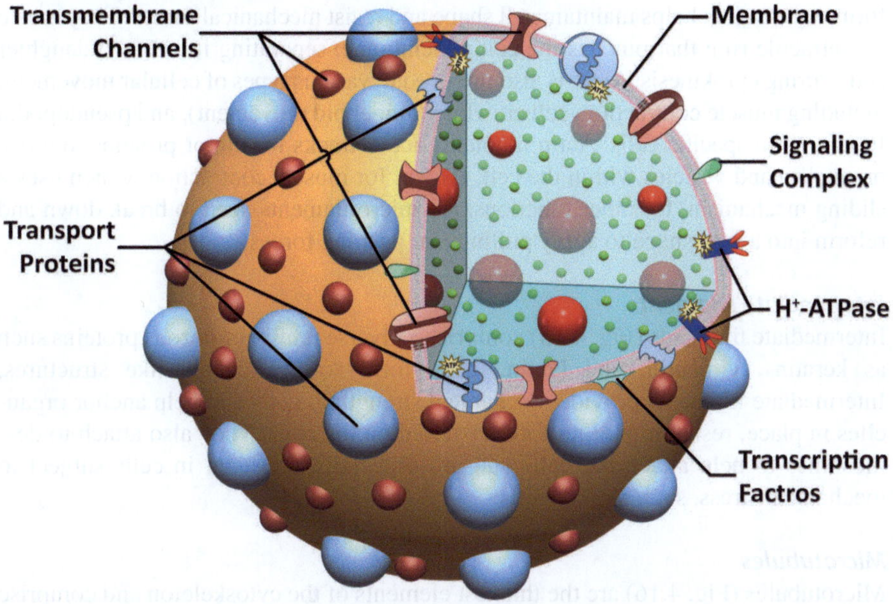

Fig. 4.13 Lysosome structure: Transmembrane channels are proteins that span the membrane and facilitate the transport of ions and molecules. H^+-ATPase is an enzyme that uses ATP to pump protons (H^+ ions) across the membrane into the lysosome, creating an acidic environment inside the lysosome necessary for its lytic function

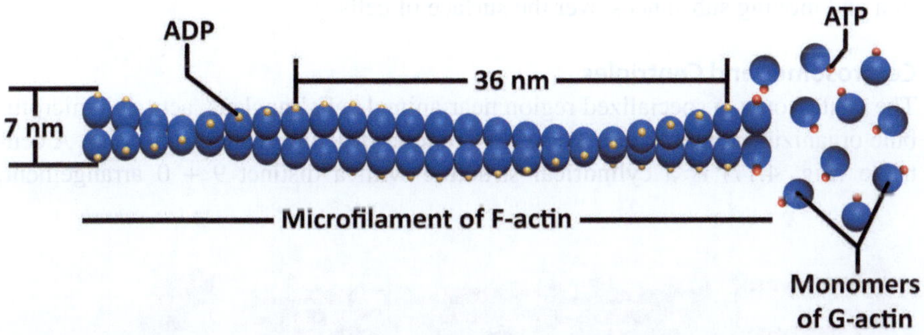

Fig. 4.14 Microfilament structure, showing its assembly mechanism, which uses the energy released from the ATP molecule attached to globular actin (G-actin) to form new filamentous actin (F-actin) extensions

Microfilaments (Actin Filaments)

Microfilaments (Fig. 4.14), the thinnest cytoskeleton elements, comprise globular actin protein monomers that polymerize to form long, thin filaments. These filaments have a double-stranded helical structure. Actin filaments provide structural support to the cell's shape. They are particularly abundant beneath the plasma membrane and form a cortex that helps maintain cell shape and resist mechanical forces. They create a contractile ring that pinches the cell's membrane, separating it into two daughter cells during cytokinesis. Actin is also involved in various types of cellular movement, including muscle contraction, cell crawling (amoeboid movement), and pseudopodia formation in specific cells. Actin filaments act as tracks for motor proteins to move organelles and vesicles within the cell. Except for muscle contraction, which uses a sliding mechanism, for other functions, the microfilaments need to break down and reform into a new shape to apply pulling and pushing forces.

Intermediate Filaments

Intermediate filaments (Fig. 4.15) comprise a diverse family of fibrous proteins such as keratins, vimentin, and lamin. They form solid and rope-like structures. Intermediate filaments provide mechanical strength to cells and help anchor organelles in place, resisting pulling forces exerted on the cells. They also attach to desmosomes to help attach the cells and are especially prevalent in cells subject to mechanical stress, such as skin cells.

Microtubules

Microtubules (Fig. 4.16) are the thickest elements of the cytoskeleton and comprise tubulin protein dimers, which assemble into long, hollow cylinders. Each microtubule is made of 13 parallel protofilaments. Microtubules help maintain cell shape and provide structural support. Microtubules form the mitotic spindle during cell division, segregating chromosomes into daughter cells. Microtubules are the core structural components of cilia and flagella, which are involved in cellular locomotion and moving substances over the surface of cells.

Centrosomes and Centrioles

The centrosome, a specialized region near animal cells' nucleus, acts as a microtubule organizing center. It consists of two unique structures called centrioles. A centriole (Fig. 4.17) is a cylindrical structure with a distinct 9 + 0 arrangement,

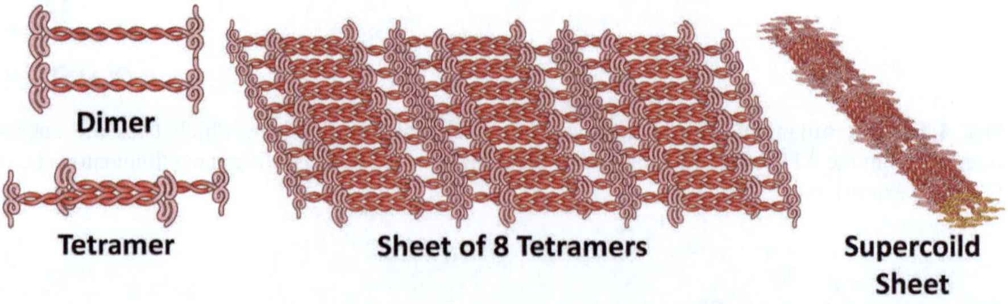

Fig. 4.15 Intermediate filament structure: A dimer is the basic building block of intermediate filaments, consisting of two polypeptide chains coiled together. The antiparallel arrangement of two dimers forms a tetramer. Eight tetramers align side by side to form a larger, sheet-like structure, and multiple sheets coiled together to form a dense and robust filament structure

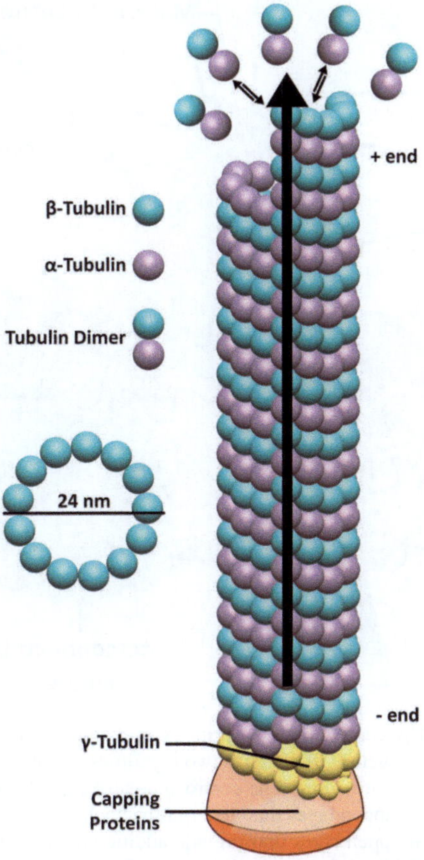

Fig. 4.16 Microtubule structure: α-Tubulin and β-tubulin are globular proteins forming heterodimers, microtubule building blocks. The heterodimers form a cylindrical structure with a diameter of 24 nm, composed of 13 protofilaments arranged in a hollow tube. The rapidly growing end of the microtubule where the addition of tubulin dimers occurs is called +end, and the more stable end, often anchored to microtubule-organizing centers (MTOCs) like the centrosome, is called -end. γ-tubulin and capping proteins, located at the microtubule-organizing center, stabilize the -end of the microtubule

comprising nine sets of triplet microtubules arranged in a ring (9 in the periphery and 0 in the center). This unique arrangement is crucial for its function. Each centriole has a proximal end, which is attached to the centrosome, and a distal end. Centrioles are vital not only to cell division but also to the formation of cilia and flagella. They duplicate during the cell cycle, ensuring each daughter cell receives a pair of centrioles. These centrioles then organize the formation of the mitotic spindle. Basal bodies derived from centrioles serve as the foundation for the formation of cilia and flagella. They anchor and organize the microtubules in these cellular appendages, allowing for their coordinated movement.

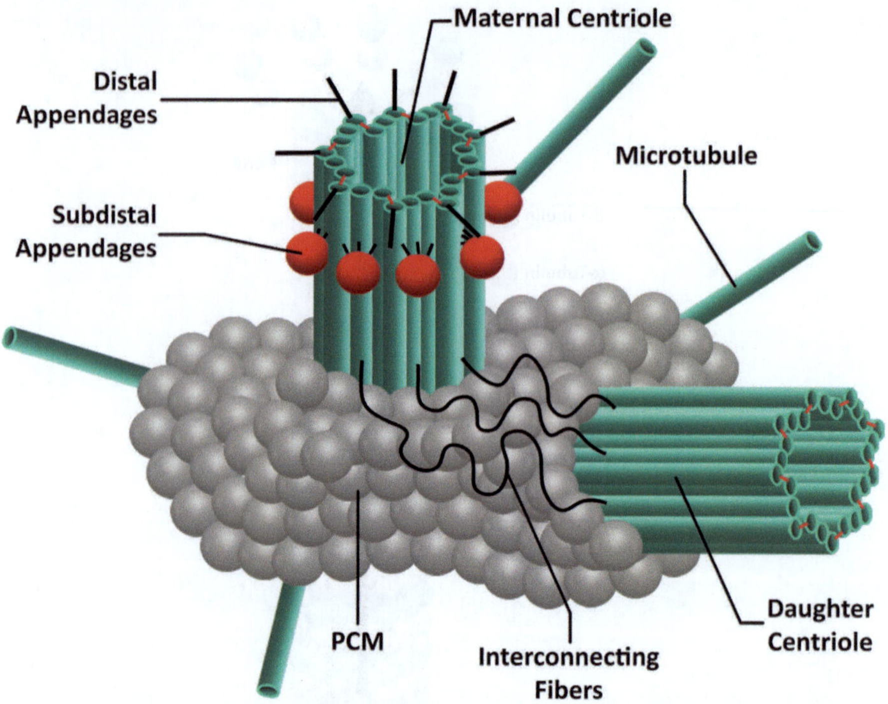

Fig. 4.17 Centrosome: The centrosome, the primary microtubule-organizing center (MTOC) in animal cells, is a complex structure. It contains two centrioles (maternal and daughter), which are the key components of the centrosome. The centrosome also includes microtubules, forming the microtubule network, distal appendages that assist in the anchoring of microtubules and cell surface interactions, subdistal appendages that provide additional anchoring sites for microtubules, pericentriolar material (PCM), a dense matrix surrounding the centrioles, rich in proteins that nucleate and anchor microtubules, and interconnecting fibers that link the two centrioles within the centrosome, facilitating their coordination and function

4.1.2.4 Cellular Appendages

Cilia, flagella, and microvilli are slender, hair-like cellular appendages in many eukaryotic organisms. Cilia and flagella serve as motion organelles, enabling cells to move through fluid environments or facilitating the movement of fluids past stationary cells. Microvilli increase the surface area of the cell surface.

Cilia

Cilia (Fig. 4.18) are motile cellular extensions located on the exposed surfaces of certain cells, typically in large numbers. Axoneme is the core structural feature of the cilium. It has a 9 + 2 arrangement, consisting of nine doublets of microtubules surrounding a central pair. Dynein arms are attached to the outer doublets of the microtubules. Dynein is a motor protein that uses energy from ATP to move along adjacent microtubules. The basal body is at the cilium's base, derived from the

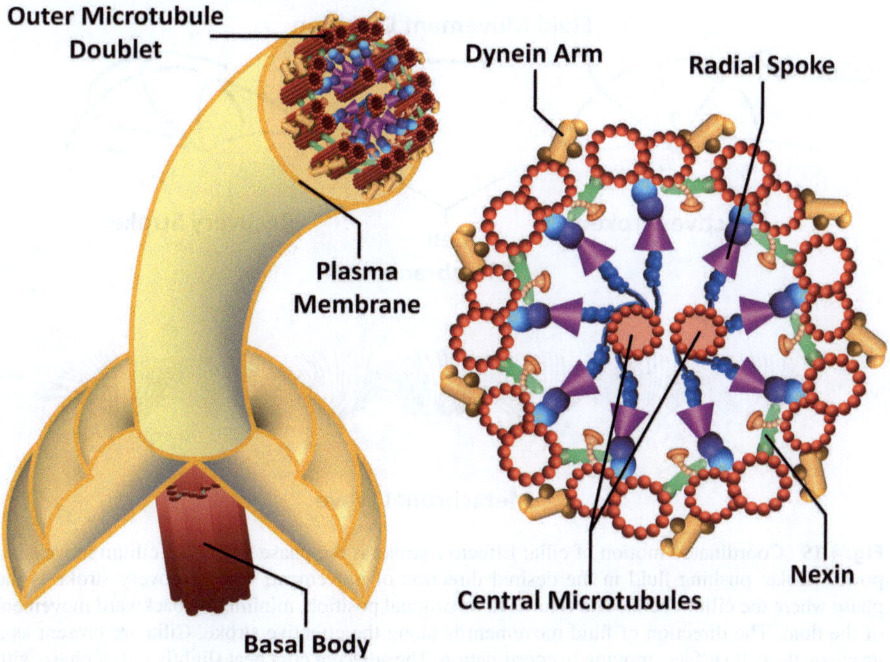

Fig. 4.18 Cilia and flagella structure: Nine pairs of microtubules arranged in a circular pattern around the periphery, and a pair of single microtubules located at the center of the structure form the basic structure of the cilium or flagellum. The dynein arm is a motor protein attached to the outer microtubule doublets, generating the sliding force that causes bending and movement. Radial spokes connect the outer microtubule doublets to the central pair, providing structural stability and aiding signal transmission. Nexin is a protein link that connects adjacent outer microtubule doublets, preventing them from sliding apart. The basal body is the base structure from which the cilium or flagellum originates, anchoring it to the cell and organizing microtubule assembly

centriole. The basal body anchors the cilium to the cell and plays a crucial role in its assembly and organization. The bending motion of cilia is achieved through a coordinated action of dynein arms. These motor proteins walk along adjacent microtubules, generating a sliding movement. This sliding causes the microtubules to bend, creating the characteristic waving motion of the cilium. The propulsive power stroke starts when the cilium is straight upward and causes the cilium to bend rapidly. When the cilium is fully bent, the recovery stroke gradually returns it to the upright position. This motion repeats 10–20 times each second (Fig. 4.19). With these two strokes, the cilium produces a pushing motion in a single direction in the fluid over the cell surface.

Flagella

Flagella are whip-like, thread-like appendages that serve primarily as organelles of locomotion. Their structure is similar to a cilium but much longer and usually present in a few numbers or even singles (Fig. 4.18). The only flagellated cell in the

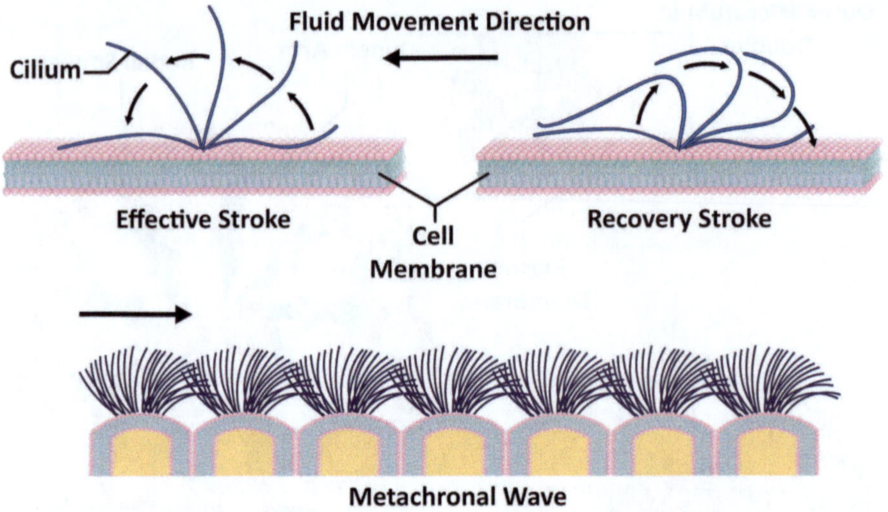

Fig. 4.19 Coordinated motion of cilia: Effective stroke is the phase where the cilium moves in a power stroke, pushing fluid in the desired direction of movement, while recovery stroke is the phase where the cilium bends and returns to its original position, minimizing backward movement of the fluid. The direction of fluid movement is along the effective stroke. Cilia are present as a group on the cell surface, moving in coordination. The adjacent cilia beat slightly out of phase with each other (metachronal), creating a smooth and continuous fluid movement

human body is a sperm, which has one propulsive flagellum, commonly called a tail. The primary function of flagella is to facilitate cellular movement.

Microvilli

Microvilli (Fig. 4.8) are tiny, finger-like projections found on the surface of certain types of eukaryotic cells. Microvilli primarily increase surface area for absorption and secretion. Microvilli are made up of actin filaments, which are thin protein fibers. These actin filaments are cross-linked and anchored to the cell's cytoskeleton. Microvilli are densely packed on the cell membrane, giving it a characteristic "brush border" appearance under a microscope.

4.1.2.5 Nucleus

The nucleus (Fig. 4.20) is a defining feature of eukaryotic cells, distinguishing them from prokaryotes. It is often called the "control center" of the cell due to its critical role in genetic information storage, regulation, and expression. The nucleus contains the instructions (as the genetic library) required to build nearly all the body's proteins. It also dictates the kinds and amounts of proteins to be synthesized at any time in response to the cell's requirements and the signals acting on it. DNA replication, a fundamental cellular reproduction and growth process, occurs within the nucleus. Additionally, the nucleus is the site of DNA repair mechanisms that rectify various types of DNA damage, ensuring the integrity of the genetic material. All body cells, except for mature red blood cells, have nuclei.

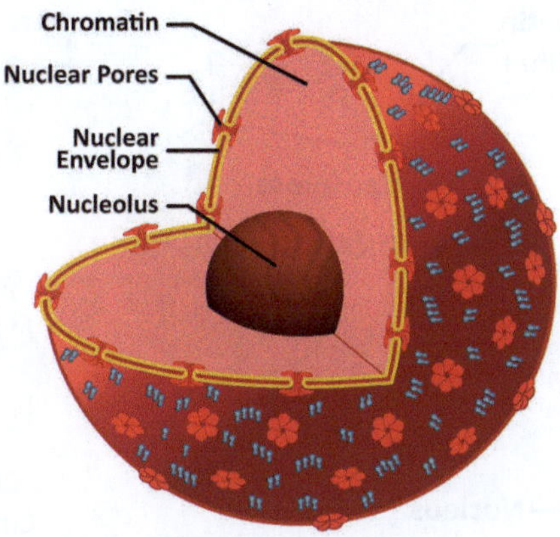

Fig. 4.20 The nucleus

The nucleus, averaging 5 μm in diameter, is larger than any cytoplasmic organelles. It is formed by three structures: the nuclear envelope, nucleolus, nucleoplasm, and chromatin.

Nuclear Envelope

A double-membraned nuclear envelope encloses the nucleus (Fig. 4.20). This system comprises an outer and inner membrane separated by a perinuclear space. The nuclear envelope is studded with nuclear pores, complex protein structures that facilitate the exchange of molecules like RNA, proteins, and ribosomal subunits between the nucleus and the cytoplasm. The outer nuclear membrane is continuous with the cytoplasm's rough ER and studded with ribosomes on its external face.

Nucleoplasm

The nucleoplasm is a semi-fluid, jellylike matrix within the nucleus. It contains various components, including chromatin (DNA and associated proteins), nucleolus, and soluble molecules.

Nucleoli

The nucleolus is a prominent, dark-staining spherical structure within the nucleus and is not enclosed by a membrane (Fig. 4.20). The nucleolus contains ribosomal proteins, ribosomal RNA (rRNA) molecules, and DNA that encodes rRNA. It is primarily responsible for the assembly of ribosomal subunits.

Chromatin

Chromatin is the material comprising the chromosomes, which consist of DNA and associated histone proteins (Page 72). It exists in two states: euchromatin (loosely packed, transcriptionally active) and heterochromatin (densely packed, transcriptionally inactive) (Fig. 4.21). The chromatin structure is dynamic and plays a vital role in gene expression.

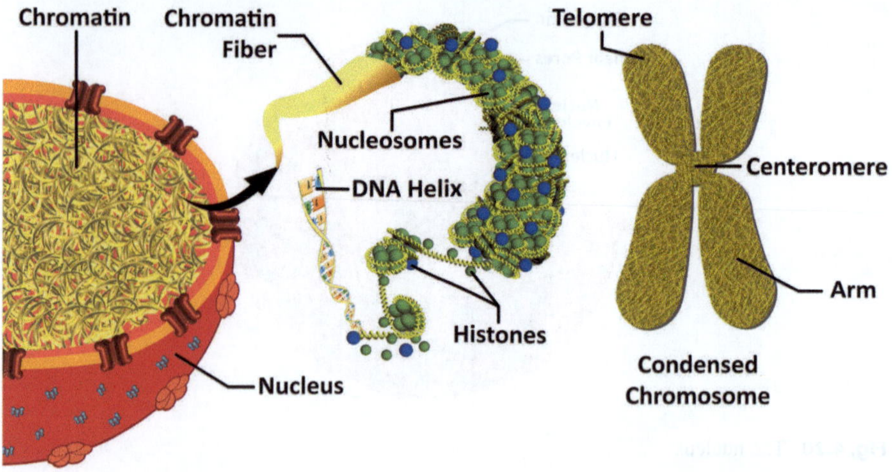

Fig. 4.21 Chromatin: the hierarchical organization of chromatin and the formation of condensed chromosomes

Chromatin is composed of approximately 30% DNA (the genetic material), 60% globular histone proteins (which package and regulate the DNA), and 10% RNA chains (which are newly formed). Histones provide a physical means for packing the very long DNA molecules in a compact, orderly way, but they also play an essential role in gene regulation.

4.1.3 Cell Cycle

The cell cycle (Fig. 4.22) is the series of changes a cell goes through from the time it is formed until it reproduces. The eukaryotic cell cycle is a highly regulated process that governs cell growth, replication, and division. It ensures the faithful transmission of genetic information from one generation of cells to the next. The cell cycle has two major periods: interphase and cell division.

4.1.3.1 Interphase

Interphase (Fig. 4.22) is the period from cell formation to cell division. It is the longest phase of the eukaryotic cell cycle, constituting about 90% of a cell's life. Interphase is a dynamic and essential period in the cell cycle. In addition to carrying on its life-sustaining activities, it prepares cells adequately for division. Interphase can be divided into three phases: G1, S, and G2.

G1 (Gap 1 Subphase)

The first gap phase, G1 (Fig. 4.22), marks the period after cell birth. During this phase, the cell is metabolically active, grows, and synthesizes RNA. This phase is the most variable in terms of length. For most of G1, virtually no activities directly related to cell division occur. As G1 ends, the centrioles start to replicate in preparation for cell division.

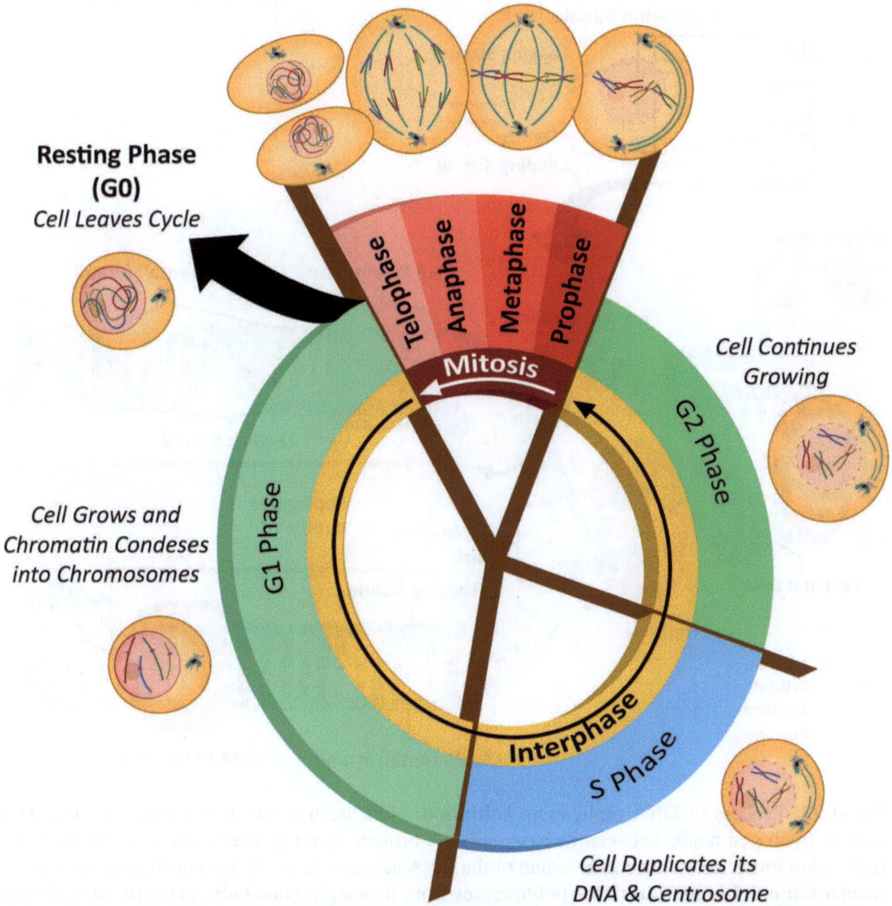

Fig. 4.22 The cell cycle, showing the stages of cell division (mitosis) and interphase, during which the cell prepares for division. Some cells exit the cell cycle and enter a quiescent state known as G0. In this phase, cells are not actively preparing to divide and can remain in this state for an extended period. Cells in the G0 phase carry out their normal functions without progressing through the cell cycle

S (Synthesis Subphase)

The synthesis phase, S (Fig. 4.22), is when DNA replication occurs. The cell duplicates its genetic material, ensuring each daughter cell receives identical genetic material. New histones are also made and assembled into chromatin.

DNA replication: DNA (deoxyribonucleic acid) is a double-stranded, helical molecule composed of nucleotide units. DNA replication (Fig. 4.23) follows a semiconservative model, meaning that each newly formed DNA molecule consists of one strand from the original DNA molecule (the template strand) and one newly synthesized strand. Since human DNA molecules are very long, replication begins at

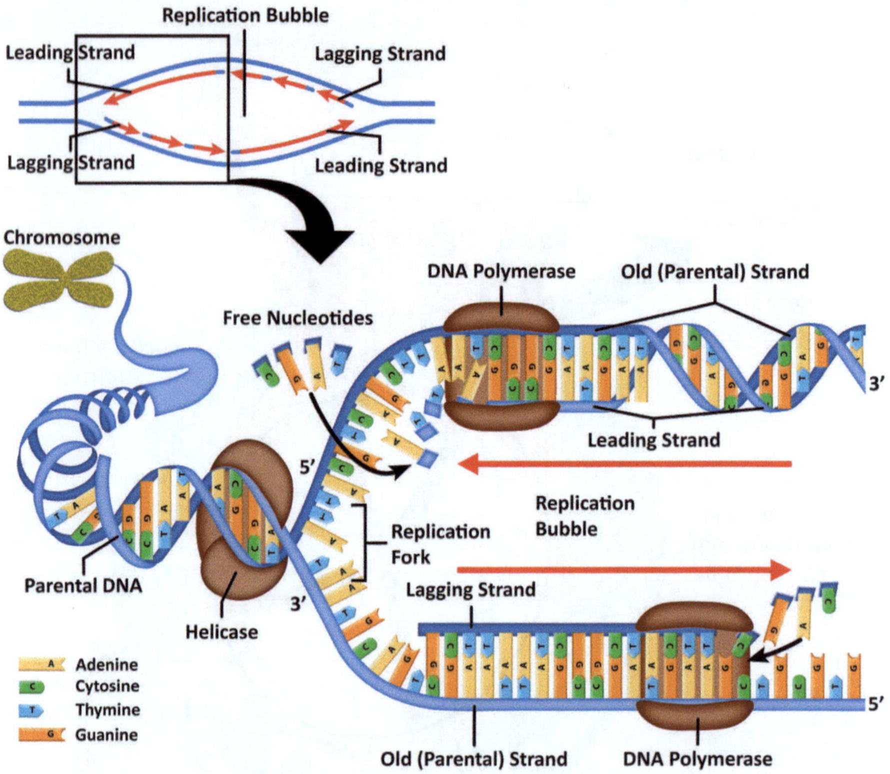

Fig. 4.23 Process of DNA replication: **Initiation:** Helicase unwinds the double-stranded DNA and the hydrogen bonds between its base pairs are broken, forming a replication bubble with two replication forks. Each nucleotide strand of the DNA acts as a template for constructing a complementary strand. **Elongation:** DNA polymerases work in one direction only, so the two new strands (leading and lagging) are synthesized in opposite directions. DNA polymerase synthesizes the leading strand continuously in the 5′ to 3′ direction, adding nucleotides complementary to the template strand. DNA polymerase synthesizes the lagging strand discontinuously in short segments called Okazaki fragments. Each fragment is initiated by an RNA primer (not shown in the image), and DNA polymerase adds nucleotides in the 5′ to 3′ direction. **Joining:** Okazaki fragments on the lagging strand are later joined together by DNA ligase (not shown in the image) to form a continuous strand. (**Parental DNA:** The double-stranded DNA molecule that is being replicated. **Replication bubble:** The localized region where the two strands of DNA have separated, allowing replication to occur. It contains two replication forks. **Replication fork:** The Y-shaped region where the DNA is split into two separate strands for replication to proceed. **Leading strand:** The strand of DNA that is synthesized continuously in the 5′ to 3′ direction. **Lagging strand:** The strand of DNA that is synthesized discontinuously in short segments called Okazaki fragments, also in the 5′ to 3′ direction. **Helicase:** The enzyme responsible for unwinding the double-stranded DNA, creating the replication fork by separating the two strands. **DNA polymerase:** The enzyme that adds nucleotides to the growing DNA strand during replication. There are two DNA polymerases shown, one for the leading strand and one for the lagging strand.)

several locations along the DNA length to increase the replication speed. Replication includes the following sequence of events:

- **Initiation:** Replication begins at specific DNA sites, known as replication origins. In eukaryotes, these origins are numerous and distributed across the genome. Each origin consists of a short sequence of DNA called the origin recognition complex (ORC), which recruits other proteins needed to initiate replication.
- **Unwinding:** Once the ORC is bound, it recruits a helicase enzyme. Helicases are responsible for unwinding the DNA double helix by breaking the hydrogen bonds between the base pairs. This creates a replication fork, separating the DNA strands and providing access to replication machinery.
- **Assembly:** With the DNA strands unwound and separated, the next step is assembly. DNA polymerases are enzymes responsible for catalyzing the formation of phosphodiester bonds between adjacent nucleotides, effectively synthesizing a new DNA strand. During and after replication, DNA polymerases have proofreading capabilities. They can detect and correct errors in base pairing.
- **Termination:** Replication forks continue moving along the DNA template until they converge (replication fork convergence), meaning the entire DNA molecule has been duplicated. Eukaryotic linear chromosomes possess particular sequences at their ends known as telomeres. These sequences do not code for any specific protein but help protect the ends of chromosomes from deterioration and ensure complete replication.

During replication, histone proteins, made in the cytoplasm and imported into the nucleus, are associated with the DNA, forming two new chromatin strands. Two identical chromatin strands remain attached by a button-like structure called a centromere (Fig. 4.21) until the cell enters the anaphase stage of mitotic cell division.

G2 (Gap 2 Subphase)

The second gap phase, G2 (Fig. 4.22), is brief and follows the S phase. During this time, the cell synthesizes proteins necessary for mitosis and undergoes final preparations for cell division. By the end of G2, centriole replication (begun in G1) is complete, signifying the beginning of cell division.

4.1.3.2 Division Phase

Cell division includes nuclear division, either mitosis (asexual division) or meiosis (sexual division), and cytokinesis (cytoplasmic division). Cell division is essential for body growth, tissue repair, and reproduction. Almost all cells in the body undergo continuous division to produce new identical cells to replace the old or worn-away cells. The cells of nervous, skeletal, and heart muscle cells are exceptionally unable to divide when fully mature, and repairs are made with fibrous tissue.

Mitosis

Mitosis is a process of nucleus division in which a parent cell divides into two identical daughter cells with the same number of chromosomes as the parent cell. Although mitosis is a continuous process, it can be considered in four stages: prophase, metaphase, anaphase, and telophase.

Prophase: Mitosis begins with prophase (Fig. 4.24), where chromatin condensation and nuclear envelope breakdown happen. During chromatin condensation, DNA coils tightly and forms visible chromosomes. The nuclear envelope disintegrates, allowing interaction between chromosomes and the mitotic spindle.

Metaphase: During metaphase (Fig. 4.24), the condensed chromosomes align at the cell's equator, the metaphase plate.

Anaphase: In anaphase (Fig. 4.24), sister chromatids (identical copies of a chromosome) separate and move toward opposite poles of the cell. This separation ensures that each daughter cell receives a complete set of chromosomes.

Telophase: Telophase (Fig. 4.24) marks the end of nuclear division. The nuclear envelope reforms around the two sets of chromosomes, beginning to decondense into chromatin.

Cytokinesis: Cytokinesis (Fig. 4.24) occurs concurrently with telophase or immediately after. A contractile ring made of actin filaments forms and pinches the cell membrane, dividing the cell into two daughter cells.

Meiosis

Meiosis is a specialized type of cell division that occurs in eukaryotic organisms. It produces gametes (sex cells) with half the parent cell's genetic material. This process is crucial for sexual reproduction as it ensures genetic diversity in offspring. Before meiosis happens, the cell undergoes interphase, which is the same as the interphase in mitosis and prepares the cell for division. The result of meiosis is the production of four daughter cells with n chromosomes from a parent cell with 2n chromosomes. Meiosis consists of two sequential divisions: meiosis I and meiosis II.

Meiosis I (Reductional Division) (Fig. 4.25)

Prophase I Meiosis I initiates with prophase I, during which the chromatin condenses into visible chromosomes. Homologous chromosomes, pairs containing similar genetic information but potentially different versions of genes (alleles), come together in a process known as synapsis. This forms structures called tetrads, which facilitate genetic recombination.

Metaphase I: The cell progresses to metaphase I following prophase I. Tetrads align along the cell's equatorial plane, the metaphase plate. Microtubules from opposite poles of the cell attach to each homologous chromosome, preparing for their separation.

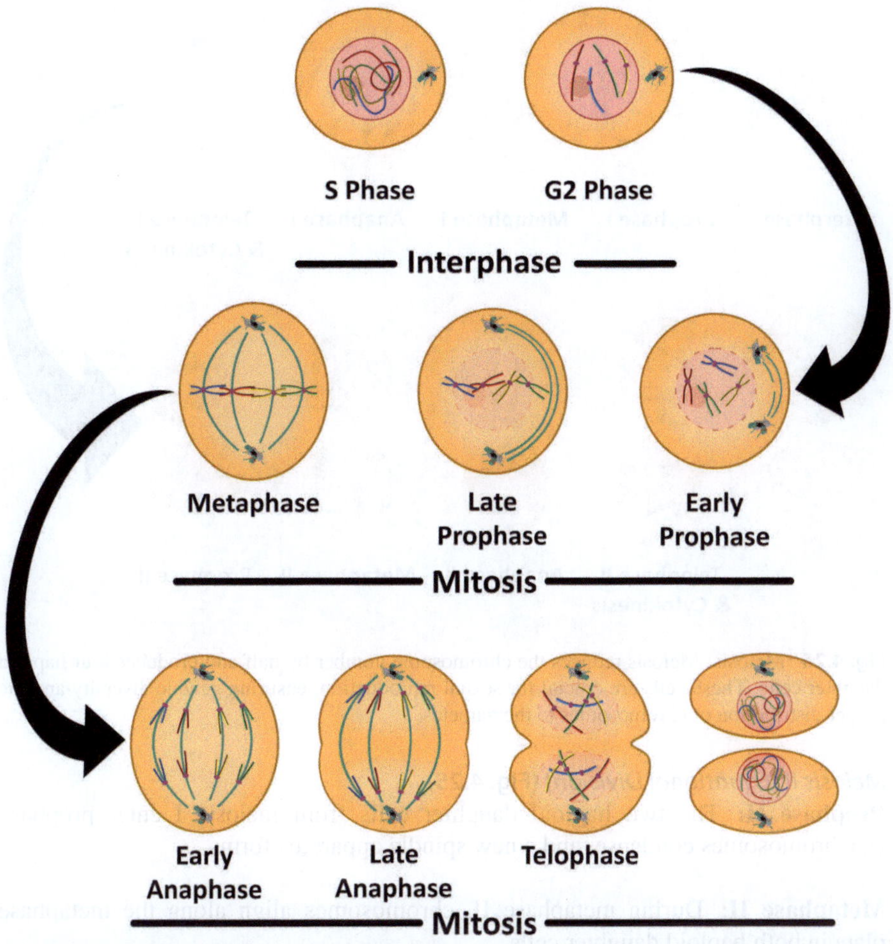

Fig. 4.24 Mitosis: mitosis ensures that genetic material is accurately and equally distributed to daughter cells, creating two daughter cells similar to the parent cell and maintaining the continuity and function of multicellular organisms

Anaphase I: During anaphase I, homologous chromosomes are pulled apart and move toward opposite poles of the cell. Unlike mitosis, in which sister chromatids separate, entire chromosomes separate, leading to the distribution of genetic variation.

Telophase I: In telophase I, the separated chromosomes reach their respective poles and begin to decondense, and the nuclear envelope may reform.

Cytokinesis I: After telophase, the cell undergoes cytokinesis, resulting in two daughter cells, each with a haploid set of chromosomes (half the number of chromosomes of the parent cells).

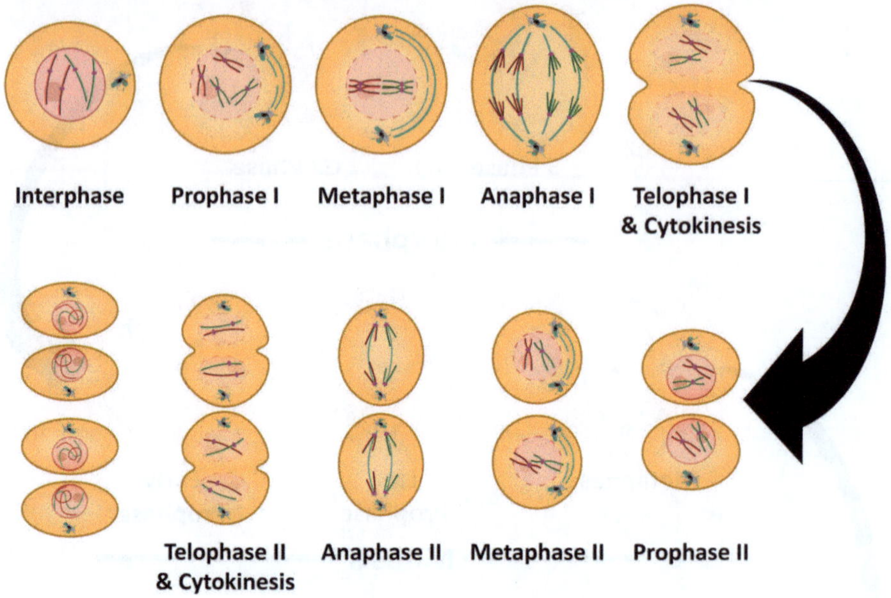

Fig. 4.25 Meiosis: Meiosis reduces the chromosome number by half and produces four haploid daughter cells. These cells are crucial for sexual reproduction, ensuring genetic diversity and the proper distribution of chromosomes to the gametes

***Meiosis II (Equational Division)* (Fig. 4.25)**

Prophase II: The two haploid daughter cells from meiosis I enter prophase II. Chromosomes condense, and a new spindle apparatus forms.

Metaphase II: During metaphase II, chromosomes align along the metaphase plate in both haploid daughter cells.

Anaphase II: Anaphase II is marked by separating sister chromatids. The chromatids, now individual chromosomes, are pulled toward opposite poles of the cell.

Telophase II: In telophase II, chromosomes arrive at opposite poles and decondense, and the nuclear envelopes reform around the chromosomes.

Cytokinesis II: Cytokinesis II then follows telophase II, resulting in four unique haploid cells, each with a distinct combination of genetic material.

4.1.3.3 Protein Synthesis

Protein synthesis is a fundamental cellular process that involves translating genetic information encoded in DNA molecules into functional proteins. The genetic code in DNA is formed by four nucleotide bases (A, G, T, and C). The sequence of these bases provides the genetic information. Each sequence of three bases, a triplet or a codon (Fig. 4.27), specifies a particular amino acid. (e.g., the DNA codon TAC specifies the amino acid methionine, and CTT specifies leucine) (Table 4.1). The sequence of codons in each gene tells exactly how a polypeptide will be made: It specifies the number, kinds, and order of amino acids needed to build a particular polypeptide.

Variations in the arrangement of A, T, C, and G allow our cells to make all the different kinds of proteins needed.

DNA, containing the genetic code, is confined inside the nucleus, but the ribosomes, which are the protein synthesis machinery, are located in the cytoplasm, outside the nucleus. To deliver the genetic information to ribosomes, the codes in DNA must be transcripted into an mRNA to pass the nucleus membrane pores. Protein synthesis occurs in several stages.

Transcription

Transcription (Fig. 4.26) involves the synthesis of messenger RNA (mRNA) from DNA. It begins with binding RNA polymerase, an enzyme, to the promoter region of a gene on the DNA. This binding signals the initiation of transcription, and the RNA polymerase unwinds the DNA helix. Then, the elongation of RNA happens. During elongation, RNA polymerase moves along the DNA template strand, synthesizing a complementary mRNA strand. The RNA strand grows in the 5′ to 3′ direction. Translation is terminated when the RNA polymerase encounters a termination signal, causing it to release both the newly synthesized mRNA and the DNA template strand. The mRNA molecule is then ready for processing and translation.

The completed mRNA molecule, which carries the genetic information from the DNA, is now ready for processing (in eukaryotes) or translation into a protein (in prokaryotes).

Table 4.1 Correlation of amino acids with their corresponding genetic codons

Amino acid	Codons	Amino acid	Codons
Alanine	GCA, GCC, GCG, GCU	**Lysine**	AAA, AAG
Arginine	AGA, AGG, CGA, CGC, CGG, CGU	**Methionine (also the start codon)**	AUG
Asparagine	AAC, AAU	**Phenylalanine**	UUC, UUU
Aspartic acid	GAC, GAU	**Proline**	CCA, CCC, CCG, CCU
Cysteine	UGC, UGU	**Serine**	UCA, UCC, UCG, UCU, AGC, AGU
Glutamic acid	GAA, GAG	**Threonine**	ACA, ACC, ACG, ACU
Glutamine	CAA, CAG	**Tryptophan**	UGG
Glycine	GGA, GGC, GGG, GGU	**Tyrosine**	UAC, UAU
Histidine	CAC, CAU	**Valine**	GUA, GUC, GUG, GUU
Isoleucine	AUA, AUC, AUU	**Stop Codon**	UAA, UAG, UGA
Leucine	UUA, UUG, CUA, CUC, CUG, CUU		

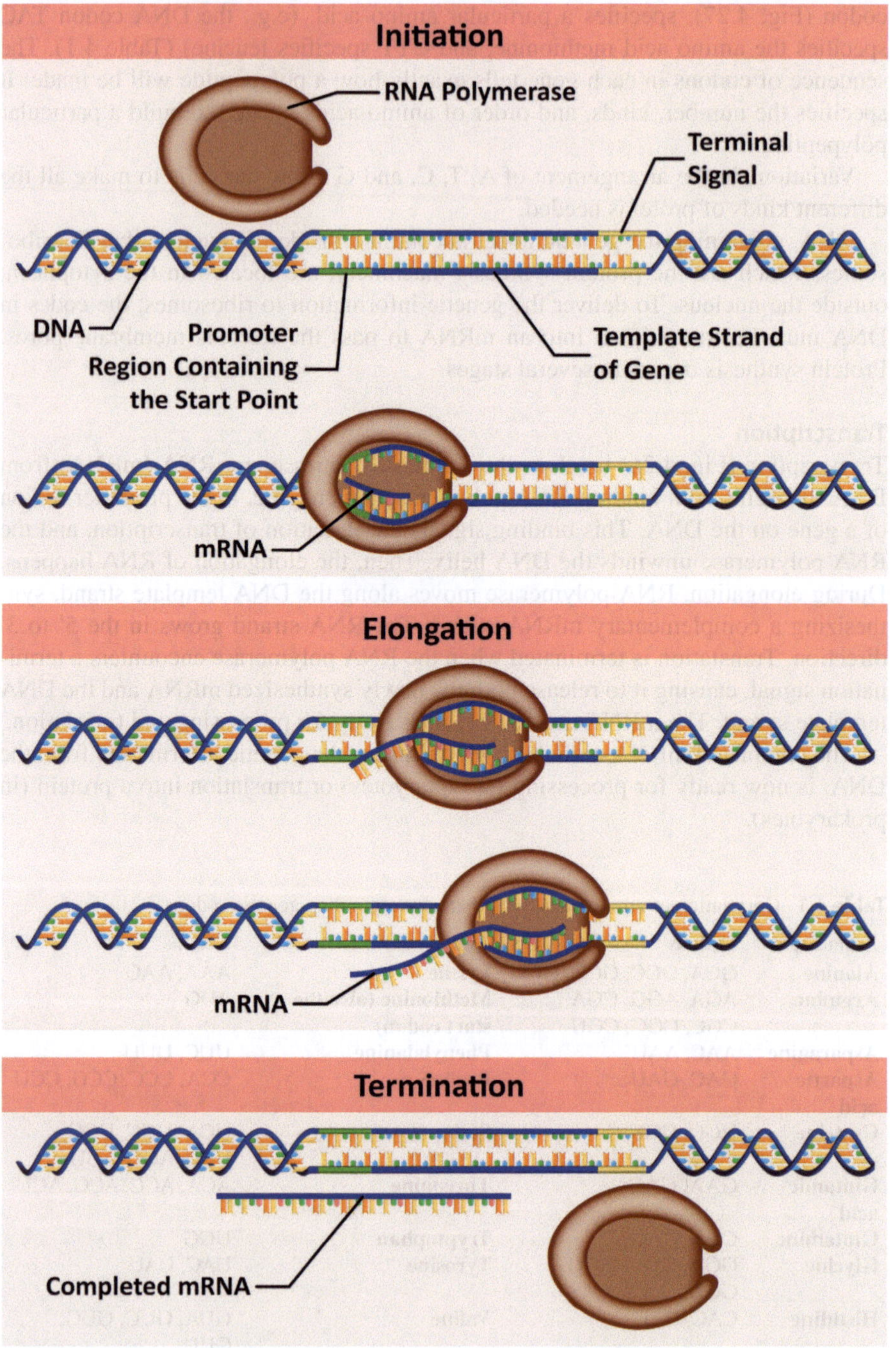

Fig. 4.26 Stages of transcription: **Initiation:** Transcription begins when RNA polymerase binds to the promoter region of the gene. This promoter contains specific DNA sequences that signal the start point of transcription. The promoter region is upstream of the gene and is essential for accurate transcription initiation. It contains specific nucleotide sequences recognized by RNA

mRNA Processing

Following transcription and before leaving the nucleus, the newly formed mRNA undergoes processing to be ready for the next step, translation. Before leaving the nucleus, the 5′ end of the mRNA molecule is modified with a 5′ cap. This cap, consisting of a modified guanine nucleotide, protects the mRNA from degradation and aids ribosome binding during translation. A poly-A tail is added at the 3′ end. This tail, composed of adenine nucleotides, protects the mRNA from degradation and assists in nuclear export and translation. Introns, noncoding regions within the pre-mRNA (containing nonsense or nontranslatable genetic codes), are removed through splicing. The splicing leaves only the exons (translatable genetic codes) containing the protein-coding information. This spliced mRNA is then ready for export to the cytoplasm.

Translation

Translation (Fig. 4.27) begins with binding the small ribosomal subunit to the mRNA molecule, followed by the attachment of an initiator tRNA carrying the amino acid methionine (or formylmethionine in bacteria). The large ribosomal subunit then joins, forming a complete ribosome. Then, the polypeptide chain starts to form, which is called elongation. During elongation, the ribosome moves along the mRNA, reading the codons (three-nucleotide sequences) and matching them with complementary tRNAs carrying the corresponding amino acids. Peptide bonds form between adjacent amino acids, resulting in the growing polypeptide chain. Translation is terminated when a stop codon is encountered on the mRNA. Release factors bind to the ribosome, causing it to dissociate into its subunits and releasing the completed polypeptide chain.

Post-translation Modifications

After translation, the polypeptide chain may undergo various modifications, including folding into its final three- dimensional structure and adding functional groups or other molecules, such as lipids or carbohydrates.

Fig. 4.26 (continued) polymerase and other transcription factors. RNA polymerase unwinds a short segment of the DNA double helix, exposing the template strand of the gene. RNA polymerase synthesizes mRNA by adding RNA nucleotides complementary to the DNA template strand. The synthesis occurs in the 5′ to 3′ direction, meaning nucleotides are added to the 3′ end of the growing mRNA strand. **Elongation:** During elongation, RNA polymerase moves along the DNA template strand, unwinding the DNA ahead of it and rewinding it behind. The enzyme adds RNA nucleotides to the 3′ end of the mRNA strand, complementary to the DNA template strand. As RNA polymerase progresses, the mRNA strand elongates. **Termination:** Transcription continues until RNA polymerase encounters a termination signal in the DNA sequence. This signal indicates the end of the gene. Once the termination signal is reached, RNA polymerase releases the newly synthesized mRNA strand and detaches from the DNA template

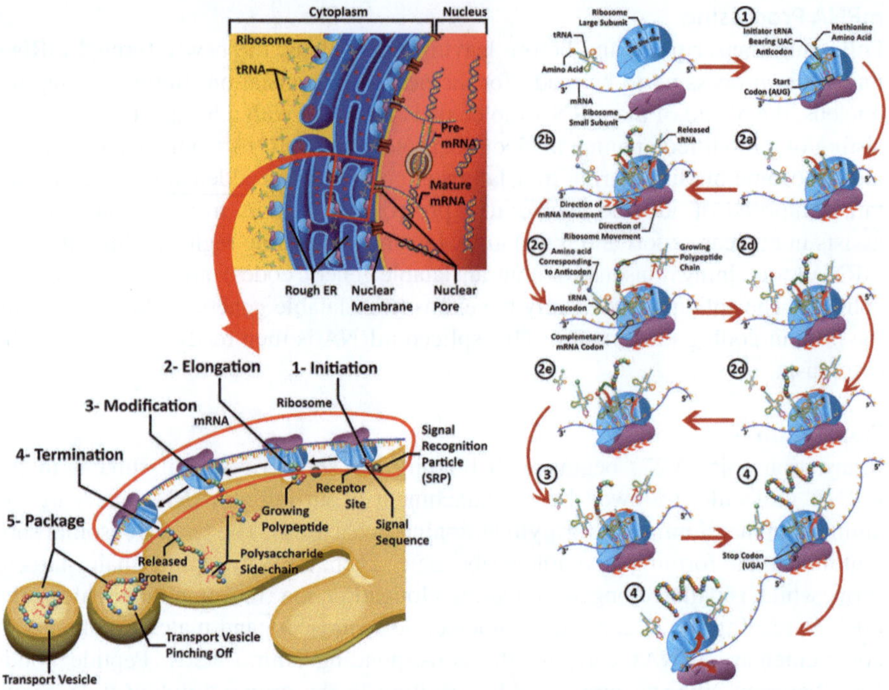

Fig. 4.27 Stages of translation: Protein synthesis needs four components: a ribosome large subunit, a ribosome small subunit, the mRNA, and tRNA carrying amino acid. **1. Initiation:** The small ribosomal subunit binds to the mRNA at the start codon (AUG), which codes for amino acid methionine. The initiator tRNA molecule, carrying the methionine amino acid, also binds to the start codon. The large ribosomal subunit joins the small subunit to form a functional ribosome. The SRP recognizes and targets specific signal sequences on newly synthesized polypeptide chains translated by ribosomes in the cytoplasm. These signal sequences are short amino acid sequences, usually found at the N-terminus of the polypeptide. They target the protein for translocation into or across the ER membrane. The SRP directs the mRNA- ribosome complex to the rough ER. There, the SRP binds to a receptor site. **2. Elongation:** During elongation, the ribosome moves along the mRNA in a 5′ to 3′ direction, reading each codon and bringing in the appropriate aminoacyl- tRNA molecule. The ribosome catalyzes the formation of peptide bonds between adjacent amino acids, forming a growing polypeptide chain. As the ribosome advances along the mRNA, it shifts one codon at a time, moving from the A site (aminoacyl site) to the P site (peptidyl site) to the E site (exit site). Once attached to the ER, the SRP is released, and the growing polypeptide snakes through the ER membrane pores into the cistern. **2a. Codon recognition:** As the ribosome advances, it encounters the next mRNA codon in the ribosomal A site (aminoacyl site). The anticodon of a transfer RNA (tRNA) molecule, carrying the corresponding amino acid, base-pairs with the mRNA codon via complementary base pairing. This ensures the correct amino acid is added to the growing polypeptide chain. **2b. Peptide bond formation:** Once the tRNA molecule with the correct anticodon is in the A site, the ribosome catalyzes the formation of a peptide bond between the amino acid carried by the tRNA in the A site and the amino acid on the tRNA in the P site (peptidyl site). A ribozyme within the ribosome, known as peptidyl transferase, carries this bond formation. It joins the carboxyl group of the amino acid in the P site with the amino group of the amino acid in the A site, forming a peptide bond. This step transfers the growing polypeptide chain from the P site's tRNA to the A site's tRNA. **2c. Entry of a new tRNA:** With the ribosome's new position, the A site is empty and ready to accept a new tRNA molecule. The process repeats as the

4.2 Part 2: Prokaryotic Cells

Prokaryotic cells represent some of Earth's oldest and most abundant life forms. They are fundamental to the planet's ecosystems and are crucial in various biological processes. All prokaryotes are single-celled organisms, and all are bacteria; hence, in the rest of this chapter, we call all prokaryotic cells bacteria.

4.2.1 Size, Shape, and Arrangement

Bacteria are incredibly diverse in size, shape, and arrangement, and these variations play crucial roles in their ecological niches and biological functions.

4.2.1.1 Size

Bacterial cells are typically much smaller than eukaryotic cells. Their size can range from about 0.5 to 5 micrometers (μm) in diameter. This small size starkly contrasts eukaryotic cells, which are generally larger, ranging from 10 to 100 μm in diameter. However, some spiral bacteria are long and have a length of about 60 μm. Since the bacteria have no intracellular route for nutrients, their size must remain small to increase the surface-to-volume ratio (to absorb more nutrients through the surface) and to keep all the internal structures easily and quickly accessible to the nutrients.

4.2.1.2 Shape

Typically, bacteria appear in three basic shapes: spherical (coccus), rodlike (bacillus), and spiral (Fig. 4.28). Some bacteria do not have a fixed shape and can change depending on environmental conditions, so they are called pleomorphic bacteria.

- Coccus (plural: cocci): These are spherical or roughly spherical bacteria. They can occur as single cells, pairs (diplococci), chains (streptococci), or clusters (staphylococci).
- Bacillus (plural: bacilli): These are rod-shaped bacteria. They can occur singly, in pairs (diplobacilli), or in chains (streptobacilli).

Fig. 4.27 (continued) ribosome recognizes the next mRNA codon and a new tRNA carrying the appropriate amino acid binds to the A site. **2d. Translocation:** After the peptide bond formation, the ribosome moves one codon along the mRNA molecule in the 5' to 3′ direction. The tRNA in the P site, which no longer carries the growing polypeptide chain, is moved to the E site (exit site) and is released from the ribosome. The tRNA in the A site, now carrying the growing polypeptide chain, is moved to the P site. Steps 2a–2d are repeated iteratively as the ribosome continues to move along the mRNA, adding amino acids to the growing polypeptide chain. **3. Termination:** Translation continues until the ribosome encounters a stop codon (UAA, UAG, or UGA) on the mRNA. There are no tRNA molecules with anticodons complementary to stop codons. Instead, release factors bind to the A site of the ribosome, causing the release of the completed polypeptide chain. The ribosomal subunits dissociate, and the mRNA molecule is released

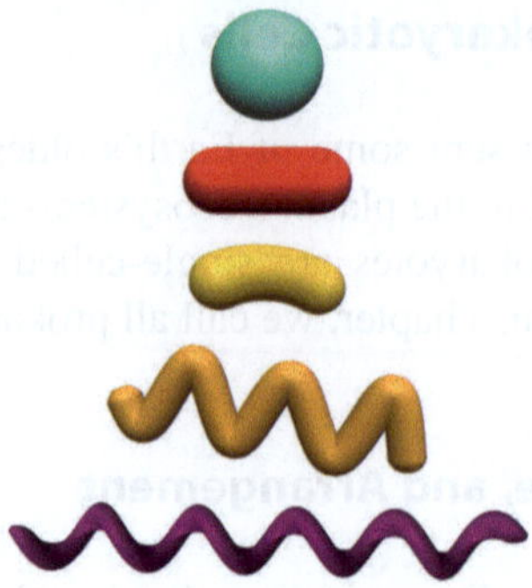

Fig. 4.28 Bacteria shapes

- Spiral bacteria have long, curved shapes and appear as:
  - Spirillum (plural: spirilla): these are spiral-shaped bacteria with a rigid helical structure.
  - Spirochete: these are spiral-shaped bacteria with a flexible, corkscrew-like structure.
  - Vibrio: these are curved rod-shaped bacteria that resemble a comma.

4.2.1.3 Arrangement

Bacteria can exist in various arrangements based on how they divide and adhere to each other (Fig. 4.29):

- **Single:** Individual bacterial cells exist independently, with no particular attachment to one another.
- **Diplo:** Bacteria appear in pairs after division. For example, cocci, like *Neisseria gonorrhoeae*, are often found in diplo-arrangements.
- **Strepto:** Bacteria form chains due to repeated divisions in one plane. This arrangement is commonly seen in cocci, such as *Streptococcus pneumoniae*.
- **Staphylo:** Bacteria form clusters resembling a bunch of grapes. This arrangement is typical of cocci like *Staphylococcus aureus*.
- **Tetrads:** Bacteria divide in two planes, creating groups of four cells in one plane. This arrangement is characteristic of cocci, such as *Micrococcus* species.
- **Sarcinae:** Bacteria divide into three planes, forming groups of eight cells. This pattern is less common and is seen in cocci, like *Sarcina ventriculi*.
- **Palisade:** Bacilli arrange themselves side by side, like a row of matchsticks. This arrangement is seen in bacteria like *Corynebacterium diphtheriae*.

It is important to note that bacteria are incredibly adaptable. Their size, shape, and arrangement can be influenced by environmental conditions such as temperature, pH, nutrient availability, and more. This adaptability is key to their success as a diverse and widespread group of microorganisms.

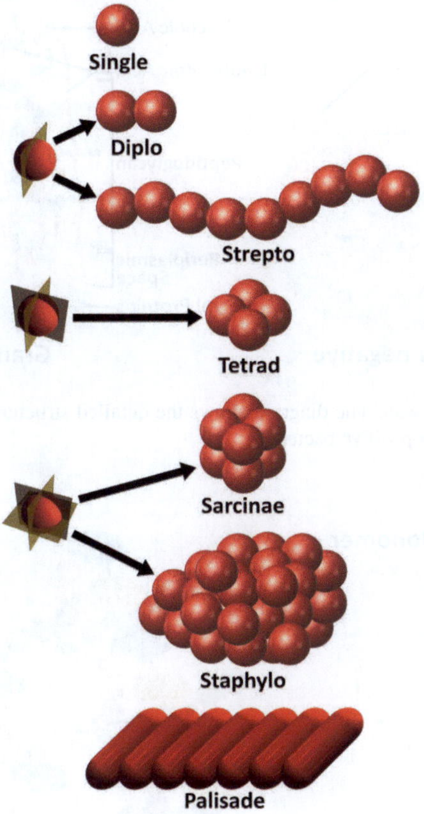

Fig. 4.29 Bacterial arrangement

4.2.2 Bacterial Structure

4.2.2.1 Cell Wall

The bacterial cell wall (Fig. 4.30) is a defining feature of prokaryotic cells. It provides structural integrity and protection from osmotic stress and is vital in bacterial physiology, pathogenicity, and antibiotic resistance. Its composition and structure vary between different bacterial species.

Peptidoglycan

The main component of bacterial cell walls is peptidoglycan, also known as murein. Peptidoglycan consists of linear chains of alternating N-acetylglucosamine (NAG) and N-acetylmuramic acid (NAM) residues, connected by β-1,4- glycosidic bonds and cross-linked by short peptides (Fig. 4.31). This structure forms a mesh-like framework around the cell.

The bacterial cell wall's composition varies significantly between Gram-positive and Gram-negative bacteria (Fig. 4.30).

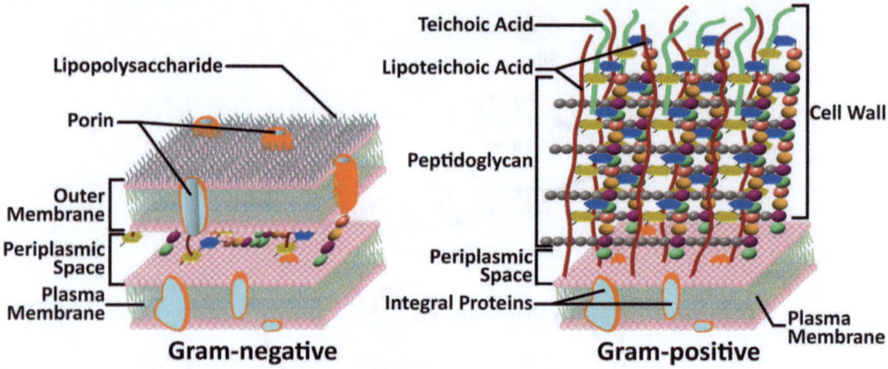

Fig. 4.30 Bacterial cell wall. The diagram shows the detailed structure and differences between Gram-negative and Gram-positive bacteria

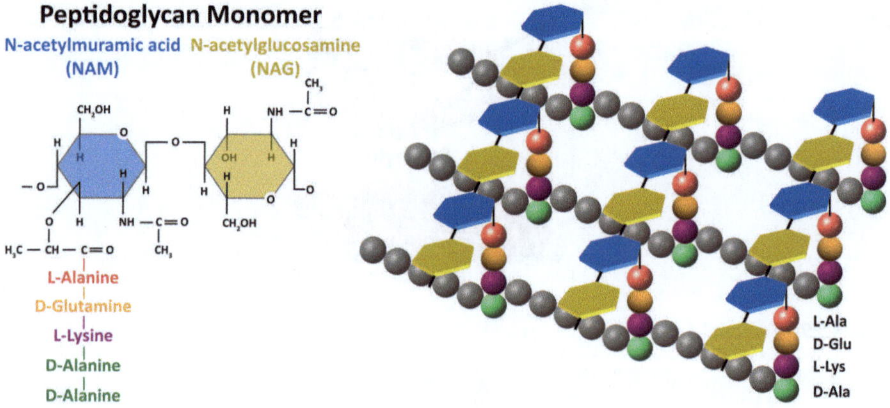

Fig. 4.31 Peptidoglycan structure: Peptidoglycan provides structural strength to the cell wall and helps maintain the cell's shape, countering osmotic pressure. Peptidoglycan is essential for the survival of bacteria, providing a robust framework that protects the cell from lysis

In Gram-positive bacteria, the cell wall is thick and consists of multiple layers of peptidoglycan. This dense structure contributes to the characteristic purple staining of these bacteria in Gram staining. Many Gram-positive bacteria have teichoic acids integrated into their cell walls. These are polymers of glycerol or ribitol phosphate that extend from the cell surface, contributing to the overall charge and stability of the cell wall.

In Gram-negative bacteria, the cell wall consists of a thin peptidoglycan layer surrounded by an outer membrane. This thin peptidoglycan layer results in the characteristic red staining of these bacteria in Gram staining.

Outer Membrane

The outer membrane is a unique and crucial component of Gram-negative bacteria (Fig. 4.30). It is an additional protective layer surrounding the peptidoglycan layer attached to it and is absent in Gram-positive bacteria. It comprises lipids, lipoproteins, and complex molecules known as lipopolysaccharides (LPS). The lipoproteins are embedded in the outer membrane and covalently bonded to the peptidoglycan. The outer membrane acts as a coarse sieve, having little control over many molecules except for some proteins. The outer surface of the outer membrane has surface antigens and receptors. The outer membrane is a barrier against many antibiotics, including some beta-lactam antibiotics, which cannot penetrate this layer.

Lipid Bilayer

The outer membrane primarily comprises lipids, including phospholipids, lipopolysaccharides, and lipoproteins. Unlike the inner membrane, which is mainly phospholipids, the outer membrane's lipid composition is asymmetric (Fig. 4.30). The outer leaflet of the outer membrane primarily contains LPS, while the inner leaflet is mainly composed of phospholipids. This asymmetry provides unique properties to the outer membrane, including its impermeability to many small hydrophobic molecules.

Lipopolysaccharides

Lipopolysaccharide (LPS) (Fig. 4.30), also called endotoxin, has several components:

- Lipid A is the innermost portion of LPS and anchors the LPS in the outer membrane. It is composed of a phosphorylated glucosamine disaccharide attached to long-chain fatty acids. Lipid A is an endotoxin and a potent activator of the host immune response.
- Core oligosaccharide consists of short sugar chains that connect lipid A to the O-antigen. It provides stability and structure to the LPS molecule.
- O-antigen is the outermost portion of LPS, which consists of a repetitive sugar polymer. It varies between bacterial species and provides antigenic diversity. It is an essential virulence factor for many pathogenic bacteria and helps identify them.

Porins

Porin proteins are embedded in the outer membrane. These form channels that allow the passive diffusion of small hydrophilic molecules, such as ions and nutrients, across the outer membrane. Porins can be selective in the types of molecules they allow to pass, contributing to the outer membrane's semi-permeable nature.

Lipoproteins

Lipoproteins are proteins covalently linked to lipids. They anchor the outer membrane to the underlying peptidoglycan layer and inner membrane, providing structural stability.

Periplasmic Space

Periplasmic space is a gap between the inner (cell) membrane and the cell wall and is a very active area of cell metabolism. It is primarily seen in Gram-negative bacteria but also as a narrow space in Gram-positive bacteria (Fig. 4.30). The equivalent of periplasmic space in most Gram-positive bacteria is called periplasms—not periplasmic spaces—where metabolic digestion occurs, and new cell wall peptidoglycan is attached. Thus, the periplasm in Gram-positive bacteria is part of the cell wall.

The periplasmic space contains a diverse array of molecules, including:

- **Peptidoglycan:** this mesh-like polymer provides structural support to the bacterial cell envelope.
- **Proteins:** enzymes, transport proteins, and other functional proteins are found in the periplasmic space and carry out various cellular processes.
- **Small molecules:** these include nutrients, metabolites, ions, and signaling molecules.
- **Cellular machinery:** the periplasmic space can harbor components of the bacterial flagellum, pili, and other structures.

4.2.2.2 Cell Membrane

The inner cell membrane (cytoplasmic or plasma membrane) (Fig. 4.30) is a vital component of bacterial cells. It is a semipermeable barrier that separates the cytoplasm from the external environment. It is composed of:

- **Phospholipid bilayer:** The primary structural element of the inner membrane is a phospholipid bilayer. The bipolar phospholipids arrange themselves into a bilayer, with the hydrophilic heads facing outward toward the aqueous environment and the hydrophobic tails aligning in the interior of the membrane. The specific types of phospholipids present in the membrane can vary among different bacterial species. Common types include phosphatidylethanolamine (PE), phosphatidylglycerol (PG), and cardiolipin (CL).
- **Proteins:** The inner membrane is studded with proteins embedded within or associated with its surface. These proteins have diverse functions, including transporting molecules across the membrane, enzymatic activities, signal transduction, and structural support.

4.2.2.3 Cytoplasm

The cytoplasm (Fig. 4.32) is the gel-like, semi-fluid substance that fills the interior of bacterial cells. The inner cell membrane surrounds it and is the site of numerous cellular activities. Water is the primary component of the cytoplasm, making up about 70–80% of its volume. It also contains organic molecules (including carbohydrates, lipids, amino acids, nucleotides, and metabolites) and inorganic ions [including potassium (K^+), sodium (Na^+), calcium (Ca^{2+}), and magnesium (Mg^{2+})]. The cytoplasm is the primary site for most of the metabolic activities in bacteria. It also houses the bacterial metabolic enzymes and organelles, including ribosomes, DNA, and RNA.

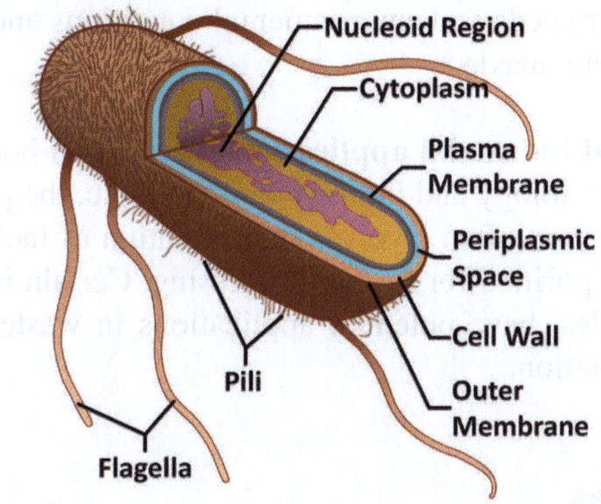

Fig. 4.32 Bacterial cell structure

4.2.2.4 Ribosomes

Bacterial ribosomes are complex molecular structures that work as machinery for protein synthesis. The bacterial ribosomes are formed by components (proteins and tRNA) similar to the eukaryotic ribosomes (Page 61) and consist of two subunits. However, the bacterial ribosomes have different proteins and tRNAs and are smaller in size than the eukaryotic ribosomes, having a sedimentation rate of 70S, compared to 80S for eukaryotic ribosomes. The sedimentation rate of the small subunit is 30S, and the large subunit is 50S (compared to 40S and 60S for eukaryotic ribosomes). (The "S" stands for Svedberg units, a measure of sedimentation rate used to describe the size of biological molecules in a centrifugal field [Page 62].) This difference in the size and the components between prokaryotic and eukaryotic ribosomes enables us to use some antibiotics targeting the bacterial ribosomes while sparing the eukaryotic ones.

4.2.2.5 Nucleoid (Nuclear Region)

Bacteria do not possess a true nucleus, as found in eukaryotic cells. Instead, they have a nucleoid or nuclear region, which serves as the bacterial cell's central genetic repository. The nucleoid is located near the cell center, but its precise positioning can vary based on the species and cell cycle stage (Fig. 4.32). It is an irregularly shaped, dense cell region where genetic material is concentrated. The nucleoid comprises deoxyribonucleic acid (DNA) and associated proteins. The DNA in the nucleoid forms a single, continuous circular chromosome in most bacteria. Nucleoid-associated proteins (NAPs) are small, positively charged proteins that bind to DNA, helping to condense and organize it within the nucleoid. Some bacteria contain smaller circular molecules of DNA called plasmids, which provide supplemental genetic information for the bacteria.

4.2.2.6 Inclusions

Inclusion bodies are dense, granular aggregates of macromolecules such as proteins, lipids, or minerals and serve various storage functions. Inclusion bodies form in response to cellular needs and environmental conditions and are degraded when the stored molecules are needed.

Biotechnological and industrial applications: Inclusion bodies have significant applications in biotechnology and industry. For example, the production of recombinant proteins in bacteria often involves the formation of inclusion bodies, which can be harvested and purified for further processing. Certain inclusion bodies, like polyphosphate granules, have potential applications in wastewater treatment and environmental remediation.

4.2.3 Endospores

Endospores are highly resistant, dormant structures formed by some bacteria in response to adverse environmental conditions. They allow the bacteria to survive extreme conditions that would typically be lethal to the vegetative form of the cell. Endospores can endure environmental conditions for extended periods, with some surviving over 10,000 years. For instance, spores from bacteria can remain dormant for at least 10,000 years while being frozen at a temperature of 14 °C within ice layers as deep as 430 meters. Furthermore, certain spores can withstand prolonged exposure to boiling temperatures.

Endospores can be easily dispersed through the environment by air, water, or animal vectors. This allows the bacteria to colonize new habitats when conditions improve, as in some diseases like tetanus. Endospores can undergo germination when environmental conditions become more favorable (such as the return of nutrients). This process involves the activation of metabolic pathways, rehydrating the core, and resuming active growth as a vegetative cell.

4.2.3.1 Structure

Structurally, an endospore consists of (Fig. 4.33):

- **Core:** The core is the central and essential part of the endospore. It contains the bacterial chromosome, ribosomes, and enzymes required for germination (resuming active growth). The core of an endospore contains high levels of dipicolinic acid (DPA), which plays a crucial role in heat resistance. DPA binds to calcium ions and helps stabilize the core's structure. Endospores have very low water content, which contributes to their resistance to heat, desiccation, and other harsh conditions.

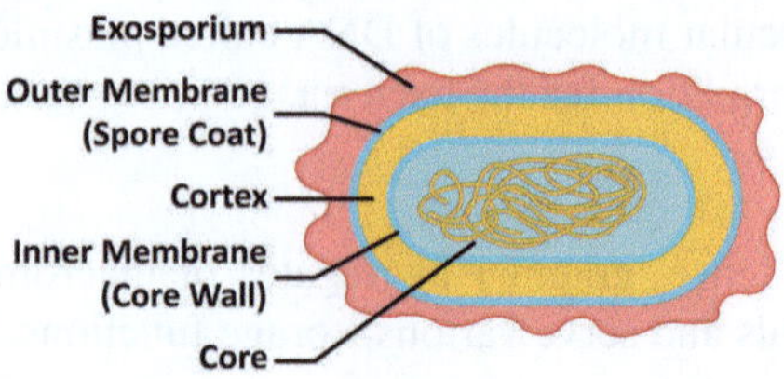

Fig. 4.33 Endospore structure

- **Inner membrane:** Surrounding the core is the inner membrane (core wall), a thin layer acting as a defense against external factors.
- **Cortex:** Outside the core wall lies the cortex, a thick, protective layer composed primarily of peptidoglycan. This layer is crucial in maintaining spore dormancy and resistance, as it helps to dehydrate the core and contributes to heat resistance.
- **Outer membrane:** Encapsulating the cortex is the outer membrane (spore coat), a multi-layered shell made of various proteins. This coat is highly resistant to chemicals, enzymes, and physical damage, safeguarding the spore's core.
- **Exosporium:** Some endospores have an additional outermost layer called the exosporium. This thin, delicate layer may contain proteins or lipids.

4.2.3.2 Endospore Formation (Sporulation)

Endospore formation (Fig. 4.34) is initiated in response to adverse environmental conditions, such as nutrient limitation, high temperatures, or exposure to toxins.

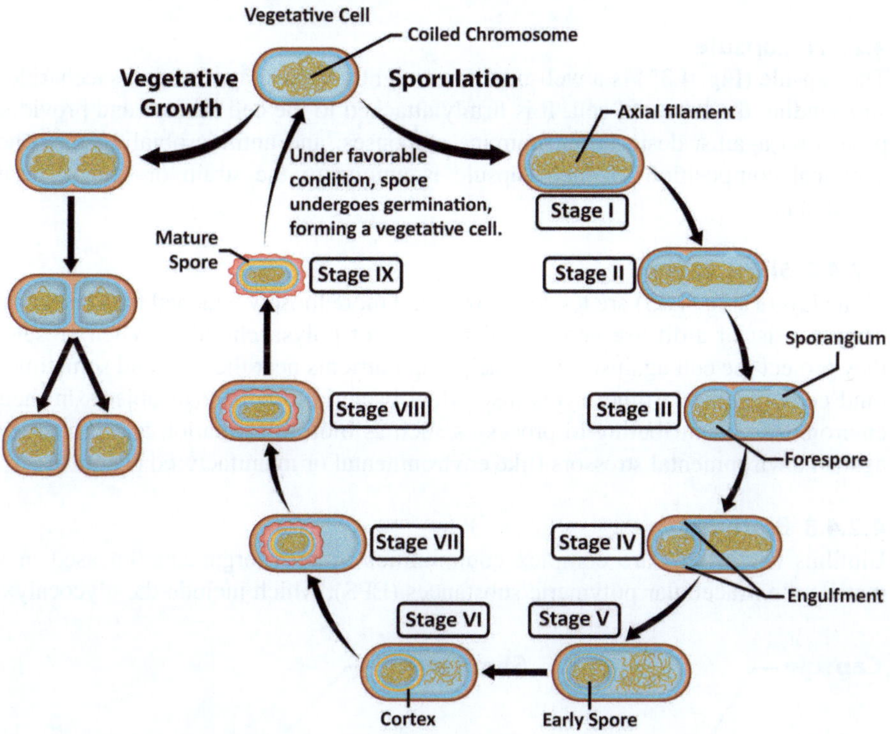

Fig. 4.34 Bacterial sporulation cycle: **Stage I:** Chromosome is duplicated and extends into an axial filament. **Stage II:** Duplicated chromosomes are separated, and the septum starts to form near one pole. **Stage III:** Septum completes, separating a forespore from the sporangium (mother cell). Each one of them receives a copy of the chromosome. **Stage IV:** Sporangium engulfs the forespore for further development. **Stage V:** Sporangium begins to synthesize spore layers around the forespore actively. The chromosome of sporangium starts to disintegrate. **Stage VI:** Cortex and outer coat layers are deposited. Sporangium chromosome disintegration continues. **Stage VII:** Forespore develops a layer of peptidoglycan (exosporium) between the original forespore membrane and sporangium. **Stage VIII:** Dipicolinic acid is synthesized, and calcium is incorporated into the spore coat, forming the mature spore. **Stage IX:** Mature spore is released with loss of the sporangium

These conditions signal to the bacteria that it is necessary to form an endospore for survival. Endospore formation begins with the bacterial cell undergoing asymmetric cell division. One of the resulting daughter cells becomes the endospore, while the other remains as the mother cell. Then, the endospore undergoes a complex maturation process involving the deposition of layers, including the cortex and sporangium. The core's DNA becomes highly condensed, and DPA accumulates.

4.2.4 Glycocalyx

The bacterial glycocalyx is a complex extracellular structure that is a gel-like layer of polysaccharides, glycoproteins, and glycolipids surrounding the outer surface of bacterial cells. It forms an essential part of the extracellular matrix and contributes significantly to the bacterium's overall physiology. Bacterial glycocalyces may be present as a capsule or slime layer.

4.2.4.1 Capsule

The capsule (Fig. 4.35) is a well-organized, tightly bound layer of polysaccharides surrounding the bacterial cell. It is firmly attached to the cell surface and provides protection against desiccation, immune responses, and antimicrobial agents. The chemical composition of each capsule is unique to the strain of bacteria that secreted it.

4.2.4.2 Slime Layer

Slime layers (Fig. 4.35) are less organized and more loosely attached than capsules. They consist of a diffuse and irregular matrix of polysaccharides. When present, they protect the cell against drying, help trap nutrients near the cell, and sometimes bind cells together. Slime layers also allow bacteria to adhere to objects in their environments, contributing to processes such as biofilm formation and protection against environmental stressors (like environmental or manufactured chemicals).

4.2.4.3 Biofilm

Biofilms (Fig. 4.36) are complex communities of microorganisms encased in a matrix of extracellular polymeric substances (EPS), which include the glycocalyx.

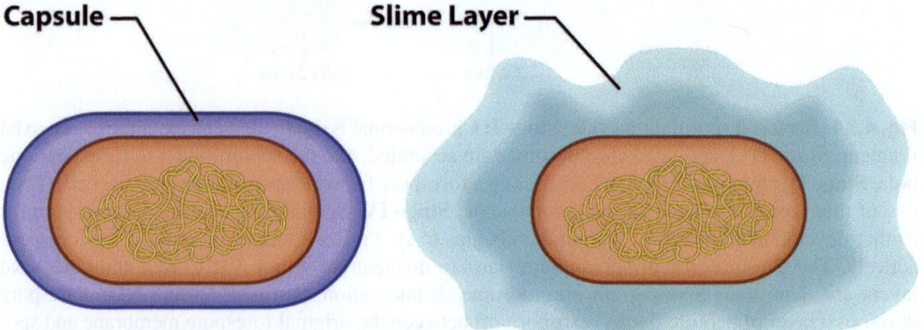

Fig. 4.35 Bacterial glycocalyx: capsule (left) and slime layer (right)

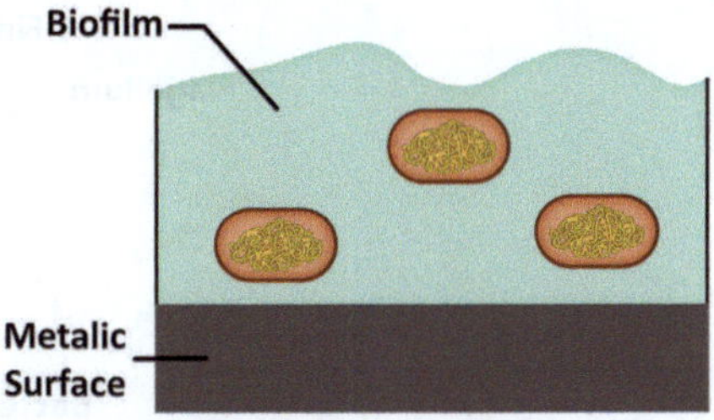

Fig. 4.36 Biofilm formed on a metallic surface

The glycocalyx provides the adhesive matrix necessary for biofilm formation, allowing bacteria to adhere to surfaces and each other. Biofilms confer protection and survival advantages in hostile environments and protect the bacteria against antibiotics.

4.2.5 Motility Appendages

About 50% of the bacteria discovered are motile and able to move. These bacteria move fast and with purpose. Bacterial motility is a critical trait that enables bacteria to navigate diverse environments, seek nutrients, avoid toxins, and colonize new niches. The motility appendages in bacteria include flagella, axial filaments, and pili.

4.2.5.1 Flagella

Flagella are long, whip-like appendages found in various microorganisms, including bacteria, that enable them to move through liquid environments. The filament is the visible, whip-like portion of the flagellum located outside the bacteria (Fig. 4.37). It is composed primarily of a protein called flagellin. Multiple flagellin subunits arrange themselves in a helical pattern to form the filament (Fig. 4.38). Located at the base of the filament, the hook is a short, flexible segment that connects the filament to the basal body. It is made up of a different protein, called hook protein, which has a different structure than flagellin. The complex structure of the basal body anchors the flagellum to the cell wall or membrane. It consists of several rings and a central rod-like structure. The basal body has one ring and spans the cytoplasmic membrane peptidoglycan layer in Gram-positive bacteria. In contrast, Gram-negative bacteria have four rings that span the outer membrane, periplasmic space, and inner membrane (Fig. 4.38).

Bacteria without flagella are atrichous. The number and arrangement of flagella are different among bacteria:

- **Monotrichous:** a single flagellum is present at one end of the bacteria.
- **Amphitrichous:** a single flagellum is present at each end of the bacteria (the bacteria has two flagella).
- **Lophotrichous:** multiple flagella are present at one or both ends of the cell.
- Peritrichous: multiple flagella are distributed over the entire surface of the cell.

Fig. 4.37 Bacterial appendages: Fimbriae are short and numerous, while sex pili are fewer in number and longer. The flagellum is the longest appendage and the fewest in number

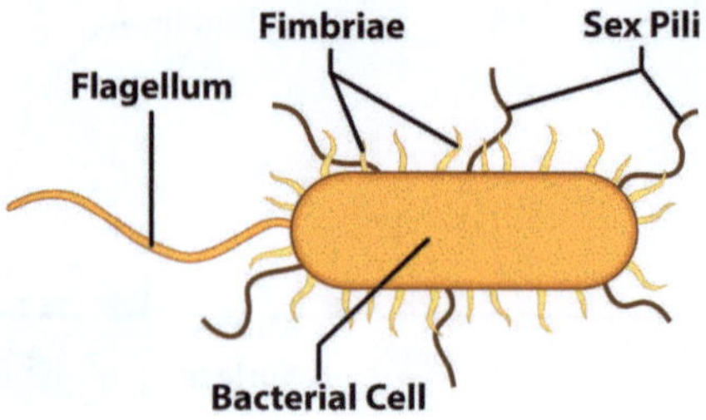

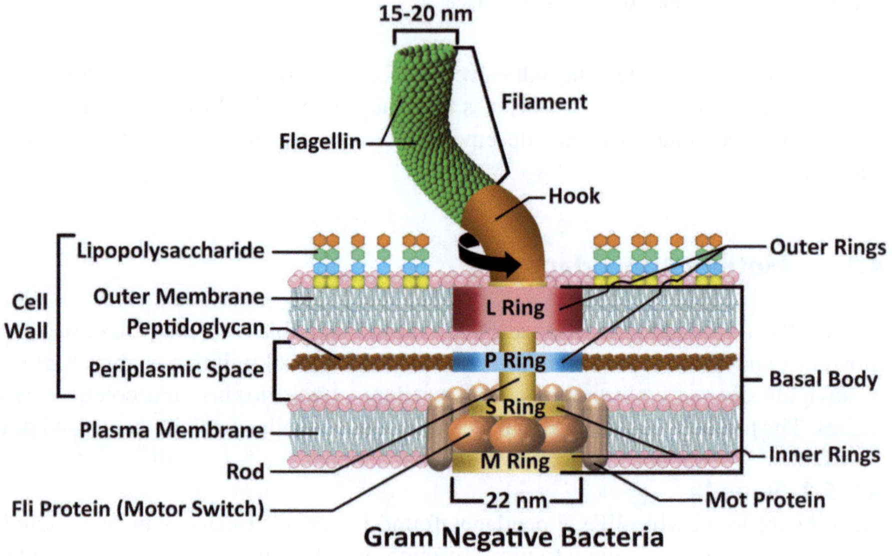

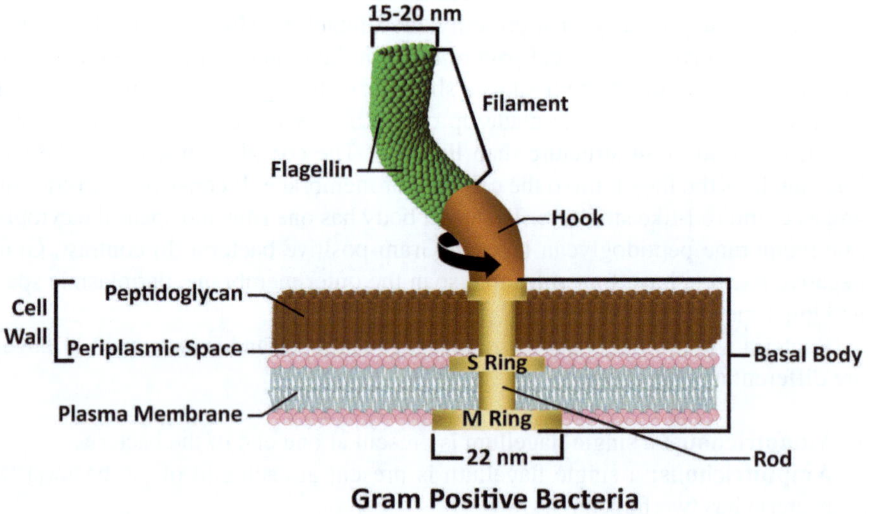

Fig. 4.38 Basal body of flagellum in Gram-negative and Gram-positive bacteria

Flagellar motion is driven by a rotary motor embedded in the cell membrane at the base of the flagellum. The motor is powered by the protons (H^+) flowing across the membrane, using the proton motive force generated by the electron transport chain or other energy-yielding reactions. Bacteria can control the direction and frequency of flagellar rotation through a range of sensory systems that respond to environmental cues, such as chemical gradients or light.

Chemotaxis is the ability of bacteria to move in response to gradients of chemicals in their environment. It involves receptors that bind specific molecules, triggering a signal transduction pathway that ultimately influences flagellar rotation to move the bacteria toward (positive chemotaxis) or away from (negative chemotaxis) the chemical gradient.

Phototaxis is the ability of bacteria to sense light and use it to guide their movement toward (positive phototaxis) or away from (negative phototaxis) the light.

4.2.5.2 Axial Filaments

Axial filaments (endoflagella or periplasmic flagella) (Fig. 4.39) are unique structures in spirochetes. They are composed primarily of a specialized protein called flagellin, with a different structural arrangement from the flagella found in other bacteria. Axial filaments are located within the periplasmic space, between bacteria's inner and outer membranes. Each axial filament is anchored at one of its ends to an end of the cytoplasmic cylinder that forms the body of the spirochete (the attachment site is equal to the basal body). Spirochetes typically have multiple axial filaments distributed along the cell's length. The helical structure of axial filaments, by coiling and uncoiling, allows them to generate a corkscrew-like motion, enabling spirochetes to move in a characteristic spiral pattern.

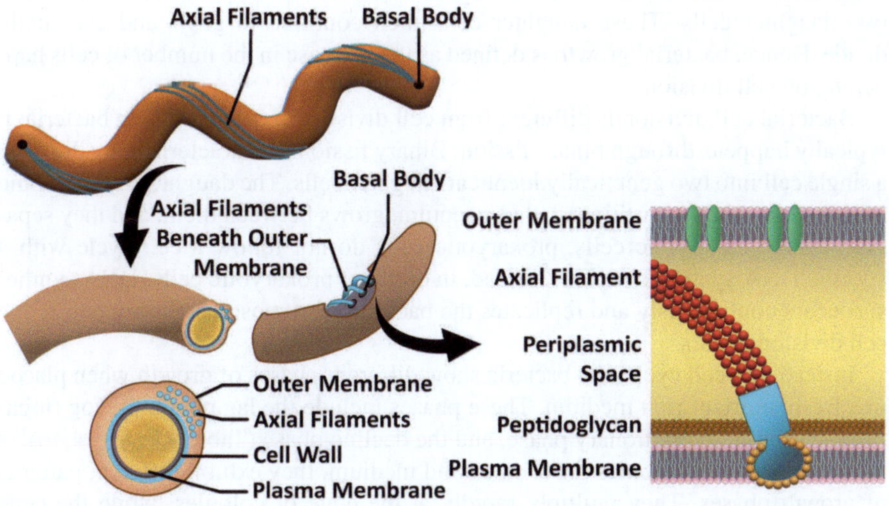

Fig. 4.39 Axial filaments in a spirochete bacterium

4.2.5.3 Pili

Pili (singular: pilus) are tiny, hair-like, hollow, filamentous appendages found on the surface of many bacteria. They serve various functions, including adhesion, motility, and genetic exchange. Pili are primarily composed of protein subunits called pilin. These pilin proteins are arranged in a spiral or linear fashion, forming the structure of the pilus. There are two types of pili in bacteria.

Fimbriae

Fimbriae (Fig. 4.37) are short, numerous pili mediating surface adhesion. They help bacteria attach to specific host tissues or to inanimate surfaces like medical devices through specialized proteins known as tip adhesins. Fimbriae are usually nonmotile. They contribute to the pathogenicity of certain bacteria—their ability to produce disease—by enhancing colonization (the development of colonies) on the surfaces of the cells of other organisms.

Sex Pili (Conjugative Pili)

Sex pili (conjugative pili) (Fig. 4.37) are longer and fewer in number compared to fimbriae. They are involved in bacterial conjugation, a process where genetic material (plasmids) is transferred from one bacterium to another. Such transfers among bacteria cause problems for humans because antibiotic resistance can be passed on with the DNA transfer.

4.2.6 Prokaryotic Growth

When we talk about growth, we usually refer to an increase in size. Even unicellular organisms experience growth in this context. Once a cell, known as the mother or parent cell, reaches about double its size, it duplicates its contents and divides into two daughter cells. These daughter cells then continue to grow and eventually divide. Hence, bacterial growth is defined as the increase in the number of cells happening by cell division.

Bacterial cell division is different from cell division in eukaryotes. In bacteria, it typically happens through binary fission. Binary fission is characterized by dividing a single cell into two genetically identical daughter cells. The daughter cells become independent when a partition, called a septum, grows between them, and they separate. Unlike eukaryotic cells, prokaryotic cells do not follow a cell cycle with a specific DNA synthesis period. Instead, in dividing prokaryotic cells, DNA synthesis occurs continuously and replicates the bacterial chromosome shortly before the cell division occurs.

Instead of a cell cycle, the bacteria show different phases of growth when placed in a fresh, nutrient-rich medium. These phases include the lag phase, the log (logarithmic) phase, the stationary phase, and the decline phase (Table 4.2; Fig. 4.40).

When bacterial colonies grow on a solid medium, they exhibit different patterns of growth phases. They multiply rapidly at the edge of colonies, while the cells

Table 4.2 Overview of bacterial growth phases: descriptions, characteristics, and influential factors

Phase	Description	Characteristics	Key factors
Lag phase	A period of adaptation after bacteria are introduced into a new environment or medium	Bacteria acclimatize to the new environment Synthesis of enzymes and other molecules occurs No significant increase in cell number The duration depends on the history of the inoculum, medium composition, and environmental conditions	Prior growth conditions Medium composition Environmental factors (e.g., temperature, pH)
Log (exponential) phase	Rapid cell division and population growth occur	Bacteria divide at a constant and maximum rate Population doubles at regular intervals Growth is limited only by the environment, not by nutrient depletion or waste accumulation The rate of growth is characteristic of the species and conditions	Nutrient availability Absence of toxic waste products Optimal growth conditions
Stationary phase	Growth rate slows and stabilizes as the number of new cells equals the number of dying cells	Nutrient levels decrease, and waste products accumulate Metabolic activity changes, often with increased production of secondary metabolites Bacteria may become more resistant to harmful conditions Phase can last for varying lengths of time	Nutrient depletion Waste accumulation Increased cell density
Death (decline) phase	The number of dying cells exceeds the number of new cells being produced	Accumulation of waste products and depletion of nutrients Cells die exponentially, but some cells may enter a dormant state and survive The death rate can vary and is influenced by some cells' persistence and harsh environment	Exhaustion of nutrients Accumulation of toxic waste products Adverse environmental conditions

closer to the center show slower growth or start to die because there is a shortage of nutrients. All phases of bacterial growth take place simultaneously within a colony, indicating that the growth is not synchronized. Bacterial growth is a dynamic process influenced by many physical, chemical, biological, and environmental factors.

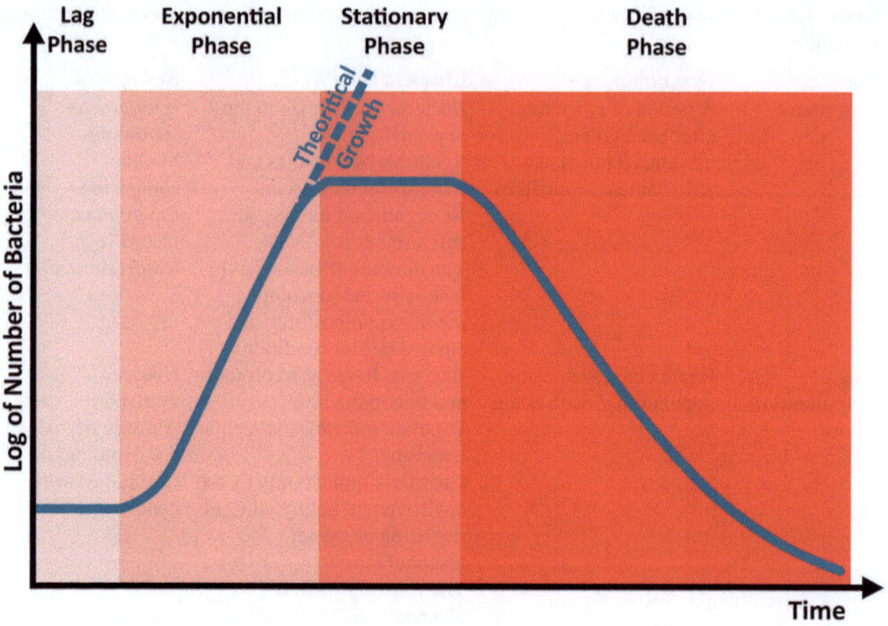

Fig. 4.40 Bacterial growth curve, representing the number of living cells in a population over time

4.2.6.1 Physical Factors Influencing Bacterial Growth

pH: pH levels profoundly affect bacterial metabolism and growth. Extremes in pH can inhibit bacterial growth. No species can tolerate the entire pH range; however, many can tolerate a pH range that overlaps two categories. Acidophiles thrive in acidic (pH of 0.1–5.4) environments, alkaliphiles in basic (pH 7.0–11.5) conditions, and neutrophiles in neutral (pH 5.4–8.0) pH ranges. Changes in pH can lead to denaturing enzymes and other proteins and interfere with cell metabolism. In many bacteria, acids are a by-product of cell metabolism produced in sufficient quantities, eventually interfering with their growth.

Temperature: Bacterial growth is highly temperature-dependent. The optimal temperature range for most bacteria, which are called mesophiles, is between 20 °C and 40 °C. Psychrophiles thrive in cold temperatures (15° to 20 °C), while thermophiles prefer high temperatures (50° to 60 °C). Human pathogens are included in the mesophile category, and most grow best in temperatures near the human body (37 °C).

Oxygen: The presence or absence of oxygen influences bacterial growth based on their species. Aerobes require oxygen for respiration, while anaerobes thrive in oxygen-limited environments. Facultative anaerobes can switch between aerobic and anaerobic metabolism.

Osmotic pressure: Bacteria have varying tolerance levels for salt concentrations, impacting their growth in saline environments. Halophiles (constituting most bacteria) thrive in high salt conditions, while osmotolerant bacteria can withstand elevated solute concentrations. In hyperosmotic environments, halophiles lose water and undergo plasmolysis, or shrinking of the cell. Using salt to preserve hams, bacon, and pickles or the sugar in jellies and jams exerts sufficient osmotic pressure to kill or inhibit microbial growth.

4.2.6.2 Chemical Factors Influencing Bacterial Growth

Nutrient availability: The availability of essential nutrients like carbon, nitrogen, phosphorus, and trace elements is crucial for bacterial growth. Limitations in these factors can lead to extended lag phases or growth inhibition.

Toxic substances: Bacteria are sensitive to toxic compounds like heavy metals, antibiotics, and chemical disinfectants. Some bacteria have developed resistance mechanisms to counteract the effects of harmful substances.

4.2.6.3 Biological Factors Influencing Bacterial Growth

Interactions with other microorganisms: Competition for resources, predation by bacteriophages, and interactions with other microorganisms can profoundly impact bacterial growth.

Quorum sensing: Bacteria communicate through quorum sensing, regulating gene expression based on population density. This communication can influence behaviors like biofilm formation, antibiotic resistance, and virulence factor production.

Genetic factors: Genetic mutations, horizontal gene transfer between cells through conjugate pili, and acquisition of plasmids can confer advantages or disadvantages in different environments.

4.2.6.4 Environmental Factors Influencing Bacterial Growth

Light and radiation: Some bacteria, like phototrophs, use light as an energy source for growth. Some bacteria may have mechanisms to withstand exposure to ionizing radiation.

Moisture levels: Water availability is crucial for bacterial growth, with some species thriving in dry conditions (xerophiles) and others in moist environments.

4.3 Part 3: Cell Destruction

Cell function is a seemingly never-ending job. Hence, there must be a process to eliminate outdated cells and organelles and recycle cellular nutrients for reuse. Cells accomplish this through autophagy, or apoptosis.

4.3.1 Autophagy

The term "autophagy"is derived from the Greek words "auto" (meaning self) and "phagy" (meaning eating), indicating a process of self-eating. Autophagy involves degrading and recycling cellular components, such as damaged organelles, misfolded proteins, and other cytoplasmic constituents. This process is orchestrated by a complex interplay of cellular machinery, including specialized proteins and membrane structures. The first step in autophagy involves forming a double-membraned structure known as the phagophore or isolation membrane. This structure is derived from a specialized compartment of the endoplasmic reticulum (ER) called the omegasome. The phagophore expands and elongates, eventually encapsulating the targeted cellular material. The sealed autophagosome then fuses with a lysosome containing hydrolytic enzymes. This fusion event forms an autolysosome, creating an acidic and enzymatically active environment necessary for degradation.

4.3.2 Apoptosis

Cell apoptosis, also known as programmed cell death, is a highly regulated and controlled process vital for maintaining tissue homeostasis and development and eliminating damaged or unwanted cells in multicellular organisms. Unlike necrosis, a form of cell death resulting from acute injury or trauma, apoptosis is a carefully orchestrated series of events that occur in a controlled manner. Various internal and external signals can initiate apoptosis. These signals can include DNA damage, cellular stress, withdrawal of growth factors, or activation of specific death receptors on the cell membrane. Apoptosis helps maintain the balance of cell populations within tissues, ensuring the removal of excess or damaged cells. It is also crucial for shaping developing tissues and organs. It allows for sculpting structures, such as fingers and toes, by eliminating cells between them.

4.3.3 Cell Aging

Cell aging, also known as cellular senescence, refers to the process by which individual cells undergo irreversible changes that limit their ability to divide and function over time. This is distinct from organismal aging, which involves the decline of entire organisms over time. Cellular aging is a complex and multifaceted phenomenon influenced by genetic, environmental, and physiological factors. Telomeres are protective caps at the ends of chromosomes, composed of repetitive DNA sequences. With each cell division, telomeres shorten due to the "end-replication problem," where the DNA polymerase enzyme cannot fully replicate the end of linear chromosomes. Eventually, critically short telomeres trigger cell cycle arrest and senescence. Accumulation of DNA damage due to factors such as radiation, oxidative stress, and chemical agents can also lead to cellular senescence. Excessive reactive oxygen species (ROS) can cause damage to cellular components, leading to senescence. Antioxidant systems help mitigate oxidative stress. Senescent cells cease to divide and enter a state of irreversible cell cycle arrest. Senescent cells often display enlarged and flattened morphology compared to actively dividing cells.

4.4 Part 4: Cell Metabolism

Cellular metabolism encompasses all the chemical reactions within a cell to maintain life. It is a highly regulated and intricate pathway network that manages the flow of energy and materials necessary for cell growth, maintenance, and reproduction. The primary currency of cellular energy is adenosine triphosphate (ATP). When ATP is hydrolyzed (broken down), it releases energy, which can be used to drive cellular processes. The metabolic pathways are either catabolic or anabolic.

Catabolic pathways break complex molecules into simpler ones, releasing energy. Common catabolic processes include glycolysis, the citric acid cycle (Krebs cycle), and oxidative phosphorylation. These pathways break down carbohydrates, fats, and proteins to generate ATP. Anabolic pathways involve synthesizing complex molecules from simpler ones, requiring energy input. Examples include gluconeogenesis (synthesis of glucose), fatty acid, and protein synthesis (Section 4.1.3.3). Anabolism is essential for building and maintaining cellular structures and molecules.

4.4.1 Anaerobic Metabolism

Anaerobic metabolism, also known as anaerobic respiration, is a cellular metabolic process without oxygen. It includes glycolysis and fermentation.

4.4.1.1 Glycolysis

Glycolysis, or the Embden–Meyerhof pathway (Fig. 4.41), is a metabolic pathway in virtually all living organisms (aerobes and anaerobes). It is the initial step in cellular energy production, converting glucose into smaller molecules, capturing energy in adenosine triphosphate (ATP), and reducing equivalents like NADH (nicotinamide adenine dinucleotide).

Energy investment phase: In the first half of glycolysis, two molecules of ATP are invested to prepare glucose for cleavage (substrate-level phosphorylation). Hexokinase phosphorylates glucose, trapping it inside the cell. This is followed by isomerization and phosphorylation steps, culminating in forming fructose-1,6-bisphosphate.

Cleavage and energy generation phase: Fructose-1,6-bisphosphate is cleaved into two three-carbon molecules: dihydroxyacetone phosphate (DHAP) and glyceraldehyde-3-phosphate (GAP). This phase leads to the subsequent steps where GAP is oxidized, yielding NADH and ATP and eventually producing two pyruvate molecules.

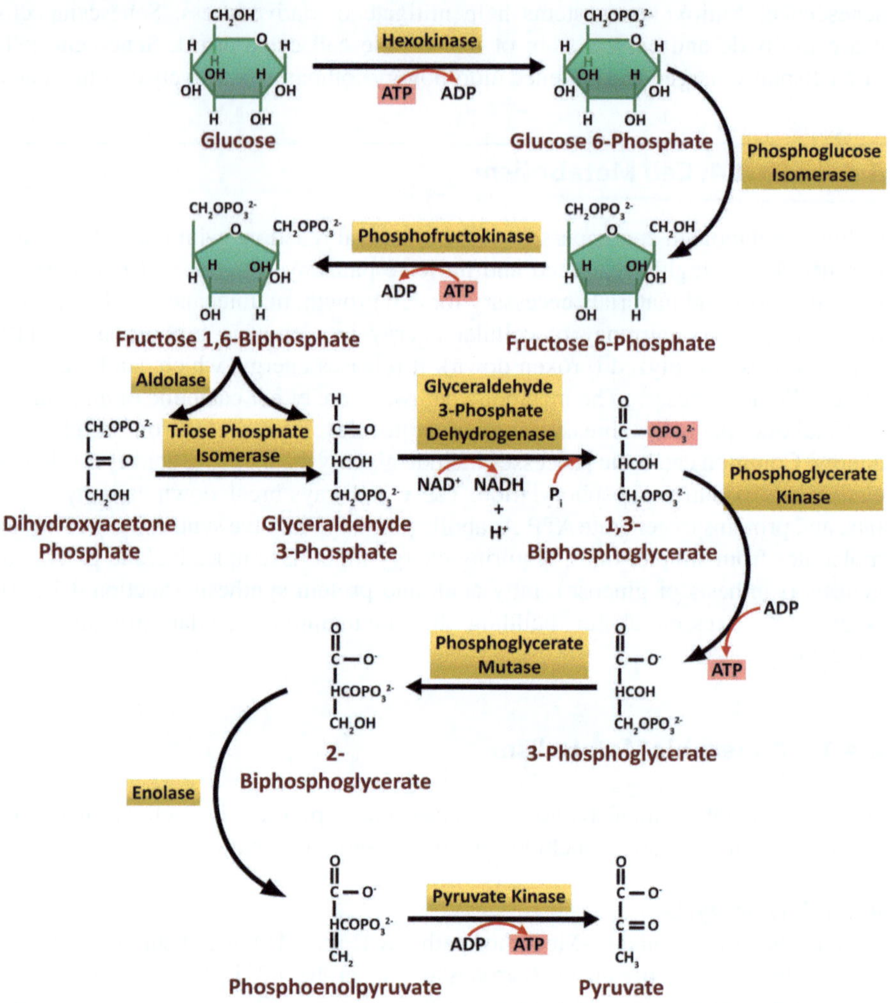

Fig. 4.41 Glycolysis pathway

Glycolysis provides cells with a relatively small amount of energy. The energy released during glycolysis of a single glucose molecule is captured in four molecules of ATPs. However, two ATPs were used in the initial phosphorylations, so glycolysis results in a net energy capture of only two ATPs per glucose molecule. Glycolysis product (pyruvate molecule) is required for the oxidative phosphorylation pathways, so glycolysis functions as the first step of the aerobic metabolism of the cells.

4.4.1.2 Fermentation

Fermentation (Fig. 4.42) is another pathway for cell anaerobic metabolism. Fermentation can appear as several pathways, including lactic acid fermentation,

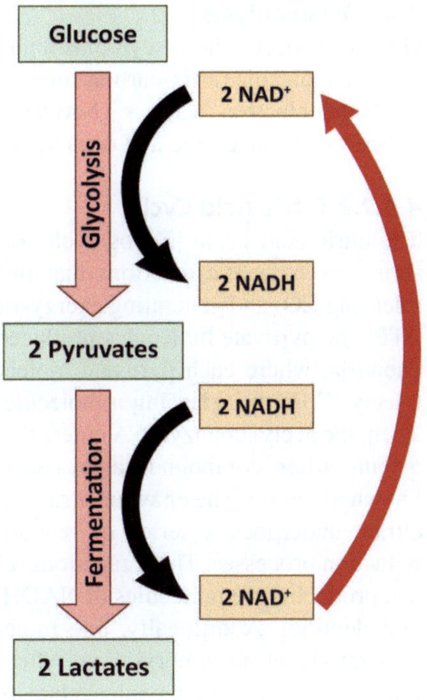

Fig. 4.42 Fermentation pathway, replenishing the NAD^+ content of the cell

alcoholic fermentation, mixed-acid fermentation, butyric acid fermentation, and propionic acid fermentation. Only lactic acid fermentation happens in animals, including human cells, among these different fermentation pathways. These pathways do not capture energy in ATP, but they remove electrons from reduced NAD (nicotinamide adenine dinucleotide) so that it can continue to act as an electron acceptor. Fermentation is a necessary cellular mechanism to provide oxidized NAD by removing the electrons from the reduced NAD produced during other metabolic pathways, primarily oxidative phosphorylation. Lactic acid fermentation is a common pathway employed by various microorganisms and muscle cells under anaerobic conditions. This pathway directly converts the pyruvic acid to lactic acid, using electrons from reduced NAD and producing oxidized NAD.

4.4.2 Aerobic Metabolism

Cellular aerobic metabolism (cellular respiration) is the process by which cells generate energy in the presence of oxygen. It is a set of metabolic reactions in eukaryotic cells' mitochondria. Cellular respiration completely breaks down glucose molecules into carbon dioxide (CO_2), water (H_2O), and energy. The energy released is harnessed in the form of ATP. Cellular respiration is divided into three main stages: glycolysis, the citric acid cycle (also known as the Krebs cycle), and oxidative phosphorylation.

4.4.2.1 Glycolysis

Glycolysis occurs in the cytoplasm and is the first step of aerobic metabolism. One glucose molecule (a six-carbon sugar) is broken into two pyruvate molecules. Some ATP and electron carriers (NADH) are generated along the way (Fig. 4.41). Glycolysis is anaerobic and does not require oxygen.

4.4.2.2 Citric Acid Cycle

The citric acid cycle [Krebs cycle, or tricarboxylic acid cycle (TCA cycle)] is a series of enzymatic reactions that further oxidize the carbon atoms in pyruvate, releasing CO_2 and generating energy-rich molecules (NADH and $FADH_2$) and some ATP. The pyruvate molecules produced in glycolysis are transported into the mitochondria, where each pyruvate molecule is converted to one acetyl coenzyme A (Acetyl CoA) by releasing a molecule of CO_2 and reducing one NAD (Fig. 4.43). Then, the acetyl coenzyme A enters the citric acid cycle. Acetyl CoA combines with a four-carbon compound called oxaloacetate, producing a six-carbon compound known as citrate. The enzyme citrate synthase catalyzes this reaction. Subsequently, citrate undergoes a series of reactions involving isomerization and oxidation-reduction processes. These reactions release two molecules of carbon dioxide (CO_2) and produce three molecules of NADH and one molecule of $FADH_2$ (flavin adenine dinucleotide). Additionally, one molecule of ATP or GTP is generated through substrate-level phosphorylation. After several intermediate reactions, the original four-carbon compound, oxaloacetate, is regenerated. This step is crucial as it ensures the continuous operation of the cycle. Oxaloacetate can then combine with another molecule of Acetyl CoA to start the cycle anew (Fig. 4.43). For every one round of the Krebs cycle (per Acetyl CoA molecule), the following are produced:

- 3 NADH molecules: these carry high-energy electrons to the electron transport chain.
- 1 $FADH_2$ molecule: another electron carrier for the electron transport chain.
- 1 ATP/GTP molecule: directly synthesized through substrate-level phosphorylation.
- 2 CO_2 molecules: released as waste products.

Additionally, since two molecules of Acetyl CoA are produced from one glucose molecule (since glucose generates two pyruvate molecules in glycolysis), each glucose molecule yields double the above outputs.

4.4.2.3 Oxidative Phosphorylation

Oxidative phosphorylation is the final stage of cellular respiration, taking place in the inner mitochondrial membrane of eukaryotic cells and the inner membrane of the prokaryotic cells. This process couples the transfer of electrons along the electron transport chain (ETC) to the synthesis of ATP. Various factors, including the availability of oxygen, the concentration of electron carriers, and the ETC complexes' activity, can control the oxidative phosphorylation rate.

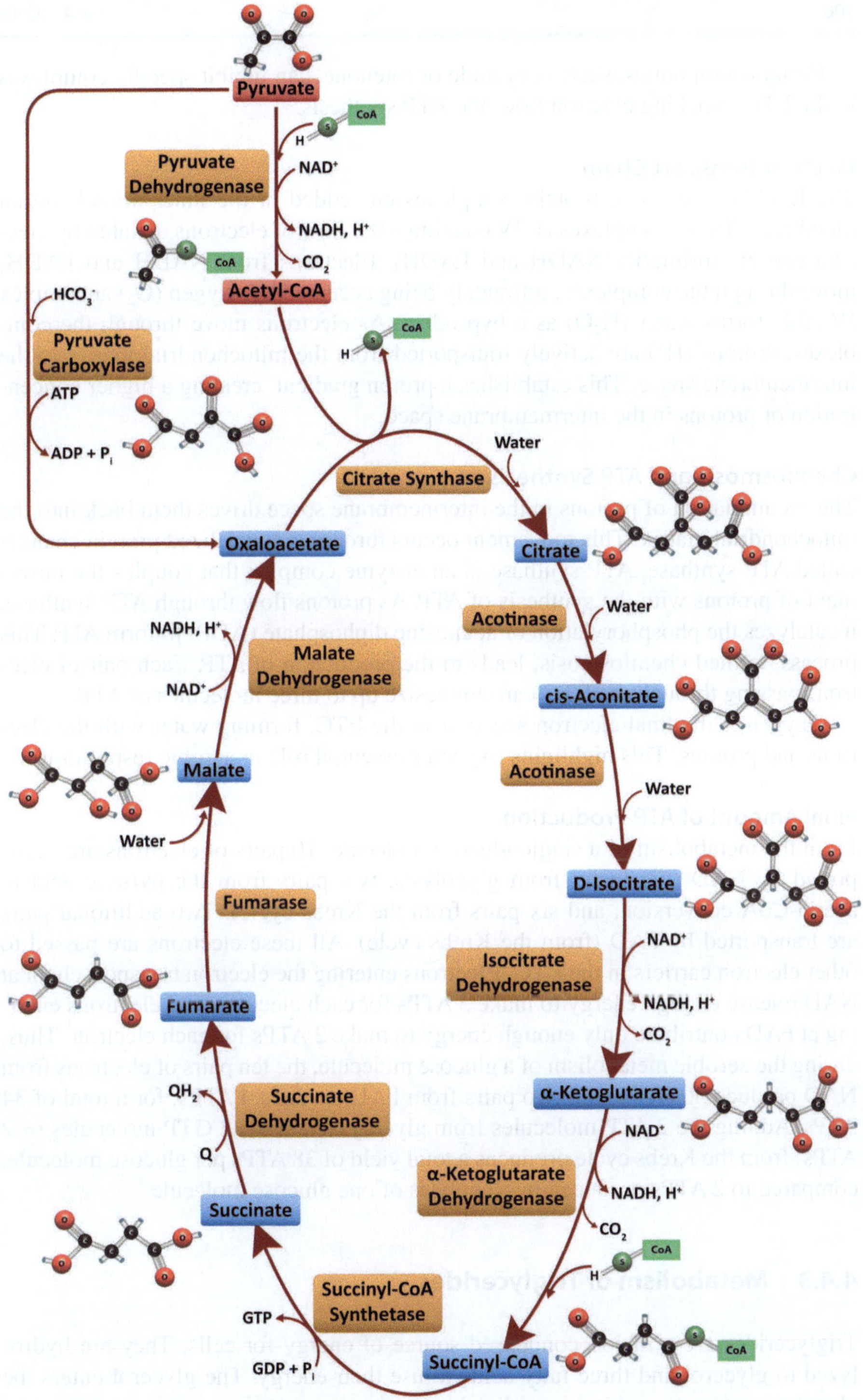

Fig. 4.43 Citric acid cycle (Krebs cycle)

Certain compounds, such as cyanide or rotenone, can inhibit specific complexes in the ETC, blocking electron flow and ATP synthesis.

Electron Transport Chain

The ETC is a series of protein complexes embedded in the inner mitochondrial membrane. These complexes (I–IV) facilitate the flow of electrons donated by electron carriers, primarily NADH and $FADH_2$. Electrons from NADH and $FADH_2$ move through the complexes, ultimately being accepted by oxygen (O_2) at complex IV. This forms water (H_2O) as a byproduct. As electrons move through the complexes, protons (H^+) are actively transported from the mitochondrial matrix to the intermembrane space. This establishes a proton gradient, creating a higher concentration of protons in the intermembrane space.

Chemiosmosis and ATP Synthesis

The accumulation of protons in the intermembrane space drives them back into the mitochondrial matrix. This movement occurs through a specialized protein channel called ATP synthase. ATP synthase is an enzyme complex that couples the movement of protons with the synthesis of ATP. As protons flow through ATP synthase, it catalyzes the phosphorylation of adenosine diphosphate (ADP) to form ATP. This process, termed chemiosmosis, leads to the production of ATP. Each pair of electrons passing through the ETC can synthesize up to three molecules of ATP.

Oxygen is the final electron acceptor in the ETC, forming water with the electrons and protons. This highlights oxygen's essential role in aerobic respiration.

Final Amount of ATP Production

From the metabolism of a single glucose molecule, 10 pairs of electrons are transported by NAD (two pairs from glycolysis, two pairs from the pyruvic acid to acetyl-CoA conversion, and six pairs from the Krebs cycle). Two additional pairs are transported by FAD (from the Krebs cycle). All these electrons are passed to other electron carriers in the ETC. Electrons entering the electron transport chain at NAD release enough energy to make 3 ATPs for each electron, and electrons entering at FAD contribute only enough energy to make 2 ATPs for each electron. Thus, during the aerobic metabolism of a glucose molecule, the ten pairs of electrons from NAD produce 30 ATPs, and two pairs from FAD produce 4 ATPs, for a total of 34 ATPs. Adding the 2 ATP molecules from glycolysis and the 2 GTP molecules (= 2 ATPs) from the Krebs cycle produces a total yield of 38 ATPs per glucose molecule, compared to 2 ATPs produced in glycolysis of one glucose molecule.

4.4.3 Metabolism of Triglycerides

Triglycerides are a major, condensed source of energy for cells. They are hydrolyzed to glycerol and three fatty acids to use their energy. The glycerol enters the glycolysis pathway and is metabolized into pyruvic acid.

The fatty acids, which usually have an even number of carbons (16, 18, or 20), are broken down into 2-carbon pieces by a metabolic pathway called β-oxidation, which happens inside the mitochondria. Before β-oxidation, fatty acids must be activated in the cytoplasm. This involves the attachment of a coenzyme A (CoA) molecule to the fatty acid, forming acyl-CoA.

The first step of β-oxidation involves the oxidation of the β-carbon of the acyl-CoA molecule by an enzyme called acyl- CoA dehydrogenase. This results in the formation of a trans-Δ_2-enoyl-CoA and the reduction of FAD to $FADH_2$. In this step, water is added across the double bond of the trans-Δ_2-enoyl-CoA molecule by an enzyme known as enoyl-CoA hydratase. This leads to the formation of L-β-hydroxyacyl-CoA. Another oxidation step occurs, catalyzed by β- hydroxyacyl-CoA dehydrogenase. This results in the formation of a ketoacyl-CoA molecule and the reduction of NAD^+ to NADH. The final step involves the cleavage of the ketoacyl-CoA molecule by thiolase, resulting in the formation of acetyl-CoA and a new acyl-CoA chain, shorter by two carbons, which re-enters the cycle for further β-oxidation.

Each produced acetyl-CoA enters a citric acid cycle to be oxidized. Hence, each triglyceride molecule containing three fatty acids with many carbons can activate several citric acid cycles, producing many ATPs. If the cell needs glucose instead of ATP, like when the blood glucose level is low, the acetyl-CoA produced from fatty acids enters a glucose- production pathway called gluconeogenesis.

4.4.4 Metabolism of Proteins

Proteins can also be used as a source of energy. They undergo hydrolysis, where proteolytic enzymes split the peptide bond between the amino acids and break the proteins into amino acids. These amino acids undergo deamination, which involves the removal of their amino groups. The resulting deaminated molecules enter either glycolysis, fermentation, or the Krebs cycle for further metabolic activities.

Suggested Reading

Campbell Biology, 12th Ed. Lisa A. Urry, et al. Pearson, 2021

Cell and Molecular Biology, 3rd Ed. Nalini Chandar, Susan M. Viselli, Lippincott Williams & Wilkins, 2023

Developmental Biology, 10th Ed. Michael Barresi, Scott Gilbert, Oxford University Press, 2023

Essential Cell Biology, 6th Ed. Bruce Alberts, et al. W. W. Norton & Company, 2023

Jawetz, Melnick, & Adelberg's Medical Microbiology, 28th Eds. Stefan Riedel, et al. McGraw-Hill Education, 2019

Molecular Cell Biology, 9th Ed. Harvey Lodish, et al. W. H. Freeman, 2021

Sherris & Ryan's Medical Microbiology, 8th Ed, Kenneth J. Ryan. McGraw-Hill, 2022

Viruses

5

In this chapter, you will learn

- Nature of viruses as agents straddling the boundary between living and nonliving entities
- General characteristics of viruses, including their lack of cellular structure
- Definition and structural composition of viruses, including their genetic material and protein coat
- Diversity of viruses affecting various life forms, from humans to bacteria
- Various shapes and sizes of viruses and their implications for transmission and infection
- Classification of viruses based on genetic material and replication methods
- Specific replication strategies of animal viruses and bacteriophages
- Concept of virus-like agents, including satellites, viroids, and prions

Viruses operate at the boundary between the living and the nonliving and have profoundly impacted health, society, and even history. From common colds to devastating pandemics, viruses are a testament to the power wielded by the world's tiniest inhabitants.

Viruses lack a cell nucleus, organelles, or cytoplasm. They can only multiply within a living host cell, and they are known as intracellular parasites.

5.1 Definition

A virus is a submicroscopic infectious agent lacking cellular structure and cannot carry out metabolic processes. It straddles the line between living and nonliving entities. A virus is incapable of independent replication and is composed of genetic material—either DNA or RNA—enclosed in a protein coat known as a capsid and sometimes also enveloped in a lipid membrane. Instead, it must infect a suitable host cell to hijack its machinery to reproduce and assemble new viral particles.

H. Hosseinzadeh, *Fundamentals of Medicine for Biomedical Engineering*,
https://doi.org/10.1007/978-3-031-74085-5_5

Viruses are incredibly diverse, infecting nearly every form of life, from animals and plants to microorganisms like bacteria (where they are known as bacteriophages) and even other viruses. Their sizes can vary, ranging from 20 to 300 nanometers in diameter. They may have simple or complex shapes, including icosahedral, helical, or more complicated multipartite structures. Their genomes are also highly varied, allowing various strategies to infect host cells, including attachment, penetration, uncoating, replication, assembly, and release.

They are of significant interest in biotechnology, where they can be used as vectors for gene delivery, and in medicine, where viral diseases remain a major research focus, especially in developing antiviral drugs and vaccines.

Despite their minute size and seeming simplicity, viruses have a profound and far-reaching impact on life on Earth, both as agents of disease and as drivers of ecological and evolutionary change.

5.2 Components

The structure of viruses is simple compared to that of cells. A viral particle (virion) includes a nucleic acid core, a surrounding protein coat called a capsid, and, in some viruses, a lipid bilayer membrane (Fig. 5.1).

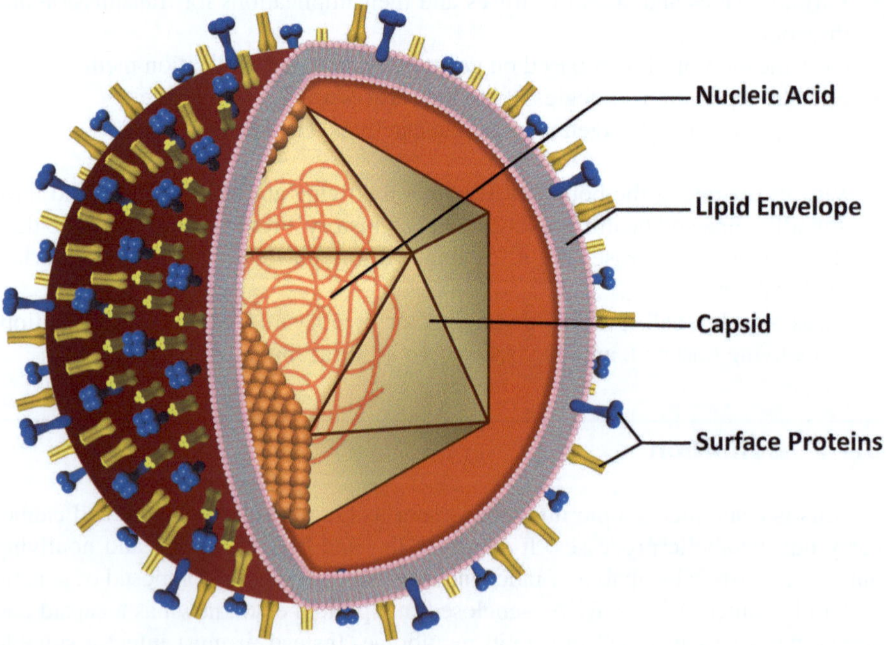

Fig. 5.1 General structure of a membraned virus. The membrane is absent in most viruses, and the capsid has various shapes

5.2.1 Nucleic Acids

The core of every virus is its genetic material, which can be either DNA (deoxyribonucleic acid) or RNA (ribonucleic acid). This genetic information is the blueprint for creating new viral particles and often encodes the strategy for hijacking the host's cellular machinery. DNA viruses usually have a double-stranded DNA genome. This genetic material is more stable and generally larger, allowing for a more complex set of instructions for the viral life cycle. RNA viruses possess a genome made up of RNA. These viruses are usually more prone to mutations due to the lack of error-correcting mechanisms during replication. This characteristic confers a high degree of adaptability but also poses challenges for treatment and prevention.

5.2.2 Capsids (the Protective Armor)

The capsid (Fig. 5.1) is a protein shell that encloses the viral genome. Although it may appear to be a simple protective casing, the viral capsid is a complex and highly organized structure that plays a multifaceted role in the virus's life cycle. The capsid is composed of protein subunits known as capsomeres. These are often self-assembling and can form the capsid without requiring external energy input. Capsids can be categorized based on their geometry into three primary types (Fig. 5.2):

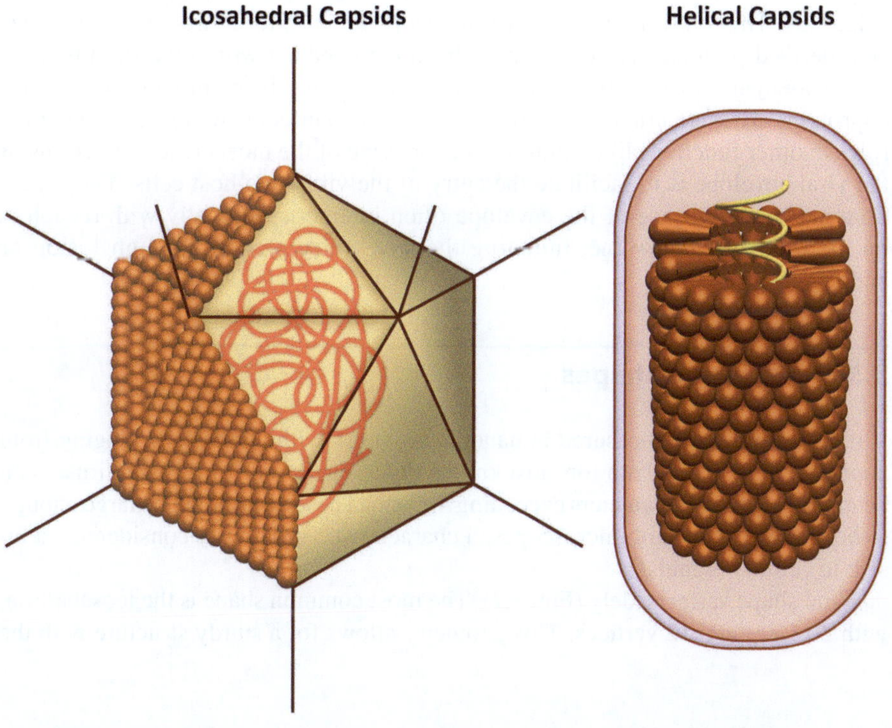

Fig. 5.2 Capsid: icosahedral and helical are among the most common shapes of capsid

Icosahedral capsids: These are the most common type and consist of 20 faces with 12 vertices resembling soccer balls. Human papillomavirus (HPV) and poliovirus are examples.

Helical capsids: These are cylindrical, rod-shaped structures found in viruses like the tobacco mosaic virus (TMV). The capsomeres are arranged in a helical pattern around the central axis.

Complex capsids: These are more elaborate and may have additional structures like tails, as seen in bacteriophages. They do not fit neatly into either the icosahedral or helical category.

The primary function of the capsid is to protect the viral genome from degradation by nucleases and to withstand harsh environmental conditions that the virus may encounter outside a host cell. Capsids often have specific protein sequences or domains that interact with receptors on the surface of potential host cells. This specificity is crucial for the virus's ability to infect particular types of cells.

5.2.3 Envelopes

While not all viruses possess an envelope, those that do often exhibit increased complexity and a more remarkable ability to adapt to various challenges the host organism presents. The viral envelope primarily comprises a lipid bilayer (Fig. 5.1), usually derived from the host cell membrane during viral budding (Fig. 5.3). This bilayer serves as a framework for inserting viral and sometimes host-derived proteins. Viral glycoproteins are embedded within the lipid bilayer, which are synthesized in the host cell and inserted into the membrane. These glycoproteins are often crucial for the virus's ability to infect new cells and may play roles in other functions like immune evasion. One of the most critical functions of the viral envelope is to facilitate the entry of the virus into host cells. The glycoproteins on the surface of the envelope often interact specifically with receptors on the host cell membrane, initiating the viral entry process through fusion or endocytosis.

5.3 Sizes and Shapes

Viruses are typically measured in nanometers (nm), with dimensions ranging from approximately 20–300 nm for most known types. However, some giant viruses can be much larger with diameters exceeding 400 nm. These viruses can be large enough to be visible under light microscopes, a characteristic previously considered exclusive to cellular organisms.

Viral shape varies widely (Fig. 5.2). The most common shape is the icosahedron, with 20 faces and 12 vertices. This geometry allows for a sturdy structure with the

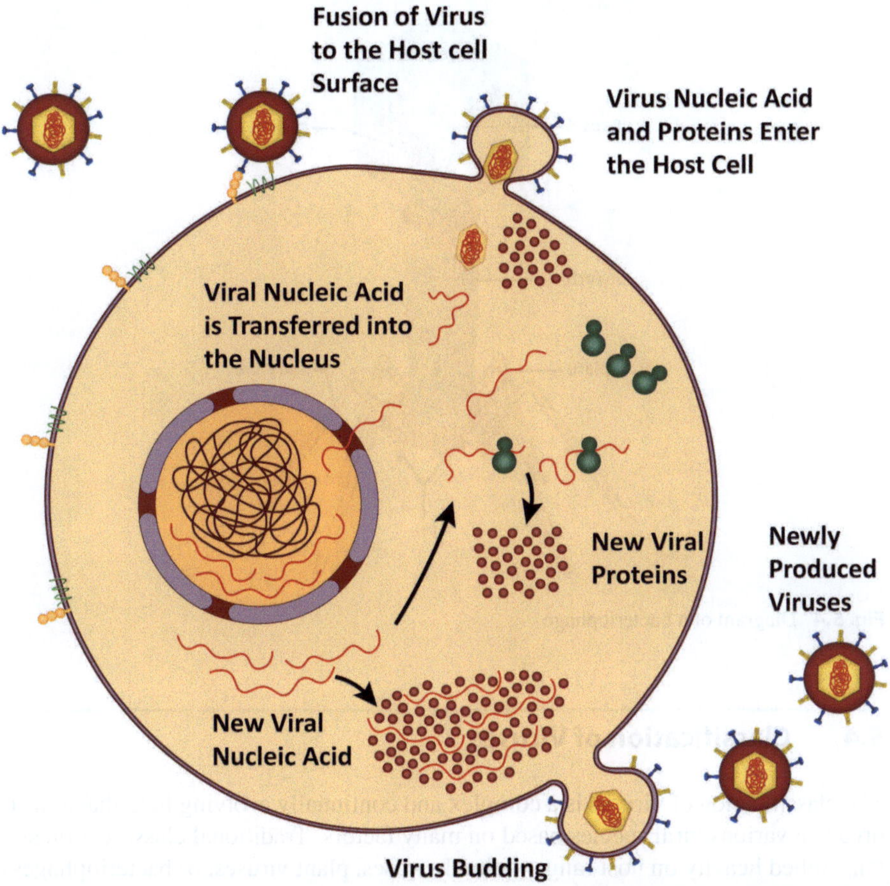

Fig. 5.3 Replication cycle of animal viruses. When the virus capsid with its nucleic acid enters the animal cell, the nucleic acid and the capsid proteins multiply and start the assembly process. The newly formed viruses bud out of the cell through the cell plasma membrane, keeping the plasma membrane as the virus's membrane

minimum use of materials, making it highly efficient. Some viruses possess helical symmetry. The genome is typically surrounded by a helical arrangement of protein subunits in these viruses. Complex viruses do not fit neatly into either category and often have additional structures like tails or layered protein coats. The most well-known examples are bacteriophages (Fig. 5.4), which often have icosahedral "heads" attached to helical "tails." Some viruses are pleomorphic, meaning they can exist in multiple shapes. Their lipid envelope allows for this variability, making them more flexible in their interactions with host cells.

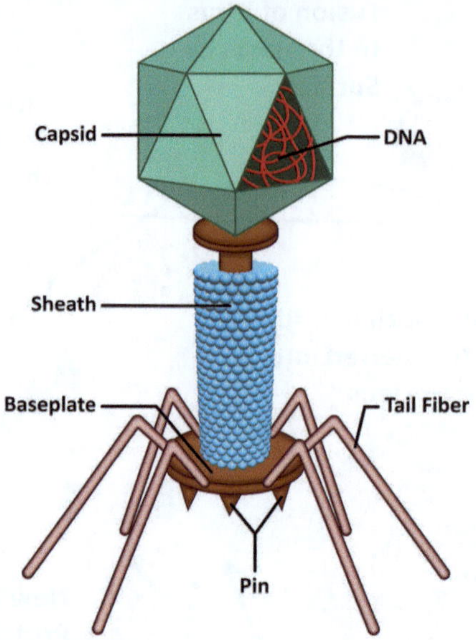

Fig. 5.4 Diagram of a bacteriophage

5.4 Classification of Viruses

The classification of viruses is a complex and continually evolving field that aims to organize various viral species based on many factors. Traditional classification systems relied heavily on host range (animal viruses, plant viruses, or bacteriophages), morphology, and biochemical properties [type of nucleic acid (DNA or RNA), polarity (positive-sense or negative-sense), and replication mechanisms].

However, the advent of molecular biology techniques, particularly whole-genome sequencing, has revolutionized our understanding and categorization of viruses. The Baltimore classification classifies viruses into seven groups based on their genome structure and the pathway used to synthesize messenger RNA (mRNA):

- Group I: double-stranded DNA viruses (e.g., herpesviruses)
- Group II: single-stranded DNA viruses (e.g., parvoviruses)
- Group III: double-stranded RNA viruses (e.g., reoviruses)
- Group IV: single-stranded RNA viruses with positive-sense RNA (e.g., coronaviruses)
- Group V: single-stranded RNA viruses with negative-sense RNA (e.g., influenza viruses)
- Group VI: single-stranded RNA retroviruses with a DNA intermediate in their life cycle (e.g., HIV)
- Group VII: double-stranded DNA retroviruses with an RNA intermediate (e.g., Hepadnaviruses like hepatitis B)

5.5 Viral Replication

The viral replication cycle is a complex, multi-step interplay between the virus and its host cell. It involves attachment (adsorption), entry (penetration), genome replication, assembly, and eventual release of new viral progeny. In eukaryotic cells, acute infections generally lead to rapid replication and cell death, whereas persistent infections can be chronic or latent, involving periods of reduced activity.

5.5.1 Replication of Animal Viruses

Animal virus replication can be broadly categorized into steps (Fig. 5.3):

(a) Attachment: The viral particle, or virion, attaches to specific receptors on the host cell surface. The specificity of this interaction determines the virus's host range.
(b) Entry: The virion or its genome enters the host cell following attachment. Mechanisms include membrane fusion, endocytosis, and direct penetration.
(c) Uncoating: The viral genome is released from its protein coat, making it accessible for replication and transcription.
(d) Genome replication: DNA viruses often use the host cell's replication machinery, while RNA viruses usually carry their own polymerases. Some viruses replicate in the nucleus, while others do in the cytoplasm.
(e) Transcription and translation: Viral genes are transcribed into messenger RNA (mRNA), then translated into viral proteins.
(f) Capsid formation: Newly synthesized viral proteins form the capsid.
(g) Genome packaging: The viral genome is packaged into the newly formed capsid. Some viruses encapsulate additional elements like enzymes necessary for the next round of replication.
(h) Release: Enveloped viruses often exit the host cell by budding, acquiring their envelope from the host cell membrane.

Nonenveloped viruses and some bacteriophages often rupture the host cell membrane to release new virions. Some viruses are released in a controlled manner via the host's secretory pathway (exocytosis).

5.5.2 Replication of Bacteriophages

The replication of bacteriophages takes the same steps, similar to animal viruses, but with some differences in the process. Bacteriophages exhibit different types of life cycles, primarily the lytic and lysogenic cycles (Fig. 5.5). In the lytic cycle, the phage replicates rapidly and lyses the bacterial cell to release progeny phages. In the lysogenic cycle, the phage genome integrates into the bacterial chromosome or exists as a plasmid, becoming a prophage. No progeny virions are produced until specific conditions trigger the lytic cycle induction.

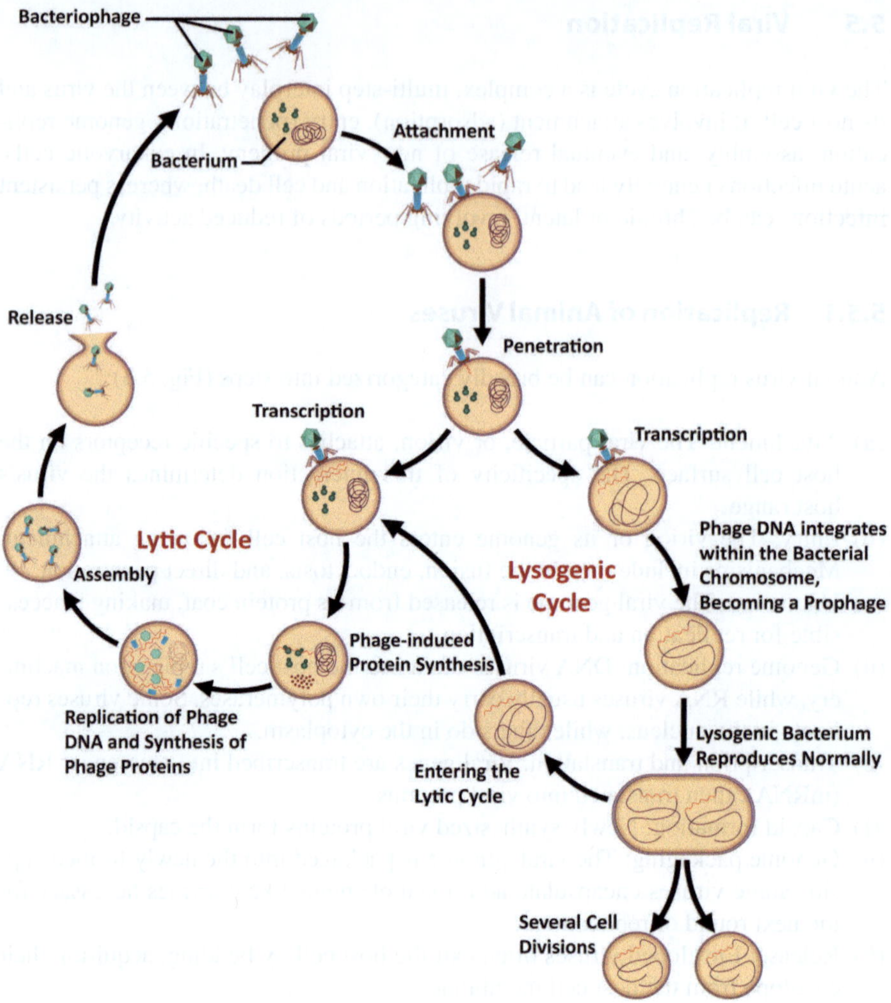

Fig. 5.5 Replication cycle of bacteriophages

5.6 Virus-Like Agents

Virus-like agents (VLAs) are a diverse group of infectious entities that share certain similarities with viruses but are not universally classified as such. These include viroids, prions, and satellite viruses, among others.

5.6.1 Satellites

The structure of satellite viruses may include a protein coat and a genome composed of RNA or DNA. They are much smaller than most viruses, often ranging from 17 to 24 nm in diameter. Satellite viruses depend on co-infection with a helper virus for replication and transmission. They possess their genome but lack essential genes for replication. Satellite viruses may modulate the pathology of diseases caused by their helper viruses.

5.6.2 Viroids

Viroids are small, circular, single-stranded RNA molecules that lack a protein coat. Viroids typically range from 250 to 400 nucleotides and adopt highly stable secondary and tertiary structures, often resembling a rod-like or branched configuration. Viroids exist inside cells, usually inside the nuclei. Viroids infect plant cells and hijack the host's transcriptional and translational machinery for their replication, often leading to plant diseases.

5.6.3 Prions

Prions (proteinaceous infectious) are infectious protein particles devoid of nucleic acids. Prions are made solely of proteins and may exist in different conformations. The infectious form of the prion protein is β-sheet rich and aggregates to form amyloid fibrils. Prions are resistant to inactivation by heating to 90 °C, which will inactivate viruses, radiation treatment that damages virus genomes, and enzymatic digestion that digests DNA or RNA but are sensitive to protein denaturing agents, such as phenol and urea. Normal prion protein has amino acids in the alpha helix with less than 5% of beta sheets. Abnormal prion protein is a misfolded protein, where most alpha helices are converted into beta sheets, resulting in more than 50% of beta sheets.

Prions cause neurodegenerative diseases in animals and humans, such as Creutzfeldt–Jakob disease (mad cow disease) and bovine spongiform encephalopathy. They induce conformational changes in normal cellular proteins, leading to their aggregation and cell death, predominantly in neural tissue.

Suggested Reading

Basic Virology, 4th Ed. Martinez J. Hewlett, et al. Wiley-Blackwell, 2021

Flexible Viruses: Structural Disorder in Viral Proteins, 1st Ed. Vladimir Uversky, Sonia Longhi, Wiley, 2011

Jawetz, Melnick, & Adelberg's Medical Microbiology, 28th Eds. Stefan Riedel, et al. McGraw-Hill Education, 2019

Principles of Virology, 5th Ed. Theodora Hatziioannou, et al. ASM Press, 2020

Sherris & Ryan's Medical Microbiology, 8th Ed, Kenneth J. Ryan. McGraw-Hill, 2022

Virology: Principles and Applications, 2nd Ed. John Carter, Venetia Saunders, Wiley, 2013

6 Tissues

In this chapter, you will learn

- Definition and function of tissues in the human body
- Types of tissues: epithelial, connective, muscular, and nervous
- Characteristics and functions of epithelial tissue
- Structure and function of the basement membrane in epithelial tissue
- Types of epithelial tissue: coverage and glandular
- Characteristics and types of simple and stratified epithelium
- Role and structure of glandular epithelium
- Types of glandular epithelium
- Overview of connective tissue and its functions in the body
- Components of connective tissue: cells, fibers, and ground substance
- Types of fibers in connective tissue: collagen, elastic, and reticular
- Classification of connective tissues based on structure and function
- Specialized connective tissues: cartilage, bone, and blood
- Overview of muscle tissue and its types: skeletal, cardiac, and smooth
- Structure and function of nervous tissue
- Phases of tissue repair: hemostasis, inflammatory response, proliferative phase, and remodeling phase
- Regenerative capacity of different tissues

Tissues (tissu = woven) are collections of cells that may be similar in structure and perform a common or related function. Tissues are intermediary organizations of complexity that bridge the gap between cells, the fundamental units of life, and organs, which are sophisticated arrangements of multiple tissue types functioning in a coordinated manner.

Furthermore, tissues are not merely aggregates of cells but also comprise a complex milieu of extracellular components—such as fibers, glycoproteins, and fluids—known collectively as the extracellular matrix (ECM).

H. Hosseinzadeh, *Fundamentals of Medicine for Biomedical Engineering*,
https://doi.org/10.1007/978-3-031-74085-5_6

Tissues are traditionally classified into four fundamental types based on their structural characteristics and functions: epithelial tissue, connective tissue, muscular tissue, and nervous tissue. These four basic tissues form the human body.

6.1 Part 1: Epithelial Tissue

Epithelial tissue, also known as the epithelium, serves as both guardian and regulator of the internal milieu against the external environment. It comprises a cell layer covering body surfaces, lines body cavities, or forms glands.

6.1.1 Characteristics

Epithelial cells are closely packed, manifesting minimal intercellular spaces and a scant extracellular matrix. The cells are generally polygonal in shape and exhibit strong lateral adhesions through specialized junctional complexes like tight junctions, adherens junctions, and desmosomes. These complexes afford mechanical strength and regulate paracellular transport (Fig. 6.1). One of the hallmark features of epithelial cells is their inherent polarity, characterized by an apical surface facing a lumen or external environment and a basal surface anchored to the underlying basement membrane. The apical and basal surfaces possess distinct proteins and lipids, thus enabling specialized functions like secretion, absorption, and filtration.

Basement membrane: The basement membrane (Fig. 6.2) is typically a thin, sheet-like structure that ranges from 50 to 100 nanometers in thickness, serving as the foundation upon which epithelial cells rest. It consists of two primary layers: the basal lamina, which is in direct contact with epithelial cells, and the reticular lamina, closer to the underlying connective tissue. The basement membrane is composed of a complex assembly of macromolecules that include:

- Collagens: Primarily type IV collagen (Table 6.1), which forms a network that provides tensile strength.
- Laminins: These are glycoproteins vital for the basement membrane's initial formation and provide anchoring sites for cells and other ECM components.
- Proteoglycans: These include heparan sulfate proteoglycans (e.g., perlecan, and agrin), which contribute to the filtration characteristics and structural stability.
- Entactins/nidogens: These are linking proteins that connect laminin and collagen networks, facilitating the structural integrity of the membrane.
- Fibronectins: Though not a core component, fibronectin links the basement membrane to the underlying connective tissue.

The basement membrane provides the essential scaffolding to maintain the spatial orientation and structural integrity of epithelial tissues. Through integrin receptors, epithelial cells anchor themselves to laminin and collagen molecules in the basement membrane, facilitating stable attachment. As a selective barrier, the

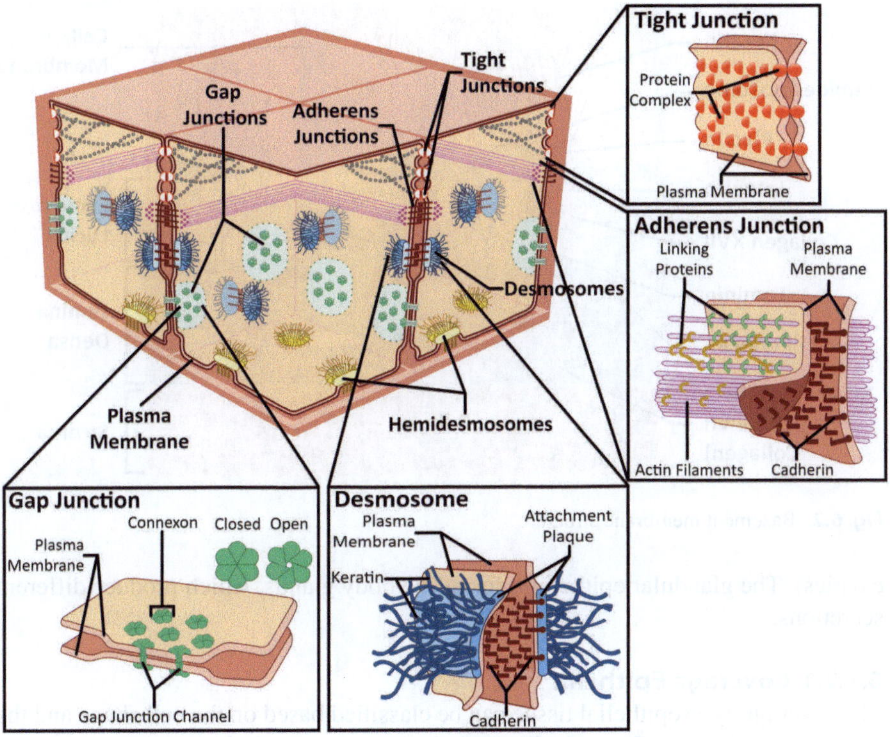

Fig. 6.1 Cell junctional complexes. **Tight junctions** are protein complexes forming a seal between adjacent cells. They prevent the passage of molecules and ions through the space between cells, creating a barrier to regulate the flow of substances in the intercellular space. **Adherens junctions** comprise cadherin proteins linked to actin filaments inside the cell. They provide strong mechanical attachments between adjacent cells, supporting tissue integrity and maintaining cell shape. **Desmosomes** comprise cadherins linked to intermediate filaments (keratin) through attachment plaques. They provide mechanical strength to tissues by anchoring cells to each other, allowing tissues to withstand mechanical stress. **Gap junctions** are formed by connexon protein complexes that create channels between adjacent cells. They facilitate direct communication between cells by allowing the passage of ions, nutrients, and other small molecules, enabling coordinated cellular activities

basement membrane regulates the passage of cells and macromolecules between the epithelium and the underlying tissues.

6.1.2 Types

The epithelial tissue is either coverage type or glandular type. The coverage epithelium covers all the body's external surfaces and the internal cavities inside the surface (gastrointestinal tract, urogenital system, abdominopelvic, and thoracic

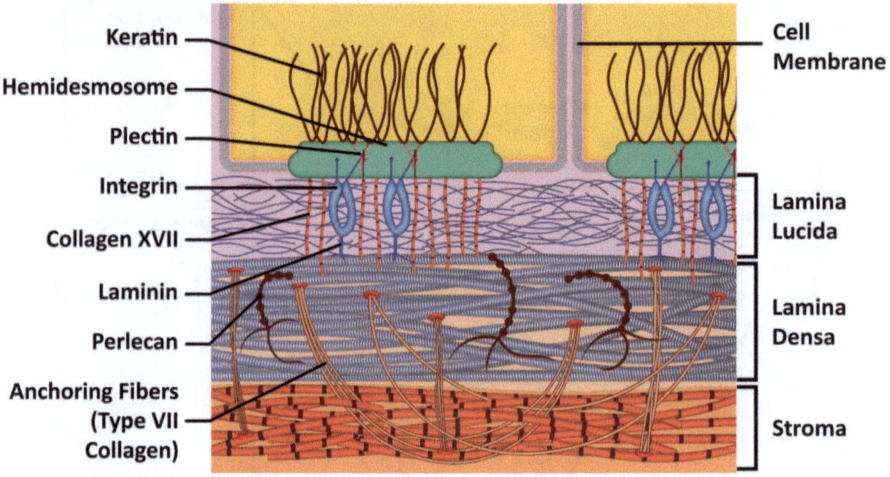

Fig. 6.2 Basement membrane structure

cavities). The glandular epithelium forms the body glands, which produce different secretions.

6.1.2.1 Coverage Epithelia

The coverage-type epithelial tissue can be classified based on the cell shape and the number of cell layers. The cell shape can be squamous, cuboid, or columnar; based on the number of cell layers, it can be simple or stratified.

Simple Epithelium

Simple epithelium (Fig. 6.3) is characterized by a single layer of similar cells directly resting upon a basement membrane. Given its monolayer organization, it is particularly specialized for filtration, absorption, and secretion processes. Often lining the internal surfaces of organs and body cavities, simple epithelium plays a pivotal role in metabolic exchange, barrier function, and sensing. Based on the cell shape, simple epithelium can be categorized into the following subtypes:

- Simple squamous epithelium (Fig. 6.3): Composed of flat, scale-like cells, this subtype is particularly suited for passive diffusion and filtration. It is commonly found in areas where rapid exchange of substances is required, such as the alveoli in the lungs, Bowman's capsule in the kidneys, and the lining of blood vessels (endothelium).
- Simple cuboidal epithelium (Fig. 6.3): This subtype consists of cube-shaped cells specialized for secretion and absorption. It is typically found in glands, the thyroid, and segments of the renal tubules.
- Simple columnar epithelium (Fig. 6.3): This subtype is characterized by tall columnar cells and is adept at secretion and absorption. It predominantly lines the gastrointestinal tract and may feature specialized structures like microvilli on the apical surface for increased functional efficiency.

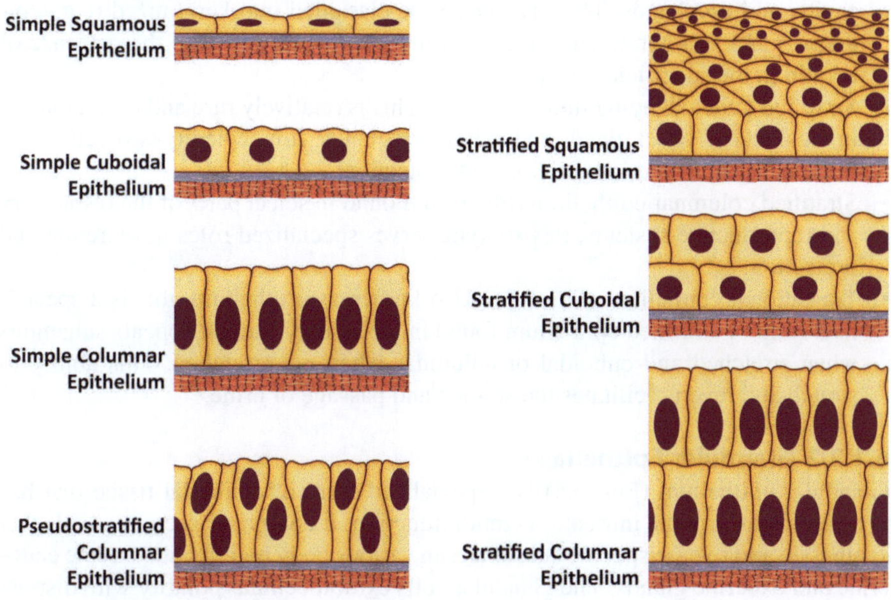

Fig. 6.3 Classification of epithelial tissues

- Pseudostratified columnar epithelium (Fig. 6.3): Although technically a form of simple epithelium, this subtype gives a false impression of being stratified due to the varying heights of its cells. Commonly ciliated, it is primarily involved in secretion and the movement of mucus and is found in the respiratory tract.

Stratified Epithelium

Stratified epithelium (Fig. 6.3), distinguished by its multi-layered cellular organization, is a form of epithelial tissue primarily geared toward protecting against physical and chemical insults. This tissue type is predominantly found in areas of the body subjected to wear and tear, such as the skin, mouth, and esophagus.

The layers of cells of the stratified epithelium can range from a few to several, depending on the location and functional requirements. While the cells in the basal layer are in contact with the basement membrane, only the most superficial layer interacts with the external environment or lumen. Hence, cellular polarity is not as evident as in simple epithelium. Cells in the basal layer are typically mitotically active, replenishing cells lost or damaged in the superficial layers. Like simple epithelia, stratified epithelia also contain various intercellular junctions like tight junctions, desmosomes, and gap junctions that confer structural integrity and enable cellular communication. Based on the cell shape in the most superficial layer, stratified epithelium can be categorized into the following subtypes:

- Stratified squamous epithelium (Fig. 6.3): It is the most common type, primarily found in areas subjected to mechanical stress and friction like the skin, oral cavity, and esophagus. This can further be classified into keratinized (e.g., epidermis) and non-keratinized (e.g., oral mucosa) types based on the presence of keratin in the superficial layers.

- Stratified cuboidal epithelium (Fig. 6.3): This is relatively rare and found mainly in the ducts of sweat glands and mammary glands. It is typically two-cell layers thick and is involved in secretion and protection.
- Stratified columnar epithelium (Fig. 6.3): Found in select parts of the respiratory and reproductive systems, this subtype serves specialized roles in secretion and protection.
- Transitional epithelium (Fig. 6.4): Also known as urothelium, this is a specialized form of stratified epithelium found in the urinary tract. It appears squamous when stretched and cuboidal or columnar when not stretched. This ability to stretch and revert facilitates the storage and passage of urine.

6.1.2.2 Glandular Epithelia

Glandular epithelium (Fig. 6.5) is a specialized form of epithelial tissue that has evolved to perform the intricate secretion function. It comprises secretory cells that synthesize and release specific substances in various body regions, such as the endocrine and exocrine glands. The glandular cells exhibit cellular polarity with distinct apical, lateral, and basal domains, facilitating secretion and interactions with surrounding tissues. Their cellular machinery is geared toward high synthetic and secretory activity, characterized by a well-developed endoplasmic reticulum and Golgi apparatus.

Endocrine: Endocrine glands (Fig. 6.5) comprise glandular epithelial cells. They lack ducts and secrete hormones directly into the bloodstream. Examples include the thyroid, adrenal, and pituitary glands (Section 12.2).

Exocrine: Unlike endocrine glands, exocrine glands (Fig. 6.5) have specialized ducts that secrete substances like enzymes, sweat, or mucus directly into specific target areas, such as bodily cavities or the skin's surface. These glands play crucial

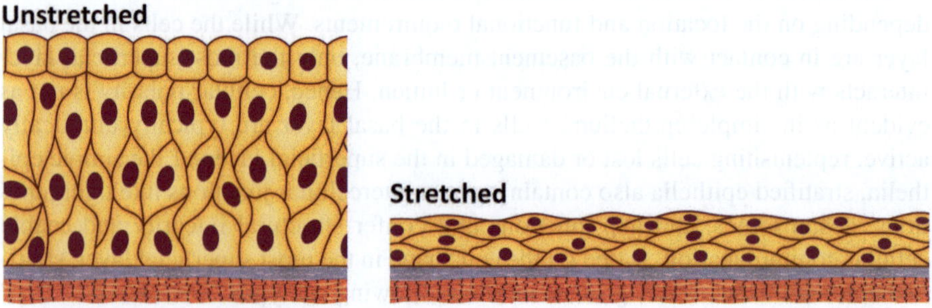

Fig. 6.4 Pseudostratified columnar tissue

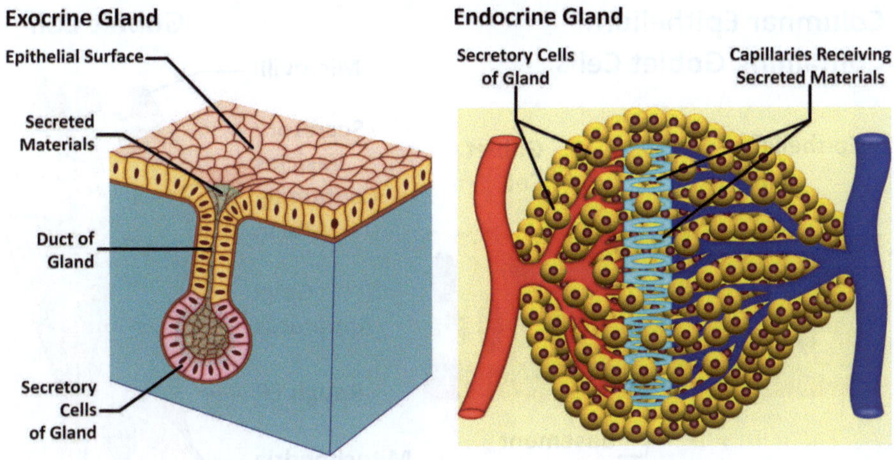

Fig. 6.5 Glandular epithelia, exocrine versus endocrine

roles in physiological processes ranging from digestion to thermoregulation and protection against microbial invaders. The basic architecture of exocrine includes

- Ductal system: One of the defining features of exocrine glands is the presence of ducts that transport secretory products from the glandular cells to the target site.
- Secretory units: These can be acini, tubules, or alveoli, and they are where the secretory products are actually synthesized and stored.
- Myoepithelial cells: Often found surrounding the secretory units, by contracting, these cells aid in expelling the glandular secretions.

Unicellular Glands

Unicellular glands represent a subclass of glandular epithelium characterized by individual secretory cells dispersed within a larger epithelial structure rather than organized as distinct glandular formations. The most widely cited example of a unicellular gland is the goblet cell (Fig. 6.6). Unicellular glands are commonly found in the respiratory tract (lining the airways to trap foreign particles), digestive system (present in the lining of the gastrointestinal tract, aiding in food passage and nutrient absorption), and reproductive systems (found in parts of the reproductive tracts, facilitating the passage of gametes).

Multicellular Glands

Unlike unicellular glands, multicellular glands manifest as discrete organ-like structures with dedicated functional units. They are ubiquitous across various body systems, serving essential roles from digestion and metabolism to reproduction and immune responses. The multicellular exocrine glands include the liver, pancreas, breast, prostate, seminal vesicle, sweat and sebaceous glands, and salivary glands.

The exocrine glands are classified (Fig. 6.7) based on their duct structure into simple (possess a single, unbranched duct) and compound lands (have branched

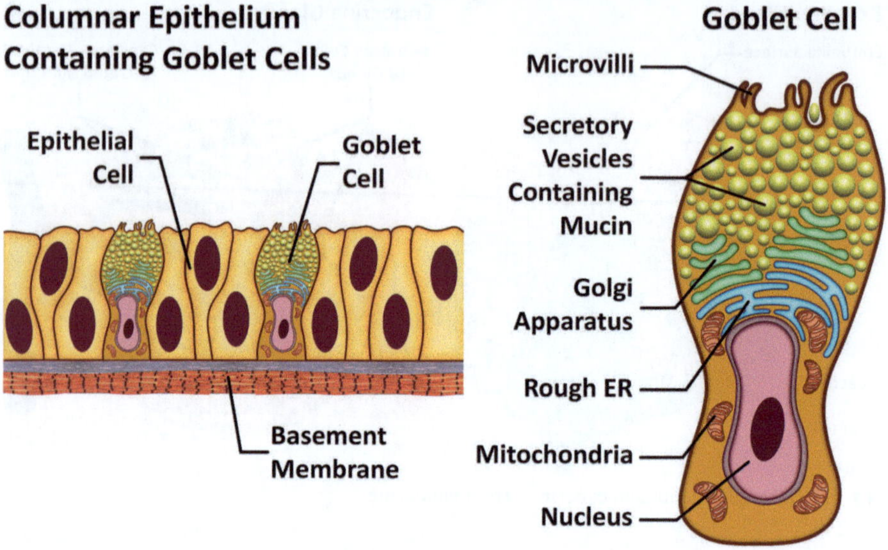

Fig. 6.6 Unicellular exocrine gland (Goblet cell)

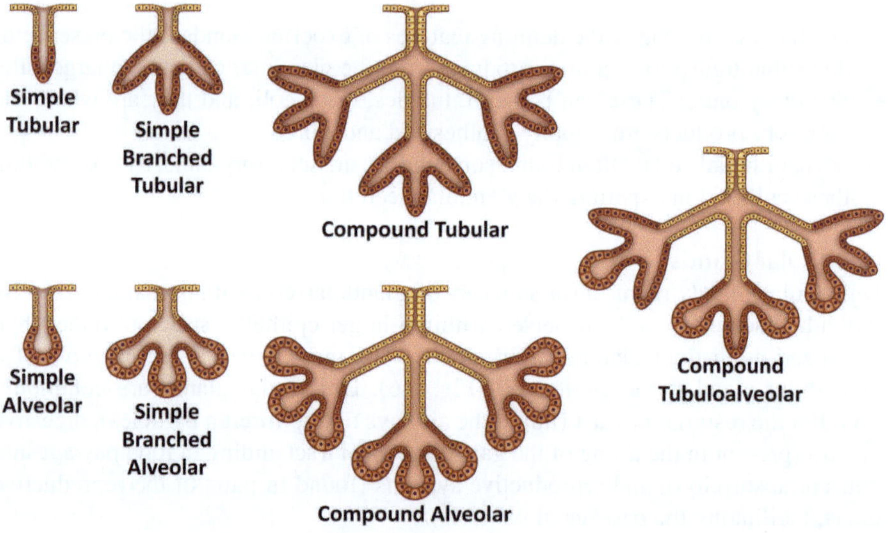

Fig. 6.7 Types of multicellular exocrine glands

ducts, often leading to multiple secretory units). The exocrine glands are classified based on their secretory units into tubular (the secretory units are tube-like in shape, e.g., intestinal glands), alveolar (acinar) (the secretory units resemble tiny sacs, e.g., mammary glands), and tubuloalveolar glands (these have tubular and alveolar components, e.g., salivary glands).

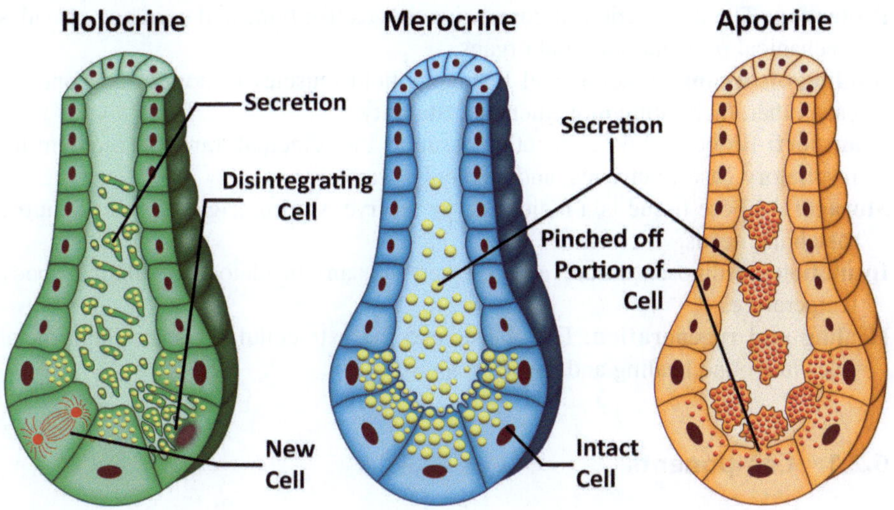

Fig. 6.8 Modes of secretion in human exocrine glands

Based on the mode of secretion (Fig. 6.8), the exocrine glands can be classified as:

- Merocrine: Secretion occurs through exocytosis without losing cell components, for example, sweat glands.
- Apocrine: Secretion involves the loss of apical portions of the cell, for example, mammary glands.
- Holocrine: The entire cell disintegrates to release its secretory products, for example, sebaceous glands.

Based on the type of secretion, they are classified into serous (produce thin, watery secretions, e.g., parotid glands), mucous (produce mucus-rich secretions, e.g., sublingual glands), and mixed (contain both serous and mucous components, e.g., submandibular glands).

6.2 Part 2: Connective Tissue

Connective tissues form a pivotal component of multicellular organisms, primarily as the supporting framework for various organs and systems. They can be found all over the body, with varying quantities in different organs (e.g., a small percentage of the brain, a more significant percentage in the skin, and almost all the tissues in a bone). There are four categories and various subcategories of connective tissue.

Although distinctive in their versatility and ubiquity, they perform a multitude of functions that include:

Structural support: Connective tissues provide a framework and support for organs and the body, contributing to their mechanical properties.

Protection: The distribution of connective tissues like bone and cartilage provides mechanical protection to vital organs.

Binding of organs: Tendons and ligaments hold muscles to bones and bones to each other, facilitating movement and stability.

Transport: Blood, a fluid connective tissue, is the principal transport medium for respiratory gases, nutrients, and metabolic waste.

Storage: Adipose tissue is a major energy reserve, storing triglycerides for future metabolic needs.

Insulation: Adipose tissue also serves as an insulator, regulating body temperature.

Healing and regeneration: Fibroblasts and the extracellular matrix play crucial roles in wound healing and tissue repair.

6.2.1 Components

Connective tissues contain three elements: ground substance, fibers, and cells (Fig. 6.9). The extracellular matrix is composed of both ground substance and fibers. The composition and arrangement of these elements can vary significantly, leading to a range of tissues.

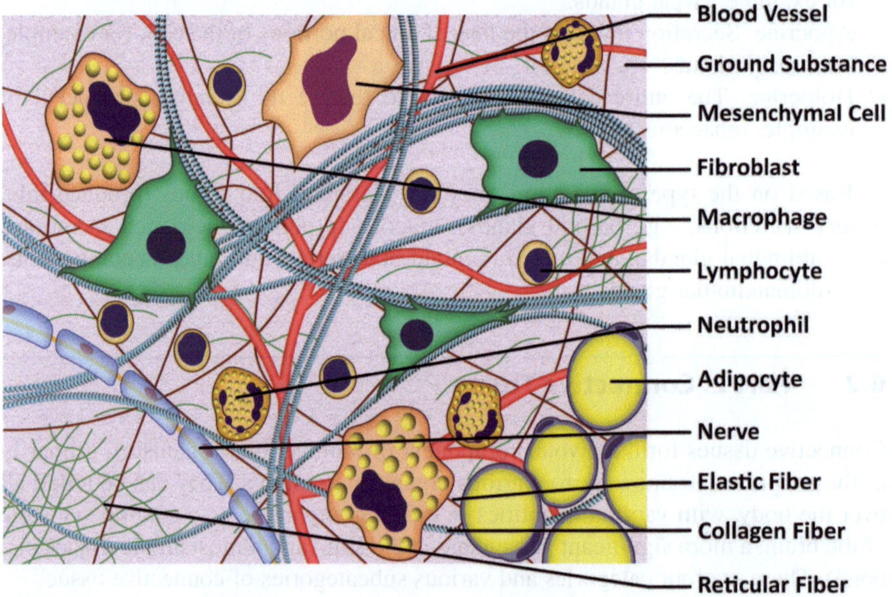

Fig. 6.9 A prototype areolar connective tissue

6.2.2 Ground Substance

The ground substance is a gel-like material that fills the spaces between cells and fibers. It contains glycosaminoglycans (GAGs), proteoglycans, and glycoproteins and acts as a medium for exchanging metabolites and signaling molecules.

6.2.2.1 Fibers

The fibers of connective tissue are proteins that provide support. The connective tissue matrix has three types of fibers: collagen, elastic, and reticular fibers. Of these, collagen fibers are by far the strongest and most abundant.

Collagen Fibers

Collagen fibers (Fig. 6.10) provide tensile strength and resist stretching. They are the most abundant fibers in most connective tissues. The basic unit of a collagen fiber is the collagen molecule, a long, rod-like structure approximately 300 nm long and 1.5 nm in diameter. Each collagen molecule is a trimer consisting of three polypeptide chains known as alpha-chains. These chains are wound together in a right-handed triple helix. Each alpha-chain consists of a sequence of amino acids following the pattern Gly-X-Y, where "Gly" stands for glycine, and X and Y are often proline and hydroxyproline, respectively. Glycine's small size allows the chains to pack closely, making the helix more stable. The stability of the collagen molecule is further enhanced by covalent cross-links that occur both within and between the triple helices. Collagen molecules associate staggered to form a larger structure called a microfibril. In this assembly, adjacent collagen molecules are displaced from each other by about 67 nm, a length often referred to as the "D-period."

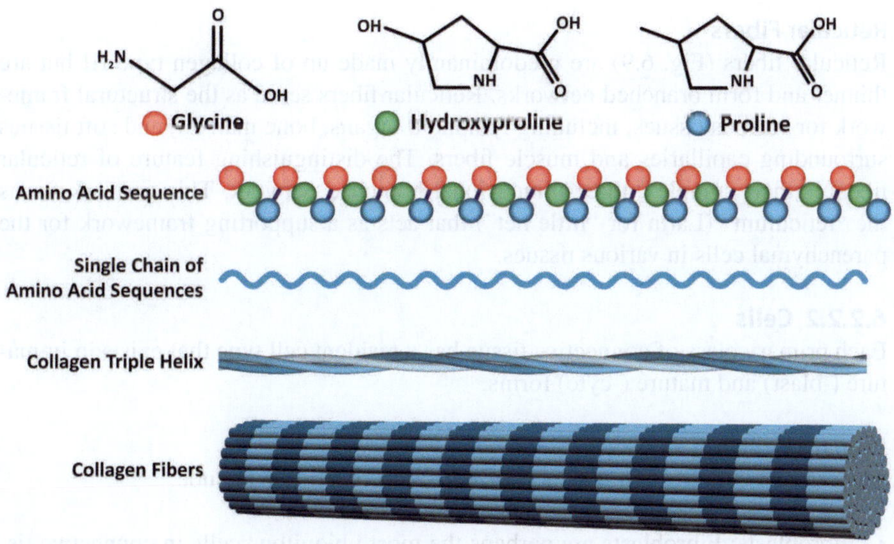

Fig. 6.10 Collagen fiber hierarchal structure

Microfibrils further aggregate to form collagen fibrils. The fibrils are cylindrical structures that vary widely in diameter, from a few nanometers to several micrometers, depending on the tissue type and physiological requirements. Collagen fibrils are bundled together to form collagen fibers, macroscopic structures visible under the light microscope. These fibers can be grouped in various orientations depending on the mechanical stresses the tissue experiences.

It is worth noting that different types of collagen (e.g., type I, II, III, etc.) may exhibit specialized structural attributes adapted for specific functions or tissue types (Table 6.1). For instance, type I collagen, prevalent in skin and bone, is incredibly strong and resists tensile force, whereas type II collagen, found in cartilage, offers more flexibility.

Elastic Fibers

Elastic fibers (Fig. 6.9) confer the critical properties of extensibility and elastic recoil to tissues, allowing them to return to their original shape after mechanical deformation. The primary components of elastic fibers are elastin and fibrillin-rich microfibrils. Elastin is a hydrophobic protein composed predominantly of glycine and proline amino acids. Unlike collagen, elastin lacks a regular secondary structure and exists in a more random coil configuration. This loose structure allows for high elasticity, enabling elastin to stretch up to five times its relaxed length without breaking. Fibrillin is a glycoprotein that forms the microfibrils' backbone surrounding the elastin core. Fibrillin provides structural support. In mature elastic fibers, a core of cross-linked elastin is encased by a mantle of fibrillin-rich microfibrils. In the walls of large arteries, they are organized into concentric sheets called lamellae. A branched network of elastic fibers in the lungs accommodates the tissue's need for frequent stretching and recoiling during breathing.

Reticular Fibers

Reticular fibers (Fig. 6.9) are predominantly made up of collagen type III but are thinner and form branched networks. Reticular fibers serve as the structural framework for various tissues, including lymphoid organs, bone marrow, and soft tissues surrounding capillaries and muscle fibers. The distinguishing feature of reticular fibers is their complex, three-dimensional branched network. This network forms the "reticulum" (Latin for "little net") that acts as a supporting framework for the parenchymal cells in various tissues.

6.2.2.2 Cells

Each primary class of connective tissue has a resident cell type that exists in immature (-blast) and mature (-cyte) forms.

Primary Cells

The primary cell types in nonspecialized connective tissue include:

- Fibroblasts: Fibroblasts are perhaps the most ubiquitous cells in connective tissue. They synthesize the extracellular matrix (ECM), including collagen, elastin,

Table 6.1 Types of collagen, their locations in the human body, and primary functions

Type of collagen	Location in the body	Primary function
Type I	Skin, tendon, vascular ligature, organs, bone	Provides structure to skin, bones, tendons, fibrous cartilage
Type II	Cartilage (including hyaline cartilage)	Provides cushioning in joints, vital for cartilage elasticity
Type III	Skin, muscle, blood vessels	Supports the structure of muscles, organs, and arteries
Type IV	Basal lamina, eye lens	Helps with filtration and is found in the layers of the skin
Type V	Cell surfaces, hair, and placenta	Aids in cell surface properties, hair, and placental integrity
Type VI	Interstitial spaces	Essential for cell adhesion and migration
Type VII	Beneath stratified squamous epithelia	Anchors fibrils to basal keratinocytes
Type VIII	Some endothelial cells	Plays a role in angiogenesis
Type IX	Cartilage	Supports and stabilizes the collagen fibrils in cartilage
Type X	Hypertrophic and mineralizing cartilage	Involved in bone formation
Type XI	Cartilage	Regulates the size of collagen fibrils
Type XII	Connective tissues associated with type I collagen	Links collagen to other matrix components
Type XIII	Widely distributed in the body	Believed to play a role in cell interaction and migration
Type XIV	Fetal tissues, other connective tissues	Associates with the surface of collagen fibrils
Type XV	Various tissues	Involves in matrix organization and vascular development
Type XVI	Connective tissues	Possibly involved in the organization of the extracellular matrix
Type XVII	Epidermal basement membrane	Necessary for the integrity and structure of the skin
Type XVIII	Vascular and visceral basement membranes	Involved in angiogenesis and cell survival
Type XIX	Various tissues	Presumed to influence cell behavior and tissue repair
Type XX	Undetermined	Function is not yet fully understood
Type XXI	Blood vessels, gastrointestinal tract, and skin	Influences cell adhesion and migration
Type XXII	Cartilage, eye, and intervertebral disc	Involved in the assembly of other collagen types
Type XXIII	Epithelial structures	Associated with cell differentiation and tissue development
Type XXIV	Undetermined	Function is not yet fully understood
Type XXV	Brain	Implicated in neurogenesis and synaptic plasticity
Type XXVI	Undetermined	Function is not yet fully understood
Type XXVII	Cartilage	Involves in the development of cartilage
Type XXVIII	Various tissues	Function is not yet fully understood

and other fibers and glycosaminoglycans (GAGs). Fibroblasts exhibit an elongated, spindle-shaped morphology, usually interspersed among collagen fibers.

- Macrophages: Functioning as tissue scavengers, macrophages are highly phagocytic cells that clear pathogens and cellular debris. They are derived from monocytes and may exist in various tissue-specific forms, such as liver Kupffer cells and central nervous system microglia.
- Mast cells: Located near vascular structures, mast cells are involved in inflammation and allergic responses. They contain cytoplasmic granules filled with histamine, heparin, and various enzymes.
- Mesenchymal stem cells: These are multipotent stem cells that can differentiate into various cell types, such as fibroblasts, adipocytes, and chondrocytes. They play a crucial role in tissue repair and regeneration.

Specialized Cells

The cells in the specialized connective tissues, like bone, cartilage, and adipose tissue, are also specialized, including:

- Osteocytes: Osteocytes synthesize and maintain bone matrix and calcium homeostasis.
- Chondrocytes: Chondrocytes produce cartilage-specific ECM components like chondroitin sulfate.
- Adipocytes: Adipocytes, or fat cells, are specialized for storing triglycerides, which serve as an energy reservoir.

6.2.3 Classification

Connective tissues are classified based on their constituent cells, extracellular matrix (ECM) composition, and specific functions.

6.2.3.1 Loose Connective Tissue

- Areolar tissue (Fig. 6.9): This is the most common type of loose connective tissue found in the subcutaneous layer, mucous membranes, and around blood vessels. It contains a variety of cells, including fibroblasts, macrophages, and mast cells. The ECM primarily comprises collagen and elastin fibers interspersed in a ground substance of glycoproteins and glycosaminoglycans (GAGs).
- Adipose tissue (Fig. 6.11): Predominantly composed of adipocytes, this tissue acts as an energy storage depot, insulator, and cushion for organs. There are two types of adipose tissue: white and brown. White adipose tissue stores triglycerides and is commonly found in the subcutaneous layer. Brown adipose tissue is specialized for thermogenesis and is mainly located around the neck and large blood vessels.

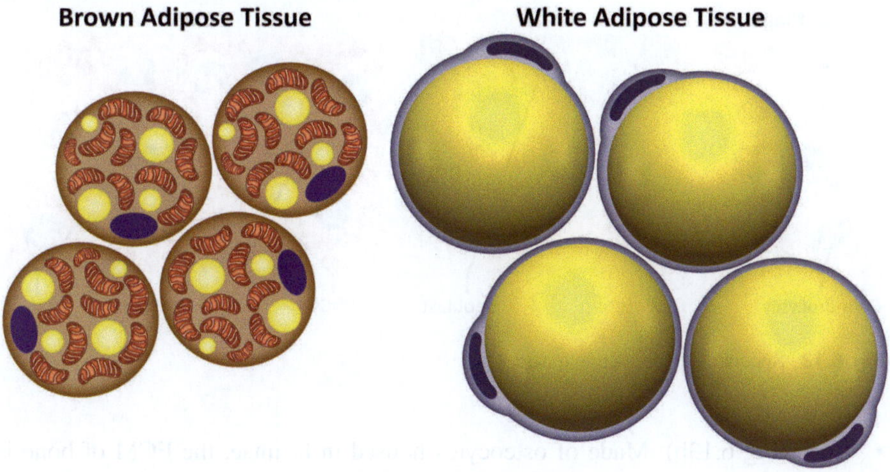

Fig. 6.11 Adipose tissue, brown versus white tissue

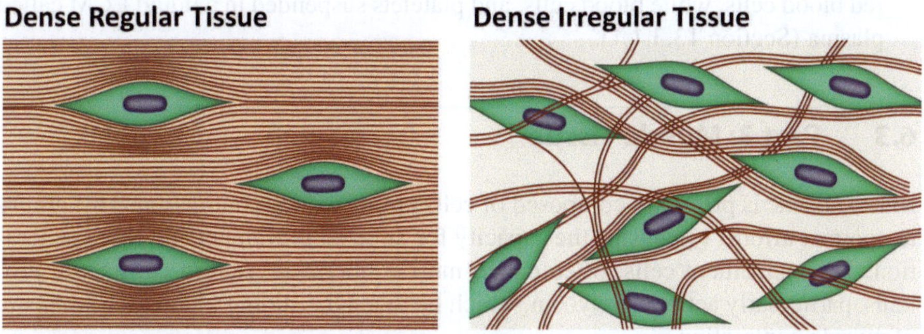

Fig. 6.12 Types of dense connective tissue

6.2.3.2 Dense Connective Tissue (Fig. 6.12)

- Dense regular tissue: Characterized by closely packed collagen fibers arranged in parallel, this tissue provides tensile strength and is found in tendons (Section 9.1.2.3) and ligaments (Section 8.2).
- Dense irregular tissue: Unlike dense regular tissue, the collagen fibers in dense irregular tissue are not organized in a specific direction. This provides strength and flexibility in multiple directions. It is commonly found in the skin's dermis and the organs' fibrous capsules.

6.2.3.3 Specialized Connective Tissues

- Cartilage (Fig 6.13a): Composed of chondrocytes embedded in an ECM rich in chondroitin sulfate and collagen fibers, cartilage provides a flexible framework for structures like the ear, nose, and intervertebral discs. There are three types: hyaline, elastic, and fibrocartilage, each with different fiber compositions and mechanical properties (Section 8.1.2).

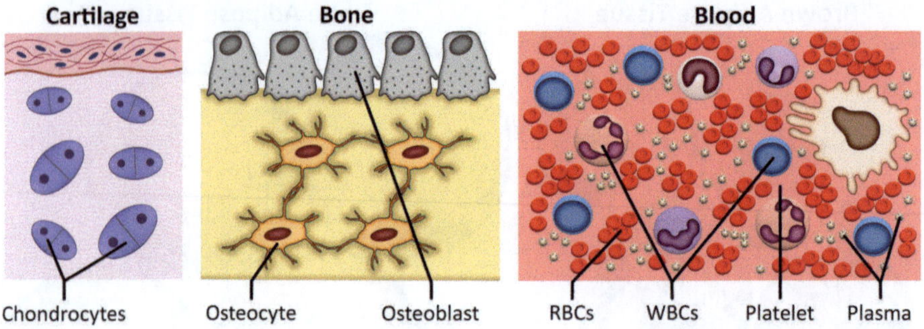

Fig. 6.13 Specialized connective tissues

- Bone (Fig 6.13b): Made of osteocytes housed in lacunae, the ECM of bone is rich in hydroxyapatite crystals and collagen fibers, providing rigidity and structural support (Section 8.3.2).
- Blood (Fig 6.13c): Often considered a fluid connective tissue, blood comprises red blood cells, white blood cells, and platelets suspended in a liquid ECM called plasma (Section 13.3.1.1).

6.3 Part 3: Muscle Tissue

Muscle tissue is primarily composed of cells optimized for contraction. This ability to contract affords organisms the capacity for directed movement and force generation. The constituent cells, often called muscle fibers, are rich in contractile proteins, particularly actin and myosin, which facilitate the sliding filament mechanism of muscle contraction.

Muscle tissue can be broadly categorized into three main types based on structural and functional characteristics.

6.3.1 Skeletal Muscle

Skeletal muscle fibers (Fig 6.14a) are elongated, multinucleated cells arranged parallelly. Due to the orderly arrangement of actin and myosin filaments, they exhibit a striated appearance under microscopic examination (Section 9.1.2.6). Skeletal muscles are primarily responsible for voluntary movements and are usually attached to bones via tendons. They also play an essential role in maintaining posture and generating heat.

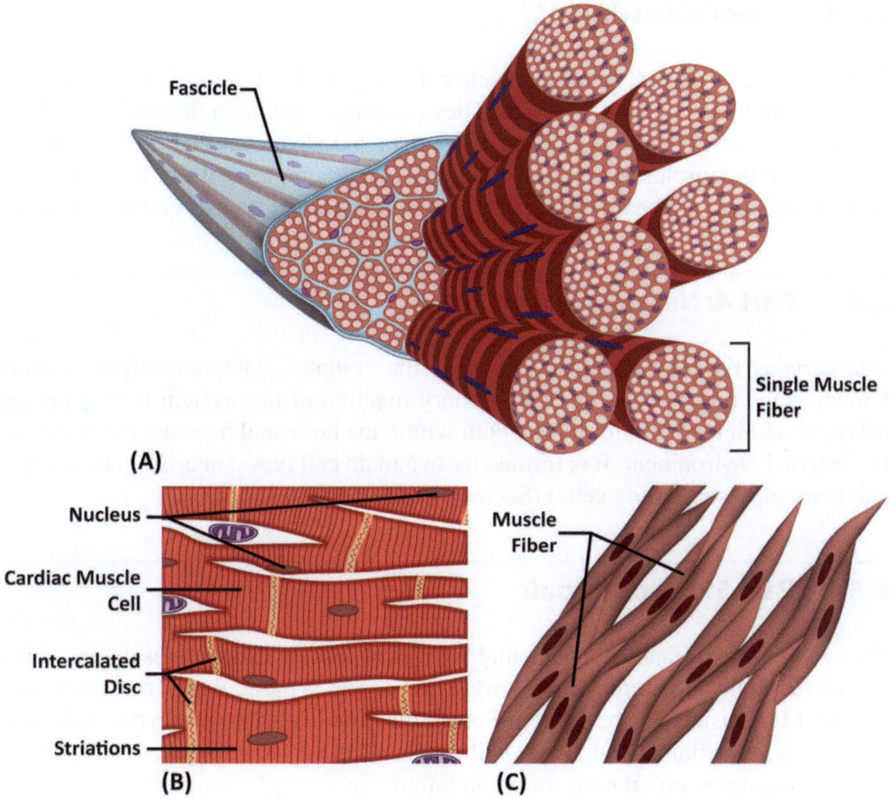

Fig. 6.14 Types of muscle tissue: (**a**) skeletal muscle; (**b**) cardiac muscle; (**c**) smooth muscle

6.3.2 Cardiac Muscle

Cardiac muscle cells (Fig 6.14b), or cardiomyocytes, are striated like skeletal muscles but are shorter, branched, and interconnected through specialized junctions known as intercalated discs, which contain gap junctions for electrical coupling (Section 13.1.1.2). Found only in the heart, cardiac muscle tissue contracts involuntarily to pump blood throughout the circulatory system. The autonomic nervous system modulates the intrinsic rhythmicity of these muscles.

6.3.3 Smooth Muscle

Unlike skeletal muscles, smooth muscle cells (Fig 6.14c) are non-striated, spindle-shaped, and contain a single nucleus. They are usually found in the walls of hollow organs and structures, such as the digestive tract, blood vessels, and respiratory passages. Smooth muscles function involuntarily, facilitating activities like the propulsion of food through the digestive tract or regulating blood flow through the arteries.

6.4 Part 4: Nervous Tissue

The nervous tissue is the cornerstone of the complex, integrative body system known as the nervous system. The primary function of this system is to generate, transmit, and integrate information both within the body and between the body and the external environment. It is formed by two main cell types: neurons (nerve cells) and neuroglia (supporting cells) (Section 10.1).

6.5 Part 5: Tissue Repair

Tissue repair represents a fundamental biological process that restores damaged tissue's structural and functional integrity. This process is particularly crucial for multicellular organisms, where the coordinated function of various cell types and organ systems is essential for survival. Failure in effective tissue repair can result in chronic inflammation, fibrosis, or organ failure.

6.5.1 Phases of Tissue Repair (Fig. 6.15)

Hemostasis: The immediate response to tissue injury is ceasing bleeding via platelet aggregation and fibrin mesh formation.

Inflammatory response: Immune cells, predominantly neutrophils and macrophages, are recruited to the injury site to clear debris and pathogens.

Proliferative phase: Includes three phenomena:

- Angiogenesis: New blood vessels form to supply nutrients and oxygen.
- Fibroblast proliferation: Fibroblasts synthesize new extracellular matrix components, primarily collagen.
- Epithelialization: Epithelial cells migrate to cover the wound, restoring the barrier function.

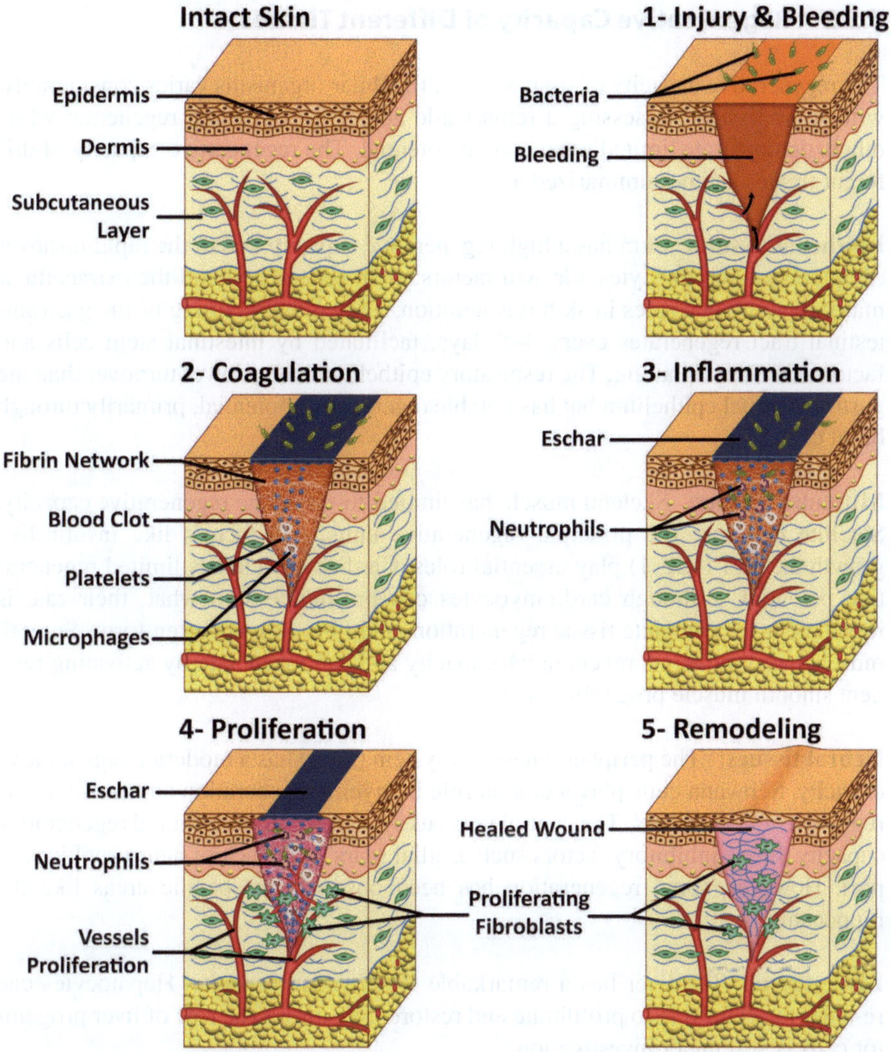

Fig. 6.15 Phases of tissue repair

Remodeling phase: Newly formed tissue changes to regain its original structure and function. Some scars may persist while excess collagen and other ECM components are degraded during remodeling.

6.5.2 Regenerative Capacity of Different Tissues

The regenerative capacity of tissues in multicellular organisms varies considerably, with some tissues possessing a remarkable ability to repair and regenerate while others demonstrate limited regenerative potential. The regenerative capacity of different tissues can be summarized as:

Epithelial tissues: Skin has a high regenerative capacity due to the rapid turnover of epidermal keratinocytes. Growth factors, stem cell niches, and the extracellular matrix play pivotal roles in skin regeneration. The epithelial lining of the gastrointestinal tract regenerates every 4–5 days, facilitated by intestinal stem cells and factors like Wnt signaling. The respiratory epithelium has a slower turnover than the gastrointestinal epithelium but has notable regenerative potential, primarily through basal cells.

Muscular tissues: Skeletal muscle has limited-to-moderate regenerative capacity. Satellite cells are the principal regenerative units, and factors like insulin-like growth factor 1 (IGF-1) play essential roles. Cardiac muscle has limited regenerative potential. Although cardiomyocytes can proliferate somewhat, their rate is insufficient for complete tissue regeneration, and fibrotic scars often form. Smooth muscle has a moderate regenerative capacity and can regenerate by activating resident smooth muscle progenitor cells.

Neural tissues: The peripheral nervous system (PNS) has a moderate regenerative capacity. Schwann cells play a crucial role in myelin regeneration and facilitate the regrowth of nerve fibers. The central nervous system (CNS) has limited regenerative capacity due to inhibitory factors such as glial scars and a lack of neurotrophic support. However, some regeneration has been observed in specific areas like the hippocampus.

Liver tissue: The liver has a remarkable regenerative capacity. Hepatocytes can re-enter the cell cycle to proliferate and restore liver mass. The role of liver progenitor cells is still under investigation.

Bone tissue: Bone tissue has an excellent regenerative capacity. Osteoblasts and osteoclasts are key players in bone remodeling, with factors like bone morphogenetic proteins (BMPs) being critical.

Suggested Reading

Advanced Physiology and Pathophysiology: Essentials for Clinical Practice, 2nd Ed. Nancy Tkacs, Linda Herrmann, Springer Publishing Company, 2024

Anatomy & Physiology, S. Meloni, et al. Independent Publisher, 2023

Anatomy & Physiology: The Unity of Form and Function, 10th Ed. Kenneth S. Saladin, McGraw Hill, 2023

Gray's Anatomy for Students, 5th Ed. Richard L. Drake, et al. Elsevier, 2020
Human Anatomy & Physiology, 11th Ed. Elaine N. Marieb and Katja Hoehn, Pearson, 2022
Junqueira's Basic Histology: Text and Atlas, 16th Ed. Anthony L. Mescher, McGraw Hill, 2021
Moore's Clinically Oriented Anatomy, 9th Ed. Keith L. Moore, et al. Lippincott Williams & Wilkins, 2020
Netter Atlas of Human Anatomy: Classic Regional Approach, 8th Ed. Frank H. Netter, Elsevier, 2022
Principles of Anatomy and Physiology

Integumentary System

7

In this chapter, you will learn

- Integumentary system's role as a biological interface between the organism and its external environment
- Skin as the largest organ, its general functions, and composition
- Structure of the epidermis and its role as a protective barrier
- Differentiation and functions of the layers of the epidermis
- Role and types of cells found in the epidermis, such as keratinocytes, melanocytes, and Langerhans cells
- Dermis, its structure, and its function in providing mechanical resilience and support
- Hair structure and function, including the growth cycle and different hair types
- Nail anatomy and the physiology of nail growth
- Types and functions of sweat glands, and the role of sweat in thermoregulation and waste elimination
- Sebaceous glands, their role in producing sebum, and the functions of sebum
- Cellular and biochemical basis of skin color, including the types of melanin and factors influencing skin pigmentation
- Skin's roles in sensation, immune function, and thermoregulation, detailing mechanisms like pain, touch, and temperature perception

The integumentary system serves as an essential biological interface between the organism and its external environment, functioning not merely as a passive barrier but as a complex, multifaceted organ system. It exemplifies an amalgamation of cellular structures, tissues, and biochemical mechanisms, all intricately coordinated to perform various functions.

H. Hosseinzadeh, *Fundamentals of Medicine for Biomedical Engineering*,
https://doi.org/10.1007/978-3-031-74085-5_7

7.1 Part 1: Skin

The skin, the most extensive organ of the human body, is the primary component of the integumentary system. It consists of a highly organized and stratified arrangement of cells and tissues. Traditionally classified into two main layers—the epidermis and dermis—skin architecture is a product of cellular specialization and complex biochemical interactions (Fig. 7.1).

7.1.1 Epidermis

The epidermis is the outermost layer of the skin and functions as the primary interface between the body and its external environment. It predominantly comprises keratinocytes but includes specialized cells such as melanocytes, Langerhans, and Merkel cells. These cellular constituents are organized into multiple strata or layers.

7.1.1.1 Layers of Epidermis

The layers, from outermost to innermost, are as follows: the stratum corneum, stratum lucidum (only in thick skin), stratum granulosum, stratum spinosum, and stratum basale (Fig. 7.2).

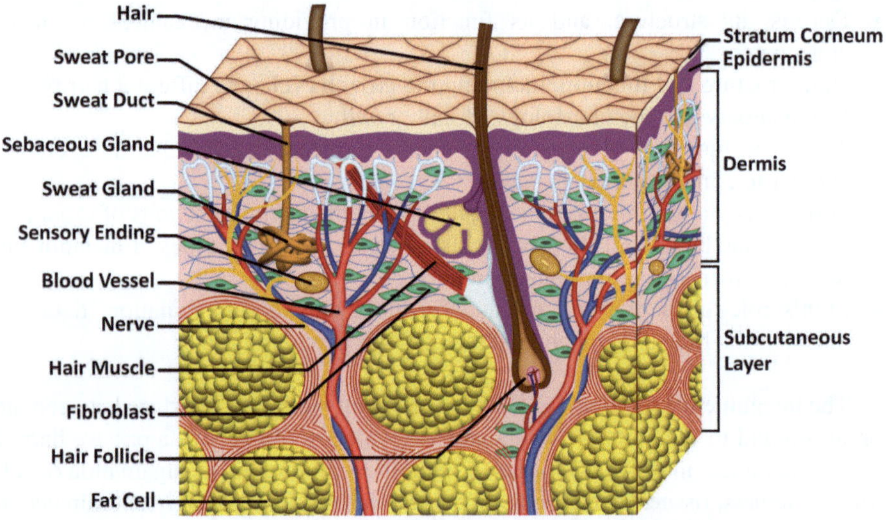

Fig. 7.1 Skin structure. Three-dimensional view of the skin and underlying subcutaneous tissue

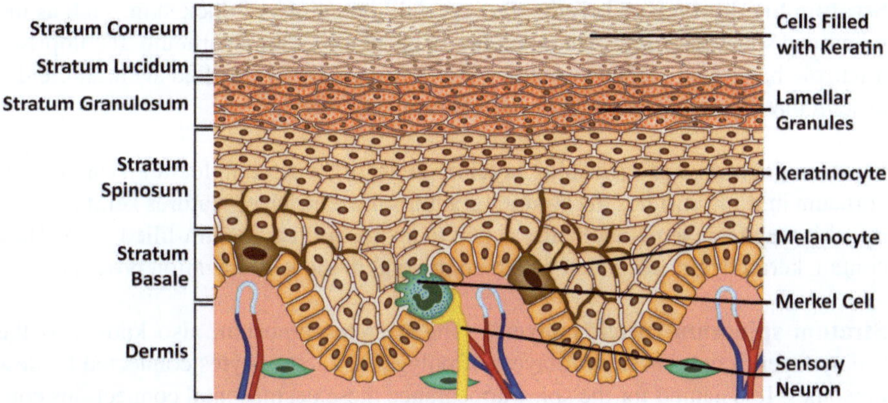

Fig. 7.2 Layers of the epidermis and epidermal cells

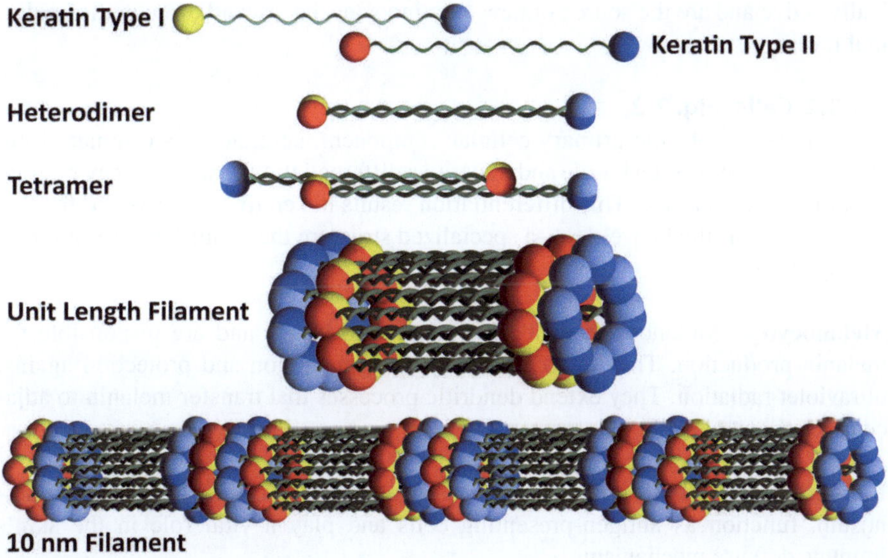

Fig. 7.3 Keratin hierarchical structure

Stratum corneum (horny layer): The stratum corneum is the most superficial layer and consists of dead, flattened, and anucleated keratinocytes filled with keratin filaments (Fig. 7.3). These cells are embedded in a lipid-rich extracellular matrix, forming a "brick and mortar" arrangement as a formidable barrier against water loss and pathogenic invasion.

Stratum lucidum (clear layer): Present only in regions of thick skin, such as the palms and soles, the stratum lucidum lies beneath the stratum corneum. It comprises a narrow band of translucent, densely packed keratinocytes filled with an eleidin protein that eventually transforms into keratin.

Stratum granulosum (granular layer): Beneath the stratum lucidum (or stratum corneum in thin skin) lies the stratum granulosum. This layer features keratinocytes transitioning from a metabolically active to an anucleated, keratin-filled state. They contain keratohyalin granules, which are pivotal in the keratinization process.

Stratum spinosum (prickly layer): The stratum spinosum, also known as the "spiny layer," contains several layers of polygonal keratinocytes connected by desmosomes. It is named for the spiny appearance these desmosomal connections confer under microscopic examination.

Stratum basale (basal layer): The stratum basale, the innermost layer, comprises a single layer of columnar or cuboidal keratinocytes. These cells are the most mitotically active and are the source of new keratinocytes that ascend through the epidermal layers.

7.1.1.2 Cells (Fig. 7.2)

Keratinocytes: As the primary cellular component, keratinocytes originate from stem cells in the stratum basale and undergo a differentiation process as they migrate toward the skin surface. This differentiation results in keratin synthesis and the formation of a cornified envelope—a specialized structure that contributes to skin barrier function.

Melanocytes: Melanocytes reside in the stratum basale and are responsible for melanin production. They contribute to skin pigmentation and protection against ultraviolet radiation. They extend dendritic processes that transfer melanin to adjacent keratinocytes.

Langerhans (dendritic) cells: Langerhans cells, located mainly in the stratum spinosum, function as antigen-presenting cells and play a vital role in the skin's immune defense mechanisms.

Tactile epithelial (Merkel) cells: These specialized cells are most commonly found in the stratum basale and are associated with nerve endings. They contribute to the skin's role in sensory perception, particularly light touch.

Various molecular components, including keratins, lipids, and enzymes, uphold the epidermis's structural integrity and functional efficacy. In particular, transglutaminases facilitate the formation of the cornified envelope. At the same time, ceramides, cholesterol, and fatty acids contribute to the lipid matrix in the stratum corneum, which is crucial for barrier function.

7.1.2 Dermis

The dermis, situated beneath the epidermis, is a robust scaffold that provides structural integrity and mechanical resilience to the skin. Its functions are far-reaching, encompassing thermoregulation, sensory perception, and immune response, among other physiological activities. The dermis comprises a complex network of collagen and elastin fibers, nerve endings, blood and lymphatic vessels, and specialized cells.

7.1.2.1 Layers of Dermis

The dermis is categorized into two primary layers based on histological characteristics and functional attributes: papillary and reticular (Fig. 7.4).

Papillary Dermis

The papillary dermis is the superficial layer, interfacing directly with the epidermis. It contains thin collagen fibers and elastin, and its primary purpose is to supply nutrients to the adjacent epidermis and regulate temperature. The layer is distinguished by its dermal papillae (Fig. 7.4)—protrusions that interdigitate with the epidermal ridges, increasing the surface area for exchanging oxygen, nutrients, and waste products. These dermal papillae also contribute to the mechanical adhesion between the dermis and the epidermis, reducing the likelihood of shear forces causing separation.

Reticular Dermis

The reticular dermis, the deeper and thicker of the two layers, comprises densely packed collagen fibers arranged parallel to the skin surface. This configuration offers the skin its tensile strength, resistance to stretching, and elasticity. The reticular layer also houses larger blood vessels, lymphatic vessels, nerve fibers, and various adnexal structures, such as hair follicles, sebaceous glands, and sweat glands (Fig. 7.4).

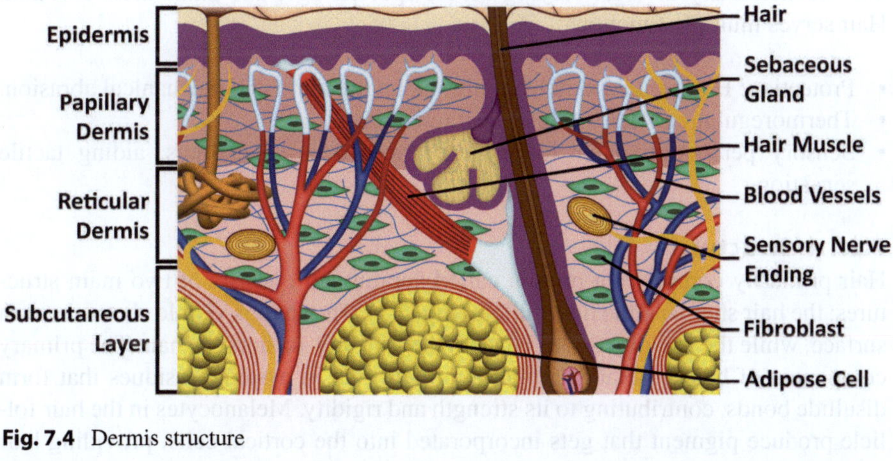

Fig. 7.4 Dermis structure

7.1.2.2 Cellular Components of Dermis

Fibroblasts: Fibroblasts are the principal cell type in the dermis and are responsible for the production and turnover of the extracellular matrix (ECM).

Mast cells: Predominantly found near blood vessels, mast cells release histamine and other inflammatory mediators, contributing to vasodilation and recruitment of immune cells to the site of injury or infection.

Macrophages: Dermal macrophages serve as phagocytic cells that ingest foreign substances and participate in antigen presentation. They are pivotal for the initiation of adaptive immune responses and tissue repair.

Other cells: Other cells in the dermis include pericytes, which stabilize blood vessels, and adipocytes, which are mainly localized in the deeper reticular dermis and hypodermis. The dermis is richly vascularized, supplying its cellular constituents and the avascular epidermis. It contains a dense network of sensory nerve endings, including mechanoreceptors like Meissner's corpuscles and thermoreceptors (Page 339), contributing to tactile and thermal sensation.

7.2 Part 2: Skin Appendages

The skin appendages—hair follicles, sebaceous glands, sweat glands, and nails (Fig. 7.1)—represent specialized structures originating from the epidermal layer and extending into the dermis and sometimes hypodermis.

7.2.1 Hair

Hair, an obvious and defining characteristic of mammals, is structurally complex and biochemically diverse, varying across species, age, and anatomical location. Hair serves multiple functions:

- Protection: Hairs shield skin from ultraviolet radiation and mechanical abrasion.
- Thermoregulation: Hairs help in heat retention.
- Sensory perception: Hair follicles are rich in nerve endings, aiding tactile sensation.

7.2.1.1 Structure

Hair primarily comprises a protein called keratin and consists of two main structures: the hair shaft and the hair follicle (Fig. 7.5). The shaft is visible above the skin surface, while the follicle resides beneath the skin and anchors the hair. The primary component of hair is keratin, a fibrous protein rich in cysteine residues that form disulfide bonds, contributing to its strength and rigidity. Melanocytes in the hair follicle produce pigment that gets incorporated into the cortical cells, providing hair its color.

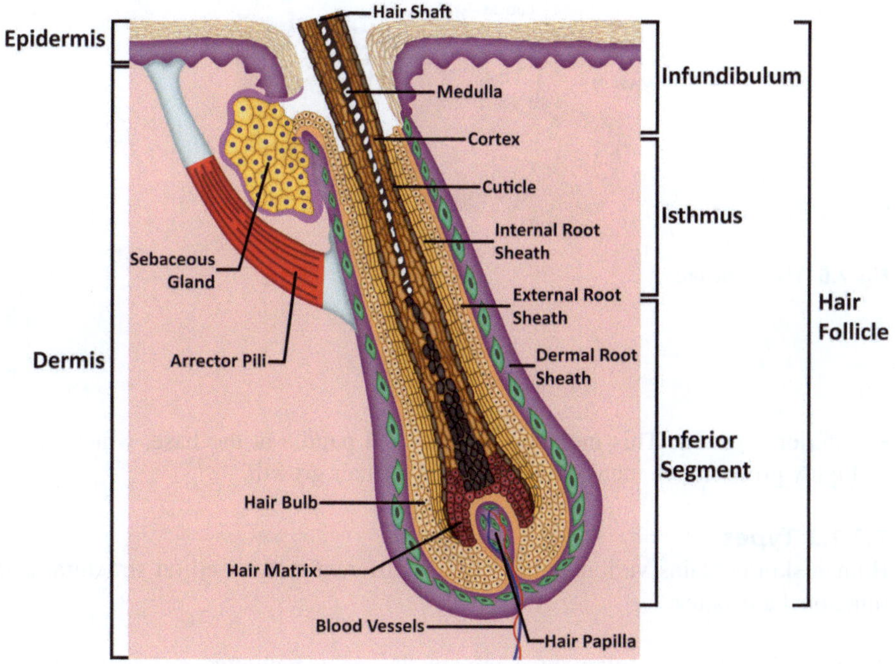

Fig. 7.5 Hair follicle structure

Shaft

The hair shaft is divided into three layers (Fig. 7.6):

- Medulla: The innermost layer comprises keratin and air spaces. It is more prominent in thicker, terminal hairs.
- Cortex: This middle layer contains spindle-shaped cells rich in keratin and pigment granules, giving the hair tensile strength and color.
- Cuticle: The outermost layer comprises overlapping keratinized cells, functioning as a protective barrier.

Follicle

The hair follicle, a complex mini-organ, extends from the epidermis into the dermis or hypodermis. It is divided into three segments (Fig. 7.5):

- Infundibulum: The upper part connected to the skin pore.
- Isthmus: The segment from the opening of the sebaceous gland duct to the insertion of the arrector pili muscle.

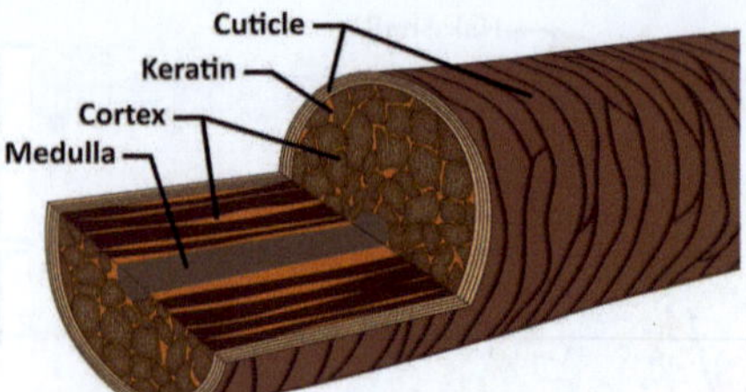

Fig. 7.6 Hair structure

- Inferior segment: This includes the bulb and papilla at the base, which house highly proliferative matrix cells essential for hair growth.

7.2.1.2 Types

Human skin contains various types of hair, differentiated based on structural and functional attributes:

Terminal hairs: Thick, pigmented hairs on the scalp, eyebrows, and eyelashes.
Vellus hairs: Thin, unpigmented hairs covering most of the body surface.
Intermediate hairs: These are transitional forms, often observed in areas like the arms and legs.

7.2.1.3 Hair Growth

Hair growth occurs cyclically, involving three main phases (Fig. 7.7):

- Anagen: The active growth phase, characterized by rapid cell division at the follicle's base.
- Catagen: The transitional phase where hair growth ceases, and the follicle undergoes structural changes.
- Telogen: The resting phase, culminating in hair shedding and renewal of the cycle.

7.2.2 Nail

Nails, comprising keratinized structures at the distal ends of the fingers and toes, offer more than just an aesthetic appendage. The main component of the nail plate is keratin, which is harder than in skin and hair due to a higher disulfide bond content between the keratinocytes. Substances like lipids and trace minerals are also present but in minute amounts.

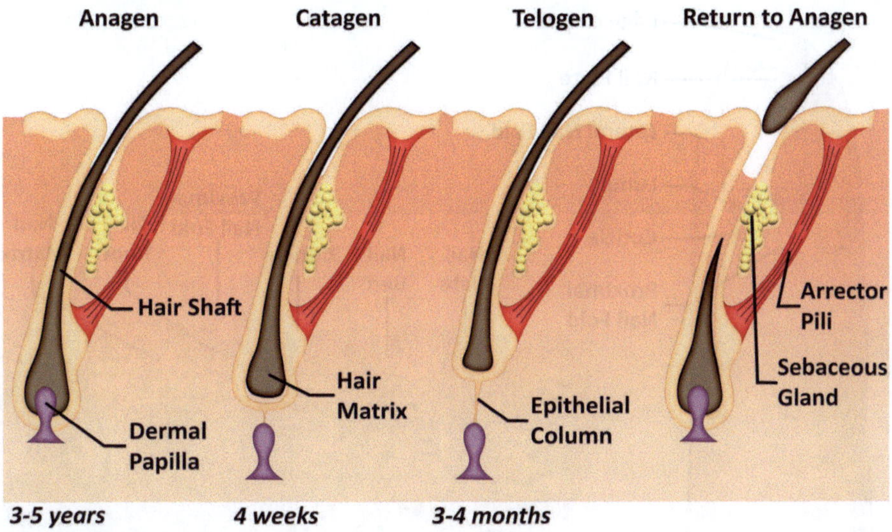

Fig. 7.7 Hair growth cycle

A nail unit includes several distinct but interrelated structures (Fig. 7.8):

- Nail plate: The hard and translucent portion composed of densely packed keratinized cells.
- Nail matrix: The tissue below the nail plate, critical for nail growth. The nail matrix, or the "keratogenous zone," contains basal cells that divide and differentiate to form the nail plate.
- Nail bed: The skin layer beneath the nail plate, adhering to it through a series of ridges and grooves. It is richly vascularized and innervated, providing the nutrients and signals for nail growth.
- Cuticle: A layer of dead, keratinized cells that serves as a seal between the nail plate and proximal nail fold.
- Nail folds: Skin folds surrounding the three sides of the nail plate.
- Lunula: The crescent-shaped, whitish area at the base of the nail plate.
- Hyponychium: The skin underneath the free edge of the nail.

7.2.3 Sweat Glands

Sweat glands (Fig. 7.1), integral components of the integumentary system, are embedded within the dermal layer of the skin and exhibit variations in their anatomical structure and biochemical secretions based on location and function. The functions of sweat glands may include:

Thermoregulation: Eccrine glands play a vital role in maintaining body temperature.

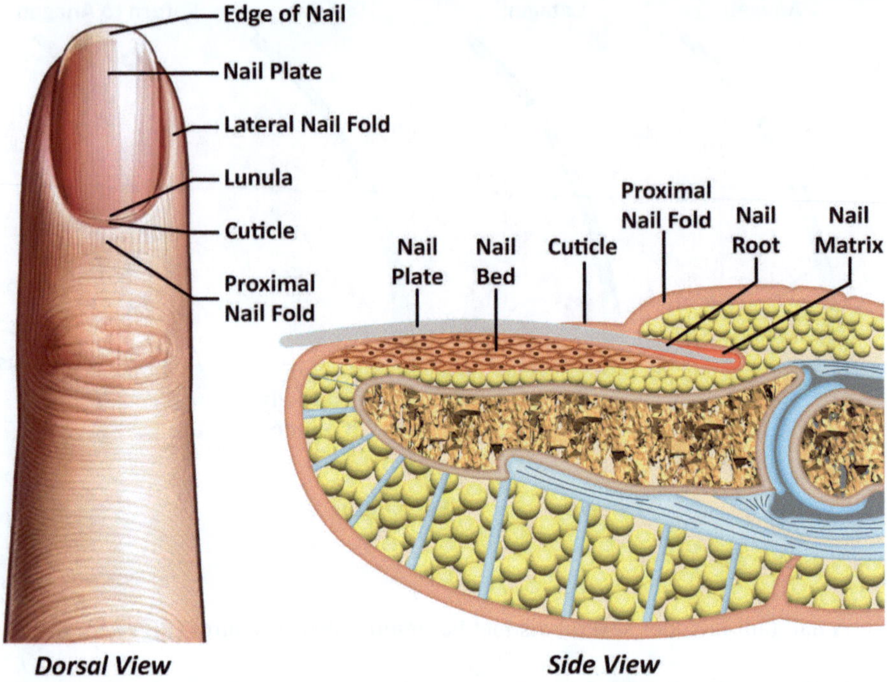

Fig. 7.8 Nail structure

Waste elimination: Sweat is a minor excretory route for excess ions and waste products.

Skin hydration: The water content of sweat contributes to skin hydration.

Sweat glands are predominantly categorized into two main types, each with its unique distribution: eccrine and apocrine sweat glands.

7.2.3.1 Eccrine Sweat Glands

The eccrine glands (Fig. 7.9) are widely distributed over the body, numerous on the palms, soles, and forehead. Eccrine sweat comprises water, sodium, potassium, chloride, and trace amounts of urea, lactate, and glucose. Eccrine sweat glands consist of two main segments:

- Secretory coil, located deep within the dermis or hypodermis, contains the glandular epithelial cells responsible for sweat production. The secretory coil is enveloped by myoepithelial cells, which facilitate the expulsion of sweat through contractions.
- Duct: A spiraling tube traverses through the dermis and epidermis, terminating at a pore on the skin surface.

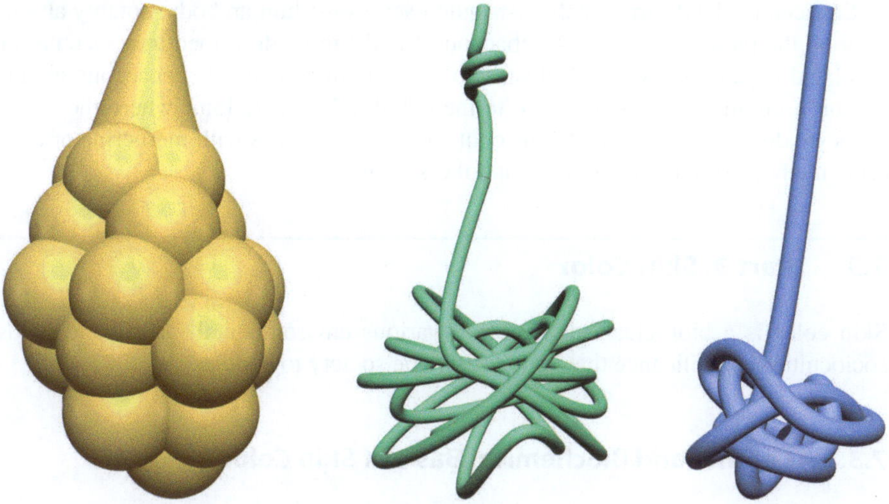

Fig. 7.9 Structure of cutaneous glands

7.2.3.2 Apocrine Sweat Glands

The apocrine glands (Fig. 7.9) are restricted to axillae, perianal region, and external genitalia. These are structurally similar to eccrine glands but larger and deeper in the dermis or hypodermis. Unlike eccrine glands, they typically release their secretions into a hair follicle rather than directly onto the skin surface. Apocrine sweat is similar to eccrine sweat but richer in proteins, lipids, and steroid compounds, contributing to its milky or yellowish color.

7.2.4 Sebaceous Glands

Sebaceous glands (Fig. 7.9) are primarily involved in lipid production and secretion. These specialized exocrine glands are located predominantly in association with hair follicles. They secrete sebum—an oily substance integral to the lubrication of skin and hair. The functional implications of sebaceous glands secretions may include:

Lubrication: Sebum coats the skin and hair, imparting moisturization and reducing water loss.

Protection: The lipid-rich sebum contributes to the skin's barrier function, preventing excessive permeability.

Antimicrobial activity: Some sebum components, like free fatty acids, have antimicrobial properties.

Sebaceous glands are widely distributed across the human body, notably absent only on the palms and soles. A sebaceous gland can be described as a saccular or lobulated organ consisting of clusters of cells known as acini. Sebaceous glands exhibit a unique form of secretion termed "holocrine secretion," where the entire sebocyte disintegrates to release its contents. This contrasts with merocrine or apocrine methods observed in other glandular systems.

7.3 Part 3: Skin Color

Skin color is a biological adaptation to various environmental factors and holds sociocultural significance that varies from one society to another.

7.3.1 Cellular and Biochemical Basis of Skin Color

Melanocytes, located in the basal layer of the epidermis, are specialized cells responsible for the synthesis of melanin, the primary pigment influencing skin color. Melanocytes form a complex network with keratinocytes, transferring melanin to these cells through structures called dendrites (Fig. 7.2). Melanin has two main types:

Eumelanin: Predominantly found in individuals with dark skin, eumelanin is brown to black and offers protection against ultraviolet (UV) radiation.
Pheomelanin: Responsible for lighter skin tones and is red to yellow.

7.3.2 Factors Governing Skin Color

Factors that govern the skin color may include:

Genetic determinants: MC1R (melanocortin 1 receptor), TYR (tyrosinase), and OCA2 and SLC45A2 genes are associated with lighter skin tones in European populations.

Epigenetic factors: Epigenetic modifications, such as DNA methylation and histone modification, can also influence the expression of genes related to skin color.

Environmental Influences Ultraviolet (UV) radiation: Melanin is a natural sunscreen. Darker skin provides an evolutionary advantage in regions with higher UV radiation by protecting against DNA damage:

- Diet: Carotenoids and flavonoids, primarily obtained from fruits and vegetables, can also contribute to skin pigmentation to some extent.
- Hormonal regulations and life stages: Fluctuations in hormones like estrogen, progesterone, and melanocyte-stimulating hormone (MSH) can affect skin pigmentation. Life stages like pregnancy and menopause can thus result in observable changes in skin color due to hormonal alterations.

7.4 Part 4: Skin Physiology

The skin is a formidable biological interface between the internal physiological system and the external environment. Its role is not restricted to mere superficial existence; it performs complex functions essential to human survival.

7.4.1 Protective Barrier Function

Physical barrier: The skin's outermost layer, the stratum corneum, is a physical barrier preventing the entry of pathogens and foreign substances. The densely packed keratinocytes, filled with the protein keratin, provide a robust yet flexible structure.

Chemical barrier: The skin's slightly acidic pH, ranging from 4.5 to 6.0, contributes to its barrier function. This acidity is maintained by the "acid mantle," a thin film on the skin's surface composed of sebum, sweat, and dead skin cells that deters the growth of harmful bacteria and fungi.

Biological barrier: The skin hosts a myriad of commensal microorganisms that form the skin microbiota. These microbes help maintain skin homeostasis and prevent pathogenic invasion by outcompeting harmful organisms for nutrients and space.

7.4.2 Sensory Perception

Mechanoreception: The skin is laden with mechanoreceptors, such as Meissner's corpuscles and Pacinian corpuscles, that are sensitive to touch, pressure, and vibration.

Thermoreception: Specialized nerve endings in the skin are sensitive to temperature changes, alerting the body to potentially harmful conditions like extreme heat or cold.

Pain perception: Nociceptors are responsible for the sensation of pain and are vital in alerting the body to potential tissue damage.

7.4.3 Thermoregulation

Insulation: The subcutaneous layer, primarily composed of adipose tissue, acts as an insulating layer, retaining body heat.

Sweat production: Sweat glands produce sweat that evaporates from the skin surface, dissipating heat and aiding in body temperature regulation.

Blood flow regulation: Blood vessels in the dermal layer can constrict or dilate, thereby regulating heat exchange with the environment.

7.4.4 Immune Function

The skin is an essential component of the innate immune system (Section 14.2.1). Langerhans cells, a specialized type of dendritic cell, capture pathogens and present them to T-cells, initiating an adaptive immune response.

7.4.5 Metabolic Functions

Synthesis of vitamin D: Upon exposure to UV radiation, the skin synthesizes vitamin D, a steroid hormone essential for calcium homeostasis and bone health.

Lipid storage: The subcutaneous layer is a storage site for lipids, which can be metabolized for energy.

7.4.6 Excretion and Absorption

While not a primary excretory organ, the skin does participate in the excretion of substances, such as sweat, which can contain electrolytes and metabolic waste. Similarly, the skin can absorb specific substances, most notably demonstrated through transdermal drug delivery systems.

Suggested Reading

Advanced Physiology and Pathophysiology: Essentials for Clinical Practice, 2nd Ed. Nancy Tkacs, Linda Herrmann, Springer Publishing Company, 2024

Anatomy & Physiology, S. Meloni, et al. Independent Publisher, 2023

Anatomy & Physiology: The Unity of Form and Function, 10th Ed. Kenneth S. Saladin, McGraw Hill, 2023

Gray's Anatomy for Students, 5th Ed. Richard L. Drake, et al. Elsevier, 2020

Human Anatomy & Physiology, 11th Ed. Elaine N. Marieb and Katja Hoehn, Pearson, 2022

Junqueira's Basic Histology: Text and Atlas, 16th Ed. Anthony L. Mescher, McGraw Hill, 2021

Moore's Clinically Oriented Anatomy, 9th Ed. Keith L. Moore, et al. Lippincott Williams & Wilkins, 2020

Netter Atlas of Human Anatomy: Classic Regional Approach, 8th Ed. Frank H. Netter, Elsevier, 2022

Principles of Anatomy and Physiology

Skeletal System

8

In this chapter, you will learn

- General overview of the human skeletal system
- Function and composition of cartilage
- Types and structures of cartilage: hyaline, elastic, and fibrocartilage
- Cells and extracellular matrix in cartilage
- Function of ligaments and their structure
- Types of collagens found in ligaments
- Basic anatomy and functions of bones
- Macroscopic and microscopic structure of bones
- Different types of bones: long, short, flat, irregular, and sesamoid
- Bone cells: osteoblasts, osteocytes, and osteoclasts
- Composition and function of the bone extracellular matrix
- Structural units of bone: osteons and trabeculae
- Process of bone formation and growth
- Detailed process of bone repair: stages and cellular activities involved
- Classification of joints: fibrous, cartilaginous, and synovial
- Structure and function of synovial joints
- Types of movements in synovial joints: gliding, angular, and rotational
- Overview of the human skeleton: axial and appendicular
- Structure and function of the skull: cranial and facial bones
- Vertebral column: structure, regions, and curvatures
- Thoracic cage structure including the sternum and ribs
- Anatomy of the upper limb bones: clavicle, scapula, humerus, radius, and ulna
- Anatomy of the lower limb bones: hip bone, femur, patella, tibia, and fibula
- Detailed description of hand bones: carpals, metacarpals, and phalanges
- Detailed description of foot bones: tarsals, metatarsals, and phalanges
- Arches of the foot: medial longitudinal, lateral longitudinal, and transverse

H. Hosseinzadeh, *Fundamentals of Medicine for Biomedical Engineering*,
https://doi.org/10.1007/978-3-031-74085-5_8

The human skeletal system (Fig. 8.1) is the foundational framework for the entire body, providing structural integrity, protection to vital organs, and facilitating locomotion. The system is an intricate arrangement of bones, cartilages, and ligaments forming an endoskeleton. The skeletal system, constituting 206 individual bones in adults, is a marvel of biological engineering that has evolved over millions of years to support a variety of functions integral to human survival and well-being.

8.1 Part 1: Cartilage

Cartilage, a semirigid yet flexible connective tissue, is essential to the vertebrate skeletal system. Notably different from bone, cartilage consists of specialized cells, extracellular fibers, and ground substances that provide structural support and biomechanical versatility.

8.1.1 Structure

Cartilage structure (Fig. 8.2) has two major components: the cellular elements and the extracellular matrix.

8.1.1.1 Cells

The predominant cell type in cartilage is the chondrocyte, which is responsible for the production and maintenance of the extracellular matrix. Chondrocytes reside in small cavities within the matrix called lacunae. As the principal cells of cartilage,

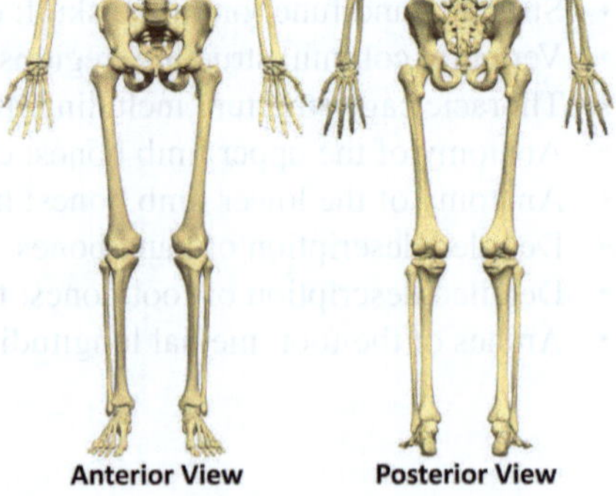

Fig. 8.1 The skeletal system of the human body

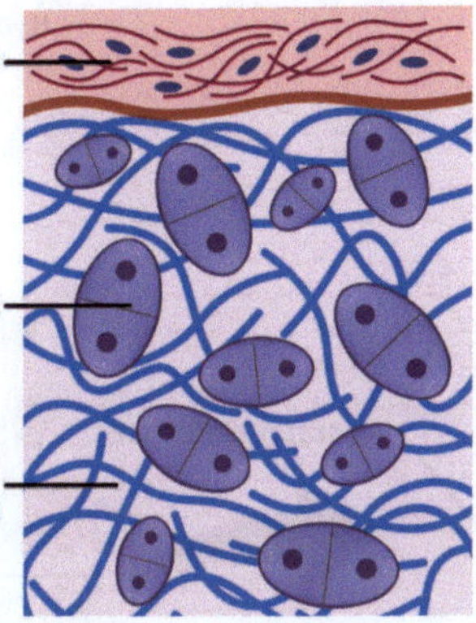

Fig. 8.2 Cartilage structure

chondrocytes secrete both the fibrous and nonfibrous elements of the matrix. Chondroblasts are the progenitor cells that differentiate into chondrocytes. During cartilage formation, known as chondrogenesis, chondroblasts actively produce the components of the extracellular matrix before they become encapsulated within the lacunae and mature into chondrocytes.

8.1.1.2 Extracellular Matrix

The extracellular cartilage matrix is a composite material comprising a dense network of fibers embedded within a hydrated gel of ground substance. This matrix is responsible for cartilage's mechanical properties, such as its tensile strength and elasticity.

Fibrous components: Collagen is the primary fibrous component of the extracellular matrix and is mainly responsible for providing tensile strength to the cartilage. While several types of collagen can be found in various types of cartilage, type II collagen is the most predominant (Section 6.2.2.1 & Table 6.1). The collagen fibers form a three-dimensional mesh that provides a scaffold for the other components of the matrix. Elastic cartilage contains elastin fibers (Page 130) in addition to collagen. Elastin provides elasticity, allowing the cartilage to return to its original shape after deforming.

Ground substance: Proteoglycans (Fig. 8.3) are large macromolecules made up of a core protein attached to long chains of repeating disaccharides, known as glycosaminoglycans (GAGs). These molecules occupy the spaces between the fibrous components and are critical for the compressive resistance of cartilage. The negative charge on the GAGs attracts water molecules, imbuing the cartilage with its characteristic resilience and resistance to compression. Glycoproteins like chondronectin and fibronectin are present to help cells adhere to the extracellular matrix, serving as a link between the chondrocytes and the matrix components.

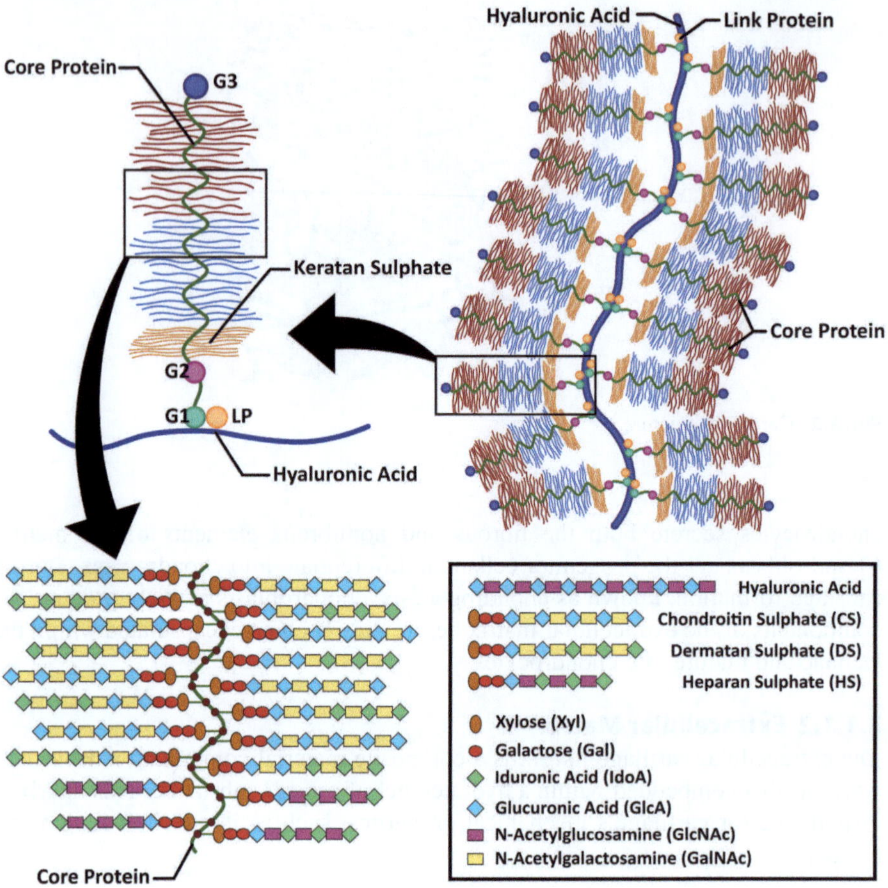

Fig. 8.3 Structural organization of proteoglycans and glycosaminoglycans (GAGs), crucial components of the extracellular matrix (ECM) in cartilage tissues

8.1.2 Types of Cartilage

Vertebrate cartilage tissue has three primary types: hyaline cartilage, elastic cartilage, and fibrocartilage (Fig. 8.4). These types differ primarily in their extracellular matrix composition and the presence or absence of particular fibers.

8.1.2.1 Hyaline Cartilage

Hyaline cartilage (Fig 8.4a) is the most prevalent type in the human body. It is translucent and appears bluish-white, with a smooth surface. The extracellular matrix primarily comprises type II collagen fibers, which are relatively delicate and disperse uniformly throughout the matrix. Chondrocytes are arranged in small clusters within their respective lacunae.

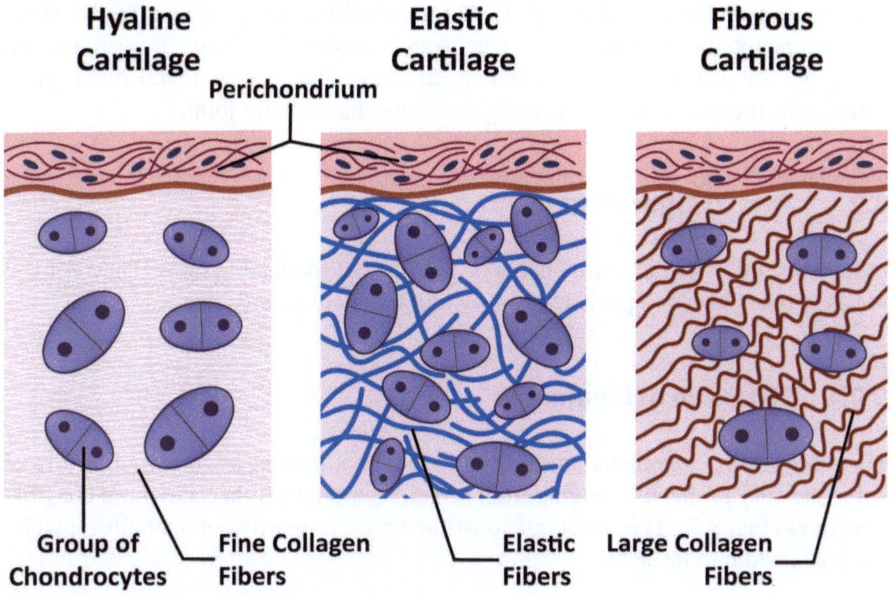

Fig. 8.4 Structural features of types of cartilage

The ground substance of hyaline cartilage contains large amounts of proteoglycans bound to hyaluronic acid. This gives hyaline cartilage its high resistance to compression. Hyaline cartilage provides smooth, low-friction surfaces for articulation in joints. It is also found in the respiratory tract, forming the structural framework for the trachea and bronchi. It is a template for bone formation during fetal development and postnatal growth.

8.1.2.2 Elastic Cartilage

Elastic cartilage (Fig 8.4b) is yellowish and characterized by its high content of elastin fibers, interwoven with collagen fibers, predominantly of type II. These elastin fibers form a mesh-like network around the chondrocytes, which reside singly within their lacunae. The abundant elastin fibers confer considerable flexibility, allowing elastic cartilage to resume its original shape after deformation. However, it still maintains rigidity due to the presence of collagen fibers. Elastic cartilage is less prone to tearing than hyaline cartilage due to its elastic properties. It is found in the external ear (auricle), epiglottis, Eustachian tubes, and laryngeal cartilages.

8.1.2.3 Fibrocartilage

Fibrocartilage (Fig 8.4c) is distinguished by its densely interwoven array of collagen fibers, mostly of type I, along with sparse chondrocytes. Unlike other types of cartilage, fibrocartilage does not possess a perichondrium, a layer of connective tissue that usually envelops cartilaginous structures. The densely packed collagen fibers give fibrocartilage high tensile strength, making it highly durable and resistant to both tensile and compressive forces. Fibrocartilage is an effective shock absorber adapted to withstand high tensile and compressive forces. It is often found in regions subjected to intense mechanical stress, like intervertebral discs, pubic symphysis, menisci of the knee, and the temporomandibular joint.

8.2 Part 2: Ligaments

Ligaments are essential components of the musculoskeletal system. They connect bones to other bones to provide stability and facilitate movement.

8.2.1 Structure of Ligaments

Macroscopically, ligaments are fibrous tissues that exhibit a whitish or yellowish color and vary in shape—ranging from broad, membranous sheets to more cord-like structures (Fig. 8.5). The structural constituents are primarily collagen fibers, which are organized into fascicles.

Microscopic structure (Fig. 8.6): The primary structural unit of ligaments is the collagen fiber. These fibers are mainly of type I collagen, with a smaller percentage being type III. The fibers are tightly packed and aligned along the axis of tension to withstand mechanical loads. Elastin fibers, although less prevalent than collagen, are present and confer a degree of elasticity to ligaments, allowing them to return to their original shape post-deformation. The extracellular matrix comprises ground substance—a gel-like material consisting of glycoproteins and proteoglycans. This structure aids in the distribution of mechanical stress and nourishes the cells. The most common cell type in ligaments is the fibroblast, responsible for the production and turnover of collagen and other extracellular matrix components.

8.2.2 Functions of Ligaments

Mechanical stability: One of the primary functions of ligaments is to provide mechanical stability to joints by limiting the range of motion and preventing dislocations.

Proprioception: Ligaments contain mechanoreceptors that aid in proprioception, allowing individuals to perceive the position and movement of their body parts.

Load distribution: By connecting various skeletal components, ligaments help distribute mechanical loads, reducing localized stress and strain.

Facilitation of movement: Although ligaments primarily restrict movement to a controlled range, their elastic properties facilitate smooth joint action.

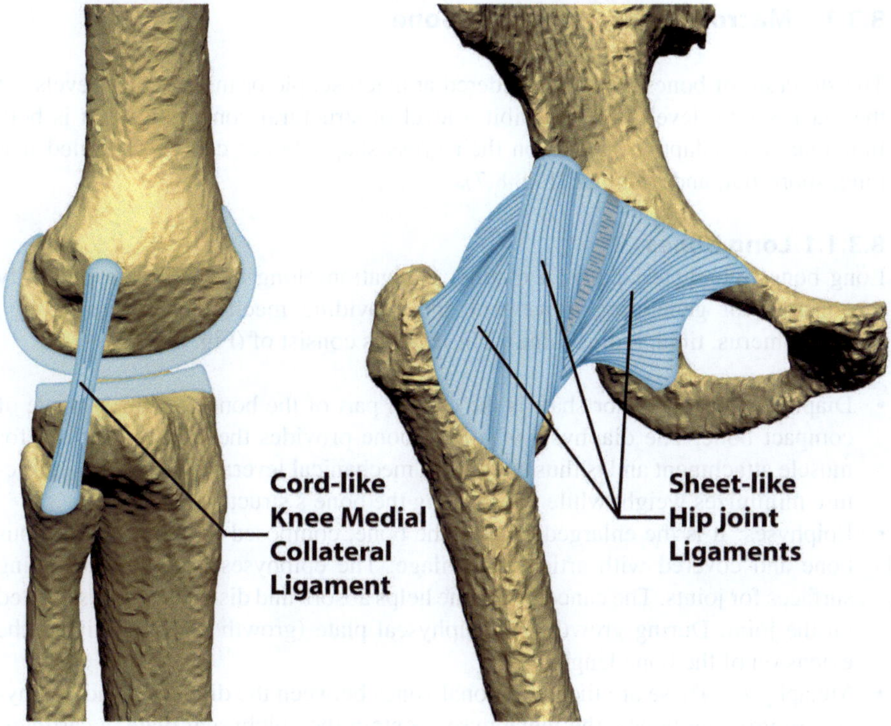

Fig. 8.5 Types of ligaments, cord-like (left) and sheet-like (right)

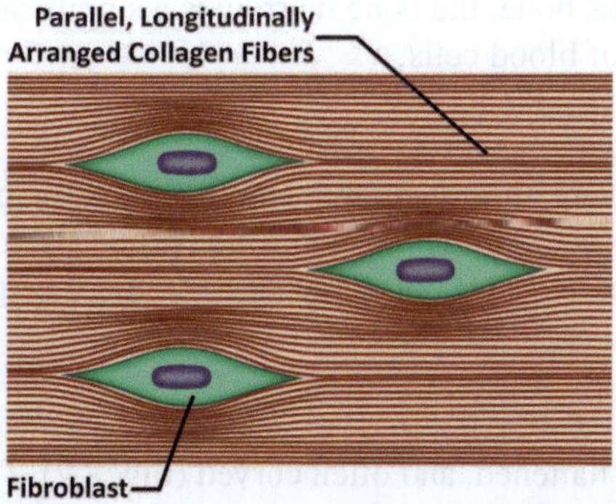

Fig. 8.6 Microscopic ligament structure as a regular dense connective tissue

8.3 Part 3: Bone

Bones are rigid organs that form the endoskeleton of vertebrate organisms. They serve myriad functions, including mechanical support, protection of vital organs, and facilitation of locomotion. Composed of an intricate architecture that incorporates both organic and inorganic components, bones are dynamic structures subject to constant remodeling.

8.3.1 Macroscopic Structure of Bone

The structure of bones can be considered at macroscopic or microscopic levels. At the macroscopic level, bones exhibit a level of structural complexity that is both functional and adaptive. Based on their gross shape, bones can be classified into long, short, flat, and irregular (Fig. 8.7).

8.3.1.1 Long Bones

Long bones, characterized by a distinct elongation along one axis, are primarily responsible for generating movement and providing mechanical leverage (e.g., femur, humerus, tibia, radius, ulna). These bones consist of (Fig. 8.8):

- Diaphysis: Diaphysis or shaft is the central part of the bone, primarily made of compact bone. The diaphysis of a long bone provides the primary surface for muscle attachment and is thus crucial for mechanical leverage. The hollow structure minimizes weight while maintaining the bone's structural integrity.
- Epiphyses: It is the enlarged ends of the bone, composed mainly of cancellous bone and covered with articular cartilage. The epiphyses provide articulating surfaces for joints. The cancellous bone helps absorb and distribute forces exerted on the joint. During growth, the epiphyseal plate (growth plate) facilitates the extension of the bone length.
- Metaphyses: These are the transitional zones between the diaphysis and epiphyses. In growing bones, the metaphyses contain the epiphyseal plate, a cartilaginous region responsible for bone elongation.
- Bone marrow cavity: Housed within the diaphysis of long bones and within the spaces in cancellous bone, the bone marrow is a significant site for hematopoiesis, the formation of blood cells.

8.3.1.2 Short Bones

Short bones (Fig. 8.7) are almost equal in length and width, providing a stable platform with limited movement. These bones are primarily made up of cancellous bone surrounded by a thin layer of compact bone [e.g., carpals (wrist bones) and tarsals (ankle bones)].

8.3.1.3 Flat Bones

These bones are thin, flattened, and often curved (Fig. 8.7). They comprise two layers of compact bone with a layer of cancellous bone sandwiched between them (e.g., skull bones, ribs, and scapulae).

8.3.1.4 Irregular Bones

Irregular bones (Fig. 8.7) have complex shapes and variable compositions of compact and cancellous bone (e.g., vertebrae, facial, and pelvic bones).

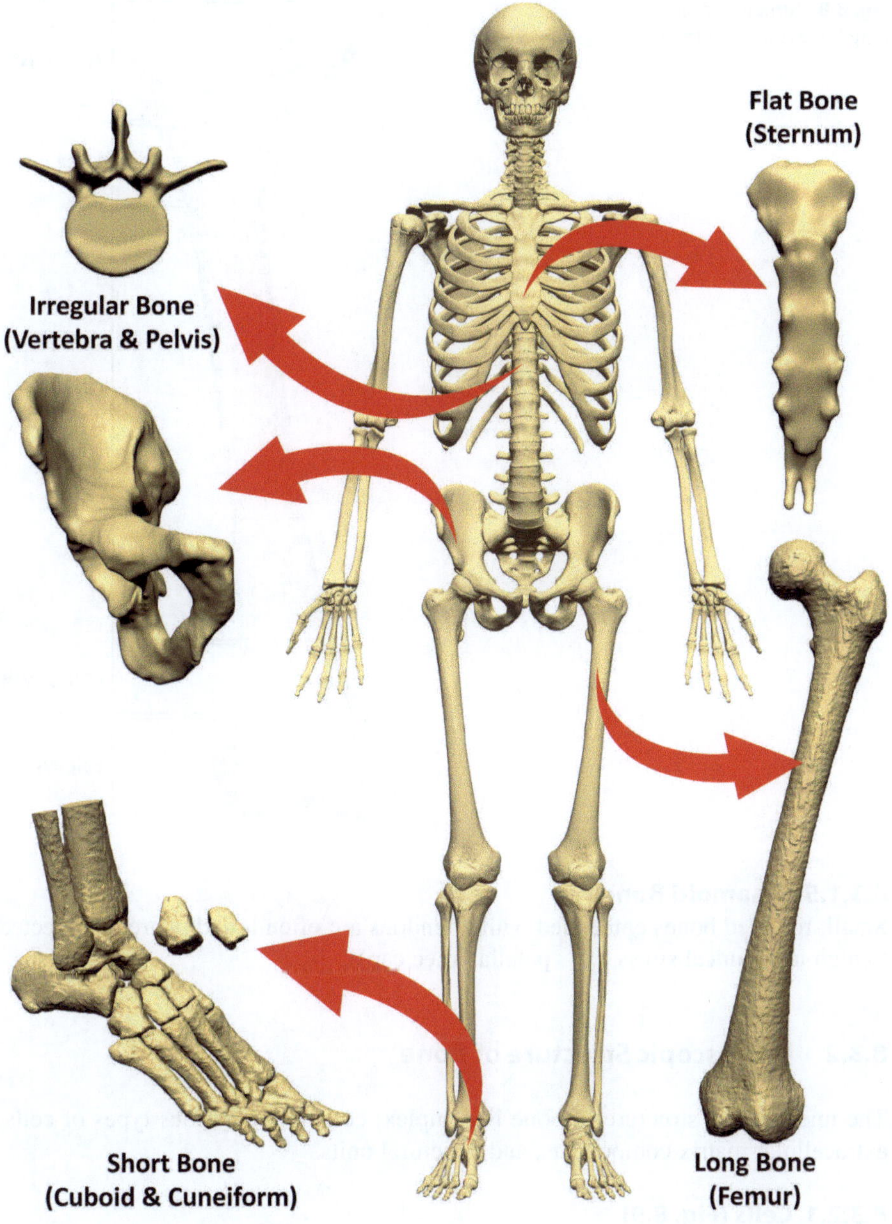

Fig. 8.7 Types of bones based on the shape

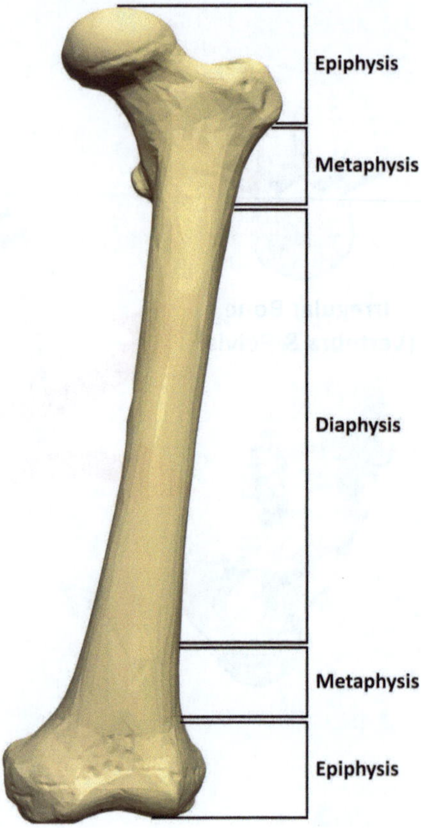

Fig. 8.8 Structure of a long bone (femur of thigh)

8.3.1.5 Sesamoid Bones

Small, rounded bones embedded within tendons are often found in areas subjected to high mechanical stress [e.g., patella (knee cap)].

8.3.2 Microscopic Structure of Bone

The microscopic structure of bone is complex, comprising various types of cells, extracellular matrix components, and structural units.

8.3.2.1 Cells (Fig. 8.9)

Osteoblasts: Osteoblasts are bone-forming cells originating from mesenchymal stem cells. A high metabolic rate characterizes these cells, producing the bone matrix's organic constituents, primarily collagen type I. They are most commonly found on the bone surface but can also be found in the periosteum and endosteum. Upon completion of bone formation, osteoblasts may undergo apoptosis, become osteocytes, or transform into lining cells.

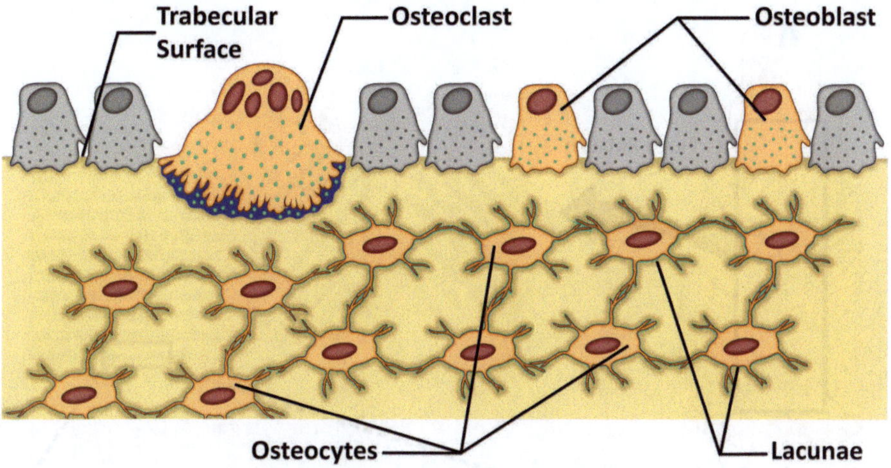

Fig. 8.9 Types of bone cells and their location on a bone trabecula

Osteocytes: Osteocytes are mature bone cells derived from osteoblasts and are the most abundant cell type in bone tissue. They are embedded in lacunae within the mineralized bone matrix and connected to each other and to surface osteoblasts through an intricate network of canaliculi. Osteocytes are thought to function as mechanosensors, detecting mechanical strains on the bone and regulating both osteoblast and osteoclast activity in response.

Osteoclasts: Osteoclasts are large, multinucleated cells responsible for bone resorption. They originate from the fusion of hematopoietic progenitors of the monocyte/macrophage lineage. Osteoclasts are highly specialized cells that adhere tightly to the bone surface, creating an isolated microenvironment wherein they secrete hydrochloric acid and enzymes like cathepsin K to dissolve the mineral and organic matrix.

8.3.2.2 Extracellular Matrix

The bone's extracellular matrix (ECM) is a composite material of organic and inorganic phases:

Organic phase: Collagen, comprising about 90% of the organic matrix, provides the bone with tensile strength and flexibility. Type I collagen is the most abundant form. Noncollagenous proteins include osteocalcin, osteonectin, and bone sialoprotein, among others, and regulate bone mineralization and cell–matrix interactions.

Inorganic phase: The inorganic phase is primarily composed of hydroxyapatite crystals ([$Ca_5(PO_4)_3OH$]), which provide hardness and compressive strength to the bone. The crystals are deposited in the gaps and pores within the collagen network (Fig. 8.10).

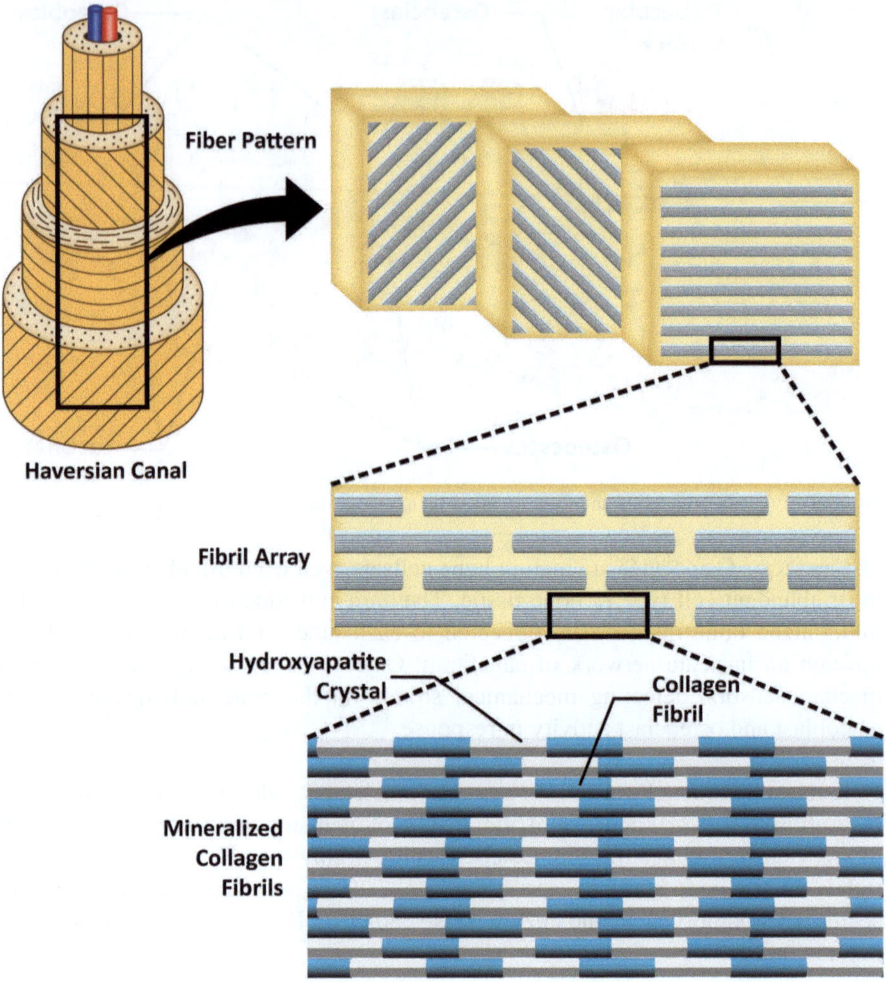

Fig. 8.10 Hierarchical organization of hydroxyapatite crystals and collagen fibers in compact bone. Each lamella has collagen fibers oriented differently, enhancing the bone's strength and resistance to stress. Collagen fibrils are organized in a staggered array within the lamellae, contributing to the structural integrity of the bone. Collagen fibrils and hydroxyapatite crystals form a composite material, giving bone unique mechanical properties

8.3.2.3 Structural Units of Bone

The structural units in compact bone are called osteon, and in cancellous bone, they are called trabeculae.

Osteons

The osteon, or the Haversian system (Fig. 8.11), is the fundamental unit of compact, or cortical, bone. Each osteon consists of concentric layers (lamellae) of calcified

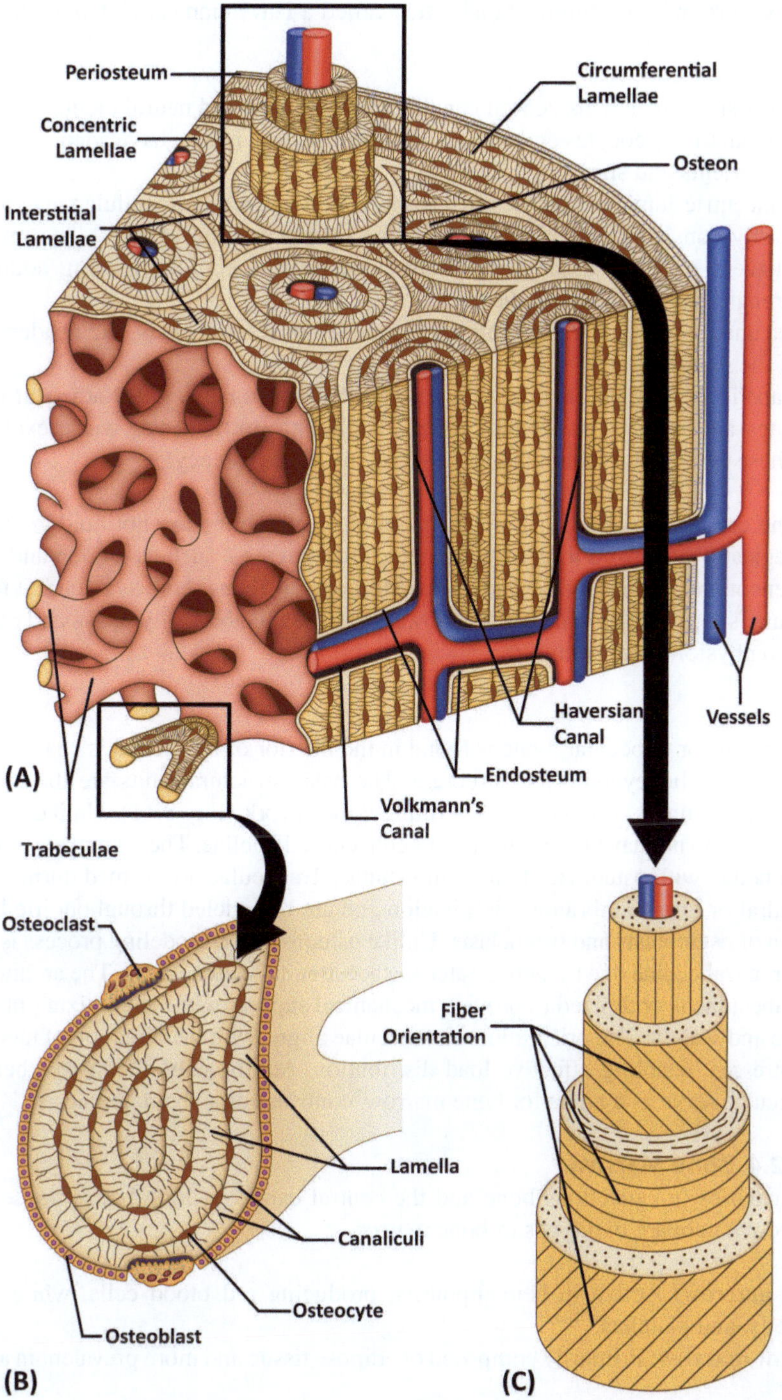

Fig. 8.11 Microscopic anatomy of long bone. (**a**) Diagram of a section of long bone diaphysis. (**b**) The structure of a trabecula in cancellous bone. (**c**) A single osteon, compact bone's fundamental functional unit, shows the alternating fiber orientation in adjacent lamellae

matrix surrounding a central canal, often called a Haversian canal. Components of an osteon include:

- Haversian canal: This central canal houses vascular and neural elements, ensuring that the osteocytes embedded in the surrounding matrix have adequate access to nutrients and signaling molecules.
- Concentric lamellae: These are layers of mineralized extracellular matrix, primarily consisting of collagen fibers and hydroxyapatite crystals. The fibers in adjacent lamellae usually run in perpendicular directions, lending additional strength and resilience to the osteon.
- Lacunae: Small cavities within the lamellae house osteocytes, the resident cells of the mature bone.
- Canaliculi: Minute channels branching off the lacunae connect adjacent osteocytes and lead to the central canal, facilitating nutrient and waste exchange. These canaliculi contain the tiny extensions of the osteocytes.

The orientation and distribution of osteons in the bone are not random but are strategically organized to withstand mechanical stresses, such as torsion and bending, encountered during daily activities. Moreover, osteons serve as a conduit for the vascular supply to the bone tissue, allowing it to be metabolically active and responsive to physiological stimuli.

Trabeculae

Cancellous, or trabecular, bone is found in the interior of bones and is characterized by a porous, honeycomb-like structure. The basic structural units are the trabeculae—thin columns or bone plates forming a meshwork (Fig. 8.11). Unlike osteons, trabeculae do not have a central canal or concentric lamellae. They are made of parallel lamellae with osteocytes housed in lacunae. Trabeculae are formed during endochondral or intramembranous ossification and are remodeled throughout life by the action of osteoblasts and osteoclasts. Unlike osteons, the remodeling process is often faster in trabeculae due to their greater surface area-to-volume ratio. The architecture of trabeculae is optimized to provide mechanical support while minimizing material usage and weight. The orientation of trabeculae aligns with the direction of mechanical stresses, enabling effective load distribution. Additionally, the spaces between trabeculae serve as a niche for bone marrow, contributing to hematopoiesis.

8.3.2.4 Bone Marrow

The cavities in cancellous bone and the central canals of long bones house bone marrow. There are two types of bone marrow:

Red marrow: Active in hematopoiesis, producing red blood cells, white blood cells, and platelets.

Yellow marrow: Primarily composed of adipose tissue and more prevalent in adults.

8.3.3 Functions of Bone

Mechanical functions: Bones provide structural support to the organism, facilitating upright posture in humans and other vertebrates. Bones are the foundational scaffolding upon which soft tissues and organs are anchored. Critical organs are encased in bony structures to protect them from mechanical damage. The bones act as a lever system set in motion by muscles. Joints allow for a range of movements.

Metabolic functions: Bone tissue is a significant reservoir of essential minerals, notably calcium and phosphorus. The bones play a vital role in maintaining mineral homeostasis through bone deposition and resorption. Bone contains alkaline mineral salts like calcium phosphate, which can act as buffers to keep the body's acid–base balance, particularly during acidosis. Adipose tissue within the marrow cavities of certain bones serves as an energy reserve, storing triglycerides that can be mobilized when needed.

Hematopoietic functions: The bone marrow, particularly in the axial skeleton and epiphyseal regions of long bones, is a primary site for hematopoiesis—the formation of blood cells. Both erythrocytes (red blood cells) and leukocytes (white blood cells) originate from pluripotent stem cells in the bone marrow.

8.3.4 Bone Formation and Growth

The origins of bone and its growth are phenomena that start in the uterus (during embryonic life) and continue until puberty when its pattern changes from growth to remodeling. Bone formation happens through intramembranous ossification (Fig. 8.12) and endochondral ossification (Fig. 8.13).

8.3.4.1 Stages of Skeletal Development

Embryonic stage: Bone formation begins around the sixth week of embryonic development.

Childhood and adolescence: Rapid bone growth occurs, mediated by growth plates. The bones lengthen and widen.

Adulthood: The epiphyseal plates close by the late teens or early twenties, and bone growth ceases. Bone remodeling continues throughout adulthood.

Aging: After reaching peak bone mass around 30, bone mass declines gradually, particularly in postmenopausal women and older men.

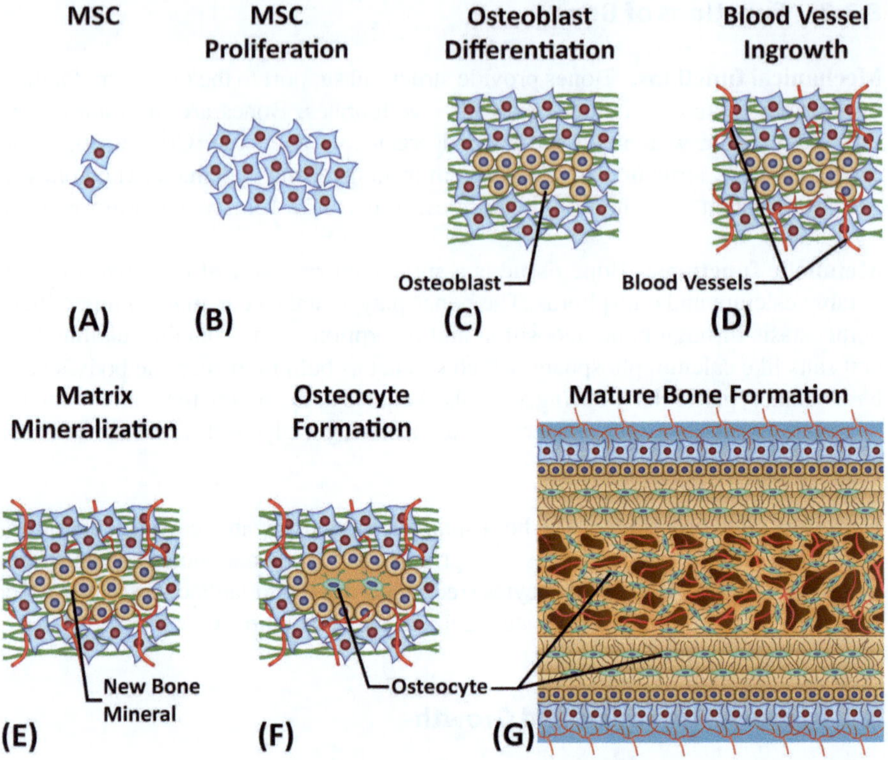

Fig. 8.12 Intramembranous ossification: (**a**) **Initial stage:** Mesenchymal stem cells (MSC), which are multipotent stromal cells, have the potential to differentiate into various cell types, including osteoblasts. (**b**) **Cell proliferation:** MSCs undergo proliferation, increasing in number and forming a dense cluster of cells that will give rise to bone-forming cells. (**c**) **Osteoblast differentiation:** MSCs differentiate into osteoblasts, the bone-forming cells responsible for synthesizing bone matrix. These cells begin to produce and secrete the organic components of the bone matrix. (**d**) **Blood vessel ingrowth:** Blood vessels grow into the developing bone tissue, providing necessary nutrients and oxygen to the proliferating cells and facilitating further bone development. (**e**) **Matrix mineralization:** Osteoblasts begin to deposit minerals, such as calcium and phosphate, into the organic matrix, leading to the formation of mineralized bone. The mineralization process hardens the bone matrix, providing strength and rigidity to the developing bone. (**f**) **Osteocyte formation:** As the matrix mineralizes, osteoblasts become entrapped within the matrix and differentiate into osteocytes. These mature bone cells reside in lacunae and maintain the bone matrix, communicating with other bone cells through canaliculi. (**g**) **Mature bone formation:** The final stage results in the formation of mature bone tissue, consisting of a well-organized structure of osteocytes embedded in a mineralized matrix with distinguishable cortical and cancellous bones and medullary cavities

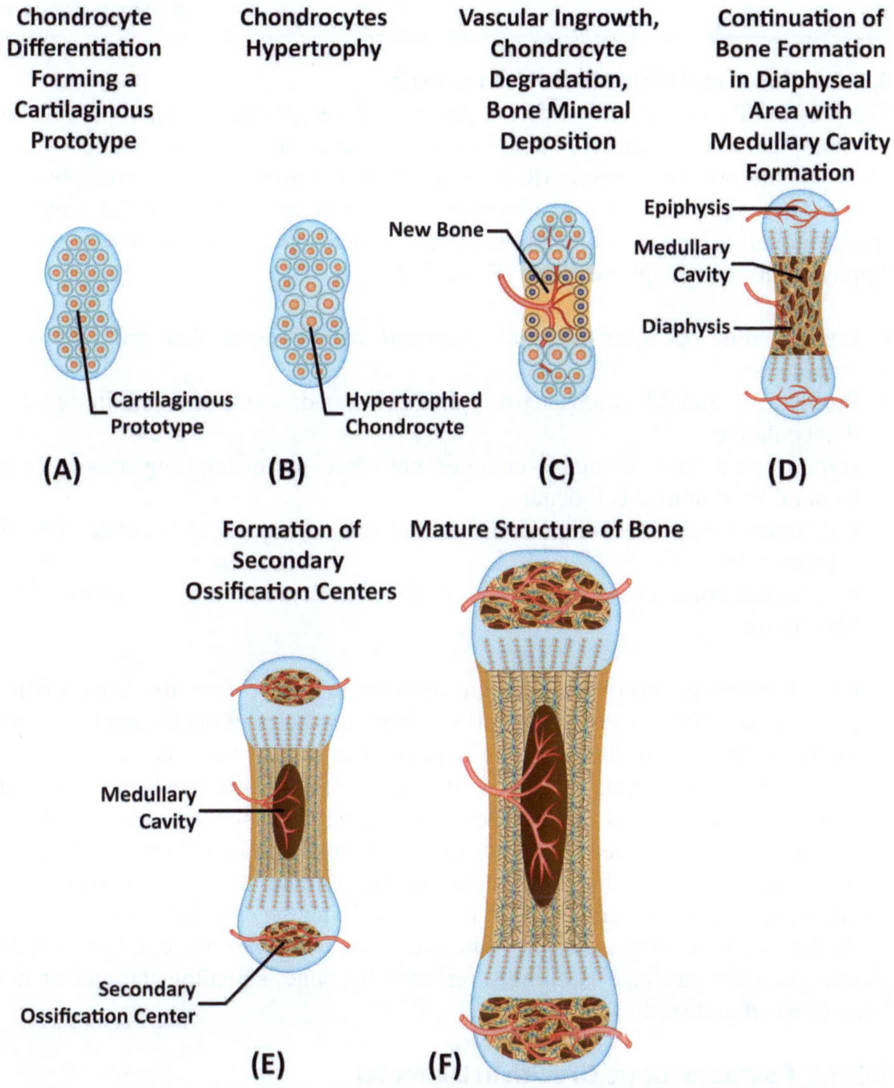

Fig. 8.13 Endochondral ossification in a long bone: (**a**) **Chondrocyte differentiation:** Mesenchymal cells differentiate into chondrocytes, forming a hyaline cartilage model that resembles the shape of the future bone (cartilaginous prototype). (**b**) **Chondrocyte hypertrophy:** Chondrocytes in the center of the cartilage model increase in size (hypertrophy), and the surrounding matrix begins to calcify. (**c**) **Vascular ingrowth, chondrocyte degradation, bone mineral deposition:** Blood vessels invade the calcified cartilage, bringing in osteoblasts and osteoclasts. Hypertrophied chondrocytes die and leave spaces that will be occupied by bone. Osteoblasts deposit bone matrix in the spaces left by the degraded chondrocytes, forming new bone. (**d**) **Formation of the diaphyseal area with medullary cavity:** Bone formation continues in the diaphysis (shaft) of the bone, and a medullary cavity forms in the center. This cavity is formed by osteoclasts breaking down the newly formed bone, creating a space filled with bone marrow. (**e**) **Formation of secondary ossification centers:** These centers form in the epiphyses (ends) of the bone, where a similar process of chondrocyte hypertrophy, vascular invasion, and bone formation occurs. (**f**) **Mature bone formation:** The mature bone structure consists of a diaphysis with a medullary cavity and epiphyses with secondary ossification centers. The remaining cartilage between the diaphysis and epiphysis forms the growth plate, which allows for continued growth in length during development

8.3.4.2 Postnatal Bone Growth in Length

The growth of bones in length during postnatal development is a highly regulated physiological process central to attaining adult stature and overall skeletal integrity. This intricate process is primarily orchestrated at the growth plates or epiphyseal plates. The epiphyseal plate is a hyaline cartilage layer sandwiched between the epiphysis and metaphysis of long bones. It is organized into different zones from the epiphyseal to the metaphyseal side (Fig. 8.14):

- Resting zone: Comprises small, scattered chondrocytes that are relatively quiescent.
- Proliferative zone: Features actively dividing chondrocytes arranged in longitudinal columns.
- Hypertrophic zone: Contains enlarged chondrocytes undergoing apoptosis, a form of programmed cell death.
- Calcification zone: Includes remnants of chondrocytes and an accumulation of calcium salts.
- Ossification zone: The site where the calcified cartilaginous matrix is replaced by bone tissue.

Chondrocytes undergo rapid mitotic division in the proliferative zone of the epiphyseal plate. This expansion in cell number contributes to the elongation of the bone. Then, they are pushed into the hypertrophic zone, where they enlarge and secrete an extracellular matrix rich in collagen and proteoglycans, which provides a scaffold for future bone formation. The extracellular matrix becomes calcified in the calcification zone, followed by the invasion of blood vessels and osteogenic cells from the metaphysis into this calcified matrix. This initiates the ossification process, where osteoblasts lay down new bone tissue, replacing the calcified cartilage.

At the end of puberty, under the influence of sex hormones, the epiphyseal plate closes when the cartilage is entirely replaced by bone, signaling the end of the bone's growth in length.

8.3.4.3 Postnatal Bone Growth in Diameter

The growth of bones in diameter or thickness, often termed "appositional growth," is critical to skeletal development and maintenance. The inner layer of the periosteum contains osteoblasts and osteogenic precursor cells. When activated, these cells differentiate into osteoblasts. Then, osteoblasts synthesize and deposit osteoid, an unmineralized organic matrix, onto the outer surface of the existing bone. This osteoid undergoes matrix mineralization, which includes the deposition of calcium and phosphate ions into the osteoid, converting it into mature bone tissue, and increasing the diameter of the bone. The endosteal surface within the medullary cavity undergoes bone resorption, primarily by osteoclasts. The combination of periosteal apposition and endosteal resorption results in the shifting or "drift" of the bone cortex, contributing to bone diameter and shape changes.

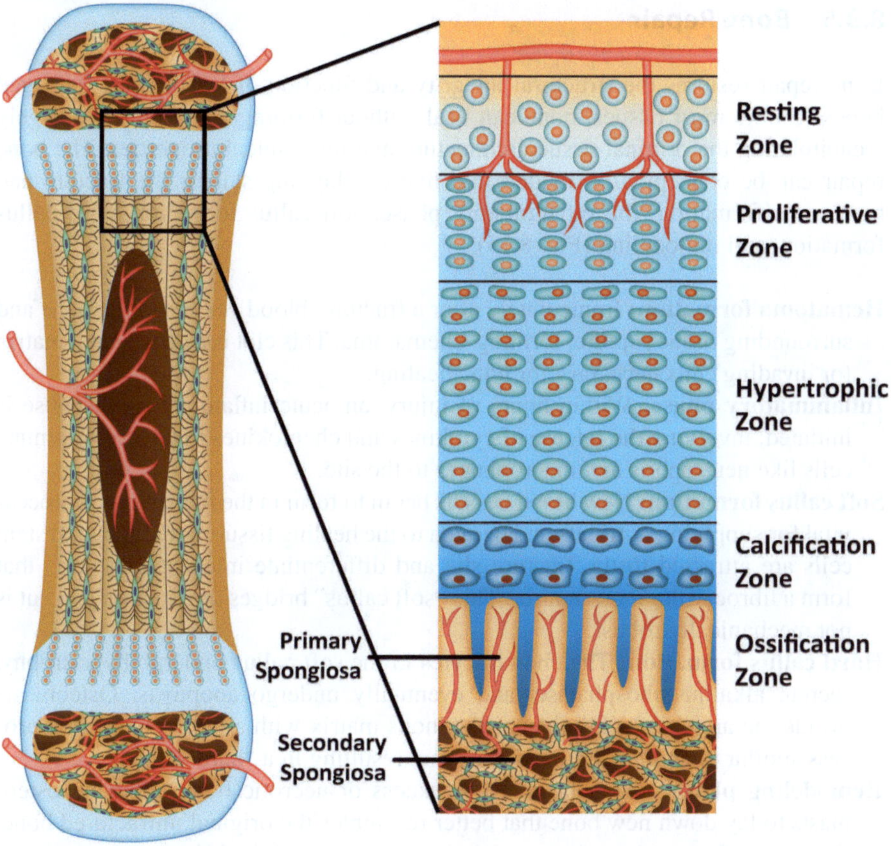

Fig. 8.14 Epiphyseal plate structure: A layer of hyaline cartilage between the epiphysis and metaphysis of long bones where new bone is produced, allowing the bone to lengthen. **Resting zone:** The closest layer to the epiphysis consists of small, inactive chondrocytes scattered within the matrix. This zone anchors the growth plate to the epiphysis **Proliferative zone:** It is located below the resting zone. Chondrocytes in this zone undergo rapid mitosis, resulting in stacked cell columns. This proliferation pushes the epiphysis away from the diaphysis, contributing to bone lengthening. **Hypertrophic zone:** This layer is the next one below the proliferative zone. Chondrocytes hypertrophy and the lacunae around them expand. This zone prepares the matrix for the next stage of calcification. **Calcification zone:** The next layer toward the diaphysis is the calcification zone. The matrix becomes calcified, and chondrocytes begin to die as they cannot survive in the calcified environment. This calcified matrix serves as a scaffold for new bone formation. **Ossification zone (primary spongiosa):** The ossification zone is the last layer toward the diaphysis. This zone is where bone tissue replaces the calcified cartilage. Osteoblasts and osteoclasts from the diaphysis invade, depositing bone matrix on the remaining cartilage framework, forming primary spongiosa

8.3.5 Bone Repair

Bone repair restores the structural integrity and function of fractured or damaged bones. Unlike most tissues, bone can heal without forming a fibrous scar, largely recapitulating the original tissue architecture and mechanical properties. The bone repair can be conceptually divided into five overlapping stages: the fracture and hematoma formation, the inflammatory phase, soft callus formation, hard callus formation, and remodeling (Fig. 8.15).

Hematoma formation: Immediately after a fracture, blood vessels in the bone and surrounding tissue rupture, forming a hematoma. This clot is a provisional matrix for invading cells necessary for bone healing.

Inflammatory phase: Within hours of injury, an acute inflammatory response is initiated, involving the release of cytokines and chemokines that recruit immune cells like neutrophils and macrophages to the site.

Soft callus formation: New blood vessels begin to form in the hematoma, a process vital for supplying nutrients and oxygen to the healing tissue. Mesenchymal stem cells are attracted to the fracture site and differentiate into chondroblasts that form a fibrocartilaginous matrix. This "soft callus" bridges the fracture gap but is not mechanically robust.

Hard callus formation: The chondrocytes in the soft callus undergo hypertrophy, secrete alkaline phosphatase, and eventually undergo apoptosis. Osteoblasts invade the area, replacing the cartilaginous matrix with a bony matrix in a process similar to endochondral ossification, resulting in a "hard callus."

Remodeling phase: Osteoclasts resorb excess or necrotic bone, allowing osteoblasts to lay down new bone that better resembles the original unfractured bone. Over time, the new bone is remodeled to restore its original shape, structure, and mechanical strength, primarily regulated by mechanical stresses, as per Wolff's law.

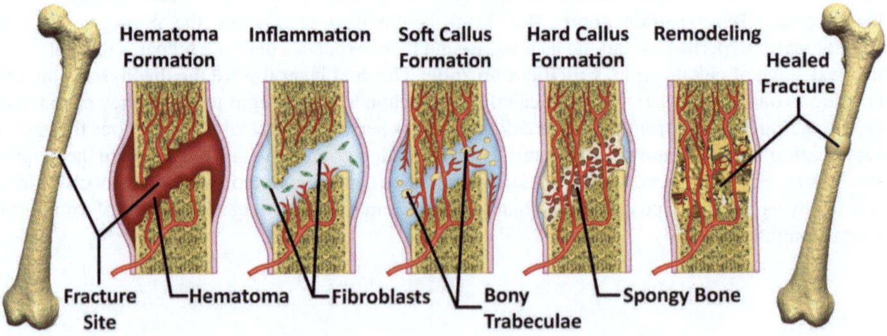

Fig. 8.15 Stages of bone repair

8.4 Part 4: Joints

Joints act as the connection between bones, facilitating various ranges of motion while ensuring structural integrity. They can have simple or complex structures involving bones, ligaments, tendons, muscles, cartilage, and synovial fluid (Table 8.1).

8.4.1 Classification

From a structural perspective, joints are classified as fibrous, cartilaginous, and synovial.

Table 8.1 Major joints in the human body: anatomy, range of motion (ROM), and functional roles

Joint	Anatomy	Range of motion	Functions
Atlantoaxial joint	Pivot joint formed by the atlas (C1) and axis (C2) with the odontoid process (dens)	Rotation primarily, limited flexion and extension	Allows for head rotation ("no" movement); crucial for cervical spine mobility
Intervertebral joints	Symphysis joints with intervertebral discs made of fibrocartilage	Limited flexion, extension, lateral flexion, and rotation	Provide stability and flexibility to the spinal column; act as shock absorbers
Glenohumeral joint	Ball-and-socket joint involving the humerus and glenoid cavity of the scapula	Flexion, extension, abduction, adduction, and rotation	Allows a wide range of arm movements; critical for lifting, pushing, and pulling
Elbow joint	Hinge joint involving the humerus, ulna, and radius	Flexion and extension	Enables bending and straightening of the arm; critical for lifting and carrying
Radioulnar joints	Pivot joints allowing the radius to rotate over the ulna	Pronation and supination	Allows for rotation of the forearm; essential for tasks requiring hand orientation
Wrist joint	Condyloid joint involving carpal bones and distal ends of the radius and ulna	Flexion, extension, abduction, and adduction	Allows complex hand movements; essential for gripping and object manipulation
Metacarpophalangeal joint	The condyloid joint connecting the metacarpal and proximal phalangeal bones	Flexion, extension, abduction, and adduction	Allows finger movement, essential for gripping and fine motor skills

(continued)

Table 8.1 (continued)

Joint	Anatomy	Range of motion	Functions
Interphalangeal joint	Hinge joints between phalanges of the fingers and toes	Flexion and extension	Allows for bending and straightening of fingers and toes; crucial for grip and balance
Hip joint	Ball-and-socket joint involving the femoral head and acetabulum of the pelvis	Flexion, extension, abduction, adduction, and rotation	Supports body weight; allows for walking, running, and jumping
Knee joint	Hinge joint involving the femur, tibia, and patella	Flexion, extension, and limited rotation when flexed	Enables walking, running, and jumping; critical for squatting and kneeling
Tibiofibular joints	Syndesmosis joints connecting the tibia and fibula	Minimal motion primarily provides stability	Provides stability to the lower leg; supports the weight-bearing role of the tibia
Ankle joint	Hinge joint involving the tibia, fibula, and talus	Dorsiflexion, plantarflexion, limited inversion, and eversion	Critical for walking, running, and jumping; allows foot movement necessary for locomotion
Midtarsal joints	Comprised of talonavicular and calcaneocuboid joints	Limited inversion, eversion, dorsiflexion, and plantarflexion	Provides flexibility and stability to the foot; helps adapt to different terrains

8.4.1.1 Fibrous Joints

Fibrous joints are a quintessential example of anatomical formations designed for stability rather than mobility. These joints are united by dense fibrous connective tissue and lack a joint cavity, which limits the range of motion. Fibrous joints can be categorized into three primary subtypes (Fig. 8.16):

Sutures: Sutures are found exclusively in the skull and tightly bind the bones, permitting strictly limited movement. The edges of the bones are interlocked or overlapped and connected by a thin layer of fibrous tissue.

Syndesmoses: Syndesmoses allow limited movement and are characterized by bones connected by a ligament or an interosseous membrane with large collagen fibers. Examples include the distal tibiofibular joint and the connection between the radius and ulna.

Gomphoses: Gomphoses are unique to dental sockets, where teeth are anchored into the jawbone. The tooth is connected to the alveolar socket of the jaw by a fibrous ligament called the periodontal ligament, which consists of collagen fibers.

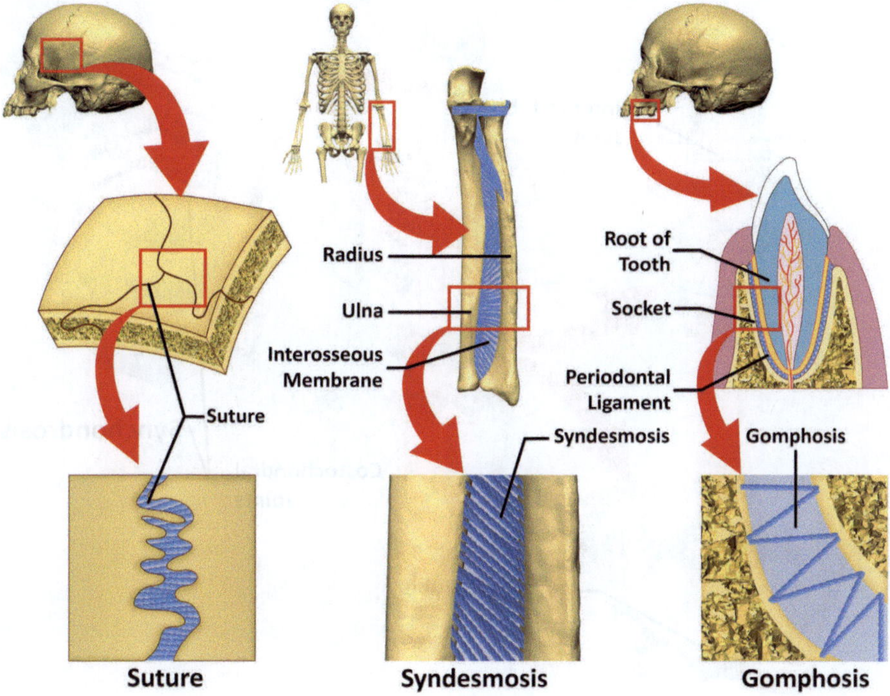

Fig. 8.16 Fibrous joints

8.4.1.2 Cartilaginous Joints

Cartilaginous joints represent an intermediary stage between the immobility of fibrous joints and the extensive mobility of synovial joints. As the name suggests, cartilaginous joints are connected by cartilage, allowing for varying degrees of movement and structural stability. Cartilaginous joints are primarily classified into two subtypes (Fig. 8.17) based on the kind of cartilage that binds them:

Synchondroses: In synchondroses, the bones are connected by hyaline cartilage. Examples include the costochondral joints of the ribs.

Symphyses: A fibrocartilaginous fusion between the articulating bones characterizes symphyses. They permit limited mobility and act as a cushion to absorb shock. Examples include the pubic symphysis and the intervertebral discs between spinal vertebral bodies.

8.4.1.3 Synovial Joints

Synovial joints (Fig. 8.18) represent the epitome of structural complexity and functional diversity among the various types of joints in the human body. Synovial joints permit a wide range of motion.

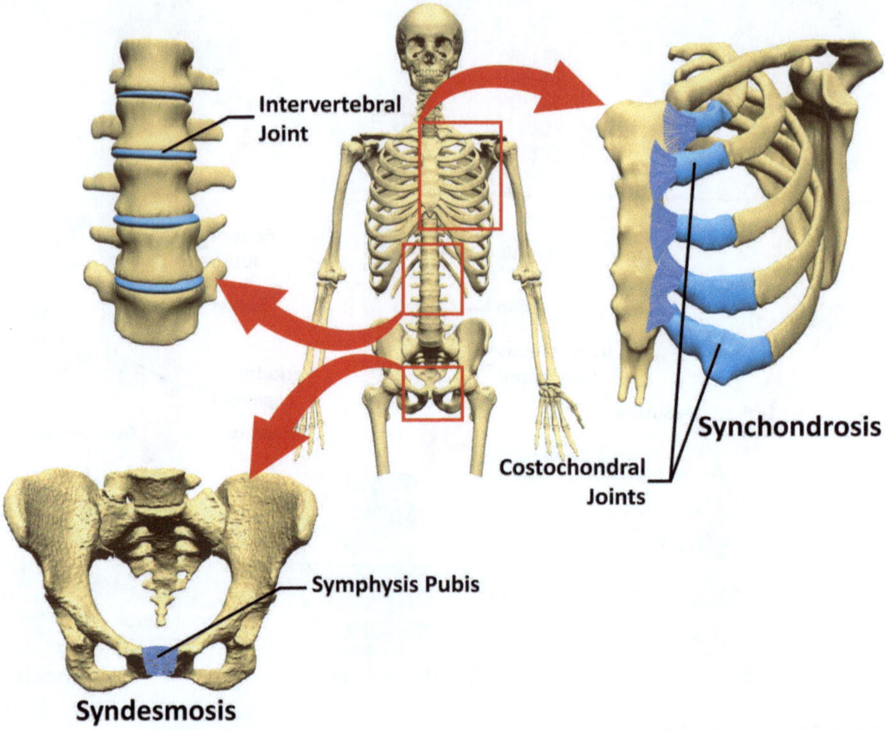

Fig. 8.17 Cartilaginous joints

Structural Components

The architecture of synovial joints is designed for both stability and mobility. The following structural components are the hallmarks of synovial joints (Fig. 8.18):

- Articular cartilage: This is a smooth, avascular layer of hyaline cartilage that covers the articulating surfaces of the bones. Its primary functions are to minimize friction and distribute load during joint movement.
- Synovial cavity: This space between the articulating bones is filled with synovial fluid, which lubricates the joint and nourishes the articular cartilage.
- Joint capsule: It provides mechanical stability and is usually made of dense irregular connective tissue.
- Synovial membrane: This specialized membrane lines the joint capsule and secretes synovial fluid. It consists of a layer of specialized cells for secretion and a sub-layer rich in blood vessels and nerves.
- Ligaments: Ligaments are dense fibrous tissue bands connecting bone to bone. In synovial joints, they are either extracapsular (outside the joint capsule), intracapsular (inside the joint capsule), or part of the joint capsule itself.
- Menisci and labra (Fig. 8.19): These fibrocartilaginous structures exist in some synovial joints to provide additional stability and shock absorption.

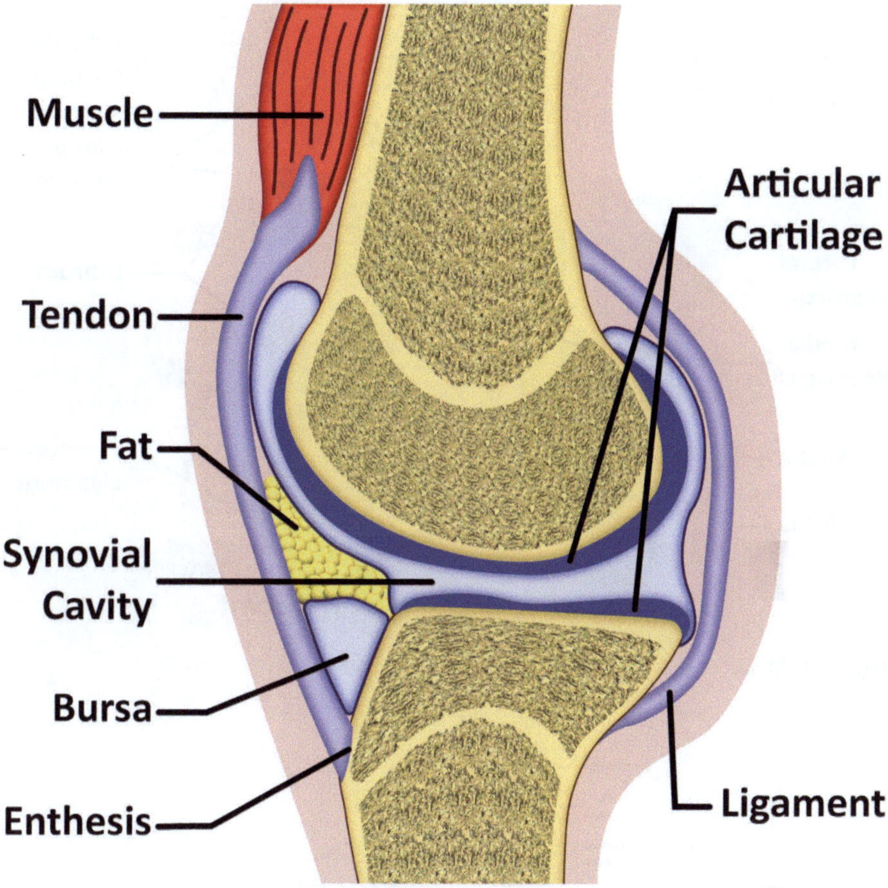

Fig. 8.18 General structure of a synovial joint

- Bursae and tendon sheaths (Fig. 8.20): These sac-like structures filled with synovial fluid reduce friction between tendons, ligaments, and bones.

Classification

Synovial joints can be classified based on the shape of their articulating surfaces and the type of movement they permit. The main types are as follows (Fig. 8.21):

- Plane joint: Plane joints, such as the intercarpal joints in the wrist, allow sliding or translational movements. The articulating surfaces are nearly flat.
- Hinge joint: Hinge joints, found in the elbow and knee, allow movement primarily along one axis, resembling the motion of a door hinge.
- Pivot joint: Pivot joints have a rounded or pointed surface fitting into a ring formed by another bone and a ligament, allowing rotational movement. Examples include the atlantoaxial joint and the proximal radioulnar joint.

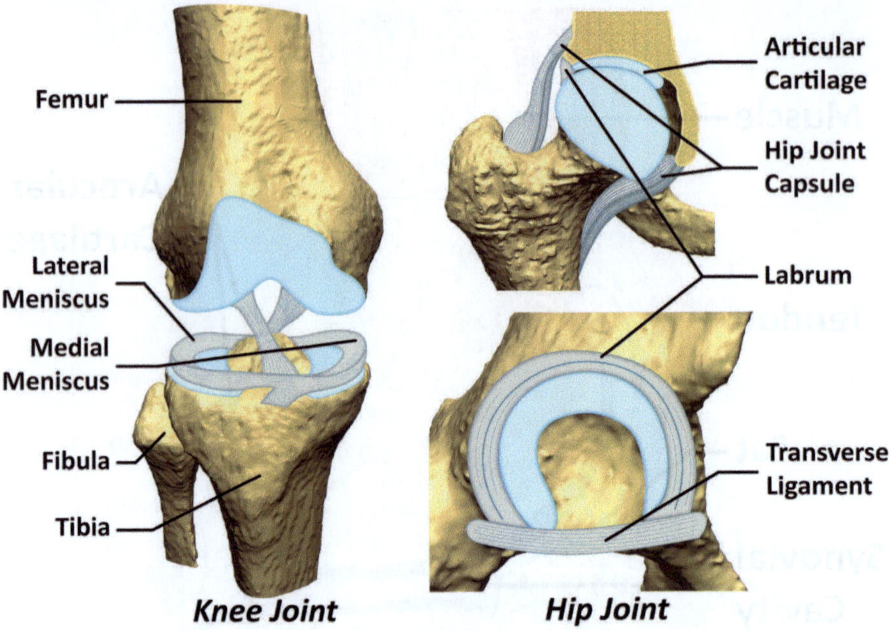

Fig. 8.19 Meniscus (left) and labrum (right) in synovial joints

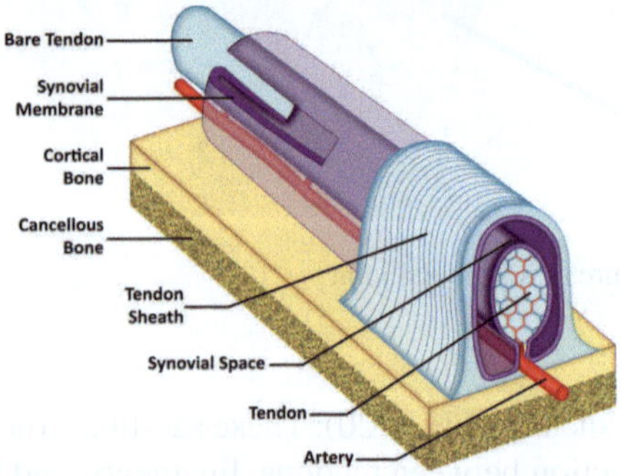

Fig. 8.20 Tendon sheath

- Condyloid (ellipsoidal) joints: Condyloid joints, such as the wrist joint, permit movement in two planes, allowing flexion, extension, adduction, abduction, and circumduction but not rotation.
- Saddle joint: In saddle joints, one of the articulating surfaces is saddle-shaped, and the other rests within it. This allows movements similar to condyloid joints but with greater freedom. An example is the carpometacarpal joint of the thumb.
- Ball-and-socket joint: Ball-and-socket joints, such as the hip and shoulder joints, offer the most extensive range of motion, allowing movement in multiple axes and planes.

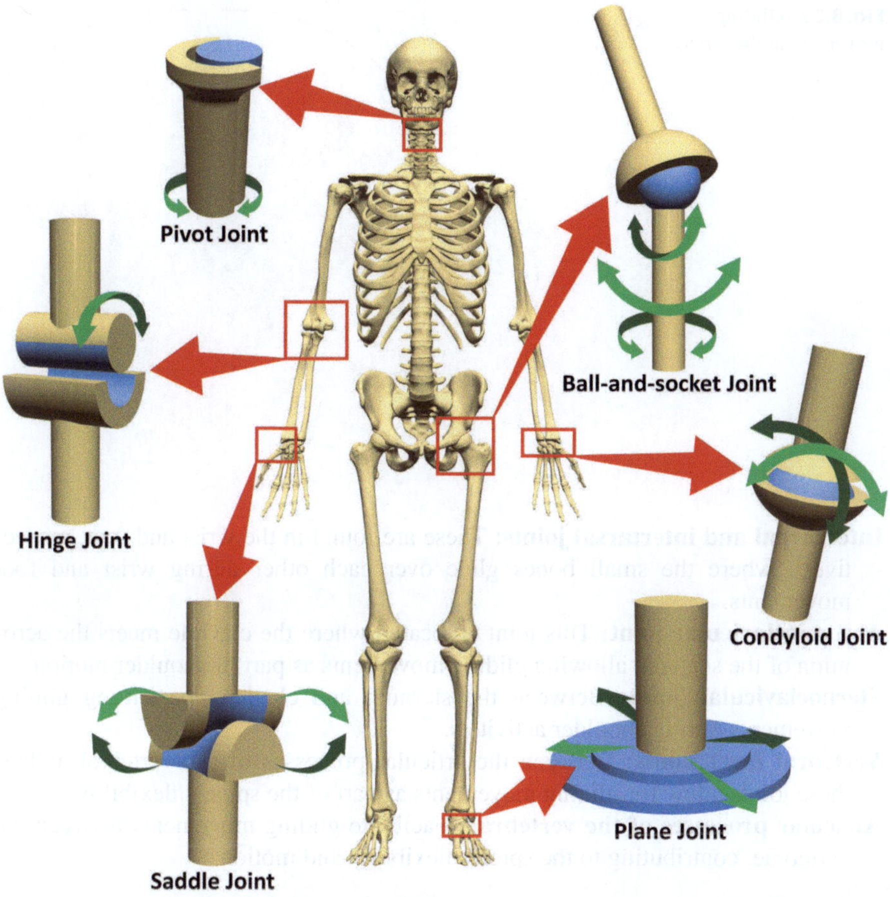

Fig. 8.21 Types of synovial joints

8.4.2 Types of Movements in Synovial Joints

Synovial joints exhibit various movements essential for the wide range of activities the human body can perform. Based on their nature, these motions can be classified as gliding, angular, or rotational.

8.4.2.1 Gliding Movements

Gliding movement (Fig. 8.22) in synovial joints is one of the fundamental types of joint movements characterized by the sliding or translational movement of articulating bone surfaces against one another. This motion occurs in joints where the articulating surfaces are nearly flat or slightly curved. This shape allows the bones to slide over each other with minimal angular change. The movement is typically linear and can occur in any direction along the plane of the joint. The range of this movement is generally small and limited by the structure of the joint and the surrounding ligaments and tendons. Examples of gliding joints include:

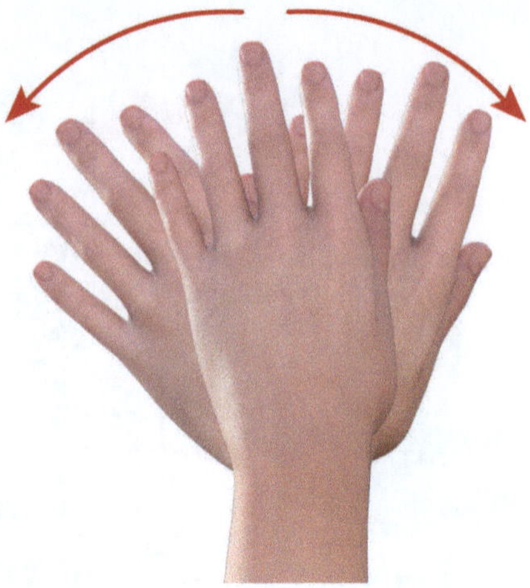

Fig. 8.22 Gliding movement at the wrist

Intercarpal and intertarsal joints: These are found in the wrist and foot, respectively, where the small bones glide over each other during wrist and foot movements.

Acromioclavicular joint: This joint is located where the clavicle meets the acromion of the scapula, allowing gliding movements as part of shoulder motion.

Sternoclavicular joint: Between the sternum and clavicle, permitting gliding movements during shoulder activities.

Vertebral facet joints: Between the articular processes of the vertebral arches, these joints allow for gliding movements as part of the spine's flexibility.

Articular processes of the vertebrae: Facilitate gliding movements between the vertebrae, contributing to the spine's flexibility and motion.

8.4.2.2 Angular Movements

Angular movement (Fig. 8.23) in synovial joints is the type of joint movement that involves a change in the angle between two bones. Key types of angular movements in synovial joints include:

Flexion: This bending movement decreases the angle between two bones. For example, bending the elbow or knee is a flexion motion.

Extension: This is the opposite of flexion, involving a straightening movement that increases the angle between body parts. Extending the elbow or knee is a typical example.

Hyperextension: This occurs when the extension goes beyond the standard anatomical position, increasing the angle between the bones to more than 180°.

Abduction: Abduction is the movement of a limb or other part away from the body's midline or from another part. Raising the arms or legs sideways away from the body demonstrates abduction.

Adduction: Adduction is the movement toward the body's midline or an adjacent body part. Lowering the arms to the sides or bringing the legs together are examples of adduction.

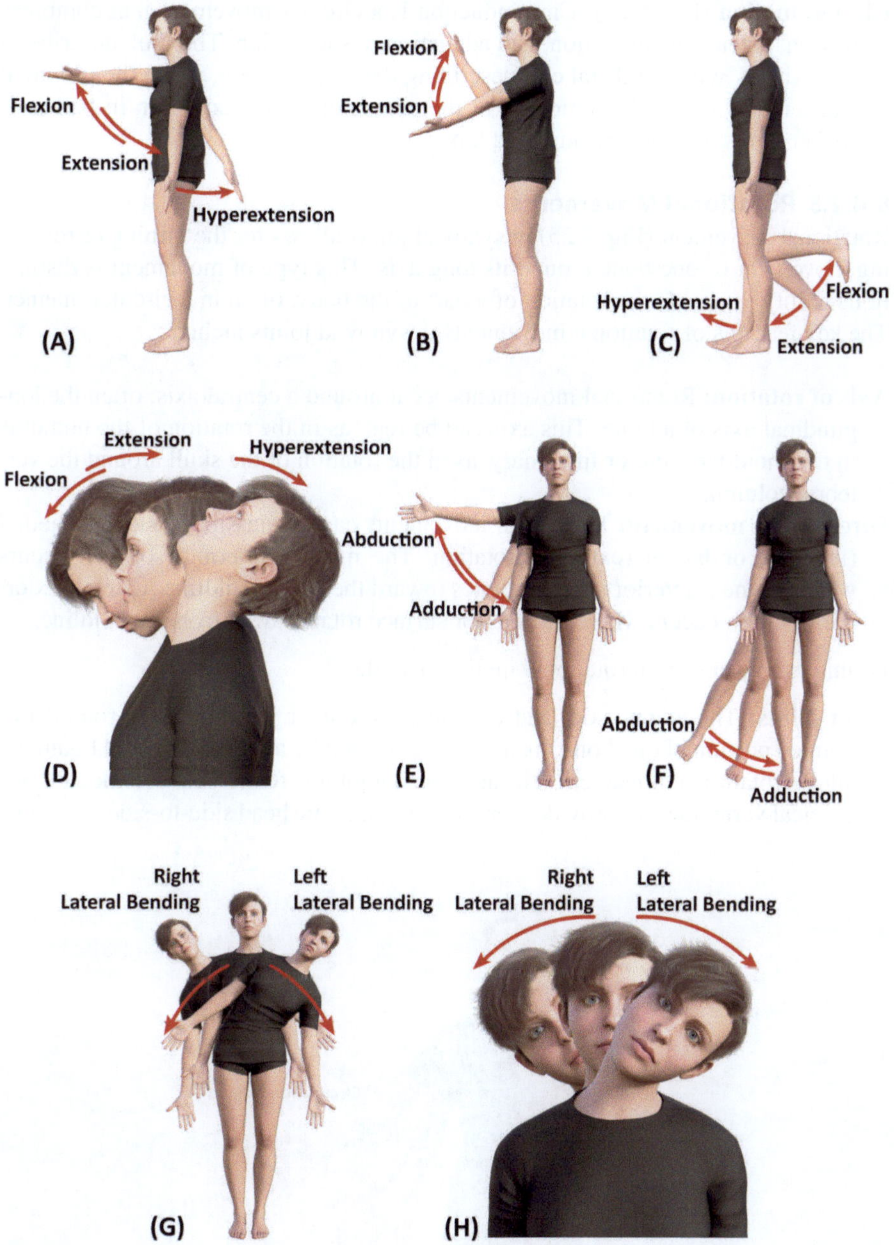

Fig. 8.23 Angular movements. (**a**) Flexion, extension, and hyperextension of the shoulder. (**b**) Flexion and extension of the elbow. (**c**) Flexion, extension, and hyperextension of the knee. (**d**) Flexion, extension, and hyperextension of the neck. (**e**) Abduction and adduction of the shoulder. (**f**) Abduction and adduction of the hip. (**g**) Lateral bending of the trunk at the lumbar spine. (**h**) Lateral bending of the neck

Circumduction (Fig. 8.24): Circumduction is a circular movement that combines flexion, extension, abduction, and adduction in succession. The limb describes a cone shape, with the distal end describing the circular base, while the proximal end remains relatively stationary at the cone's apex. It's common in ball-and-socket joints like the shoulder and hip.

8.4.2.3 Rotational Movements

Rotational movement (Fig. 8.25) in synovial joints allows for the turning or revolving movement of one bone around its long axis. This type of movement is distinct in its ability to enable the rotation of a part of the body, often in a circular manner. The key features of rotational movements in synovial joints include:

Axis of rotation: Rotational movements occur around a central axis, often the longitudinal axis of a bone. This axis can be real, as in the rotation of the humerus in the shoulder joint, or imaginary, as in the rotation of the skull around the vertebral column.

Direction of movement: Rotational movements can be classified as either medial (internal) or lateral (external) rotation. The medial (internal) rotation occurs when a bone's anterior surface rotates toward the body's midline. Lateral (external) rotation occurs when the anterior surface rotates away from the midline.

Examples of joints with rotational motion include:

Pivot joints: The atlantoaxial and proximal radioulnar joint, where a rounded or pointed portion of one bone fits into a ring formed by another bone and ligament, allows rotational movement. The atlantoaxial joint is found between the first two cervical vertebrae and provides for the rotation of the head side-to-side. Proximal

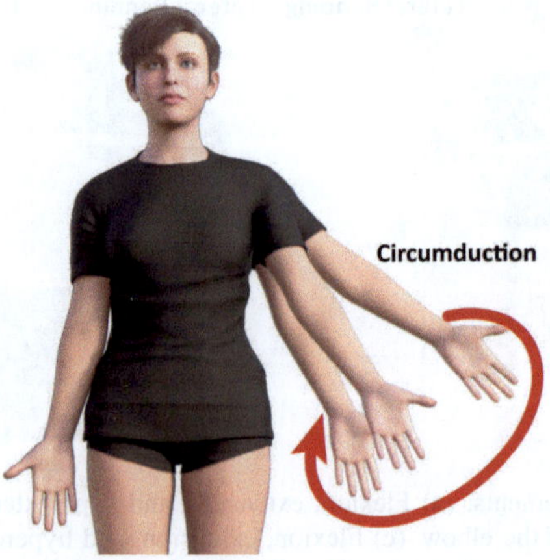

Fig. 8.24 Circumduction: a complex motion involving a combination of flexion, extension, abduction, and adduction that results in a conical movement of a limb

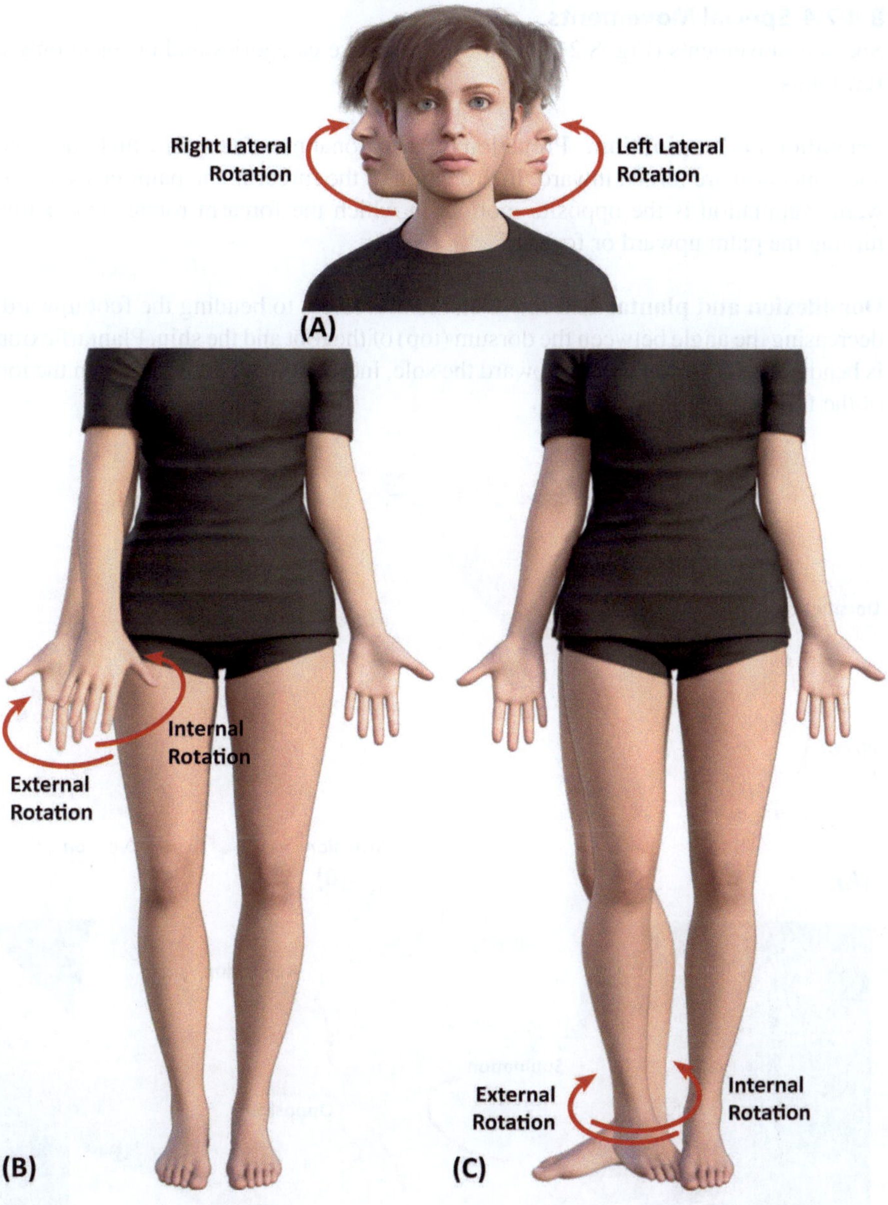

Fig. 8.25 Rotational movements. (**a**) Lateral rotation of the neck. (**b**) Internal and external rotation of the upper limb. (**c**) Internal and external rotation of the lower limb

and distal radioulnar joints enable forearm rotation; pronation and supination movements occur here.

Hip and shoulder joints: These ball-and-socket joints allow for a significant range of rotational movements, contributing to the limbs' extensive range of motion.

8.4.2.4 Special Movements

Specific movements (Fig. 8.26) do not fit the above categories and occur at only a few joints.

Pronation and supination: Pronation is a rotational movement in which the hand and upper arm are turned inward. In pronation of the forearm, the palm faces downward. Supination is the opposite motion, in which the forearm rotates externally, turning the palm upward or forward.

Dorsiflexion and plantarflexion: Dorsiflexion refers to bending the foot upward, decreasing the angle between the dorsum (top) of the foot and the shin. Plantarflexion is bending the foot downwards toward the sole, increasing the angle between the top of the foot and the shin.

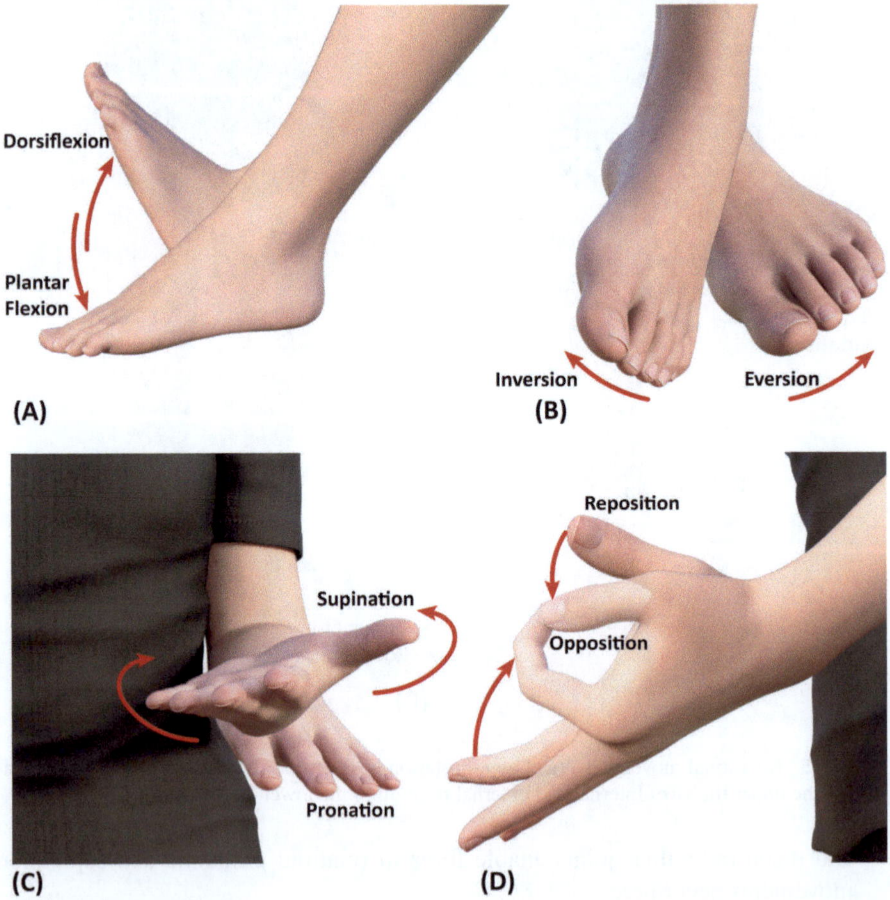

Fig. 8.26 Specific movements. (**a**) Ankle dorsiflexion and plantar flexion. (**b**) Subtalar inversion and eversion. (**c**) Forearm supination and pronation. (**d**) Thumb opposition and reposition

Inversion and eversion: Inversion involves turning the sole inward toward the body's midline. Eversion is turning the sole outwards, away from the midline.

Opposition and reposition: Opposition occurs when the thumb and a finger move toward each other, as in pinching. Reposition is the movement back to the anatomical position from the opposition.

Protraction and retraction: Protraction moves a body part forward, or anteriorly, along the horizontal plane. It's essentially a forward movement from the standard anatomical position. Retraction is the opposite of protraction, involving moving a body part backward or posteriorly along the horizontal plane. This movements are seen in scapulothoracic and temporomandibular joints. Protraction of the scapula consists of moving the shoulder blades away from the spine, as seen when reaching forward, while retraction of the scapula pulls the shoulder blades toward the spine, seen in actions like pulling the shoulders back to stand up straight. In the temporomandibular joint (jaw), protraction moves the chin forward, a motion commonly associated with jutting the jaw out, while retraction moves the chin backward, reversing the protraction motion.

Elevation and depression: Elevation refers to the movement of a body part vertically upward along the frontal plane. Depression is the opposite of elevation, involving the downward movement of a body part along the frontal plane. This movements are seen in scapulothoracic and temporomandibular joints. Elevation of the scapula involves raising the shoulder blades, as seen when shrugging the shoulders. In contrast, depression of the scapula moves the shoulder blades downward, as seen when lowering the shoulders from a shrugged position. In the jaw, elevation moves the lower jaw upward, closing the mouth, while depression lowers the lower jaw, opening the mouth.

8.5 Part 5: Human Skeleton

The human skeleton can be broadly categorized into two main subdivisions: axial and appendicular.

8.5.1 Axial Skeleton

The axial skeleton serves as the central core framework of the body, providing an anchor for the appendicular skeleton and protecting vital organs within the cranial and thoracic cavities. It consists of 80 bones organized into three primary regions: the skull, vertebral column, and thoracic cage (Fig. 8.27).

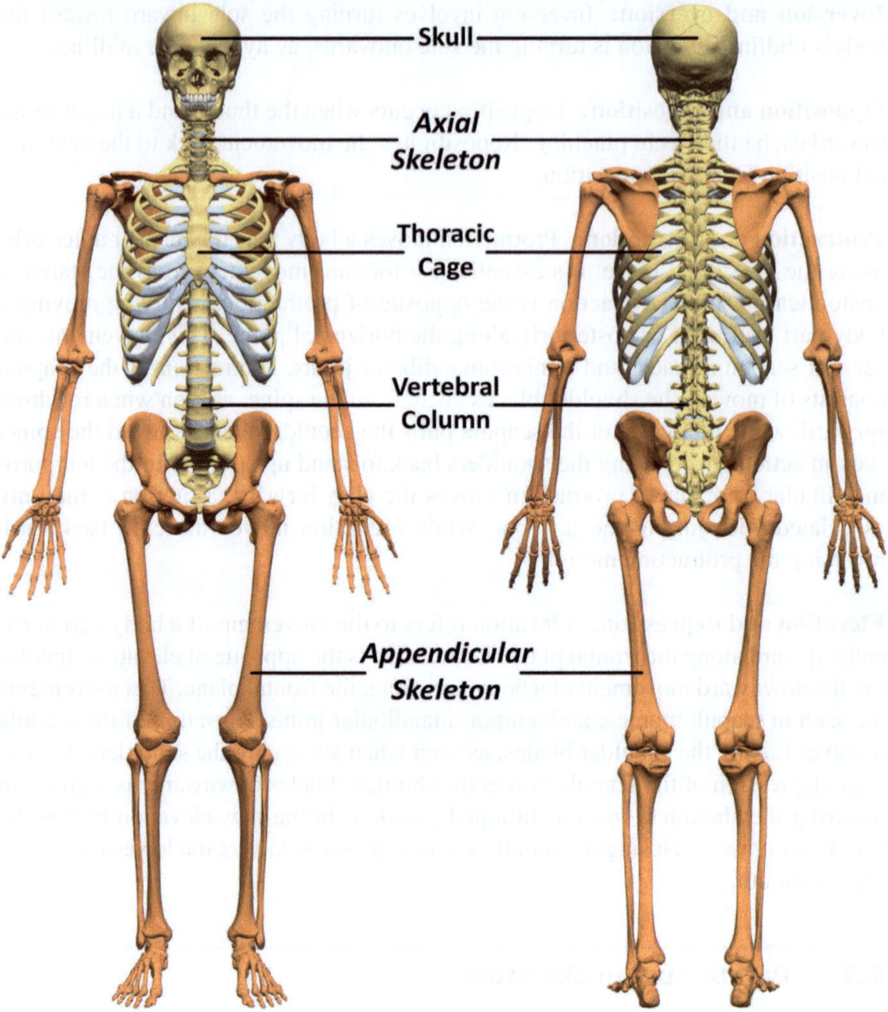

Fig. 8.27 The human skeleton. The bones of the axial skeleton are colored light gold, and the bones of the appendicular skeleton are sandy brown

8.5.1.1 Skull

The skull (Fig. 8.28) encases and protects the brain. It is comprised of 22 individual bones (excluding the ossicles of the middle ear). The skull is subdivided into the cranium, which contains eight bones, and the facial skeleton, which consists of 14 bones. The mandible is the only bone in the skull capable of independent motion.

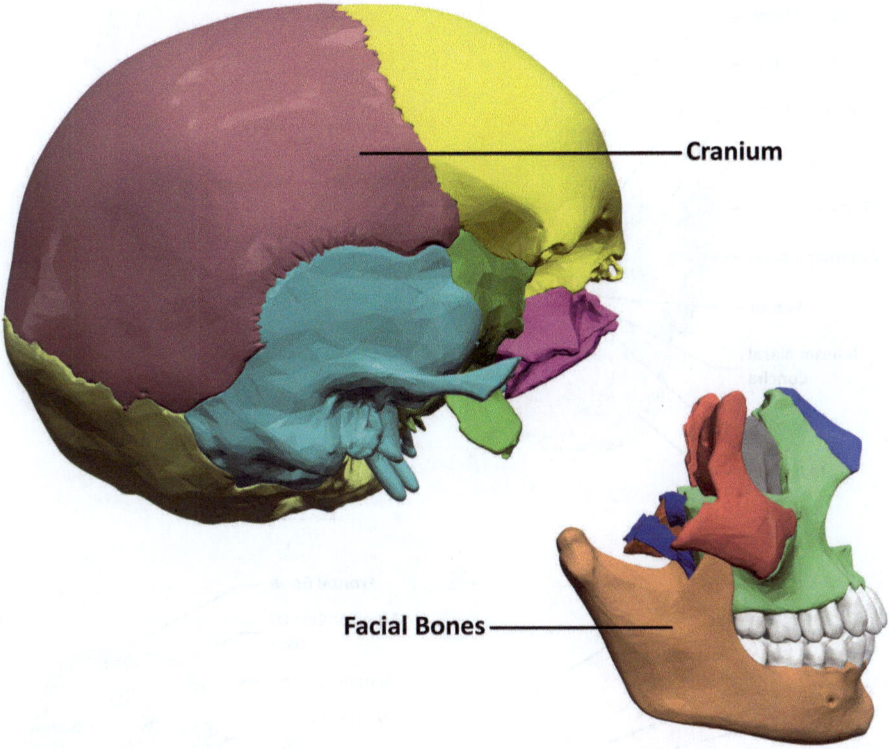

Fig. 8.28 The skull: cranial and facial divisions

Cranium

The cranium is a complex bony structure that serves as the protective encasement for the brain and sensory organs such as the eyes and ears. It is a part of the axial skeleton and consists of eight distinct bones joined together via sutures, which are fibrous joints that allow for strictly limited movement. These bones include the frontal bone, two parietal bones, two temporal bones, the occipital bone, the sphenoid bone, and the ethmoid bone (Figs. 8.29, 8.30, 8.31, and 8.32). Table 8.2 provides a detailed description of the individual bones and their particular characteristics.

Face

The facial bony structure, or skeleton (Fig. 8.33), is an intricately designed construct comprising 14 individual bones, separate from the eight cranial bones. These bones serve multiple functions, including forming the facial contours, housing

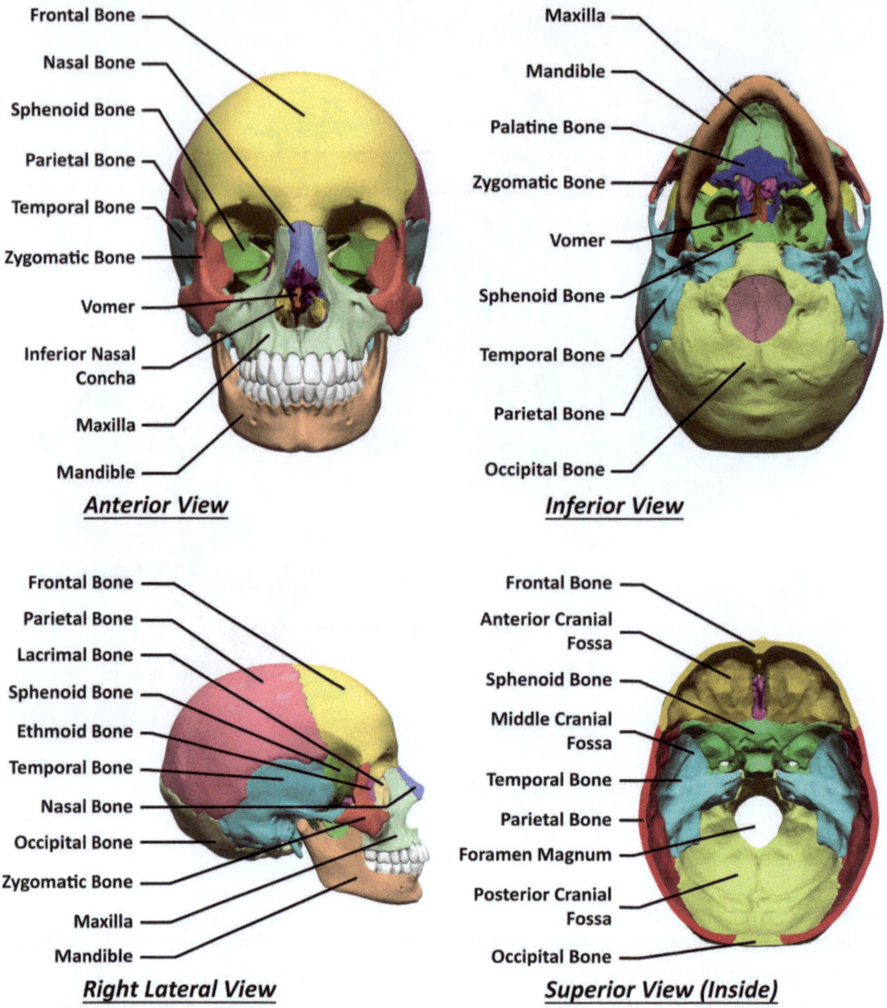

Fig. 8.29 Bones of the skull

sensory organs, and providing the framework for soft tissue attachments (Figs. 8.33 and 8.34). The bones also facilitate essential physiological functions such as mastication and food and air passage. Table 8.3 provides a detailed description of the individual bones and their particular characteristics.

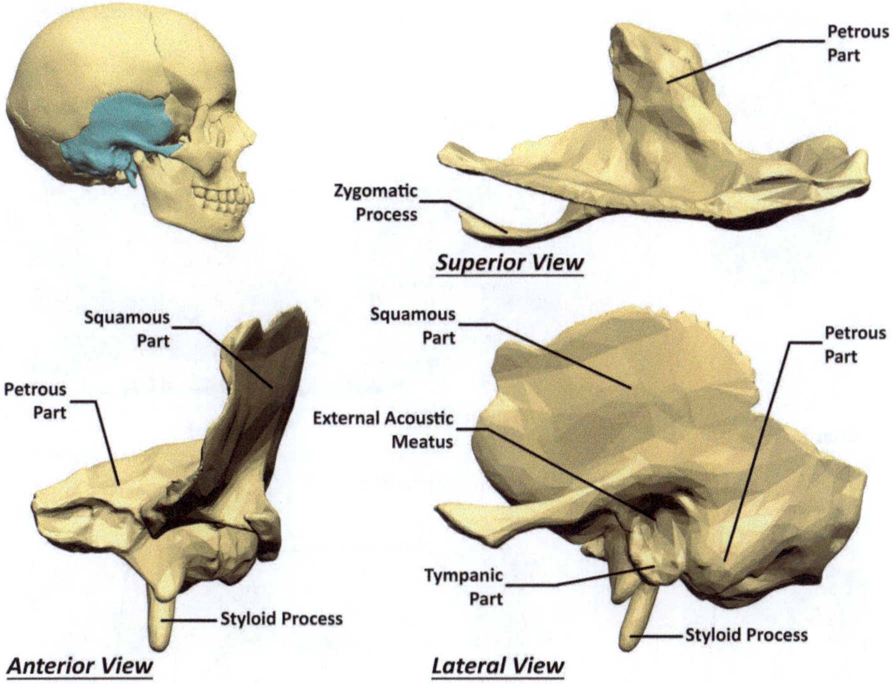

Fig. 8.30 The temporal bone

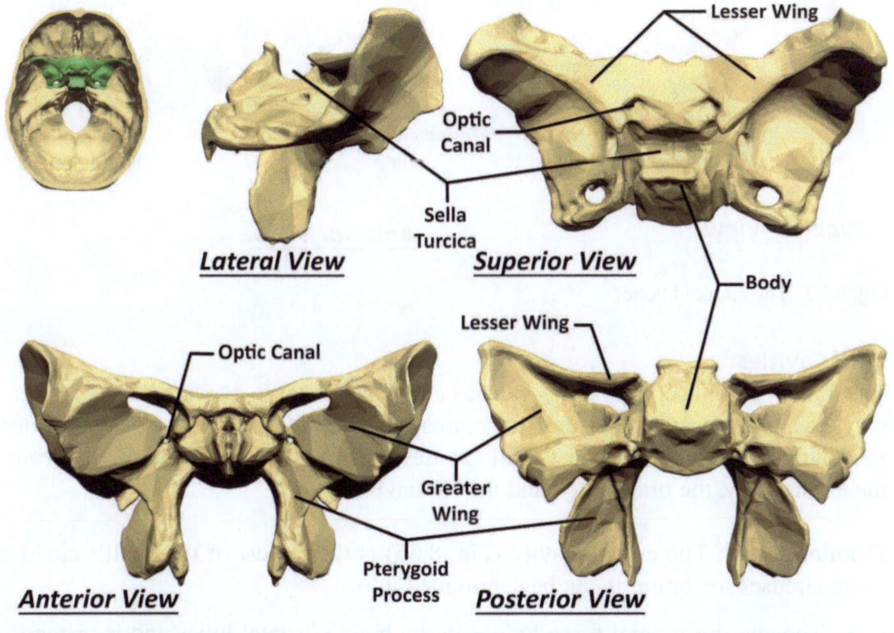

Fig. 8.31 The sphenoid bone

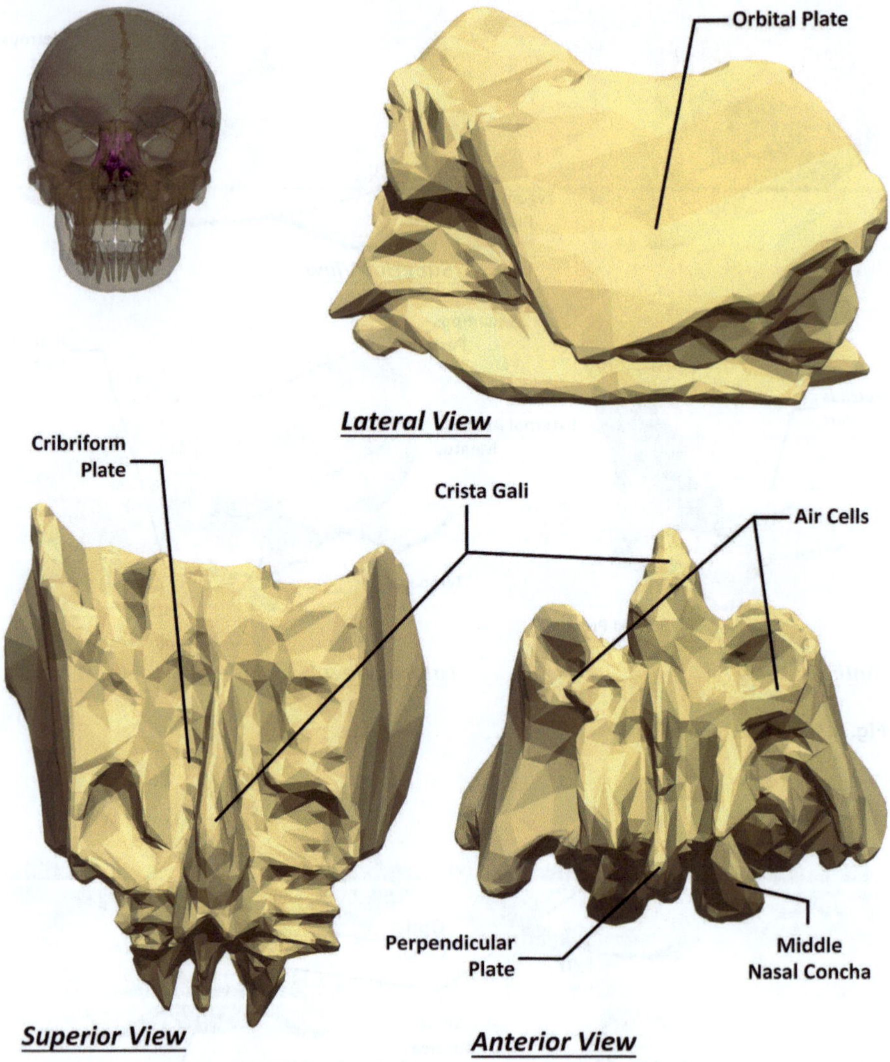

Fig. 8.32 The ethmoid bone

Skull Cavities

The human skull accommodates various cavities that serve as repositories for sensory organs, protect vital neural structures, and provide pathways for neural and vascular conduits. The primary skull cavities include the cranial cavity, the orbits, the nasal cavity, the oral cavity, and the ear cavities.

Cranial cavity: The cranial cavity (Fig. 8.35) is the largest of the skull's cavities and houses the brain. It can be subdivided into:

- The anterior cranial fossa lodges in the brain's frontal lobes and is primarily formed by the frontal bone and parts of the sphenoid bone.

Table 8.2 Cranial bones: locations, attachments, markings, and important features

Name	Location	Attachments	Markings and notable structures	Important features
Frontal bone	Forehead, anterior part of the skull	Articulates with the two parietal bones via coronal suture	Frontal sinuses, glabella, supraorbital foramen	Protection of frontal lobes of the brain
Parietal bone	Sides and roof of the cranial cavity	Frontal bone via coronal suture; opposing parietal bone via sagittal suture; temporal bone via squamous suture; occipital bone via lambdoid suture	Temporal lines for muscle attachment	Forms the main side walls of the cranial cavity
Temporal bone (Fig. 8.30)	Lateral sides of the skull, adjacent to the ear region	Parietal bones via squamous suture; occipital bone via occipitomastoid suture; sphenoid bone via greater wings	External acoustic meatus, mastoid process, styloid process, zygomatic process, mandibular fossa	Houses structures of the ear; attachment point for jaw muscles
Occipital bone	Posterior and inferior aspects of the cranial cavity	Parietal bones via lambdoid suture; temporal bones via occipitomastoid suture; first cervical vertebra (atlas) via occipital condyles	Foramen magnum, occipital condyles, external occipital protuberance	Protects the cerebellum; the attachment point for ligaments and muscles of the neck
Sphenoid bone (Fig. 8.31)	The central base of the skull	Articulates with all other cranial bones	Sella turcica, greater and lesser wings, optic canal, foramen rotundum, foramen ovale	Forms part of the eye socket; houses the pituitary gland
Ethmoid bone (Fig. 8.32)	Between the orbits, forms part of the nasal cavity	Frontal bone via ethmoidal notch; sphenoid bone via sphenoidal body; nasal bones; maxillary bones	Cribriform plate, perpendicular plate, ethmoidal sinuses, superior and middle nasal conchae	Partitions nasal cavity; contributes to orbits

- The middle cranial fossa houses the temporal lobes and the hypophysis (pituitary gland) and mainly comprises the sphenoid and temporal bones.
- The posterior cranial fossa accommodates the cerebellum and the brainstem. The occipital bone and the posterior parts of the temporal and sphenoid bones form it.

Orbital cavities: The orbits are bony sockets that contain the eyes, associated muscles, nerves, and vessels. The orbits are pyramidal with an apex that points posteriorly and are formed by seven bones, including the frontal, sphenoid, zygomatic, maxillary, palatine, lacrimal, and ethmoid.

Nasal cavity: The nasal cavity is a structure between the orbits and above the oral cavity. It is separated by the nasal septum, primarily formed by the perpendicular plate of the ethmoid bone and the vomer. The cavity also includes three conchae

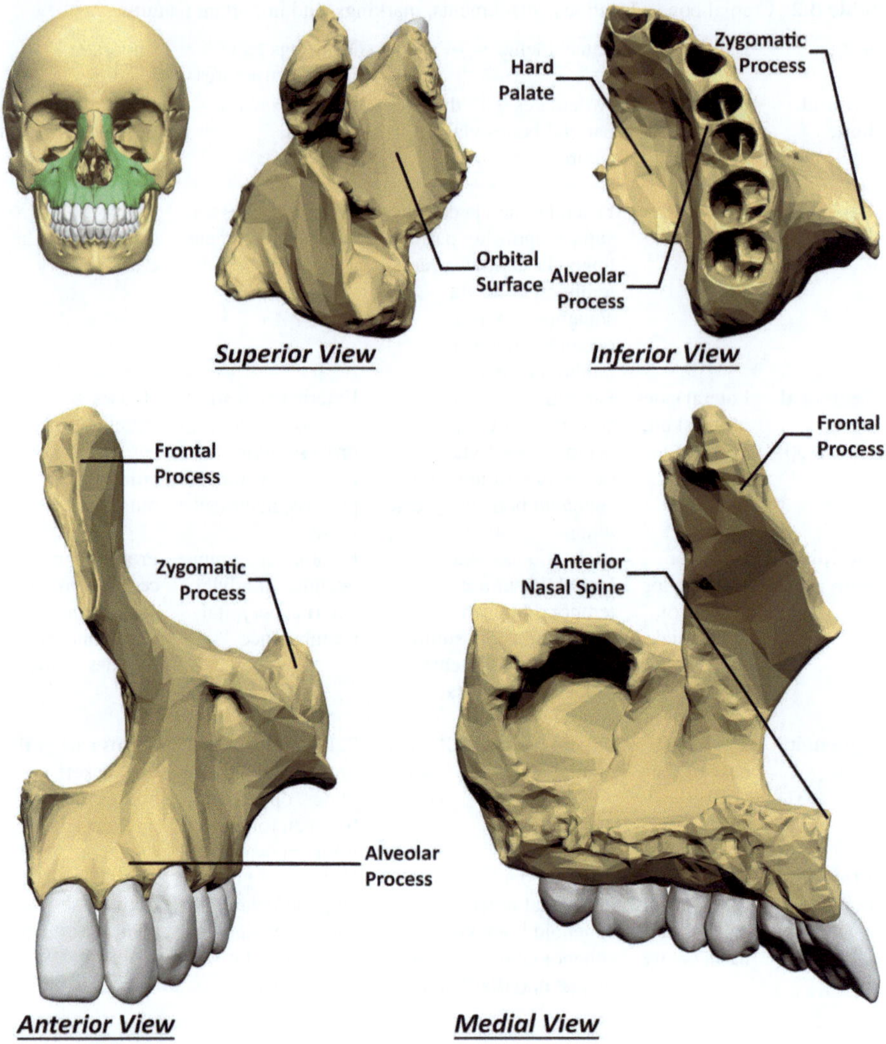

Fig. 8.33 The maxilla

that humidify and warm the air. The bones contributing to this cavity include the ethmoid, sphenoid, maxillae, palatines, and nasal bones.

Paranasal sinuses: Paranasal sinuses (frontal, maxillary, sphenoidal, and ethmoidal sinuses) are connected to the nasal cavity (Page 563).

Ear cavities: The ear cavities are located in the petrous part of the temporal bones and can be divided into the external auditory meatus, middle ear, and inner ear (Section 11.2).

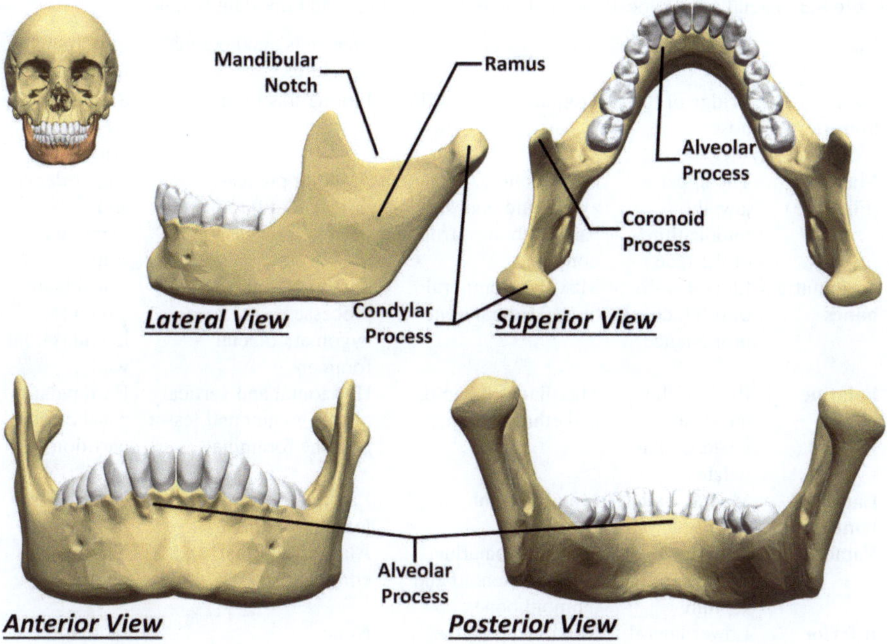

Fig. 8.34 The mandible

Oral cavity: The oral cavity extends from the lips to the oropharyngeal isthmus and includes the mouth and the beginning of the throat. The maxilla and mandible form the bony framework of this cavity, and it serves primarily for ingestion, chewing, and speech (Section 16.2.1.1).

8.5.1.2 Hyoid Bone

The hyoid bone is a unique structure in the human skeletal system, distinguished by its isolated position in the neck and its absence of articulation with any other bone. Located between the mandible and the thyroid cartilage, it is an anchoring structure for many muscles and ligaments involved in functions such as swallowing, speech, and respiration. It has a horseshoe-like shape.

8.5.1.3 Vertebral Column

The vertebral column, or the spinal column or backbone (Fig. 8.36), is the primary axial support for the human body. It is a complex, segmented structure that houses and protects the spinal cord, supports the skull, and provides a scaffold for the attachment of the rib cage and musculature of the back. Comprising 33 vertebrae in its embryonic state, it eventually fuses to comprise 26 vertebrae in the adult human due to the fusion of certain segments. The vertebral column is not a straight line but exhibits several curves that confer both strength and flexibility.

Table 8.3 Facial bones: locations, attachments, markings, and important features

Name	Location	Attachments	Markings and notable structures	Important features
Nasal bones	Bridge of the nose	Frontal bone, maxillae	Internasal suture	Support for the nose bridge
Maxillae (Fig. 8.33)	The upper jaw, the middle third of the face	Nasal bones, zygomatic bones, palatine bones, frontal bone	Alveolar process, infraorbital foramen, palatine process	Upper dental arch, maxillary sinus
Zygomatic bones	Lateral walls of orbit, cheek prominence	Maxillae, temporal bones, frontal bone	Temporal and frontal processes, zygomaticofacial foramen	Cheekbone structure, lateral orbital wall
Palatine bones	Back of the nasal cavity, posterior hard palate	Maxillae, sphenoid, and ethmoid bones	Horizontal and vertical plates, greater and lesser palatine foramina	Hard palate, nasal cavity partition
Lacrimal bones	Medial walls of orbit	Maxillae, ethmoid bones	Lacrimal groove, lacrimal foramen	Housing for lacrimal sac
Vomer	The lower part of the nasal septum	Maxillae, palatine bones, sphenoid and ethmoid bones	Alae, sphenoidal, and ethmoidal crests	Nasal cavity partition
Inferior nasal conchae	Lower lateral walls of the nasal cavity	Maxillae, palatine bones	None	Increase nasal epithelial surface area
Mandible (Fig. 8.34)	Lower jaw	Temporal bones (at temporomandibular joints)	Mental foramen, coronoid process, condyloid process, mandibular notch, ramus, body, alveolar process	Lower dental arch, muscle attachment

Regions

The vertebral column is categorized into five regions based on anatomical and functional differences (Fig. 8.36):

Cervical region: This is the uppermost region of the vertebral column, comprising seven vertebrae (C1-C7). The first vertebra, known as the atlas, directly supports the skull and allows for the head's nodding motion. The second vertebra, the axis, has a peg-like projection known as the odontoid process (or dens) around which the atlas rotates, allowing the head to turn from side to side.

Thoracic region: This region consists of 12 vertebrae (T1-T12) and is the primary point of articulation for the ribs. The presence of costal facets for rib attachment characterizes thoracic vertebrae. This region is less mobile than the cervical and lumbar regions but provides robust structural support for the chest cavity.

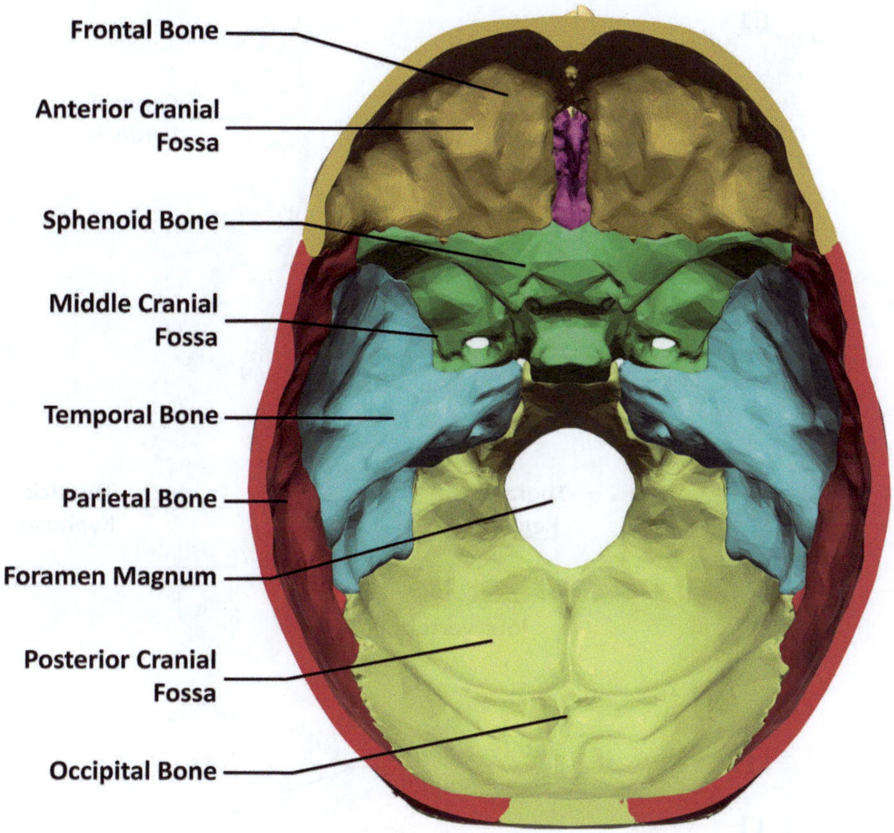

Fig. 8.35 Cranial cavity. Calvaria removed

Lumbar region: The lumbar region consists of five large, robust vertebrae (L1-L5) that support most of the body's weight. These vertebrae are built for strength, have large bodies, and have relatively short, thick processes.

Sacral region: The sacral region comprises five fused vertebrae (S1-S5) to form a single, triangular-shaped bone called the sacrum. The sacrum articulates with the iliac bones on either side to create the sacroiliac joints, critical elements in transferring weight from the spinal column to the lower limbs.

Coccygeal region: The coccyx, commonly called the tailbone, typically consists of four fused vertebrae. This vestigial structure is an attachment site for certain ligaments and muscles of the pelvic floor.

Curvatures

The vertebral column presents four curves (Fig. 8.36):

- Cervical lordosis: A forward (anterior) curve in the cervical region.
- Thoracic kyphosis: A backward (posterior) curve in the thoracic region.

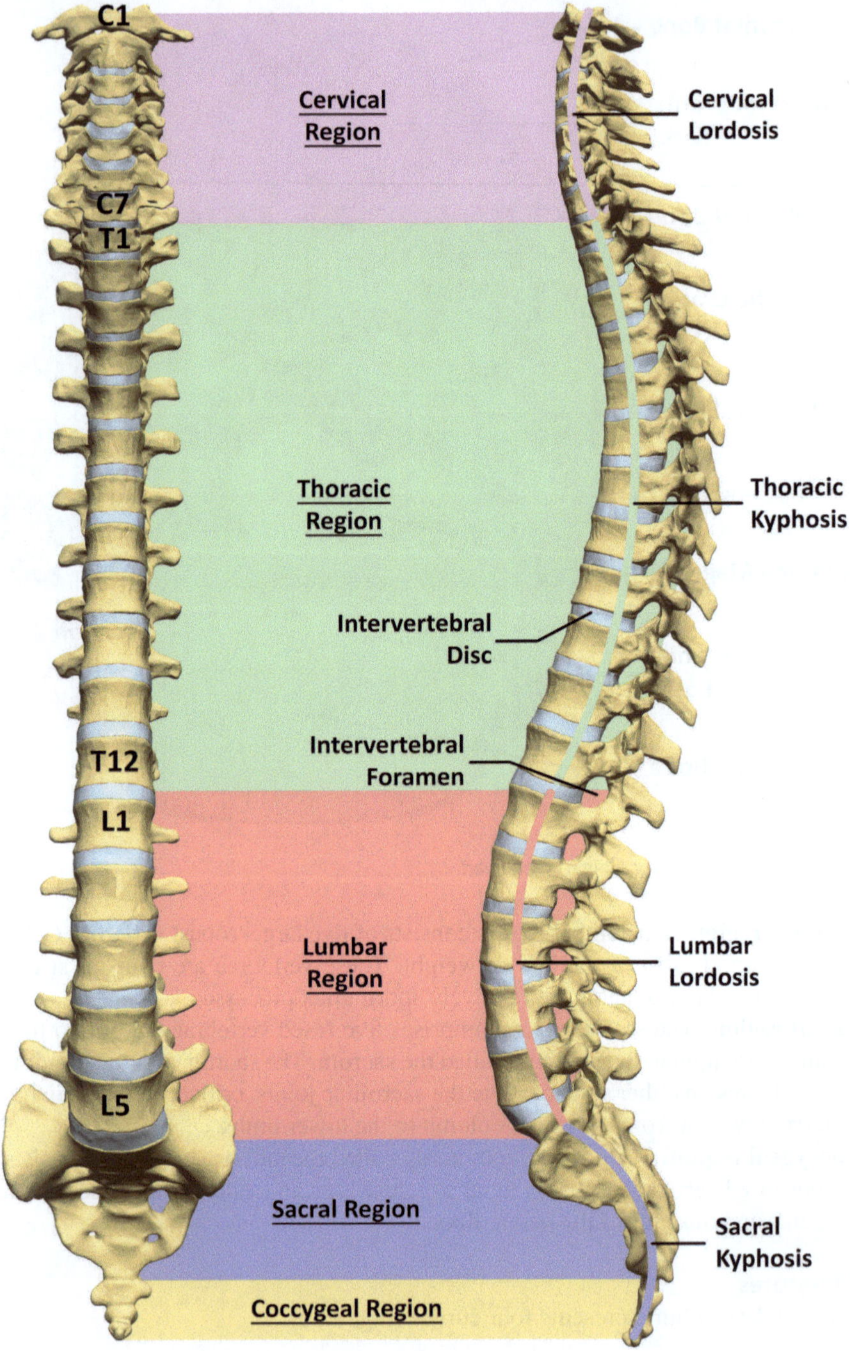

Fig. 8.36 The vertebral column

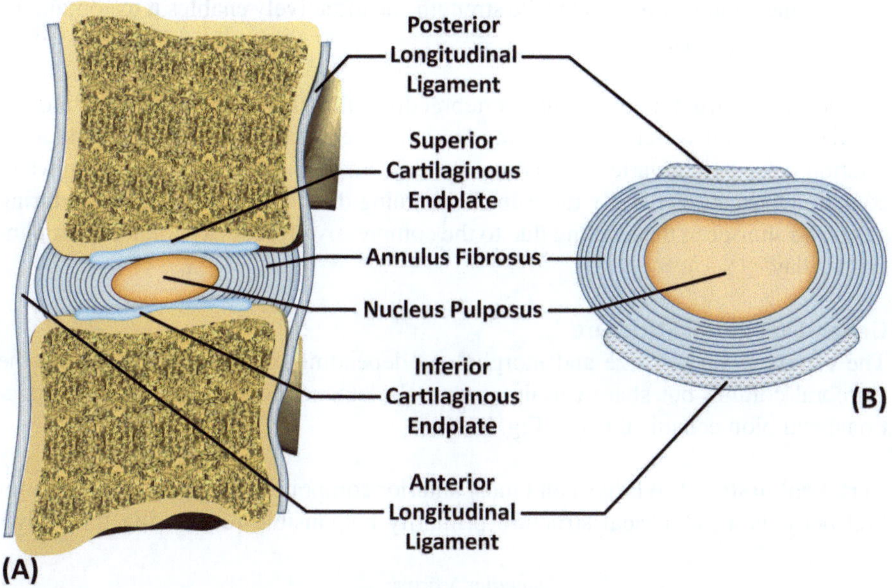

Fig. 8.37 Intervertebral disc. (**a**) Sagittal section. (**b**) Axial section

- Lumbar lordosis: A forward (anterior) curve in the lumbar region.
- Sacral kyphosis: A backward (posterior) curve in the sacral region.

These curves enhance the column's resilience and mechanical strength, enable balanced bipedal locomotion, and distribute axial loads during movement and at rest.

Intervertebral Discs

The intervertebral discs are specialized, fibrocartilaginous structures between adjacent vertebrae along the vertebral column (Fig. 8.37). They serve as the spine's shock-absorbing system, facilitate a range of movements, and contribute to the overall stability and flexibility of the vertebral column. Comprising approximately 25% of the spine's total length, these discs are essential for distributing mechanical stress during various activities such as lifting, twisting, and bending. Each intervertebral disc is composed of two main components (Fig. 8.37):

- Nucleus pulposus: This is the inner, gel-like core primarily made of water, collagen fibers, and proteoglycans. It has high elasticity and is the main shock absorber, distributing compressive loads applied to the spine. The high water content helps the disc to withstand compression.
- Annulus fibrosus: This outer layer consists of multiple concentric rings of fibrocartilage known as lamellae. These lamellae are rich in type I collagen and are organized in a parallel fashion but at oblique angles to other layers. This structure

gives the annulus fibrosus tensile strength and effectively enables it to contain the nucleus pulposus.

The unique structure of the intervertebral discs allows for various kinds of motion between adjacent vertebrae, including flexion, extension, lateral bending, and axial rotation. Disc height varies throughout the day due to the diurnal variation in water content. Discs are generally taller in the morning due to fluid accumulation during sleep and shorter in the evening due to the compressive forces experienced throughout the day.

General Vertebral Structure

The vertebrae vary in size and morphology depending on their location along the vertebral column but share certain structural elements contributing to their functional and biomechanical roles (Fig. 8.38).

Vertebral body: The largest and most anterior component of a vertebra, the vertebral body, is a cylindrical structure primarily responsible for bearing the body's

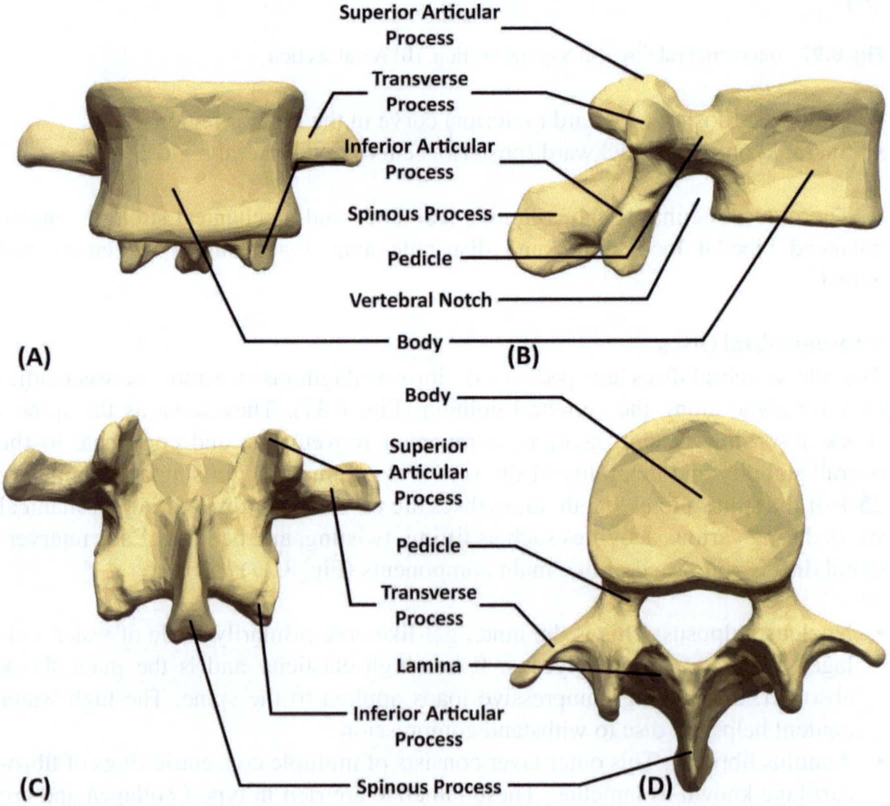

Fig. 8.38 Typical vertebral structure (L1). (**a**) Anterior view. (**b**) Lateral view. (**c**) Posterior view. (**d**) Superior view

weight and providing a surface for articulation with intervertebral discs. It is composed of cancellous bone enveloped by a thin layer of cortical bone.

Vertebral arch: This bony arch encloses the vertebral foramen and is formed by two pedicles and two laminae.

- Pedicles: These short, stout processes project posteriorly from the vertebral body to meet the laminae.
- Laminae: These flat, broad plates extend medially from the pedicles to complete the vertebral arch.

Vertebral foramen: The hollow space enclosed by the vertebral body and vertebral arch through which the spinal cord passes.

Spinous process: A singular, posteriorly projecting process that arises at the junction of the two laminae. It serves as an attachment point for muscles and ligaments.

Transverse processes: These are lateral projections arising from the junction between the pedicles and laminae, serving as additional muscle and ligament attachment sites.

Articular processes: There are four articular processes per vertebra: two superior and two inferior. These processes contain articular facets coated with hyaline cartilage and articulate with corresponding facets on adjacent vertebrae, thereby forming facet (zygapophyseal) joints, which are synovial.

Variations in Vertebral Morphology

There are significant regional variations in the morphology of the individual vertebral components.

Cervical vertebrae (Fig. 8.39): The vertebral body is smaller and broader than other regions, often concave superiorly and convex inferiorly. The transverse foramina, unique to cervical vertebrae, allows for the passage of the vertebral arteries, which supply blood to the brain. Spinous processes, generally bifid (split) in shape, except for C1 and sometimes C7, are shorter and project posteriorly and slightly upward. The C1 (atlas) and C2 (axis) vertebrae have unique structures:

Atlas (C1) (Fig. 8.39) is a unique ring-like structure without a vertebral body or spinous process. It articulates with the skull's occipital bone and allows for the head's "nodding" motion.

Axis (C2) (Fig. 8.39) is characterized by a unique, tooth-like projection called the dens or odontoid process, which articulates with the atlas to allow for rotational movement of the head.

Thoracic vertebrae (Fig. 8.40): The vertebral body is heart-shaped and larger than the cervical but smaller than the lumbar vertebrae. Bodies bear facets or demifacets for rib articulation. Transverse processes are long and robust, with facets for rib

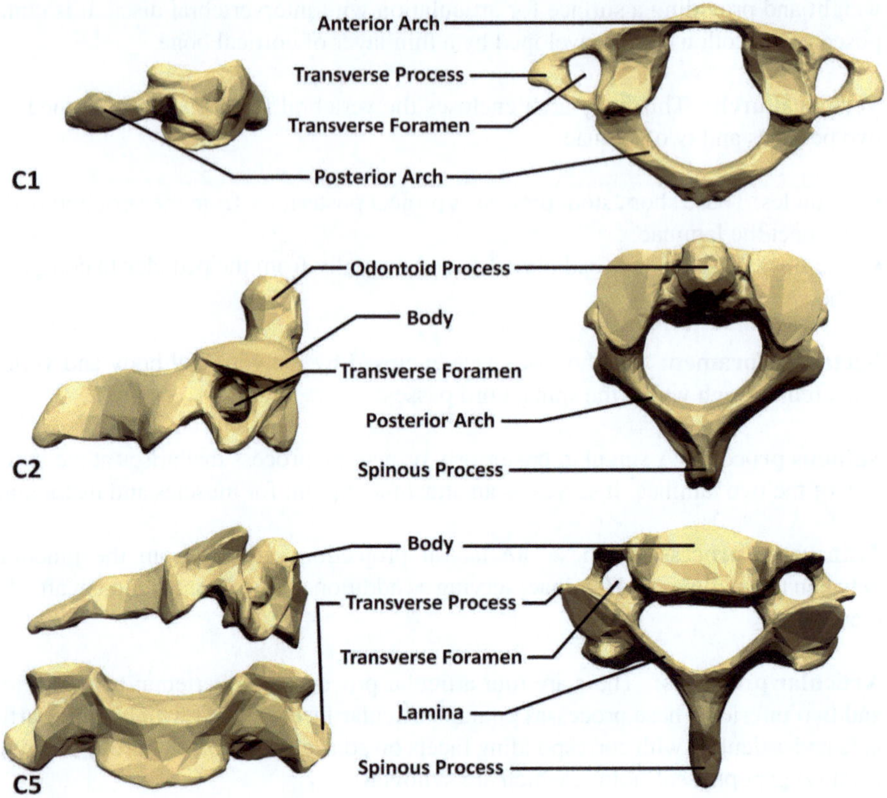

Fig. 8.39 Cervical vertebrae: atlas (C1), axis (C2), and C5

articulation (except T11 and T12). Spinous processes are long, slender, and pointed, generally directed inferiorly. They serve as attachment sites for numerous back muscles.

Lumbar vertebrae (Fig. 8.40): The vertebral body is the largest and most robust among all vertebrae. It is kidney-shaped and adapted to bear most of the body's weight. Transverse processes are shorter but bulkier than thoracic vertebrae, without facets for rib articulation. Spinous processes are short, flat, and hatchet-shaped, extending directly backward and providing significant surface area for the attachment of powerful back muscles.

Sacrum (Fig. 8.41): Five fused vertebrae form a single, triangular-shaped bone called the sacrum. The sacrum articulates with the iliac bones to form the sacroiliac joints, forming the pelvic girdle. Anterior and posterior sacral foramina serve as passageways for sacral nerves. Promontory, the upper anterior portion of the sacrum, is prominent and used as an anatomical landmark.

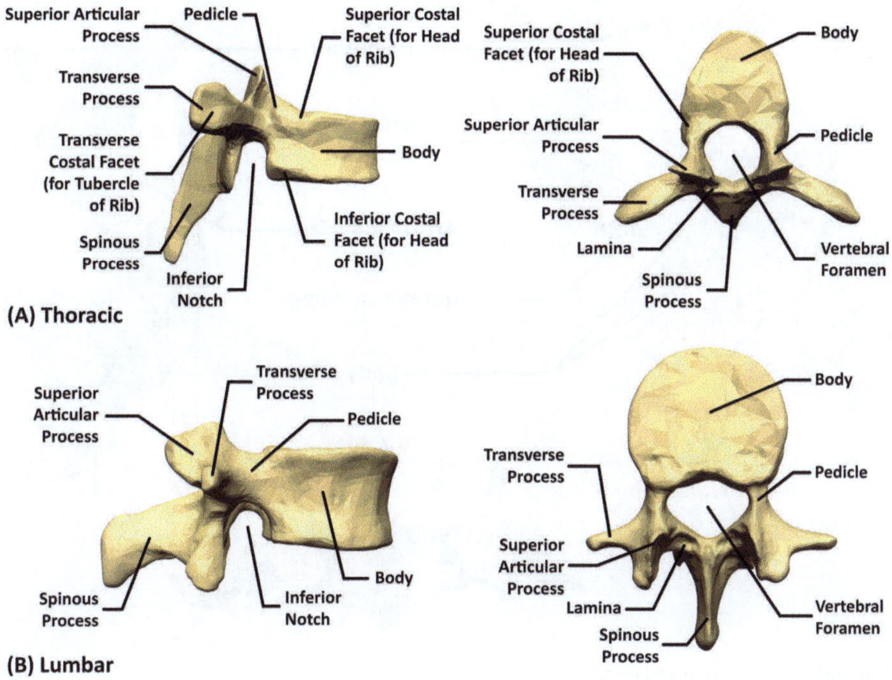

Fig. 8.40 Thoracic and lumbar vertebrae

8.5.1.4 Thoracic Cage

The thoracic cage, also known as the rib cage (Fig. 8.42), provides protective encasement for the thoracic organs, including the heart and lungs. It also serves as a rigid support for the shoulder girdle and upper limbs while offering flexible anchorage for many muscles involved in respiration, positioning of the vertebral column, and movements of the pectoral girdle. The thoracic cage is formed anteriorly by the sternum, posteriorly by the thoracic vertebrae, and laterally and anterolaterally by the ribs and their cartilages. The thoracic cage is conical, wider at the base, and narrower at the top. The costal margin, formed by the costal cartilages of ribs 7–10, is an inverted V-shaped structure that provides a lower boundary for the thoracic cage and is an attachment site for abdominal muscles.

Sternum

The sternum or breastbone (Fig. 8.43) is a flat, elongated bone located in the midline of the anterior thoracic wall. It consists of three parts: manubrium, body, and xiphoid process.

Manubrium: The uppermost part of the sternum is called the manubrium. It is a flattened, triangular-shaped bone that articulates with the clavicles (collarbones) at its superior border and the first rib and part of the second rib at its lateral edges.

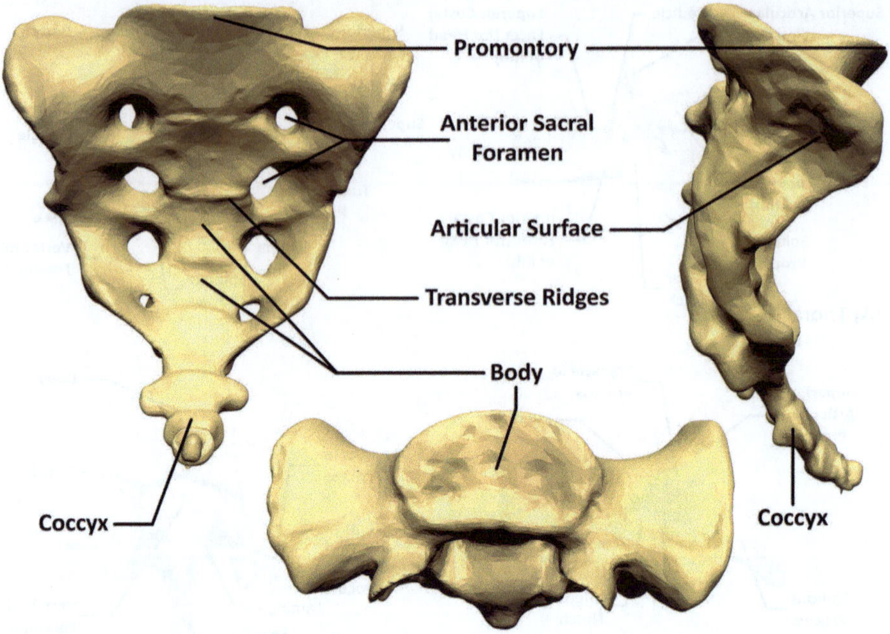

Fig. 8.41 The sacrum and coccyx

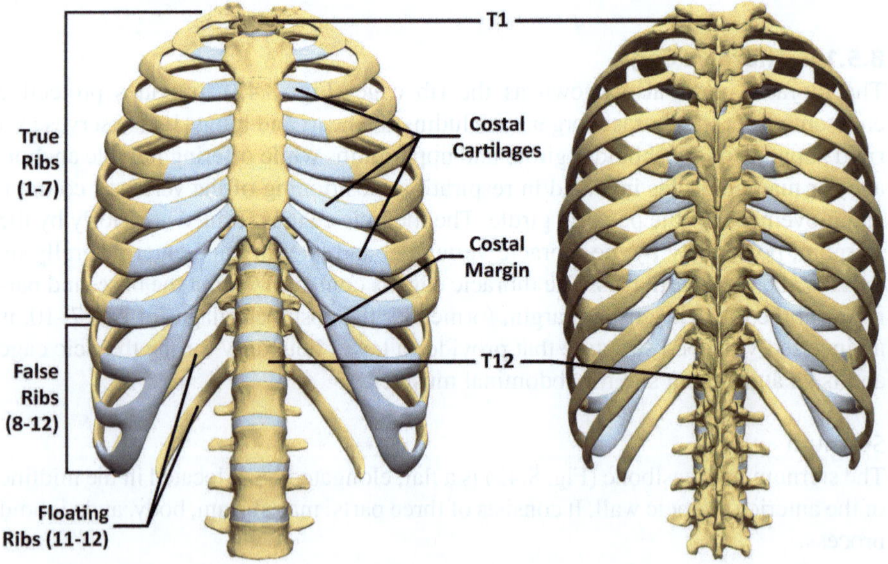

Fig. 8.42 Thoracic cage: anterior view (right) and posterior view (left)

Body (or gladiolus): Below the manubrium lies the largest portion of the sternum, known as the body or gladiolus. It is long and flat, resembling a blade or sword, and forms the central part of the sternum. The body of the sternum articulates with the costal cartilages of the second through seventh ribs.

Xiphoid process: The xiphoid process is the sternum's smallest and most inferior part. It is a thin, pointed extension that projects downward from the body of the sternum. The xiphoid process is made of cartilage during childhood and gradually ossifies into bone as a person matures.

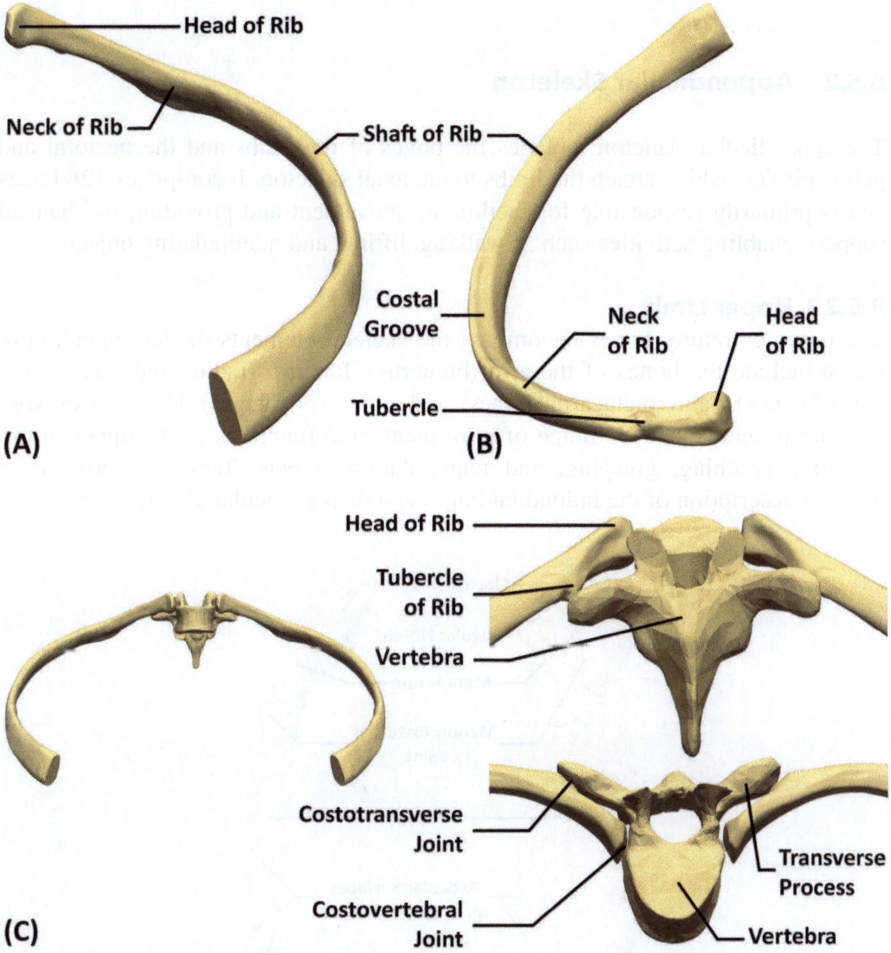

Fig. 8.43 A typical rib structure. (**a**) Left rib 6, superior view. (**b**) Left rib 6, inferior view. (**c**) Costovertebral articulation: anterior view (left), posterior view (upper right), and superior view (lower right)

Ribs

Ribs are 12 pairs of elongated, curved bones that articulate posteriorly with the thoracic vertebrae. There are three types of ribs based on their attachment to the sternum (Fig. 8.44): Ribs 1–7 are termed "true ribs" and connect directly to the sternum through their costal cartilages. Ribs 8–10 are termed "false ribs" and are connected to the sternum indirectly via the costal cartilage of the rib above. Ribs 11 and 12 are termed "floating ribs" and do not connect to the sternum; instead, their costal cartilages terminate within the musculature of the abdominal wall. Intercostal spaces between adjacent ribs are filled with intercostal muscles, which assist in respiratory mechanics.

8.5.2 Appendicular Skeleton

The appendicular skeleton includes the bones of the limbs and the pectoral and pelvic girdles, which attach the limbs to the axial skeleton. It comprises 126 bones and is primarily responsible for facilitating movement and providing mechanical support, enabling activities such as walking, lifting, and manipulating objects.

8.5.2.1 Upper Limb

The upper extremity bones encompass the skeletal elements of the upper limbs, which include the bones of the arm (humerus), forearm (radius and ulna), wrist (carpal bones), palm (metacarpal bones), and fingers (phalanges). These bones work together to enable a wide range of movements and functions in the upper limbs, including reaching, grasping, and manipulating objects. Table 8.4 provides a detailed description of the individual bones and their particular characteristics.

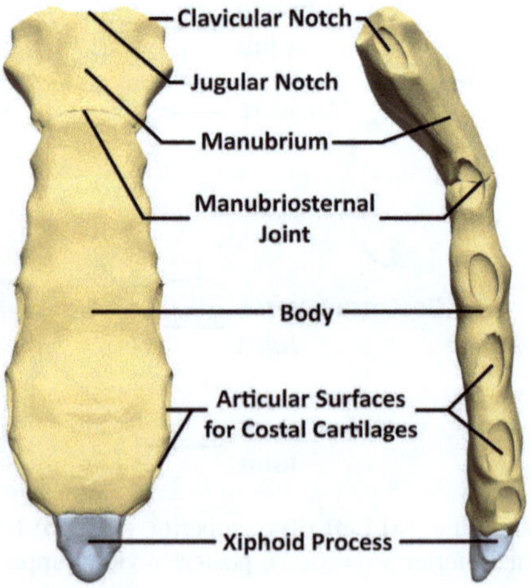

Fig. 8.44 The sternum

Table 8.4 Upper limb bones: articulations, parts, markings, and important features (Figs. 8.45, 8.46, 8.47, 8.48, and 8.49)

Bone name	Articulations	Parts	Markings	Important features
Clavicle	Sternoclavicular joint with sternum; acromioclavicular joint with scapula	Sternal end, acromial end, shaft	Sternal end, acromial end, conoid tubercle	S-shaped; supports the shoulder anteriorly; acts as a brace for the arm
Scapula	Glenohumeral joint with humerus; acromioclavicular joint with clavicle	Spine, acromion, coracoid process, body, glenoid cavity	Glenoid cavity, coracoid process, acromion, suprascapular notch, infraglenoid and supraglenoid tubercles	Triangular flat bone; serves as the point of attachment for many shoulder muscles
Humerus	Glenohumeral joint with scapula; elbow joint with radius and ulna	Head, neck, shaft, distal end	Greater and lesser tubercles, deltoid tuberosity, surgical neck, anatomical neck, capitulum, trochlea, medial and lateral epicondyles	The longest bone of the upper limb; has a spherical head that fits into the glenoid cavity
Radius	Proximal radioulnar joint with ulna; elbow joint with humerus; distal radioulnar joint with ulna; radiocarpal joint with carpals	Head, neck, shaft, distal end	Radial head, radial tuberosity, styloid process, ulnar notch	Lateral bone of the forearm; permits pronation and supination
Ulna	Elbow joint with humerus; proximal and distal radioulnar joints with radius	Olecranon, trochlear notch, coronoid process, shaft, distal end	Olecranon, coronoid process, trochlear notch, radial notch, ulnar styloid process	The medial bone of the forearm; forms the major portion of the elbow joint
Carpals	Radiocarpal joint with radius; Intercarpal joints	Scaphoid, lunate, triquetrum, pisiform, trapezium, trapezoid, capitate, hamate	None prominent; may contain accessory ossicles	Comprise the wrist; arranged in two rows of four

(continued)

Table 8.4 (continued)

Bone name	Articulations	Parts	Markings	Important features
Metacarpals	Carpometacarpal joints with carpals; metacarpophalangeal joints with proximal phalanges	Base, shaft, head	Base, shaft, head	Five bones numbered I–V medially to laterally; provide structure to the palm
Phalanges	Interphalangeal joints between phalanges; metacarpophalangeal joints with metacarpals	Proximal, middle, distal phalanges for fingers; proximal, distal for thumb	Base, shaft, head	Articulate with each other and the metacarpals to form the fingers

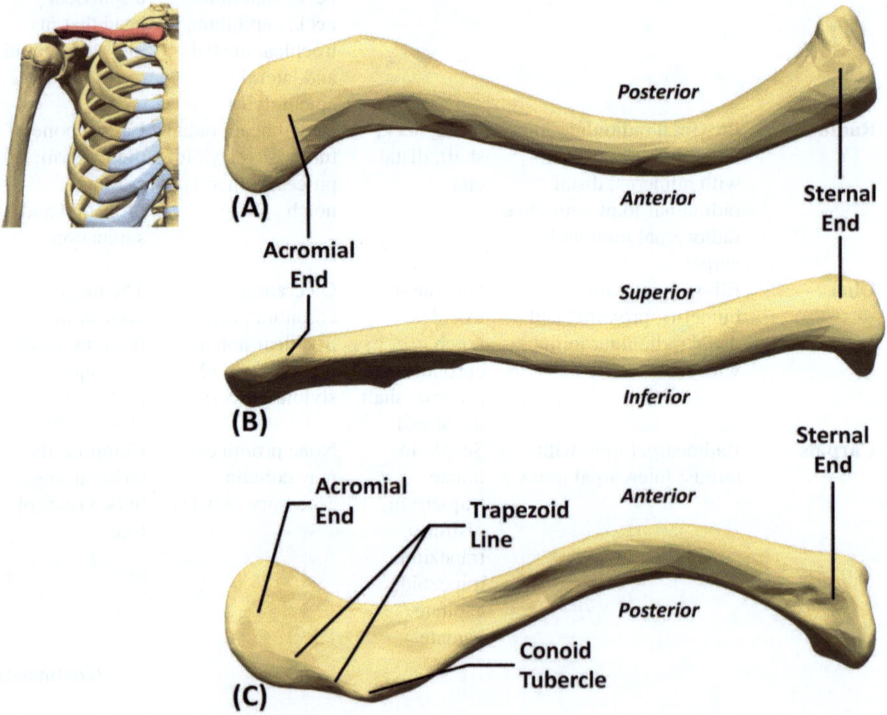

Fig. 8.45 The clavicle. (**a**) Superior view. (**b**) Anterior view. (**c**) Inferior view

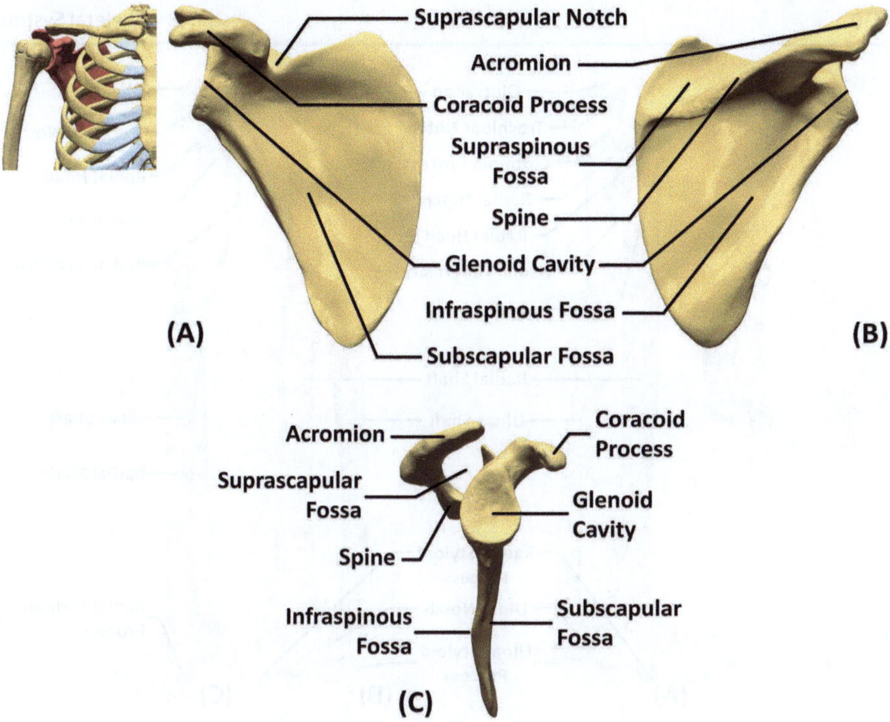

Fig. 8.46 The scapula. (**a**) Anterior view. (**b**) Posterior view. (**c**) Lateral view

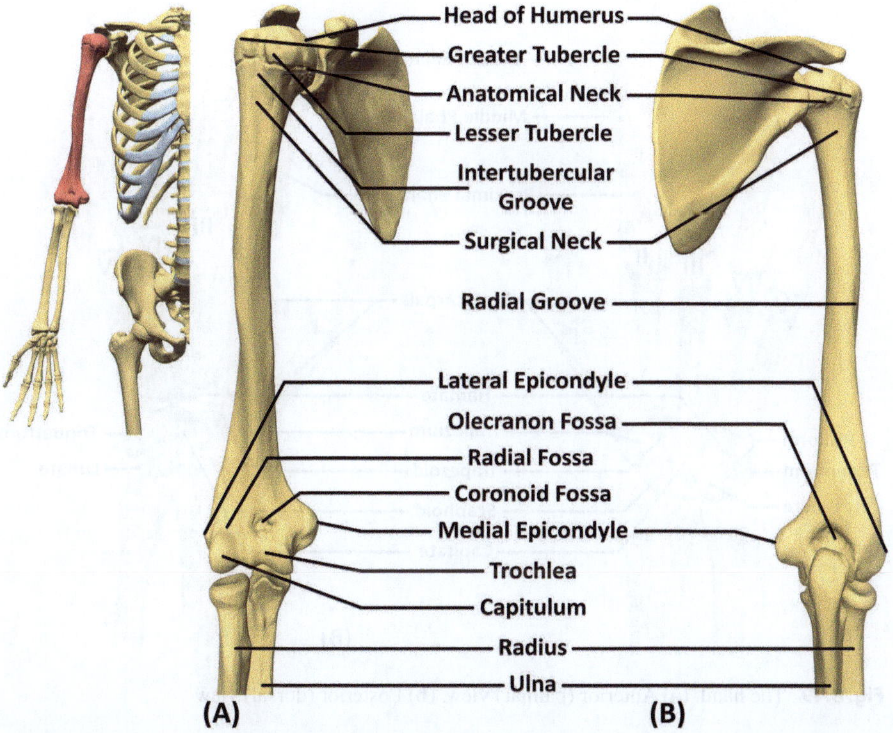

Fig. 8.47 The humerus, the glenohumeral, and the elbow joints. (**a**) Anterior view. (**b**) Posterior view

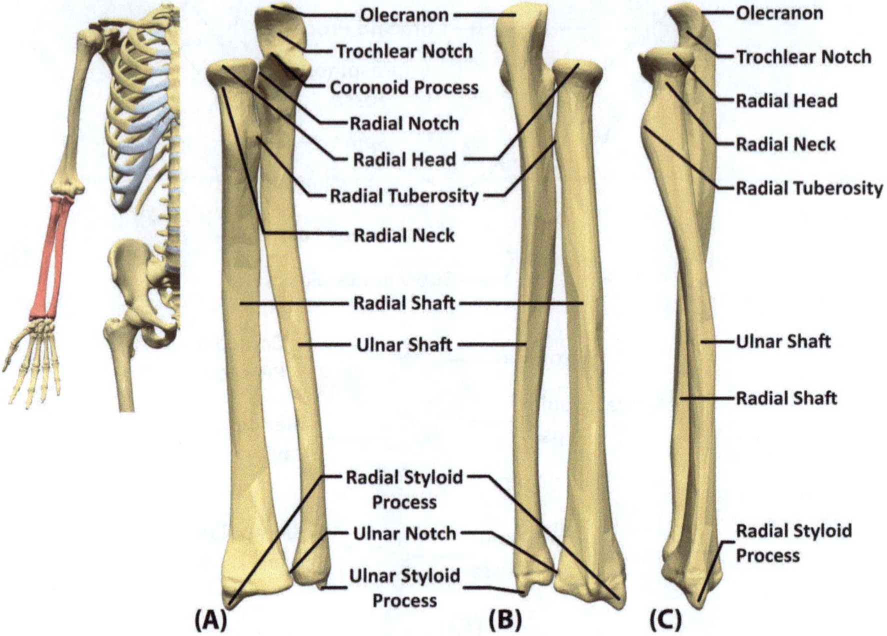

Fig. 8.48 Radius and ulna: (**a**) Anterior view. (**b**) Posterior view. (**c**) Lateral view

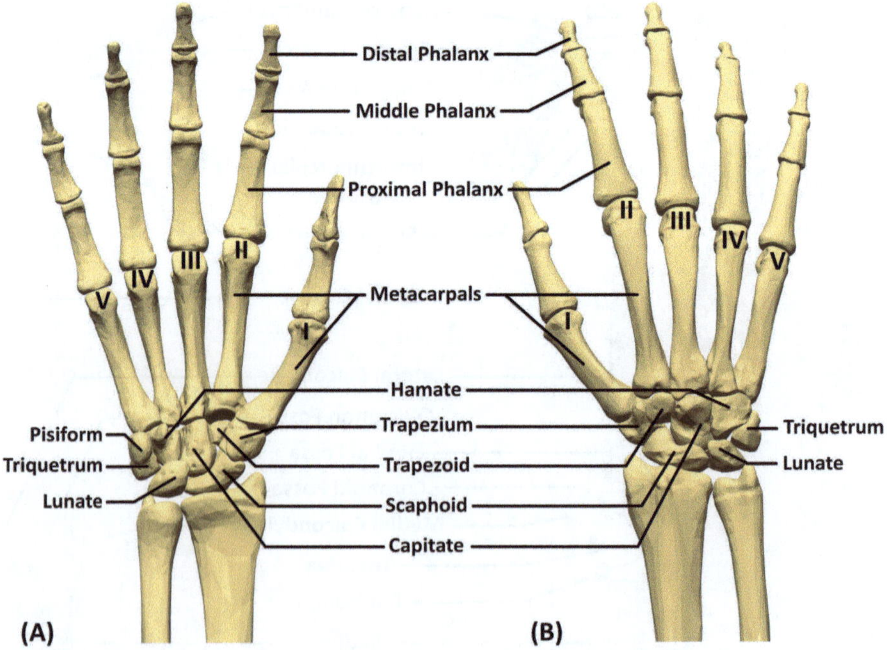

Fig. 8.49 The hand. (**a**) Anterior (palmar) view. (**b**) Posterior (dorsal) view

8.5.2.2 Lower Limb

The lower extremity bones work with various muscles, ligaments, and joints to facilitate bipedal locomotion, support the body's weight, and perform a wide range of activities such as walking, running, jumping, and maintaining balance. Table 8.5 provides a detailed description of the individual bones and their particular characteristics.

8.5.2.3 Arch of the Foot

The arch of the foot is a complex anatomical structure designed for both stability and mobility, crucial for bipedal locomotion, shock absorption, and weight distribution. Comprising bones, ligaments, muscles, and tendons, the foot arch is anatomically categorized into three primary types: the medial longitudinal arch, the lateral longitudinal arch, and the transverse arch (Fig. 8.56).

Medial longitudinal arch: The most prominent and easily recognizable arch extends from the calcaneus (heel bone) to the metatarsal heads of the first through the third toes. It comprises the calcaneus, talus, navicular, three cuneiforms, and the first three metatarsals. The medial longitudinal arch is maintained and supported by various anatomical components, such as the plantar aponeurosis (plantar fascia), the tibialis posterior tendon, and several intrinsic foot muscles. The primary function is shock absorption. It flattens slightly upon weight-bearing and springs back during the push-off phase of gait, acting much like a spring.

Lateral longitudinal arch: This arch is flatter than its medial counterpart and extends from the calcaneus to the metatarsal heads of the fourth and fifth toes. It consists of the calcaneus, cuboid, and fourth and fifth metatarsals. The peroneus longus and peroneus brevis tendons, along with intrinsic foot muscles and interconnecting ligaments, are significant contributors to the maintenance and function of this arch. Though not as flexible as the medial arch, the lateral longitudinal arch distributes body weight to the lateral side of the foot.

Transverse arch: The transverse arch runs perpendicular to the length of the foot, forming a half-dome over the midfoot region. It is formed by the cuboid, cuneiform bones, and the bases of all five metatarsals. The transverse arch is primarily maintained by the ligaments connecting the tarsal bones and some of the foot's intrinsic muscles. This arch aids in weight distribution across the width of the foot and provides added stability, particularly during dynamic activities that require lateral movement.

Table 8.5 Lower limb bones: articulations, parts, markings, and important features (Figs. 8.50, 8.51, 8.52, 8.53, 8.54, and 8.55)

Bone name	Articulations	Parts	Markings	Important features
Hip bone	Sacroiliac joint with sacrum; pubic symphysis	Ilium, ischium, pubis	Iliac crest, anterior superior iliac spine, posterior superior iliac spine, ischial tuberosity, pubic tubercle	Provides support for the vertebral column and connects the trunk to the lower limbs
Femur	Hip joint with pelvis; knee joint with tibia and patella	Head, neck, shaft, distal end	Greater and lesser trochanters, linea aspera, medial and lateral condyles, medial and lateral epicondyles	The longest and strongest bone in the body; articulates with hip and knee
Patella	Patellofemoral joint with femur	Singular sesamoid bone	Anterior and posterior surfaces, base, apex	Protects the knee joint and improves leverage of the quadriceps femoris
Tibia	Knee joint with femur; ankle joint with talus; syndesmosis with fibula	Medial and lateral condyles, shaft, distal end	Tibial tuberosity, medial malleolus, interosseous border	Medial and larger of the two lower leg bones; bears most of body weight
Fibula	Syndesmosis with tibia; ankle joint with talus	Head, neck, shaft, distal end	Head, lateral malleolus, interosseous border	Lateral and smaller of the two lower leg bones; non-weight bearing
Tarsals	Ankle joint with tibia and fibula; intertarsal joints	Talus, calcaneus, navicular, cuboid, and three cuneiforms	Tuberosity of calcaneus, trochlea of talus	Seven bones that form the ankle and heel
Metatarsals	Tarsometatarsal joints with tarsals; metatarsophalangeal joints with proximal phalanges	Base, shaft, head	Base, shaft, head	Five long bones in the foot numbered I–V from medial to lateral
Phalanges	Interphalangeal joints between phalanges; metatarsophalangeal joints with metatarsals	Proximal, middle, distal phalanges for toes; proximal, distal for big toe	Base, shaft, head	Articulate with each other and the metatarsals to form the toes

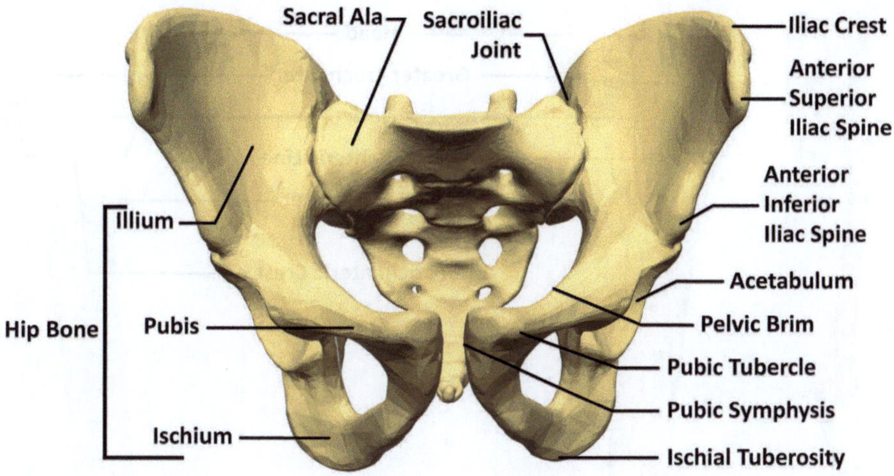

Fig. 8.50 Pelvis. The pelvis consists of the two hip bones and the sacrum

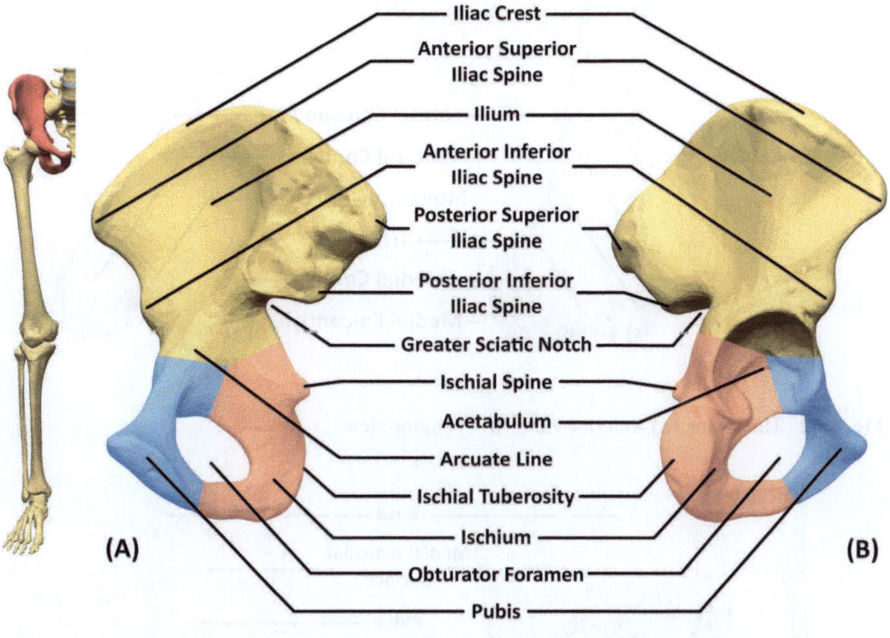

Fig. 8.51 The right hip bone. (**a**) Lateral view. (**b**) Medial view of the right hip bone. The acetabulum is the point of fusion of the ilium (gold), ischium (red), and pubic (blue) bones

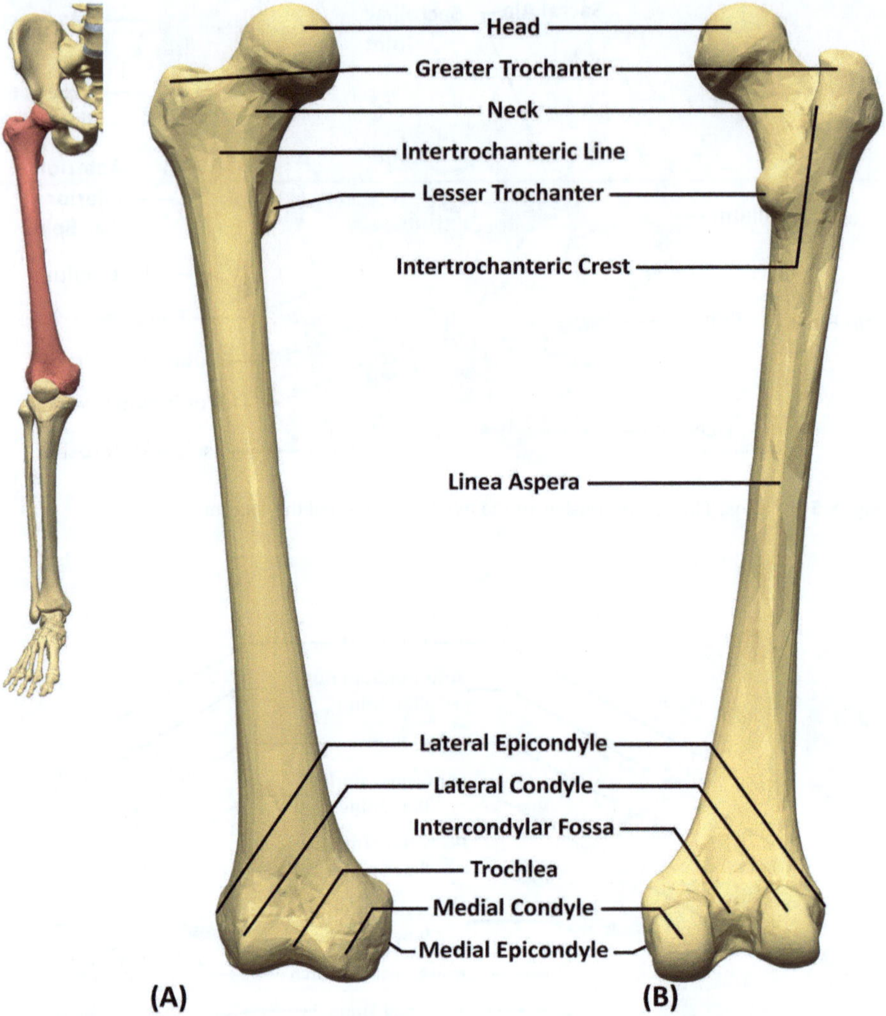

Fig. 8.52 The femur. (**a**) Anterior view. (**b**) Posterior view

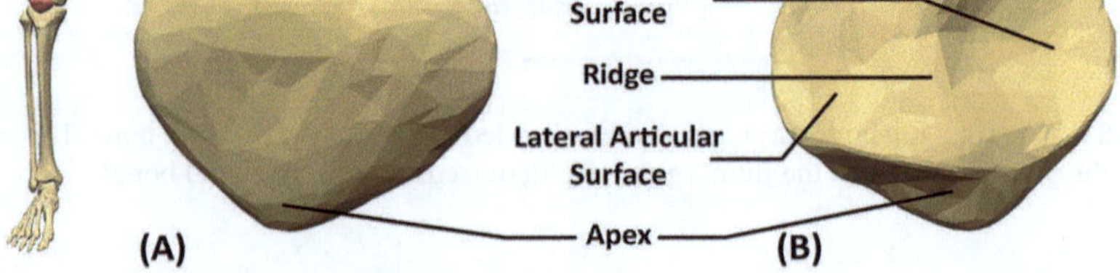

Fig. 8.53 The patella. (**a**) Anterior view. (**b**) Posterior view

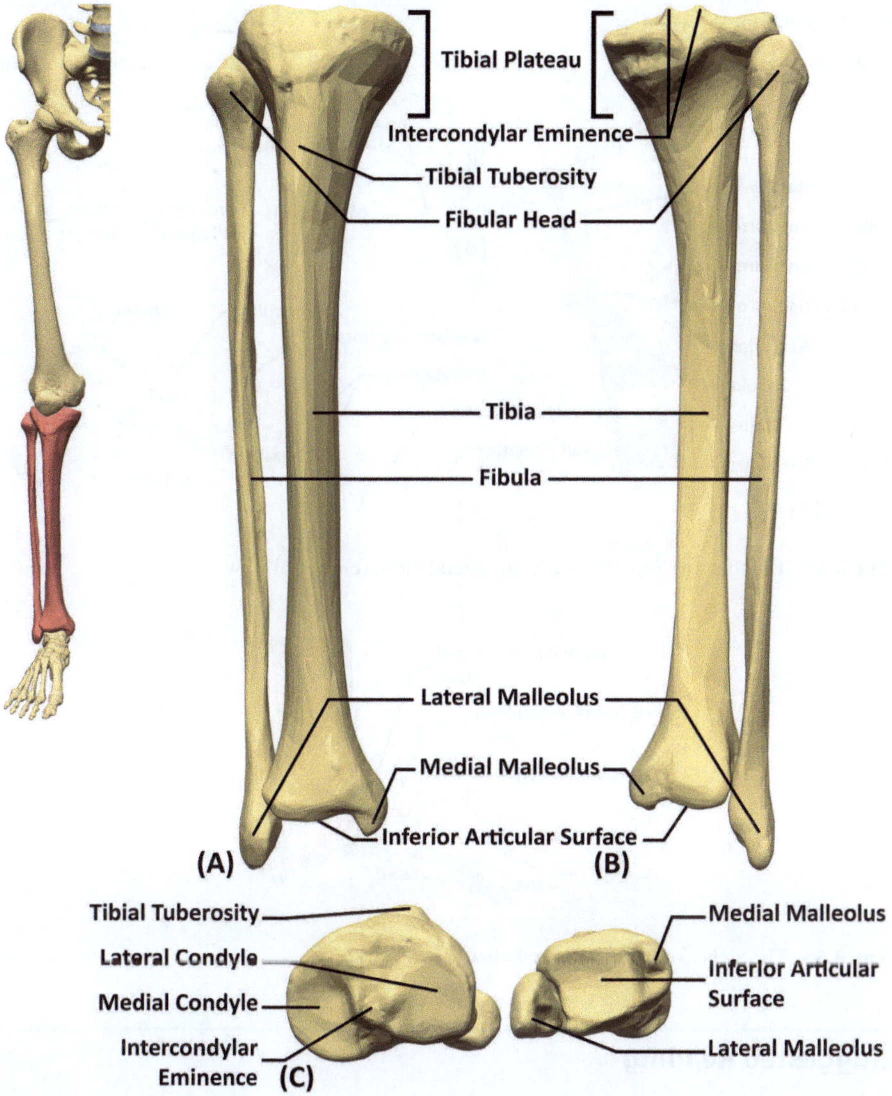

Fig. 8.54 The tibia and fibula. (**a**) Anterior view. (**b**) Posterior view. (**c**) Superior articular surface, or tibial plateau (left), inferior articular surface, or tibial plafond (right)

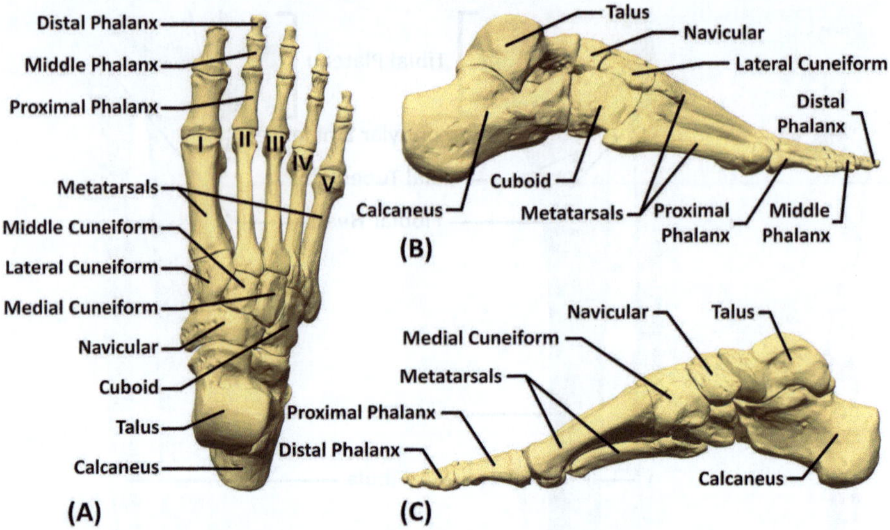

Fig. 8.55 The foot. (**a**) Superior view. (**b**) Lateral view. (**c**) Medial view

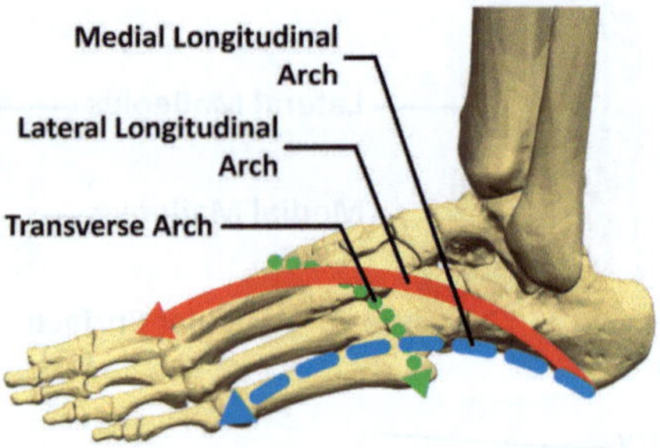

Fig. 8.56 The arches of the foot (lateral view)

Suggested Reading

Advanced Physiology and Pathophysiology: Essentials for Clinical Practice, 2nd Ed. Nancy Tkacs, Linda Herrmann, Springer Publishing Company, 2024

Anatomy & Physiology, S. Meloni, et al. Independent Publisher, 2023

Anatomy & Physiology: The Unity of Form and Function, 10th Ed. Kenneth S. Saladin, McGraw Hill, 2023

Gray's Anatomy for Students, 5th Ed. Richard L. Drake, et al. Elsevier, 2020

Human Anatomy & Physiology, 11th Ed. Elaine N. Marieb and Katja Hoehn, Pearson, 2022

Moore's Clinically Oriented Anatomy, 9th Ed. Keith L. Moore, et al. Lippincott Williams & Wilkins, 2020

Netter Atlas of Human Anatomy: Classic Regional Approach, 8th Ed. Frank H. Netter, Elsevier, 2022

Principles of Anatomy and Physiology, 16th Ed. Gerard J. Tortora, Bryan H. Derrickson, Wiley, 2020

Muscular System

9

In this chapter, you will learn

- Overview of the muscular system and its functions
- Types of muscle tissue: skeletal, smooth, and cardiac
- Skeletal muscle functions
- Skeletal muscle structure at macroscopic and microscopic levels
- Detailed structure and function of the neuromuscular junction
- Definition and role of motor units in muscle function
- Types of muscle attachments
- Basic structure of muscle fiber, myofibrils, and sarcomeres
- Types of myofilaments: actin and myosin
- Sarcoplasmic reticulum and its function in calcium storage and release
- Transverse tubules (T-tubules) and their role in muscle contraction
- Mechanism of muscle contraction
- Different types of ion channels in muscle fibers and their functions
- Role of voltage-gated calcium channels in muscle excitation and contraction
- Generation and maintenance of resting membrane potential in muscle cells
- Mechanism of action potential propagation in muscle fibers
- Overview of excitation–contraction coupling in skeletal muscle
- Description of a muscle twitch and its phases
- Concept of graded muscle contractions
- Concept of muscle tone and its physiological basis
- Types of muscle contractions: isotonic and isometric
- Energy sources for muscle contraction: ATP, creatine phosphate, and metabolic pathways
- Anaerobic and aerobic pathways in muscle metabolism
- Types of muscle fibers and their characteristics
- Detailed structure and function of smooth muscle tissue

H. Hosseinzadeh, *Fundamentals of Medicine for Biomedical Engineering*,
https://doi.org/10.1007/978-3-031-74085-5_9

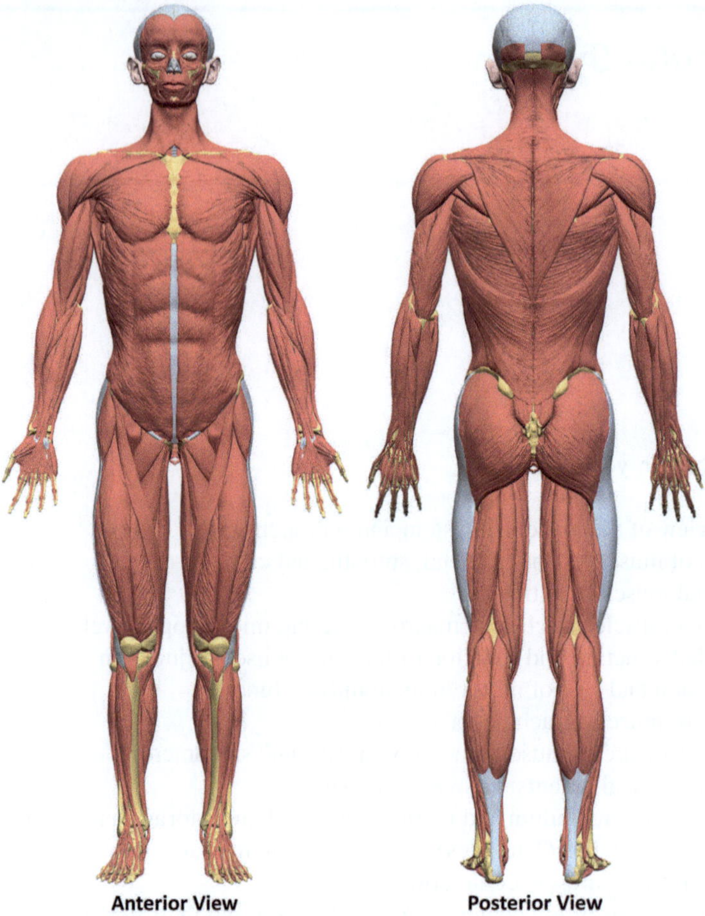

Fig. 9.1 Superficial muscles of the body

The muscular system is essential as a conductor and performer, orchestrating various biological functions for survival and well-being. Comprising an intricate network of muscles, tendons, and associated tissues, it facilitates body movement and governs critical internal processes ranging from cardiovascular circulation to digestion (Fig. 9.1).

While skeletal muscles enable locomotion and voluntary actions, smooth and cardiac muscles unconsciously labor to sustain physiological equilibrium.

The muscle tissue is divided into three primary types: skeletal, smooth, and cardiac. Each type exhibits distinct structural configurations, functions, and regulatory mechanisms, thus serving unique roles within the broader physiological framework.

9.1 Part 1: Skeletal Muscles

The skeletal muscle tissue, characterized by its striated appearance, multinucleated cellular structure, and a high degree of contractility, serves as the primary effector tissue for the somatic nervous system, enabling a wide range of voluntary physical activities. The skeletal muscle tissue is distinguished by its responsiveness to conscious neural input, unlike its smooth and cardiac counterparts, which are regulated mainly by involuntary autonomic controls.

9.1.1 Function

Skeletal muscle tissue has manifold functions, extending beyond facilitating bodily movement to encompass a diverse array of physiological processes:

Locomotion and volitional movement: Skeletal muscle tissue is integral to executing a wide range of voluntary movements, from gross motor skills like walking and running to fine motor activities like writing. Skeletal muscles produce the mechanical forces necessary to move bones and joints, enabling locomotion.

Postural maintenance: Skeletal muscles contribute significantly to maintaining upright posture and positional stability. Specific muscles, often termed "postural muscles," work in semi-continuous contraction to counteract gravitational forces and maintain structural alignment.

Thermoregulation: The metabolic activities intrinsic to muscle contraction generate heat, contributing to overall body thermoregulation.

Energy metabolism: Skeletal muscles act as significant metabolic organs, regulating glucose homeostasis and lipid metabolism. They store glycogen and triglycerides, serving as energy reservoirs that can be rapidly mobilized during increased physiological demand.

Protection and support: Skeletal muscles provide mechanical protection to internal organs. For instance, the abdominal muscles shield the digestive organs, while the muscles surrounding the ribcage protect the heart and lungs.

Communication: Specialized groups of skeletal muscles facilitate facial expressions, gestures, and vocalization. These muscles enable communication, which is critical for social interaction and the expression of emotions.

9.1.2 Structure

The anatomy and physiology of skeletal muscle tissue are inextricably linked to its structure at macroscopic and microscopic levels. From its hierarchical arrangement of fascicles and muscle fibers to its microscopic constituents such as sarcomeres, myofibrils, and associated proteins, skeletal muscle tissue presents a complex yet elegantly orchestrated structure.

9.1.2.1 Neurovascular Structure

Skeletal muscles'functionality is intricately tied to their neurovascular supply, which facilitates the transduction of neural impulses for contraction and the delivery and removal of essential biochemical substrates and waste products.

Neuromuscular Junctions

The neuromuscular junction (NMJ) (Fig. 9.2) interfaces the nervous system and skeletal muscle fibers. This specialized synapse is a site of tremendous physiological importance, acting as the locus where neural impulses are transduced into mechanical action through muscle contraction. The NMJ is formed by:

- Presynaptic terminal: The presynaptic terminal, or synaptic bouton, is the bulbous end of the motor neuron axon that contains synaptic vesicles filled with neurotransmitters, predominantly acetylcholine (ACh). These vesicles are localized near the presynaptic membrane and ready for exocytosis upon the arrival of an action potential.
- Synaptic cleft: The synaptic cleft is the narrow extracellular gap (~20–50 nm) that separates the presynaptic terminal from the postsynaptic muscle membrane. The cleft contains extracellular matrix proteins and acetylcholinesterase, an enzyme that breaks down acetylcholine.
- Postsynaptic membrane: The postsynaptic membrane is a specialized sarcolemma (muscle cell membrane) region directly opposite the presynaptic terminal. It is characterized by deep infoldings called junctional folds, which increase the surface area for neurotransmitter binding. The membrane is densely packed with nicotinic acetylcholine receptors (nAChRs), ion channels that are opened by the binding of acetylcholine.

Physiology of the Neuromuscular Junction

The arrival of a neural action potential (Section 10.1.2.4) at the presynaptic terminal triggers the opening of voltage-gated calcium channels. The influx of calcium ions into the presynaptic terminal catalyzes the fusion of synaptic vesicles with the presynaptic membrane, facilitating the release of acetylcholine into the synaptic cleft via exocytosis. Acetylcholine diffuses across the synaptic cleft and binds to the nicotinic acetylcholine receptors on the postsynaptic membrane. Each receptor requires the binding of two acetylcholine molecules to induce a conformational change, leading to the opening of the ion channel. The activated nAChRs are ligand-gated cation channels, allowing the passage of sodium ions (Na^+) into the muscle cell and potassium ions (K^+) out of it. However, the influx of sodium ions is substantially larger than the efflux of potassium ions, resulting in a localized depolarization known as the endplate potential (EPP). Acetylcholine is rapidly hydrolyzed by acetylcholinesterase into choline and acetate, terminating the signal. The choline is reabsorbed into the presynaptic terminal for the resynthesis of acetylcholine, ensuring the system's readiness for subsequent neural inputs.

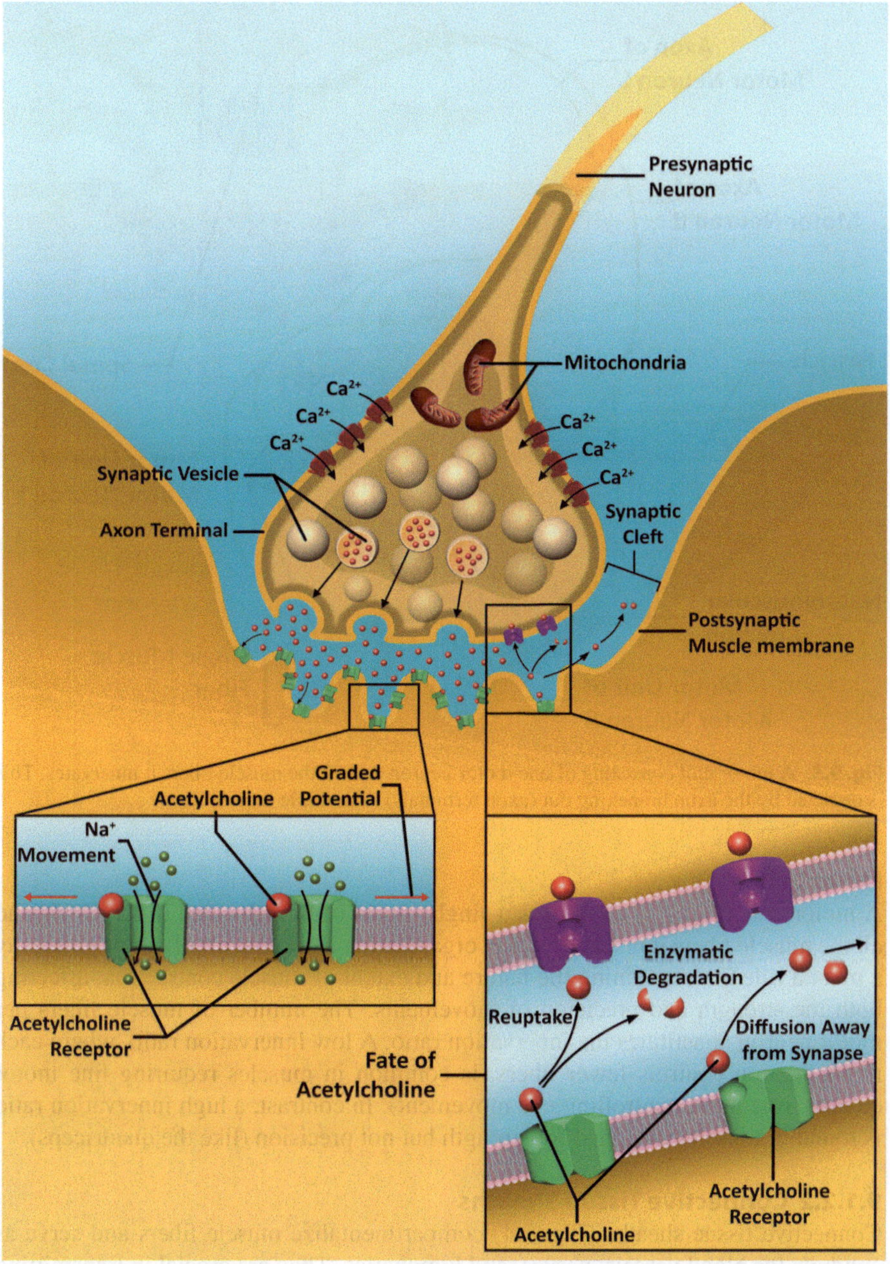

Fig. 9.2 Neuromuscular junction, showing the process of synaptic transmission at a neuromuscular junction, highlighting the interactions between the presynaptic neuron, synaptic cleft, and postsynaptic muscle membrane. The right inset illustrates acetylcholine binding to its receptor, causing Na^+ to cross the postsynaptic muscle membrane, depolarization, and generating a graded potential. The left inset shows the signal termination mechanisms: the breakdown of acetylcholine by acetylcholinesterase, the reuptake of choline into the presynaptic neuron, and acetylcholine diffusion away from the synapse

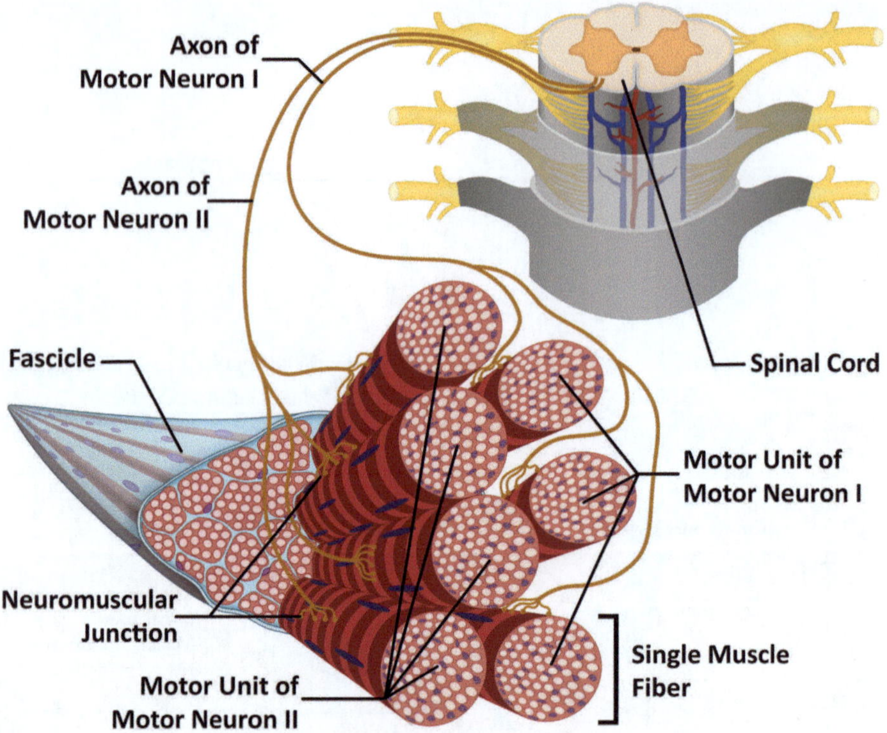

Fig. 9.3 A motor unit consisting of one motor neuron and all the muscle fibers it innervates. This is depicted by the axon branching out (axon terminals) to multiple muscle fibers

Motor Unit

A motor unit (Fig. 9.3) comprises a single motor neuron (located in the CNS) and all the muscle fibers it innervates. The organization and behavior of motor units play a pivotal role in determining the nature and extent of muscle contraction, affecting both the strength and precision of movements. The number of muscle fibers per motor neuron constitutes the innervation ratio. A low innervation ratio, where each motor neuron controls fewer fibers, is common in muscles requiring fine motor control (like those controlling eye movement). In contrast, a high innervation ratio is found in muscles designed for strength but not precision (like the quadriceps).

9.1.2.2 Connective Tissue Sheaths

Connective tissue sheaths (Fig. 9.4) compartmentalize muscle fibers and serve as conduits for blood vessels, nerves, and lymphatics. They are crucial in transmitting mechanical forces generated during muscle contraction to the skeleton.

Epimysium: The epimysium is the outermost layer of connective tissue surrounding a muscle. Composed primarily of dense irregular connective tissue, it envelops the entire muscle organ and aids in maintaining its structural integrity. The

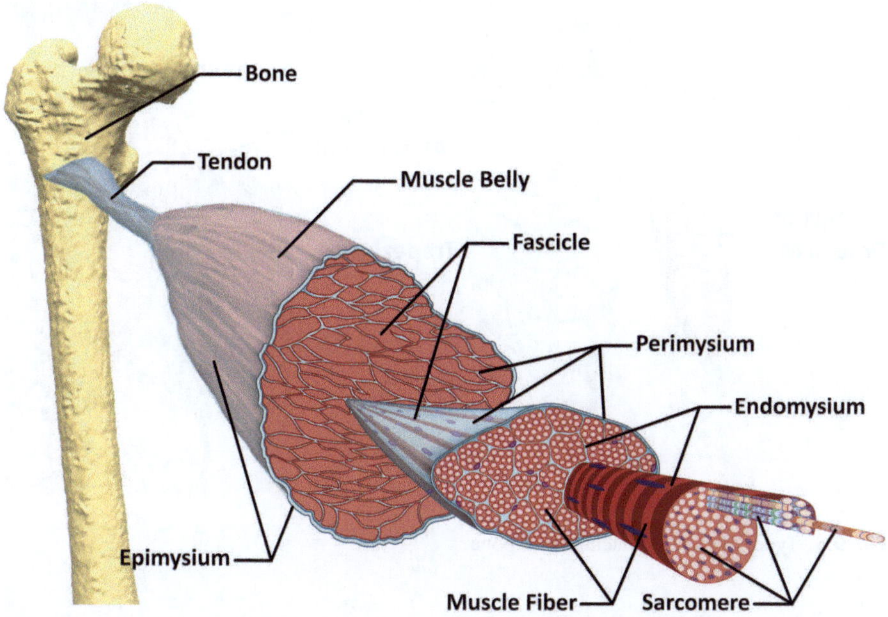

Fig. 9.4 Hierarchical structure of connective tissue sheaths of skeletal muscle, from the attachment to the bone down to the muscle fibers

epimysium is continuous with the tendons or aponeuroses that connect the muscle to bones or other muscles, facilitating force transmission.

Perimysium: Beneath the epimysium lies the perimysium, a layer of dense connective tissue that segregates the muscle fibers into fascicles. By doing so, the perimysium plays a vital role in organizing the muscle fibers, allowing for more effective force generation and transmission.

Endomysium: The innermost layer of connective tissue is the endomysium, primarily composed of reticular fibers and surrounds each muscle fiber. The endomysium houses capillaries, nerve fibers, and lymphatic vessels, serving as a conduit for essential nutrients and signaling molecules.

9.1.2.3 Muscle Attachments

Muscle attachments represent the anatomical and functional interfaces between skeletal muscles and bones, facilitating the transmission of force that enables movement and stabilization of joints.

Tendons: Tendons (Fig. 9.5) are the most common form of muscle attachment and are composed predominantly of dense regular connective tissue. They are strong, fibrous cords that attach muscle to bone, consisting mainly of collagen fibers

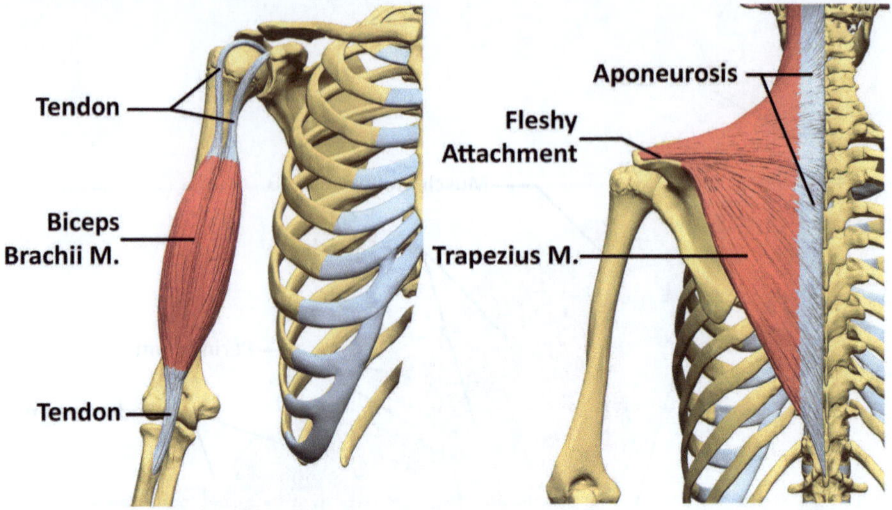

Fig. 9.5 Types of muscle attachment to bone

arranged in parallel bundles. Tendons are continuous with the epimysium, the outermost connective tissue sheath of the muscle, and are embedded into the periosteum of the bone.

Aponeuroses: Aponeuroses (Fig. 9.5) are flat, sheet-like tendons that attach muscles to bones or other muscles. Composed of dense regular connective tissue like tendons, aponeuroses offer a broader area of attachment, which can be beneficial for muscles with a flat shape or for joining muscles in a functional complex.

Fleshy attachments (Fig. 9.5): While tendons and aponeuroses represent the majority of muscle attachments, some muscles attach directly to bone through fleshy attachments. In these cases, short collagen fibers extend directly from the muscle's connective tissue sheaths into the bone's periosteum.

9.1.2.4 Musculotendinous Junction

The musculotendinous junction (MTJ) (Fig. 9.6) serves as the interface between muscle fibers and their associated tendons, representing a zone of transition where contractile elements of the muscle are anchored to noncontractile, fibrous tendons. The MTJ is characterized by a specialized region of transitional tissue that gradually changes from the muscle's connective tissue sheaths to tendon tissue. This transition is not abrupt but occurs through a gradient that helps distribute force and minimize the risk of injury at the interface. The muscle fibers and tendon tissue at the MTJ often exhibit an interdigitating pattern, resembling fingers interlocking. This increases the surface area for attachment, distributes the mechanical load, and provides additional strength to the junction.

The MTJ is the weakest part of the connective tissue structures connecting the muscle to the bone. Hence, it is a common site for sports-related injuries.

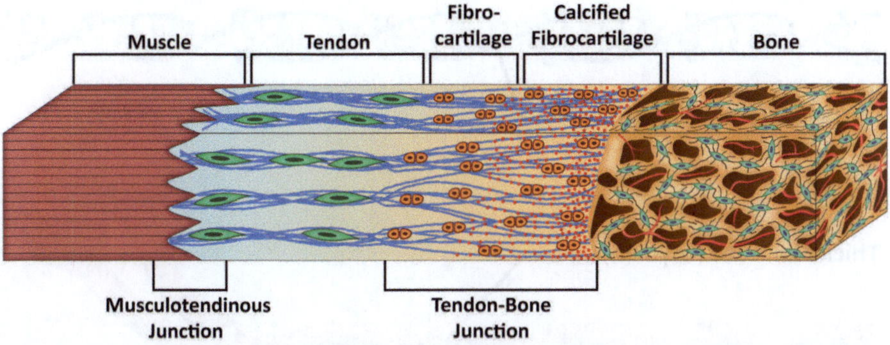

Fig. 9.6 Tendon junctions, musculotendinous and tendon–bone, showing the gradual changes in tissue composition and structure essential for effectively transmitting forces generated by muscle contractions to the skeletal system

9.1.2.5 Tendon–Bone Junction

The tendinoosseous junction (Fig. 9.6), called the tendon–bone junction or enthesis, is a structural and functional transition zone between the compliant tendon and the rigid bone. This unique junction is essential for effectively transferring mechanical forces from muscle to skeletal elements, facilitating movement and providing stability. The tendon–bone junction is not a simple point of attachment but a gradual transition where soft tissue (tendon) interfaces with hard tissue (bone). This gradient comprises four zones: tendon tissue, fibrocartilage, mineralized fibrocartilage, and bone. The gradual transition is vital for minimizing stress concentrations at the interface. A layer of fibrocartilage often exists at the tendon–bone interface. Fibrocartilage is rich in type I and II collagen and acts as a buffer to absorb compressive forces, thus facilitating a smoother force transition between the tendon and bone. Sharpey's fibers are collagen fibers that penetrate the bone matrix and provide mechanical anchorage of the tendon to the bone. Sharpey's fibers are more mineralized than other tendon fibers and have a greater type I collagen content, making them particularly strong and capable of withstanding high tensile loads. The ECM at the tendon–bone junction is rich in proteoglycans and glycoproteins, contributing to its unique mechanical properties.

9.1.2.6 Muscle Fibers

Individual skeletal muscle cells are called muscle fibers. These fibers are surrounded by a connective tissue layer known as the endomysium. Each muscle fiber (Fig. 9.4) is elongated, cylindrical, and multinucleated, with the nuclei situated peripherally. Within each muscle fiber are myofibrils, which are long, cylindrical structures that run parallel to the length of the fiber. These myofibrils are the contractile elements of the muscle cell and are composed of repeating units of sarcomeres.

The myofibrils are cylindrical structures, with diameters ranging between 1 and 2 micrometers, and can extend several centimeters in length, essentially spanning the entire length of the muscle cell. The myofibrils occupy a substantial portion of the cytoplasmic space within the muscle fiber and run parallel to the fiber's long

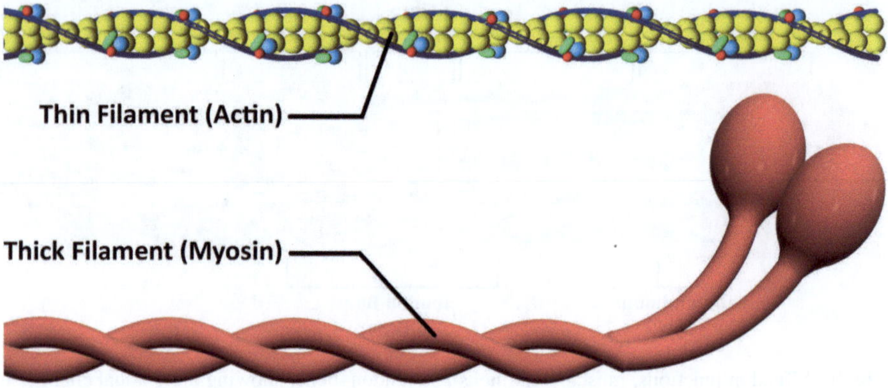

Fig. 9.7 Contractile filaments: thin (actin) and thick (myosin) filaments

axis. Myofibrils are composed of repeating units called sarcomeres, the smallest functional units of the muscle cell. A sarcomere is delineated by two adjacent Z-discs (or Z-lines) and contains thick and thin myofilaments (Fig. 9.7).

Striation

Striations (Fig. 9.8) in skeletal muscle tissue represent a highly organized, repetitive pattern of bands visible under the microscope, resulting from the precise arrangement of contractile and structural proteins within the myofibrils.

- A-band (anisotropic band): The darker of the two bands in the striation pattern, the A-band is characterized by its anisotropic properties under polarized light. It corresponds to the length of the thick myosin filaments within the sarcomere and includes regions where thick and thin filaments overlap. It is in this region that cross-bridges form during muscle contraction.
- I-band (isotropic band): Appearing lighter than the A-band, the I-band is isotropic under polarized light. The I-band comprises thin actin filaments and spans two adjacent sarcomeres bisected by a Z-disc. During muscle contraction, the I-band shortens as the actin filaments slide over the myosin filaments.
- Z-disc (Z-line): The Z-discs form the boundaries between adjacent sarcomeres and appear as dark lines in the center of the I-bands. These Z-discs anchor the thin filaments and transmit the force generated by contracting sarcomeres along the length of the myofibril.
- H-zone: This is a lighter region within the middle of the A-band where only thick myosin filaments are present. The H-zone disappears during maximal muscle contraction when the actin filaments overlap entirely with the myosin filaments.
- M-line: Located in the middle of the H-zone, the M-line is a complex of proteins that holds the thick filaments in place and aids in sarcomere organization.

The visible striations are not static but dynamically change during muscle contraction and relaxation.

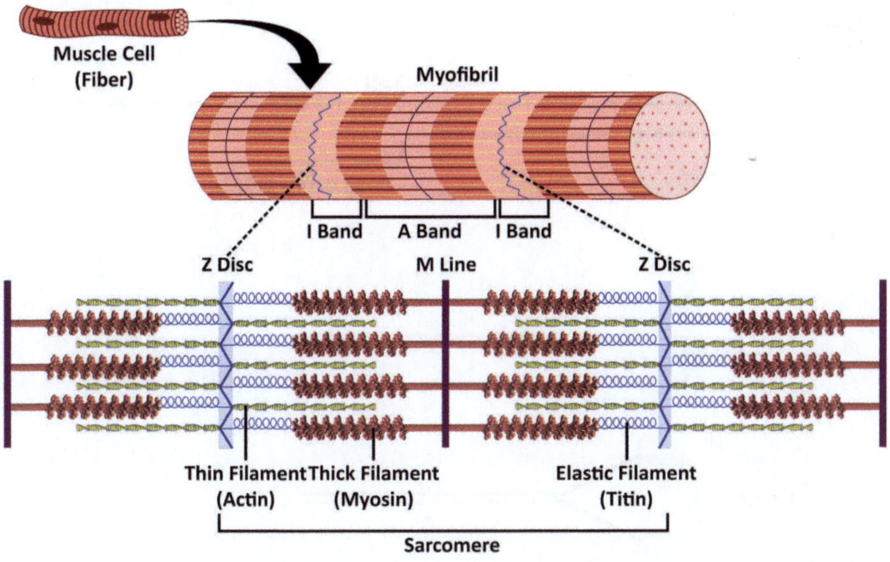

Fig. 9.8 Microscopic structure of the fundamental functional unit of a skeletal muscle fiber, a sarcomere

Sarcomere

The sarcomere (Fig. 9.8) represents the fundamental contractile unit of striated muscle tissue, such as skeletal and cardiac muscle. It is the segment of a myofibril bounded by two Z-discs (also known as Z-lines) and comprises an intricate arrangement of protein filaments.

Myofilaments

Myofilaments in skeletal muscles are the primary constituents of the sarcomeres and are responsible for the process of muscle contraction. They are of two principal types: thick and thin filaments. Thick filaments comprise the protein myosin, and the protein actin primarily forms the thin filaments.

Thick filaments: The myosin myofilaments, commonly referred to as "thick filaments" (Fig. 9.9), are central to the process of muscle contraction, serving as molecular motors that generate force by interacting with actin filaments. Each thick filament consists of approximately 200–500 myosin molecules. Each myosin molecule contains two heavy chains, forming a coiled-coil structure constituting the molecule's head, neck, and tail. Each myosin molecule is also associated with two pairs of light chains. Essential light chain (ELC) and regulatory light chain (RLC) are wrapped around the neck region between the myosin head and tail to stabilize the myosin head.

The myosin molecules are organized bipolarly, with the tails centered in the filament and the heads projecting outward. The projecting myosin heads are organized to interact with actin filaments in cross-bridges, enabling the sliding

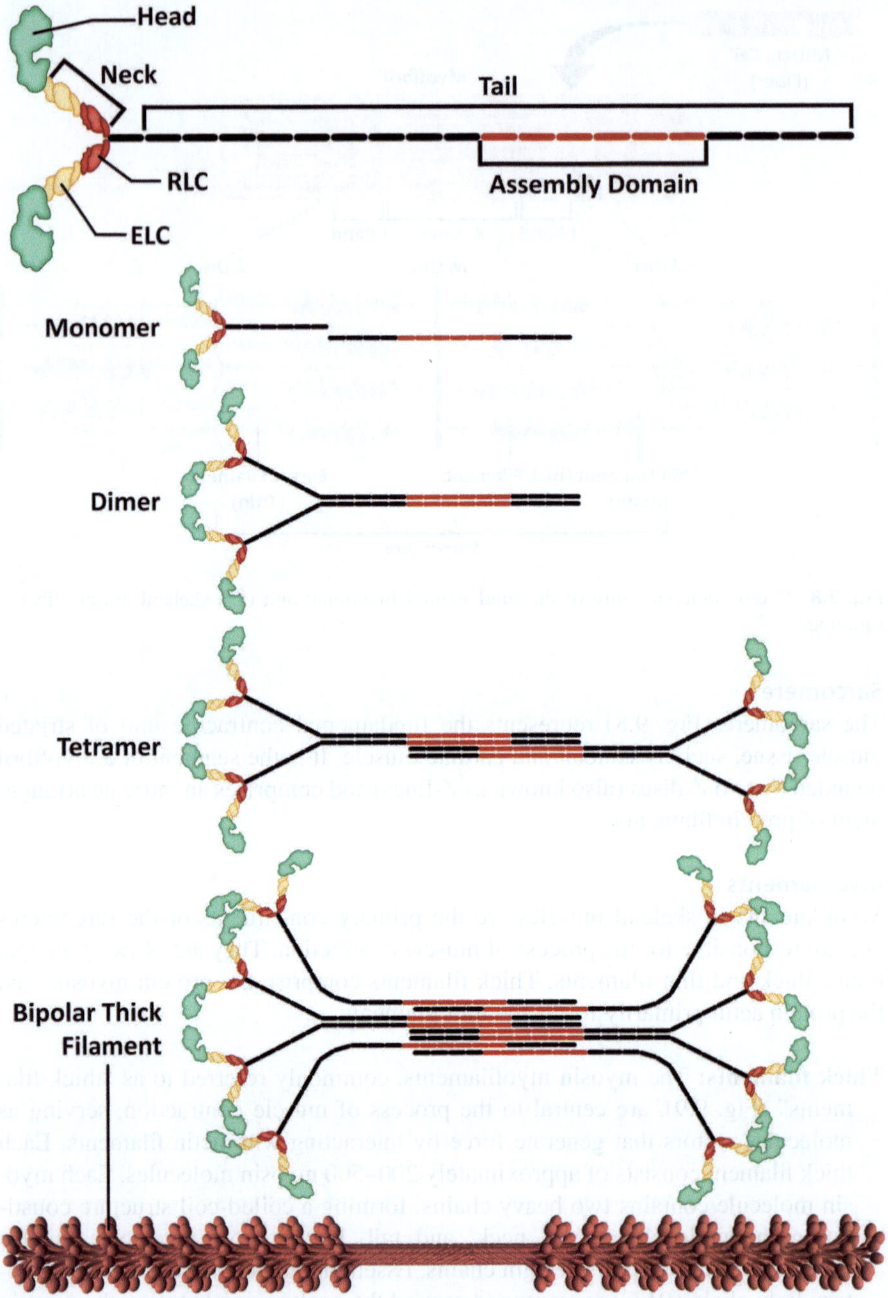

Fig. 9.9 The composition and molecular structure of the thick filament

mechanism central to muscle contraction. The ATPase site on the myosin head hydrolyzes ATP to ADP and inorganic phosphate (Pi), providing the energy for the mechanical work of contraction. The myosin heads have specific binding sites for actin, forming cross-bridges that allow for the sliding of the thin filaments during muscle contraction.

Thin filaments: Thin filaments (Fig. 9.10) are intricate structures primarily comprised of the protein actin. These filaments also contain the regulatory proteins tropomyosin and troponin. The polymerized form of actin is a double-stranded helix with a helical pitch of approximately 36 nm. Each strand is a linear chain of actin monomers, and the double helix comprises about 13 monomers per turn. Tropomyosin is a two-chained coiled protein that lies in the groove of the actin helix. In relaxed muscles, it blocks the myosin-binding sites on actin. Tropomyosin molecules are arranged end-to-end along the actin filament, with each tropomyosin molecule associated with a troponin complex. Troponin is a complex of three subunits—troponin T (TnT), troponin I (TnI), and troponin C (TnC); TnT binds to tropomyosin, anchoring the troponin complex, TnI inhibits actin–myosin

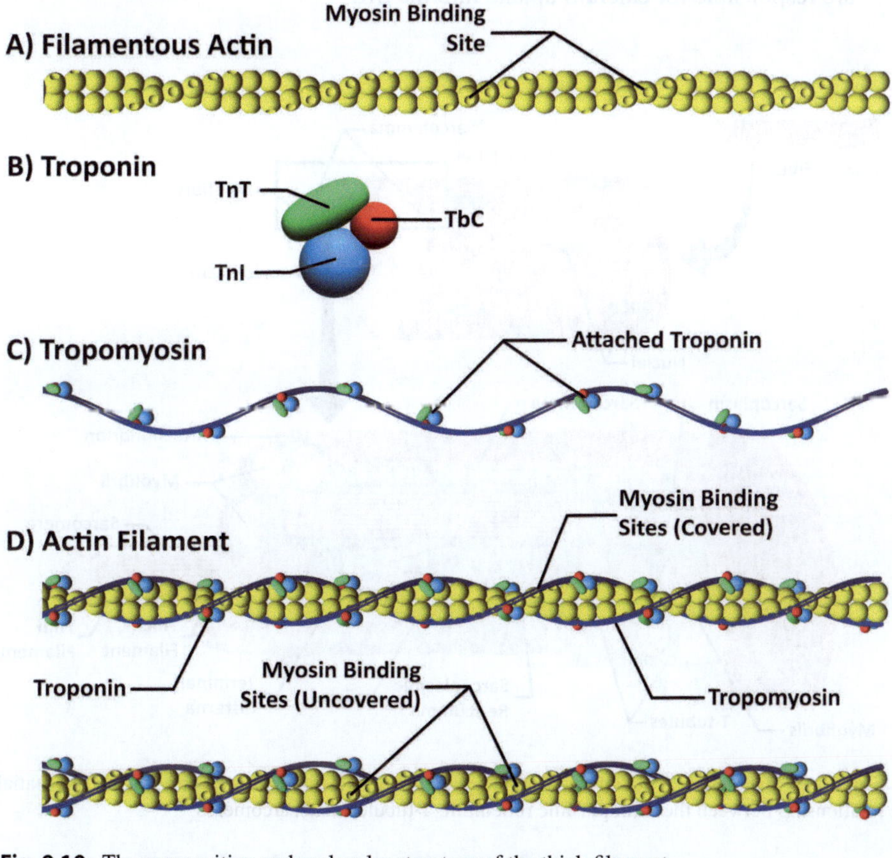

Fig. 9.10 The composition and molecular structure of the thick filament

interaction by stabilizing the tropomyosin blockade, and TnC is the calcium-binding subunit that undergoes a conformational change upon calcium binding, which moves tropomyosin, exposing the myosin-binding sites on actin.

Sarcoplasmic Reticulum

The sarcoplasmic reticulum (SR) (Fig. 9.11) is an elaborate network of membrane-bound tubules and cisternae that envelop the myofibrils.

- Terminal cisternae: These are large, flattened sacs at the ends of the SR, which lie adjacent to the T-tubules, collectively forming the triad junction in skeletal muscle.
- Longitudinal tubules: These are elongated, interconnected tubules parallel to the myofibrils. The sarcoplasmic reticulum contains the following specialized molecules:
- Calcium channels: Ryanodine receptors (RyR1 in skeletal muscle) are the primary calcium release channels located in the SR membrane.
- Calcium pumps: Sarco/endoplasmic reticulum Ca^{2+}-ATPase (SERCA) pumps are responsible for calcium uptake into the SR.

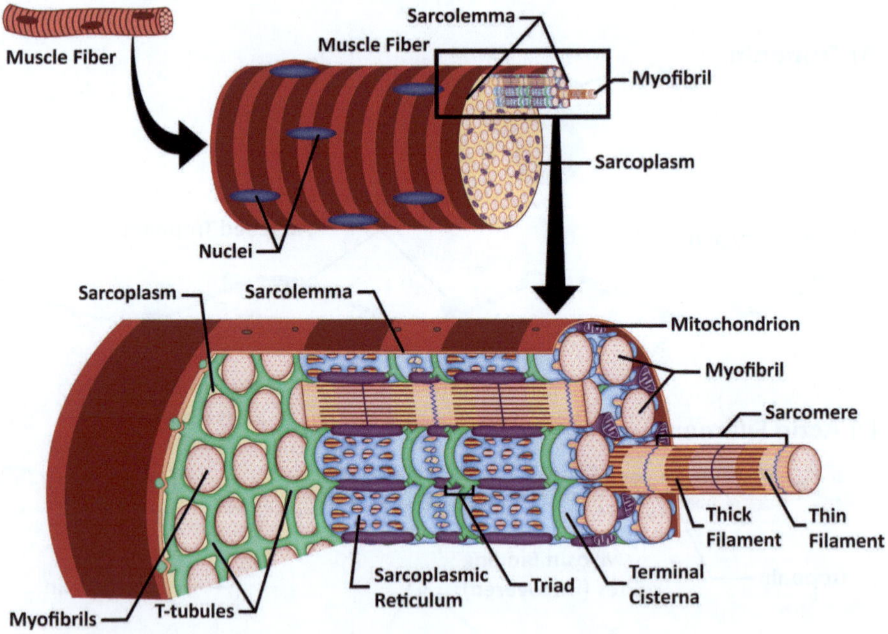

Fig. 9.11 The hierarchical and microscopic structure of skeletal muscle fiber, showing the spatial relationship between the sarcoplasmic reticulum, T-tubules, and sarcomeres

- Calcium-binding proteins: Calsequestrin is the main calcium-binding protein within the SR's lumen, which helps store high concentrations of calcium ions.

The SR serves as the main intracellular storage site for calcium ions, which are critical for initiating muscle contraction. Upon depolarization of the T-tubules, RyR1 channels open, releasing calcium ions into the sarcoplasm, enabling actin–myosin cross-bridge formation. SERCA pumps transport calcium ions back into the SR post-contraction, facilitating muscle relaxation.

Transverse Tubules (T-Tubules)

T-tubules (Fig. 9.11) are invaginations of the sarcolemma (muscle cell membrane) that penetrate the muscle cell, running perpendicular to the myofibrils. Each T-tubule is closely associated with two terminal cisternae of the SR, forming a "triad." The T-tubules possess various ion channels, including voltage-gated sodium and potassium channels, which are crucial for action potential propagation. T-tubules transmit action potentials from the sarcolemma deep into the muscle fiber, ensuring uniform excitation. They have a vital role in the excitation–contraction coupling. T-Tubules' architecture allows for rapid, localized changes in calcium concentration, ensuring that a small electrical signal can result in a robust mechanical response.

9.1.3 Muscle Contraction

Skeletal muscle fiber contraction underscores the intricate coordination between neural and muscular systems and involves a sophisticated sequence of electrochemical and molecular interactions. The stimulus for muscle contraction is initiated at the neuromuscular junction, where a motor neuron interfaces with a skeletal muscle fiber. The process then proceeds through an array of well-coordinated steps, including action potential propagation, excitation–contraction coupling, and the eventual triggering of the contractile machinery within the muscle cell.

9.1.3.1 Ion Channels in Muscle Fibers

Ion channels (Fig. 9.12) in skeletal muscle cells are transmembrane proteins that selectively allow ions to pass through the membrane, often in a voltage-dependent or ligand-gated manner.

Leak Channels

Leak channels (Fig. 9.12) are characterized by their continuous or "leaky" conductance of ions. Unlike voltage-gated or ligand-gated channels, they lack specific gating mechanisms that open or close in response to changes in membrane potential or the binding of ligands. Leak channels are typically formed by integral membrane proteins that create a pore or channel through which ions can pass. Different leak channels are selective for specific ions, such as potassium (K^+), sodium (Na^+), or chloride (Cl^-). Leak ion channels are essential for establishing and maintaining the

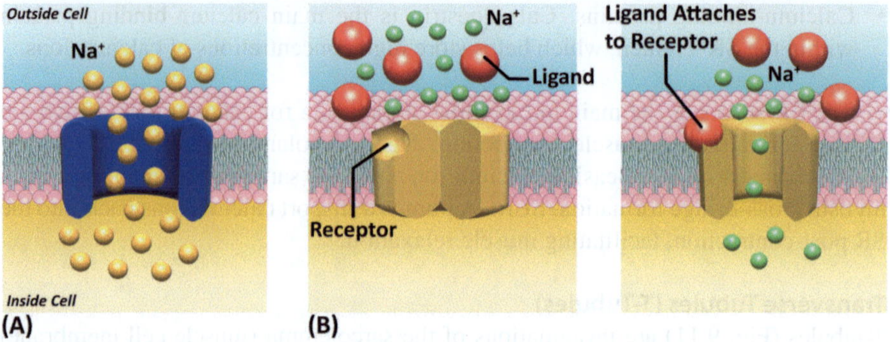

Fig. 9.12 Ion channels. (**a**) Leak ion channel. (**b**) Ligand-gated ion channel

resting membrane potential in muscle cells. Leak channels also maintain ionic homeostasis by allowing a small, continuous flux of ions across the cell membrane. This helps prevent excessive buildup or depletion of specific ions within the cell. While leak channels do not contribute significantly to action potential generation, they help set the baseline membrane potential. This baseline potential determines the starting point from which voltage-gated channels, such as voltage-gated sodium and potassium channels, initiate action potentials in response to stimuli.

Ligand-Gated Sodium Channels

Ligand-gated sodium channels (Fig. 9.12) in muscles, also known as nicotinic acetylcholine receptors (nAChRs) or simply acetylcholine receptors, are specialized ion channels that play a fundamental role in muscle physiology, particularly in skeletal muscle contraction. These receptors are activated by the neurotransmitter acetylcholine (ACh) released from motor neurons at the neuromuscular junction. Ligand-gated sodium channels in muscles are pentameric structures composed of five subunits. Each subunit consists of multiple transmembrane domains. Two alpha (α) subunits, one beta (β) subunit, one gamma (γ) subunit, and one delta (δ) subunit make up the typical composition of the muscle nAChR. The transmembrane regions of the subunits create a central pore or channel. The extracellular regions of the α subunits contain the binding sites for acetylcholine. When ACh binds to these sites, it induces conformational changes in the receptor. Ligand-gated sodium channels are predominantly found at the neuromuscular synapses, where motor neurons communicate with muscle fibers. When an action potential arrives at the neuromuscular junction, it triggers the release of acetylcholine from motor neuron terminals. Acetylcholine binds to the ligand-binding sites on the α subunits of the nAChR, inducing a conformational change. The conformational change triggered by acetylcholine binding leads to the opening of the sodium channel pore. This allows an influx of sodium ions into the muscle fiber. The influx of sodium ions through the nAChR channels causes a rapid and localized depolarization of the muscle cell membrane (sarcolemma). This depolarization propagates along the sarcolemma and initiates an action potential.

Voltage-Gated Sodium Channels (NaV)

Voltage-gated sodium channels (Fig. 9.13) are responsible for the rapid influx of Na^+ ions, leading to the action potential's depolarization phase. The core of the voltage-gated sodium channel is a large transmembrane protein (alpha subunit). The alpha subunit contains four homologous domains (I, II, III, and IV), each comprising six transmembrane segments (S1–S6). The S4 segments in each domain serve as voltage sensors, detecting changes in membrane potential. The S5 and S6 segments in each domain of the alpha subunit form the pore-lining region of the channel. This pore is responsible for the selective passage of sodium ions across the cell membrane during channel activation. The S6 segments from domains III and IV and a short P-loop segment form the activation gate. The activation gate controls the opening and closing of the channel in response to changes in membrane potential. The inactivation gate is formed by a short sequence called the "IFM motif" between domains III and IV in the intracellular loop. This gate is responsible for the rapid inactivation of the sodium channel after it opens in response to depolarization. Inactivation is crucial for preventing sustained sodium influx and allowing for repolarization. At the neuromuscular junction, sodium channels are closely associated with acetylcholine receptors, facilitating the initiation of the action potential.

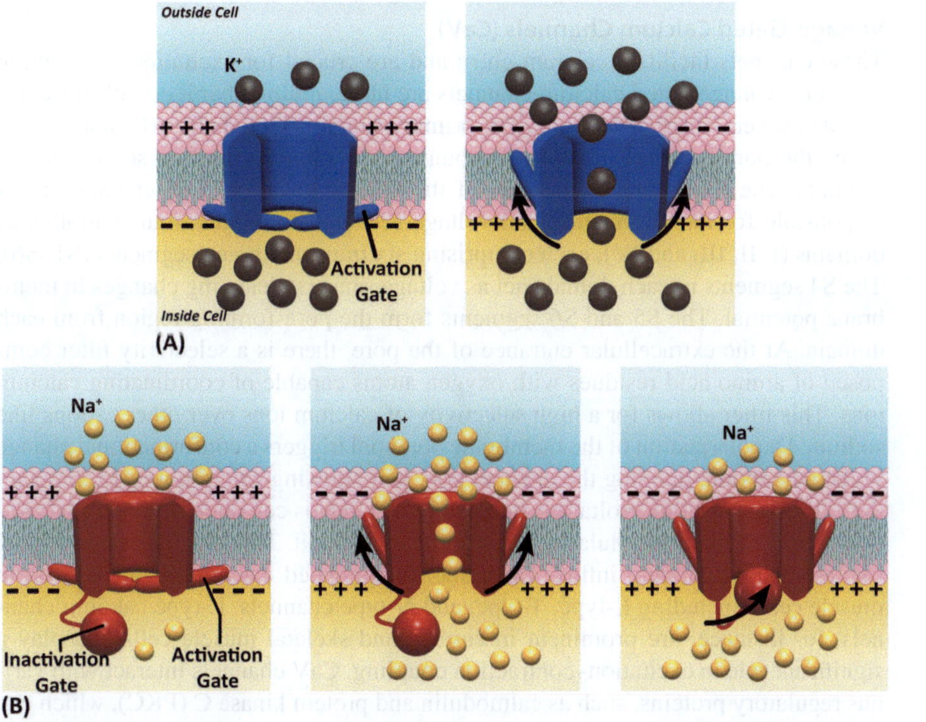

Fig. 9.13 Voltage-gated potassium (**a**) and sodium (**b**) channels

Voltage-Gated Potassium Channels (KV)

Voltage-gated potassium channels (Fig. 9.13) allow the efflux of K^+ ions, contributing to the repolarization and hyperpolarization phases of the action potential. Voltage-gated potassium channels consist of four identical subunits arranged around a central ion-conducting pore. Each subunit contains six transmembrane segments (S1–S6) and a reentrant P-loop that lines the ion-conducting pore. The S1–S4 segments in each subunit constitute the voltage-sensing domain. The S4 segment is vital and contains positively charged amino acids (arginine or lysine) at regular intervals, which act as voltage sensors. Changes in membrane potential alter the position of these charged residues in response to depolarization or hyperpolarization. The S5 and S6 segments from each of the four subunits come together to create the pore-forming region of the channel. The P-loop, formed by parts of the S5 and S6 segments, lines the ion-conducting pore and plays a critical role in ion selectivity and permeation. Voltage-gated potassium channels possess two main gating mechanisms: activation and inactivation. The channel undergoes conformational changes in response to membrane depolarization, opening the channel pore and allowing potassium ions to flow out of the cell. Some potassium channels exhibit inactivation mechanisms, similar to sodium channels, where an intracellular domain or ball-and-chain mechanism can occlude the channel pore to terminate ion flow.

Voltage-Gated Calcium Channels (CaV)

These channels facilitate calcium entry and are crucial for excitation–contraction coupling. Voltage-gated calcium channels are large, multi-subunit complexes composed of several distinct subunits. The primary subunits include the $\alpha 1$ subunit (also called the pore-forming subunit), β subunits, $\alpha 2\delta$ subunits, and, in some cases, γ subunits. The $\alpha 1$ subunit is the core of the voltage-gated calcium channel and is responsible for ion permeation and voltage sensing. It contains four homologous domains (I, II, III, and IV), each comprising six transmembrane segments (S1–S6). The S4 segments in each domain act as voltage sensors, detecting changes in membrane potential. The S5 and S6 segments form the pore-forming region from each domain. At the extracellular entrance of the pore, there is a selectivity filter composed of amino acid residues with oxygen atoms capable of coordinating calcium ions. This filter allows for a high selectivity of calcium ions over other cations like sodium. Depolarization of the membrane potential triggers a conformational change in the $\alpha 1$ subunit, opening the channel pore and allowing calcium ions to flow into the cell. After opening, voltage-gated calcium channels can undergo rapid inactivation mediated by intracellular regions of the $\alpha 1$ subunit. This inactivation is crucial for terminating calcium influx. Different voltage-gated calcium channels exist in muscle cells, including L-type, T-type, and N-type channels. L-type calcium channels, for instance, are prominent in cardiac and skeletal muscle cells and play a significant role in excitation–contraction coupling. CaV channels interact with various regulatory proteins, such as calmodulin and protein kinase C (PKC), which can modulate channel activity.

Ryanodine Receptors (RyR)

They function as calcium-release channels on the sarcoplasmic reticulum (RyR1 in skeletal muscle). They are controlled by voltage-sensitive dihydropyridine receptors (DHPR) on the T-tubules and by intracellular calcium levels. Ryanodine receptors exist as homotetramers, composed of four identical subunits that assemble to form a functional channel. Each subunit comprises over 5000 amino acids, making RyRs some of the largest ion channels known. The RyR is an integral membrane protein in the sarcoplasmic reticulum (SR), an organelle responsible for storing and releasing calcium ions in muscle cells. It spans the SR membrane, forming a pore that allows the passage of calcium ions. The amino-terminal domain is located on the cytoplasmic side of the RyR. It contains a calmodulin-binding region, which can interact with calcium-bound calmodulin to regulate RyR activity. The central core domain forms the main body of each RyR subunit. The transmembrane domain contains the ion-conducting pore formed by each subunit's S5 and S6 segments. This pore allows calcium ions to move from the SR lumen into the cytoplasm. Ryanodine receptors possess several cytoplasmic domains crucial for their regulation and interaction with other proteins. These domains include the "handle," "clamp," and "feet" regions, which are involved in channel gating and anchoring to the SR membrane. RyRs are closely associated with dihydropyridine receptors (DHPRs) in skeletal muscle, forming a complex known as the DHPR-RyR couplon. The DHPR acts as a voltage sensor that detects membrane depolarization and communicates this signal to the RyR to initiate calcium release during excitation–contraction coupling. The opening of RyRs in response to a calcium signal rapidly releases calcium ions from the SR, which triggers muscle contraction. This release of calcium ions is a fundamental step in excitation–contraction coupling in skeletal and cardiac muscles.

Chloride Channels

Muscle cells can contain different types of chloride channels, such as voltage-gated chloride channels (ClC channels), calcium-activated chloride channels (CaCCs), and cystic fibrosis transmembrane conductance regulator (CFTR) channels. Most chloride channels, including ClC channels and CaCCs, are integral membrane proteins that span the cell membrane. Chloride channels have pore-forming domains that determine their ion selectivity and conductance properties. Chloride channels can have different gating mechanisms that control their opening and closing. Voltage-gated chloride channels, for example, open or close in response to changes in membrane potential, while intracellular calcium levels activate calcium-activated chloride channels. Chloride channels often have regulatory regions that can be phosphorylated, leading to changes in channel activity. Chloride channels in muscle cells are involved in maintaining membrane potential, regulating cell volume, and influencing the excitability of muscle cells. In some cases, chloride channels may contribute to muscle membrane hyperpolarization.

9.1.3.2 Resting Membrane Potentials

The resting membrane potential (Fig. 9.14) is the steady electrical potential difference across the cell membrane when the muscle cell is quiescent. This is typically around −80 to −90 millivolts (mV) for skeletal muscle cells. Various ion channels primarily govern the resting potential of muscle cells.

Leak channels: Leak channels, also known as background channels, are constitutively open and allow the passive movement of ions across the cell membrane. K^+ leak channels are essential in determining the resting potential of muscle cells. They allow potassium ions to move out of the cell, creating a negative resting potential. Na^+ leak channels are also present but in a much lower concentration.

Sodium-potassium (Na^+/K^+) pump: While not an ion channel, the Na^+/K^+ pump plays a crucial role in maintaining the resting potential. It actively transports three Na^+ out of the cell and two K^+ into the cell against their respective concentration gradients. This action helps establish a negative membrane potential by maintaining a higher concentration of potassium ions inside the cell.

Molecular mechanisms and ion movements: Potassium leak channels are typically open at rest, allowing a continuous efflux of K^+ ions out of the cell along their chemical gradient. As K^+ ions exit the cell, they carry a positive charge, leaving the interior more negatively charged. This contributes to the negative resting potential. The resting membrane potential of muscle cells is determined by the equilibrium potential for K^+ ions (EK) due to the high permeability of the membrane to potassium through leak channels. At the resting potential, the net movement of potassium ions out of the cell equals the net movement of potassium ions into the cell due to the sodium-potassium pump. This results in a stable equilibrium. Na^+ leak channels

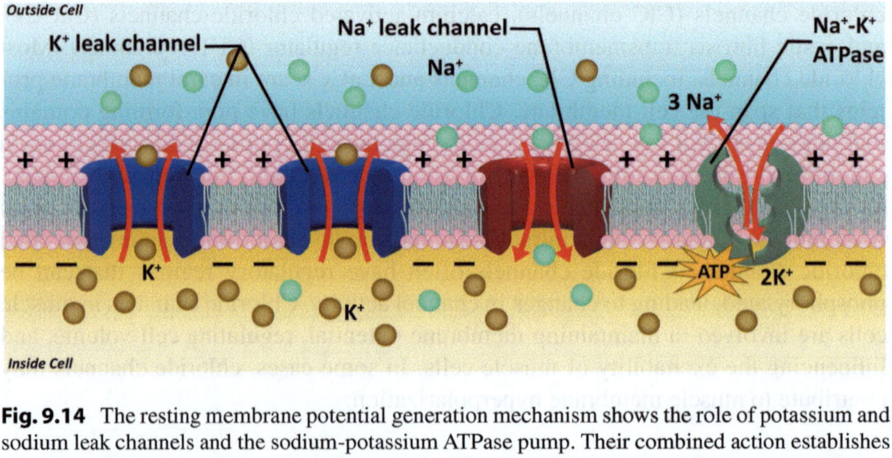

Fig. 9.14 The resting membrane potential generation mechanism shows the role of potassium and sodium leak channels and the sodium-potassium ATPase pump. Their combined action establishes a resting membrane potential, typically around −70 mV, which is essential for the excitability of muscle cells

are also present, leading to an influx of Na^+ ions into the muscle fibers along their electrochemical gradient, making inside the cells more positive, thus opposing the resting membrane potential. Since the number of Na^+ leak channels is much lower than K^+ leak channels, its contribution to resting membrane potential is less significant. Intracellular anions like Cl^-, proteins, and phosphates contribute to a more negative intracellular environment. Maintaining the resting potential is an energy-dependent process, requiring ATP for the activity of the Na^+/K^+ pump, which rectifies any slight deviations from the resting potential. The final resting membrane potential is an equilibrium between all these factors.

9.1.3.3 Action Potentials

An action potential (Fig. 9.15) is a rapid, transient change in the electrical potential across the sarcolemma, to +20 to +30 mV, initiated by external stimuli and propagated along the membrane. It consists of depolarization and repolarization phases. Action potential generation in muscle cells begins with depolarization, the rapid change of the membrane potential from its resting state to a more positive value. This depolarization is initiated by the arrival of a nerve impulse at the neuromuscular junction, where the neurotransmitter acetylcholine (ACh) is released. Upon binding of ACh to receptors on the muscle cell membrane, there is a localized influx

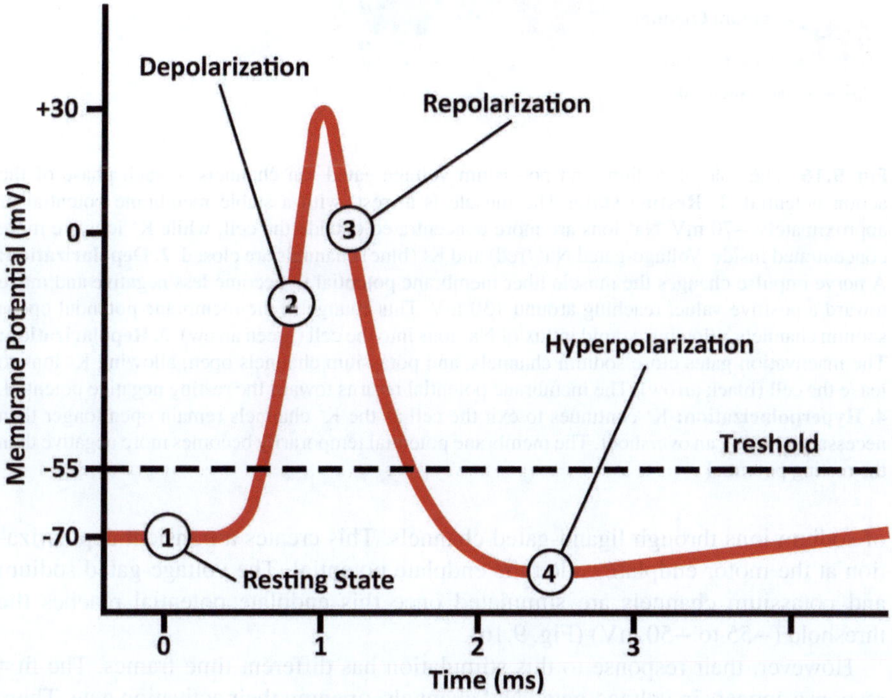

Fig. 9.15 Recording of an action potential (AP) in a muscle fiber, depicting changes in membrane potential (mV) over time (ms). An AP is a brief change in membrane potential

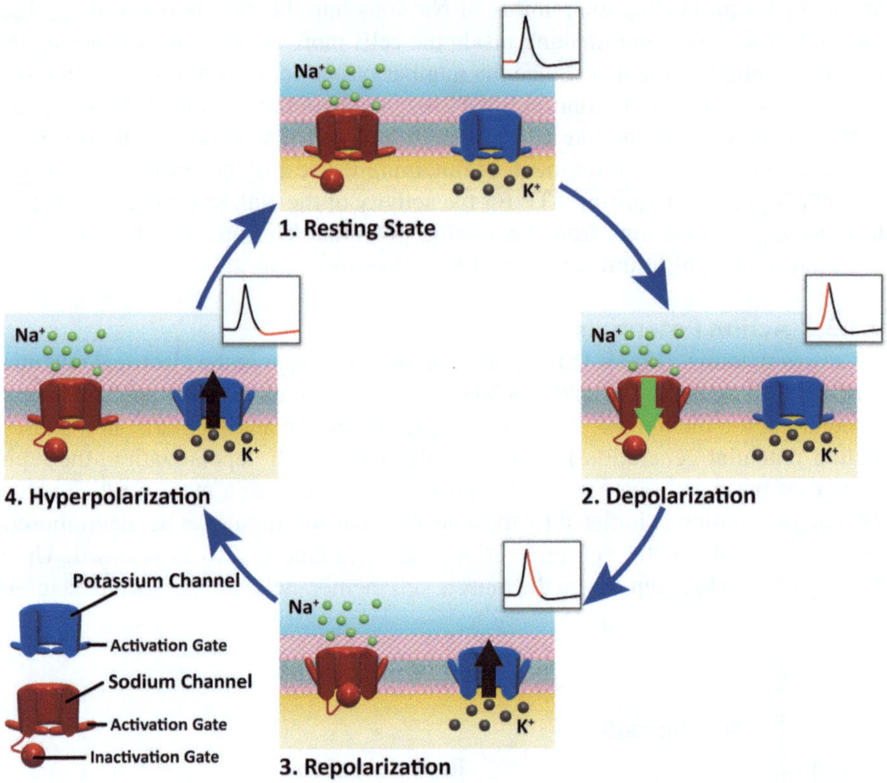

Fig. 9.16 The role of sodium and potassium voltage-gated ion channels in each phase of the action potential. **1. Resting state:** The muscle is at rest, with a stable membrane potential of approximately −70 mV. Na^+ ions are more concentrated outside the cell, while K^+ ions are more concentrated inside. Voltage-gated Na^+ (red) and K^+ (blue) channels are closed. **2. Depolarization:** A nerve impulse changes the muscle fiber membrane potential to become less negative and move toward a positive value, reaching around +30 mV. This change in the membrane potential opens sodium channels, allowing a rapid influx of Na^+ ions into the cell (green arrow). **3. Repolarization:** The inactivation gates close sodium channels, and potassium channels open, allowing K^+ ions to leave the cell (black arrow). The membrane potential returns toward the resting negative potential. **4. Hyperpolarization:** K^+ continues to exit the cell as the K^+ channels remain open longer than necessary, causing an overshoot. The membrane potential temporarily becomes more negative than the resting potential

of sodium ions through ligand-gated channels. This creates a transient depolarization at the motor endplate called the endplate potential. The voltage-gated sodium and potassium channels are stimulated once this endplate potential reaches the threshold (−55 to −50 mV) (Fig. 9.16).

However, their response to this stimulation has different time frames. The first response appears in voltage-gated Na^+ channels, opening their activation gate. Then, the inactivation gates of voltage-gated Na^+ channels close, and the activation gates of voltage-gated K^+ channels open. Opening the voltage-gated sodium channels along the sarcolemma allows a massive influx of Na^+ ions into the cell, causing a rapid depolarization of the membrane. This depolarization spreads along the sarcolemma in a wave-like manner. When the action potential reaches its peak (+20 to +30 mV), the inactivation gates of voltage-gated sodium channels close, stopping the sodium influx, and voltage-gated potassium channels open, allowing potassium ions to exit the cell. This helps restore the negative resting membrane potential (repolarization). The last change in voltage-gated channels is the inactivation (closure) of K^+ channels. This late voltage-gated K^+ channel inactivation leads to prolonged K^+ ions efflux outside the muscle fibers, creating more negative potential across the membrane (hyperpolarization) (−100 mV). Meanwhile, the inactivated voltage-gated Na^+ channels, with the activation gates open and inactivation gates closed, reset (inactivation gates open and activation gate closes) and prepare for another activation cycle. Hyperpolarization gradually returns to resting membrane potential (−90 mV) through the function of the Na^+-K^+ ATPase pump.

The depolarization and repolarization processes continue to propagate along the sarcolemma, triggering the release of Ca^{2+} ions from the sarcoplasmic reticulum. The calcium ions play a pivotal role in muscle contraction by binding to troponin and allowing myosin and actin to interact. This results in the contraction of sarcomeres, leading to muscle shortening. This contraction is sustained as long as the action potential propagates along the sarcolemma.

The inactivation gates of the voltage-gated sodium channels ensure the transient nature of the action potential. These inactivation gates also produce an absolute refractory period during which the muscle fiber does not respond to any stimulus. This period ensures that another action potential cannot be initiated immediately, allowing for signal directionality and preventing continuous firing.

9.1.3.4 Excitation–Contraction Coupling

Excitation–contraction coupling transforms an electrical stimulus—an action potential—into a mechanical response, namely muscle contraction. The excitation–contraction coupling events include a motor neuron stimulus initiates an action potential at the neuromuscular junction, which then propagates along the sarcolemma. The action potential travels down the T-tubules, bringing the depolarization into proximity to the SR. The DHPR, a voltage-sensitive L-type calcium channel in the T-tubules, senses the change in membrane potential. Upon sensing the voltage change, DHPR undergoes a conformational change and mechanically interacts with RyR1. This mechanical interaction induces the opening of RyR1 channels, rapidly releasing Ca^{2+} ions from the SR into the cytosol. The released Ca^{2+} ions bind to the troponin complex on the thin filament of the sarcomere. This binding prompts a change in the conformation of the associated tropomyosin, exposing myosin-binding sites on actin filaments. Myosin heads in the thick filament bind to the exposed sites on actin, forming cross-bridges. Upon cross-bridge formation, the myosin heads pivot, pulling the actin filaments toward the center of the sarcomere. This process is fueled by the hydrolysis of ATP to adenosine diphosphate (ADP) and inorganic phosphate (P_i). ADP molecules are detached, and new ATP molecules bind to the myosin heads, detaching the myosin heads from the actin filaments. ATP bound to myosin is hydrolyzed, which re-cocks the myosin heads back to their original position, ready for another round of cross-bridge formation and power stroke. As the action potential ceases, Ca^{2+} ions are pumped back into the sarcoplasmic reticulum by sarcoplasmic/endoplasmic reticulum Ca^{2+}-ATPase (SERCA). As Ca^{2+} ions are removed from the cytosol, the myosin-binding sites on actin are re-concealed, resetting the system for the next contraction.

9.1.3.5 Muscle Twitch

A muscle twitch is a physiological phenomenon that represents the mechanical response of a skeletal muscle fiber to a single action potential from a motor neuron. A single, brief electrical stimulus elicits this transient contraction. A muscle twitch is generally subdivided into three primary phases: the latent period, the contraction phase, and the relaxation phase (Fig. 9.17).

Latent period: This brief delay occurs immediately following the initiation of the action potential and precedes the onset of muscle contraction. During this phase, which typically lasts a few milliseconds, the excitation–contraction coupling process is initiated, but no observable mechanical activity occurs. The action potential travels along the sarcolemma and down the T-tubules, releasing Ca^{2+} ions from the sarcoplasmic reticulum.

Contraction phase: Following the latent period, the muscle fiber undergoes a period of contraction, during which it generates force and, depending on the load, may shorten. This phase is mediated by cross-bridges forming between actin and myosin filaments, facilitated by the elevated cytosolic Ca^{2+} concentration. The length of the contraction phase can vary depending on several factors, including the type of muscle fiber, the load against which it is contracting, and the intracellular Ca^{2+} concentration.

Relaxation phase: After reaching peak tension, the muscle fiber relaxes and returns to its resting length. The active reuptake of Ca^{2+} ions into the sarcoplasmic reticulum via the sarcoplasmic/endoplasmic reticulum Ca^{2+}-ATPase (SERCA) pumps

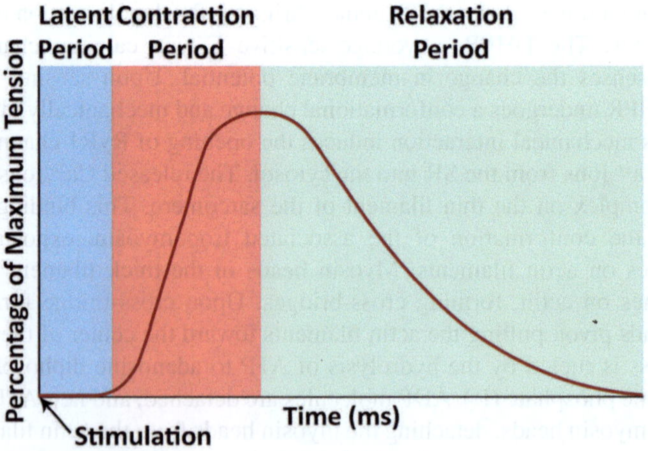

Fig. 9.17 A myogram showing a typical muscle twitch

characterizes this phase. The decrease in intracellular Ca^{2+} concentration leads to the dissociation of Ca^{2+} from troponin, facilitating the repositioning of tropomyosin and thereby inhibiting further actin–myosin interactions.

9.1.3.6 Graded Muscle Contractions

Graded muscle contractions refer to the varying degrees of muscle tension or force generated by a muscle in response to different levels of stimuli. This modulation is vital for fine-tuning movement control and executing various types of physical activity.

Temporal Summation (Frequency Summation)

Temporal summation (Fig. 9.18) in skeletal muscle contraction is a mechanism that amplifies force output by the successive and rapid delivery of action potentials to the muscle fibers. Temporal summation is fundamental for the fine-tuned regulation of skeletal muscle force. When a second action potential is generated before the first twitch has had time to relax, the two responses summate temporally. This results in a contraction of greater force than a single twitch.

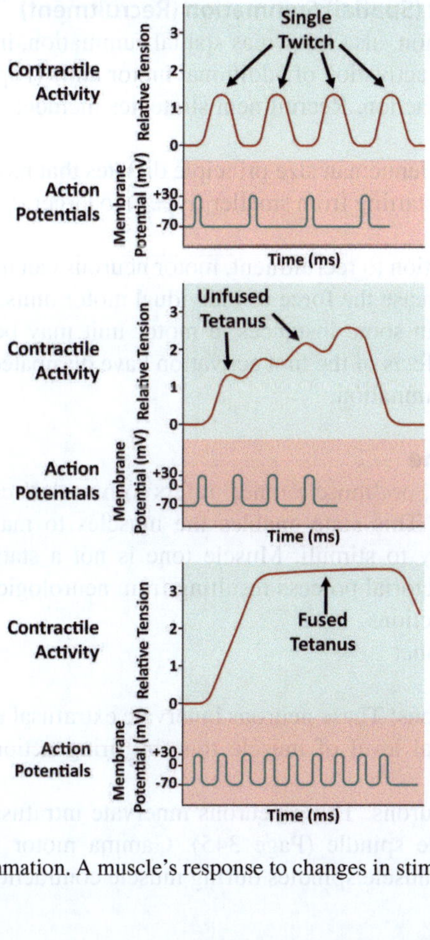

Fig. 9.18 Temporal summation. A muscle's response to changes in stimulation frequency

- Types of temporal summation
  - Incomplete (unfused) tetanus (Fig. 9.18): If the frequency of action potentials is high but allows for partial relaxation between individual twitches, the muscle exhibits a pattern known as incomplete tetanus, characterized by fluctuating tension levels.
  - Complete (fused) tetanus (Fig. 9.18): At even higher frequencies that allow no relaxation, the muscle tension plateaus maximally, leading to complete tetanus.
- Influencing factors
  - Stimulus frequency: The degree of temporal summation highly depends on the frequency of incoming action potentials. Higher frequencies are more likely to produce tetanic contractions.
  - Muscle fiber type: Slow-twitch fibers are less prone to temporal summation than fast-twitch fibers due to different calcium handling and myosin ATPase activities.
  - Fatigue: Muscle fatigue can alter the effectiveness of temporal summation by affecting intracellular ion concentrations and metabolic status.

Multiple Motor Unit (Spatial) Summation (Recruitment)

Recruitment summation, also known as spatial summation, in skeletal muscle contraction, refers to the activation of additional motor units (Page 222) to increase the force of muscle contraction. Recruitment strategies include:

Size principle: The Henneman size principle dictates that motor units are generally recruited orderly, starting from smaller (type I) to larger (type II) units, as greater force is required.

Rate coding: In addition to recruitment, motor neurons can modulate the firing rate to increase or decrease the force of individual motor units.

Double discharge: In some instances, a motor unit may be activated more than once before the effects of the first activation have dissipated, providing a form of intramotor unit summation.

9.1.3.7 Muscle Tone

Skeletal muscle tone, or "muscle tone," refers to muscles' continuous contraction and passive tension. This state enables the muscles to maintain posture and to respond more rapidly to stimuli. Muscle tone is not a static phenomenon but a dynamic and multifactorial process resulting from neurological, physiological, and biomechanical interactions.

Basis of muscle tone:

- Alpha motor neurons: These neurons innervate extrafusal muscle fibers and contribute to the basal level of muscle tone by firing action potentials at a low, steady rate.
- Gamma motor neurons: These neurons innervate intrafusal muscle fibers comprising the muscle spindle (Page 345). Gamma motor neurons maintain the sensitivity of the muscle spindles during muscle contraction.

- Cross-bridge formation: Even when a muscle appears relaxed, minimal cross-bridge formation between actin and myosin filaments contributes to resting muscle tone.
- Ionic homeostasis: The maintenance of ionic gradients, particularly of sodium and potassium ions, across the muscle membrane affects the excitability of the muscle and, hence, its tone.
- Stretch reflex: This involuntary motor response occurs when a muscle is stretched. It resists stretching and contributes to muscle tone by activating muscle spindles and Golgi tendon organs (Page 342).
- Reciprocal inhibition: When one muscle contracts, its antagonist is inhibited, which is essential for coordinated movement and plays a role in modulating muscle tone.

9.1.3.8 Isotonic Contractions

Isotonic contractions (Fig. 9.19) are a specific type of muscle contraction in which the muscle changes length while maintaining constant tension. A muscle's ability to generate force during an isotonic contraction is highly dependent on its length before the contraction (force-length relationship). At an optimal length, maximal force can be produced.

Isotonic contractions can be divided into concentric and eccentric categories (Fig. 9.19):

Concentric contractions: In concentric contractions, the muscle fibers shorten as they generate force, pulling the muscle's origin and insertion points closer together. For instance, the biceps brachii undergoes concentric contraction during the upward phase of a bicep curl.

Eccentric contractions: The muscle fibers lengthen in eccentric contractions while generating force. Eccentric contractions decelerate the movement initiated by the concentric contraction of the opposing muscle. An example would be the controlled weight lowering during a bicep curl, which elongates the biceps brachii.

9.1.3.9 Isometric Contractions

Isometric contractions refer to tension generation without a corresponding change in the muscle's length. In isometric contractions, the muscle remains at a constant length, which means that the muscle's initial length greatly influences the tension generated. At the optimal length, the maximal number of myosin–actin cross-bridges can be formed, generating maximal tension. Isometric contractions often stabilize joints, providing a counterforce to an external load without moving.

9.1.3.10 Energy for Contraction

The energetic demands of skeletal muscle contractions are met through a complex interplay of metabolic pathways that provide the necessary adenosine triphosphate (ATP), the cell's primary energy currency. Each source of energy is utilized to varying degrees depending on factors such as the duration and intensity of muscle activity, the availability of substrates, and the muscle fiber type.

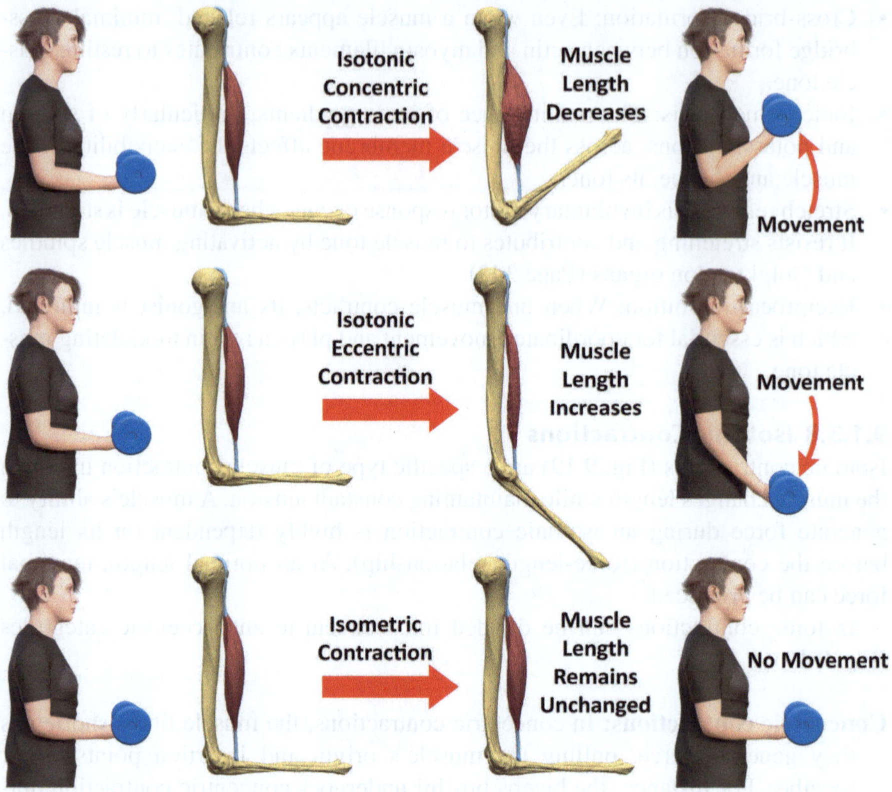

Fig. 9.19 Types of skeletal muscle contraction: isotonic concentric, isotonic eccentric, and isometric contractions

Direct Phosphorylation of ADP by Creatine Phosphate

Creatine phosphate (phosphocreatine) is an immediate, high-energy reserve in skeletal muscle contractions. Its function is particularly essential during the initial phase of high-intensity, short-duration activities, when it acts as a rapid energy supplier by donating its phosphate group to replenish adenosine triphosphate (ATP).

Creatine kinase catalyzes the transfer of the phosphate group from creatine phosphate to adenosine diphosphate (ADP), regenerating ATP. The reaction can be summarized as:

$$\text{Creatine Phosphate} + \text{ADP} \rightleftharpoons \text{ATP} + \text{Creatine}$$

The creatine phosphate system acts as a buffer, storing high-energy phosphate groups that can be rapidly mobilized. This enables the muscle to meet the sudden demand for energy, especially in the absence of adequate ATP reserves. During low activity or rest periods, creatine phosphate reserves are replenished from ATP generated primarily through oxidative metabolism.

Creatine phosphate is vital in activities requiring a quick burst of energy, such as sprinting or lifting a heavy weight. During the first few seconds of these activities, the creatine phosphate system is the primary energy source. As high-intensity exercise continues beyond a few seconds, other energy systems, like anaerobic glycolysis and aerobic metabolism, take over.

Anaerobic Pathway: Glycolysis and Lactic Acid Formation

This pathway breaks down glucose derived from blood or stored intracellularly as glycogen into pyruvate while generating 2 ATP molecules per glucose molecule (Section 4.4.1). In anaerobic conditions, pyruvate is converted into lactate (Fig. 9.20). Anaerobic glycolysis can fuel moderate to high-intensity muscle activity for about 2 min.

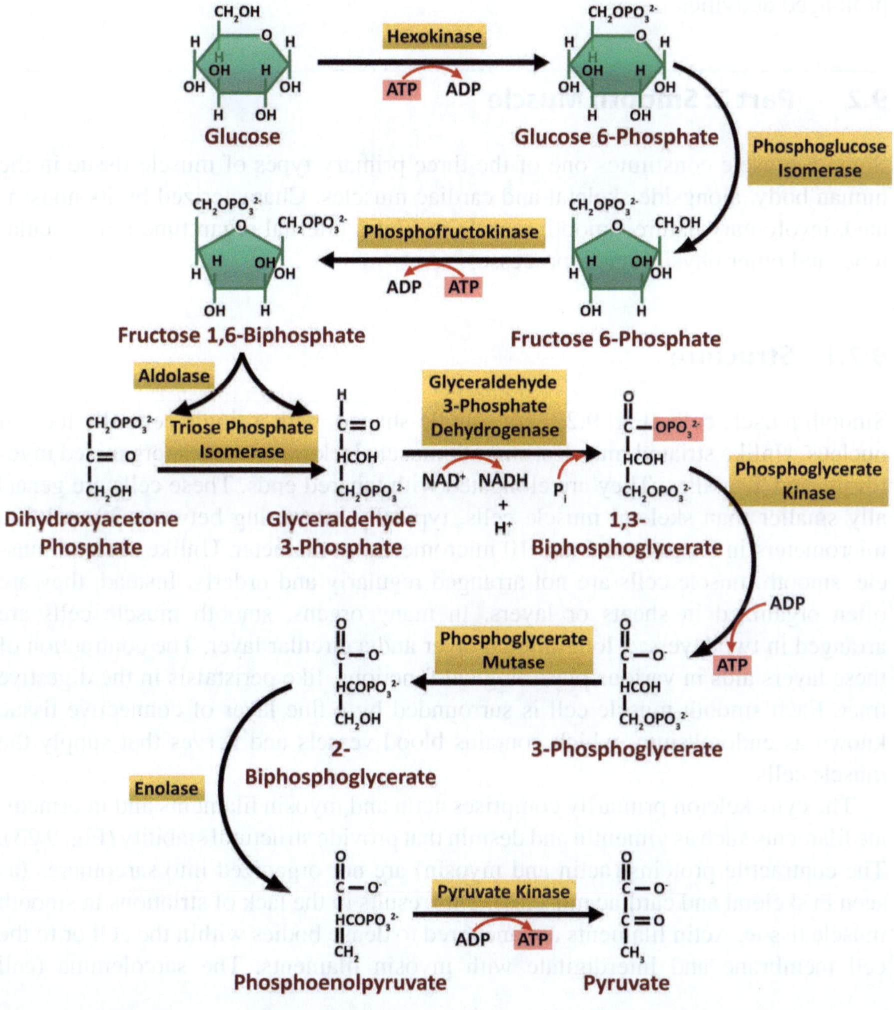

Fig. 9.20 Glycolysis (anaerobic) pathway

Aerobic Respiration

Aerobic respiration (Fig. 9.21), occurring in the mitochondria, is the most efficient pathway for ATP production (Section 4.4.2). It requires oxygen to oxidize carbohydrates, fats, and proteins, typically in the form of pyruvate, fatty acids, and amino acids, respectively. The electron transport chain and chemiosmotic coupling produce a large amount of ATP, making this pathway suitable for prolonged, low-intensity activities.

9.1.3.11 Muscle Fiber Type

Skeletal muscle fiber types (Table 9.1) are classified based on morphological, biochemical, and functional characteristics influencing their contractile properties and metabolic efficiencies. Multiple fiber types enable the skeletal muscle to accommodate diverse functional demands, from rapid, explosive movements to sustained, prolonged activities.

9.2 Part 2: Smooth Muscle

Smooth muscle constitutes one of the three primary types of muscle tissue in the human body, alongside skeletal and cardiac muscles. Characterized by its nonstriated, involuntary nature, smooth muscle regulates internal organ function, vascular tone, and other physiological processes.

9.2.1 Structure

Smooth muscle cells (Fig. 9.22) are spindle-shaped, with a single centrally located nucleus. Unlike striated muscles, smooth muscle lacks sarcomeres, organized myofibrils, and T-tubules. They are elongated with tapered ends. These cells are generally smaller than skeletal muscle cells, typically measuring between 30 and 200 micrometers in length and 5 and 10 micrometers in diameter. Unlike skeletal muscle, smooth muscle cells are not arranged regularly and orderly. Instead, they are often organized in sheets or layers. In many organs, smooth muscle cells are arranged in two layers: a longitudinal layer and a circular layer. The contraction of these layers aids in various physiological functions, like peristalsis in the digestive tract. Each smooth muscle cell is surrounded by a fine layer of connective tissue known as endomysium, which contains blood vessels and nerves that supply the muscle cells.

The cytoskeleton primarily comprises actin and myosin filaments and intermediate filaments such as vimentin and desmin that provide structural stability (Fig. 9.23). The contractile proteins (actin and myosin) are not organized into sarcomeres (as seen in skeletal and cardiac muscles). This results in the lack of striations in smooth muscle tissue. Actin filaments are anchored to dense bodies within the cell or to the cell membrane and interdigitate with myosin filaments. The sarcolemma (cell

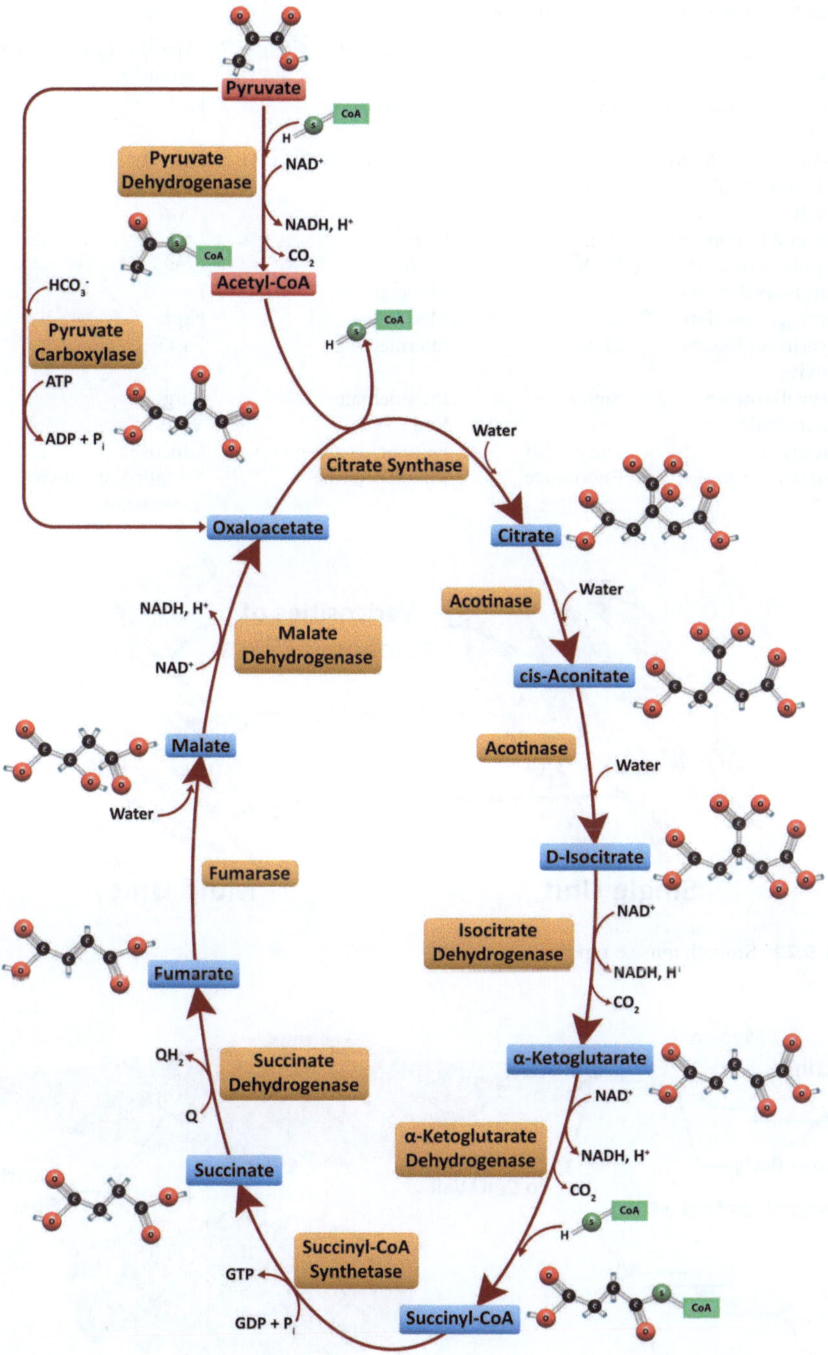

Fig. 9.21 Aerobic (oxidative) pathway

Table 9.1 Types of skeletal muscle fibers

Skeletal muscle fiber type	Type I (slow-twitch)	Type IIa (fast-twitch oxidative)	Type IIx (fast-twitch glycolytic)
Muscle contraction speed	Slow	Intermediate	Fast
Metabolic pathways	Oxidative	Oxidative and glycolytic	Glycolytic
Mitochondrial density	High	High	Low
Myoglobin content	High	High	Low
Capillary density	High	High	Low
Fatigue resistance	High	Moderate	Low
Force production	Low	Moderate	High
Myosin ATPase activity	Slow	Intermediate	Fast
Fiber diameter	Small	Intermediate	Large
Color (staining)	Red	Red	White
Energy source	Fatty acids	Fatty acids and glucose	Glucose
Activity examples	Endurance activities	Mixed activities	Sprinting, explosive movements

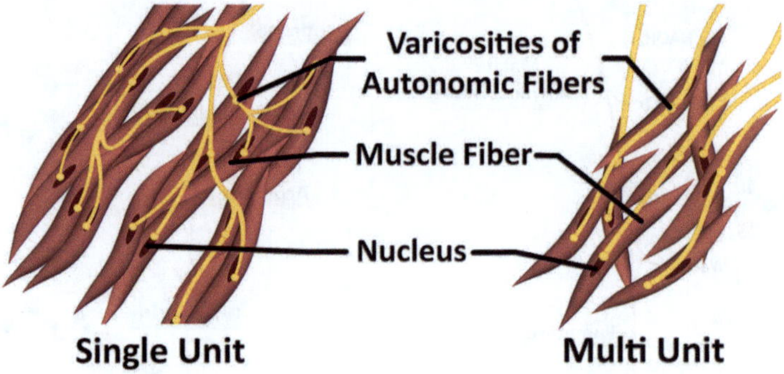

Fig. 9.22 Smooth muscle types and structure

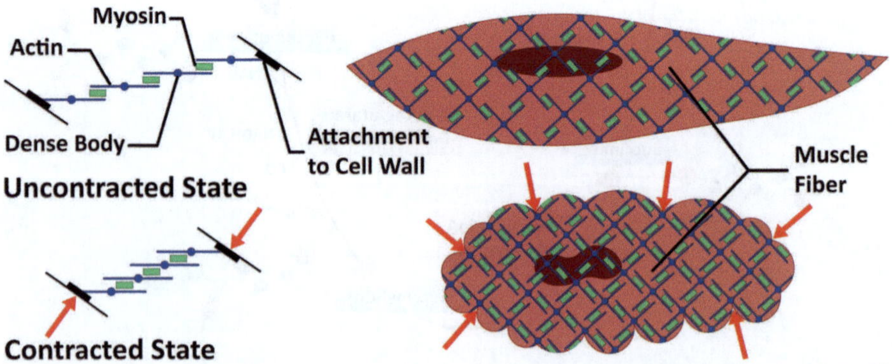

Fig. 9.23 The arrangement of contractile filaments (actin and myosin) in smooth muscle fiber and their interaction pattern during contraction

membrane) contains specialized invaginations known as caveolae that serve as reservoirs for calcium ions and signaling molecules, contributing to contraction regulation.

9.2.2 Functional Aspects

The autonomic nervous system primarily controls smooth muscle. It does not have the neuromuscular junctions characteristic of skeletal muscle. Instead, neurotransmitters are released from varicosities of autonomic nerve fibers and diffuse into the muscle tissue (Fig. 9.22). Smooth muscle cells have a greater capacity for regeneration and repair than skeletal and cardiac muscle cells. Smooth muscle contraction is generally slower and more sustained than skeletal muscle contraction. It can maintain tension over a more extended period without fatigue. Smooth muscle can maintain tension over various lengths, a property known as plasticity.

9.2.3 Types (Fig. 9.22)

Multiunit smooth muscle: Found primarily in the eye's iris, the ductus deferens, and certain small arterioles, multiunit smooth muscle cells operate largely independently of each other. They receive direct innervation from autonomic neurons and exhibit minimal electrical coupling.

Unitary (or single-unit) smooth muscle: Commonly found in the digestive tract, urinary bladder, and certain large arteries, unitary smooth muscle cells are electrically coupled through gap junctions and often contract as a coordinated group. These cells demonstrate pacemaker activity and can initiate contractions independently of the neural input.

Vascular smooth muscle: This subtype primarily lines the walls of blood vessels and is crucial for regulating vascular tone. Depending on the specific vessel and physiological conditions, vascular smooth muscle can display characteristics of both multiunit and unitary types.

9.2.4 Contraction Mechanism

Intracellular calcium ion concentration regulates smooth muscle contraction (Fig. 9.23). Calcium ions activate calmodulin, which activates myosin light-chain kinase (MLCK). The activation of MLCK leads to the phosphorylation of myosin, enabling it to form cross-bridges with actin and initiate contraction.

The contraction cascade in smooth muscle cells includes:

Receptor activation: Various signaling molecules like acetylcholine, norepinephrine, or hormones can activate receptors on the smooth muscle cell membrane.

Signal transduction: Activation of G-protein-coupled receptors (GPCRs) can increase intracellular calcium levels via the inositol triphosphate (IP3) pathway or calcium influx via ion channels.

Calmodulin activation: Elevated calcium levels bind to calmodulin, activating MLCK.

Cross-bridge formation: Activated MLCK phosphorylates myosin, facilitating interaction with actin to form cross-bridges.

Power stroke and sliding filament mechanism: The myosin heads undergo a power stroke, pulling the actin filaments and causing muscle contraction.

Relaxation: Contraction ceases when myosin light-chain phosphatase (MLCP) dephosphorylates myosin, an action regulated by intracellular calcium levels.

9.3 Part 3: Skeletal Muscle System

The skeletal muscle system is the principal mediator of mechanical force and movement. Constituting approximately 40% of total body weight, it is an extensive network of muscle tissues that confers mobility and contributes to posture, thermoregulation, and various physiological functions. It interacts intricately with the skeletal, nervous, and circulatory systems to generate coordinated actions ranging from fine motor skills to gross locomotor activities.

9.3.1 How Are Skeletal Muscles Named?

Accurate nomenclature allows clinicians, anatomists, and researchers to communicate clearly and efficiently, avoiding misunderstandings that could lead to clinical or scholarly errors. Primary determinants of skeletal muscle naming include:

Location in the body: Muscles are often named based on anatomical location. For instance, the "rectus abdominis" refers to a muscle in the abdominal region, while "tibialis anterior" denotes a muscle at the front of the tibia.

Shape: The shape of the muscle can also inform its name. For example, the "deltoid" muscle roughly resembles a triangle, akin to the Greek letter delta.

Size: Muscle names sometimes reflect the muscle's size. The terms "major" and "minor" or "longus" and "brevis" are often used to distinguish muscles of different sizes but similar functions or locations. For instance, the "gluteus maximus" is larger than the "gluteus minimus."

Direction of muscle fibers: The orientation of muscle fibers can be a defining characteristic. For example, "rectus" refers to muscles with fibers that run parallel to the midline, whereas "oblique" refers to muscles with fibers that run diagonally.

Number of origins: Muscles may be named according to their number of origins, known as "heads." For instance, the "biceps brachii" has two points of origin, while the "triceps brachii" has three.

Function: Functional attributes often serve as a basis for naming muscles. For example, "flexor" and "extensor" muscles are named for their primary function of flexing or extending a joint, respectively.

Points of attachment: The origin and insertion points can be descriptive factors in muscle naming. For instance, the "sternocleidomastoid" originates from the sternum and clavicle and inserts onto the mastoid process.

9.3.2 Fascicle Arrangement

"Fascicle arrangement" refers to the spatial organization of muscle fibers within a given muscle. The angle at which the fascicles insert into the tendons and their overall orientation about the muscle's line of action can vary significantly.

These variations lead to different categories of fascicle arrangements (Fig. 9.24).

9.3.2.1 Circular

The fascicles are arranged in a circular pattern, usually surrounding an opening or orifice, as seen in the orbicularis oris muscle around the mouth.

9.3.2.2 Convergent

In convergent arrangements, the muscle fibers are spread over a large area but converge to a single point. An example is the pectoralis major. Functional characteristics include:

- Versatile, capable of applying force in different directions depending on activation pattern
- Moderate force production

9.3.2.3 Parallel

In parallel arrangements, the fascicles run parallel to the longitudinal axis of the muscle. This arrangement is most common and is found in muscles like the biceps brachii and rectus abdominis.

9.3.2.4 Pennate

In pennate arrangements, the fascicles are short and attach obliquely to a central tendon that runs the length of the muscle. They can be:

- Unipennate: One side of the central tendon has fascicles attaching at an angle, as in the extensor digitorum longus.
- Bipennate: Both sides of the central tendon have fascicles attaching at an angle, as in the rectus femoris.
- Multipennate: Multiple rows of fascicles are present, as in the deltoid muscle.

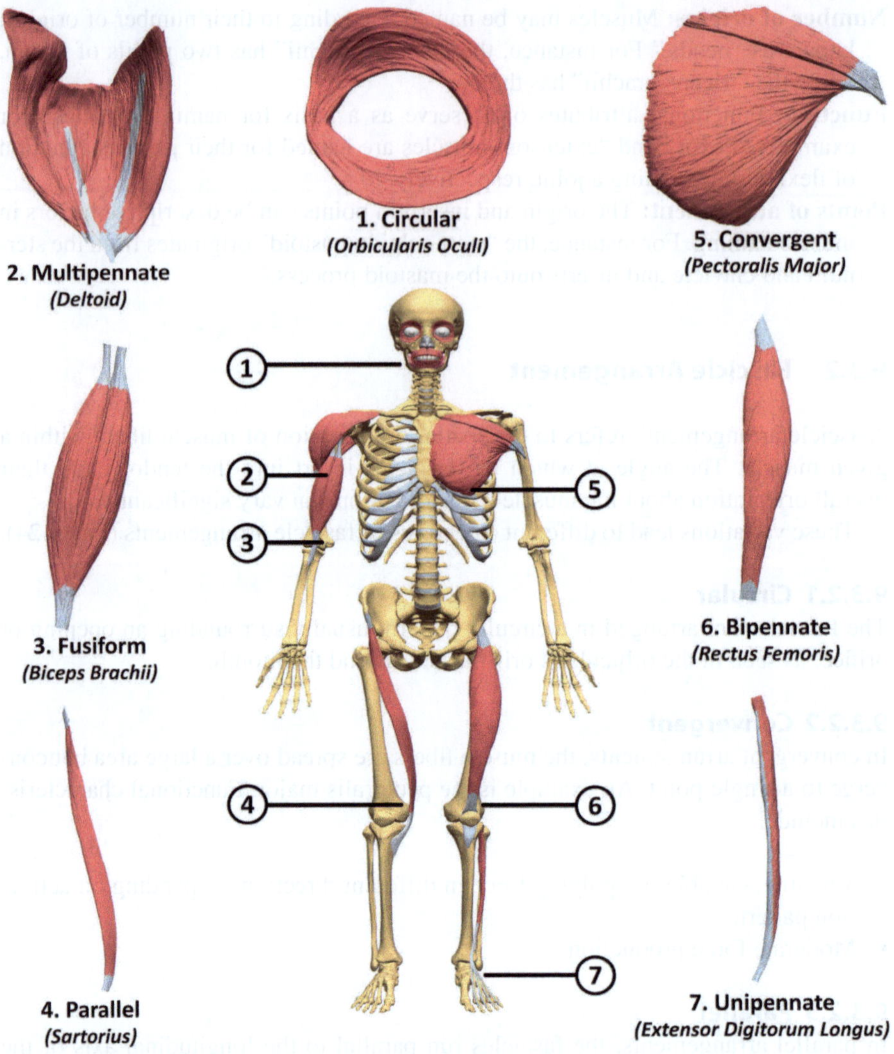

Fig. 9.24 Patterns of fascicle arrangement in skeletal muscles

9.3.3 Mechanism of Muscle Action

The functional capability of muscle tissue is not solely dependent on its intrinsic properties but is also profoundly influenced by its interaction with skeletal and joint structures. One of the principal frameworks that elucidates this interaction is the lever system.

9.3.3.1 Lever System

In mechanical terms, a lever system (Fig. 9.25) is an arrangement of a rigid bar (bone), fulcrum (joint), resistance (load), and force (muscular effort). The lever

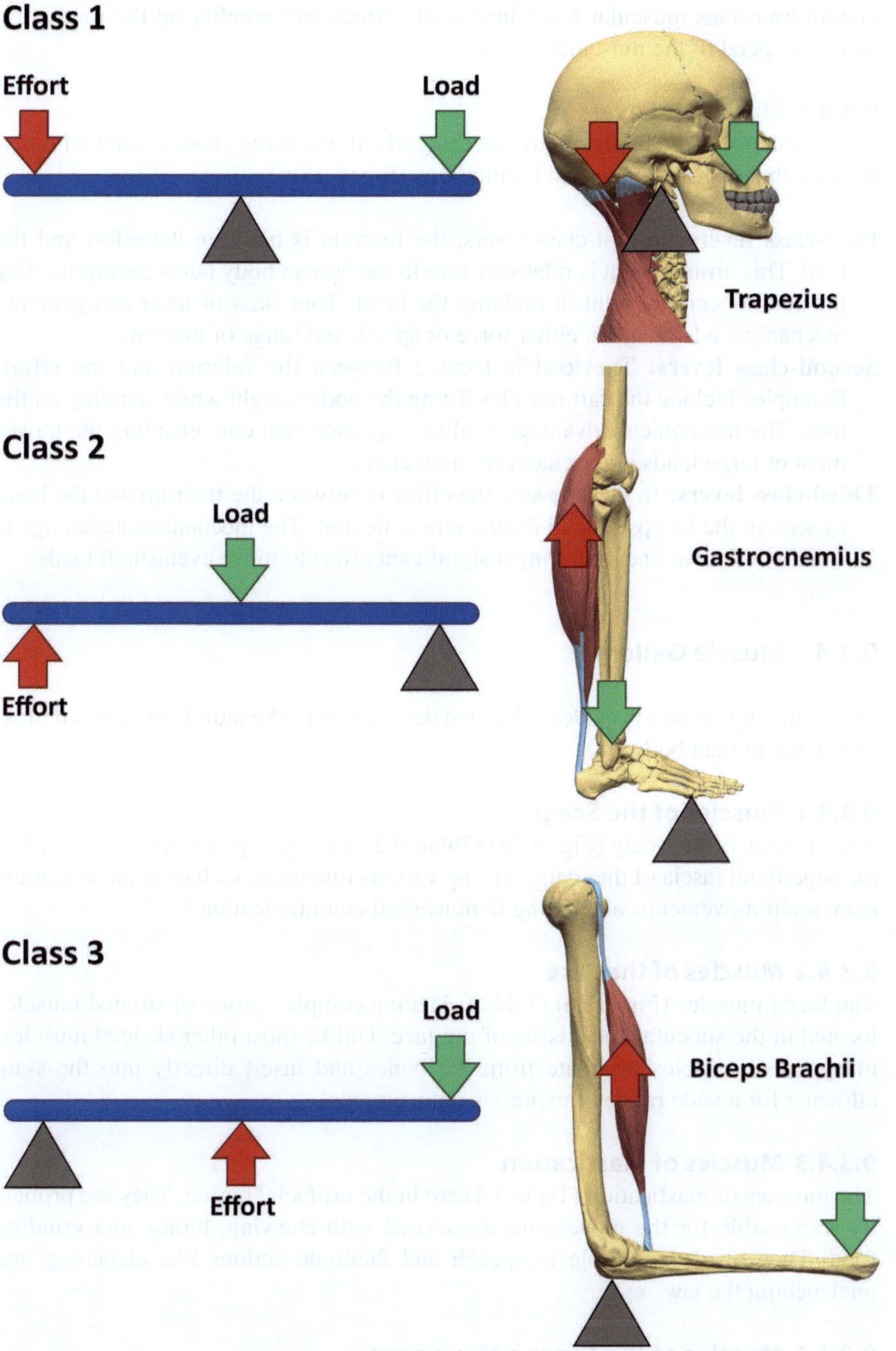

Fig. 9.25 Lever systems operating at a mechanical or displacement advantage

system translates muscular force into work, effectively amplifying the force, distance, or speed of the muscular action.

9.3.3.2 Classes of Levers

Lever systems in the human body are categorized into three classes, each with distinct mechanical advantages and limitations (Fig. 9.25):

First-class levers: In first-class levers, the fulcrum is between the effort and the load. This arrangement is relatively rare in the human body but is exemplified by the atlantooccipital joint in nodding the head. This class of lever can generate mechanical advantage in either force or speed, and range of motion.

Second-class levers: The load is located between the fulcrum and the effort. Examples include the calf muscles lifting the body weight while standing on the toes. The mechanical advantage is always greater than one, enabling the movement of large loads with relatively small effort.

Third-class levers: In these levers, the effort is between the fulcrum and the load, as seen in the biceps brachii during elbow flexion. The mechanical advantage is usually less than one, requiring a significant effort to move even small loads.

9.3.4 Muscle Gallery

The following section provides a detailed description of the individual skeletal muscles in the human body.

9.3.4.1 Muscles of the Scalp

The muscles of the scalp (Fig. 9.26) (Table 9.2) are a group of muscles situated in the superficial fascia of the scalp, serving various functions, including facial expression, scalp movements, and aiding in nonverbal communication.

9.3.4.2 Muscles of the Face

The facial muscles (Fig. 9.26) (Table 9.3) are a complex group of striated muscles located in the subcutaneous tissue of the face. Unlike most other skeletal muscles, many facial muscles originate from the bones and insert directly into the skin, allowing for a wide range of expressive movements.

9.3.4.3 Muscles of Mastication

The muscles of mastication (Table 9.4) are in the orofacial region. They are primarily responsible for the movements associated with chewing, biting, and grinding food. They also play a role in speech and facilitate actions like clenching and unclenching the jaw.

9.3.4.4 Muscles of the Tongue Movement

The muscles responsible for tongue movement (Table 9.5) are an intricate set of muscular tissues that play pivotal roles in mastication and deglutition (swallowing), articulation during speech, and other oral functions. These muscles are broadly

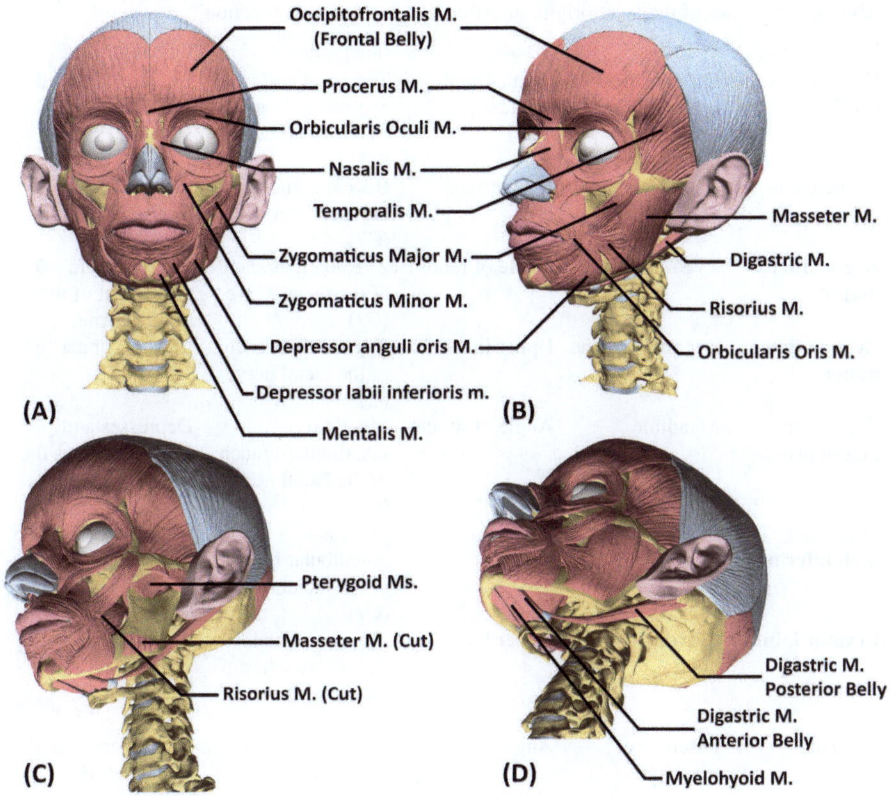

Fig. 9.26 Muscles of the scalp and face. (**a**) Anterior view. (**b**) Left oblique view. (**c**) Masseter and risorius muscles were removed to see the pterygoid muscles. (**d**) Muscles of the mouth floor

Table 9.2 Muscles of scalp: origin, insertion, innervation, and function

Muscle name	Origin	Insertion	Innervation	Function
Occipitofrontalis	Occipital belly: Occipital bone	Frontal belly: Skin of eyebrows and root of nose	Occipital belly: Posterior auricular nerve (C2, C3) Frontal belly: Temporal branches of the facial nerve (C7)	Elevates eyebrows, wrinkles forehead, and assists in retracting the scalp
Temporoparietalis	Fascia overlying temporal region	Galea aponeurotica	Auriculotemporal nerve (branch of V3)	Assists in moving the auricle of the ear and the scalp
Auricularis anterior	Fascia overlying the temporal region	Helix of ear	Temporal branches of the facial nerve (C7)	Helps in moving the auricle of the ear forward
Auricularis superior	Galea aponeurotica	Helix of ear	Temporal branches of the facial nerve (C7)	Helps in elevating the auricle of the ear
Auricularis posterior	Mastoid part of the temporal bone	Concha of auricle	Posterior auricular nerve (branch of facial nerve, C7)	Helps in retracting the auricle of the ear

Table 9.3 Muscles of the face: origin, insertion, innervation, and function

Muscle name	Origin	Insertion	Innervation	Function
Orbicularis oculi	Orbital rim, frontal bone, maxilla	Skin around eyelids	Temporal and zygomatic branches of the facial nerve (C7)	Closure of eyelids; squinting
Orbicularis oris	Maxilla, mandible	Skin around lips	Buccal branch of the facial nerve (C7)	Closure and protrusion of lips
Zygomaticus major	Zygomatic bone	Angle of mouth	Zygomatic branch of the facial nerve (C7)	Elevation and lateral movement of the mouth corner
Zygomaticus minor	Zygomatic bone	Upper lip	Zygomatic branch of the facial nerve (C7)	Elevates upper lip
Depressor anguli oris	Mandible	Angle of mouth	Marginal mandibular branch of the facial nerve (C7)	Depresses and laterally moves the mouth corner
Depressor labii inferioris	Mandible	Lower lip	Marginal mandibular branch of the facial nerve (C7)	Depresses lower lip
Levator labii superioris	Maxilla, frontal bone	Upper lip	Zygomatic and buccal branches of the facial nerve (C7)	Elevates upper lip
Risorius	Fascia over masseter	Angle of mouth	Buccal branch of the facial nerve (C7)	Retraction of angle of the mouth
Buccinator	Alveolar process of maxilla and mandible	Orbicularis oris	Buccal branch of the facial nerve (C7)	Assists in mastication by pushing food against molars
Mentalis	Mandible	Skin of chin	Marginal mandibular branch of the facial nerve (C7)	Elevates and protrudes lower lip; wrinkles chin
Corrugator supercilii	Frontal bone	Skin of eyebrow	Temporal branch of the facial nerve (C7)	Draws eyebrows medially and downward
Procerus	Nasal bone	Skin between eyebrows	Buccal branch of the facial nerve (C7)	Pulls down medial angle of eyebrows; wrinkles nose bridge
Nasalis	Maxilla	Aponeurosis over bridge of nose	Buccal branch of the facial nerve (C7)	Compresses nasal cartilages; flares nostrils

Table 9.4 Muscles of mastication: origin, insertion, innervation, and function

Muscle name	Origin	Insertion	Innervation	Function
Masseter	Zygomatic arch	Angle and ramus of mandible	Mandibular branch of trigeminal nerve (V3)	Elevation of mandible; chewing
Temporalis	Temporal fossa	Coronoid process of mandible	Deep temporal nerves (V3)	Elevation and retraction of mandible
Medial pterygoid	Maxilla, palatine bone, pterygoid fossa	Medial surface of mandible's angle	Medial pterygoid nerve (V3)	Elevation and protrusion of mandible; helps in side-to-side movements
Lateral pterygoid	Greater wing of sphenoid, lateral pterygoid plate	Condylar process of mandible	Lateral pterygoid nerve (V3)	Protrusion and side-to-side movement of mandible
Digastric (anterior belly)	Digastric fossa of mandible	Body of hyoid bone via a fibrous loop	Mylohyoid nerve (V3)	Depression of mandible; elevation of hyoid bone during swallowing
Digastric (posterior belly)	Mastoid notch of temporal bone	Body of hyoid bone via a fibrous loop	Facial nerve (C7)	Depression of mandible; elevation of hyoid bone during swallowing
Mylohyoid	Mylohyoid line of mandible	Midline raphe and body of hyoid bone	Mylohyoid nerve (V3)	Elevation of floor of mouth and tongue; assists in swallowing

Table 9.5 Muscles of tongue movement: origin, insertion, innervation, and function

Muscle name	Origin	Insertion	Innervation	Function
Genioglossus	Superior mental spine of mandible	Inferior aspect and body of tongue, hyoid bone	Hypoglossal nerve (CN XII)	Protrusion, depression, and retraction of the tongue
Hyoglossus	Body of hyoid bone	Inferior aspect of tongue	Hypoglossal nerve (CN XII)	Retraction and depression of the tongue
Styloglossus	Styloid process of the temporal bone	Inferior lateral aspect of the tongue	Hypoglossal nerve (CN XII)	Retraction and elevation of the tongue
Palatoglossus	Palatine aponeurosis	Lateral aspect of the tongue	Pharyngeal plexus, Vagus nerve (CN X)	Elevation of the posterior tongue, depression of soft palate
Intrinsic muscles (superior longitudinal, inferior longitudinal, transverse, vertical)	Within the substance of the tongue	Within the substance of the tongue	Hypoglossal nerve (CN XII)	Alter the shape of the tongue for speech and swallowing

categorized into intrinsic and extrinsic muscle groups, each with specific responsibilities in controlling tongue shape, position, and motion.

9.3.4.5 Muscles of the Neck

The neck muscles represent a complex anatomical assembly that provides structural support, movement, and protection for the cervical spine, trachea, esophagus, and various blood vessels and nerves. These muscles are categorized into anterior (Fig. 9.27) (Table 9.6) and posterior groups, with functions ranging from head and neck movement to swallowing and speech. The muscles in the posterior group are part of the spinal muscles (Section 9.3.4.8 & Fig. 9.32).

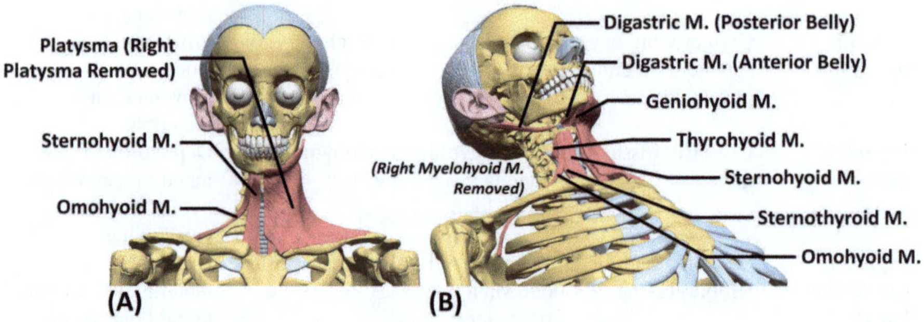

Fig. 9.27 Muscles of the anterolateral neck and throat

Table 9.6 Anterior neck muscles: origin, insertion, innervation, and function

Muscle name	Origin	Insertion	Innervation	Function
Sternohyoid	Manubrium of sternum, clavicle	Body of hyoid bone	Ansa cervicalis (C1– C3)	Depresses hyoid bone
Sternothyroid	Manubrium of sternum	Oblique line of thyroid cartilage	Ansa cervicalis (C2– C3)	Depresses thyroid cartilage
Thyrohyoid	Oblique line of thyroid cartilage	Body of hyoid bone	C1 via hypoglossal nerve	Elevates thyroid cartilage; depresses hyoid bone
Omohyoid	Superior border of scapula	Body of hyoid bone	Ansa cervicalis (C1– C3)	Depresses and retracts hyoid bone
Digastric (anterior belly)	Digastric fossa of mandible	Body of hyoid bone via a fibrous loop	Mylohyoid nerve (V3)	Depresses mandible; elevates hyoid bone
Digastric (posterior belly)	Mastoid notch of temporal bone	Body of hyoid bone via a fibrous loop	Facial nerve (C7)	Depresses mandible; elevates hyoid bone
Mylohyoid	Mylohyoid line of mandible	Midline raphe and body of hyoid bone	Mylohyoid nerve (V3)	Elevation of floor of mouth and tongue; assists in swallowing
Geniohyoid	Inferior mental spine of mandible	Body of hyoid bone	C1 via hypoglossal nerve	Depresses mandible; elevates and draws hyoid bone anteriorly
Platysma	Fascia covering pectoralis major and deltoid	Mandible, skin below mouth	Cervical branch of the facial nerve (C7)	Depresses mandible; tenses skin of neck and lower face

9.3.4.6 Muscles of the Thorax

The thorax muscles facilitate vital functions such as respiration, postural support, and upper-body movement. These muscles are broadly categorized into those involved in respiration (respiratory muscles) (Fig. 9.28) (Table 9.7) and those contributing to the structure and stability of the thoracic wall (Fig. 9.29) (Table 9.8).

9.3.4.7 Muscles of the Abdominal Wall

The abdominal wall muscles provide mechanical support to the abdominal organs, participate in the mechanics of respiration, facilitate postural adjustments, and are involved in various movements such as flexion and rotation of the trunk. They are categorized primarily into anterolateral and posterior muscles (Fig. 9.31) (Table 9.9).

9.3.4.8 Muscles of the Spinal Column

The muscles associated with the spinal column (Fig. 9.32) (Table 9.10) are complex structures that maintain posture, provide stability, and facilitate vertebral column movements.

9.3.4.9 Muscles of the Pelvic Diaphragm

The pelvic diaphragm (Fig. 9.33) (Table 9.11) is a muscular and fascial structure that spans the pelvic cavity, serving as its inferior boundary. Composed of two

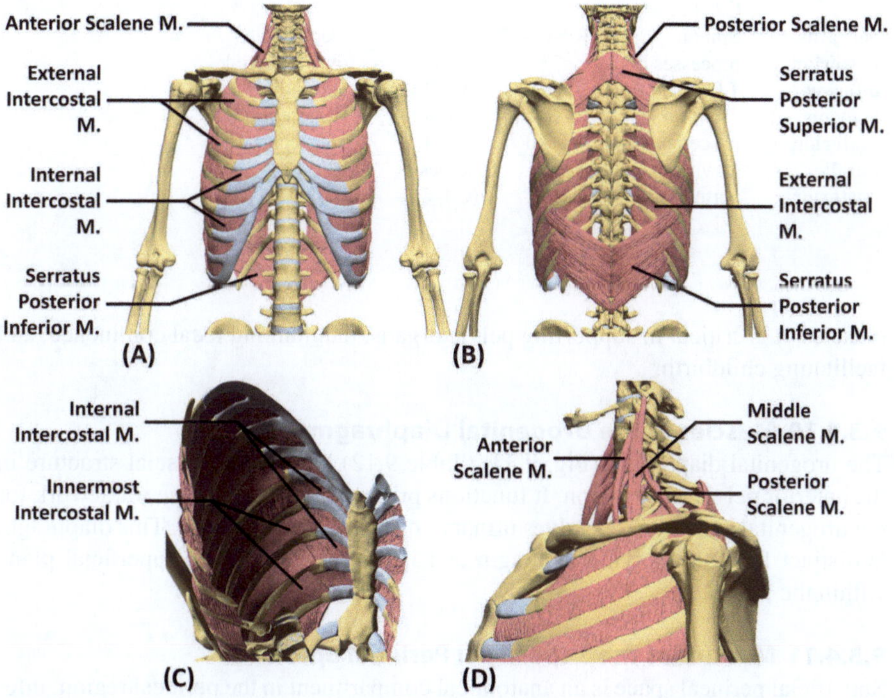

Fig. 9.28 Muscles of respiration, excluding the diaphragm (Fig. 9.30). (**a**) Anterior view. (**b**) Posterior view. (**c**) Internal view of the chest wall. (**d**) Lateral view of deep neck area

Table 9.7 Muscles of respiration: origin, insertion, innervation, and function

Muscle name	Origin	Insertion	Innervation	Function	Important features
Diaphragm (Fig 9.30)	Xiphoid process, lower ribs, lumbar vertebrae	Central tendon	Phrenic nerve (C3–C5)	Principal muscle of inspiration; separates thoracic and abdominal cavities	Dome-shaped; has three openings for the passage of aorta, esophagus, and vena cava
External intercostal	Inferior border of the rib above	Superior border of the rib below	Intercostal nerves (T1–T11)	Elevate ribs during inspiration	Run inferoanteriorly from rib to rib
Internal intercostal	Superior border of the rib below	Inferior border of the rib above	Intercostal nerves (T1–T11)	Depress ribs during forced expiration	Run inferoposteriorly from rib to rib
Innermost intercostal	Inner aspect of ribs	Inner aspect of adjacent ribs	Intercostal nerves (T1–T11)	Similar to internal intercostals in function	Deepest layer of intercostal muscles
Serratus posterior superior	Ligamentum nuchae, spinous processes of C7–T3	Ribs 2-5	Intercostal nerves (T2–T5)	Elevate ribs during forced inspiration	
Serratus posterior inferior	Spinous processes of T11–L2	Ribs 9-12	Intercostal nerves (T9–T12)	Depress ribs during forced expiration	
Scalene (anterior, middle, posterior)	Transverse processes of cervical vertebrae	First and second ribs	Cervical spinal nerves (C3–C8)	Elevate first and second ribs during forced inspiration	

muscles, it is critical in supporting pelvic organs, maintaining fecal continence, and facilitating childbirth.

9.3.4.10 Muscles of the Urogenital Diaphragm

The urogenital diaphragm (Fig. 9.33) (Table 9.12) is a musculofascial structure in the anterior pelvic floor region. It functions primarily as a supportive framework for the urogenital organs and involves urinary continence mechanisms. This diaphragm is distinct from the pelvic diaphragm and is oriented in a more superficial plane within the pelvis.

9.3.4.11 Muscles of the Superficial Perineal Space

Superficial perineal space is an anatomical compartment in the perineal region, inferior to the pelvic diaphragm and superior to the perineal membrane. This space

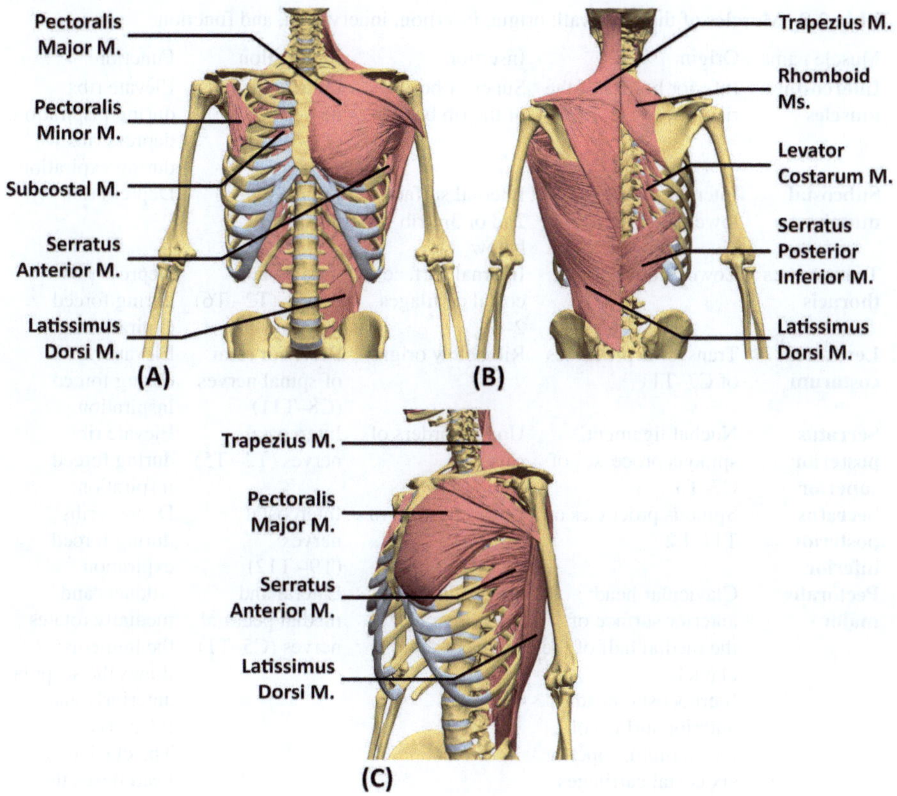

Fig. 9.29 Muscles of the thoracic wall. (**a**) Anterior view. (**b**) Posterior view. (**c**) Oblique lateral view

contains a collection of muscles (Fig. 9.33) (Table 9.13) that serve primarily sexual and supportive functions.

9.3.4.12 Muscles of the Shoulder Joint

The shoulder represents an anatomical and functional unit that facilitates a broad range of motions and provides stability to the upper limb. The shoulder is essential for lifting, pushing, and pulling. Several muscles contribute to the stability and movement of the shoulder joint (Fig. 9.34) (Table 9.14). These muscles can be broadly categorized into two main groups: the extrinsic and intrinsic muscles of the shoulder.

Table 9.8 Muscles of thoracic wall: origin, insertion, innervation, and function

Muscle name	Origin	Insertion	Innervation	Function
Intercostal muscles	Inferior border of the rib above	Superior border of the rib below	Intercostal nerves (T1– T11)	Elevate ribs during inspiration, depress ribs during expiration
Subcostal muscles	Internal surface of lower ribs	Internal surface of 2nd or 3rd rib below	Intercostal nerves (T12)	Depress ribs
Transversus thoracis	Lower sternum	Internal surface of costal cartilages 2–6	Intercostal nerves (T2– T6)	Depress ribs during forced expiration
Levatores costarum	Transverse processes of C7–T11	Rib below origin	Posterior rami of spinal nerves (C8–T11)	Elevate ribs during forced inspiration
Serratus posterior superior	Nuchal ligament, spinous processes of C7–T3	Upper borders of ribs 2–5	Intercostal nerves (T2– T5)	Elevate ribs during forced inspiration
Serratus posterior inferior	Spinous processes of T11–L2	Lower borders of ribs 9–12	Intercostal nerves (T9– T12)	Depress ribs during forced expiration
Pectoralis major	Clavicular head: anterior surface of the medial half of the clavicle Sternocostal head: anterior surface of the sternum, superior six costal cartilages, aponeurosis of external oblique muscle	Lateral lip of the intertubercular sulcus of the humerus	Lateral and medial pectoral nerves (C5–T1)	Adducts and medially rotates the humerus; draws the scapula anteriorly and inferiorly The clavicular head flexes the humerus, and the sternocostal head extends it from the flexed position
Pectoralis minor	3rd to 5th ribs near their costal cartilages	Medial border and superior surface of the coracoid process of the scapula	Medial pectoral nerve (C8, T1)	Stabilizes the scapula by drawing it inferiorly and anteriorly against the thoracic wall
Serratus anterior	External surfaces of lateral parts of 1st to 8th ribs	Anterior surface of the medial border of the scapula	Long thoracic nerve (C5, C6, C7)	Protracts and holds the scapula against the thoracic wall; rotates the scapula

(continued)

Table 9.8 (continued)

Muscle name	Origin	Insertion	Innervation	Function
Latissimus dorsi	Spinous processes of T7-L5 vertebrae, thoracolumbar fascia, iliac crest, inferior 3 or 4 ribs, inferior angle of scapula	Floor of the intertubercular sulcus of the humerus	Thoracodorsal nerve (C6, C7, C8)	Extends, adducts, and medially rotates the humerus; raises the body toward arms during climbing
Trapezius	Medial third of the superior nuchal line, external occipital protuberance, nuchal ligament, and spinous processes of C7-T12 vertebrae	Lateral third of the clavicle, acromion, and spine of the scapula	Accessory nerve (CN XI) and C3, C4 spinal nerves	Elevates, retracts, and rotates the scapula; superior fibers elevate, middle fibers retract, and inferior fibers depress the scapula
Rhomboid major	Spinous processes of T2-T5 vertebrae	Medial border of the scapula from the level of the spine to the inferior angle	Dorsal scapular nerve (C4, C5)	Retracts the scapula and rotates it to depress the glenoid cavity; fixes the scapula to the thoracic wall
Rhomboid minor	Nuchal ligaments and spinous processes of C7-T1 vertebrae	Smooth triangular area at the medial end of the scapular spine	Dorsal scapular nerve (C4, C5)	Retracts the scapula and rotates it to depress the glenoid cavity; fixes the scapula to the thoracic wall

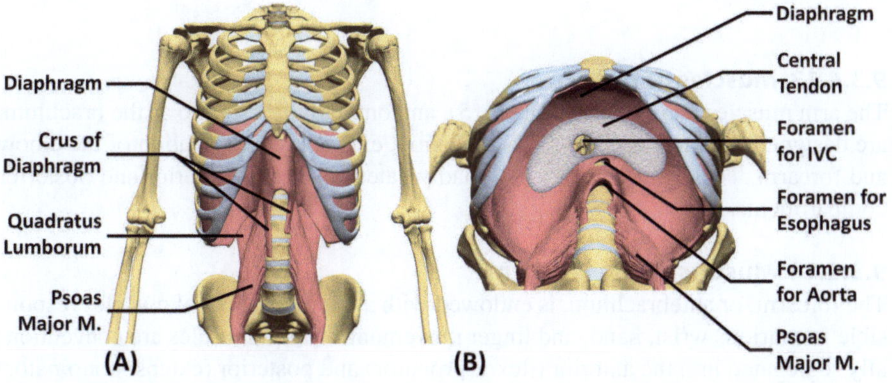

Fig. 9.30 Diaphragm. (**a**) Anterior view. (**b**) Inferior view

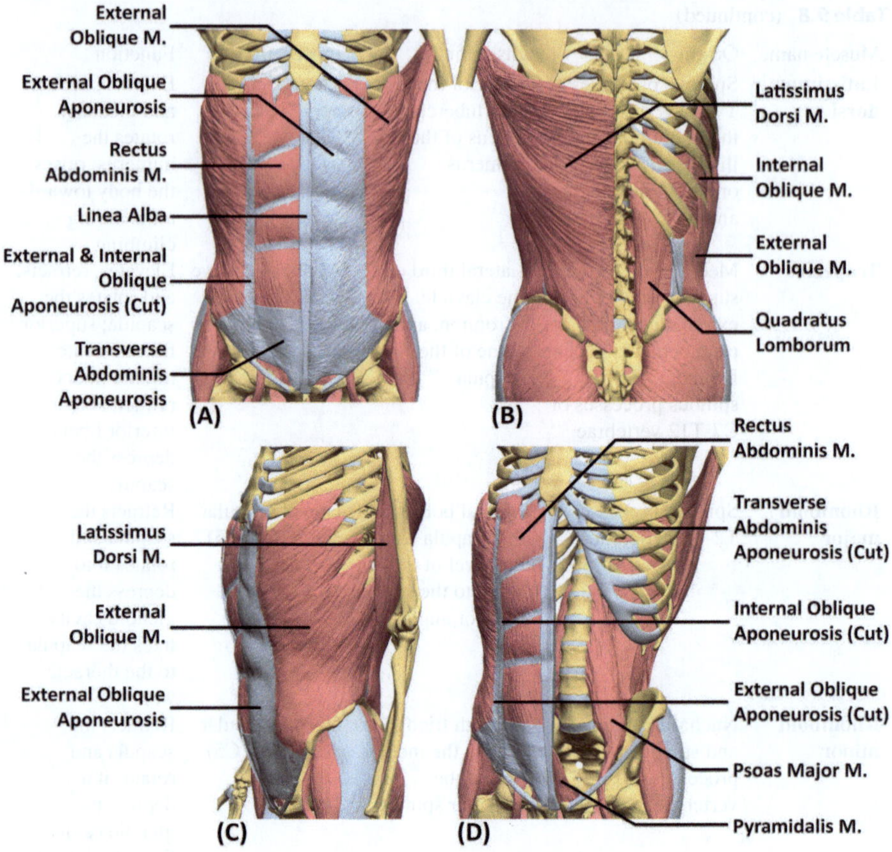

Fig. 9.31 Muscles of the abdominal wall. (**a**) Anterior view. (Aponeuroses of the right external and internal oblique muscles are removed.) (**b**) Posterior view. (The right latissimus dorsi muscle is removed.) (**c**) Left lateral oblique view. (**d**) Left anterior oblique view. (Left abdominal wall muscles are removed.)

9.3.4.13 Muscles of the Arm

The arm muscles (Fig. 9.35) (Table 9.15), anatomically referred to as the brachium, are designed for various tasks, such as flexion, extension, and rotation of the elbow and forearm. These muscles can be broadly categorized into anterior and posterior compartments.

9.3.4.14 Muscles of the Forearm

The forearm, or antebrachium, is endowed with a complex array of muscles responsible for various wrist, hand, and finger movements. These muscles are conventionally segregated into the anterior (flexor/pronator) and posterior (extensor/supinator) compartments (Fig. 9.36) (Table 9.16).

Table 9.9 Muscles of abdomen: origin, insertion, innervation, and function

Muscle name	Origin	Insertion	Innervation	Function
Anterolateral muscles				
Rectus abdominis	Pubic symphysis and pubic crest	Xiphoid process and costal cartilages of ribs 5–7	T7–T12 intercostal nerves, L1	Flexion of trunk, assists in forced expiration
External oblique	Ribs 5–12	Iliac crest, inguinal ligament, linea alba	T7–T12 intercostal nerves, iliohypogastric, ilioinguinal	Trunk flexion, lateral flexion, rotation
Internal oblique	Thoracolumbar fascia, iliac crest, inguinal ligament	Inferior borders of 10th–12th ribs, linea alba	T7–T12 intercostal nerves, iliohypogastric, ilioinguinal	Trunk flexion, lateral flexion, rotation
Transversus abdominis	Inguinal ligament, iliac crest, thoracolumbar fascia, cartilages of lower ribs	Linea alba, pubic crest	T7–T12 intercostal nerves, iliohypogastric, ilioinguinal	Compresses abdominal contents, assists in forced expiration
Pyramidalis	Pubic bone	Linea alba	Subcostal nerve (T12), occasionally L1	Tenses linea alba
Posterior muscles				
Psoas major	Lumbar vertebrae (bodies, discs, and transverse processes of L1–L5)	Lesser trochanter of femur	Lumbar plexus (L1–L3)	Hip flexion, lateral flexion of vertebral column
Psoas minor	Bodies of T12 and L1	Pectineal line and iliopubic eminence	Lumbar plexus (L1)	Weak trunk flexor
Quadratus lumborum	Iliac crest, iliolumbar ligament	12th rib, transverse processes of L1–L4	Lumbar plexus (T12, L1–L4)	Lateral flexion of vertebral column, depresses 12th rib

9.3.4.15 Muscles of the Hand

The hand is an intricate anatomical structure equipped with diverse muscles that enable fine motor skills, including grasping, pinching, and manipulating objects. These muscles are typically categorized into two groups: the extrinsic and intrinsic muscles. The extrinsic muscles in the forearm control gross movements of the hand and fingers, whereas the intrinsic muscles (Fig. 9.37) (Table 9.17), situated within the hand itself, control more nuanced movements.

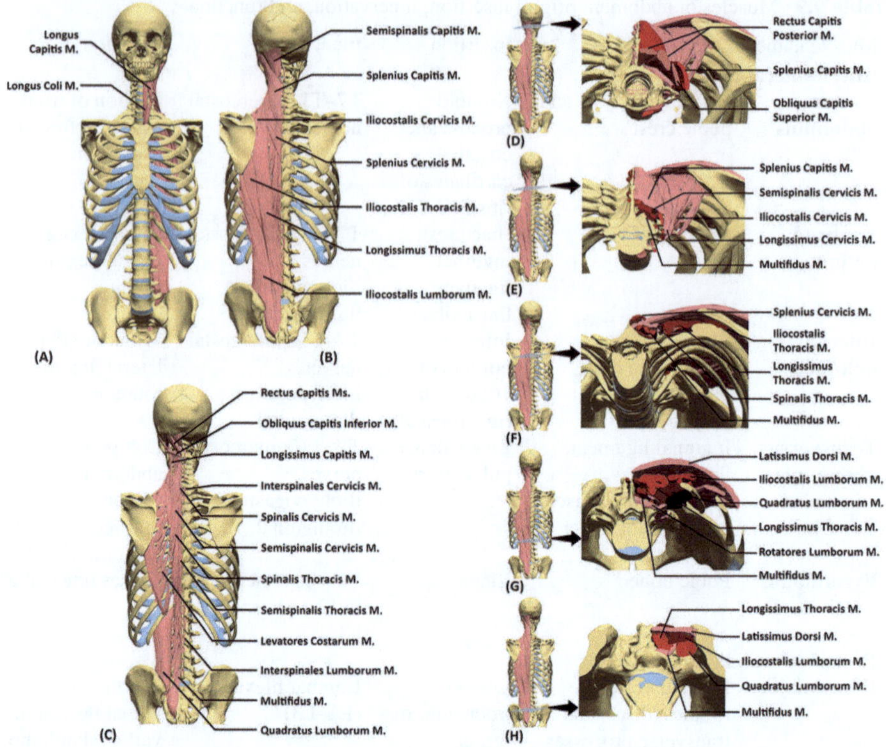

Fig. 9.32 Muscles of the spinal column. (**a**) Anterior view. (**b**) Posterior view: superficial layer. (**c**) Posterior view: deep layer. (**d**) Axial cross-section at the level of C1. (**e**) Axial cross-section at the level of C6. (**f**) Axial cross-section at the level of T5. (**g**) Axial cross-section at the level of L1. (**h**) Axial cross-section at the level of L5

9.3.4.16 Muscles of the Hip

The hip muscles (Fig. 9.38) (Table 9.18) are responsible for many functions, most notably facilitating locomotion, maintaining postural equilibrium, and enabling a range of movements such as flexion, extension, abduction, adduction, and rotation.

9.3.4.17 Muscles of the Thigh

The thigh muscles (Fig. 9.39) (Table 9.19) are robust structures central to lower limb movement and stability. They are primarily responsible for walking, running, and maintaining an upright posture. These muscles are categorized into three main groups based on their anatomical location and functions: the anterior (quadriceps), posterior (hamstrings), and medial (adductors) thigh muscles.

Table 9.10 Muscles of the spinal column: origin, insertion, innervation, and function

Muscle name	Origin	Insertion	Innervation	Function
Erector spinae (iliocostalis, longissimus, spinalis)	Iliac crest, sacrum, lumbar and lower thoracic vertebrae	Ribs and various vertebrae	Dorsal rami of spinal nerves	Extension and lateral flexion of the vertebral column
Multifidus	Sacrum, ilium, and thoracic and lumbar vertebrae	Spinous processes of vertebrae above origin	Dorsal rami of spinal nerves	Stabilizes vertebrae during movement
Semispinalis	Transverse processes of vertebrae	Spinous processes of vertebrae above origin	Dorsal rami of spinal nerves	Extension and rotation of the vertebral column
Rotatores	Transverse processes of vertebrae	Spinous processes of vertebrae above origin	Dorsal rami of spinal nerves	Stabilization and rotation of the vertebral column
Interspinales	Spinous processes of vertebrae	Spinous processes of adjacent vertebrae	Dorsal rami of spinal nerves	Extension and stabilization of the vertebral column
Intertransversarii	Transverse processes of vertebrae	Transverse processes of adjacent vertebrae	Dorsal rami of spinal nerves	Lateral flexion and stabilization of the vertebral column
Levatores costarum	Transverse processes of C7 to T11 vertebrae	Rib below origin	Dorsal rami of spinal nerves	Elevate ribs, assist in respiration

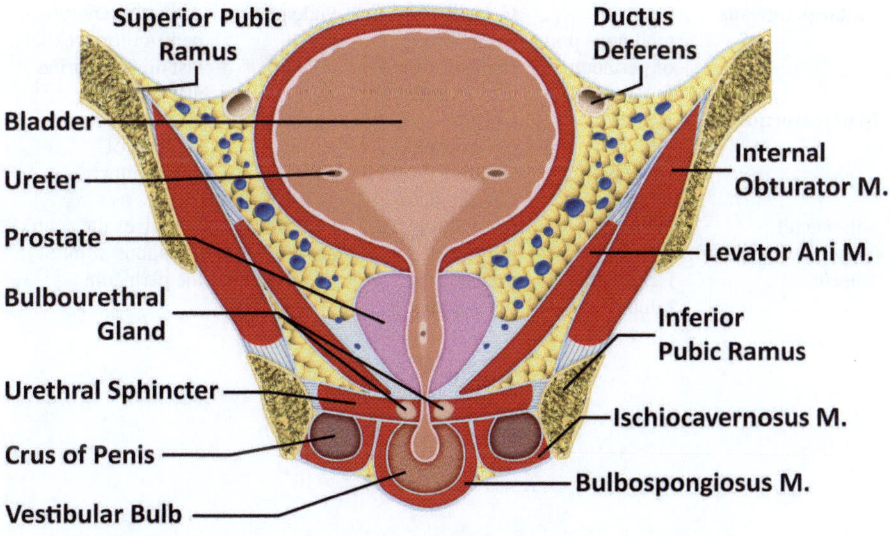

Fig. 9.33 Pelvic floor muscles

Table 9.11 Muscles of pelvic diaphragm: origin, insertion, innervation, and function

Muscle name	Origin	Insertion	Innervation	Function
Levator ani (iliococcygeus, pubococcygeus, puborectalis)	Ischial spine, posterior part of the obturator internus fascia	Coccyx, anococcygeal ligament, walls of pelvis	Pudendal nerve (S3–S4), sacral plexus	Supports pelvic viscera, aids in urinary and fecal continence, assists in childbirth
Coccygeus (ischiococcygeus)	Ischial spine	Lower sacrum and upper coccyx	S4, S5, and coccygeal nerves	Supports pelvic viscera, pulls coccyx anteriorly following defecation and childbirth

Table 9.12 Muscles of urogenital diaphragm: origin, insertion, innervation, and function

Muscle name	Origin	Insertion	Innervation	Function
Deep transverse perineal muscle	Ischiopubic ramus	Midline raphe or perineal body	Pudendal nerve (S2–S4)	Supports pelvic viscera, stabilizes perineal body
Sphincter urethrae	Ischiopubic ramus	Merges with the other side around urethra	Pudendal nerve (S2–S4)	Constricts urethra, maintains urinary continence
Compressor urethrae	Ischiopubic ramus	Female urethra	Pudendal nerve (S2–S4)	Compresses and stabilizes female urethra

Table 9.13 Muscles of superficial perineal space: origin, insertion, innervation, and function

Muscle name	Origin	Insertion	Innervation	Function
Bulbospongiosus	Central tendinous point of perineum	Corpus spongiosum (male)/Clitoral crus (female)	Pudendal nerve (S2-S4)	Assists in erection of penis/clitoris, expels last drops of urine and semen
Ischiocavernosus	Ischial tuberosity and ischiopubic ramus	Corpus cavernosum (male)/Clitoral crus (female)	Pudendal nerve (S2-S4)	Aids in sustaining erection of penis/clitoris
Superficial transverse perineal muscle	Ischial tuberosity or ischiopubic ramus	Central tendinous point of perineum or perineal body	Pudendal nerve (S2-S4)	Stabilizes the central tendinous point of the perineum

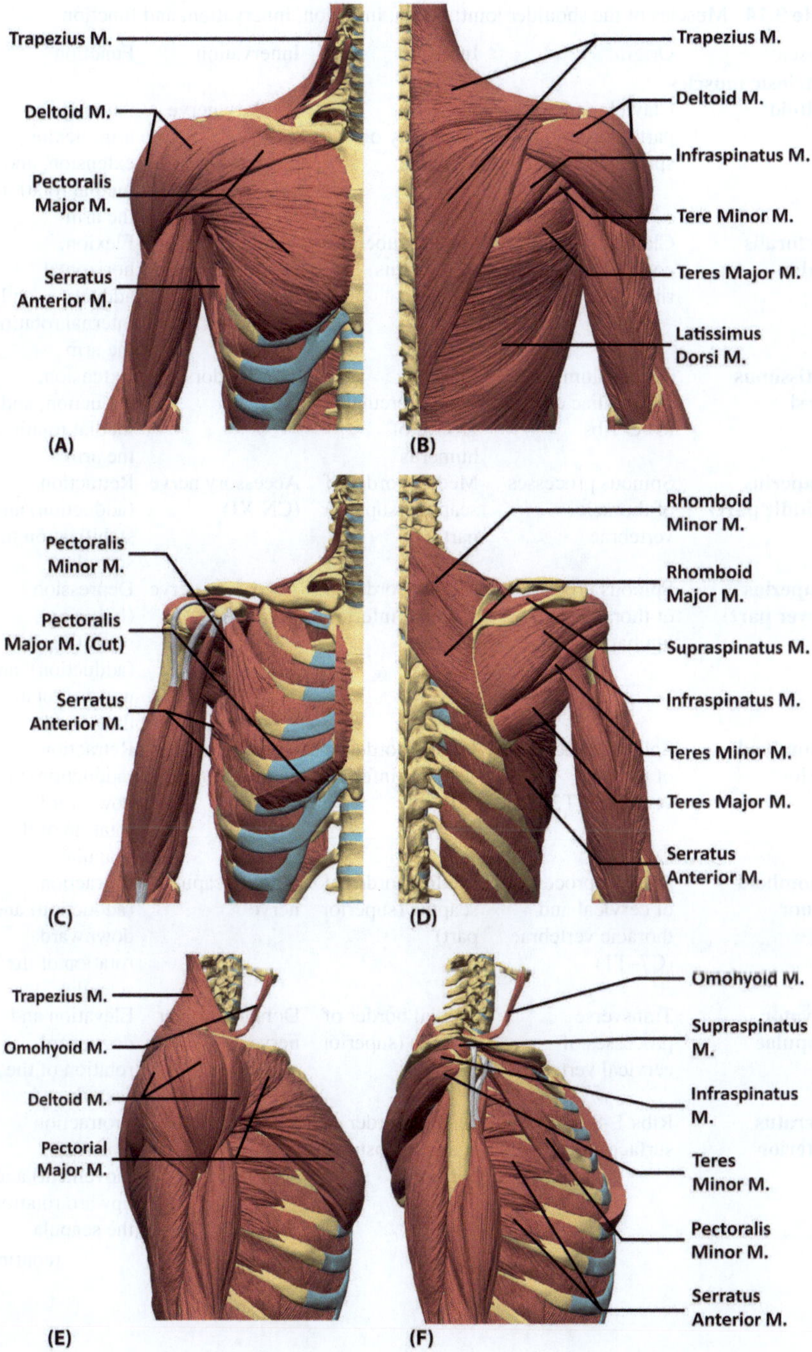

Fig. 9.34 Muscles of the shoulder. (**a**) Anterior view. (**b**) Posterior view. (**c**) Anterior view. (Deltoid and pectoralis major muscles are removed.) (**d**) Posterior view. (Deltoid and pectoralis major muscles are removed.) (**e**) Lateral view. (**f**) Lateral view. (Deltoid and pectoralis major muscles are removed)

Table 9.14 Muscles of the shoulder joint: origin, insertion, innervation, and function

Muscle	Origin	Insertion	Innervation	Function
Extrinsic muscles				
Deltoid	Clavicle (anterior part), acromion, spine of scapula	Deltoid tuberosity of humerus	Axillary nerve	Abduction of the arm, flexion, extension, and medial rotation of the arm
Pectoralis major	Clavicle, sternum, costal cartilages of ribs 1–6	Greater tubercle of humerus	Medial and lateral pectoral nerves	Flexion, horizontal adduction, and internal rotation of the arm
Latissimus dorsi	Thoracolumbar fascia, iliac crest, lower ribs	Floor of intertubercular groove of humerus	Thoracodorsal nerve	Extension, adduction, and medial rotation of the arm
Trapezius (middle part)	Spinous processes of thoracic vertebrae	Medial border of scapula (superior part)	Accessory nerve (CN XI)	Retraction (adduction) and stabilization of the scapula
Trapezius (lower part)	Spinous processes of thoracic and lumbar vertebrae	Medial border of scapula (inferior part)	Accessory nerve (CN XI)	Depression (lowering), retraction (adduction), and upward rotation of the scapula
Rhomboid major	Spinous processes of thoracic vertebrae (T2–T5)	Medial border of scapula (inferior part)	Dorsal scapular nerve	Retraction (adduction) and downward rotation of the scapula
Rhomboid minor	Spinous processes of cervical and thoracic vertebrae (C7–T1)	Medial border of scapula (superior part)	Dorsal scapular nerve	Retraction (adduction) and downward rotation of the scapula
Levator scapulae	Transverse processes of cervical vertebrae	Medial border of scapula (superior part)	Dorsal scapular nerve	Elevation and downward rotation of the scapula
Serratus anterior	Ribs 1–8 (anterior surface)	Medial border of scapula (costal surface)	Long thoracic nerve	Protraction (forward movement) and upward rotation of the scapula

(continued)

Table 9.14 (continued)

Muscle	Origin	Insertion	Innervation	Function
Intrinsic muscles				
Supraspinatus	Supraspinous fossa of scapula	Greater tubercle of humerus	Suprascapular nerve	Initiates shoulder abduction and stabilizes the shoulder joint
Infraspinatus	Infraspinous fossa of scapula	Greater tubercle of humerus	Suprascapular nerve	External rotation of the arm and shoulder joint stabilization
Teres minor	Lateral border of scapula (superior part)	Greater tubercle of humerus	Axillary nerve	External rotation and stabilization of the shoulder joint
Subscapularis	Subscapular fossa of scapula	Lesser tubercle of humerus	Upper and lower subscapular nerves	Internal rotation and stabilization of the shoulder joint
Teres major	Inferior angle of scapula	Medial lip of intertubercular sulcus of humerus	Lower subscapular nerve (C5, C6)	Adduction and internal rotation of the arm

9.3.4.18 Muscles of the Leg

The leg muscles (Fig. 9.40) (Table 9.20) provide support, stability, and mobility to the knee, ankle, and foot. They can be broadly categorized into three main groups based on their location and primary functions: the anterior leg muscles (dorsiflexors and invertors), the posterior leg muscles (plantar flexors), and the lateral leg muscles (evertors).

9.3.4.19 Muscles of the Foot

The foot muscles (Fig. 9.41) (Table 9.21) are a group of intricate structures contributing to the foot's fine motor control, stability, and flexibility.

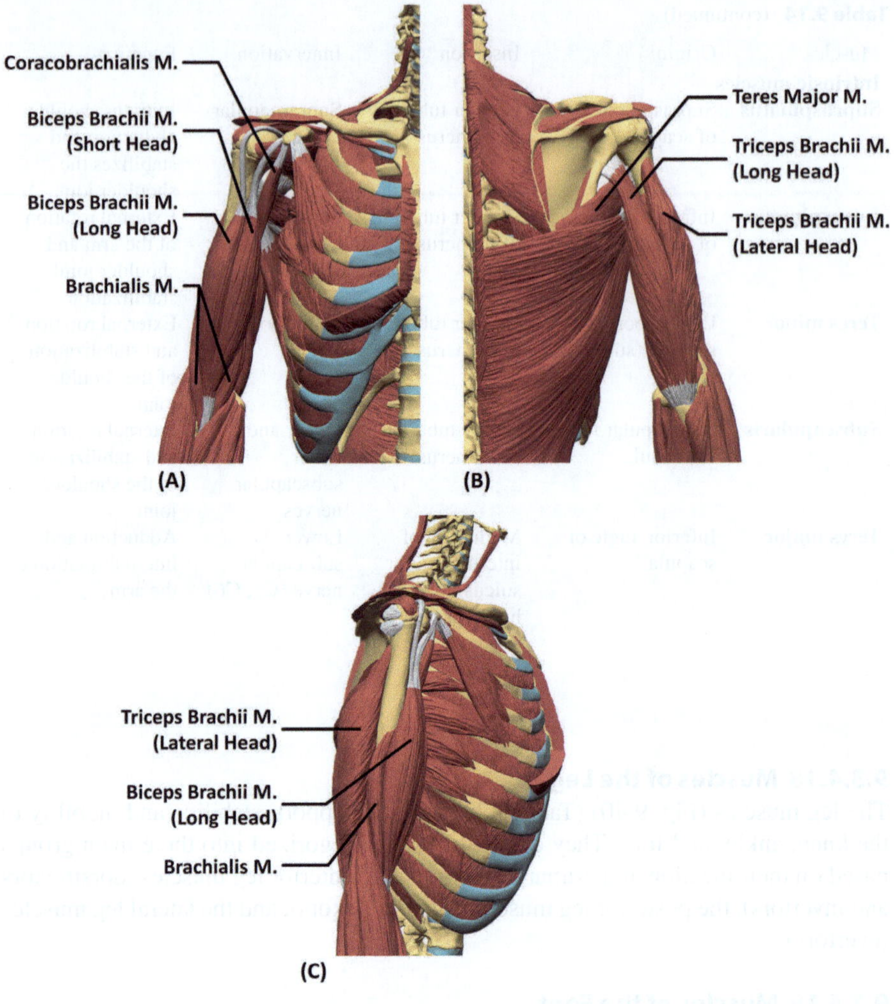

Fig. 9.35 Muscles of the shoulder. (**a**) Anterior view. (**b**) Posterior view. (**c**) Lateral view. (In all views, the deltoid, pectoralis major, infraspinatus, and teres major and minor muscles are removed)

Table 9.15 Muscles of arm: origin, insertion, innervation, and function

Muscle name	Origin	Insertion	Innervation	Function
Biceps brachii	Long head: supraglenoid tubercle of scapula; short head: coracoid process of scapula	Radial tuberosity and fascia of forearm	Musculocutaneous nerve (C5, C6)	Flexion and supination of the forearm
Brachialis	Distal half of anterior humerus	Coronoid process and tuberosity of ulna	Musculocutaneous nerve (C5, C6), radial nerve (C7)	Flexion of the forearm
Coracobrachialis	Coracoid process of scapula	Middle third of humeral shaft	Musculocutaneous nerve (C5, C6, C7)	Assists in flexion and adduction of the arm
Triceps brachii	Long head: infraglenoid tubercle of scapula; lateral head: upper half of posterior humerus; medial head: lower half of posterior humerus	Olecranon process of ulna	Radial nerve (C6, C7, C8)	Extension of the forearm
Anconeus	Lateral epicondyle of humerus	Lateral surface of olecranon and proximal ulna	Radial nerve (C7, C8, T1)	Assists triceps in forearm extension, stabilizes elbow joint

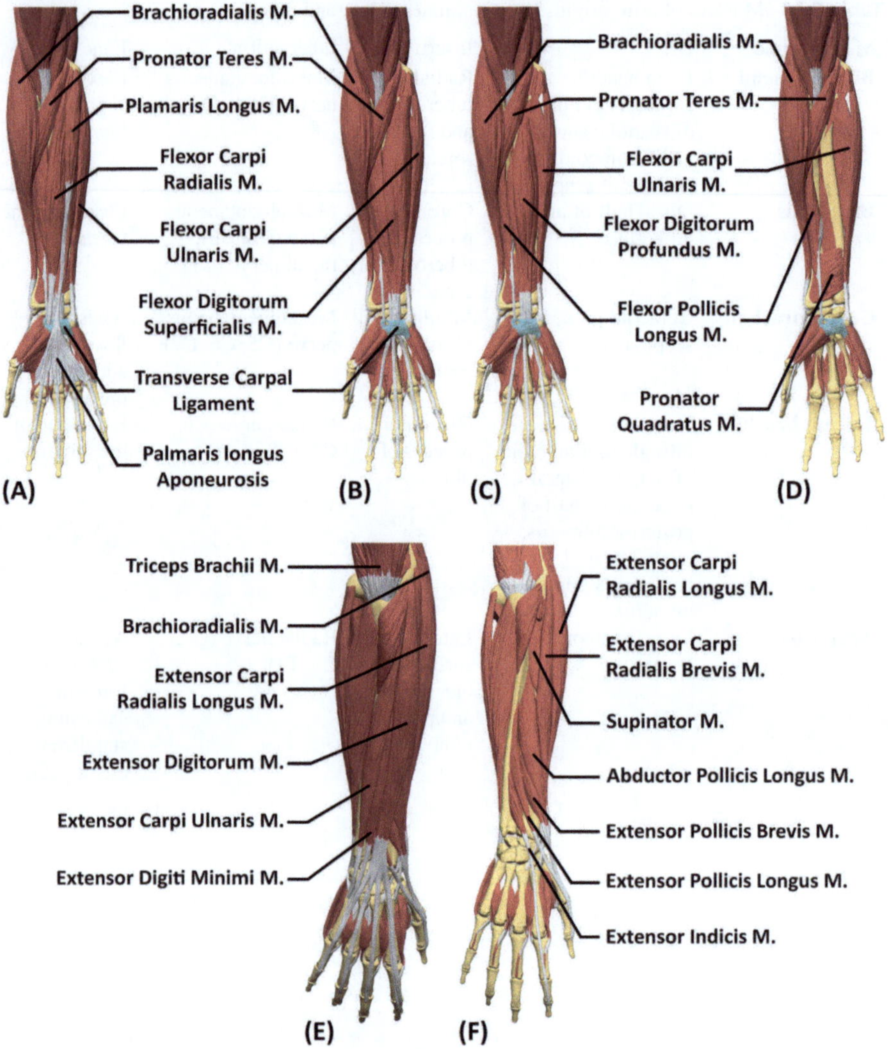

Fig. 9.36 Muscles of the forearm. (**a**) Anterior (flexor) view. (**b**) Anterior (flexor) view. (The palmaris longus and flexor carpi radialis muscles are removed.) (**c**) Anterior (flexor) view. (Flexor digitorum superficialis muscle is also removed.) (**d**) Anterior (flexor) view. (Flexor digitorum profundus muscle is also removed.) (**e**) Posterior (extensor) view. (**f**) Posterior (extensor) view. (Extensor digitorum, extensor carpi ulnaris, and extensor digiti minimi muscles are removed.)

Table 9.16 Muscles of the forearm: compartment, origin, insertion, innervation, and function

Muscle name	Origin	Insertion	Innervation	Function
Anterior compartment (superficial)				
Pronator teres	Humeral head: medial epicondyle of humerus; ulnar head: coronoid process of ulna	Middle of the lateral surface of the radius	Median nerve (C6, C7)	Pronates and flexes forearm
Flexor carpi radialis	Medial epicondyle of humerus	Bases of second and third metacarpals	Median nerve (C6, C7)	Flexes and abducts the wrist
Palmaris longus	Medial epicondyle of humerus	Flexor retinaculum and palmar aponeurosis	Median nerve (C7, C8)	Tenses palmar fascia, weak wrist flexor
Flexor carpi ulnaris	Humeral head: medial epicondyle of humerus; ulnar head: olecranon and posterior border of ulna	Pisiform, hamate, and base of fifth metacarpal	Ulnar nerve (C7, C8)	Flexes and adducts the wrist
Flexor digitorum superficialis	Medial epicondyle of the humerus, ulna, and radius	Middle phalanges of the four fingers (index to ring)	Median and ulnar nerves (C7, C8, T1)	Flexes the proximal interphalangeal joints of the fingers and assists in wrist flexion
Anterior compartment (deep)				
Flexor digitorum profundus	Upper 3/4 of the volar and medial surfaces of ulna and interosseous membrane	Distal phalanges of fingers 2–5	Median (C8, T1) (lateral half) and ulnar (C7, C8) (medial half) nerves	Flexes distal phalanges of fingers 2–5
Flexor pollicis longus	Anterior surface of the radius and interosseous membrane	Distal phalanx of the thumb	Anterior interosseous nerve (branch of median nerve) (C8, T1)	Flexes distal phalanx of the thumb
Pronator quadratus	Distal quarter of the anterior ulna	Distal quarter of the anterior radius	Anterior interosseous nerve (branch of median nerve) (C8, T1)	Pronates forearm
Posterior compartment				
Brachioradialis	Lateral supracondylar ridge of humerus	Lateral surface of distal radius	Radial nerve (C5, C6)	Flexes the elbow when the forearm is in mid-pronation

(continued)

Table 9.16 (continued)

Muscle name	Origin	Insertion	Innervation	Function
Extensor carpi radialis longus	Lateral supracondylar ridge of humerus	Base of the second metacarpal	Radial nerve (C6, C7)	Extends and abducts the wrist
Extensor carpi radialis brevis	Lateral epicondyle of humerus	Base of the third metacarpal	Deep branch of the radial nerve (C6, C7)	Extends and abducts the wrist
Extensor digitorum	Lateral epicondyle of humerus	Extensor expansions of the second to fifth fingers	Posterior interosseous nerve (C7, C8)	Extends the second to fifth fingers and the wrist
Extensor digiti minimi	Lateral epicondyle of humerus	Extensor expansion of the fifth finger	Posterior interosseous nerve (C7, C8)	Extends the fifth finger
Extensor carpi ulnaris	Lateral epicondyle of humerus and posterior border of the ulna	Base of the fifth metacarpal	Posterior interosseous nerve (C7, C8)	Extends and adducts the wrist
Abductor pollicis longus	Posterior surface of ulna and radius, interosseous membrane	Base of the first metacarpal	Posterior interosseous nerve (C7, C8)	Abducts and extends the thumb
Extensor pollicis brevis	Posterior surface of radius, interosseous membrane	Base of proximal phalanx of thumb	Posterior interosseous nerve (C7, C8)	Extends the proximal phalanx of thumb
Extensor pollicis longus	Posterior surface of ulna, interosseous membrane	Base of distal phalanx of thumb	Posterior interosseous nerve (C7, C8)	Extends the distal phalanx of thumb
Extensor indicis	Posterior surface of ulna, interosseous membrane	Extensor expansion of the index finger	Posterior interosseous nerve (C7, C8)	Extends the index finger
Supinator	Lateral epicondyle of humerus, proximal ulna	Lateral surface of proximal radius	Deep branch of the radial nerve (C5, C6)	Supinates forearm

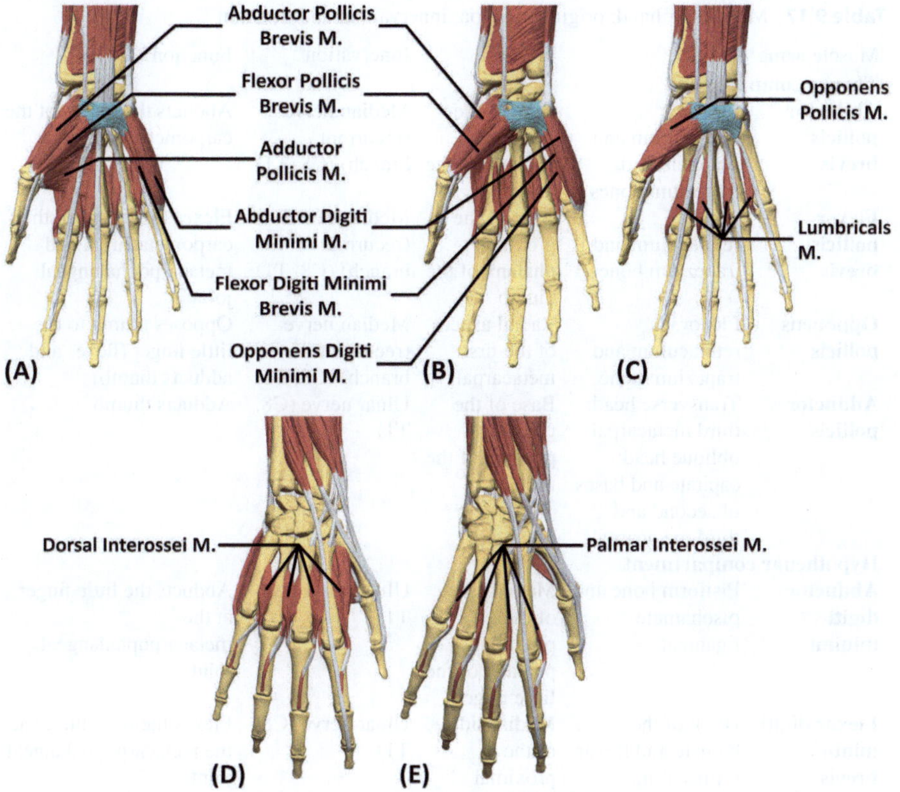

Fig. 9.37 Muscles of the hand. (**a**) Palmar (anterior) view. (**b**) Palmar (anterior) view. (Adductor pollicis muscle is removed.) (**c**) Palmar (anterior) view. (The hypothenar and superficial thenar muscles are also removed.) (**d**) Dorsal (posterior) view. (**e**) Dorsal (posterior) view. (Dorsal interossei muscles and extensor tendons are removed)

Table 9.17 Muscles of hand: origin, insertion, innervation, and function

Muscle name	Origin	Insertion	Innervation	Function
Thenar compartment				
Abductor pollicis brevis	Flexor retinaculum and scaphoid and trapezium bones	Base of the proximal phalanx of the thumb	Median nerve (recurrent branch) (C8, T1)	Abducts the thumb at the carpometacarpal joint
Flexor pollicis brevis	Flexor retinaculum and trapezium bone	Base of the proximal phalanx of the thumb	Median nerve (recurrent branch) (C8, T1)	Flexes the thumb at the carpometacarpal and metacarpophalangeal joints
Opponens pollicis	Flexor retinaculum and trapezium bone	Radial aspect of the first metacarpal	Median nerve (recurrent branch) (C8, T1)	Opposes thumb to the little finger (flexes and adducts thumb)
Adductor pollicis	Transverse head: third metacarpal; oblique head: capitate and bases of second and third metacarpals	Base of the proximal phalanx of the thumb	Ulnar nerve (C8, T1)	Adducts thumb
Hypothenar compartment				
Abductor digiti minimi	Pisiform bone and pisohamate ligament	Medial side of the proximal phalanx of the little finger	Ulnar nerve (C8, T1)	Abducts the little finger at the metacarpophalangeal joint
Flexor digiti minimi brevis	Hook of the hamate and flexor retinaculum	Medial side of the proximal phalanx of the little finger	Ulnar nerve (C8, T1)	Flexes the little finger at the metacarpophalangeal joint
Opponens digiti minimi	Hook of the hamate and flexor retinaculum	Medial aspect of the fifth metacarpal	Ulnar nerve (C8, T1)	Opposes the little finger to the thumb
Midpalmar compartment				
Palmaris brevis	Flexor retinaculum and palmar aponeurosis	Skin of ulnar border of palm	Ulnar nerve (C8, T1)	Wrinkles skin of ulnar palm, improving grip
Lumbricals (1st to 4th)	Tendons of flexor digitorum profundus	Extensor expansions of digits 2–5	Median nerve (C8, T1) (lumbricals 1st and 2nd), ulnar nerve (C8, T1) (lumbricals 3rd and 4th)	Flex the metacarpophalangeal joints and extend interphalangeal joints
Interossei (dorsal)	Adjacent sides of metacarpal bones	Bases of proximal phalanges and extensor expansions	Ulnar nerve (C8, T1)	Abduct fingers away from the midline (third digit)
Interossei (palmar)	Adjacent sides of metacarpal bones	Bases of proximal phalanges and extensor expansions	Ulnar nerve (C8, T1)	Adduct fingers toward the midline (third digit)

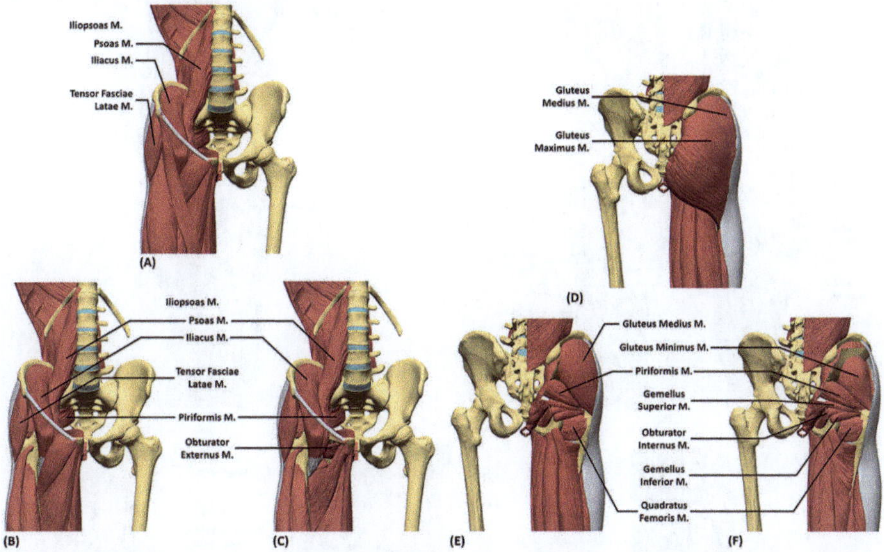

Fig. 9.38 Muscles of the hip. (**a**) Anterior view. (**b**) Anterior view. (Sartorius and rectus femoris muscles are removed). (**c**) Anterior view. (Pectineus muscle is also removed.) (**d**) Posterior view. (**e**) Posterior view. (Gluteus maximus muscle is removed.) (**f**) Posterior view. (Gluteus medius muscle is also removed.)

Table 9.18 Muscles of the hip: origin, insertion, innervation, and function

Muscle name	Region	Origin	Insertion	Innervation	Function
Iliopsoas	Anterior	Iliac fossa and lumbar vertebrae	Lesser trochanter of femur	Femoral nerve (L1–L3)	Hip flexion
Gluteus maximus	Posterior	Dorsal ilium, sacrum, and coccyx	Gluteal tuberosity and iliotibial tract	Inferior gluteal nerve (L5–S2)	Extension and lateral rotation
Gluteus medius	Lateral	Outer surface of ilium	Greater trochanter of femur	Superior gluteal nerve (L4–S1)	Abduction and medial rotation
Gluteus minimus	Lateral	Outer surface of ilium	Greater trochanter of femur	Superior gluteal nerve (L4–S1)	Abduction and medial rotation
Tensor fasciae latae	Lateral	ASIS and anterior iliac crest	Iliotibial tract	Superior gluteal nerve (L4–S1)	Abduction and medial rotation
Gemellus superior	Lateral	Ischial spine	Greater trochanter of femur	Nerve to obturator internus (L5–S1)	Lateral rotation
Gemellus inferior	Lateral	Ischial tuberosity	Greater trochanter of femur	Nerve to quadratus femoris (L4–S1)	Lateral rotation
Obturator internus	Lateral	Obturator membrane and obturator foramen	Greater trochanter of femur	Nerve to obturator internus (L5–S1)	Lateral rotation
Obturator externus	Lateral	Margin of obturator foramen	Trochanteric fossa of femur	Obturator nerve (L3–L4)	Lateral rotation
Piriformis	Lateral	Anterior surface of sacrum	Greater trochanter of femur	Nerves from sacral plexus (S1–S2)	Lateral rotation
Quadratus femoris	Posterior	Ischial tuberosity	Intertrochanteric crest of the femur	Nerve to quadratus femoris (L4–S1)	Lateral rotation

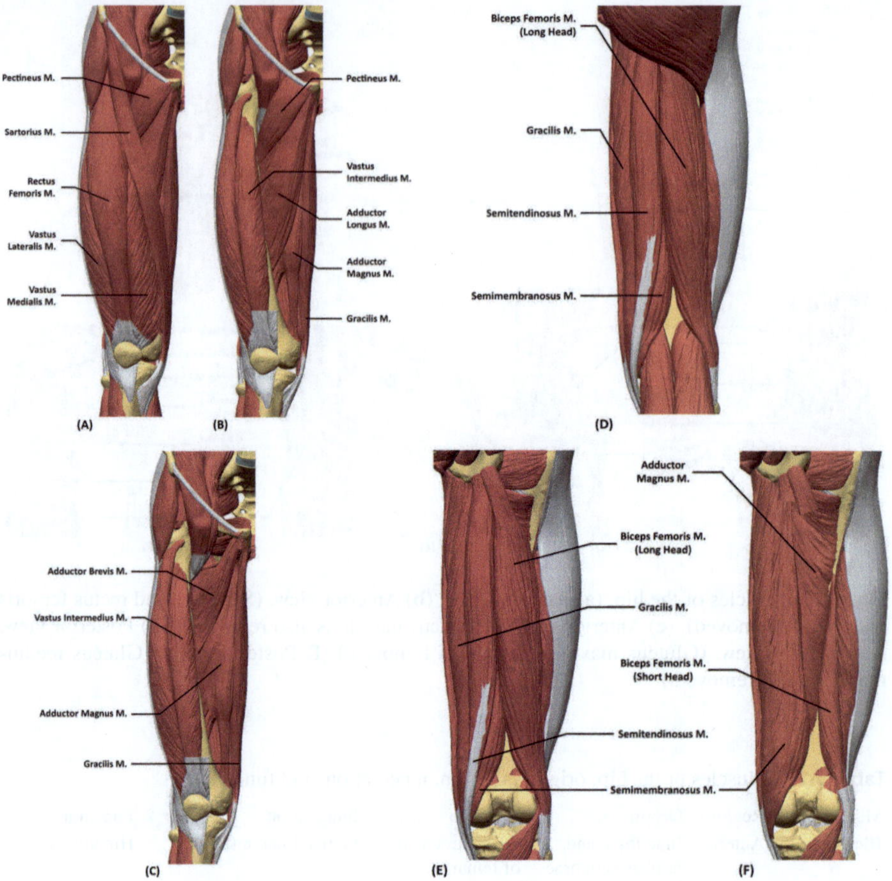

Fig. 9.39 Muscles of the thigh. (**a**) Anterior view. (**b**) Anterior view. (Rectus femoris and vastus medialis muscles are removed.) (**c**) Anterior view. (Pectineus and adductor longus muscles are also removed.) (**d**) Posterior view. (**e**) Posterior view. (Gluteus maximus muscle is removed.) (**f**) Posterior view. (The long head of the biceps femoris muscle is also removed)

Table 9.19 Muscles of the thigh: compartment, origin, insertion, innervation, and function

Muscle name	Origin	Insertion	Innervation	Function
Anterior compartment				
Rectus femoris	Anterior inferior iliac spine (AIIS)	Quadriceps tendon to patella to tibial tuberosity	Femoral nerve (L2–L4)	Extension of knee, flexion of hip
Vastus lateralis	Greater trochanter, intertrochanteric line, linea aspera	Quadriceps tendon to patella to tibial tuberosity	Femoral nerve (L2–L4)	Extension of knee
Vastus medialis	Intertrochanteric line, linea aspera	Quadriceps tendon to patella to tibial tuberosity	Femoral nerve (L2–L4)	Extension of knee
Vastus intermedius	Anterior and lateral surface of femur	Quadriceps tendon to patella to tibial tuberosity	Femoral nerve (L2–L4)	Extension of knee
Sartorius	Anterior superior iliac spine (ASIS)	Medial side of proximal tibia	Femoral nerve (L2–L3)	Flexion and lateral rotation at hip, flexion at knee
Pectineus	Superior ramus of pubis	Pectineal line of femur	Femoral nerve (L2–L3)	Adduction and flexion of hip
Posterior compartment				
Biceps femoris	Long head: ischial tuberosity; short head: linea aspera, lateral supracondylar line	Head of fibula and lateral condyle of tibia	Long head: Tibial part of sciatic nerve (S1–S2); short head: Common fibular part of sciatic nerve (L5–S2)	Extension of hip, flexion, and lateral rotation of knee
Semitendinosus	Ischial tuberosity	Medial surface of upper tibia	Tibial part of sciatic nerve (S1–S2)	Extension of hip, flexion, and medial rotation of knee
Semimembranosus	Ischial tuberosity	Medial condyle of tibia	Tibial part of sciatic nerve (S1–S2)	Extension of hip, flexion, and medial rotation of knee

(continued)

Table 9.19 (continued)

Muscle name	Origin	Insertion	Innervation	Function
Medial compartment				
Adductor longus	Body of pubis near the pubic symphysis	Middle third of linea aspera	Obturator nerve (L2–L4)	Adduction and medial rotation of hip
Adductor brevis	Inferior ramus of the pubis	Pectineal line and linea aspera	Obturator nerve (L2–L4)	Adduction and medial rotation of hip
Adductor magnus	Ischiopubic ramus and ischial tuberosity	Linea aspera and adductor tubercle	Obturator and tibial nerves (L2–L4)	Adduction, extension, and medial rotation
Gracilis	Inferior ramus of the pubis	Upper medial shaft of tibia	Obturator nerve (L2–L3)	Adduction and flexion of hip, flexion of knee
Obturator externus	Margin of obturator foramen	Trochanteric fossa of the femur	Obturator nerve (L3–L4)	Lateral rotation of hip

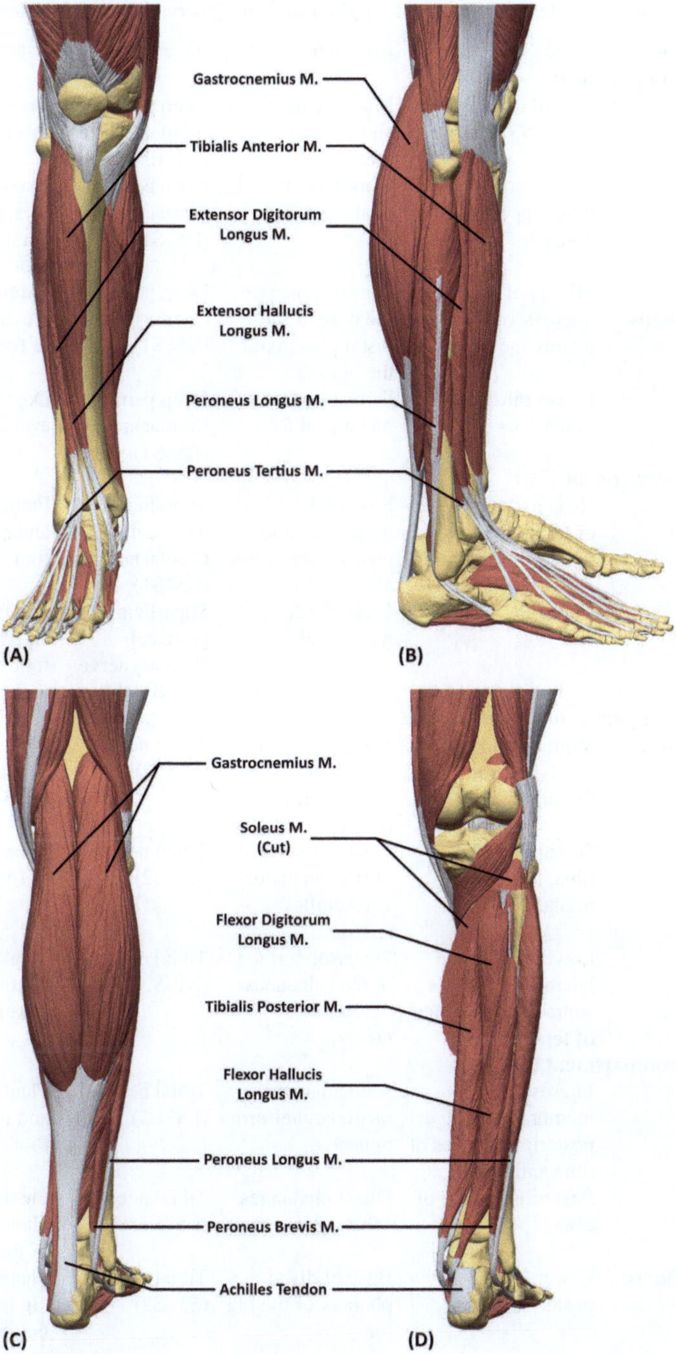

Fig. 9.40 Muscles of the leg. (**a**) Anterior view. (**b**) Lateral view. (**c**) Posterior view. (**d**) Posterior view. (Gastrocnemius and soleus muscles are removed)

Table 9.20 Muscles of leg: compartment, origin, insertion, innervation, and function

Muscle name	Origin	Insertion	Innervation	Function
Anterior compartment				
Tibialis anterior	Lateral condyle and upper 2/3 of tibial shaft	Medial cuneiform and base of 1st metatarsal	Deep peroneal (fibular) nerve (L4–L5)	Dorsiflexion and inversion of foot
Extensor digitorum longus	Lateral condyle of tibia, upper 3/4 of fibula	Dorsal surface of phalanges 2–5	Deep peroneal (fibular) nerve (L5–S1)	Extension of toes, dorsiflexion of foot
Extensor hallucis longus	Middle of fibula and interosseous membrane	Dorsal aspect of the base of the distal phalanx of the big toe	Deep peroneal (fibular) nerve (L5–S1)	Extension of big toe, dorsiflexion of foot
Fibularis (peroneus) tertius	Lower third of fibula	Dorsal surface of metatarsal 5	Deep peroneal (fibular) nerve (L5–S1)	Dorsiflexion and eversion of foot
Lateral compartment				
Fibularis (peroneus) longus	Head and upper 2/3 of fibula	Base of 1st metatarsal and medial cuneiform	Superficial peroneal (fibular) nerve (L5–S1)	Plantar flexion and eversion of foot
Fibularis (peroneus) brevis	Lower 2/3 of fibula	Base of 5th metatarsal	Superficial peroneal (fibular) nerve (L5–S1)	Plantar flexion and eversion of foot
Posterior compartment (superficial)				
Gastrocnemius	Medial and lateral condyles of the femur	Posterior surface of the calcaneus via Achilles tendon	Tibial nerve (S1–S2)	Plantar flexion of foot, flexion of knee
Soleus	Soleal line of the tibia, head of the fibula	Posterior surface of the calcaneus via Achilles tendon	Tibial nerve (S1–S2)	Plantar flexion of foot
Plantaris	Lower part of the lateral supracondylar ridge of femur	Posterior surface of the calcaneus via Achilles tendon	Tibial nerve (S1–S2)	Weak plantar flexor and knee flexor
Posterior compartment (deep)				
Tibialis posterior	Interosseous membrane, posterior surfaces of tibia and fibula	Navicular and medial cuneiform bones	Tibial nerve (L4–L5)	Plantar flexion and inversion of foot
Flexor digitorum longus	Posterior surface of tibia	Distal phalanges of toes 2–5	Tibial nerve (L5–S1)	Flexion of toes, plantar flexion of foot
Flexor hallucis longus	Lower 2/3 of posterior fibula	Base of distal phalanx of the big toe	Tibial nerve (S1–S2)	Flexion of the big toe, plantar flexion of foot

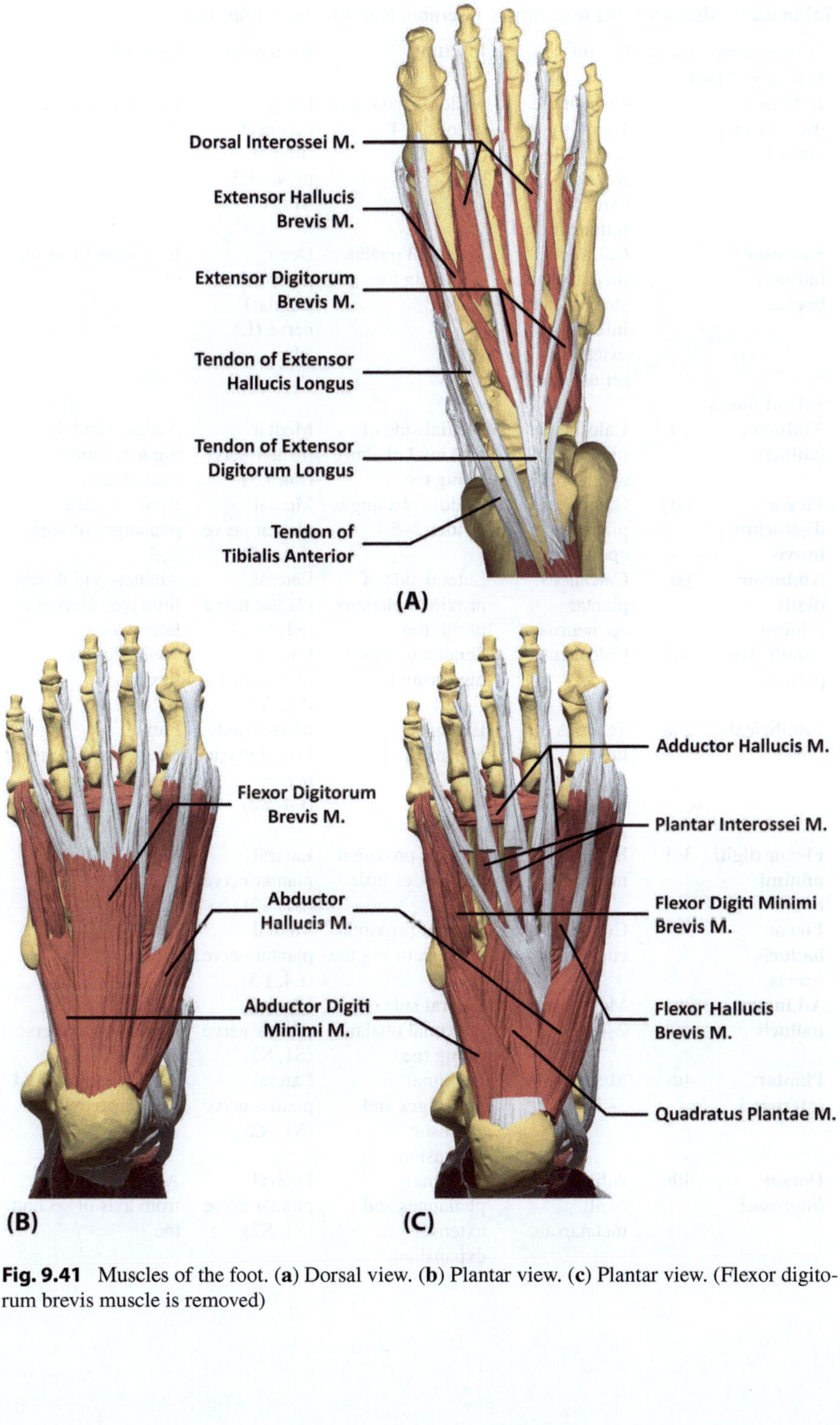

Fig. 9.41 Muscles of the foot. (**a**) Dorsal view. (**b**) Plantar view. (**c**) Plantar view. (Flexor digitorum brevis muscle is removed)

Table 9.21 Muscles of the foot: origin, insertion, innervation, and function

Muscle name	Layer	Origin	Insertion	Innervation	Function
Dorsum of foot					
Extensor digitorum brevis	–	Calcaneus, lateral to the stem of the inferior extensor retinaculum	Middle phalanges of toes 2–4	Deep peroneal (fibular) nerve (L5, S1)	Extension of toes 2–4
Extensor hallucis brevis	–	Calcaneus, medial to the stem of the inferior extensor retinaculum	Proximal phalanx of the big toe	Deep peroneal (fibular) nerve (L5, S1)	Extension of the big toe
Sole of foot					
Abductor hallucis	1st	Calcaneus, plantar aponeurosis	Medial side of proximal phalanx of big toe	Medial plantar nerve (L4, L5)	Abducts and flexes big toe, supports medial arch
Flexor digitorum brevis	1st	Calcaneus, plantar aponeurosis	Middle phalanges of toes 2–5	Medial plantar nerve (L4, L5)	Flexes middle phalanges of toes 2–5
Abductor digiti minimi	1st	Calcaneus, plantar aponeurosis	Lateral side of proximal phalanx of 5th toe	Lateral plantar nerve (S1, S2)	Abducts and flexes little toe, supports lateral arch
Quadratus plantae	2nd	Calcaneus	Tendon of flexor digitorum longus	Lateral plantar nerve (S1, S2)	Assists in toe flexion
Lumbricals	2nd	Tendons of flexor digitorum longus	Extensor expansions of toes 2–5	Medial and lateral plantar nerves (L4–S2)	Flex metatarsophalangeal joints, extend interphalangeal joints
Flexor digiti minimi brevis	3rd	Base of 5th metatarsal	Base of proximal phalanx of little toe	Lateral plantar nerve (S1, S2)	Flexes little toe
Flexor hallucis brevis	3rd	Cuboid, lateral cuneiform	Base of proximal phalanx of big toe	Medial plantar nerve (L4, L5)	Flexes big toe
Adductor hallucis	3rd	Metatarsals 2–4	Lateral side of proximal phalanx of big toe	Lateral plantar nerve (S1, S2)	Adducts big toe, supports transverse arch
Plantar interossei	4th	Metatarsals 3–5	Proximal phalanges and extensor expansions	Lateral plantar nerve (S1, S2)	Adduct toes toward axis of second toe
Dorsal interossei	4th	Adjacent sides of all metatarsals	Proximal phalanges and extensor expansions	Lateral plantar nerve (S1, S2)	Abduct toes away from axis of second toe

Suggested Reading

Advanced Physiology and Pathophysiology: Essentials for Clinical Practice, 2nd Ed. Nancy Tkacs, Linda Herrmann, Springer Publishing Company, 2024

Anatomy & Physiology, S. Meloni, et al. Independent Publisher, 2023

Anatomy & Physiology: The Unity of Form and Function, 10th Ed. Kenneth S. Saladin, McGraw Hill, 2023

Gray's Anatomy for Students, 5th Ed. Richard L. Drake, et al. Elsevier, 2020

Human Anatomy & Physiology, 11th Ed. Elaine N. Marieb and Katja Hoehn, Pearson, 2022

Moore's Clinically Oriented Anatomy, 9th Ed. Keith L. Moore, et al. Lippincott Williams & Wilkins, 2020

Netter Atlas of Human Anatomy: Classic Regional Approach, 8th Ed. Frank H. Netter, Elsevier, 2022

Principles of Anatomy and Physiology, 16th Ed. Gerard J. Tortora, Bryan H. Derrickson, Wiley, 2020

Nervous System

10

In this chapter, you will learn:

- Overview of the nervous system.
- Description of neuroglia in the CNS and PNS.
- Description of neuron structure: cell body, dendrites, axon, and synaptic terminals.
- Structure and function of the myelin sheath and nodes of Ranvier.
- Classification of neurons based on structure and function.
- Electrophysiological properties of neurons, including different types of ion channels.
- Mechanisms establishing the resting membrane potential in neurons.
- Detailed explanation of action potential generation and propagation.
- Discussion on refractory periods: absolute and relative.
- Saltatory conduction and its efficiency compared to continuous conduction in unmyelinated axons.
- Overview of synapses, including electrical and chemical synapses.
- Role and types of neurotransmitters in neural communication.
- Discussion of neural processing: serial and parallel processing.
- Structural and functional division of the CNS: brain and spinal cord.
- Discussion on the brain's gray and white matter, their organization, and functions.
- Detailed anatomy of the brain hemispheres and their functional areas.
- Overview of the basal nuclei and their roles in motor control and cognitive functions.
- Discussion on the structure and function of diencephalon.
- Detailed anatomy and function of the cerebellum.
- Overview of the ventricular system and cerebrospinal fluid (CSF) dynamics and function.
- Detailed structure and function of the meninges.
- Overview of the spinal cord and its anatomical features.
- Role of the autonomic nervous system in involuntary functions.

H. Hosseinzadeh, *Fundamentals of Medicine for Biomedical Engineering*,
https://doi.org/10.1007/978-3-031-74085-5_10

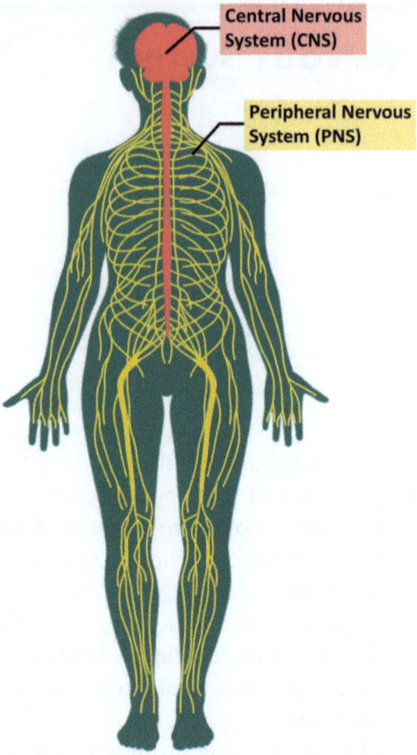

Fig. 10.1 General structure of nervous system

The nervous system is a complex, intricate network of neurons and associated cells that functions as an organism's primary messaging and control center. It orchestrates many physiological processes, from basic reflexes to higher-order cognitive activities like learning, memory, and emotion. Fundamentally, the nervous system is subdivided into two major components: the central nervous system (CNS) and the peripheral nervous system (PNS) (Fig. 10.1).

10.1 Part 1: Cells in the Nervous System

The nervous system's cellular architecture is remarkably complex. It consists primarily of two types of cells: neurons and glial cells. These cells collaborate intricately to facilitate communication, modulation, and maintenance functions that underpin the system's extensive range of activities.

10.1.1 Neuroglia

Neuroglia, often called glial cells, are auxiliary yet indispensable nervous system elements that provide structural and functional support to neurons.

10.1.1.1 Neuroglia in CNS (Fig. 10.2)

Astrocytes: Astrocytes are star-shaped cells predominantly found in the CNS. Their processes interact with neurons, blood vessels, and other glial cells, thus playing a multifaceted role in neural activity. Astrocytes form the blood-brain barrier by enveloping capillaries with their end feet, regulating the exchange of substances. They maintain ionic balance in the extracellular space, particularly concerning potassium ions. Astrocytes release neurotrophic factors that support neuron survival and function. Astrocytes can modulate neuronal activity through the release and uptake of neurotransmitters.

Oligodendrocytes: Oligodendrocytes are smaller than astrocytes and characterized by fewer processes. They are mainly located in the white matter of the CNS. Their primary role is to produce the myelin sheath, a fatty insulating layer that wraps around axons, enhancing electrical signal transmission. They also provide nutritional support to the axons they myelinate.

Microglia: Microglia are small, irregularly shaped cells found throughout the CNS and possess phagocytic abilities. Microglia act as resident immune cells in the CNS, monitoring for pathological conditions. They are responsible for clearing cellular debris, pathogens, and damaged neurons.

Ependymal Cells: Ependymal cells are ciliated cells lining the ventricular system of the CNS. Their cilia help circulate cerebrospinal fluid (CSF) and form a barrier between the ventricular system and the parenchyma of the CNS.

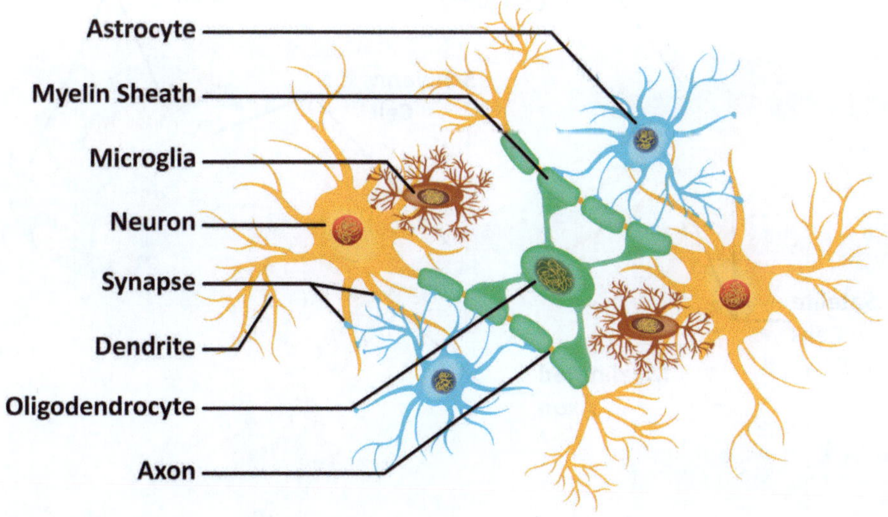

Fig. 10.2 Neuroglia in the central nervous system

10.1.1.2 Neuroglia in PNS (Fig. 10.3)

Schwann Cells Schwann cells are the primary glial cells in the PNS, and each cell is associated with a single axon or a group of smaller axons. Like oligodendrocytes, Schwann cells form the myelin sheath around axons in the PNS, facilitating rapid signal transmission. They also provide trophic support to axons.

Satellite Cells: Satellite cells are flat cells that surround neuron cell bodies in the peripheral ganglia. They provide physical support to neurons and help regulate the chemical environment around them.

10.1.2 Neurons

Neurons, the principal cells of the nervous system, serve as the foundational elements for neural communication and information processing. Their unique structural features facilitate the rapid transmission and reception of electrochemical signals.

10.1.2.1 Neuron Structure

Neurons have three primary structural components: the cell body or soma, dendrites, and the axon (Fig. 10.4).

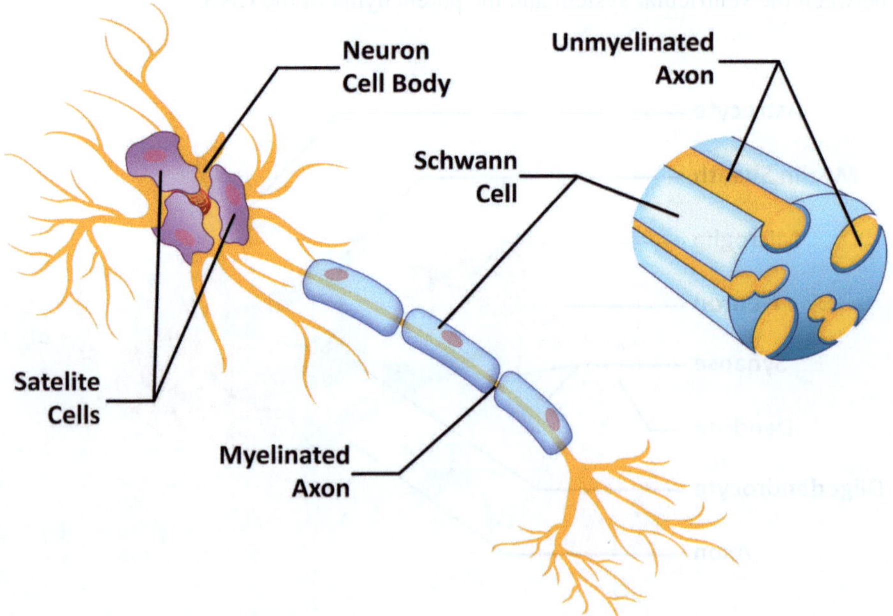

Fig. 10.3 Neuroglia in the peripheral nervous system

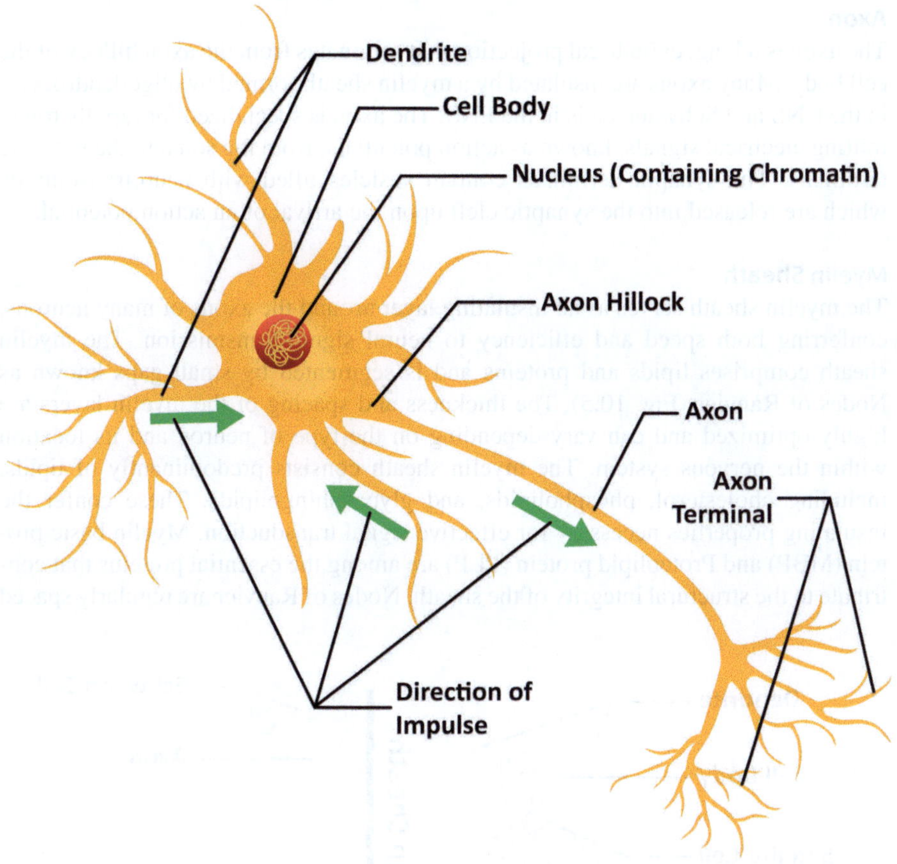

Fig. 10.4 Structure of a typical neuron

Cell Body (Soma)

The cell body houses the nucleus, endoplasmic reticulum, Golgi apparatus, mitochondria, and other cellular organelles vital for metabolic activities. It maintains the cell's structural and functional integrity. The cell body contains the nucleus, which regulates gene expression and influences neuronal function. It serves as the initial segment where incoming signals from dendrites may be summated before being transmitted down the axon.

Dendrites

Dendrites are branching, tree-like structures emanating from the cell body. They often contain specialized protrusions called dendritic spines that increase surface area and serve as synaptic contact points. Dendrites act as the primary sites for receiving signals from other neurons. They assimilate multiple incoming signals and transmit the integrated signals toward the cell body. Changes in the morphology of dendritic spines are associated with synaptic plasticity, which is crucial for learning and memory.

Axon

The axon is a long, cylindrical projection that originates from the axon hillock of the cell body. Many axons are insulated by a myelin sheath formed by oligodendrocytes in the CNS and Schwann cells in the PNS. The axon is specialized for rapidly transmitting electrical signals, known as action potentials, from the soma to the synaptic terminals. The synaptic terminals contain vesicles filled with neurotransmitters, which are released into the synaptic cleft upon the arrival of an action potential.

Myelin Sheath

The myelin sheath serves as an insulating layer around the axons of many neurons, conferring both speed and efficiency to neural signal transmission. The myelin sheath comprises lipids and proteins and is segmented by small gaps known as Nodes of Ranvier (Fig. 10.5). The thickness and spacing of the myelin layers are highly optimized and can vary depending on the type of neuron and its location within the nervous system. The myelin sheath consists predominantly of lipids, including cholesterol, phospholipids, and glycosphingolipids. These confer the insulating properties necessary for effective signal transduction. Myelin basic protein (MBP) and Proteolipid protein (PLP) are among the essential proteins that contribute to the structural integrity of the sheath. Nodes of Ranvier are regularly spaced

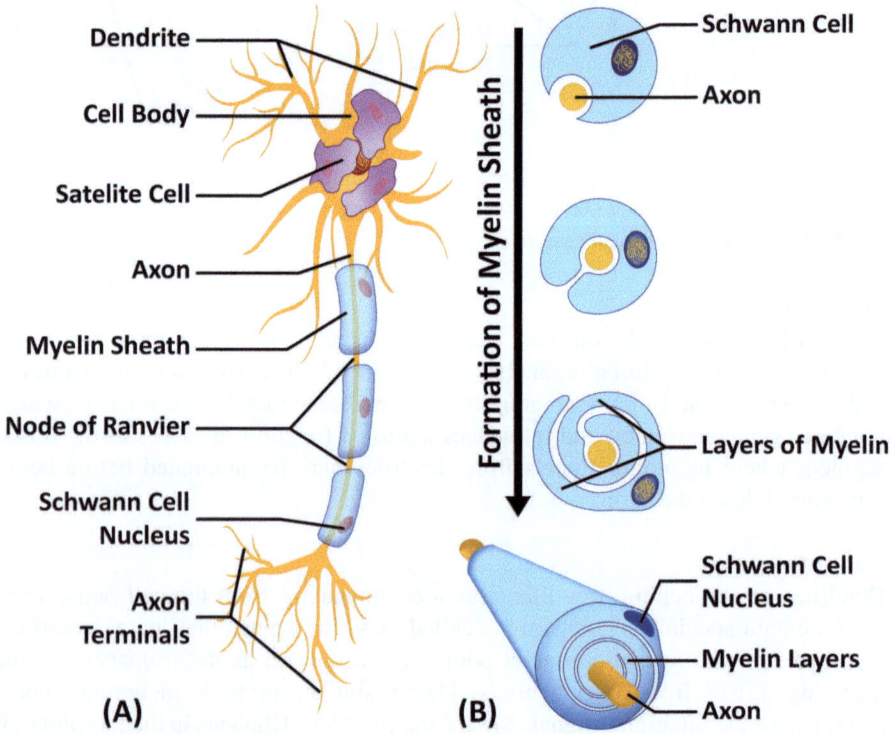

Fig. 10.5 (**a**) A myelinated axon. (**b**) Myelination process in the peripheral nervous system

gaps in the myelin sheath where the axon is exposed. These nodes contain high concentrations of voltage-gated sodium and potassium channels, essential for action potential propagation.

In the CNS, the myelin sheath is formed by oligodendrocytes, which extend their processes to wrap around the axons multiple times. A single oligodendrocyte can myelinate several axonal segments (Fig. 10.3). Schwann cells are.

responsible for myelination in the PNS. Unlike oligodendrocytes, a Schwann cell typically wraps around a single axon segment. The innermost layer of the Schwann cell membrane forms the myelin sheath in the PNS (Fig. 10.5).

One of the most prominent functions of the myelin sheath is to increase the velocity of action potential propagation. The insulation provided by the sheath allows for saltatory conduction (Page 306), where the action potential "jumps" from one node of Ranvier to the next, thereby speeding up the signal transmission. By enabling saltatory conduction and reducing the need for ion channels along the entire length of the axon, the myelin sheath enhances energy efficiency in neural signaling. The myelin sheath protects the axon, reducing its vulnerability to mechanical stress and chemical insults. It also provides trophic support to the axon, helping to maintain its structural and functional integrity.

10.1.2.2 Neurons Classification

Neurons display incredible diversity in structure and function, allowing for complex CNS and PNS operations. Their heterogeneity can be classified through various lenses, including structural features, electrochemical properties, and functional roles.

Structural Classification

One of the most traditional methods of classifying neurons is based on their morphological characteristics, particularly the number and arrangement of their cellular processes (Fig. 10.6).

- Multipolar Neurons: They are characterized by one axon and multiple dendrites emanating from the cell body. They are most commonly found in the CNS, including the cerebral cortex and spinal cord. These neurons are involved in complex information processing and integration.
- Bipolar Neurons: These neurons have two processes—an axon and a dendrite—that extend from opposite poles of the cell body. They are primarily found in sensory systems, such as the eye's retina and olfactory (smelling) epithelium.
- Unipolar Neurons: They possess a single process that bifurcates, serving as both an axon and a dendrite. They are commonly found in the dorsal root ganglia of the spinal cord, part of the PNS. These neurons are mainly involved in transmitting sensory information.
- Anaxonic Neurons: These neurons lack a distinguishable axon and feature multiple dendrites. They are found in some areas of the CNS, such as the retina and the brain cortex, and are primarily involved in local processing. They do not generate action potentials.

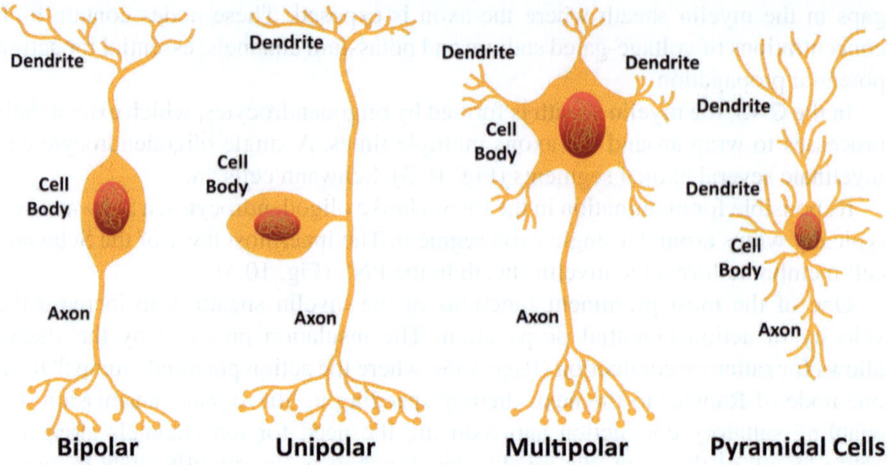

Fig. 10.6 Structural classes of neurons

Functional Classification
Neurons can also be categorized based on their functional roles within neural circuits:

Sensory Neurons: They are responsible for transducing environmental stimuli into neural signals. They are primarily found in sensory organs and convey information to the CNS.

Motor Neurons: These neurons transmit signals from the CNS to effector organs like muscles and glands, initiating movement or physiological responses.

Interneurons: Primarily found within the CNS, interneurons act as connectors between sensory and motor neurons and are involved in complex computational tasks.

Electrophysiological Classification
Based on their electrophysiological properties, neurons can be classified into various types, such as:

- Spiny vs. Aspiny: Depending on the presence or absence of dendritic spines, neurons may be termed spiny or aspiny. Spiny neurons are generally excitatory, whereas aspiny neurons are often inhibitory.
- Firing Patterns: Based on how neurons generate action potentials, they can be categorized as fast-spiking, regular-spiking, or bursting, among other types.

10.1.2.3 Resting Membrane Potential
The resting membrane potential is the electrical voltage difference across the cell membrane when the neuron is non-signaling. The resting membrane potential arises due to the differential distribution of ions across the neuron's plasma membrane and

the membrane's selective permeability to specific ions. The intracellular and extracellular milieus contain different concentrations of essential ions, primarily sodium (Na^+), potassium (K^+), chloride (Cl^-), and calcium (Ca^{2+}).

Typically, K^+ concentration is higher inside the cell, while Na^+, Cl^-, and Ca^{2+} are more concentrated outside. The resting membrane potential sets the stage for action potential generation and propagation, serving as the "baseline" from which depolarization and hyperpolarization occur.

Membrane Ion Channels

Ion channels are integral membrane proteins that form pores in the lipid bilayer, enabling selective passage of ions across the cellular membrane. Pertinent to the context of resting membrane potential, they can be primarily categorized into:

Leak Channels (Fig. 10.7): These channels are always open, allowing ions to move down their electrochemical gradient. Leak channels for potassium (K^+) are highly abundant, rendering the membrane significantly more permeable to K^+ ions. This contributes to a resting membrane potential that is close to the equilibrium potential for K^+.

Voltage-Gated Channels (Fig. 10.7): These channels open or close in response to changes in membrane potential. Although not directly involved in maintaining the resting state, closed voltage-gated sodium (Na^+) and potassium (K^+) channels ensure these ions are primed for activation should depolarization occur.

Ligand-Gated Channels (Fig. 10.7): These channels open or close in response to the binding of specific ligands. Ligand-gated channels, such as GABA_A receptors, may influence the resting membrane potential when inhibitory neurotransmitters are present but are not integral to its essential establishment.

Pump Channels (Fig. 10.8): Utilize energy from ATP to transport ions against their electrochemical gradient. The Na^+/K^+ ATPase pump maintains the steep gradient for sodium and potassium ions across the membrane, which is crucial for sustaining the resting membrane potential.

Generating the Resting Membrane Potential

The resting membrane potential arises from a complex interplay between ion concentrations and selective permeability of the neuronal membrane. Key ions include sodium (Na^+), potassium (K^+), chloride (Cl^-), and calcium (Ca^{2+}). The concentration and electrical gradients constitute electrochemical gradients for each ion. These gradients drive ions toward or away from their equilibrium potential, defined as the membrane potential at which no net movement of a particular ion exists. The neuronal membrane's selective permeability arises from various ion channels, which can be gated or constitutively open, allowing specific ions to pass through.

The most critical players in setting the resting membrane potential are the potassium (K^+) leak channels. These channels are open at rest, allowing K^+ ions to move out of the cell and down their electrochemical gradient. This outward movement of positively charged ions renders the inside of the cell more negative, contributing significantly to the resting membrane potential. Sodium (Na^+) leak channels are far less prevalent than their potassium counterparts but are crucial for maintaining a

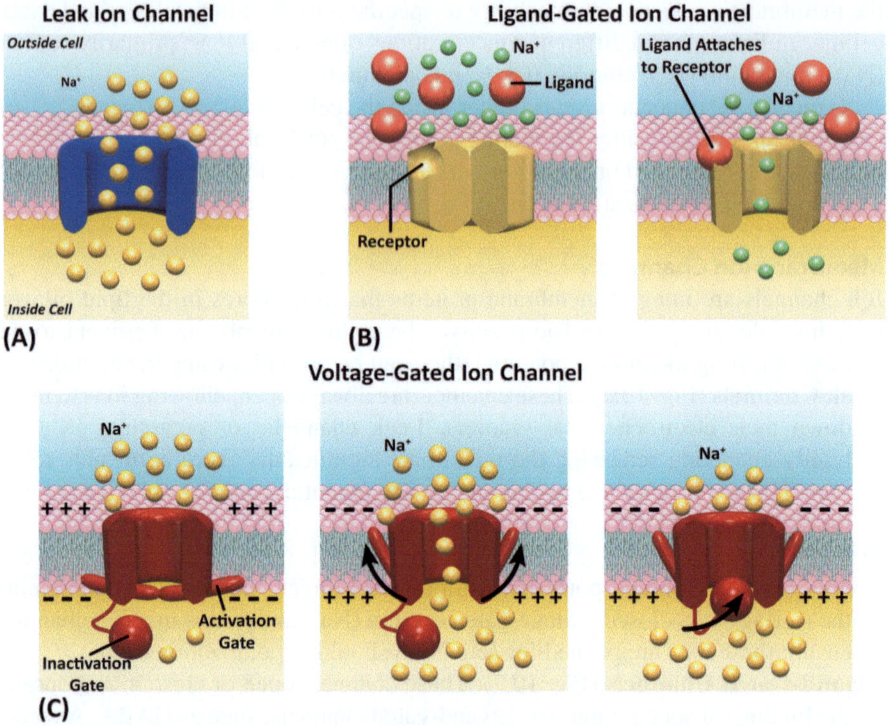

Fig. 10.7 Ion channels in neural membranes create the resting and action potentials

balance. These channels allow a slow influx of Na^+ ions, slightly counterbalancing the hyperpolarizing effect of K^+ outflow. The Na^+/K^+ ATPase pump uses energy from ATP hydrolysis to exchange three Na^+ ions out of the cell for two K^+ ions into the cell (Fig. 10.9). While not directly responsible for setting the resting membrane potential, the pump is essential for maintaining the ionic gradients that permit it.

Fig. 10.8 (continued) cell. **Step 2. Na^+ ion binding:** Three Na^+ ions can bind to the pump inside the cell. **Step 3. ATP binding:** ATP binds to the pump**. Step 4. Phosphorylation:** ATP is hydrolyzed to ADP, transferring a phosphate group to the pump (phosphorylation), causing a conformational change, which moves the Na^+ ions across the membrane to the extracellular space. **Step 5. Na^+ ion release:** The conformational change decreases the pump's affinity for Na^+ ions, releasing them outside the cell. The pump releases the three bound Na + ions outside the cell. **Steps 6–7. K^+ ion binding:** The K^+ binding sites are exposed, and this conformation of the pump has a high affinity for K^+ ions. Hence, two K^+ ions bind to the pump from the extracellular fluid. **Step 8. Dephosphorylation:** The binding of K^+ ions triggers the dephosphorylation of the pump. The phosphate group is detached from the pump, causing the reverse confrontational change and bringing the K^+ ions into the intracellular space. At the same time, this change decreases the pump affinity for K^+ ions and increases the affinity for Na^+ ions. **Step 9. K^+ release:** The two K^+ ions are released inside the cell. **Step 10. Cycle repeats:** The pump is now ready to bind three more Na^+ ions, and the cycle continues

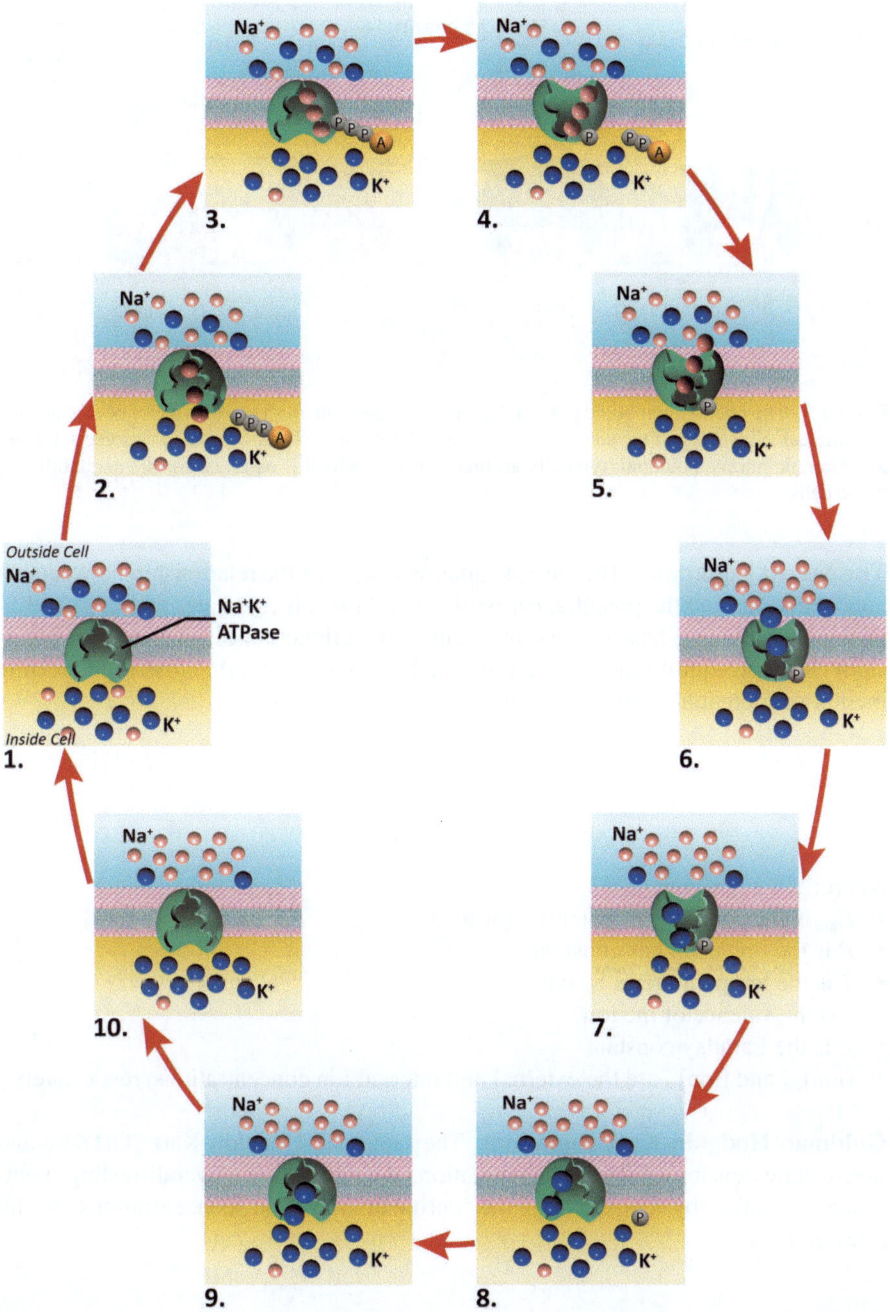

Fig. 10.8 Na^+-K^+ ATPase pump: It maintains the electrochemical gradient across the plasma membrane. The pump cycle involves the active transport of sodium (Na^+) and potassium (K^+) ions against their concentration gradients using energy derived from ATP hydrolysis. **Step 1. Resting state:** The Na^+/K^+ ATPase is in its default conformation with a high affinity for Na^+ ions inside the

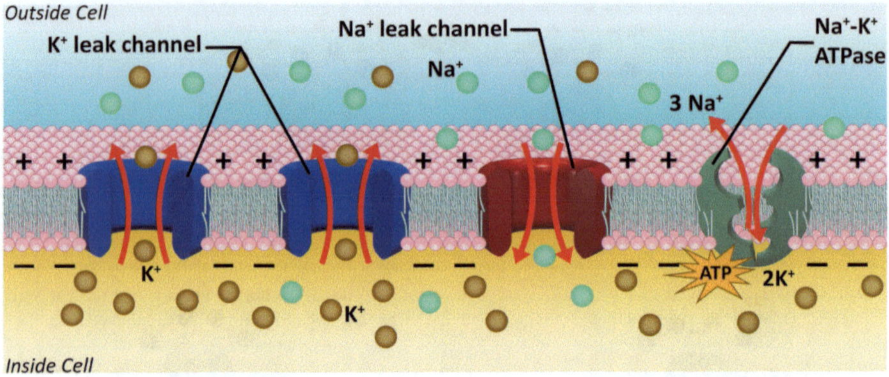

Fig. 10.9 The resting membrane potential generation mechanism shows the role of potassium and sodium leak channels and the sodium-potassium ATPase pump. Their combined action establishes a resting membrane potential, typically around −70 mV, which is essential for the excitability of nerve cells

The Nernst Equation: The Nernst equation describes the relationship between ion concentrations and the potential across the membrane. It calculates an ion's equilibrium potential (E_{ion}) based on its intra- and extracellular concentrations. For potassium (K^+), the equilibrium potential is usually around −75 mV, close to the resting membrane potential of many neurons.

$$E_{ion} = \frac{RT}{zF} \ln\left(\frac{[ion]_{\text{out}}}{[ion]_{\text{in}}}\right)$$

where:

- E_{ion} is the equilibrium potential for the ion
- R is the universal gas constant
- T is the temperature in Kelvin
- z is the valence of the ion
- F is the Faraday constant
- $[\text{ion}]_{\text{out}}$ and $[\text{ion}]_{\text{in}}$ are the external and internal ion concentrations, respectively

Goldman-Hodgkin-Katz Equation: The Goldman-Hodgkin-Katz (GHK) equation is an extension of the Nernst equation. It calculates the overall resting membrane potential by considering the permeabilities and concentrations of all relevant ions.

$$V_m = \frac{RT}{F} \ln\left(\frac{P_{\text{Na}}\left[\text{Na}^+\right]_{\text{out}} + P_{\text{K}}\left[\text{K}^+\right]_{\text{out}} + P_{\text{Cl}}\left[\text{Cl}^-\right]_{\text{out}}}{P_{\text{Na}}\left[\text{Na}^+\right]_{\text{in}} + P_{\text{K}}\left[\text{K}^+\right]_{\text{in}} + P_{\text{Cl}}\left[\text{Cl}^-\right]_{\text{in}}}\right)$$

10.1.2.4 Change in the Resting Membrane Potential

The change in the resting membrane potential refers to a deviation from the neuron's stable, non-signaling state of electrical polarization across its membrane. This alteration can occur due to various factors, such as changes in ion concentration, altered ion channel activity, or external stimuli. Typically, changes in the resting membrane potential can be classified as either depolarization, where the membrane potential becomes less negative, or hyperpolarization, where it becomes more negative. These shifts are preparatory steps for neuronal signaling events like generating graded or action potentials.

Graded Potentials

Graded potentials (Fig. 10.10) are localized, transient changes in membrane potential that serve as the fundamental units of subthreshold neuronal signaling. Unlike action potentials, which are all-or-nothing events, graded potentials vary in amplitude and duration. Graded potentials arise from the movement of ions across the neuronal membrane. The specific ions involved depend on the ion channels that are activated. Common ions include sodium (Na^+), potassium (K^+), chloride (Cl^-), and calcium (Ca^{2+}). Graded potentials predominantly involve voltage-gated and ligand-gated ion channels. Graded potentials are often localized to specific neuron regions, such as the dendrites or cell bodies. Their effects diminish with distance from the point of origin due to the resistive and capacitive properties of the neuronal membrane.

Temporal and Spatial Summation of Graded Potentials

Temporal summation refers to the additive effect of multiple graded potentials occurring quickly at the same synaptic site. Spatial summation involves the simultaneous occurrence of graded potentials at multiple sites on the neuron, which combine to affect the overall membrane potential. Graded potentials serve as the primary means of integrating multiple incoming signals, enabling a neuron to make a "decision" about whether or not to generate an action potential.

Action Potentials

The action potential in neurons is a rapid, transient change in membrane potential (becoming less negative) that serves as the fundamental unit of electrical signaling within the nervous system. The action potential is propagated along the axon towards the axon terminals, which triggers neurotransmitters' release into the synaptic cleft, facilitating communication with other neurons or effector cells.

Stages of Action Potential Generation (Fig. 10.11)

Action potentials are initiated when the membrane potential reaches a threshold level, generally around −55 mV. Stimuli that result in depolarization but do not reach the threshold will not generate an action potential and are considered subthreshold stimuli. When the threshold is reached, voltage-gated Na^+ channels open rapidly (Fig. 10.12). Sodium ions rush into the cell, causing a rapid depolarization of the membrane. The entry of Na^+ causes more voltage-gated Na^+ channels to open,

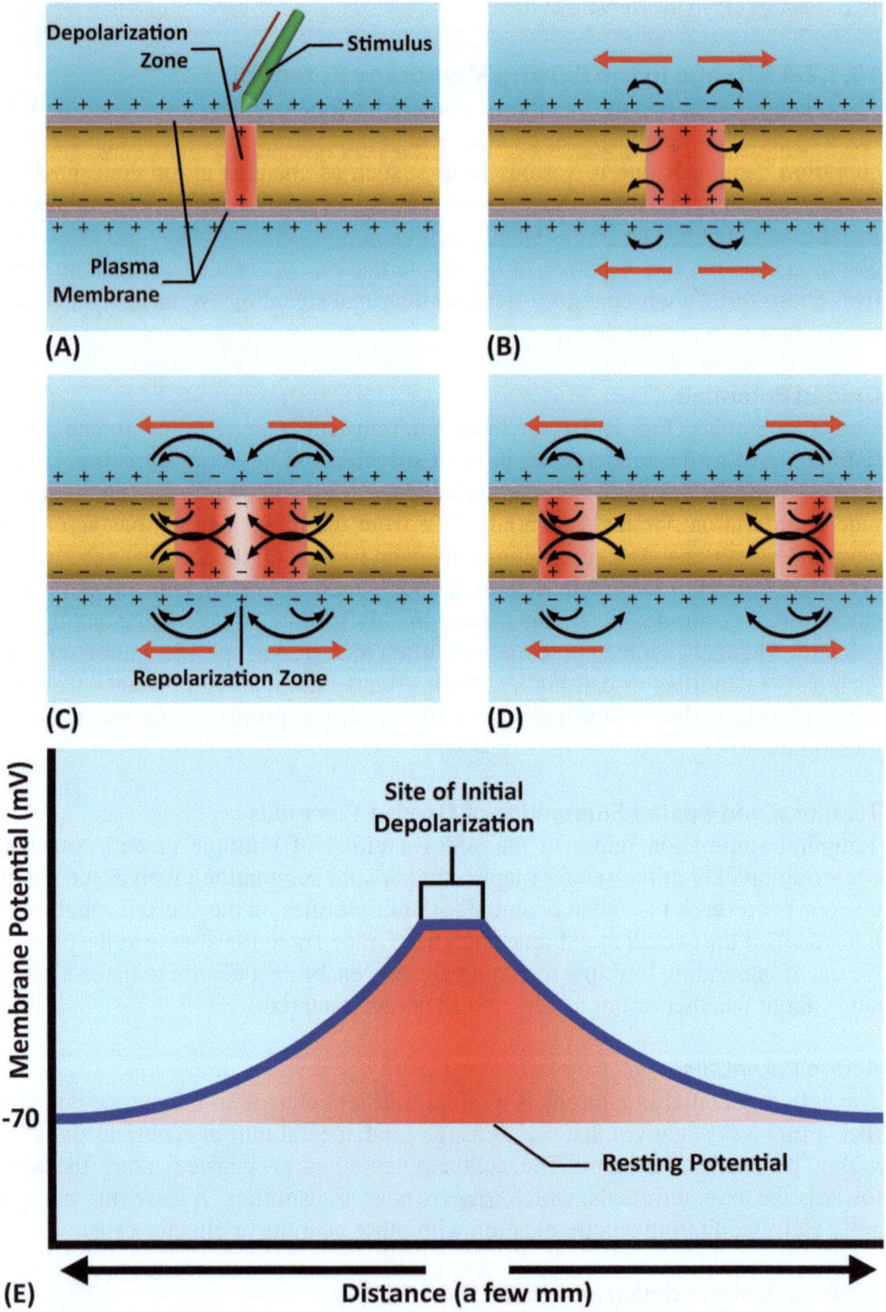

Fig. 10.10 The spread and decay of a graded potential. (**a**) **Depolarization:** A small patch of the membrane (red area) depolarizes. (**b–d**) **Depolarization spreads:** Opposite charges attract each other. This creates local currents (black arrows) that depolarize adjacent membrane areas, spreading the depolarization wave (red arrows). Depolarization spreads in both directions and as it propagates, the previously depolarized membrane patch repolarizes. (**e**) The voltage declines with distance from the stimulus, resulting in membrane potential decay. Consequently, graded potentials are short-distance signals

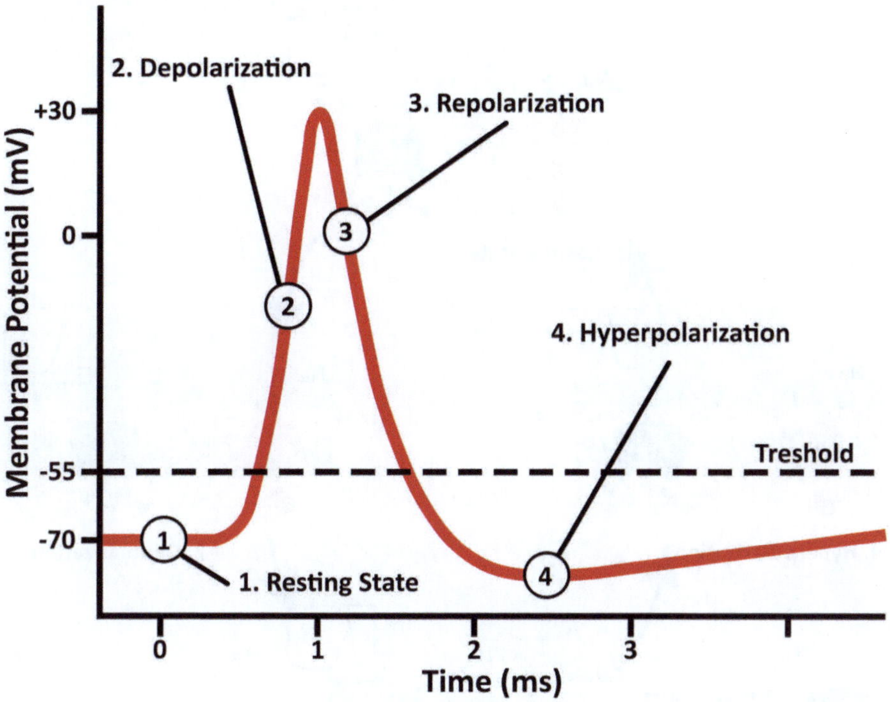

Fig. 10.11 Recording of an action potential (AP) in a nerve fiber, depicting changes in membrane potential (mV) over time (ms). An AP is a brief change in membrane potential

accelerating the depolarization process. During the peak, the membrane potential may reach +30 to +40 mV. Na^+ channels possess an inactivation gate that closes after a brief period, halting the influx of Na^+. Voltage-gated K^+ channels open more slowly than Na^+ channels in response to the depolarization. Once they open, potassium ions exit the cell, rapidly repolarizing and restoring the negative membrane potential. K^+ channels close more slowly, leading to an overshoot and hyperpolarization, making the internal environment more negative than the resting membrane potential.

The membrane potential eventually returns to the resting state, aided by the action of ion pumps and leak channels.

Refractory Periods

The refractory period is a time window during which the neuron's excitability is either completely suppressed or significantly reduced after initiating an action potential. It governs the frequency at which a neuron can fire and plays a critical role in the directional propagation of action potentials along axons. This concept is subdivided into two primary categories: absolute and relative refractory periods (Fig. 10.13).

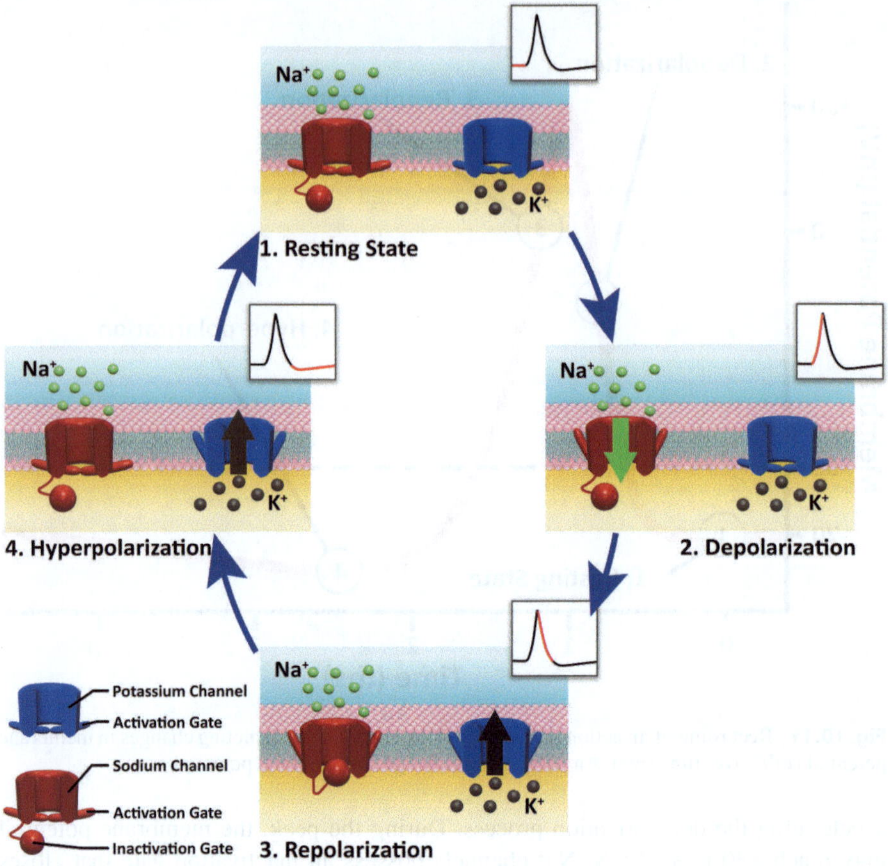

Fig. 10.12 The role of sodium and potassium voltage-gated ion channels in each phase of the action potential. **1. Resting state:** The neuron is at rest, with a stable membrane potential of approximately −70 mV. Na^+ ions are more concentrated outside the cell, while K^+ ions are more concentrated inside. Voltage-gated Na^+ (red) and K^+ (blue) channels are closed. **2. Depolarization:** A nerve impulse from a presynaptic neuron changes the membrane potential to become less negative and moves towards a positive value, reaching around +30 mV. This change in the membrane potential opens sodium channels, allowing a rapid influx of Na^+ ions into the cell (green arrow). **3. Repolarization:** The inactivation gates close sodium channels, and potassium channels open, allowing K^+ ions to leave the cell (black arrow). The membrane potential returns toward the resting negative potential. **4. Hyperpolarization:** K^+ continues to exit the cell as the K^+ channels remain open longer than necessary, causing an overshoot. The membrane potential temporarily becomes more negative than the resting potential

Absolute Refractory Period: The absolute refractory period (ARP) is the interval following the initiation of an action potential during which it is impossible to elicit a second action potential in the neuron, irrespective of the strength of the stimulus applied. After opening, voltage-gated Na^+ channels enter an inactivated state characterized by the closure of the inactivation gate (Fig. 10.12). In this state, the channels are refractory to further opening until they return to their

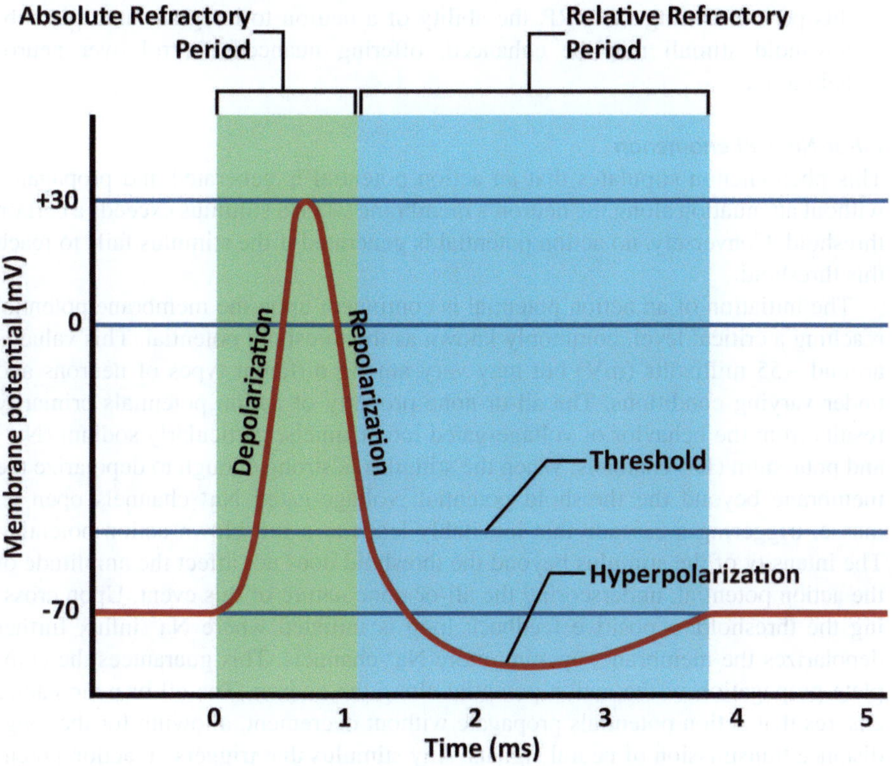

Fig. 10.13 Absolute and relative refractory periods in an action potential

resting state. The transition from the inactivated state to the resting state is sequential and time-dependent, rendering the membrane incapable of supporting another depolarizing event until sufficient numbers of Na^+ channels have returned to their resting state. The ARP typically lasts approximately 1–2 milliseconds. The ARP ensures that action potentials propagate in one direction along the axon, from the axon hillock to the axon terminals, since the membrane behind the advancing action potential front is refractory. It also sets an upper limit on the frequency of action potentials that a neuron can generate, thereby regulating the rate of information transmission.

Relative Refractory Period: The relative refractory period (RRP) follows the absolute refractory period and is characterized by a time during which a stronger-than-normal stimulus is required to initiate a second action potential. During the RRP, a fraction of voltage-gated Na^+ channels have returned to their resting state but may not be sufficient for a standard threshold stimulus to initiate an action potential. Voltage-gated K^+ channels that opened during the repolarization phase may still be open, contributing to a hyperpolarized state and requiring a more potent stimulus to reach the threshold for action potential initiation. The RRP allows neurons to be more selective in responding to incoming stimuli, ensuring that only more powerful or sustained inputs can initiate action potentials during this period. During the RRP, the ability of a neuron to integrate multiple subthreshold stimuli may be enhanced, offering nuanced control over neuronal output.

All-or-None Phenomenon

This phenomenon stipulates that an action potential is generated and propagated without attenuation along the neuron's membrane when a stimulus exceeds a certain threshold. Conversely, no action potential is generated if the stimulus fails to reach this threshold.

The initiation of an action potential is contingent upon the membrane potential reaching a critical level, commonly known as the threshold potential. This value is around −55 millivolts (mV) but may vary among different types of neurons and under varying conditions. The all-or-none property of action potentials primarily results from the behavior of voltage-gated ion channels, particularly sodium (Na^+) and potassium (K^+) channels. When the stimulus is strong enough to depolarize the membrane beyond the threshold potential, voltage-gated Na^+ channels open en masse, triggering a cascade that inevitably leads to a full-blown action potential. The intensity of the stimulus beyond the threshold does not affect the amplitude of the action potential, underscoring the all-or-none nature of this event. Upon crossing the threshold, a positive feedback loop is initiated where Na^+ influx further depolarizes the membrane, opening more Na^+ channels. This guarantees the complete propagation of the action potential along the neuron. The all-or-none nature ensures that action potentials propagate without decrement, allowing for the long-distance transmission of neural signals. Any stimulus that triggers an action potential at the axon hillock will lead to an impulse that reaches the axon terminals with the same amplitude. An all-or-none action potential is metabolically efficient for the neuron because the same amount of ionic rebalancing via the Na^+/K^+ pump is required regardless of the intensity of the triggering stimulus as long as it exceeds the threshold.

Propagation of Action Potential

The action potential is a dynamic event that, once initiated, must reliably travel from the neuron's cell body to the axon terminals to ensure the transmission of information to subsequent neurons or effector cells. When an action potential begins, Na^+ ions rush into the cell locally. The local inward current depolarizes adjacent membrane regions, opening more Na^+ channels in these areas and thus initiating a new action potential. The process continues sequentially along the axon, propagating the action potential. The absolute refractory period ensures that the action potential travels in one direction, preventing backward propagation (Fig. 10.14). The all-or-none principle ensures that the action potential retains its amplitude and form as it propagates.

In unmyelinated axons, the action potential is propagated continuously (Fig. 10.15), with every axon segment undergoing the entire cycle of depolarization and repolarization. Propagation speed is slower in unmyelinated axons and is directly proportional to the axon's diameter. In myelinated axons, the action

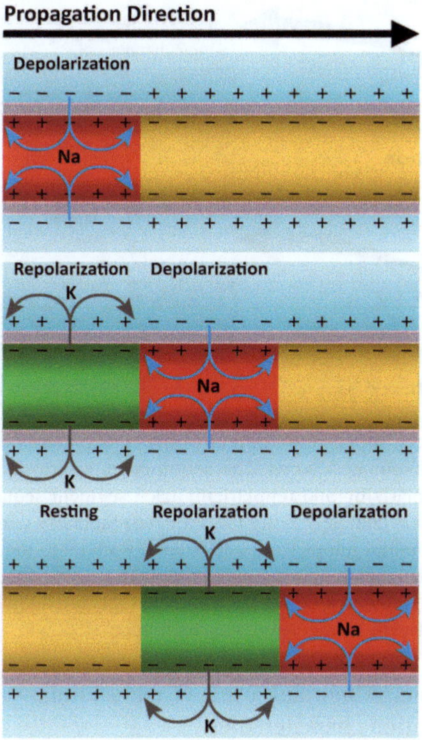

Fig. 10.14 Propagation of an action potential (AP) along an axon. The action potential begins with depolarization, where Na^+ channels open, allowing Na^+ ions to rush into the cell. This influx of positive ions causes the membrane potential to become more positive. Due to ion channels' sequential opening and closing, the action potential propagates along the axon in one direction. Following depolarization, K^+ channels open, allowing K^+ ions to exit the cell. This movement of positive ions out of the cell restores the negative membrane potential, repolarizing the membrane. After repolarization, the membrane returns to its resting state but briefly enters a refractory period where it cannot fire another action potential. This refractory period is the basis for unidirectional propagation of the AP along the axon

potential appears to "jump" between the nodes of Ranvier (saltatory conduction) (Fig. 10.14). Saltatory conduction is faster and more energy-efficient than continuous conduction. Voltage-gated Na^+ channels are concentrated at the nodes of Ranvier, and depolarization occurs only at these nodes, making the conduction more efficient.

Modulation of Propagation

- Axon Diameter: Larger-diameter axons conduct action potentials more rapidly due to lower internal resistance.
- Temperature: Lower temperatures can slow enzymatic activities and ion channel kinetics, decreasing propagation speed.

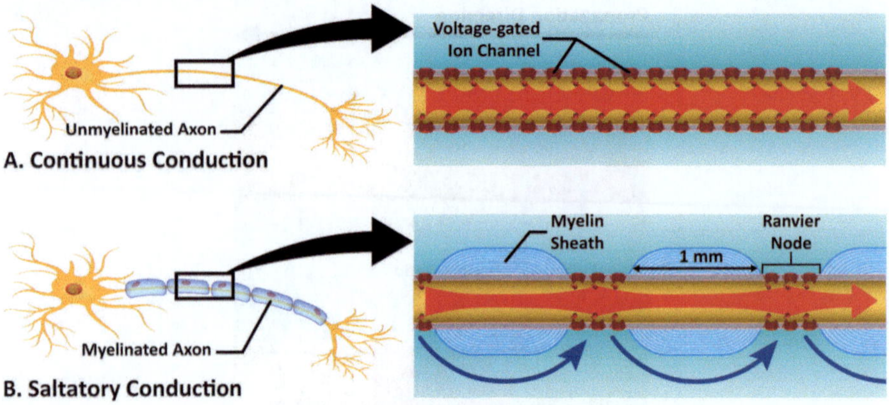

Fig. 10.15 Action potential propagation in unmyelinated and myelinated axons. (**a**) Continuous conduction occurs in unmyelinated axons. The action potential travels along the entire length of the axon membrane, depolarizing and repolarizing each segment sequentially, thus opening and closing all voltage-gated ion channels along the axon. (**b**) Saltatory conduction occurs in myelinated axons. The action potential jumps from one node of Ranvier to the next, skipping the myelinated segments. The myelin sheath acts as an insulator, increasing the speed of conduction. The nodes of Ranvier are gaps in the myelin sheath where voltage-gated ion channels are concentrated, allowing the action potential to be regenerated at each node

- Ionic Concentrations: Abnormal levels of extracellular or intracellular ions can modify the propagation characteristics.

10.1.3 Synapses

Neural synapses are the neural network's primary information exchange and signal modulation sites. They are the junctions where neurons communicate with each other or with effector cells. Neural synapses can be classified based on their structure or function.

10.1.3.1 Structural Classification of Synapses

Electrical Synapses

Two neurons connect through gap junctions in electrical synapses (Fig. 10.16). Gap junctions are aggregates of intercellular channels that establish cytoplasmic continuity between adjacent cells. Gap junctions allow for the passage of ions, leading to the depolarization or hyperpolarization of the postsynaptic cell almost instantaneously. Transmission speed is substantially higher than chemical synapses because it skips the delay associated with neurotransmitter release and reception. Electrical synapses are generally symmetrical, allowing for bidirectional information flow. This is critical in networks requiring synchronized activity, such as cardiac muscle and specific neural circuits.

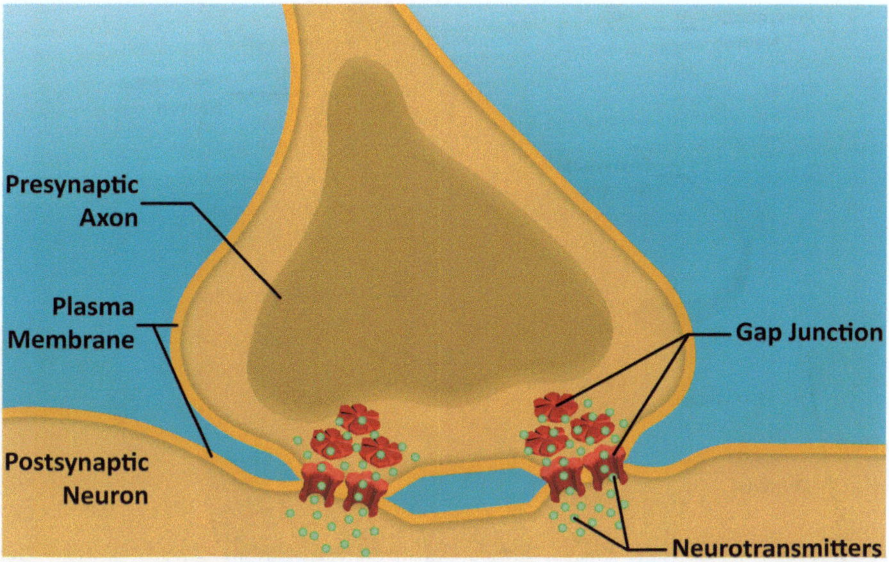

Fig. 10.16 Electrical synapse: Gap junction allows direct communication between the cytoplasm of adjacent neurons through channel proteins, facilitating the rapid transmission of electrical signals

Chemical Synapses

Chemical synapses (Fig. 10.17) serve as the primary sites for transmitting information between neurons in the nervous system. Unlike electrical synapses, which allow for direct ionic flow between adjoining cells, chemical synapses employ neurotransmitters as signaling molecules. Chemical synapses have several types:

Axodendritic Synapses: Occur between the axon of one neuron and the dendrite of another.

Axosomatic Synapses: Involve the axon of one neuron and the cell body of another.

Axoaxonic Synapses: Found between two axons.

Dendrodendritic Synapses: Exist between dendrites of different neurons.

The general structure of a chemical synapse includes the following components:

- Presynaptic Terminal: This is the specialized portion of the neuron that contains synaptic vesicles filled with neurotransmitters. It also houses the molecular machinery necessary for vesicle fusion and neurotransmitter release.
- Synaptic Cleft: This extracellular space separates the pre- and postsynaptic membranes, typically spanning a distance of 20–30 nanometers.
- Postsynaptic Density: A specialized region of the postsynaptic membrane that contains neurotransmitter receptors and various signaling molecules. This structure is vital for initiating downstream signal transduction pathways.

The signal transduction mechanism (Fig. 10.17) in chemical synapses is by second messenger systems, allowing for more complex signal transduction pathways

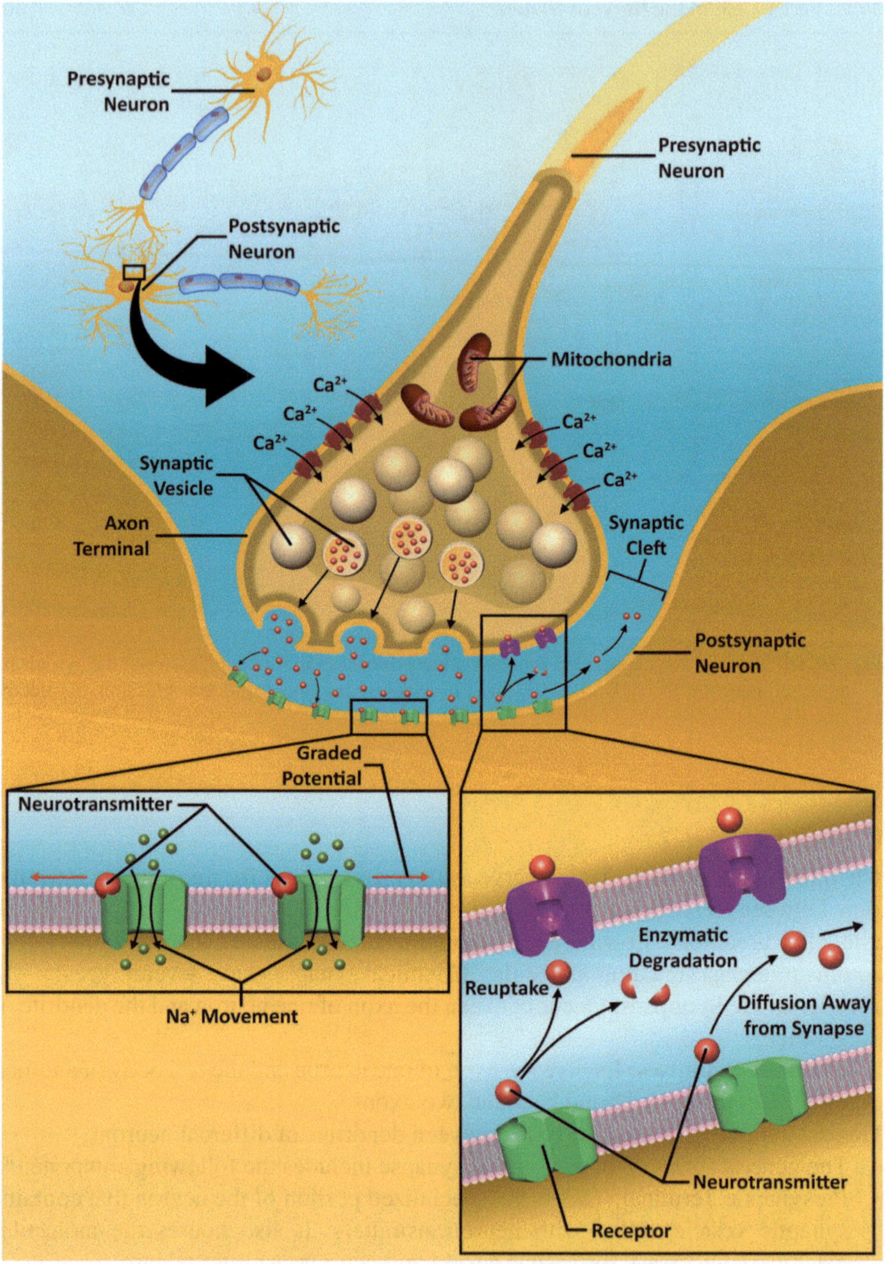

Fig. 10.17 Chemical synapse: The sequence of events in synaptic transmission. **Action potential arrival:** An action potential arrives at the axon terminal of the presynaptic neuron. **Calcium influx:** Voltage-gated calcium channels open, allowing calcium ions (Ca^{2+}) to enter the axon terminal. **Neurotransmitter release:** The influx of calcium triggers synaptic vesicles to fuse with the presynaptic membrane and release neurotransmitters into the synaptic cleft. **Binding to receptors:** Neurotransmitters diffuse across the synaptic cleft and bind to receptors on the postsynaptic membrane. **Postsynaptic potential:** This binding causes ion channels to open or close, leading to changes in the postsynaptic cell's membrane potential (graded potential). **Signal termination:** Neurotransmitters are either taken back into the presynaptic neuron through reuptake mechanisms or degraded by enzymes in the synaptic cleft

and greater modulation. An incoming action potential depolarizes the presynaptic terminal. This depolarization opens voltage-gated calcium channels, allowing calcium ions (Ca^{2+}) to enter the presynaptic cell. Calcium ions bind to synaptotagmin, a protein on the vesicle membrane, facilitating the fusion of synaptic vesicles with the presynaptic membrane. This fusion results in the release of neurotransmitters into the synaptic cleft by exocytosis. Released neurotransmitters diffuse across the synaptic cleft and bind to specific receptors on the postsynaptic density. Depending on the receptor type, this can result in either a postsynaptic depolarization (excitatory) or hyperpolarization (inhibitory). Neurotransmitters are often taken back into the presynaptic terminal for recycling. Some neurotransmitters are enzymatically broken down. A minor portion might diffuse away from the synaptic cleft.

Mixed Chemoelectrical Synapses

Mixed chemoelectrical synapses are unique structures with chemical and electrical synapse features, enabling them to engage in more complex and diversified modes of intercellular communication. These hybrid synapses are particularly relevant in specialized neural circuits requiring rapid signaling and signal modulation.

Synaptic Delay

Synaptic delay is the temporal interval between the arrival of an action potential at the presynaptic terminal and the onset of the postsynaptic response. Although often measured in milliseconds, this temporal feature significantly shapes neural dynamics and information processing. Typically, synaptic delays can range from approximately 0.3 to 5 milliseconds in chemical synapses. Due to their direct ionic coupling, electrical synapses experience minimal delay, often 0.1 milliseconds or less.

10.1.3.2 Functional Classification of Synapses

Excitatory Synapses: Facilitate the generation of an action potential in the postsynaptic cell.

Inhibitory Synapses: Reduce the likelihood of an action potential in the postsynaptic cell.

10.1.3.3 Neurotransmitters

Neurotransmitters are endogenous chemicals that mediate signaling across synapses (Table 10.1). The receptors they bind to are specialized proteins that facilitate the transduction of these signals into cellular responses. Neurotransmitters can be classified by their chemical structure: amino acids, monoamines, acetylcholine, peptides, purines, and gases.

10.1.4 Neural Processing

Neural processing refers to the computational methods employed by the nervous system to receive, integrate, and transmit information. Neural processing forms the core of all cognitive and behavioral functions and is characterized by various

Table 10.1 Neurotransmitters: structure and function

Neurotransmitter	Structure	Function	Inhibitory/ Stimulatory
Glutamate	Amino acid	Excitatory neurotransmission	Stimulatory
GABA	Amino acid	Inhibitory neurotransmission	Inhibitory
Glycine	Amino acid	Inhibitory neurotransmission	Inhibitory
Dopamine	Monoamine	Reward, motivation, motor control, regulation of prolactin secretion	Both, depending on receptor type
Norepinephrine	Monoamine	Arousal, alertness, blood pressure regulation	Stimulatory
Epinephrine	Monoamine	Hormonal and neurotransmitter roles, stress response	Stimulatory
Serotonin	Monoamine	Mood regulation, appetite, sleep, memory, sexual desire	Mostly inhibitory
Histamine	Monoamine	Wakefulness, appetite control, learning, memory	Stimulatory
Acetylcholine	Ester	Muscle activation, arousal, attention, memory	Both, depending on receptor type
Substance P	Peptide	Pain perception, stress response	Stimulatory
Endorphins	Peptides	Pain relief, reward	Inhibitory
Neuropeptide Y	Peptide	Appetite regulation, energy balance, stress response	Stimulatory
Adenosine	Purine	Modulates sleep and cardiovascular function	Inhibitory
Nitric oxide	Gas	Vasodilation, neurotransmission	Stimulatory
Anandamide	Fatty acid amide	Pain, mood, appetite, memory	Inhibitory
Tyramine	Trace amine	Blood pressure regulation	Stimulatory
Phenethylamine	Trace amine	Mood, focus	Stimulatory
ATP	Nucleoside triphosphate	Cellular energy transfer, neurotransmission	Both, depending on receptor type

patterns of neural activity and connectivity. Serial and parallel processing architectures represent two fundamental approaches in the organization and functioning of neural networks. While these are often viewed as discrete entities, they frequently operate in tandem, contributing to the robustness and flexibility of the nervous system.

Serial Processing: Serial processing often follows a linear or hierarchical model where information flows sequentially from one stage to another. In this design, the connections between neurons or neural assemblies are strictly unidirectional, thus facilitating linear information flow. Examples include reflex arcs, where sensory input is processed and sequentially leads to a motor output.

Parallel Processing: Here, multiple pathways or circuits simultaneously process different aspects of the same information. Parallel pathways often diverge from a common source and may eventually converge to a common output, but the data is processed independently in between. Examples include visual processing, different pathways process color, motion, and shape in parallel.

Parallel processing allows for the simultaneous maintenance and manipulation of multiple information units, while serial processing might be employed for tasks that require focused attention. While selecting and initiating a specific motor action might occur serially, controlling different muscle groups happens in parallel. Emotional responses often require both serial processing (e.g., evaluating the significance of a stimulus) and parallel processing (e.g., physiological responses, facial expressions).

10.1.5 Neural Circuits

Distinct types of neural circuits employ various configurations of neuronal connections to fulfill specific computational tasks.

Classification of Neural Circuits
Simple Circuits

- Chain Circuit: This circuit involves a linear sequence of neurons, often found in reflex arcs.
- Loop Circuit: This circuit consists of neurons connected in a closed loop, allowing for feedback mechanisms.

Complex Circuits

- Parallel Circuit: Multiple pathways allow simultaneous processing of the same information.
- Reverberating Circuit: A circuit with feedback loops that can sustain activity after removing the initiating stimulus.
- Diverging Circuit: A single neuron synapses onto multiple downstream neurons, amplifying the signal.
- Converging Circuit: Multiple neurons synapse onto a single neuron, often to integrate information.

Specialized Circuits

- Oscillatory Circuits: Generate rhythmic patterns of activity commonly found in the production of brain waves.
- Polysynaptic Circuits: Involve multiple synaptic connections and are often implicated in complex computational tasks.

10.2 Part 2: Nervous System Structures

This section will discuss the nervous system's architectural framework and function (anatomical configuration, functional roles, and interconnections). We will examine the central nervous system, comprising the brain and spinal cord, and the peripheral nervous system, including nerves and ganglia.

10.2.1 General Organization

The nervous system can be divided into two principal components:

- The central nervous system (CNS) comprises the brain and spinal cord and serves as the primary processing center for the nervous system.
- The peripheral nervous system (PNS) consists of nerves and ganglia that extend beyond the CNS and functions as the communication lines between the CNS and the rest of the body.

10.2.2 Central Nervous System

The central nervous system (CNS), consisting of the brain and spinal cord, is differentiated into two types of neural tissue: white matter and gray matter. These tissues serve distinct yet interrelated functions, and their organization is pivotal for the functional integrity of the CNS. Gray matter comprises neuronal cell bodies, dendrites, and unmyelinated axons, while white matter consists mainly of myelinated axons. Gray matter is the primary site for processing and integrating information and houses various specialized regions, each responsible for specific functions such as sensory perception, voluntary muscle movement, and cognition. White matter is primarily responsible for transmitting neural signals and facilitates rapid communication between different gray matter areas within the spinal cord and between the spinal cord and the brain.

The brain consists of cerebral hemispheres (cerebrum), diencephalon, brainstem, cerebellum, limbic system, and ventricular system, and is located inside the cranial cavity (Table 10.2) (Fig. 10.18). The spinal cord is inside the spinal column and extends from the foramen magnum to the first or second lumbar vertebrae level (Fig. 10.18).

10.2.2.1 White and Gray Matter (Fig. 10.19)

Organization of White and Gray Matter in the Cerebrum

Gray Matter: Cortical gray matter is the cerebrum's outer layer and is responsible for higher-order brain functions such as perception, voluntary muscle movement, and thought processes. Subcortical gray matter includes clusters of gray matter within the white matter, known as nuclei, such as the basal ganglia, thalamus, hypothalamus, and amygdala.

White Matter: Found beneath the cortical gray matter, white matter forms the internal structure of the cerebrum. It is organized into tracts, or bundles of axons, that allow communication between different brain parts.

Organization of White and Gray Matter in the Diencephalon

The thalamus, hypothalamus, and epithalamus are predominantly gray matter structures but are surrounded and penetrated by white matter tracts connecting them to the cerebrum and other CNS structures.

Table 10.2 Brain components: location, structure, and function

Name	Location	Structure	Function
Cerebrum	Largest part of the brain, occupying most of the skull's upper part	Divided into two hemispheres (left and right), each divided into four lobes (frontal, parietal, temporal, occipital)	Responsible for higher brain functions, including thought, action, and emotion
Cerebellum	Located under the cerebrum at the back of the skull	Smaller than the cerebrum but highly folded; divided into two hemispheres	Coordinates muscle movements, maintains posture and balance
Brainstem	Situated beneath the cerebrum and in front of the cerebellum	Comprises the midbrain, pons, and medulla oblongata	Controls essential life functions like breathing, heartbeat, and blood pressure
Hippocampus	Located in the medial temporal lobe	A curved structure resembling a seahorse	Key role in memory formation and spatial navigation
Amygdala	Deep within the temporal lobes, adjacent to the hippocampus	Almond-shaped	Involved in emotion processing, particularly fear and pleasure
Hypothalamus	Located below the thalamus, just above the brainstem	Small, comprising various nuclei and neural networks	Regulates various homeostatic processes, including temperature control, thirst, hunger, and sleep
Thalamus	Positioned at the top of the brainstem, between the cerebral cortex and the midbrain	Composed of two symmetrical structures	Acts primarily as a relay station for sensory and motor signals to the cerebral cortex
Corpus callosum	Lies beneath the cerebral cortex at the longitudinal fissure	A wide, flat bundle of neural fibers	Connects the left and right hemispheres of the brain, allowing for communication between them
Basal ganglia	A group of nuclei lying deep within the cerebral hemispheres	Includes structures like the striatum, globus pallidus, and substantia nigra	Involved in the control of voluntary motor movements, procedural learning, routine behaviors, and emotion
Ventricular system	Distributed throughout the brain	Consists of four interconnected ventricles	Circulates cerebrospinal fluid (CSF) to cushion the brain, remove waste, and provide nutrients

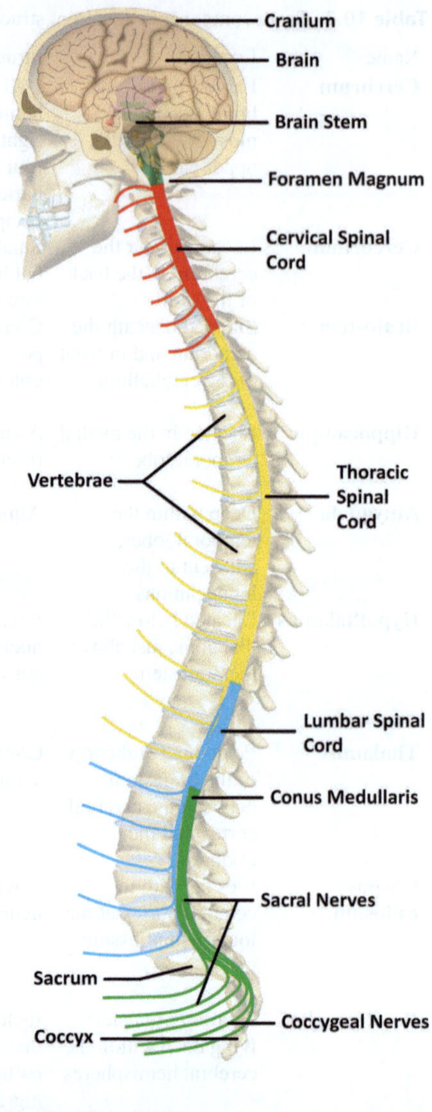

Fig. 10.18 Central nervous system structure

Organization of White and Gray Matter in the Cerebellum

Like the cerebrum, the cerebellum contains an outer layer of gray matter (the cerebellar cortex) and an inner layer of white matter.

Organization of White and Gray Matter in the Spinal Cord

Gray Matter: Located centrally, the gray matter in the spinal cord is shaped like a butterfly or the letter "H." It contains dorsal and ventral horns, which respectively handle sensory and motor functions.

White Matter: Surrounds the gray matter and is divided into anterior, posterior, and lateral columns or funiculi. These columns contain ascending and descending tracts that relay information between the spinal cord and the brain.

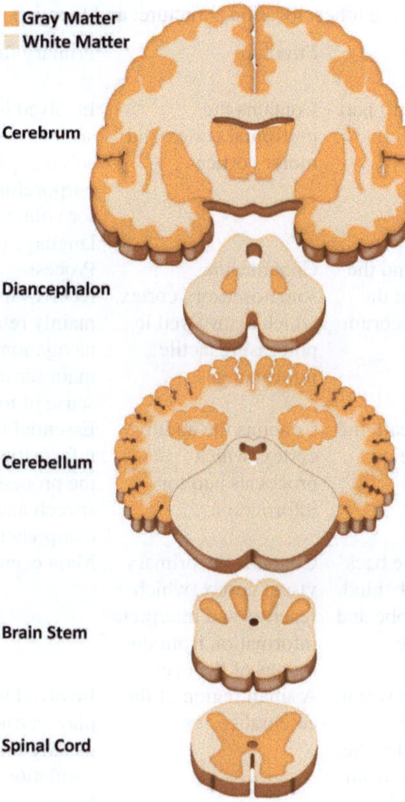

Fig. 10.19 Cross-sectional views of the central nervous system showing the distribution of gray and white matter. **Cerebrum:** The gray matter is primarily located in the outer cortex, while the white matter lies beneath it. **Diencephalon:** The gray matter is compact in the center, surrounded by a large area of white matter. **Cerebellum:** The cerebellar cortex comprises gray matter, while white matter is located internally. **Brain stem:** The gray matter in the brain stem is found in discrete nuclei, while white matter forms tracts passing through. **Spinal cord:** In the spinal cord, the gray matter forms an H-shaped structure in the center, surrounded by white matter on the outside

10.2.2.2 Brain Hemispheres

The cerebral hemispheres, the most prominent components of the human brain, serve as the principal substrates for higher cognitive functions, including perception, voluntary motor actions, language, and abstract reasoning. Comprising the frontal, parietal, temporal, occipital, and insular lobes (Table 10.3), the hemispheres manifest an intricate architecture of gyri and sulci and house various specialized areas for complex tasks. Gyri (singular: Gyrus) are the elevated folds of the cerebral cortex (Table 10.4). Sulci (singular: Sulcus) are the shallow grooves that separate gyri. A fissure is a deep groove that delineates major regions or lobes of the brain (Table 10.5) (Fig. 10.20).

Table 10.3 Brain hemisphere lobes: location, structure, and function

Name	Location in cerebrum	Structure	Primary function
Frontal lobe	Anterior (front) part of the cerebrum	Contains the prefrontal cortex and motor cortices	Involved in executive functions such as decision-making, problem-solving, planning, and control of purposeful behaviors; also crucial for voluntary movement and language processing
Parietal lobe	Located behind the frontal lobe at the top of the cerebrum	Contains the somatosensory cortex, which is involved in processing tactile information	Processes sensory information it receives from the outside world, mainly relating to spatial sense and navigation (proprioception), the main sensory receptive area for the sense of touch
Temporal lobe	Located beneath the lateral fissure on both cerebral hemispheres	Contains the auditory cortex, which processes auditory information	Essential for processing auditory information and is also crucial for the processing of semantics in both speech and vision, including the comprehension of language
Occipital lobe	Located at the back of the brain, behind the parietal lobe and temporal lobe	Contains the primary visual cortex, which receives and interprets information from the retinas of the eyes	Main center for visual processing
Insular cortex (insula)	Located deep within the lateral sulcus, which separates the temporal lobe from the frontal and parietal lobes	A small region of the cerebral cortex	Involved in consciousness and plays a role in diverse functions usually linked to emotion or the regulation of the body's homeostasis, including perception, motor control, self-awareness, cognitive functioning, and interpersonal experience
Cingulate cortex	Positioned immediately above the corpus callosum	Divided into an anterior and posterior cingulate	Plays a role in emotion formation and processing, learning, and memory; also involved in autonomic functions, such as regulating blood pressure and heart rate

Cerebral Cortex (Gray Matter)

The cerebral cortex, often deemed the "seat of consciousness," is broadly organized into functional areas that manage an array of cognitive, sensory, and motor functions. The functional areas of the cerebral cortex are typically categorized into three primary types (Table 10.6) (Fig. 10.21):

- Motor Areas: Responsible for controlling voluntary movements.
- Sensory Areas: Involved in the perception and interpretation of sensory stimuli.
- Association Areas: Engage in complex processing, integrating information from sensory and motor areas to facilitate higher cognitive functions.

Table 10.4 Major gyri of brain: location, structure, and function

Name	Location	Structure	Function
Precentral gyrus	Located in the frontal lobe, anterior to the central sulcus	A prominent gyrus running parallel to the central sulcus	Primary motor cortex: responsible for voluntary movement
Postcentral gyrus	Located in the parietal lobe, posterior to the central sulcus	Runs parallel to the central sulcus	Primary somatosensory cortex processes tactile and proprioceptive information
Superior frontal gyrus	Located in the frontal lobe, superior to the inferior frontal gyrus	Extends from the front to the back of the frontal lobe	Involved in self-awareness, coordination of complex movements, and social behavior
Middle frontal gyrus	Located in the frontal lobe, between the superior and inferior frontal gyri	Occupies the middle portion of the frontal lobe	Involved in attention, working memory, and executive functions
Inferior frontal gyrus	Located in the frontal lobe, inferior to the middle frontal gyrus	Divided into three parts: pars opercularis, pars triangularis, and pars orbitalis	Contains Broca's area (in the dominant hemisphere), essential for language production
Superior temporal gyrus	Located in the temporal lobe, above the middle temporal gyrus	Runs along the lateral side of the temporal lobe	Contains the primary auditory cortex and Wernicke's area (in the dominant hemisphere), crucial for language comprehension
Middle temporal gyrus	Located in the temporal lobe, between the superior and inferior temporal gyri	Extends along the length of the temporal lobe	Involved in processing auditory and visual information and in semantic memory
Inferior temporal gyrus	Located in the temporal lobe, below the middle temporal gyrus	Situated at the lower part of the temporal lobe	Involved in visual processing, particularly complex stimuli like faces and scenes
Cingulate gyrus	Located above the corpus callosum in the medial aspect of the brain	A curved structure following the contour of the corpus callosum	Part of the limbic system, involved in emotional processing, learning, and memory
Parahippocampal gyrus	Located in the medial temporal lobe	Wraps around the hippocampus	Involved in memory encoding and retrieval

Table 10.5 Major sulci and fissures of the brain: location, structure, and features

Name	Location	Structure	Important features
Longitudinal fissure	Extends from the front to the back of the brain, dividing it into left and right hemispheres	Deep groove running along the midline of the brain	Contains the falx cerebri, a fold of dura mater, and is the largest fissure in the brain
Lateral (sylvian) fissure	Separates the frontal and parietal lobes from the temporal lobe on each side of the brain	Runs diagonally from the bottom of the frontal lobe to the top of the temporal lobe	Also reveals the insula when its lips are pulled apart and contains branches of the middle cerebral artery
Transverse cerebral fissure	Separates the cerebrum from the cerebellum	Lies between the underside of the occipital lobes and the top of the cerebellum	Contains the tentorium cerebelli, another fold of dura mater, and provides a passage for veins and the great cerebral vein
Parieto-occipital fissure	Located on the medial surface of the hemisphere, separating the parietal lobe from the occipital lobe	Not as prominent as other fissures and is mainly seen on the medial surface of the brain	Important for distinguishing between areas responsible for sensory processing and visual processing
Calcarine fissure	Located in the occipital lobe	Runs horizontally across the occipital lobe on the medial side	Crucial for processing visual information, with different parts of the fissure processing different parts of the visual field
Central sulcus (sulcus of Rolando)	Divides the frontal lobe from the parietal lobe	Prominent sulcus running roughly from the top to the bottom of the lateral side of the brain	Acts as a crucial landmark for mapping cerebral functions
Lateral sulcus (sylvian fissure)	Separates the temporal lobe from the frontal and parietal lobes	One of the earliest-developing and most prominent features of the human brain	Contains essential structures, including the insula and parts of the auditory cortex
Parieto-occipital sulcus	Separates the parietal lobe from the occipital lobe	Visible on the medial surface of the brain	Essential for dividing visual processing regions in the occipital lobe from sensory processing regions in the parietal lobe
Calcarine sulcus	Located in the occipital lobe	Prominent sulcus on the medial surface of the brain	The location of the visual cortex means this sulcus is vital in processing visual information
Cingulate sulcus	Located just above the corpus callosum	Runs parallel to the curve of the corpus callosum on the medial surface of the brain	Essential for separating areas involved in cognitive functions and emotional processing

(continued)

Table 10.5 (continued)

Name	Location	Structure	Important features
Precentral sulcus	Located in the frontal lobe, anterior to the central sulcus	Runs parallel and anterior to the central sulcus	Separates the precentral gyrus (primary motor cortex) from the rest of the frontal lobe
Postcentral sulcus	Located in the parietal lobe, posterior to the central sulcus	Runs parallel and posterior to the central sulcus	Separates the postcentral gyrus (primary somatosensory cortex) from the rest of the parietal lobe
Superior temporal sulcus	Located on the lateral surface of the temporal lobe	Runs parallel to the lateral sulcus, along the length of the temporal lobe	Critical for language comprehension and social perception

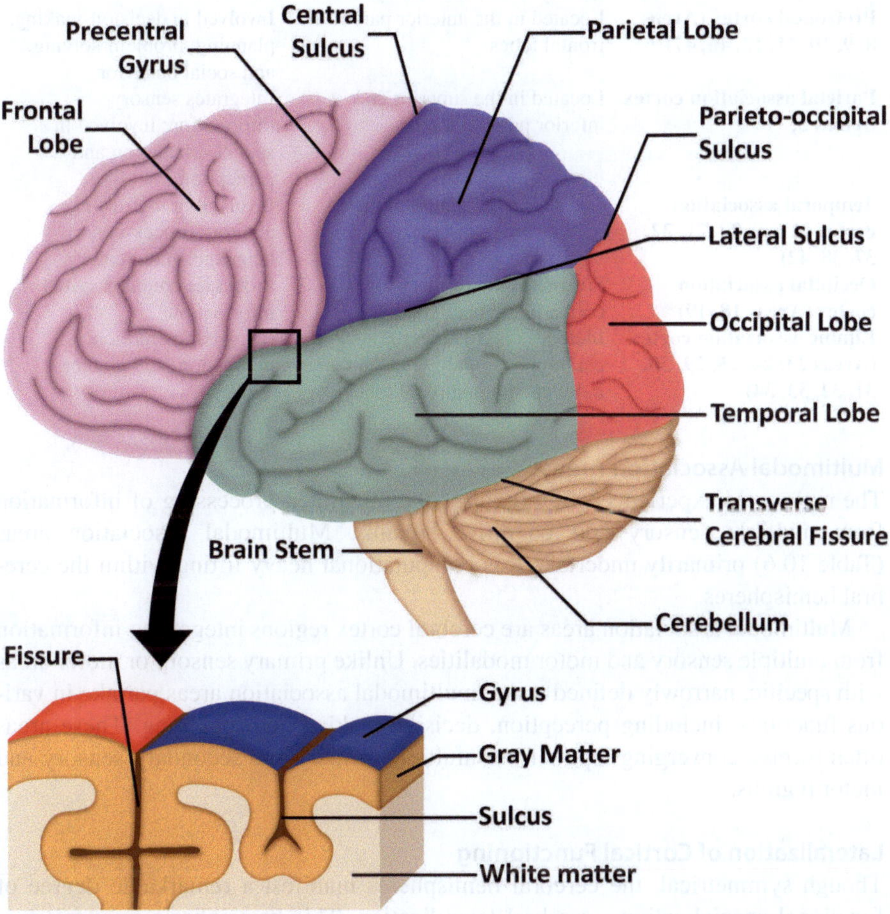

Fig. 10.20 Anatomical brain features: lobes, major sulci, and fissures of the cerebral hemispheres

Table 10.6 Major areas of cerebral cortex: location, function

Name	Location	Function
Primary motor cortex (Area 4)	Located in the precentral gyrus of the frontal lobe	Controls voluntary movements
Primary somatosensory cortex (Areas 1, 2, 3)	Located in the postcentral gyrus of the parietal lobe	Processes somatic sensations (touch, pressure, pain, temperature)
Primary auditory cortex (Area 41)	Located in the superior temporal gyrus	Processes auditory information
Primary visual cortex (Area 17)	Located in the calcarine sulcus of the occipital lobe	Processes visual information
Broca's area (Areas 44, 45)	Located in the inferior frontal gyrus of the frontal lobe, usually in the left hemisphere	Involved in speech production and language processing
Wernicke's area (Area 22)	Located in the posterior part of the superior temporal gyrus, usually in the left hemisphere	Involved in language comprehension
Prefrontal cortex (Areas 8, 9, 10, 11, 12, 46, 47)	Located in the anterior part of the frontal lobes	Involved in decision-making, planning, problem-solving, and social behavior
Parietal association cortex (Areas 5, 7)	Located in the superior and inferior parietal lobules	Integrates sensory information; involved in spatial orientation and perception
Temporal association cortex (Areas 20, 21, 22, 37, 38, 42)	Located in the temporal lobes	Involved in high-level auditory processing, language, and memory
Occipital association cortex (Areas 18, 19)	Surrounds the primary visual cortex in the occipital lobe	Processes complex visual information
Limbic association cortex (Areas 23, 24, 25, 29, 30, 31, 32, 33, 34)	Includes cingulate gyrus, parahippocampal gyrus, and areas in the medial temporal lobe	Involved in emotion, motivation, and memory

Multimodal Association Areas

The real-world experience often requires the integrated processing of information from multiple sensory and cognitive streams. Multimodal association areas (Table 10.6) primarily undertake this computational heavy lifting within the cerebral hemispheres.

Multimodal association areas are cerebral cortex regions integrating information from multiple sensory and motor modalities. Unlike primary sensory or motor areas with specific, narrowly defined tasks, multimodal association areas partake in various functions, including perception, decision-making, and planning. These areas often receive converging inputs from multiple primary and secondary sensory and motor regions.

Lateralization of Cortical Functioning

Though symmetrical, the cerebral hemispheres manifest a remarkable degree of functional specialization—cerebral lateralization. This hemispheric asymmetry is crucial in determining the range of abilities, behaviors, and cognitive processes

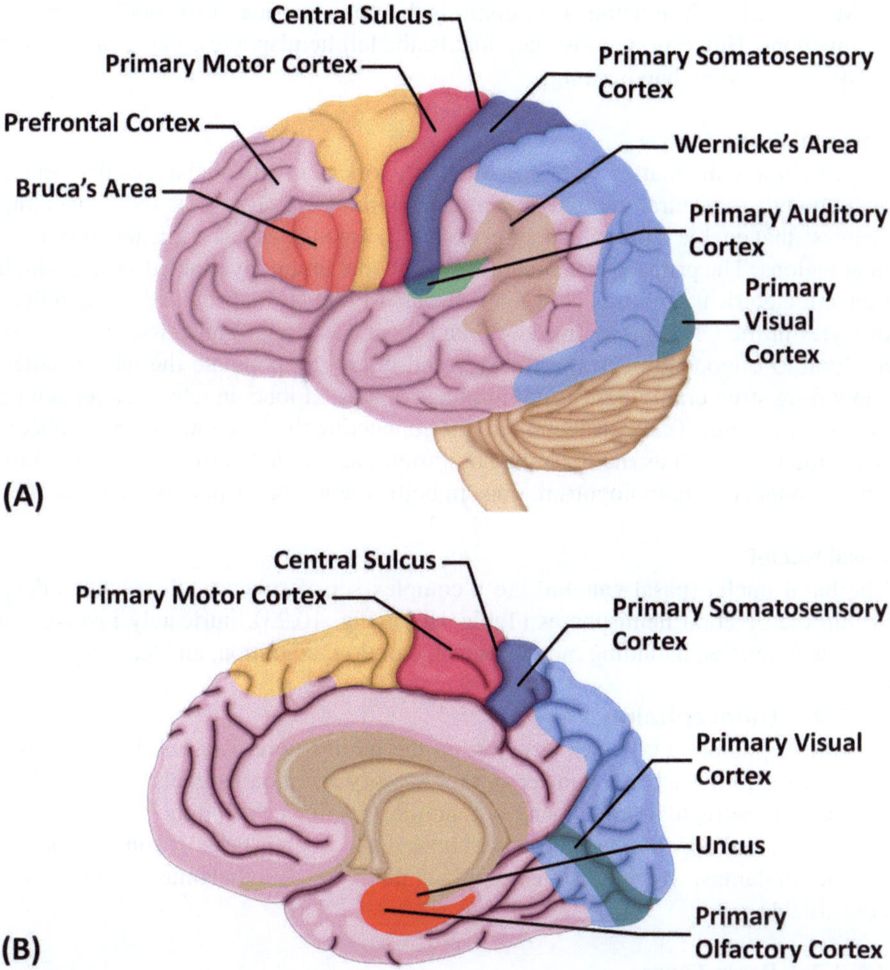

Fig. 10.21 Major functional areas of the cerebral cortex: (**a**) Lateral view. (**b**) Parasagittal medial view

humans can undertake. Research indicates that males might have greater lateralization than females, particularly for language processing and spatial reasoning. Core areas of lateralization include:

- Language Processing: The left hemisphere is specialized for language abilities in most right-handed and left-handed individuals. Broca's area for speech production and Wernicke's area for language comprehension are predominantly found in the left hemisphere.
- Spatial Abilities: The right hemisphere specializes in spatial reasoning, face recognition, and visual perception.
- Emotional Processing: The right hemisphere is predominantly responsible for processing emotions and controlling emotional expression.

- Motor Skills: Each hemisphere controls the opposite side of the body for motor functions. However, in most individuals, the left hemisphere controls fine motor skills, including handwriting.

Cerebral White Matter

The cerebral white matter is an intricate meshwork of myelinated axons that serves as a conduit for neural communication across the CNS. It acts as the connecting “cables” that enable rapid information exchange among various cortical and subcortical regions. The primary constituents of white matter are myelinated axons, which transmit electrical impulses between neurons. Myelination, provided by oligodendrocytes in the CNS, offers insulation and increases signal transmission speed. In addition to oligodendrocytes, astrocytes and microglia populate the white matter, supporting structural integrity and immune response. Fibers in white matter can be classified as (Fig. 10.22) projection fibers, connecting the cerebral cortex to subcortical structures such as the thalamus, brainstem, and spinal cord, and commissural fibers, connecting homologous regions in both cerebral hemispheres (Table 10.7).

Basal Nuclei

The basal nuclei (basal ganglia) are a complex set of subcortical structures deep within the cerebral hemispheres (Table 10.8) (Fig. 10.23), intricately involved in various functions, including motor control, emotion, cognition, and learning.

10.2.2.3 Diencephalon

The diencephalon, a complex structure in the brain’s core, is a hub for numerous crucial functions, such as a relay center for sensory and motor signals to the cerebral cortex and the regulation of many autonomic functions, including body temperature, sleep, and appetite, hormonal regulation, and homeostasis. It consists mainly of the thalamus, hypothalamus, epithalamus, and subthalamus (Table 10.9) (Fig. 10.24).

10.2.2.4 Brain Stem

The brain stem, located at the base of the brain, is a nexus between the cerebral hemispheres, the cerebellum, and the spinal cord. It maintains essential physiological functions, including respiration, heart rate, blood pressure, digestion, and mediating sensory and motor pathways. The brain stem is traditionally divided into three primary regions: midbrain, pons, and medulla oblongata (Table 10.10) (Fig. 10.24). The brain stem is the origin point for ten of the twelve cranial nerves, which perform sensory and motor functions for the head and neck. It also houses a complex mix of white and grey matter, including ascending and descending tracts and many nuclei.

10.2.2.5 Cerebellum

The cerebellum, a Latin term translating to “little brain,” is often considered the “auto-pilot” of the brain. It plays a vital role in motor control, coordination, posture, and balance. Although it accounts for approximately 10% of the brain’s total

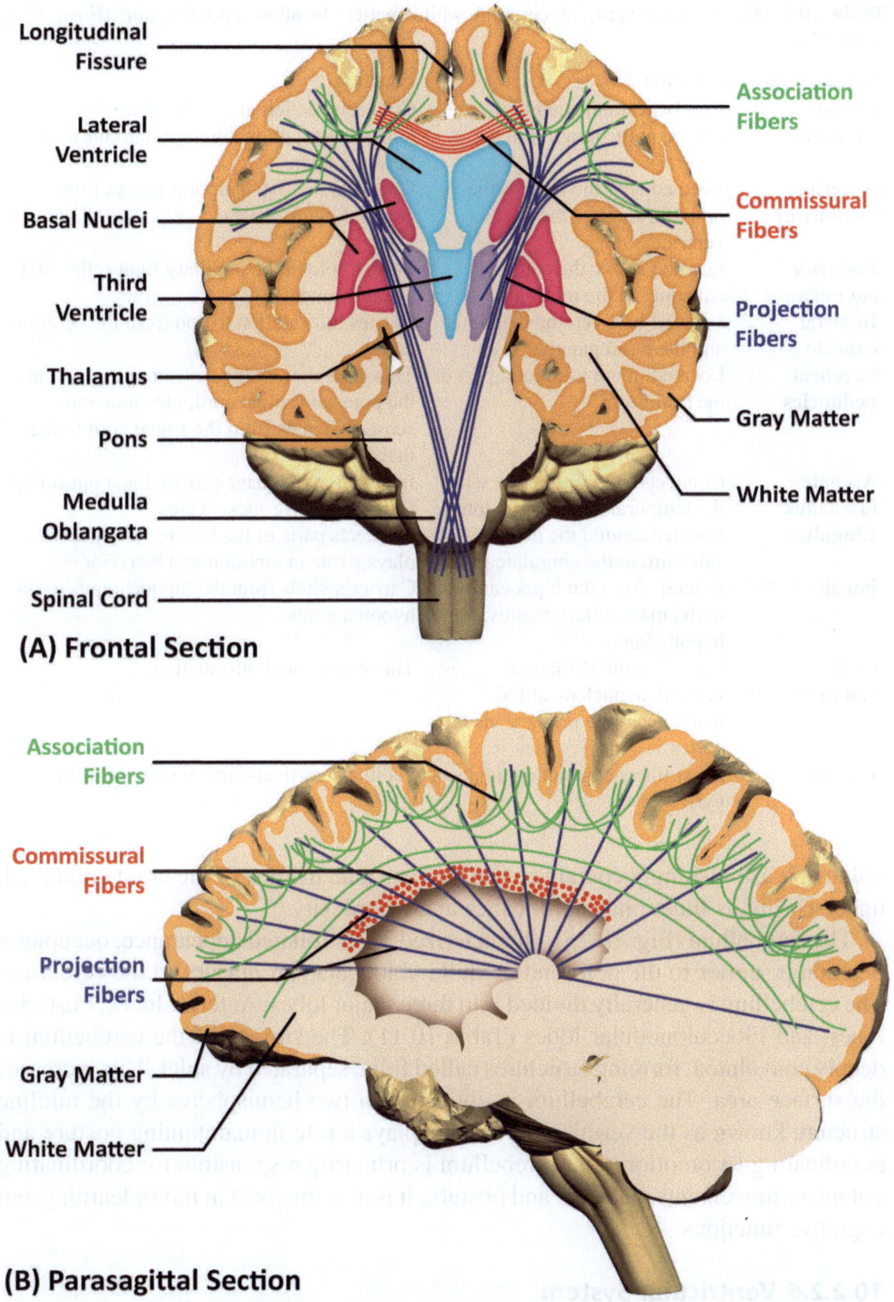

Fig. 10.22 Anatomical illustration of white fiber tracts of the cerebral hemispheres. Commissural, association, and projection fibers run within the cerebrum and between the cerebrum and lower CNS centers. Association fibers (green) connect different parts of the same hemisphere, commissural fibers (red) link corresponding regions of the two hemispheres, and projection fibers (blue) connect the cerebral cortex with lower brain structures and the spinal cord

Table 10.7 Major components of cerebral white matter: location, and function (Figs. 10.22 and 10.23)

Name	Location	Function
Corpus callosum	Lies beneath the cerebral cortex, at the longitudinal fissure	Connects the left and right cerebral hemispheres, allowing for communication between them
Anterior commissure	Located near the base of the brain, in front of the third ventricle	Connects the two temporal lobes of the cerebral hemispheres
Posterior commissure	Located above the cerebral aqueduct in the midbrain	Plays a role in the pupillary light reflex and eye movement
Internal capsule	Located between the thalamus and the basal ganglia	Carries information to and from the cerebral cortex
Cerebral peduncles	Located in the midbrain, part of the brainstem	Transmits motor signals from the cortex to the pons and spinal cord; also transmits sensory signals from the spinal cord to the thalamus
Arcuate fasciculus	Connects the frontal lobe with the temporal and parietal lobes	Involved in language processing, connecting Broca's and Wernicke's areas
Cingulum	Located around the corpus callosum in the cingulate gyrus	Connects parts of the limbic system and plays a role in emotion and behavior
Fornix	Extends from the hippocampus to the mammillary bodies of the hypothalamus	Carries signals from the hippocampus to the hypothalamus
Optic radiation	Extends from the lateral geniculate nucleus of the thalamus to the primary visual cortex	Transmits visual information
U-fibers	Short fibers connecting adjacent gyri	Facilitate cortical-cortical communication

volume, it astonishingly contains more neurons than the rest of the brain combined, underscoring its functional significance and complexity.

The cerebellum (Fig. 10.25), characterized by its foliated appearance, occupies a position posterior to the pons and medulla oblongata and inferior to the cerebrum. The cerebellum is generally divided into three major lobes: Anterior lobes, Posterior lobes, and Flocculonodular lobes (Table 10.11). The surface of the cerebellum is deeply convoluted, forming structures called folia, separated by sulci. This increases the surface area. The cerebellum is divided into two hemispheres by the midline structure known as the vermis. The vermis plays a role in maintaining posture and coordinating locomotion. The cerebellum is primarily responsible for coordinating voluntary movements, balance, and posture; it is also involved in motor learning and cognitive functions.

10.2.2.6 Ventricular System

The ventricular system (Fig. 10.26) is an intricate network of interconnected cavities and channels within the central nervous system (CNS), comprising the brain and spinal cord. It serves primarily as a conduit for cerebrospinal fluid (CSF) and

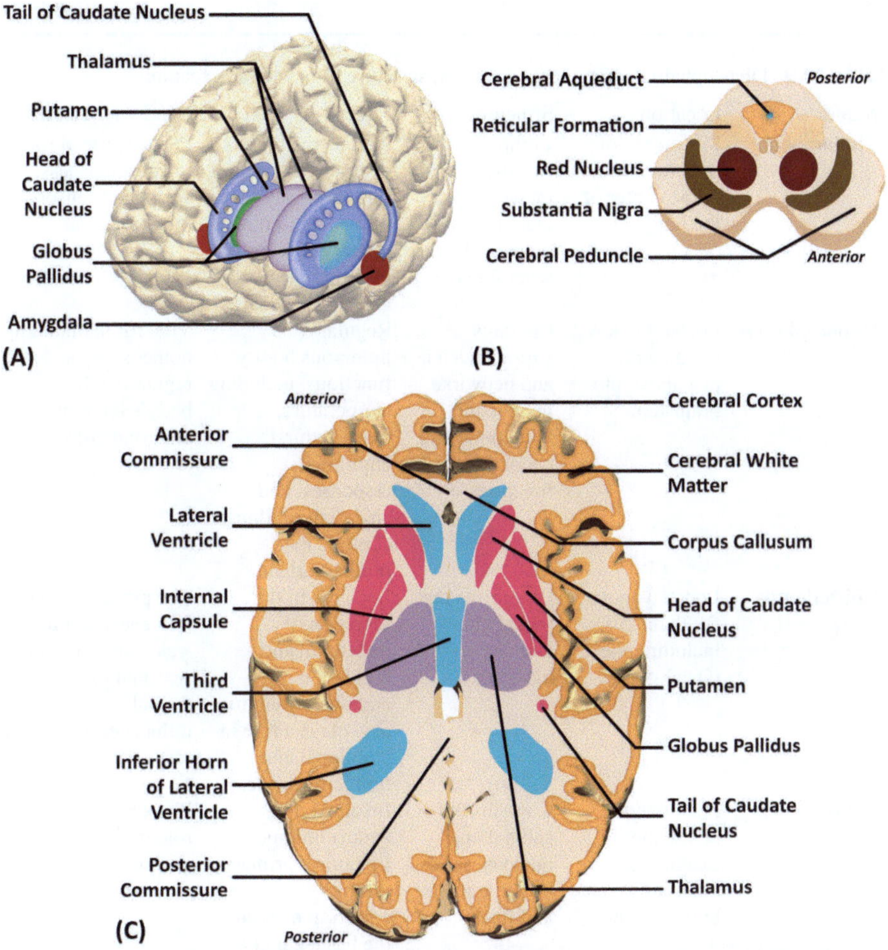

Fig. 10.23 Basal nuclei. (**a**) Three-dimensional view of the basal nuclei, showing their position in the cerebrum. (**b**) Cross section of the midbrain, showing the location of substantia nigra. (**c**) The transverse section of the cerebrum and diencephalon shows the relationship of the basal nuclei to the thalamus and the lateral and third ventricles

Table 10.8 Basal nuclei: location and function

Name	Location	Function
Caudate nucleus	Alongside the lateral ventricles of the brain	Involved in motor processes and has roles in learning, memory, and associative functions
Putamen	Lies at the base of the forebrain, lateral to the caudate nucleus and thalamus	Works in conjunction with the caudate nucleus to regulate movements and influence various types of learning
Globus pallidus	Located medial to the putamen	Involved in the regulation of voluntary movement
Substantia nigra	In the midbrain, adjacent to the ventral tegmental area	Plays an important role in reward, addiction, and movement
Subthalamic nucleus	Located in the diencephalon, below the thalamus	Involved in regulating movements produced by the striatum and globus pallidus

Table 10.9 Diencephalon components: location, structure, function, and features

Name	Location	Structure	Function	Important features
Thalamus	Sits on top of the brainstem, in the center of the brain	Composed of two oval masses of gray matter organized into several nuclei	Acts as the brain's central relay station for sensory information; directs sensory signals to the appropriate areas of the cortex	Plays a crucial role in consciousness, sleep, and sensory interpretation
Hypothalamus	Located below the thalamus, just above the brainstem	Contains various nuclei and networks involved in autonomic and endocrine functions	Regulates numerous bodily functions, including temperature, hunger, thirst, emotional responses, and circadian rhythms; controls the pituitary gland	Vital for maintaining homeostasis and regulating the body's internal environment
Epithalamus	Located above the thalamus, including the pineal gland	Comprises the pineal gland and the habenula	Involved in the secretion of melatonin and regulation of circadian rhythms; also plays a role in emotional and limbic responses	The pineal gland influences sleep-wake patterns and seasonal functions; the habenula influences emotional responses to odors
Subthalamus	Positioned below the thalamus, adjacent to the hypothalamus	Contains the subthalamic nucleus, as well as nerve tracts and fibers	Involved in controlling motor functions, primarily through connections with the basal ganglia	Plays a significant role in the regulation of movements

consists of four interconnected chambers (ventricles) and several connecting ducts (Table 10.12).

10.2.2.7 Meninges

The meninges are specialized connective tissue layers that encapsulate the central nervous system (CNS), consisting of the brain and spinal cord. These protective sheaths function as mechanical shielding, chemical stability maintenance, and fluid and blood circulation facilitation. The meninges comprise three primary layers: the dura mater, the arachnoid mater, and the pia mater (Table 10.13) (Fig. 10.27). These layers collectively form a structural continuum from the skull's inner surface to the neural tissue, creating separate compartments and potential spaces.

10.2.2.8 Cerebrospinal Fluid (CSF)

Cerebrospinal fluid (CSF) (Table 10.14) is a clear, colorless physiological fluid that envelops the brain and spinal cord. The primary site for the production of CSF is the

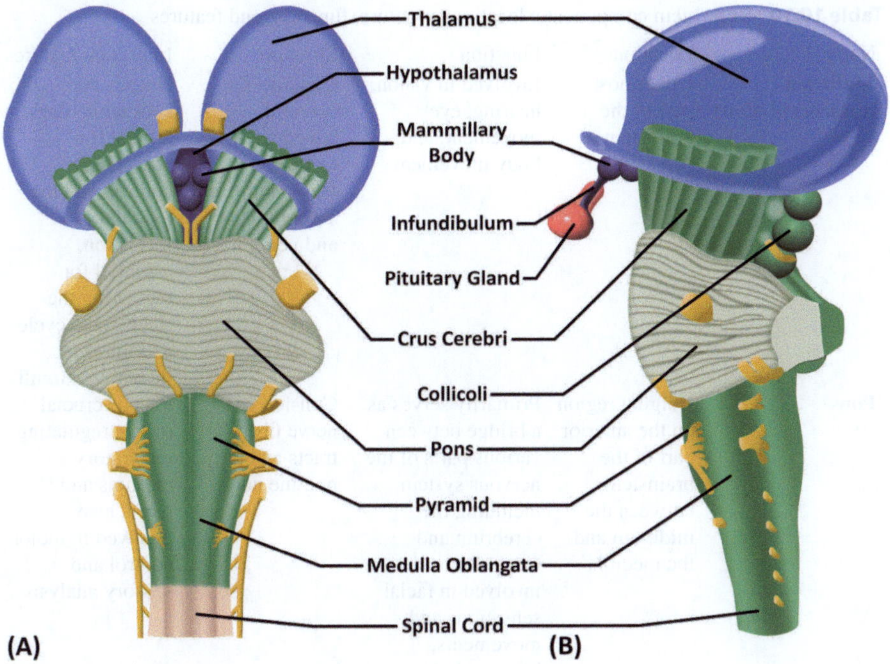

Fig. 10.24 The diencephalon (purple) and the brain stem (green): (**a**) Anterior view. (**b**) Lateral view

choroid plexus (Fig. 10.28), a specialized vascular structure located within the brain's ventricular system. Ependymal cells lining the choroid plexus filter plasma to produce CSF, secreting approximately 500–700 mL daily. The CSF initially circulates through the interconnected chambers of the ventricular system (Fig. 10.29): from the lateral ventricles into the third ventricle and then into the fourth ventricle. From the fourth ventricle, the CSF flows into the subarachnoid space that envelops both the brain and spinal cord, facilitated by the median and lateral apertures. CSF is primarily reabsorbed into the venous system via arachnoid granulations, specialized protrusions of the arachnoid mater into the dural sinuses in the upper part of the cranial cavity.

The CSF acts as a buoyant cushion that substantially reduces the brain's effective weight, protecting it from mechanical trauma. It also serves as a medium for distributing nutrients and signaling molecules and plays a crucial role in maintaining the ionic balance and pH within the CNS. CSF also facilitates the removal of metabolic waste products such as excess neurotransmitters and cellular debris.

10.2.2.9 Blood-Brain Barrier

The blood-brain barrier (BBB) is a highly selective semipermeable border that separates the circulating blood from the brain's extracellular fluid and the central nervous system (CNS). This complex biological interface performs a critical

Table 10.10 Brain stem components: location, structure, function, and features

Name	Location	Function	Structure	Important features
Midbrain (mesencephalon)	Uppermost part of the brainstem	Involved in vision, hearing, eye movement, and body movement	Contains the cerebral aqueduct, tectum, tegmentum, and several nuclei and nerve tracts	Houses the substantia nigra, critical for movement control, and the reticular formation, essential for regulating the sleep-wake cycle and filtering incoming stimuli
Pons	Bulging region on the anterior part of the brainstem, between the midbrain and the medulla	Primarily serves as a bridge between various parts of the nervous system, including the cerebrum and cerebellum; involved in facial sensations and movements, hearing, balance, and taste	Consists of nerve fiber tracts and the pontine nuclei	Plays a crucial role in regulating respiratory rhythms and sleep; also involved in motor control and sensory analysis
Medulla oblongata	Lowest part of the brainstem, continuing as the spinal cord	Regulates vital functions, such as heart rate, blood pressure, and breathing; also involved in reflexes like swallowing, coughing, vomiting, and sneezing	Contains the pyramids, which are two ridges on its ventral surface; the pyramidal decussation occurs here	Contains centers that control autonomic functions, such as the cardiac, respiratory, and vasomotor centers

gatekeeping function, allowing for the regulated passage of essential nutrients and ions while limiting the entry of toxins, pathogens, and peripheral immune cells.

The brain endothelial cell (Fig. 10.28) is the BBB's fundamental unit, which forms the walls of the cerebral microvasculature. Unlike peripheral endothelial cells, those comprising the BBB are characterized by the absence of fenestration and have reduced rates of pinocytosis. Endothelial cells are connected by tight junctions (Page 52). Astrocytic end-feet almost surround the endothelial cells and are essential in the induction and maintenance of the BBB properties. Pericytes within the basement membrane regulate endothelial cell proliferation and vascular stability.

10.2.2.10 Spinal Cord

The spinal cord (Fig. 10.30), an elongated, tubular structure encapsulated within the vertebral column, facilitates communication between the brain and the peripheral nervous system. The spinal cord spans from the foramen magnum at the base of the

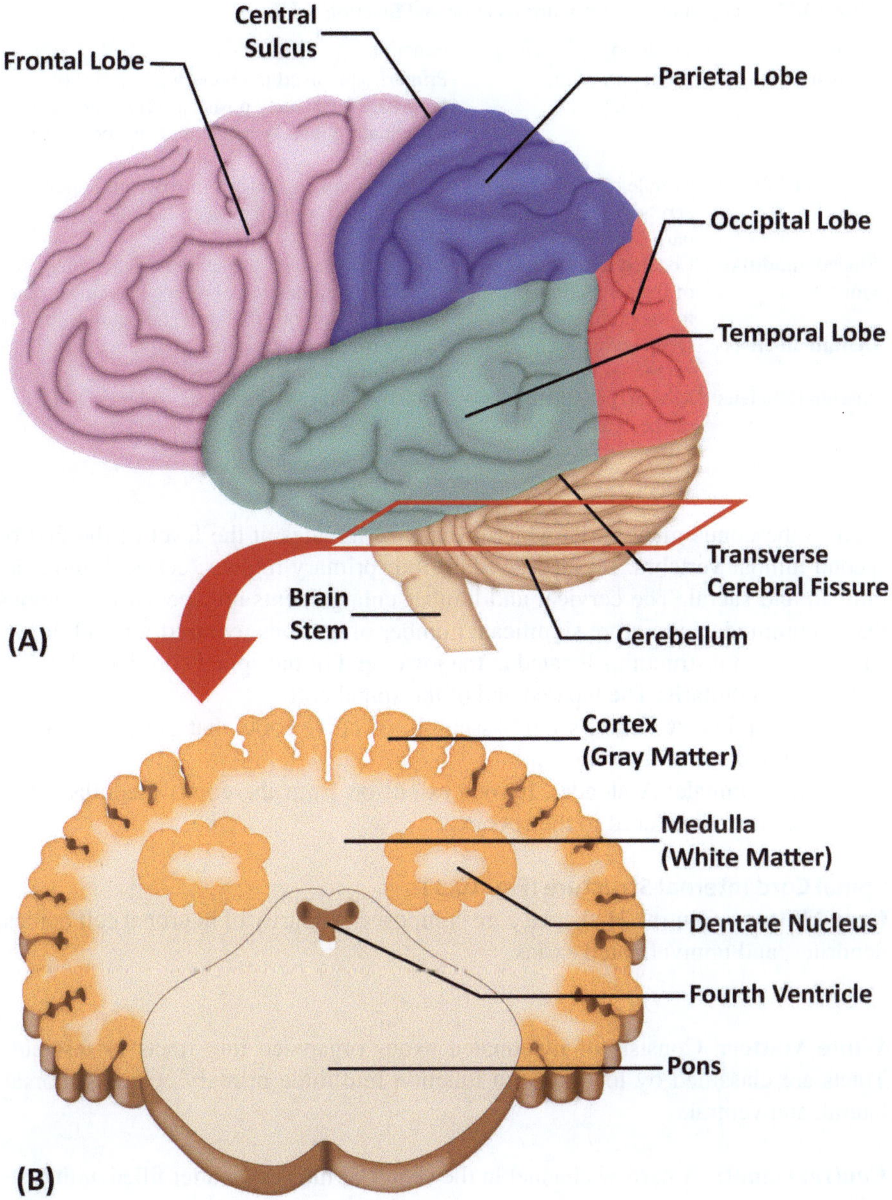

Fig. 10.25 Cerebellum: (**a**) Lateral view, showing its position below the cerebrum. (**b**) Cross section of cerebellum

Table 10.11 Cerebellum components: location and function

Name	Location	Function
Anterior lobe	Front part of the cerebellum	Primarily involved in regulating unconscious proprioceptive (body position) sensations and in coordinating muscle movements for posture and locomotion
Posterior lobe	Middle section of the cerebellum, the largest part	Involved in fine-tuning of movement, motor learning, and cognitive functions
Flocculonodular lobe	Located in the lower part of the cerebellum, close to the brainstem	Involved in balance and spatial orientation, particularly in the control of eye movements and vestibular functions (balance and equilibrium)
Dentate nucleus	Deep within each cerebellar hemisphere	Plays a role in the planning, initiation, and control of voluntary movements
Fastigial nucleus	Medial, near the center of the cerebellum	Primarily involved in balance and regulating muscle tone

skull to the conus medullaris, which usually terminates at the level of the first or second lumbar vertebra. It is divided into four primary regions: cervical, thoracic, lumbar, and sacral. The cervical and lumbar enlargements are specialized regions that accommodate the more significant number of neurons required for limb innervation. Terminal structures located at the lower end of the spinal cord include:

- Conus Medullaris: The tapered end of the spinal cord.
- Cauda Equina: A "horse's tail" assembly of nerve roots that emerge from the lumbar and sacral regions.
- Filum Terminale: A slender, fibrous extension from the conus medullaris that anchors the spinal cord to the coccyx.

Spinal Cord Internal Structure (Fig. 10.31)

Grey Matter: Central, H-shaped core composed primarily of neuronal cell bodies, dendrites, and unmyelinated axons.

White Matter: Consists of myelinated axons organized into tracts or fasciculi. Tracts are classified by location and function into three primary columns: dorsal, lateral, and ventral.

Central Canal: A narrow channel in the center of the grey matter filled with CSF.

Protection: The spinal cord is ensheathed by three layers of meninges: the dura mater, the arachnoid mater, and the pia mater (Section 10.2.2.7).

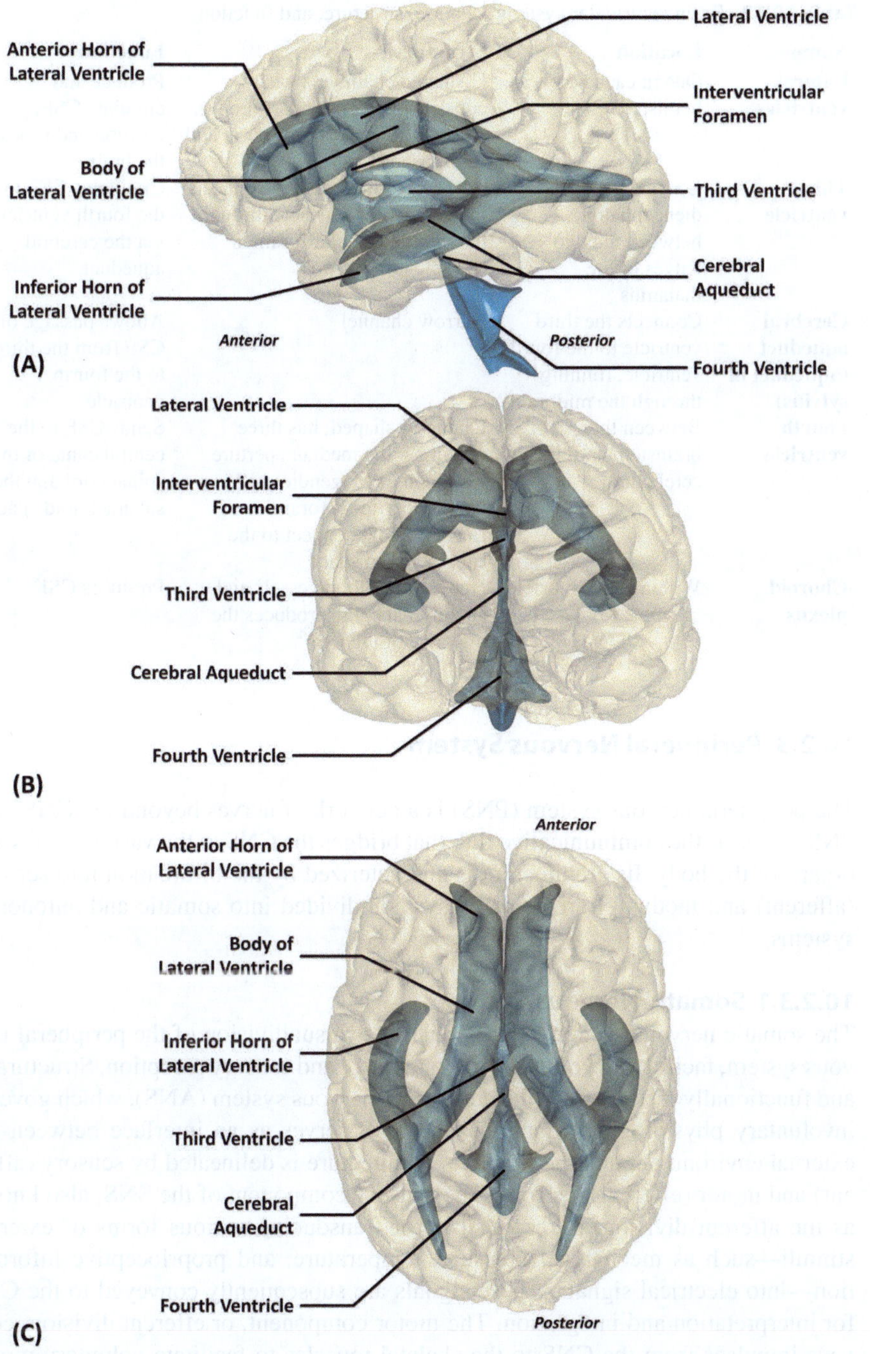

Fig. 10.26 The ventricular system of the brain: (**a**) Lateral view. (**b**) Anterior view. (**c**) Superior view

Table 10.12 Brain ventricular system: location, structure, and function

Name	Location	Structure	Function
Lateral ventricles	One in each cerebral hemisphere	C-shaped chambers; each consists of a central part (body), anterior horn, posterior horn, and inferior horn	Produce and circulate CSF; cushion and protect the brain
Third ventricle	Located in the diencephalon, between the two halves of the thalamus	Narrow, vertical cavity; connects to each lateral ventricle through the interventricular foramen (foramen of Monro)	Circulates CSF to the fourth ventricle via the cerebral aqueduct
Cerebral aqueduct (aqueduct of sylvius)	Connects the third ventricle to the fourth ventricle, running through the midbrain	Narrow channel	Allows passage of CSF from the third to the fourth ventricle
Fourth ventricle	Between the brainstem and the cerebellum	Diamond-shaped; has three openings: the median aperture (foramen of Magendie) and two lateral apertures (foramina of Luschka) that connect to the subarachnoid space	Sends CSF to the central canal of the spinal cord and the subarachnoid space
Choroid plexus	Within each ventricle	Network of blood vessels and ependymal cells; produces the majority of the CSF	Produces CSF

10.2.3 Peripheral Nervous System

The peripheral nervous system (PNS) is a network of nerves beyond the CNS. The PNS serves as the communicative link that bridges the CNS to the various limbs and organs of the body. Its architecture is characterized by the bifurcation into sensory (afferent) and motor (efferent) pathways, subdivided into somatic and autonomic systems.

10.2.3.1 Somatic Nervous System

The somatic nervous system (SNS) constitutes a subdivision of the peripheral nervous system, facilitating voluntary motor control and sensory reception. Structurally and functionally distinct from the autonomic nervous system (ANS), which governs involuntary physiological functions, the SNS serves as an interface between the external environment and the CNS. Its architecture is delineated by sensory (afferent) and motor (efferent) neurons. The sensory component of the SNS, also known as the afferent division, is responsible for transducing various forms of external stimuli—such as mechanical pressure, temperature, and proprioceptive information—into electrical signals. These signals are subsequently conveyed to the CNS for interpretation and integration. The motor component, or efferent division, conveys impulses from the CNS to the skeletal muscles to facilitate voluntary movement. Composed of alpha motor neurons, whose cell bodies reside in the spinal cord, these efferent pathways innervate skeletal muscle fibers through specialized

Table 10.13 Brain covering (meninges): location, structure, and features

Name	Location	Structure	Important features
Dura mater	Outermost layer, adhering to the skull and vertebral canal	Thick, tough, and durable membrane	Composed of two layers: The periosteal layer (attached to the skull) and the meningeal layer (deeper layer). Contains venous sinuses, which drain blood from the brain.
Arachnoid mater	Middle layer, situated between the dura and pia mater	Thin, web-like membrane	Lacks blood vessels and is separated from the pia mater by the subarachnoid space. This space is filled with cerebrospinal fluid (CSF) and contains major blood vessels.
Pia mater	Innermost layer, closely enveloping the brain and spinal cord	Delicate, highly vascularized membrane	Follows the contours of the brain, dipping into the sulci and fissures. Provides a supportive framework for blood vessels.
Subdural space	Potential space between the dura mater and arachnoid mater	Contains a thin layer of fluid	
Subarachnoid space	Actual space between the arachnoid mater and pia mater	Filled with cerebrospinal fluid (CSF)	Provides a cushion for the brain, acting as a shock absorber; also important for the exchange of nutrients and waste between the brain and blood.
Epidural space	Potential space between the dura mater and the surrounding bones	Typically absent in the cranium but present in the spinal column	In the spine, it contains fat and small blood vessels and is a site for administering epidural anesthesia.

neuromuscular synapses. The efferent neurons are activated by motor commands generated in the motor cortices of the brain, and their activation leads to muscle contraction and the generation of force and movement.

An SNS's notable feature is its conscious and reflexive control over the skeletal muscles. While voluntary movements result from complex neural circuits involving the cerebral cortex, cerebellum, and basal ganglia, the SNS also supports automatic responses, known as somatic reflexes.

Sensory Receptors

Sensory receptors serve as the frontline instruments of perception, acting as the biological interfaces that transduce various forms of environmental stimuli into electrochemical signals that the nervous system can interpret. These specialized cellular structures are distributed throughout the body and are intricately tailored to detect specific types of stimuli, such as light, pressure, temperature, and chemical concentration. The subsequent conversion of these stimuli into electrical signals initiates complex neural pathways, leading to sensory perception and, potentially,

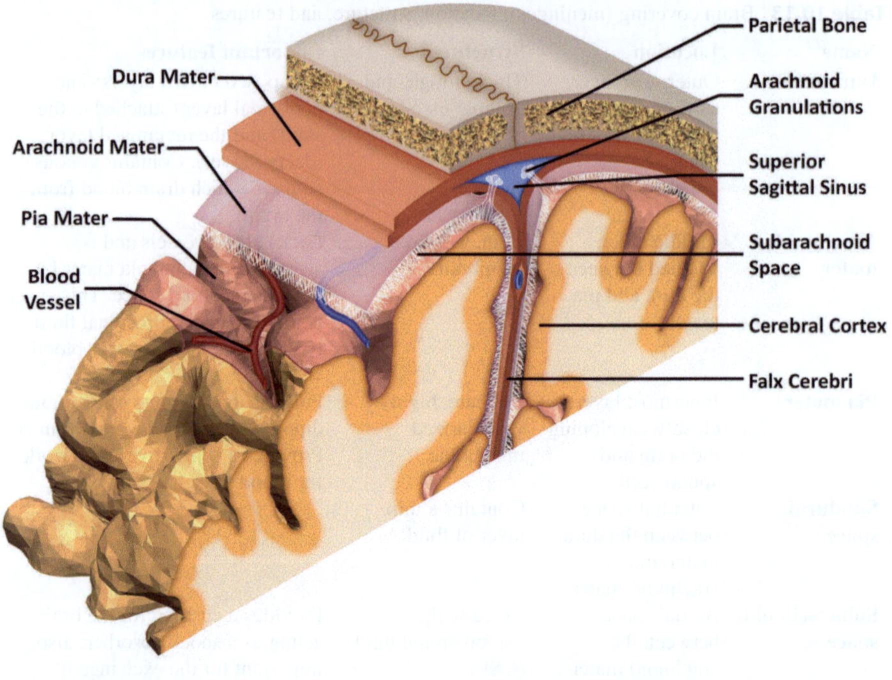

Fig. 10.27 The meninges: This illustration shows the meninges around the brain. The meninges around the spinal cord have the same structure, except that there is a space between the vertebral bone and the dura mater, called epidural space, which is absent in brain meninges

behavioral responses. Based on the stimulus, sensory receptors can be classified into:

- Photoreceptors: Sensitive to light; predominantly found in the eye's retina (Section 11.1.3.3).
- Mechanoreceptors: Detect mechanical changes, such as pressure, vibration, and stretch. They are commonly found in the skin and internal organs.
- Thermoreceptors: Respond to changes in temperature; located in the skin, mucous membranes, and hypothalamus.
- Nociceptors: Detect painful stimuli and are distributed throughout the body, including the skin and internal organs.
- Chemoreceptors: Sensitive to chemical changes, including smell and taste; they are found in the olfactory epithelium and taste buds (Sections 11.3.1.1 & 11.3.2.2).

Sensory Perception

Sensory perception is a complex physiological and cognitive process through which the body gathers and interprets information from external and internal environments. It involves detecting, transduction, transmitting, and interpreting sensory stimuli, creating conscious sensory experiences. Sensory perception begins with specialized sensory receptors that are distributed throughout the body. When

Table 10.14 Cerebrospinal fluid: physical, cellular, and chemical characteristics

Characteristic type	Component	Normal values or characteristics
Physical characteristics	Color	Clear and colorless
	Clarity	Transparent
	Volume	125–150 mL in total (adult)
	Pressure	70–180 mm H_2O (lateral decubitus position)
Cellular components	White blood cells (WBCs)	0–5 cells/μL
	Red blood cells (RBCs)	None
Chemical components	Protein	15–45 mg/dL
	Glucose	45–80 mg/dL (or 2/3 of blood glucose level)
	Lactate	10–20 mg/dL
	pH	7.31–7.34
	Chloride (cl^-)	119–126 mEq/L
	Sodium (Na^+)	Similar to plasma levels
	Potassium (K^+)	2.5–3.0 mEq/L
	Osmolality	Similar to plasma osmolality
	Immunoglobulins	Low concentration
	Biomarkers (e.g., tau, β-amyloid)	Varies depending on specific biomarkers

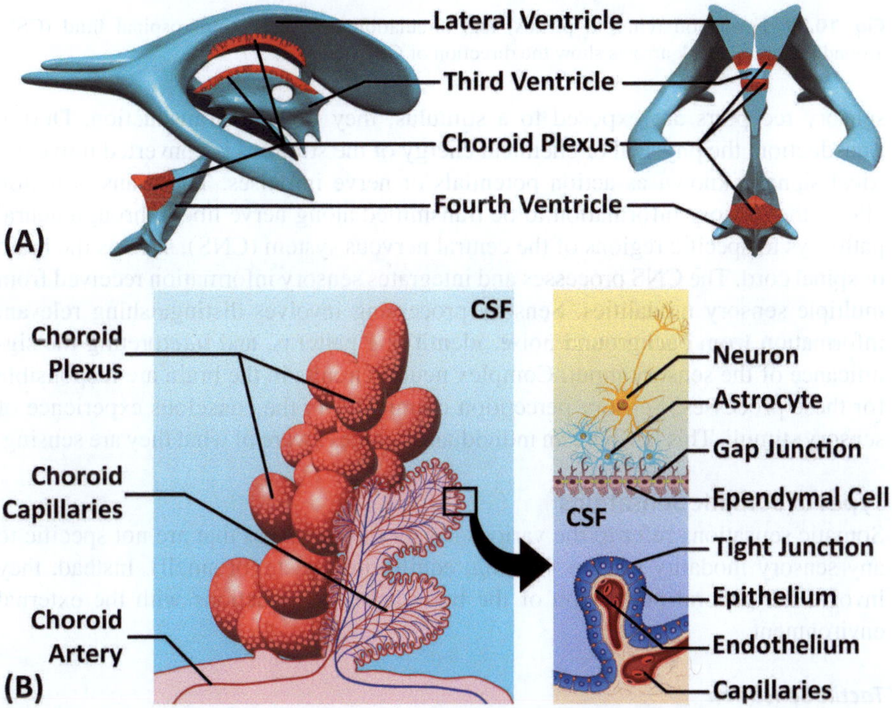

Fig. 10.28 Choroid plexus: (**a**) Anatomical location of the choroid plexus in the ventricles. (**b**) Structure of a choroid plexus. Left, macroscopic structure. Right, microscopic structure of choroid plexus, forming the blood-brain barrier (BBB)

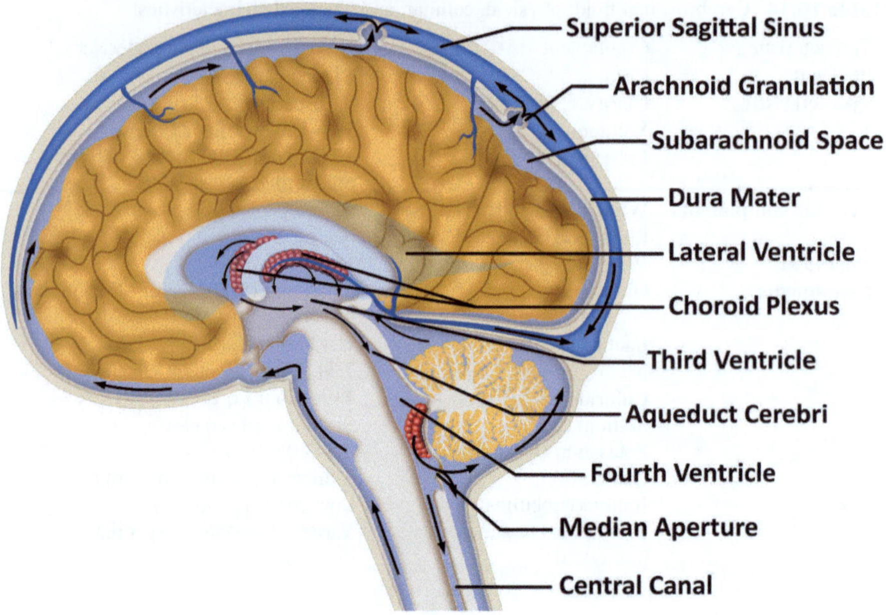

Fig. 10.29 Formation (choroid plexus) and circulation pattern of cerebrospinal fluid (CSF) around the brain: black arrows show the direction of CSF flow

sensory receptors are exposed to a stimulus, they undergo transduction. During transduction, the physical or chemical energy of the stimulus is converted into electrical signals, known as action potentials or nerve impulses. This transformation allows the sensory information to be transmitted along nerve fibers through neural pathways to specific regions of the central nervous system (CNS), such as the brain or spinal cord. The CNS processes and integrates sensory information received from multiple sensory modalities. Sensory processing involves distinguishing relevant information from background noise, identifying patterns, and interpreting the significance of the sensory input. Complex neural circuits in the brain are responsible for these processes. Sensory perception culminates in the conscious experience of sensory stimuli. This is where an individual becomes aware of what they are sensing.

Types of Somatic Sensations

Somatic sensations refer to the various sensory experiences that are not specific to any sensory modality (vision, hearing, equilibrium, taste, or smell). Instead, they involve the general perception of the body and its interactions with the external environment.

Tactile Sensation

Tactile sensation refers to the perception of touch and pressure on the skin or other body surfaces. It includes various submodalities, such as light touch, deep pressure, and vibration. Tactile sensation relies on specialized sensory receptors known as mechanoreceptors distributed throughout the skin and deeper tissues. These receptors include:

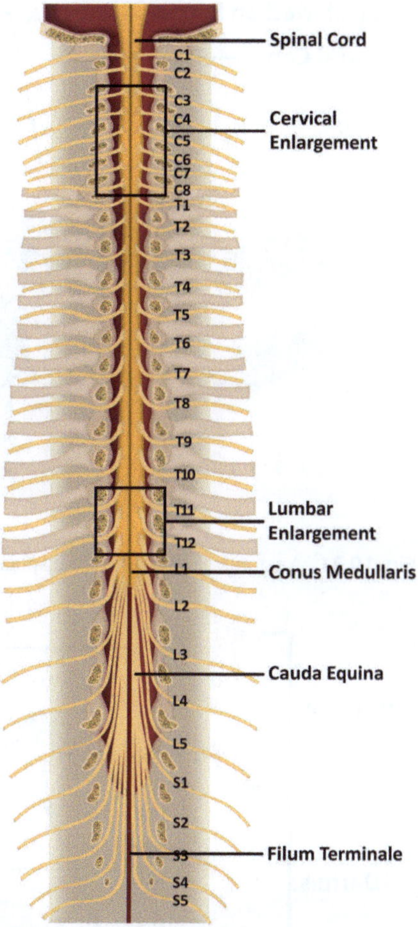

Fig. 10.30 Spinal cord and spinal nerves: dorsal view

Meissner's Corpuscles (Fig. 10.32): These mechanoreceptors are found in the superficial layers of the skin, particularly in areas sensitive to fine touch, such as the fingertips, lips, and palms. Meissner's corpuscles are responsible for detecting light touch and changes in surface texture.

Merkel Cells and Discs (Fig. 10.32): Located in the epidermis, Merkel cells and associated Merkel discs are involved in perceiving sustained pressure, shapes, and roughness of objects. They are mainly concentrated in fingertips and fingertips.

Pacinian Corpuscles (Fig. 10.32): Pacinian corpuscles are found in deeper layers of the skin and are sensitive to deep pressure and high-frequency vibrations. They detect rapid changes in pressure, such as those caused by tapping or vibration.

Ruffini Endings (Fig. 10.32): These mechanoreceptors are located in the deep layers of the skin and respond to skin stretch and deformation. Ruffini endings contribute to the perception of continuous pressure and skin distortion.

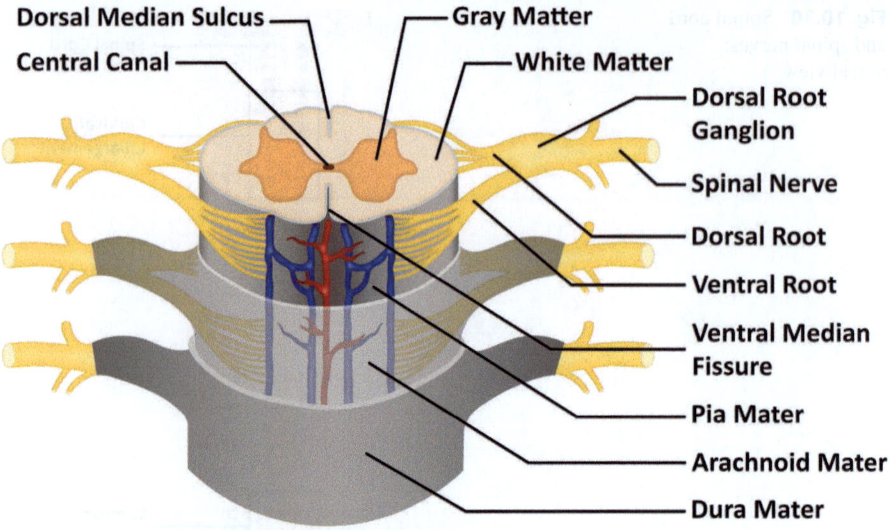

Fig. 10.31 Anatomical structure of the spinal cord

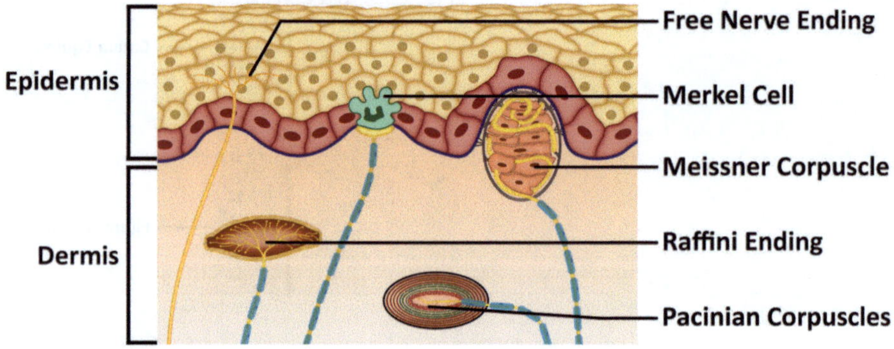

Fig. 10.32 Tactile sensory receptors

Free Nerve Endings: Free nerve endings are the most common tactile receptor type distributed throughout the skin and various tissues. They detect temperature changes, pain, and tactile sensations, including light touch and pressure.

Thermal Sensation

Thermal sensation involves the perception of temperature changes, allowing us to distinguish between hot and cold stimuli. Thermal sensation relies on specialized sensory receptors, known as thermoreceptors, distributed throughout the body, primarily in the skin and mucous membranes. Thermal sensation involves the activation of various receptors. Free nerve endings are the most common thermoreceptors and are present throughout the skin and other tissues. They respond to cold and warm stimuli, contributing to the detection of temperature changes. Cold thermoreceptors are specialized receptors that are particularly sensitive to decreases in temperature. They are more numerous in the skin's upper layers. Warm thermoreceptors are sensitive to increases in temperature and are found in the skin's deeper layers.

Pain Sensation

Pain sensation, or nociception, is the perception of tissue damage or potentially harmful stimuli. Nociceptors, specialized pain receptors, are activated by factors like intense pressure, extreme temperatures, chemical irritants, or tissue injury.

Pain is a warning signal essential for protecting the body from harm. Nociceptors are distributed throughout the body, including the skin, internal organs, and tissues. Pain sensation can be categorized into different types based on its characteristics and underlying causes. Nociceptive pain is a type of pain caused by the activation of nociceptors in response to noxious stimuli. It is typically a protective mechanism that warns of tissue damage or injury. Nociceptive pain can be further divided into two subtypes. Somatic nociceptive pain arises from the stimulation of nociceptors in the skin, muscles, bones, and connective tissues, usually caused by cuts, bruises, or fractures. Visceral nociceptive pain results from the activation of nociceptors in internal organs. This type of pain is often described as diffuse, deep, and less localized, as internal organs have fewer pain receptors. Neuropathic pain occurs due to damage or dysfunction of the nervous system itself. It is often characterized by abnormal sensations such as burning, tingling, or shooting pain.

Inflammatory pain is associated with the body's inflammatory response to injury or infection. Inflammation can sensitize nociceptors, leading to increased pain perception. Psychological factors, such as stress, anxiety, or emotional distress, primarily drive psychogenic pain. This type of pain may not have an apparent physical cause but is still experienced by the individual.

Itch Sensation

Itch sensation often begins with detecting irritants or allergens by specific receptors on sensory nerve endings in the skin. These irritants include chemicals, histamines, insect bites, allergenic substances, or certain medications. Upon exposure to irritants, cells in the skin, such as mast cells, may release a variety of itch-inducing substances, including histamines, proteases, cytokines, and neuropeptides. The released itch mediators bind to specific receptors on sensory nerve fibers, particularly C-fibers and Aδ-fibers. These receptors include histamine receptors (H1 and H4 receptors), protease-activated receptors (PARs), and other ion channels or receptors sensitive to itch mediators. The binding of itch mediators to their receptors triggers intracellular signaling cascades within the sensory nerve fibers. This generates action potentials transmitted to the central nervous system (CNS) along sensory nerve fibers. Itch sensations can vary in intensity, duration, and quality.

Tickling Sensation

Tickling begins with the application of a tactile stimulus to sensitive areas of the skin. These areas typically include regions that are more sensitive to touch, such as the soles of the feet, underarms, sides, and neck. The tactile stimulus activates mechanoreceptors in the skin. These receptors include Meissner's corpuscles, Merkel cells and discs, and free nerve endings. Meissner's corpuscles, in particular, are sensitive to gentle touch and low-frequency vibrations. After transmission to specific regions in the brain, particularly the somatosensory cortex and higher-order brain centers involved in emotional and social processing, the brain processes and integrates the tickling input.

Tickling sensations can be categorized into two main types. Knismesis is a type of tickling that involves light, gentle, and often subtle stimuli that elicit a tickling sensation. This type of tickling can provoke an itching or tingling sensation and may lead to a desire to scratch the affected area. Knismesis is often associated with small insects or other irritants on the skin's surface. Gargalesis is a more intense and playful tickling involving repetitive, vigorous, or unexpected stimuli. Gargalesis tickling is often social and can lead to uncontrollable laughter.

Stretch Sensation

Stretch receptors, such as muscle spindles (Page 345) and Golgi tendon organs, are responsible for sensing changes in muscle length and tension. This sensation contributes to the perception of muscle stretch, resistance, and tension during movements.

Proprioception

Proprioception, also known as kinesthesia or kinesthetic sense, is one of the fundamental sensory systems in the human body. It refers to the ability to sense the position, movement, and orientation of one's body and body parts without relying on visual or auditory cues. Proprioception is mediated through a complex interplay of sensory receptors, neural pathways, and the central nervous system. Key components of the proprioceptive system include muscle spindles, Golgi tendon organs, joint receptors, cutaneous receptors, and the vestibular system (Section 11.2.3.2).

Muscle spindles detect changes in muscle length and the rate of change, providing information about muscle stretch to the central nervous system.

Golgi tendon organs (Fig. 10.33) detect changes in muscle tension and force generation, helping regulate muscle contractions and prevent overloading. Golgi tendon organs are encapsulated sensory structures found within the tendons at the musculotendinous junction. Golgi tendon organs are encapsulated by a collagenous capsule, which protects and isolates them from the surrounding tissues. This capsule ensures that the tension experienced by the tendon is accurately transmitted to the receptor. Within the capsule, there are numerous sensory endings or nerve terminals. These nerve endings are intertwined among the collagen fibers of the tendon. Golgi tendon organs contain intrafusal muscle fibers, which are tiny, specialized muscle fibers. These intrafusal fibers are attached parallel to the tendon and run in series with the muscle fibers. Their contractile activity is controlled by motor neurons, ensuring that the intrafusal fibers are stretched or relaxed in response to changes in muscle tension. The sensory nerve endings within Golgi tendon organs are connected to afferent nerve fibers, which transmit the sensory information the

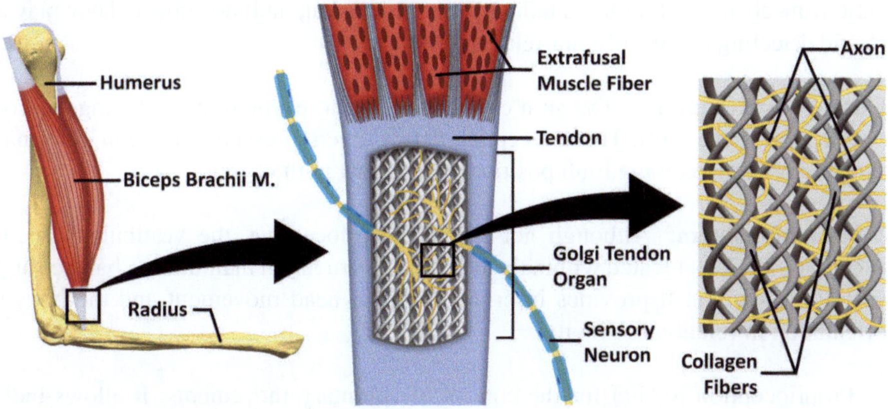

Fig. 10.33 Structure of the Golgi tendon organ

receptor generates to the central nervous system (CNS). These afferent fibers are usually type Ib afferent nerve fibers. The primary function of Golgi tendon organs is to monitor muscle tension or force generated during muscle contraction. When the muscle contracts and creates tension, it pulls on the tendon, causing it to stretch. Golgi tendon organs are highly sensitive to changes in tension within the tendon. As muscle tension increases, the tendon is stretched further, activating the sensory nerve endings within the Golgi tendon organ. The sensory information from Golgi tendon organs is transmitted to the CNS, which provides feedback about the intensity of muscle contraction. This feedback is essential for fine-tuning muscle activity during movement. Golgi tendon organs act as protective mechanisms. When muscle tension becomes too high and potentially harmful, such as during excessive muscle contraction, it triggers the Golgi tendon reflex or inverse myotatic reflex. This reflex results in muscle relaxation, preventing muscle or tendon damage.

Joint Receptors: Various mechanoreceptors are situated within the capsules and ligaments of joints. These receptors sense joint position, angle, and pressure changes, providing information about joint movement and position. There are several types of joint receptors. Ruffini endings (type I joint receptors) are encapsulated receptors in the joint capsule and ligaments. They are sensitive to static joint position and respond to sustained mechanical deformation of the joint capsule. Ruffini endings play a role in detecting joint angles and providing information about joint stability. Pacinian corpuscles (type II joint receptors) are located in the joint capsule and are sensitive to rapid changes in joint movement and acceleration. They are particularly responsive to dynamic changes in joint position and velocity. Golgi-Mazzoni corpuscles (type III joint receptors) are specialized receptors found at the ligament/tendon-bone junctions (enthesis) within some joints. They are sensitive to changes in joint tension and are thought to provide feedback about muscle contraction forces and joint loading. Free nerve endings (type IV joint receptors) are unencapsulated nerve endings scattered throughout the joint capsule, ligaments, and other joint tissues. They respond to various mechanical stimuli, including pressure, tension, and movement. They play a role in detecting a range of joint-related sensations.

Cutaneous Receptors: The skin contains mechanoreceptors contributing to proprioception (Page 342). These receptors detect touch, pressure, and skin deformation, aiding in perceiving limb position and contact with objects.

Vestibular System: Although not strictly proprioceptive, the vestibular system (Section 11.2.3.2), located within the inner ear, is crucial in maintaining balance and spatial orientation. It provides information about head movement and the body's orientation in relation to gravity.

Proprioception is vital for the control of voluntary movements. It allows individuals to make precise adjustments in muscle activity and joint positioning to perform tasks such as walking, grasping objects, and maintaining balance. Proprioception contributes significantly to postural control. It helps individuals maintain an upright posture, adjust their body position in response to external forces, and prevent falls. Proprioception enables individuals to sense their body's position in space, which is essential for navigating the environment and coordinating movements. Proprioceptive feedback from muscle spindles and Golgi tendon organs helps regulate muscle tone, ensuring muscles are at an appropriate level of contraction even at rest. Activities that require precise control, such as playing a musical instrument or typing on a keyboard, rely on accurate proprioceptive feedback for skillful performance. Proprioceptive training is often used in rehabilitation to prevent injuries and improve performance in athletes. Enhancing proprioception can help individuals respond effectively to unexpected challenges and avoid overuse injuries.

Reflex Activity

Reflex activity enables organisms to respond rapidly and appropriately to environmental stimuli without requiring higher-level cognitive intervention. The reflex arc comprises sensory receptors, neural pathways, and effector organs.

Spinal reflexes (Fig. 10.34) are a specific subset of reflexes mediated primarily through neural circuits within the spinal cord. These involuntary and stereotyped motor responses to sensory stimuli serve as essential mechanisms for rapid behavioral adaptation, homeostatic regulation, and protective actions. Unlike cranial reflexes, which involve neural pathways in the brain, spinal reflexes do not require cortical processing, making them particularly efficient for immediate responses.

Tendon (muscle) stretch reflexes, also called deep tendon reflexes (DTRs), are one of the spinal reflexes that provide motor control and proprioceptive regulation. This reflex involves the stretch of a muscle's tendon, which leads to a reflexive contraction of the muscle. The tendon stretch reflex involves several key anatomical components, including muscle and its tendon, muscle spindles, and nerves.

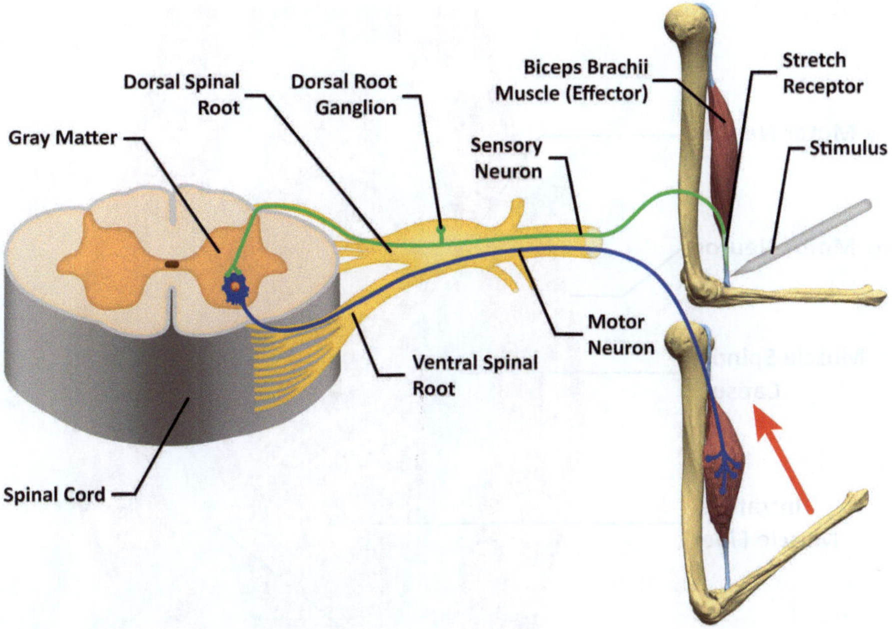

Fig. 10.34 The basic structure of a monosynaptic reflex arc

The muscle involved in the reflex is often referred to as the "target muscle." This muscle contains specialized fibers known as intrafusal fibers that are sensitive to changes in muscle length.

Within the target muscle, muscle spindles are proprioceptive sensory receptors that play a central role in the tendon stretch reflex. Muscle spindles (Fig. 10.35) are elongated, fusiform-shaped structures parallel to the regular muscle fibers within a skeletal muscle. They consist of several key components. The core of the muscle spindle is made up of intrafusal muscle fibers. These muscle fibers are distinct from the regular, contractile, extrafusal muscle fibers. Intrafusal muscle fibers are specialized for sensory function rather than force generation. They are thinner and lack the contractile apparatus in extrafusal muscle fibers. Muscle spindles contain two types of intrafusal muscle fibers: nuclear bag and nuclear chain fibers. Nuclear bag fibers have a cluster of nuclei located in a bag-like structure in the central region of the fiber. These nuclei are distributed irregularly, giving the fiber a "baggy" appearance. The other type of intrafusal muscle fiber is nuclear chain fibers. These fibers have a chain-like arrangement of nuclei in a single row, giving them a more elongated appearance than nuclear bag fibers. Muscle spindles are innervated by sensory neurons, specifically Ia afferent fibers. These sensory nerve endings wrap around the central region of the intrafusal muscle fibers, forming annulospiral endings. These endings are highly sensitive to changes in muscle length and stretch. Muscle spindles also receive motor innervation from gamma motor neurons, innervating the contractile regions of the intrafusal muscle fibers, specifically the polar regions of the nuclear bag and chain fibers. Activation of gamma motor neurons adjusts the

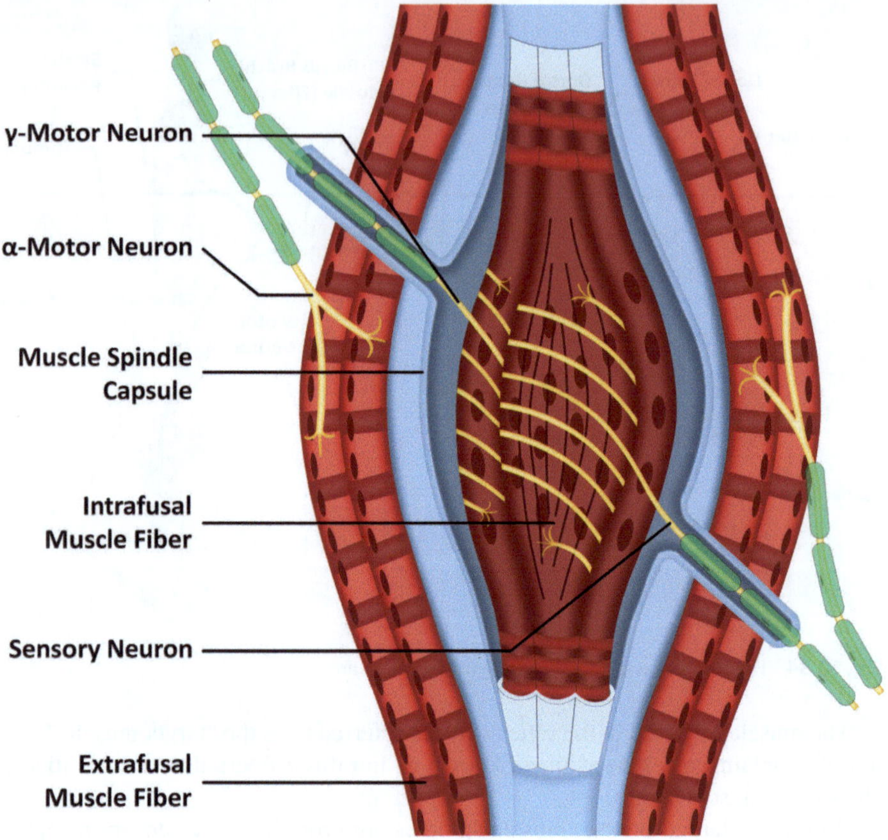

Fig. 10.35 Structure of the muscle spindle

sensitivity and tension within the muscle spindle. Muscle spindles are encapsulated by connective tissue, which provides protection and structural support. This capsule is continuous with the surrounding muscle tissue.

The tendon stretch reflex (Fig. 10.34) operates through a well-defined mechanism involving the following steps: The reflex begins when an external force stretches the target muscle. This stretch could be due to various factors, such as a tap on the tendon or a sudden increase in muscle length. The intrafusal muscle fibers within the spindles are also stretched when the muscle is stretched. As the intrafusal muscle fibers stretch, they activate the sensory nerve endings, specifically the Ia afferent fibers associated with the muscle spindle. These sensory neurons generate action potentials in response to the stretch. These action potentials travel along the Ia afferent fibers to the spinal cord, synapsing with motor neurons.

The connection between the sensory and motor neurons in the spinal cord is monosynaptic, meaning only one synapse is involved. This allows for a rapid and automatic response. The sensory input from the muscle spindle excites the alpha motor neurons that innervate the same muscle, leading to muscle contraction. This contraction opposes the initial stretch, helping to maintain muscle length and prevent overstretching. The outcome of the tendon stretch reflex is a reflexive contraction of the target muscle, which resists the stretch and maintains muscle tone.

The tendon stretch reflex helps maintain muscle tone by ensuring that muscles remain at an optimal resting length, which is critical for posture and motor control. The reflex acts as a protective mechanism by preventing excessive stretching of muscles, which could lead to muscle damage or injury.

Nerves

Nerves are essential cables transmitting electrical signals between the CNS and muscles, skin, and internal organs. The basic unit of a peripheral nerve is the nerve fiber, the axon of a neuron (Fig. 10.36), which is classified into two types based on myelination (Fig. 10.3):

- Myelinated Fibers: These fibers are covered by a myelin sheath, a protective layer of lipid and protein that insulates the nerve fiber and increases the speed of nerve impulse transmission.
- Unmyelinated Fibers: These fibers lack a myelin sheath and conduct impulses more slowly.

In the PNS, Schwann cells wrap around the axons of neurons to form the myelin sheath. Schwann cells provide support for unmyelinated fibers but do not wrap around the fibers completely.

Each nerve fiber (axon and its myelin sheath) is encased in a delicate connective tissue layer called the endoneurium, which provides mechanical and physiological support to the nerve fiber. Multiple nerve fibers are bundled together into groups called fascicles. Fascicles are surrounded by connective tissue, known as the perineurium, which provides additional protection and helps maintain the internal environment of the nerve fibers. The peripheral nerve is then wrapped in a dense connective tissue sheath called the epineurium. This outermost layer binds all the fascicles together, providing strength and protection to the nerve as a whole.

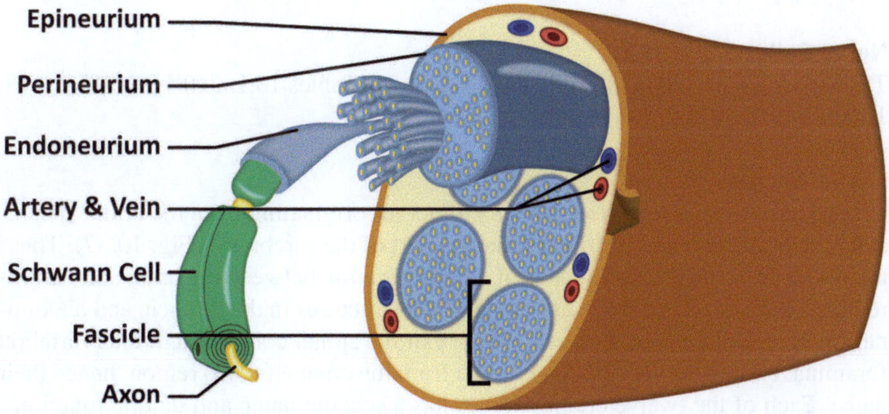

Fig. 10.36 Structure of a nerve, showing connective tissue wrappings

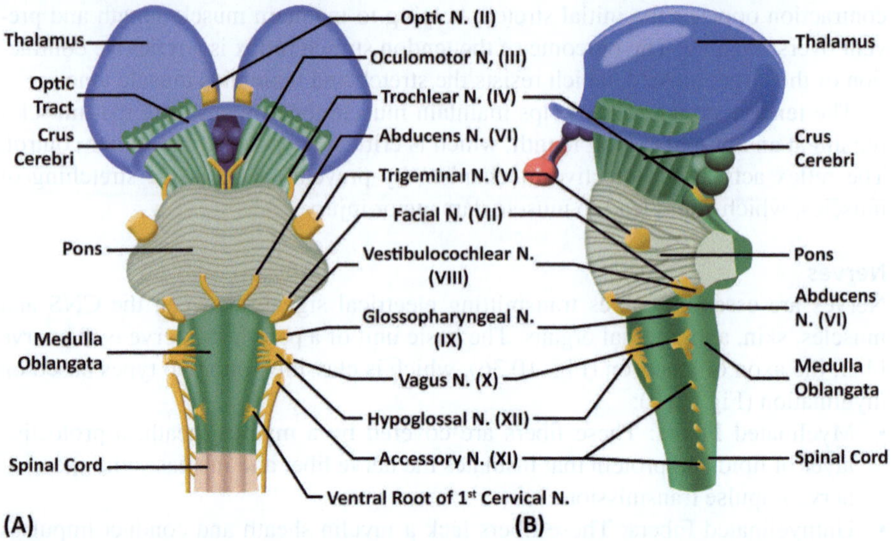

Fig. 10.37 Origin of cranial nerves (except the first cranial nerve) in the brainstem

Sensory (Afferent) Nerves: Sensory or afferent nerves are specialized to convey information from sensory receptors in the peripheral tissues to the CNS. These nerves enable the perception of various environmental stimuli, both external and internal.

Motor (Efferent) Nerves: otor or efferent nerves facilitate the transmission of impulses from the CNS to effector organs like muscles and glands, thereby enabling voluntary and involuntary actions. Motor nerves can be somatic or autonomic.

Mixed Nerves: Mixed nerves contain both sensory and motor fibers and can perform dual functions. They are typically found in spinal and some cranial nerves.

Nerve Gallery

The individual nerves in the body are discussed in Tables 10.15, 10.16, 10.17, 10.18, 10.19, and 10.20.

Cranial Nerves

Cranial nerves are a set of twelve pairs of nerves originating from the brain, specifically the brainstem and the diencephalon (part of the forebrain) (Fig. 10.37). These nerves serve as the primary means of communication between the brain and various regions of the head and neck, as well as some structures in the thoracic and abdominal cavities. Unlike spinal nerves, which exit the spinal cord through intervertebral foramina, cranial nerves directly emerge from the cranial (skull) region, hence their name. Each of the twelve cranial nerves has a specific name and unique functions,

Table 10.15 Cranial nerves: origin, course, function, and features

Cranial nerve	Origin	Course	Function	Important features
I. Olfactory	Olfactory epithelium	Through cribriform plate	Smell	Only nerve attached to the cerebrum; does not pass through the brainstem.
II. Optic	Retina	Through optic canal	Vision	Formed by axons of retinal ganglion cells; crosses at optic chiasma.
III. Oculomotor	Midbrain	Through superior orbital fissure	Eye movement, pupil constriction, eyelid elevation	Controls most eye movements; involved in pupil constriction and accommodation.
IV. Trochlear	Midbrain	Through superior orbital fissure	Superior oblique muscle movement (downward, inward eye rotation)	Smallest cranial nerve; the only nerve to emerge dorsally from the brainstem.
V. Trigeminal	Pons	Divides into three branches: Ophthalmic, Maxillary, Mandibular	Sensation of the face, motor functions of mastication	Largest cranial nerve; has both sensory and motor functions.
VI. Abducens	Pons	Through superior orbital fissure	Lateral rectus muscle movement (abduction of eye)	Primarily controls outward gaze.
VII. Facial	Pons	Through internal acoustic meatus to the stylomastoid foramen	Facial expression, taste, salivation, tear production	Provides motor innervation to facial muscles and stapedius; involved in taste (anterior 2/3 of the tongue), provides motor innervation to salivary and lacrimal glands.
VIII. Vestibulocochlear	Pons, Medulla	Through internal acoustic meatus	Hearing, balance	Composed of two parts: cochlear (hearing) and vestibular (balance).
IX. Glossopharyngeal	Medulla	Through jugular foramen	Taste, salivation, carotid body reflexes	Involved in swallowing, salivation, and monitoring carotid body and sinus; involved in taste (posterior 1/3 of the tongue).

(continued)

Table 10.15 (continued)

Cranial nerve	Origin	Course	Function	Important features
X. Vagus	Medulla	Through jugular foramen	Control of PNS efferents, taste, swallowing, speech, cough reflex	Muscles of the soft palate, pharynx, larynx, thoracic and abdominal viscera; critical role in autonomic control.
XI. Accessory	Medulla and spinal cord (C1–C5)	Through jugular foramen	Shoulder and neck movement	Has cranial and spinal roots; innervating sternocleidomastoid, and trapezius.
XII. Hypoglossal	Medulla	Through hypoglossal canal	Tongue movement	Innervates intrinsic and extrinsic tongue muscles; essential for speech, food manipulation, and swallowing.

Table 10.16 Cervical plexus branches: origin, course, and function

Component	**Origin**	**Course**	**Function**
Lesser occipital nerve (C2, C3)	C2–C3	Ascends along the posterior border of the sternocleidomastoid muscle	Provides sensory innervation to the skin of the neck and scalp posterior and superior to the ear
Great auricular nerve (C2, C3)	C2–C3	Ascends across the sternocleidomastoid muscle towards the ear	Supplies sensory innervation to the skin over the parotid gland, mastoid process, and both surfaces of the ear
Transverse cervical nerve (C2, C3)	C2–C3	Runs anteriorly across the sternocleidomastoid muscle	Provides sensory innervation to the anterior and lateral parts of the neck
Supraclavicular nerves (C3, C4)	C3–C4	Descend over the clavicle into the shoulder region	Supply sensory innervation to the skin over the clavicle, shoulder, and upper chest
Phrenic nerve (C3, C4, C5)	C3–C4 (mainly), with contributions from C5	Descends through the neck into the thorax to reach the diaphragm	Provides motor innervation to the diaphragm, essential for breathing
Ansa cervicalis (C1–C3)	C1–C3	Forms a loop in the neck	Supplies motor innervation to the infrahyoid muscles, aiding in swallowing and speech

Table 10.17 Brachial plexus branches: origin, course, and function

Component	Origin	Course	Function
Musculocutaneous nerve	Lateral cord	Through the upper arm	Innervates biceps brachii, brachialis, coracobrachialis; sensory to lateral forearm
Axillary nerve	Posterior cord	Around the surgical neck of the humerus	Innervates deltoid, teres minor; sensory to skin over deltoid
Radial nerve	Posterior cord	Down the posterior arm and forearm	Innervates triceps brachii, extensor muscles of the forearm; sensory to posterior arm and hand
Median nerve	Lateral and medial cords	Through the anterior forearm into the hand	Innervates most of the flexor muscles in the forearm; sensory to lateral palm and fingers
Long thoracic nerve	C5–C7	Runs along the lateral chest wall	Motor to serratus anterior muscle
Suprascapular nerve	C5–C6 (from upper trunk)	Passes through the suprascapular notch	Motor to supraspinatus and infraspinatus muscles; sensory from shoulder joint
Dorsal scapular nerve	C5 root	Runs along the medial border of the scapula	Motor to the rhomboids and levator scapulae muscles
Thoracodorsal nerve	C6–C8 (from posterior cord)	Descends along the posterior wall of the axilla	Motor to latissimus dorsi muscle

including sensory, motor, or both, depending on the nerve. They are numbered and named based on their location or function.

These nerves play crucial roles in controlling various sensory and motor functions, including those related to vision, hearing, balance, taste, facial expressions, swallowing, and many other vital activities of the head and neck region (Fig. 10.38).

Spinal Nerves

Each spinal nerve (Fig. 10.31) is connected to the spinal cord by two roots: A dorsal (posterior) root containing sensory nerve fibers and has a dorsal root ganglion containing the cell bodies of sensory neurons, and a ventral (anterior) root, containing motor nerve fibers that arise from the motor neurons in the spinal cord. The dorsal and ventral roots combine to form a mixed spinal nerve, which splits into the dorsal and ventral rami. The dorsal ramus supplies the muscles and skin of the posterior trunk, while the ventral ramus supplies the anterior trunk and limbs. There are eight cervical (C1–C8), twelve thoracic (T1–T12), five lumbar (L1–L5), five sacral (S1–S5), and one coccygeal (Co) pairs of spinal nerves, categorized based on their region of origin along the spinal cord (Fig. 10.30). In the cervical, lumbar, and sacral regions, the ventral rami of several spinal nerves come together to form nerve plexuses, giving rise to the major peripheral nerves of the limbs.

Table 10.18 Lumbosacral plexus branches: course and function

Component	Course	Function
Iliohypogastric (L1)	Runs along the iliac crest	Sensory to the skin of the hip and lower abdomen
Ilioinguinal (L1)	Emerges near the iliac crest	Sensory to the upper medial thigh, root of the penis, and upper part of the scrotum, or mons pubis and labia majora
Genitofemoral (L1–L2)	Pierces the psoas major	Sensory to the upper part of the anterior thigh and the genitalia
Lateral cutaneous nerve of thigh (L2–L3)	Emerges from the lateral border of the psoas major	Sensory to the skin of the lateral thigh
Femoral nerve (L2–L4)	Passes beneath the inguinal ligament into the thigh	Motor to anterior thigh muscles; sensory to anterior thigh and medial leg
Obturator nerve (L2–L4)	Passes through the obturator foramen into the medial thigh	Motor to medial thigh muscles; sensory to the skin of the medial thigh
Sciatic nerve (L4–S3)	Emerges from the pelvis and runs into the posterior thigh; at the knee level, it divides into tibial and common peroneal nerves	Motor to the posterior thigh muscles; branches to all leg and foot muscles
Superior gluteal (L4–S1)	Exits the pelvis through the greater sciatic foramen	Motor to gluteus medius, minimus, and tensor fasciae latae muscles
Inferior gluteal (L5–S2)	Exits the pelvis through the greater sciatic foramen	Motor to gluteus maximus muscle
Pudendal (S2–S4)	Leaves the pelvis through the greater sciatic foramen, reenters the pelvis through the lesser sciatic foramen	Sensory and motor to the perineum, including external anal and urethral sphincters
Posterior cutaneous nerve of thigh (S1–S3)	Descends in the posterior thigh	Sensory to the skin of the posterior thigh and part of the perineum

Cervical Plexus

The cervical plexus (Table 10.16) (Fig. 10.39) is a network of nerves formed by the anterior (ventral) rami of the first four cervical spinal nerves (C1–C4). It is located in the neck and provides motor and sensory innervation to parts of the neck, upper shoulders, and part of the head.

Brachial Plexus

The brachial plexus (Table 10.17) (Fig. 10.40) is a network of nerves that originates from the ventral rami of the lower four cervical nerves (C5–C8) and the first thoracic nerve (T1). Its structure is organized into several interconnected components: roots, trunks, divisions, cords, and branches.

Table 10.19 Nerves supplying thorax and abdomen: origin, course, and function

Nerve	Origin	Course	Function
Intercostal nerves (T1–T11)	Thoracic spinal nerves (T1–T11)	Run along the intercostal spaces	Provide motor innervation to the intercostal muscles and sensory innervation to the thoracic and abdominal wall
Subcostal nerve (T12)	Thoracic spinal nerve (T12)	Runs below the 12th rib	Similar to intercostal nerves, supplies abdominal muscles and overlying skin
Phrenic nerve (C3–C5)	Cervical spinal nerves (C3–C5)	Descends through the neck, between the lung and heart, to the diaphragm	Motor and sensory to the diaphragm; also carries sensory fibers from the pericardium and pleura
Vagus nerve (CN X)	Cranial nerve X	Descends from the neck into the thorax and abdomen	Parasympathetic innervation to the heart, lungs, and digestive tract
Splanchnic nerves	Thoracic spinal nerves (T5–T12)	Pass through the diaphragm to enter the abdominal cavity	Provide sympathetic innervation to the abdominal and pelvic viscera
Thoracic sympathetic trunk	Sympathetic ganglia in the thoracic region	Runs vertically along each side of the vertebral column	Sympathetic innervation to the heart, lungs, and upper limbs
Celiac plexus	Formed by the greater splanchnic nerve (T5–T9) and lesser splanchnic nerve (T10–T11)	Located around the celiac trunk of the abdominal aorta	Provides sympathetic innervation to the upper abdominal viscera
Superior mesenteric plexus	Formed by the lesser and least splanchnic nerves	Near the superior mesenteric artery	Supplies the small intestine and the proximal large intestine
Inferior mesenteric plexus	Lumbar splanchnic nerves	Near the inferior mesenteric artery	Innervates the distal large intestine, rectum, and urogenital organs

Roots: The brachial plexus begins with five roots derived from the anterior rami of the C5, C6, C7, C8, and T1 spinal nerves. These roots emerge from the spinal cord and travel through the neck.

Trunks: The roots converge to form three trunks:

- Upper Trunk: Formed by the union of the C5 and C6 roots.
- Middle Trunk: Continuation of the C7 root.
- Lower Trunk: Formed by the union of the C8 and T1 roots.

Divisions: Each trunk then splits into an anterior and a posterior division:

- Anterior Divisions: Generally, supply muscles that flex the limb.
- Posterior Divisions: Generally, supply muscles that extend the limb.

Table 10.20 Nerves supplying pelvis: origin, course, and function

Nerve	Origin	Course	Function
Pelvic splanchnic nerves (parasympathetic)	S2–S4 (from sacral plexus)	Descend into the pelvis	Provide parasympathetic innervation to the pelvic organs (bladder, rectum, sex organs)
Hypogastric nerves (sympathetic)	T12–L2 (from lumbar sympathetic trunk)	Descend into the pelvis, forming the superior hypogastric plexus, then the hypogastric nerves	Sympathetic innervation to pelvic organs
Inferior hypogastric plexus (pelvic plexus)	Contributions from hypogastric nerves, pelvic splanchnic nerves, and sacral sympathetic trunk	Located on either side of the rectum in males and the vagina in females	Mixed autonomic (sympathetic and parasympathetic) innervation to the bladder, uterus, prostate, and rectum
Pudendal nerve	S2–S4 (from sacral plexus)	Exits the pelvis through the greater sciatic foramen, then reenters via the lesser sciatic foramen	Sensory and motor to the perineum, including external anal and urethral sphincters
Superior gluteal nerve	L4–S1 (from sacral plexus)	Exits the pelvis through the greater sciatic foramen	Motor to gluteus medius, minimus, and tensor fasciae latae muscles
Inferior gluteal nerve	L5–S2 (from sacral plexus)	Exits the pelvis through the greater sciatic foramen	Motor to the gluteus maximus muscle
Sacral sympathetic trunk	Continuation of the lumbar sympathetic trunk	Runs on the anterior surface of the sacrum	Contributes to the inferior hypogastric plexus
Coccygeal plexus	S4–S5 and coccygeal nerves	Located near the coccyx	Provides sensory innervation to the skin over the coccyx
Uterovaginal plexus (in females)	Derived from the inferior hypogastric plexus	Located around the uterus and vagina	Autonomic innervation to the uterus and vagina
Prostatic plexus (in males)	Derived from the inferior hypogastric plexus	Located around the prostate	Autonomic innervation to the prostate, seminal vesicles, and erectile tissues

Cords: The divisions reconverge to form three cords named according to their position relative to the axillary artery:

- Lateral Cord: Formed by the convergence of the anterior divisions of the upper and middle trunks.
- Posterior Cord: Formed by the union of all three posterior divisions.
- Medial Cord: Continuation of the anterior division of the lower trunk.

The brachial plexus provides the motor and sensory innervation to the upper limbs, including the shoulder, arm, forearm, and hand.

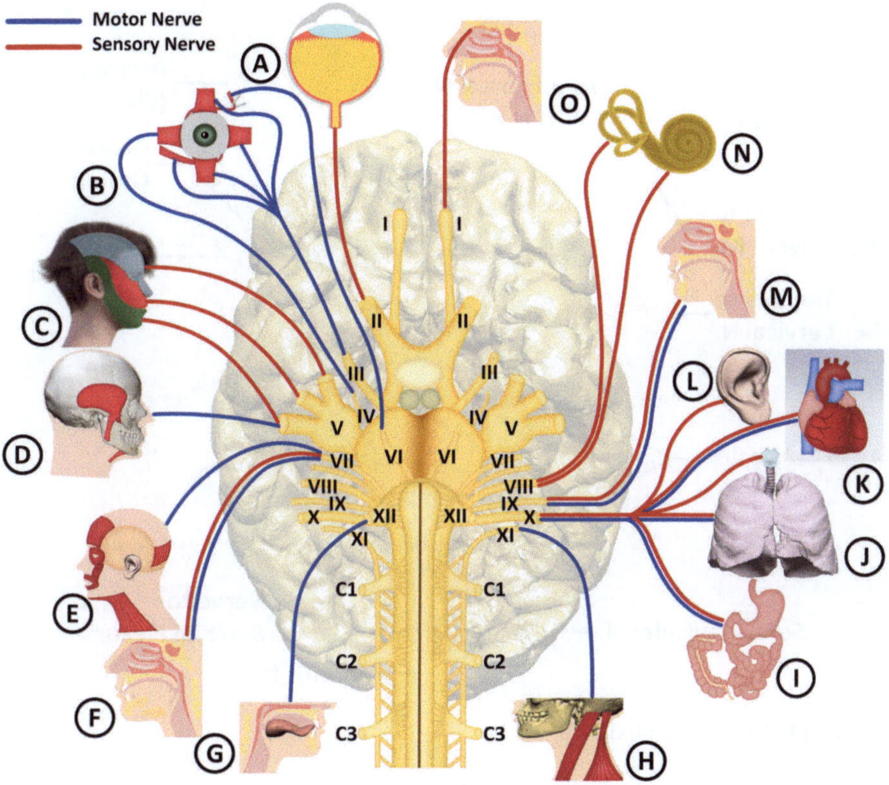

Fig. 10.38 Function of the cranial nerves: (**a**) **Optic nerve (CN II):** Carries visual information from the retina to the brain. (**b**) **The oculomotor nerve (CN III)** controls most of the eye's movements, including constriction of the pupil and maintaining an open eyelid; **the trochlear nerve (CN IV)** controls the superior oblique muscle of the eye, which is responsible for downward and lateral eye movement; and **abducens nerve (CN VI)** controls the lateral rectus muscle, which is responsible for outward gaze. (**c, d**) **Trigeminal nerve (CN V):** Through ophthalmic (V1), maxillary (V2), and mandibular (V3) branches, provides sensation to the face and motor functions such as biting and chewing. (**e, f**) **Facial nerve (CN VII):** Controls facial expression muscles and conveys taste sensations from the anterior two-thirds of the tongue and oral cavity. It also provides motor innervation to salivary glands. (**g**) **Hypoglossal nerve (CN XII):** Controls tongue movements. (**h**) **Accessory nerve (CN XI):** Controls the sternocleidomastoid and trapezius muscles. (**i–l**) **Vagus nerve (CN X):** Controls the function of the heart, lung, and digestive system by providing motor and sensory innervation. It also provides sensation to the external auditory canal. (**m**) **Glossopharyngeal nerve (CN IX):** Has a role in taste, salivation, and monitoring of the carotid body and sinus. (**n**) **Vestibulocochlear nerve (CN VIII):** It is a pure sensory nerve responsible for hearing and balance. (**o**) **Olfactory nerve (CN I):** It is responsible for the sense of smell and transmits sensory information from the nasal cavity to the brain

Lumbosacral Plexus

The lumbosacral plexus (Fig. 10.41) is a network of nerves formed from the lumbar (L1 to L5) and sacral nerves (S1 to S4). It provides motor and sensory innervation to the lower limbs, pelvis, and perineum (Table 10.18).

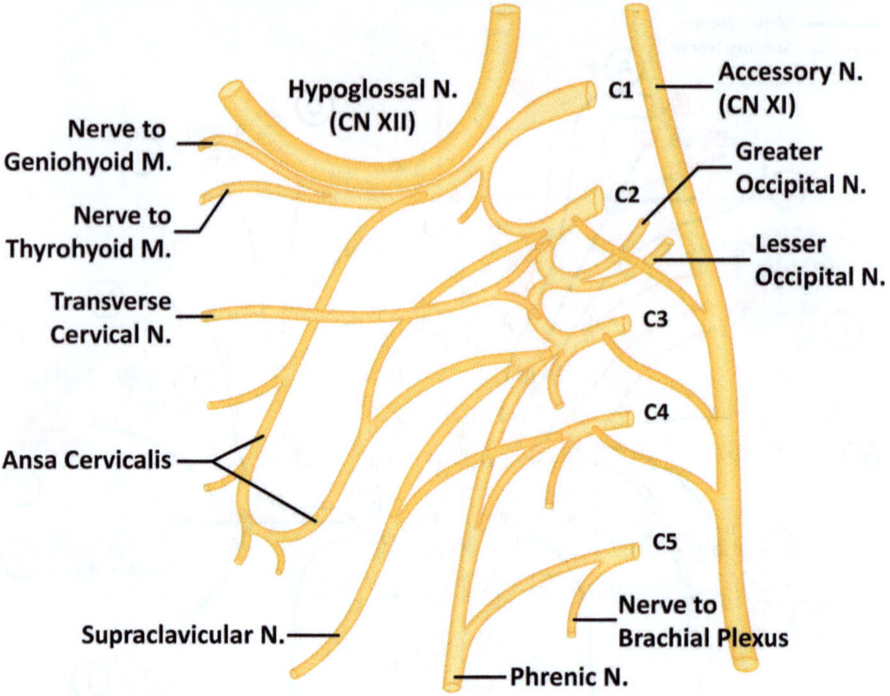

Fig. 10.39 The cervical plexus

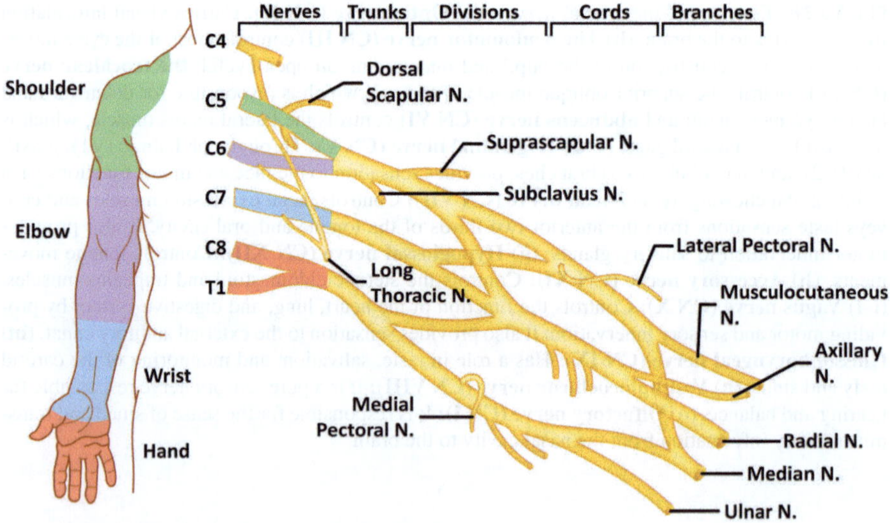

Fig. 10.40 The brachial plexus

Thorax and Abdomen Nerves

The nerves supplying the thorax and abdomen (Table 10.19) primarily consist of branches from the thoracic and lumbar spinal cord segments through the spinal nerves and the vagus nerve (tenth cranial nerve). These nerves provide sensory and motor innervation to these regions' muscles, skin, and organs.

Nerves to Pelvis

Autonomic nerves innervate the internal organs of the pelvis, while the sacral nerves provide pelvic sensation and muscle innervation (S2–S4). The pelvic area's autonomic nerves form several plexuses (Table 10.20).

Dermatomes

Dermatomes are specific skin areas innervated by single spinal nerves and their corresponding dorsal root ganglia. The body is mapped into a series of contiguous

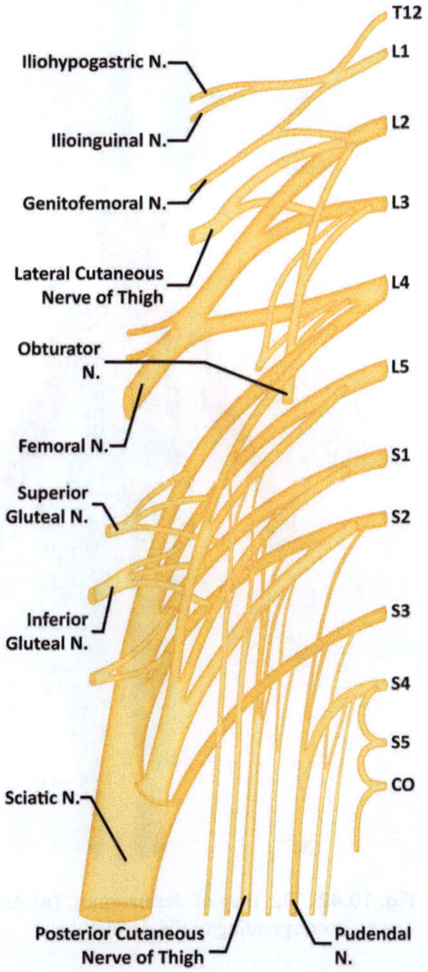

Fig. 10.41 The lumbosacral plexus

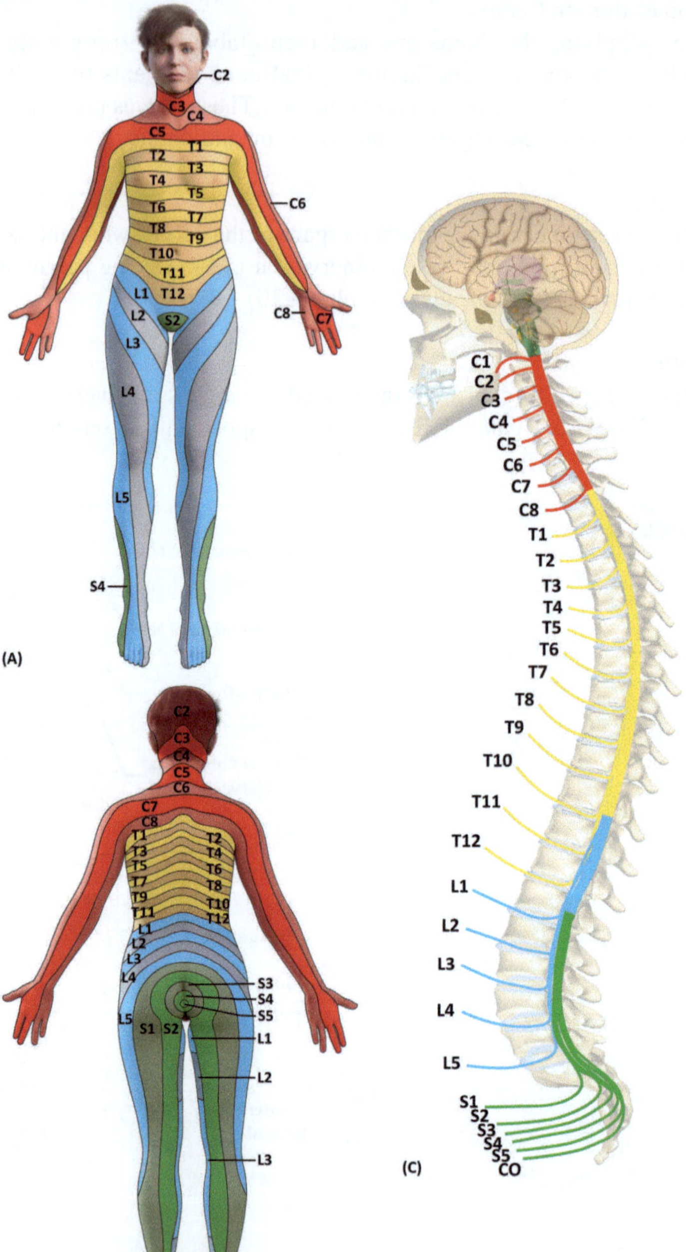

Fig. 10.42 The map of dermatomes: (**a**) Anterior view. (**b**) Posterior view. (**c**) Spinal cord and nerves, corresponding to the dermatomes

dermatomes, each linked to a specific spinal nerve root from C2 to S5 (Fig. 10.42). Notably, the C1 spinal nerve usually does not have an associated dermatome.

10.2.3.2 Autonomic Nervous System

Within the PNS, the autonomic nervous system (ANS) emerges as a complex set of efferent neurons that govern involuntary physiological processes. Unlike the somatic nervous system, the ANS plays a critical role in maintaining homeostasis by modulating the function of internal organs, glands, and blood vessels. The ANS is broadly categorized into two divisions: sympathetic and parasympathetic nervous systems.

Sympathetic Division

The sympathetic nervous system orchestrates the body's rapid involuntary response to stressful or emergency situations, commonly described as the "fight-or-flight" system. The sympathetic nervous system (Fig. 10.43) originates from the spinal cord's thoracic and upper lumbar segments, specifically between T1 and L2. Gray and white rami communicantes serve as conduits for sympathetic fibers, linking the spinal nerves with the sympathetic chain ganglia. Preganglionic neurons are located in the lateral horn of the spinal cord. These neurons have myelinated axons that exit through the ventral root. Postganglionic neurons are typically located in the sympathetic ganglia, have nonmyelinated axons, and extend to the effector organs.

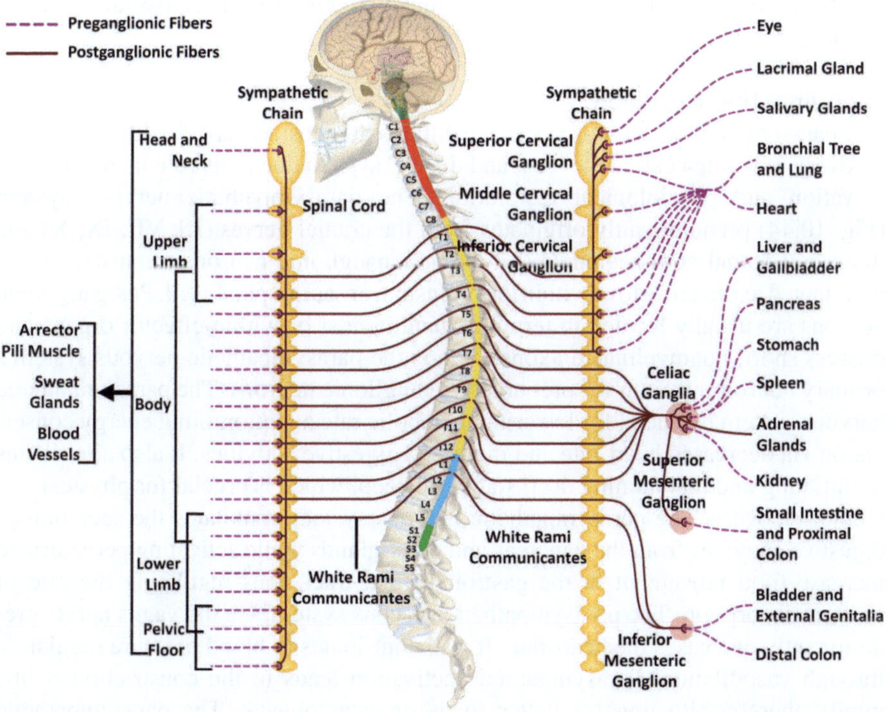

Fig. 10.43 Sympathetic nervous system

Sympathetic ganglia include paravertebral (also known as sympathetic chain ganglia), prevertebral, and terminal ganglia. Preganglionic neurons release acetylcholine (ACh), which acts on nicotinic receptors in the ganglia. Norepinephrine (NE) is the primary neurotransmitter released by postganglionic neurons; NE acts on α and β adrenergic receptors. The sympathetic nervous system is central to the activation of the fight-or-flight response, a physiological reaction that occurs in response to a perceived harmful event or threat. Upon perception of a threat, the sympathetic nervous system is activated through various pathways, primarily involving the adrenal medulla, which secretes catecholamines like adrenaline and noradrenaline (Table 12.5). The sympathetic nervous system stimulates the heart via sympathetic cardiac nerves, increasing heart rate and the force of cardiac contractions. It acts to constrict peripheral blood vessels, thereby increasing blood pressure, essential for adequate perfusion during physical activity or stress.

Through vasoconstriction, the sympathetic nervous system redistributes blood flow, directing it away from nonessential organs like the gastrointestinal tract and toward essential organs like the heart and skeletal muscles. It also stimulates the liver to convert glycogen to glucose, increasing blood glucose levels to fuel muscular activity. Through the action of hormones like adrenaline, the sympathetic nervous system promotes the breakdown of lipids in adipose tissues, releasing fatty acids into the bloodstream for energy. It causes bronchodilation, increasing airflow into the lungs and thus oxygen supply to the body, a crucial aspect of preparing the body for rapid action. The sympathetic nervous system also contributes to thermoregulation through piloerection ("goosebumps") to minimize heat loss and promote sweating for cooling.

Parasympathetic Division

The parasympathetic nervous system maintains homeostasis and facilitates rest and recovery. It is often called the "rest and digest" system due to its role in energy conservation and physiological restoration. The parasympathetic nervous system (Fig. 10.44) predominantly originates from the cranial nerves (III, VII, IX, X) and the sacral spinal cord segments (S2–S4). Preganglionic neurons often have long, myelinated axons close to or within the brainstem or sacral spinal cord. Postganglionic neurons are usually located in terminal ganglia near or within effector organs and possess short, nonmyelinated axons. ACh is the parasympathetic nervous system's primary neurotransmitter for pre- and postganglionic neurons. The parasympathetic nervous system is crucial for lowering metabolic rate and promoting energy conservation via decreased heart rate and increased digestive activities. It also contributes to initiating and maintaining deep stages of sleep, which is crucial for physical and mental recovery. The parasympathetic nervous system stimulates the secretion of digestive enzymes from the pancreas and other glands while activating peristalsis to increase food movement in the gastrointestinal tract, thus regulating the rate of nutrient absorption. The parasympathetic nervous system, via the vagus nerve, predominantly decreases the heart rate. It also contributes to blood pressure regulation through vasodilation. Parasympathetic activation leads to the constriction of the pupils, thereby allowing for better focus on near objects. The parasympathetic

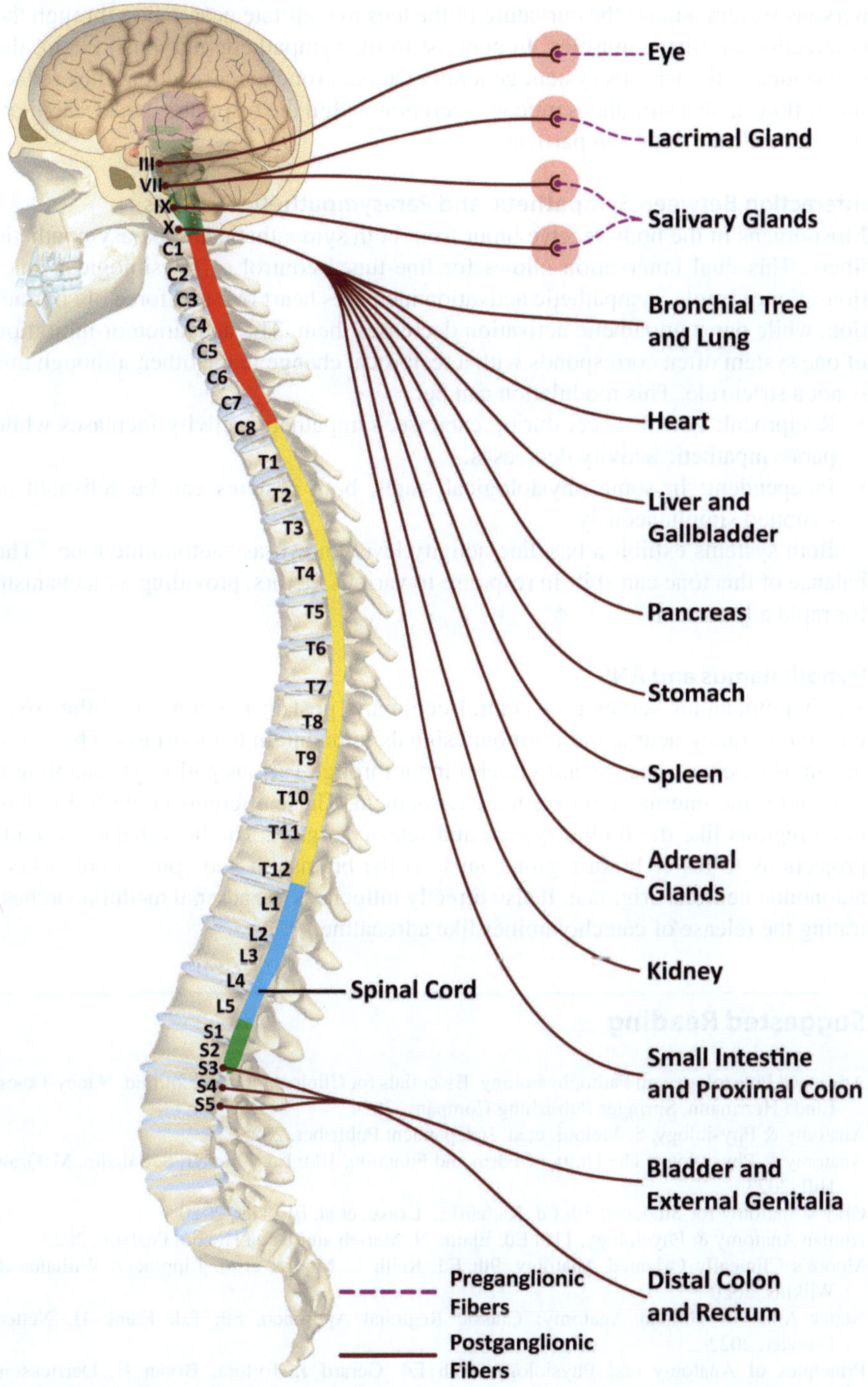

Fig. 10.44 Parasympathetic nervous system

nervous system adjusts the curvature of the lens to facilitate near vision through the contraction of ciliary muscles. In contrast to the sympathetic nervous system, the parasympathetic nervous system generally causes bronchoconstriction, thus reducing airflow. It also stimulates mucous secretion within the respiratory tract, facilitating the removal of foreign particles.

Interaction Between Sympathetic and Parasympathetic Systems

Most organs in the body receive input from both sympathetic and parasympathetic fibers. This dual innervation allows for fine-tuned control of physiological functions. For example, sympathetic activation increases heart rate and force of contraction, while parasympathetic activation decreases them. The activation or inhibition of one system often corresponds with a reciprocal change in the other, although this is not a strict rule. This modulation can be:

- Reciprocal: For instance, during exercise, sympathetic activity increases while parasympathetic activity decreases.
- Independent: In some physiological states, both systems can be activated or inhibited simultaneously.

Both systems exhibit a baseline activity level known as "autonomic tone." The balance of this tone can shift in response to various factors, providing a mechanism for rapid adjustments.

Hypothalamus and ANS

The hypothalamus serves as a central command post in the control of the ANS, integrating many neural and hormonal signals to maintain homeostasis. The hypothalamus receives sensory and visceral input through various pathways, enabling it to monitor the internal and external environment. These afferents come from other brain regions like the limbic system and sensory organs. The hypothalamus sends projections to lower brain regions, such as the brainstem and spinal cord, where autonomic neurons originate. It also directly influences the adrenal medulla, orchestrating the release of catecholamines like adrenaline.

Suggested Reading

Advanced Physiology and Pathophysiology: Essentials for Clinical Practice, 2nd Ed. Nancy Tkacs, Linda Herrmann, Springer Publishing Company, 2024

Anatomy & Physiology, S. Meloni, et al. Independent Publisher, 2023

Anatomy & Physiology: The Unity of Form and Function, 10th Ed. Kenneth S. Saladin, McGraw Hill, 2023

Gray's Anatomy for Students, 5th Ed. Richard L. Drake, et al. Elsevier, 2020

Human Anatomy & Physiology, 11th Ed. Elaine N. Marieb and Katja Hoehn, Pearson, 2022

Moore's Clinically Oriented Anatomy, 9th Ed. Keith L. Moore, et al. Lippincott Williams & Wilkins, 2020

Netter Atlas of Human Anatomy: Classic Regional Approach, 8th Ed. Frank H. Netter, Elsevier, 2022

Principles of Anatomy and Physiology, 16th Ed. Gerard J. Tortora, Bryan H. Derrickson, Wiley, 2020

Special Senses

11

In this chapter, you will learn:

- Accessory structures of the eye such as eyebrows, eyelids, and lacrimal apparatus.
- Structure and function of the eye, including detailed anatomy like the sclera, cornea, iris, lens, and retina.
- Mechanisms of phototransduction in the retina.
- Neural pathways involved in vision from the eye to the brain.
- Understanding color vision and the role of different cones sensitive to various wavelengths.
- Anatomy and function of the ear across its three main sections: external, middle, and inner ear.
- Structural and functional analysis of the cochlear duct and the organ of Corti.
- The mechanics of hearing, including the function of auditory ossicles and cochlear mechanics.
- The process of sound transduction from waves to neural signals.
- The role of the vestibular system in maintaining balance and equilibrium.
- Introduction to the olfactory system and the process of olfaction.
- Structure and function of the gustatory system, including taste buds and their role.
- Detailed discussion on the perception of sound characteristics such as pitch, loudness, and timbre.

The special senses—olfaction (smell), gustation (taste), vision (sight), audition (hearing), and equilibrium (balance)— constitute a sophisticated array of physiological mechanisms that serve as conduits between an organism and its external environment. Contrasted with somatosensory modalities like touch, temperature, and pain, which provide more general information about the body's immediate surroundings, the special senses allow nuanced interactions with external stimuli.

H. Hosseinzadeh, *Fundamentals of Medicine for Biomedical Engineering*,
https://doi.org/10.1007/978-3-031-74085-5_11

11.1 Part 1: The Eye and Vision

The eye, a remarkable organ of the visual system, is not just a biological counterpart to optical instruments. It's the key to our sense of vision, a highly specialized structure that transforms electromagnetic waves into information the brain can interpret, enabling us to interact with our environment.

11.1.1 Accessory Eye Structures

The accessory eye structures include the eyebrows, eyelids, eyelashes, conjunctiva, lacrimal apparatus, and extrinsic eye muscles.

11.1.1.1 Eyebrows

The eyebrows are composed of a region of dense, specialized hair follicles situated on the supraorbital ridge above the orbit of the eye. These hairs grow in a specific direction, usually outward and slightly upward. One of the primary functions of the eyebrows is to divert sweat, rain, and other forms of moisture away from the eyes. Eyebrows are also integral to human facial expressions, conveying emotions and intentions.

11.1.1.2 Eyelashes

Eyelashes (Fig. 11.1) are hair follicles embedded in the margins of both the upper and lower eyelids. Eyelashes possess associated sebaceous glands, specifically the glands of Moll and the glands of Zeis, which secrete oils to coat the eyelashes. Eyelashes serve as a protective barrier against dust, debris, and small insects from entering the

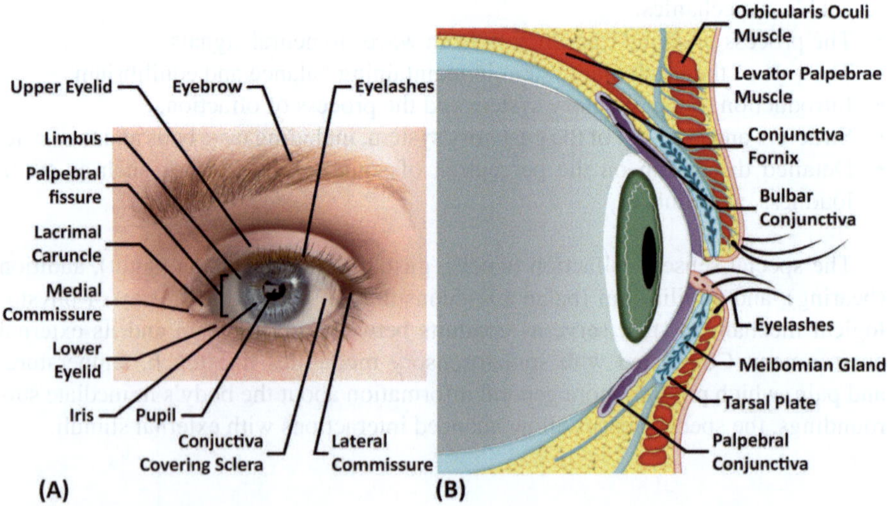

Fig. 11.1 The eye and its accessory structures: (**a**) Surface structure of the left eye. (**b**) Sagittal section of the accessory structures of the eye

eye. The eyelashes' curvature helps capture and redirect such materials away from the ocular surface. Eyelashes are highly sensitive to touch and serve a sensory function by triggering the blink reflex. When an object comes close to the eye or contacts the eyelashes, this reflex is activated, and the eyelids close protectively.

11.1.1.3 Eyelids

The eyelids are complex structures with several layers (Fig. 11.2). The outermost layer comprises thin, highly flexible skin containing sweat and sebaceous glands. Beneath the skin are muscles primarily responsible for eyelid movement, notably the orbicularis oculi and the levator palpebrae superioris. Beneath the muscles are the tarsal plates, which provide structural integrity. The eyelids house specialized glands like Meibomian glands, which secrete an oily substance to maintain tear film, as well as the glands of Moll and Zeis. The innermost layer is the palpebral conjunctiva, a mucous membrane that lines the back of the eyelids and folds back onto itself to line the front part of the eyeball up to the edge of the cornea. The lateral and medial ends where the upper and lower eyelids meet are called lateral and medial canthi. The eyelid's edge contains eyelashes and openings of the Meibomian glands. The eyelids

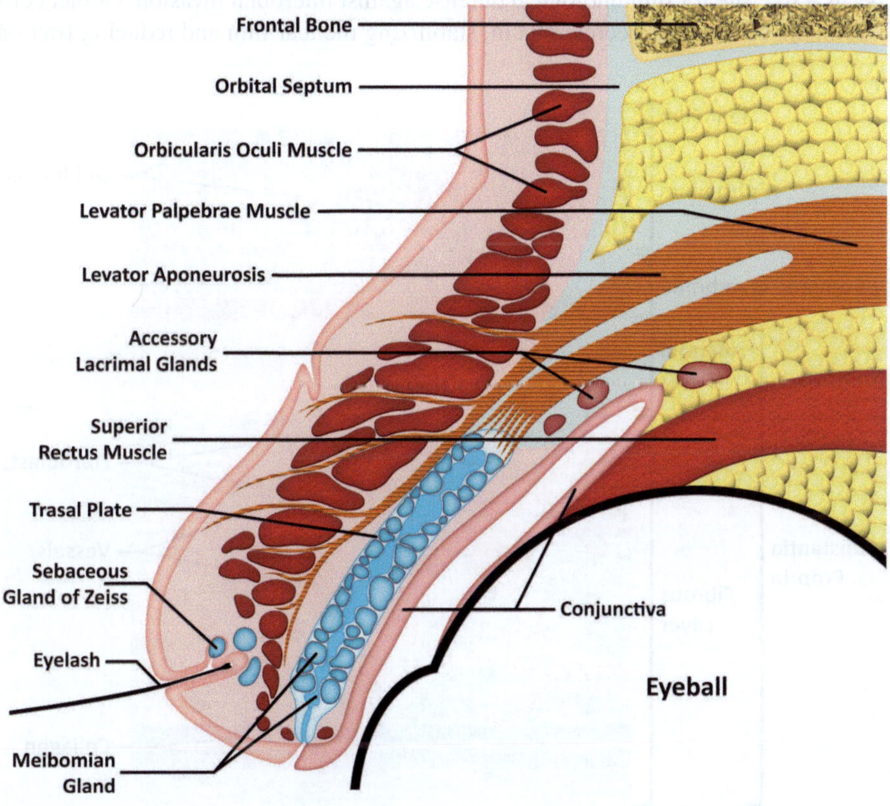

Fig. 11.2 Sagittal section of upper eyelid

are a physical barrier, protecting the eyes from environmental debris, wind, and other physical threats. By distributing the tear film across the eyes' surface, the eyelids contribute to ocular lubrication, preventing dry eyes. By adjusting the size of the palpebral aperture, the space between the open eyelids, the eyelids control the amount of light reaching the retina. Blinking facilitates the movement of tears toward the puncta, aiding in debris removal. The blink reflex, triggered by optical or mechanical stimuli, helps protect the eye from potential damage.

11.1.1.4 Conjunctiva

The conjunctiva (Fig. 11.1) is a transparent mucous membrane that lines the eyelids' interior surface and the eyeball's anterior surface. The conjunctiva comprises multiple layers (Fig. 11.3). The outermost layer consists of stratified columnar epithelial cells. Goblet cells are scattered within the epithelial layer and are responsible for mucin secretion. The inner layer, the stroma, contains connective tissue, blood vessels, and immune cells.

Owing to its mucous secretion and epithelial architecture, the conjunctiva is a physical barrier against environmental pathogens and mechanical irritants. Its extensive vascular supply and immune cells in the stroma allow the conjunctiva to be an active site for immunological defense against microbial invasion. Goblet cells within the conjunctiva secrete mucin, stabilizing the tear film and reducing friction

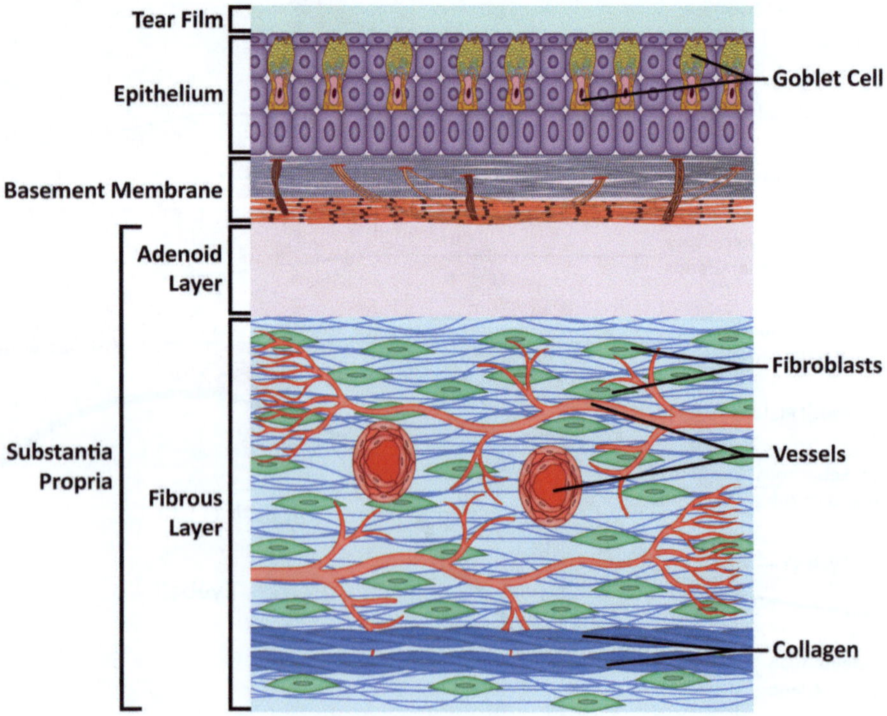

Fig. 11.3 Microscopic structure of conjunctiva

between the eyelid and the eyeball, contributing to comfort and clarity of vision. The vascular network within the conjunctiva facilitates the transfer of essential nutrients to other ocular structures.

11.1.1.5 Lacrimal Apparatus

The lacrimal apparatus (Fig. 11.4) is an intricate system of glands, ducts, and canals responsible for tears' production, distribution, and drainage. The lacrimal gland, located in the superolateral aspect of the orbit, is the primary gland responsible for tear production. It consists of acinar cells that secrete aqueous tear components.

The lacrimal caruncle is a small fleshy elevation at the medial canthus containing sebaceous and sweat glands. Its upper and lower lacrimal puncta drain tears into their respective canaliculi, emptying into the lacrimal fossa's lacrimal sac. The nasolacrimal duct extends from the lacrimal sac inferiorly and opens into the inferior meatus of the nasal cavity, completing the drainage pathway.

The lacrimal glands contribute to the constant production of tears to keep the ocular surface lubricated. The lacrimal gland significantly increases tear production in response to irritants or emotional stimuli. Each blink of the eyelids helps spread the tear film across the ocular surface, ensuring uniform distribution and preventing dry spots. The blinking action facilitates the drainage of tears by generating a pumping action that propels tears through the canaliculi into the lacrimal sac and eventually into the nasolacrimal duct. Tears contain enzymes like lysozyme and immunoglobulins that contribute to the eye's defense mechanisms against microbial invasion.

11.1.1.6 Extrinsic Eye Muscles

The extrinsic eye muscles, also known as the extraocular muscles, are a complex and finely coordinated set of muscles responsible for eyeball movement within the orbit. The extrinsic eye muscles (Table 11.1) (Fig. 11.5) can be classified into two groups based on their course and shape:

- Rectus Muscles: Comprising four muscles—superior, inferior, medial, and lateral rectus.
- Oblique Muscles: These include the superior and inferior oblique muscles and are involved in complex torsional movements of the eye.

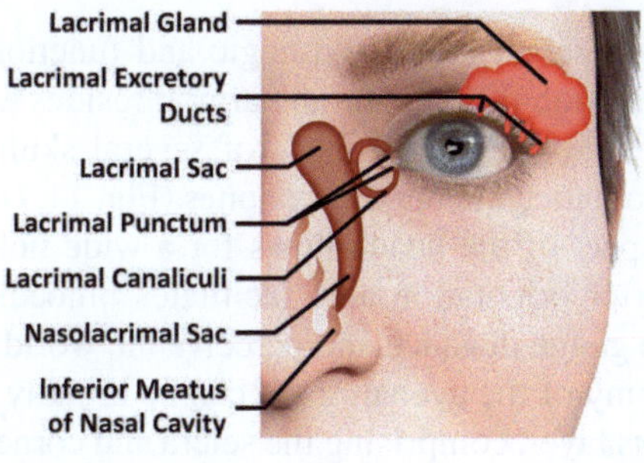

Fig. 11.4 The lacrimal apparatus

Table 11.1 Extrinsic muscles of the eye: origin, insertion, innervation, and function

Muscle	Origin	Insertion	Innervation	Function
Superior rectus	Common tendinous ring at the apex of the orbit	Superior surface of the eyeball	Oculomotor nerve (CN III)	Elevates the eye, turns it medially, and helps in intorsion
Inferior rectus	Common tendinous ring	Inferior surface of the eyeball	Oculomotor nerve (CN III)	Depresses the eye, turns it medially, and helps in extorsion
Lateral rectus	Common tendinous ring	Lateral surface of the eyeball	Abducens nerve (CN VI)	Abducts the eye (moves it laterally)
Medial rectus	Common tendinous ring	Medial surface of the eyeball	Oculomotor nerve (CN III)	Adducts the eye (moves it medially)
Superior oblique	Sphenoid bone	Travels through a pulley-like structure (trochlea) and inserts on the superior, lateral surface of the eyeball	Trochlear nerve (CN IV)	Depresses the eye, turns it laterally, and helps in intorsion
Inferior oblique	Maxillary bone (anterior aspect of the orbital Floor)	Inserts on the inferior, lateral surface of the Eyeball	Oculomotor nerve (CN III)	Elevates the eye, turns it laterally, and helps in extorsion

Most extrinsic eye muscles originate from the common tendinous ring at the orbital apex, providing a stable base for controlled movements. These muscles work in pairs to coordinate complex movements, ensuring the eyes move together. This coordination is essential for binocular vision.

11.1.2 Eyeball

The human eyeball (Fig. 11.6) is a spheroid structure that serves as the principal organ of vision, situated within the orbital cavity of the skull. Encased in a bony socket for protection, the eyeball is anchored by muscles that afford it a range of motion, and it is cushioned by fatty tissue that provides further insulation and protection.

The location of the eyeball is both strategic and functional, optimized for the field of vision and protection from potential harm. It resides within the orbit, a pyramidal structure formed by the convergence of several skull bones, including the frontal, sphenoid, zygomatic, and maxillary bones (Fig. 11.7). The placement of the eyes on the frontal aspect of the head allows for a wide field of vision and depth perception. This anterior orientation also facilitates binocular vision, which provides the capability to gauge distance and perceive the world in three dimensions.

The complex anatomy of the eyeball itself can be broadly categorized into three layers: the outer fibrous layer, comprising the sclera and cornea; the middle vascular

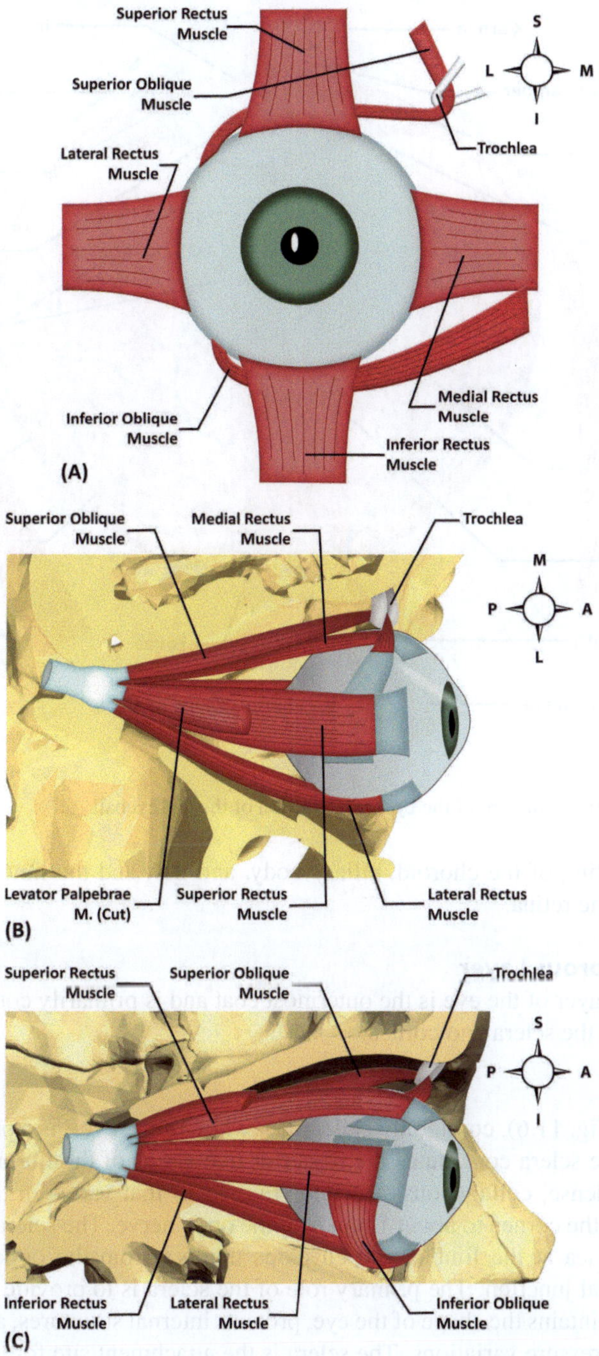

Fig. 11.5 Extrinsic eye muscles: (**a**) Anterior view of the right eye. (**b**) Superior view of the right eye. (**c**) Lateral view of the right eye

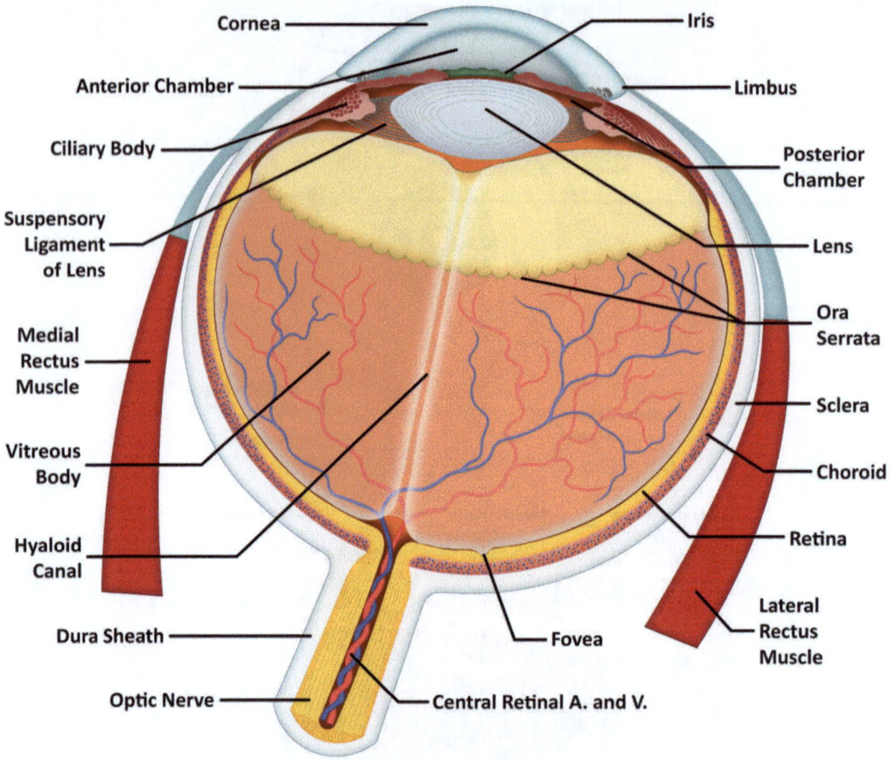

Fig. 11.6 Internal structure of the eye: axial section of the left eyeball

layer, consisting of the choroid, ciliary body, and iris; and the inner neural layer, made up of the retina.

11.1.2.1 Fibrous Layer

The fibrous layer of the eye is the outermost coat and is primarily composed of two components: the sclera and cornea.

Sclera

The sclera (Fig. 11.6), commonly called the "white of the eye," is a protective outer covering. The sclera constitutes the posterior five-sixths of the fibrous layer of the eye. It is a dense, collagenous structure (Fig. 11.8) that varies in thickness from 0.3 mm near the cornea to about 1 mm near the optic nerve. The sclera is continuous with the cornea at the limbus and envelops the eye from the optic nerve to the corneal-scleral junction. The primary role of the sclera is to provide a rigid framework that maintains the shape of the eye, protects internal structures, and withstands intraocular pressure variations. The sclera is the attachment site for the extraocular muscles, facilitating eye movements in various directions.

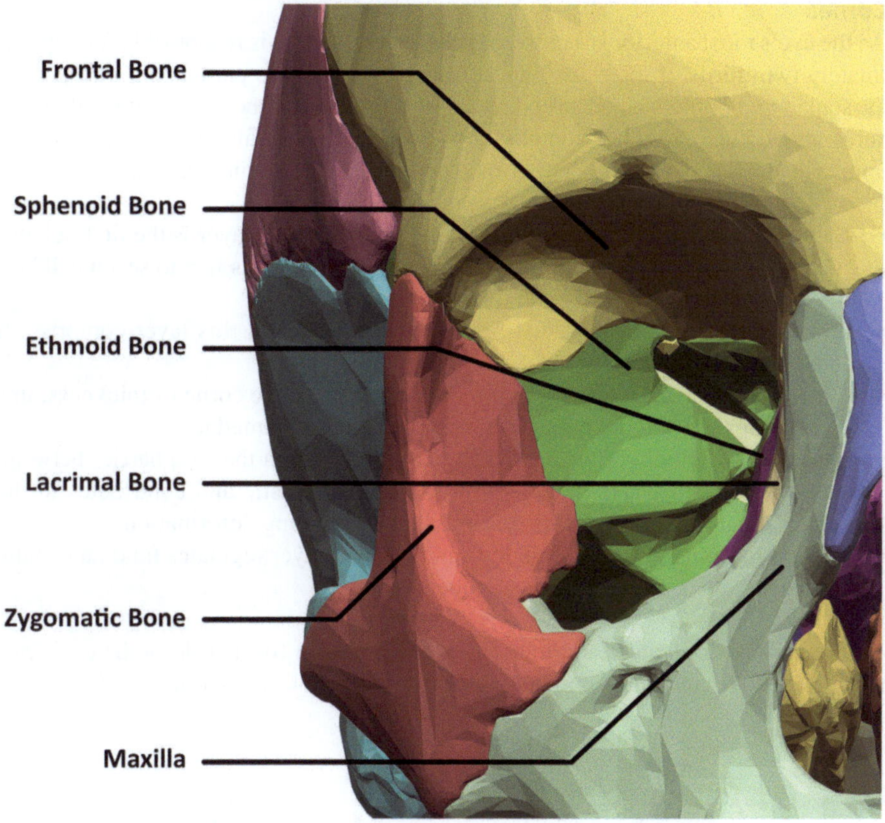

Fig. 11.7 Orbital cavity

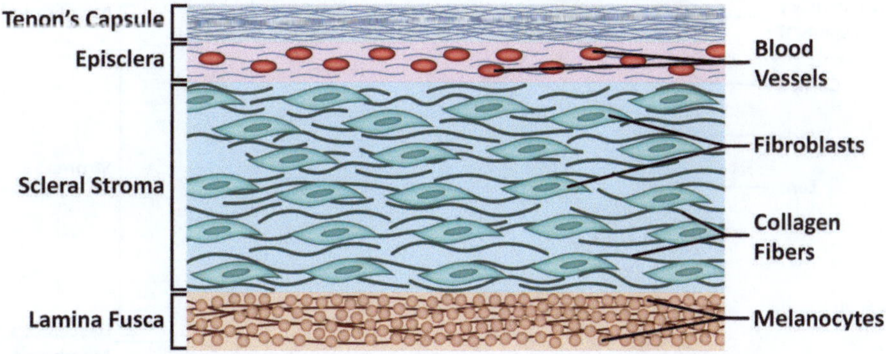

Fig. 11.8 Microscopic structure of the sclera

Cornea

As the eye's most anterior structure, the cornea (Fig. 11.9) is responsible for approximately two-thirds of the eye's total refraction power. The cornea is a transparent, avascular, and dome-shaped structure. In an adult human, the cornea typically measures about 11–12 mm horizontally and 10–11 mm vertically, with a thickness of approximately 500–600 μm. The cornea comprises five distinct layers:

- Corneal Epithelium: This stratified squamous epithelial layer is the first defense against environmental factors and pathogens. It comprises five to seven cell layers and is approximately 50 μm thick.
- Bowman's Layer: Situated just beneath the epithelium, this layer comprises a dense sheet of collagen fibers.
- Corneal Stroma: Constituting approximately 90% of the cornea's thickness, this layer comprises parallel collagen fibrils organized in lamellae.
- Descemet's Membrane: A thick, acellular basal lamina that is a barrier between the corneal stroma and endothelium. It is highly elastic and contributes to the cornea's ability to return to its original shape following deformation.
- Corneal Endothelium: This single squamous cell layer regulates fluid and solute exchange between the cornea and aqueous humor.

The cornea's curvature and transparency are crucial for its role as the eye's primary refractive medium, accounting for approximately two-thirds of its total

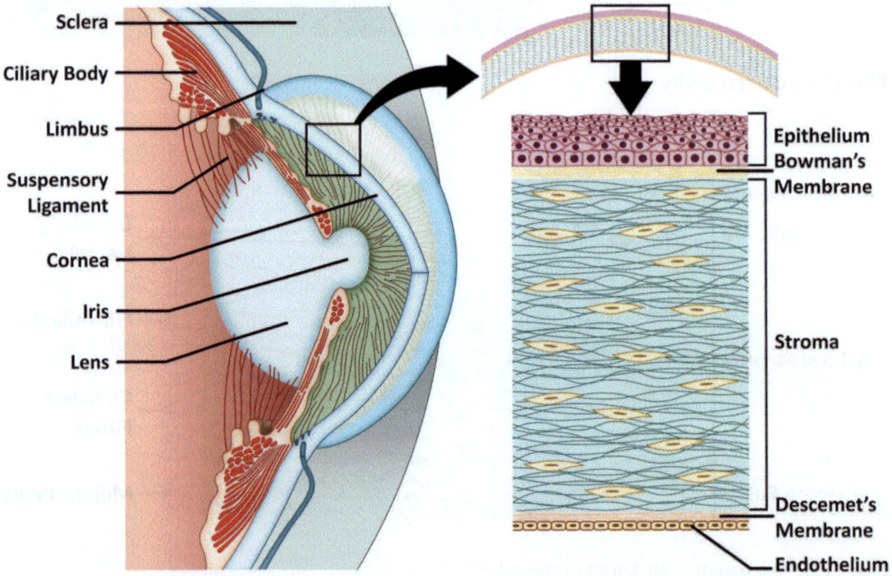

Fig. 11.9 Structure of the anterior chamber (left). (Cornea and iris are cut). Microscopic structure of the cornea (right)

refractive power. The cornea also serves as a protective barrier against physical injuries and microbial invasions, mainly due to its epithelial layer and Bowman's layer. The cornea is highly innervated by sensory nerve fibers from the ophthalmic division of the trigeminal nerve, making it one of the most sensitive tissues in the body. This sensitivity is critical for protective reflexes like blinking and tearing.

11.1.2.2 Vascular Layer

The vascular layer (Fig. 11.10), commonly known as the uvea, constitutes the middle coat of the eyeball and comprises the iris, ciliary body, and choroid.

Choroid

The choroid (Fig. 11.11) is situated between the retina and the sclera. The choroid extends from the ora serrata anteriorly to the optic nerve posteriorly, covering the posterior five-sixths of the eye. The choroid is the primary vascular supply to the outer retina, providing oxygen and essential nutrients. The melanocytes within the choroidal stroma absorb excess light, reducing intraocular scatter and improving image clarity.

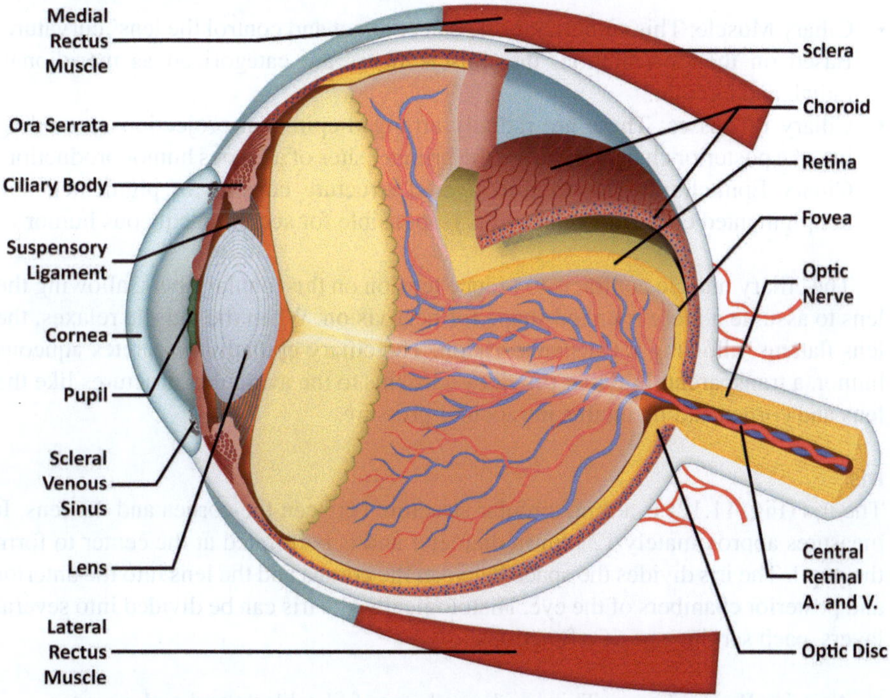

Fig. 11.10 The layers of the eye, axial cut. (The upper part of the eyeball has part of the choroid and retina inverted)

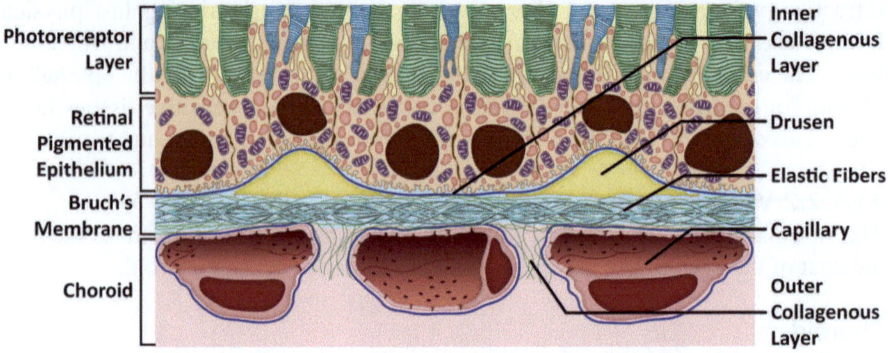

Fig. 11.11 Microscopic structure of the choroid and outer layers of the retina

Ciliary Body

The ciliary body (Fig. 11.12) is positioned between the iris anteriorly and the choroid posteriorly. The ciliary body is a ring-shaped structure that encircles the lens. It is approximately 6 mm wide and extends posteriorly to form the ora serrata, where the sensory retina ends. The ciliary body is organized into three main zones:

- Ciliary Muscle: This smooth muscle can contract and control the lens' curvature. Based on their orientation, the muscle fibers are categorized as meridional, radial, and circular.
- Ciliary Processes: These are radially arranged epithelial projections extending into the posterior chamber and are the primary sites of aqueous humor production.
- Ciliary Epithelium: This double-layered structure consists of pigmented and nonpigmented epithelial cells and is responsible for secreting aqueous humor.

The ciliary muscle contracts to reduce tension on the zonular fibers, allowing the lens to assume a more rounded shape for near vision. When the muscle relaxes, the lens flattens, allowing for distance vision. The ciliary epithelium secretes aqueous humor, a transparent fluid that provides nutrients to the avascular structures like the lens and cornea and maintains intraocular pressure.

Iris

The iris (Fig. 11.12) is a thin, circular structure between the cornea and the lens. It measures approximately 12 mm in diameter and is perforated at the center to form the pupil. The iris divides the space between the cornea and the lens into the anterior and posterior chambers of the eye. Histologically, the iris can be divided into several layers, each serving specific functions:

- Anterior Border Layer: This is a loose layer of fibroblasts and melanocytes.
- Stroma: This layer, comprised mainly of collagen fibers, fibroblasts, melanocytes, and blood vessels, contributes to the iris' rigidity and color.

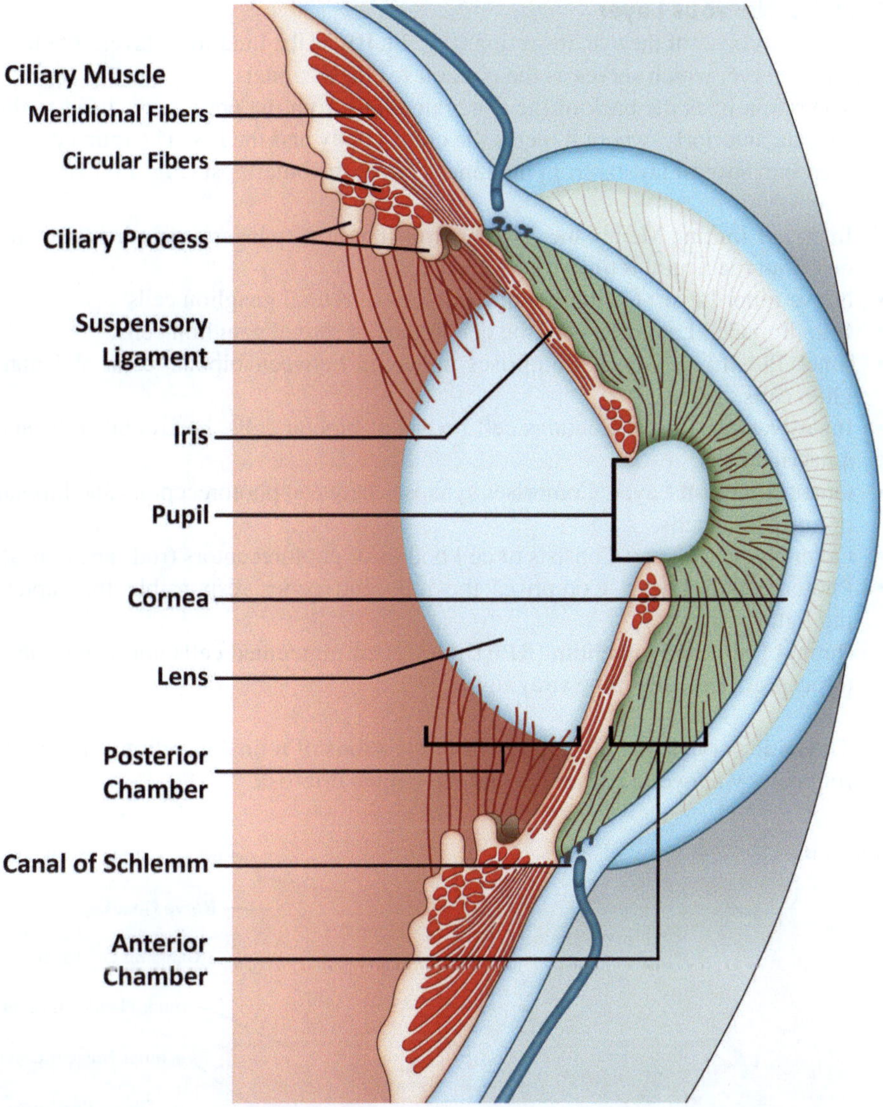

Fig. 11.12 The structure of the anterior segment of the eye, showing the anatomy of the ciliary body, iris, limbus, and anterior and posterior chambers

- Muscle Layer: This layer contains two types of smooth muscles: the sphincter pupillae, which constricts the pupil, and the dilator pupillae, which dilates it.
- Pigmented Epithelium: It consists of two layers of cuboidal cells filled with melanin granules, which absorb stray light and contribute to the eye's color.

The primary function of the iris is to regulate the amount of light entering the eye. The sphincter pupillae and dilator pupillae muscles contract or relax to adjust pupil size and modulate light intake.

11.1.2.3 Nervous Layer

The nervous layer of the eye, the retina (Fig. 11.10), is the innermost layer. It houses the optic nerve, which serves as the conduit for these signals.

The retina lines the back of the eye, extending from the optic nerve head to the ora serrata anteriorly, where it meets the ciliary body and the iris. The retina can be divided into several layers, from the innermost to the outermost (Fig. 11.13):

- Internal Limiting Membrane: The innermost boundary layer separating the vitreous from the neural retina.
- Nerve Fiber Layer: Composed of the axons of retinal ganglion cells.
- Ganglion Cell Layer: Houses the cell bodies of retinal ganglion cells.
- Inner Plexiform Layer: Comprises synapses between bipolar cells and ganglion cells.
- Inner Nuclear Layer: Contains cell bodies of bipolar cells, horizontal cells, and amacrine cells.
- Outer Plexiform Layer: Comprises synapses between photoreceptors and bipolar or horizontal cells.
- Outer Nuclear Layer: Consists of cell bodies of photoreceptors (rods and cones).
- Photoreceptor Layer: Comprises the rods and cones responsible for capturing light.
- Retinal Pigment Epithelium (RPE): A layer of pigmented cells nourishing photoreceptors and absorbing stray light.

The optic nerve is mainly composed of the axons of retinal ganglion cells and is surrounded by a myelin sheath beyond the lamina cribrosa.

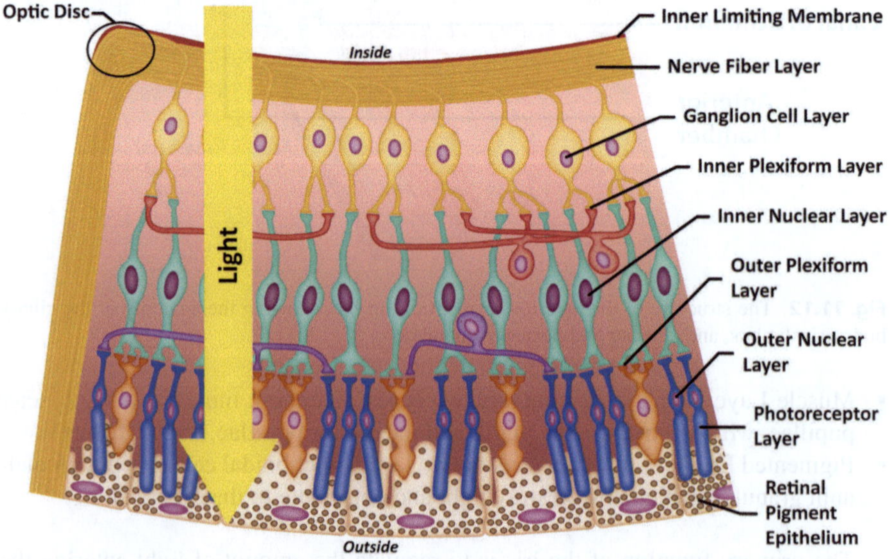

Fig. 11.13 Microscopic structure of the retina

There are three particular regions in the retina:

Fovea Centralis: The fovea is a small, central pit within the retina, approximately 1.5 mm in diameter, where the inner retinal layers are displaced to reduce light scattering (Fig. 11.14). This structural specialization allows for the most direct path of light to the photoreceptor cells, specifically cones, which are densely packed in this region. The fovea is crucial for tasks requiring high visual acuity, such as reading, driving, and any activity requiring fine detail recognition. It is the part of the retina where photopic (daylight) vision is most accurate and solely responsible for our central field of vision.

Macula Lutea: The macula is an oval-shaped, pigmented area around 5.5 mm in diameter surrounding the fovea. It contains a high density of cones and some rods (Fig. 11.14). The macula contributes to our central vision and is vital for color discrimination and detail perception.

Optic Disc: The optic disc, commonly called the "blind spot," is a circular region where the optic nerve exits the retina. It lacks photoreceptor cells and is approximately 1.5 mm in diameter. It is located nasally relative to the fovea. While the optic disc does not contribute to image formation, it serves as the point of exit for axons of the ganglion cells, which make up the optic nerve (Fig. 11.13).

Rods and cones (Section 11.1.3.3) capture photons and convert them into electrochemical signals through phototransduction. The retina incorporates multiple

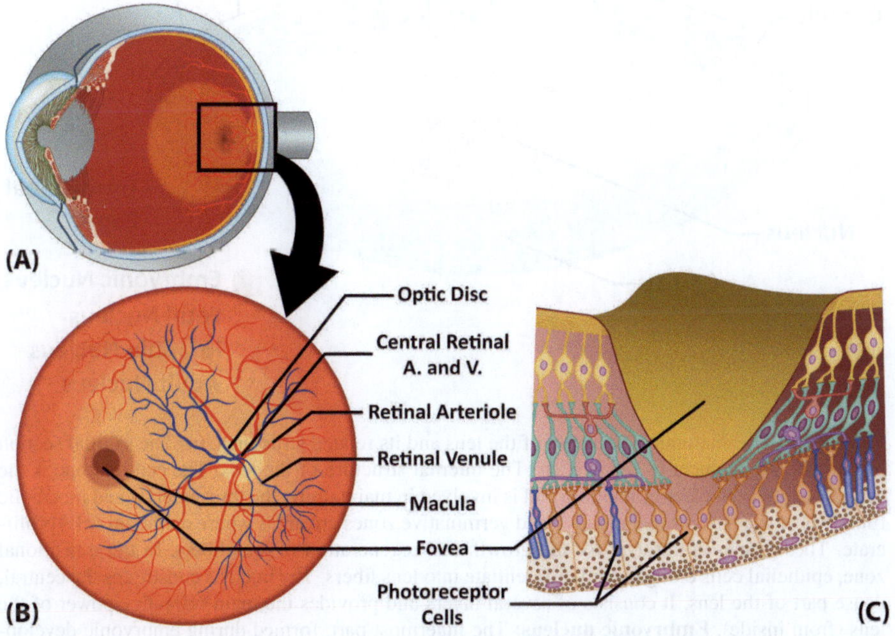

Fig. 11.14 Macula and fovea. (**a**) The parasagittal section of the right eye shows the retina's internal surface. (**b**) Anterior view of the retina's internal surface shows the location of the fovea, macula, and optic disc. (**c**) The microscopic structure of the retina in the fovea region

types of cells that participate in preliminary signal processing, including contrast and edge detection, before transmitting the signals to the brain.

11.1.2.4 Lens

Situated between the anterior and posterior chambers of the eye, the crystalline lens (Fig. 11.15) is a biconvex, transparent, and avascular structure, approximately 10 mm in diameter and 4 mm in thickness. It is suspended by zonular fibers that

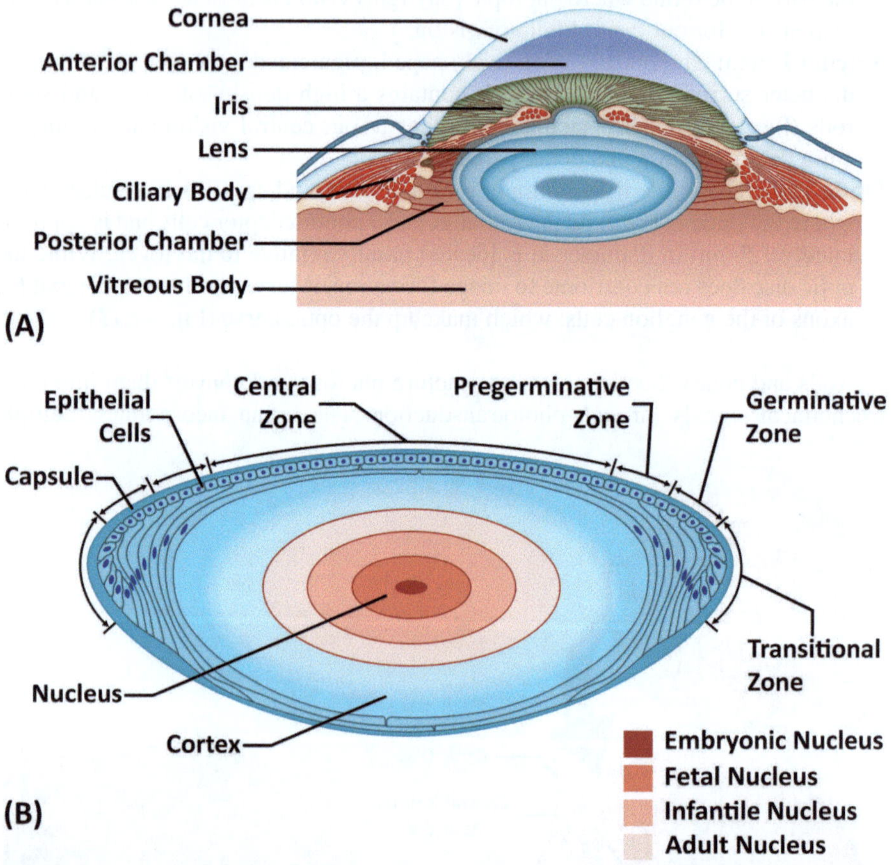

Fig. 11.15 The anatomical structure of the lens and its related structures. (**a**) The sagittal section of the anterior segment of the eye. (**b**) The internal structure of the lens. The central zone is the central region of the lens epithelium and is involved in maintaining the transparency and metabolic functions of the lens. Pregerminative and germinative zones are areas where epithelial cells proliferate. These zones are crucial for the growth and regeneration of lens fibers. In the transitional zone, epithelial cells elongate and differentiate into lens fibers. The nucleus constitutes the central, dense part of the lens. It consists of several layers and provides the main refractive power of the lens (from inside). **Embryonic nucleus:** The innermost part, formed during embryonic development. **Fetal nucleus:** Surrounds the embryonic nucleus, formed before birth. **Infantile nucleus:** Formed during infancy. **Adult nucleus:** The outermost layer of the nucleus, formed during adulthood

extend from the ciliary body, allowing the lens to alter its shape. The lens comprises primarily of:

- Capsule: This is the outermost layer of the lens, a basement membrane made of collagen and glycoproteins. It provides structural integrity and an attachment point for the zonular fibers.
- Epithelium: Located just beneath the capsule at the anterior aspect, this monolayer of cuboidal cells produces the lens fibers that comprise the bulk.
- Lens Fibers: These are highly elongated, transparent cells that constitute the bulk of the lens. They lack organelles to maintain transparency and are rich in proteins called crystallins.
- Nucleus: This is the centermost region composed of older lens fibers. Based on the density of the fibers, it can be further categorized into soft and hard nuclei, and based on the age of the fibers, it can be classified into embryonic, fetal, infantile, and adult nuclei.

The primary role of the lens is to refract incoming light and focus it onto the retina. It contributes to approximately one-third of the eye's overall refractive power. Through the accommodation process (Fig. 11.16), the lens alters its curvature to focus on objects at varying distances. When ciliary muscles contract, zonular fibers slacken, allowing the lens to adopt a more rounded shape for near vision. Conversely, ciliary muscle relaxation tightens zonular fibers, flattening the lens for distant vision.

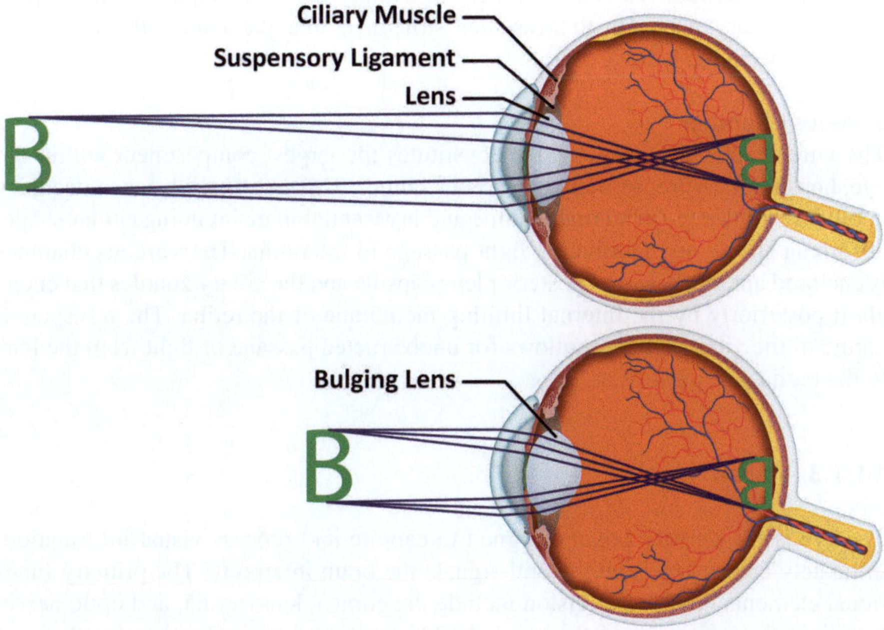

Fig. 11.16 The process of accommodation

11.1.2.5 Chambers

These chambers—the anterior chamber, the posterior chamber, and the vitreous chamber (Fig. 11.9)—are essential for maintaining the eye's shape, refractive properties, and overall functionality.

Anterior Chamber

The eye's anterior chamber is a fluid-filled space between the cornea and the iris (Fig. 11.15). The anterior chamber is bounded anteriorly by the endothelial layer of the cornea, posteriorly by the anterior surface of the iris and the part of the anterior lens surface that is visible through the pupil, and laterally by the peripheral zone of the anterior iris and the corneoscleral limbus. The trabecular meshwork, located at the angle of the anterior chamber, serves as the primary drainage route for aqueous humor. Schlemm's canal, located adjacent to the trabecular meshwork, acts as a secondary passage for aqueous humor.

Posterior Chamber

The posterior chamber (Fig. 11.15) is located between the iris's posterior surface and the lens's anterior face and its zonular fibers. It serves as a key region for producing and circulating aqueous humor. The ciliary processes of the ciliary body secrete aqueous humor into the posterior chamber. It flows through the pupil into the anterior chamber and is drained primarily via the trabecular meshwork and Schlemm's canal. The dynamic balance between aqueous humor production and drainage is crucial in maintaining a stable intraocular pressure (IOP), which typically ranges between 10 and 21 mmHg in healthy eyes. The aqueous humor provides necessary nutrients to avascular structures like the lens and cornea and removes metabolic waste.

Vitreous Chamber

The vitreous chamber (Fig. 11.15) constitutes the largest compartment within the eye, housing the vitreous body or vitreous humor. This gel-like substance occupies about 80% of the eye's internal volume and is essential in maintaining ocular shape, absorbing shock, and facilitating light passage to the retina. The vitreous chamber is enclosed anteriorly by the posterior lens capsule and the ciliary zonules that encircle it posteriorly by the internal limiting membrane of the retina. The transparent nature of the vitreous humor allows for unobstructed passage of light from the lens to the retina.

11.1.3 Vision

The eye is the sensory organ designed to capture and process visual information, ultimately converting it into neural signals the brain interprets. The primary functional elements involved in vision include the cornea, lens, retina, and optic nerve. Light's path as it traverses the eye is highly organized to facilitate optimal visual acuity, contrast, and color perception. The passage of light through the eye includes the cornea, anterior chamber, pupil, lens, vitreous humor, and retina.

Focusing light is a cornerstone of visual perception, allowing the eye to produce clear and distinct images on the retina. Refraction is the bending of light as it passes from one medium to another with differing optical densities. Light first undergoes refraction in the eye at the air-cornea interface and subsequently at the aqueous humor, lens, and vitreous humor interfaces. About two-thirds of the eye's optical power is attributed to the cornea. The curvature of the cornea is essential for its refractive function. The crystalline lens contributes to approximately one-third of the eye's refractive power. Accommodation refers to the dynamic adjustment of the lens curvature to focus on objects at varying distances.

11.1.3.1 Near Sight

Near sight (Fig. 11.17), focusing on a close object, initiates the contraction of the ciliary muscle. This contraction causes the ciliary body to move inward, releasing tension on the suspensory ligaments. With decreased zonular tension, the lens assumes a more spherical, convex shape due to its inherent elasticity. This shape alteration increases the lens' refractive power, allowing the eye to focus on near

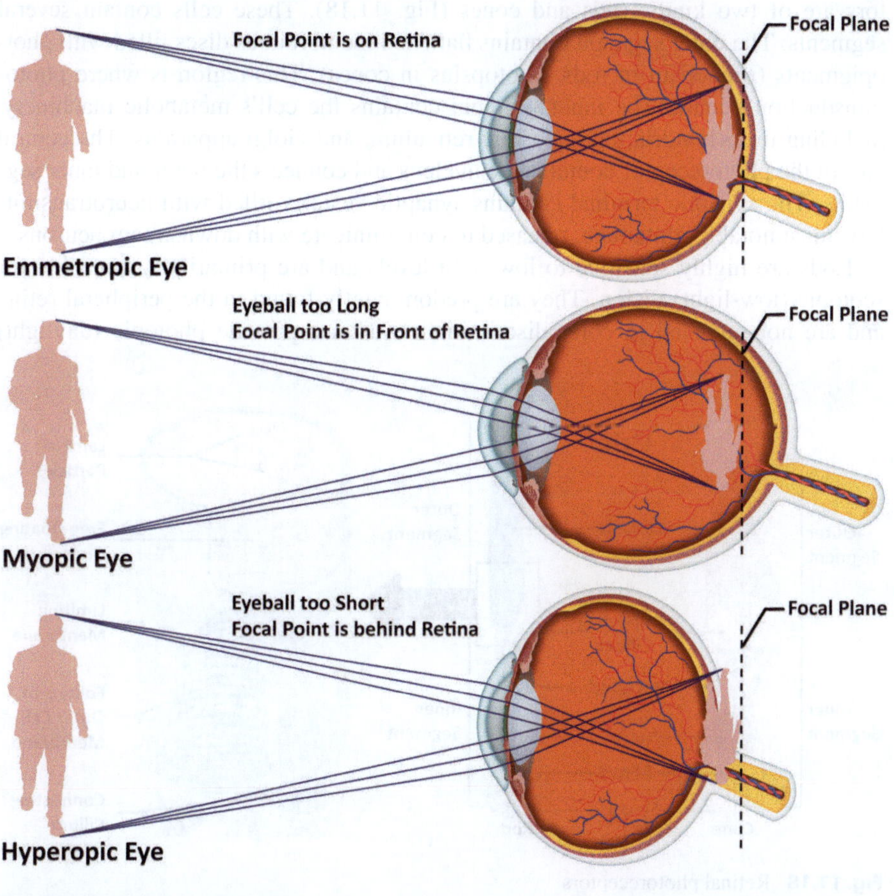

Fig. 11.17 Refractive problems

objects. When accommodating for near sight, the pupil typically constricts, limiting the amount of light entering the eye and reducing spherical and chromatic aberrations, enhancing depth of field and image quality.

11.1.3.2 Far-Sight

Far-sight (Fig. 11.17), focusing on distant objects, relaxes the ciliary muscle. This relaxation causes an outward pull on the ciliary body, increasing tension on the zonules. The increased tension exerted on the lens by the zonules when the ciliary muscle relaxes leads to a more flattened, less convex lens shape. This reduces the lens' refractive power, enabling the eye to focus on distant objects. Pupil dilation is common during far focus, as a larger aperture allows more light to enter, particularly in low-light conditions.

11.1.3.3 Phototransduction

Phototransduction is the biochemical and biophysical process through which the eye's photoreceptor cells, rods, and cones convert light stimuli into electrical signals. Photoreceptors are specialized neural cells within the retina. The photoreceptors are of two kinds: rods and cones (Fig. 11.18). These cells contain several segments: The outer segment contains flattened membranous discs filled with photopigments (rhodopsin in rods, photopsins in cones). This region is where phototransduction occurs. The inner segment contains the cell's metabolic machinery, including mitochondria, endoplasmic reticulum, and Golgi apparatus. The central part of the photoreceptor contains the nucleus and connects the outer and inner segments. The synaptic terminal contains synaptic vesicles filled with neurotransmitters, most notably glutamate, released to communicate with downstream neurons.

Rods are highly sensitive to low light levels and are primarily responsible for scotopic (low-light) vision. They are predominantly found in the peripheral retina and are not involved in color discrimination. Cones provide photopic (daylight)

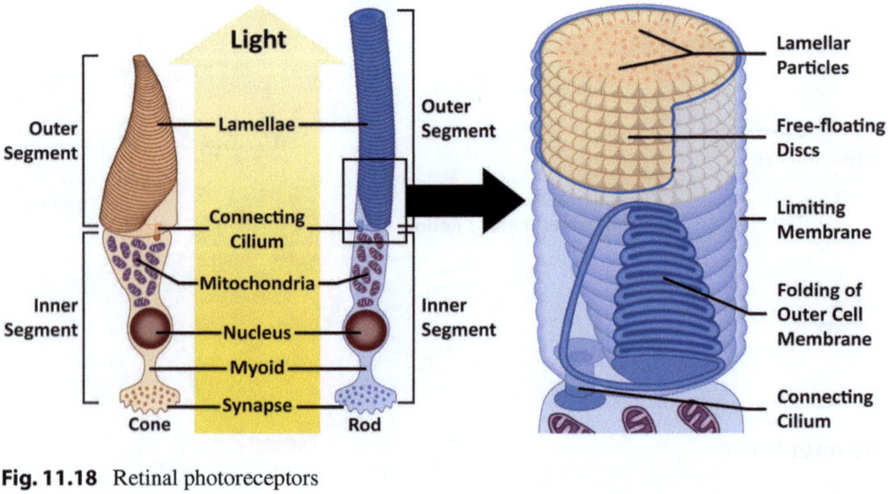

Fig. 11.18 Retinal photoreceptors

vision and are responsible for high-acuity tasks and color perception. They are concentrated in the fovea, the central region of the retina.

Visual pigments (Fig. 11.19) are specialized molecules in photoreceptor cells that can absorb photons and translate this light energy into a biochemical cascade, culminating in neural signals. Rhodopsin in the rods is highly sensitive to low light levels and primarily facilitates scotopic or nighttime vision. It is most sensitive to wavelengths around 498 nm, often in the green-blue spectrum. Cones contain three types of photopsins sensitive to different parts of the visible spectrum, enabling trichromatic (three-color) vision. The variants are:

- S-cone photopsin: Sensitive to short wavelengths (~420–440 nm), blue color.
- M-cone photopsin: Sensitive to medium wavelengths (~534–545 nm), green color.
- L-cone photopsin: Sensitive to long wavelengths (~564–580 nm), red color.

Visual pigments are G-protein-coupled receptors (GPCRs) (Fig. 11.20) that consist of two main components:

- Opsin: This is a seven-transmembrane protein that forms the structure of the visual pigment.
- Retinal: This chromophore is derived from vitamin A, attached to opsin, and is the light-sensitive component.

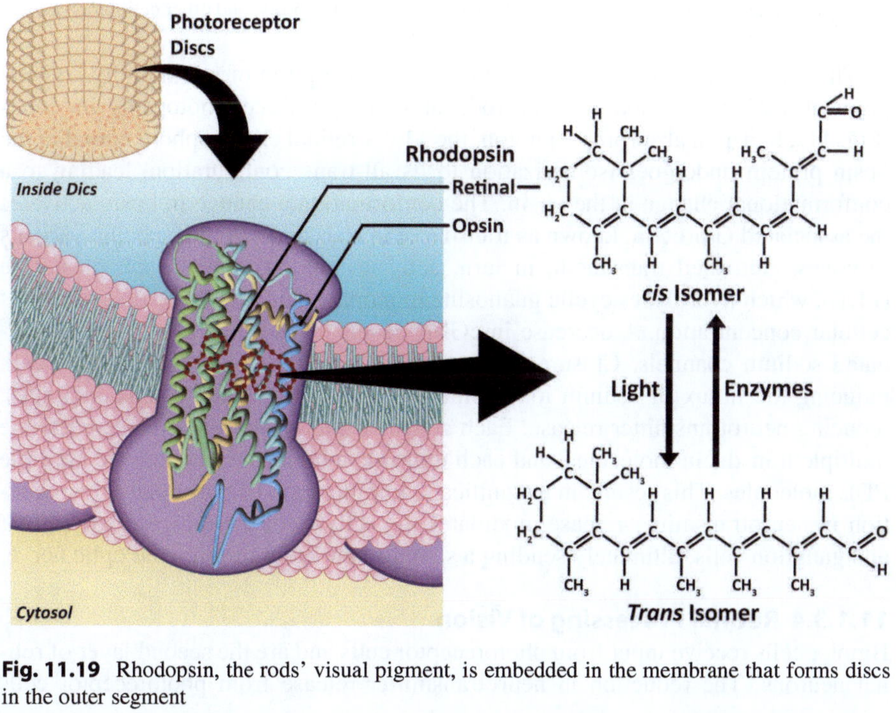

Fig. 11.19 Rhodopsin, the rods' visual pigment, is embedded in the membrane that forms discs in the outer segment

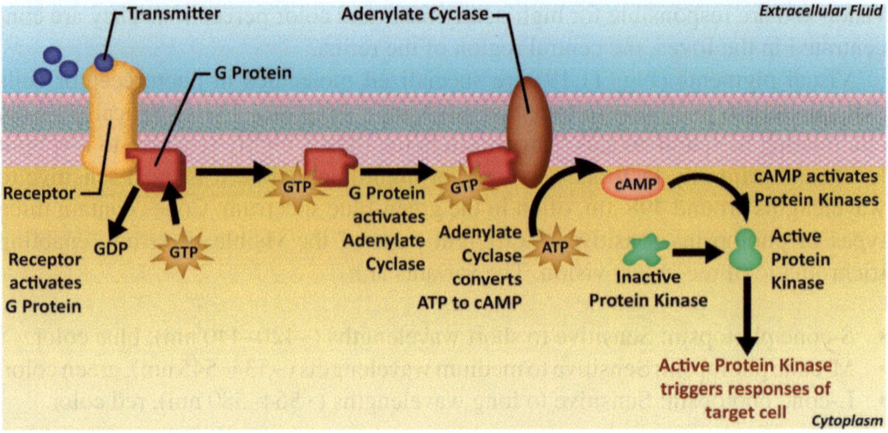

Fig. 11.20 Signal transduction pathway involving G-protein-coupled receptors (GPCRs): A hormone (ligand) binds to a specific receptor on the cell membrane. The receptor is a G-protein-coupled receptor (GPCR), which undergoes a conformational change upon hormone binding. The receptor activates an associated G protein by exchanging GDP for GTP on the alpha subunit of the G protein. The GTP-bound alpha subunit of the G protein dissociates from the beta and gamma subunits and activates adenylate cyclase, an enzyme embedded in the cell membrane. Adenylate cyclase converts ATP to cAMP, which serves as a second messenger in the cell. cAMP amplifies the signal by activating downstream targets. cAMP activates protein kinase A (PKA) by binding to its regulatory subunits, causing the release and activation of the catalytic subunits. PKA phosphorylates various target proteins, leading to changes in their activity. These responses can include changes in gene expression, enzyme activity, ion channel permeability, and other cellular processes

The initial event in phototransduction is the absorption of a photon by a visual pigment molecule—rhodopsin in rods and one of three photopsins in cones (Fig. 11.21). Upon absorbing a photon, the 11.cis-retinal chromophore bound to the opsin protein undergoes isomerization to its all-trans configuration, leading to a conformational change in the opsin. The conformational change in opsin activates the associated G protein, known as transducin in rods and its cone-specific variants in cones. Activated transducin, in turn, activates the enzyme phosphodiesterase (PDE), which hydrolyzes cyclic guanosine monophosphate (cGMP), decreasing its cellular concentration. A decrease in cGMP levels leads to the closure of cGMP-gated sodium channels. Closure of the sodium channels hyperpolarizes the cell, reducing the influx of sodium ions while the efflux of potassium ions continues, reducing neurotransmitter release. Each activated rhodopsin molecule can activate multiple transducin molecules, and each transducin molecule can activate multiple PDE molecules. This results in a significant amplification of the signal. The reduction in neurotransmitter release modulates the activity of the downstream bipolar and ganglion cells, ultimately sending a signal to the brain through the optic nerve.

11.1.3.4 Retinal Processing of Vision

Bipolar cells receive input from photoreceptor cells and are the second layer of retinal neurons. The reduction in neurotransmitter release from photoreceptor cells

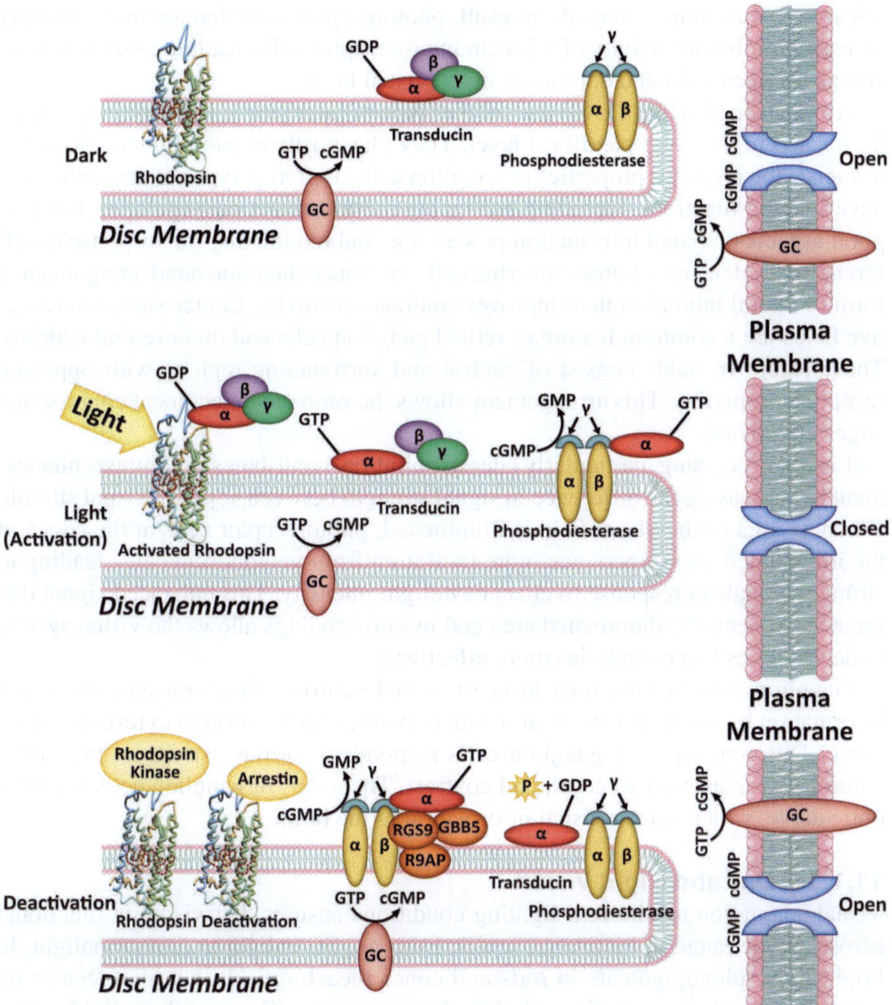

Fig. 11.21 The molecular events of phototransduction using G-protein-coupled receptors (GPCRs)

upon hyperpolarization changes bipolar cell membrane potential. Bipolar cells transmit the signal to ganglion cells, the retina's output neurons. Horizontal cells (Fig. 11.13) are specialized interneurons in the retina located in the outer plexiform layer, between photoreceptor cells (rods and cones) and bipolar cells. Their primary role is lateral inhibition, which enhances contrast by suppressing the activity of neighboring photoreceptor cells. Light stimulates a photoreceptor cell and activates neighboring horizontal cells through synaptic connections. Horizontal cells then send inhibitory signals to photoreceptor cells connected to them, reducing their release of neurotransmitters. As a result, photoreceptor cells that are more strongly activated inhibit the activity of adjacent photoreceptor cells, leading to sharper transitions between light and dark areas in the visual field.

Amacrine cells (Fig. 11.13) are another type of interneuron in the retina, primarily located in the inner plexiform layer. They play a role in lateral interactions and influence the response properties of ganglion cells. Different types of amacrine cells have specific functions, including enhancing motion detection, regulating the temporal aspects of visual information processing, and modulating the response to different light intensities. Some amacrine cells mediate center-surround antagonism, a form of lateral inhibition that improves contrast sensitivity. Center-surround receptive fields are a common feature of retinal ganglion cells and other retinal neurons. These receptive fields consist of central and surrounding regions with opposing receptive properties. This arrangement allows the retina to detect local contrast and edges effectively.

Lateral processing, particularly lateral inhibition, contributes to contrast enhancement by increasing the difference in signal strength between adjacent visual stimuli. When an area of the visual field is illuminated, photoreceptor cells at the edges of the illuminated area experience more inhibition from neighboring cells, leading to stronger signals in response to changes in light intensity. This enhanced signal difference between the illuminated area and its surroundings allows the visual system to detect edges and boundaries more effectively.

Ganglion cells are the final layer of retinal neurons. They transmit the visual information by action potentials along their axons, which converge to form the optic nerve. Different types of ganglion cells respond to various aspects of the visual stimulus, such as motion, color, and contrast. The axons of ganglion cells form the optic nerve, which carries visual information to the brain.

11.1.3.5 Day and Night Vision

Visual adaptation to different lighting conditions ensures that vision is functional across a wide range of luminance levels, from bright sunlight to dim moonlight. In bright light, photopigments in rods and cones bleach quickly but also regenerate rapidly, preventing saturation of the photoreceptors. The pupillary light reflex reduces the amount of light entering the eye, serving as an immediate mechanical response to bright light. In dim light conditions, photopigments regenerate more slowly, allowing the rods to accumulate more pigment and increase their sensitivity. In low light, the pupil dilates to allow more light to reach the retina, enhancing sensitivity.

11.1.3.6 Color Vision

The trichromatic theory suggests that color vision is based on the activation of three different types of cone cells, each sensitive to a specific range of wavelengths:

- Short-wavelength cones (S-cones) respond most strongly to short wavelengths, corresponding to the blue part of the spectrum.
- Medium-wavelength cones (M-cones) respond to medium wavelengths, corresponding to the green part of the spectrum.

- Long-wavelength cones (L-cones) are most sensitive to long wavelengths, corresponding to the red part of the spectrum.

The brain processes the relative activation of these three types of cones to perceive a wide range of colors. Color perception arises from the combination of signals from these cones in different proportions.

The opponent-process theory suggests that color vision is based on opposing pairs of color-sensitive cells responding to complementary pairs of colors. The primary opponent pairs are red vs. green, blue vs. yellow, and black vs. white (brightness). In this theory, the perception of color results from the differential firing of these opponent pairs. For example, if red-sensitive cells are activated, green-sensitive cells are inhibited, leading to the perception of red.

In the retina, information from cones is processed and transmitted to ganglion cells, which send signals to the brain via the optic nerve. Ganglion cells with receptive fields sensitive to specific colors respond differentially to the activation of cones. These signals are then processed in the thalamus's lateral geniculate nucleus (LGN) and the visual cortex, where color perception is refined and integrated.

11.1.3.7 Optic Chiasm

The optic nerves from each eye meet at the optic chiasm, where fibers corresponding to the nasal half of each retina cross over to the opposite hemisphere. At the same time, the temporal (outer) portions remain uncrossed. The optic chiasm is responsible for dividing the visual field into two hemifields: the left and right visual fields. Consequently, each brain hemisphere processes information from the left and right visual fields. Thus, the brain maintains the lateralization of visual information, meaning that the left hemisphere primarily processes visual input from the right visual field. In contrast, the right hemisphere primarily processes visual input from the left visual field.

11.1.3.8 Depth Perception

Depth perception allows for accurate gauging of spatial relationships between objects in the environment. It is a multidimensional phenomenon mediated by various optical, neural, and psychological factors. Stereopsis, or binocular disparity, is the most direct method for perceiving depth. Each eye receives a slightly different image due to their horizontal separation. The brain processes these disparities to perceive depth. Even with one eye closed, depth perception is possible through monocular cues such as size, overlap, relative height, texture gradient, and linear perspective.

11.2 Part 2: Hearing and Balance—Ear

The auditory system, colloquially referred to as the ear, is an intricate and multifaceted anatomical and physiological structure responsible for the perception of sound. The ear also plays an essential role in maintaining balance and spatial orientation. It can be anatomically divided into three primary regions: the external ear, the middle ear, and the inner ear.

11.2.1 External Ear

The external ear, an introductory component of the auditory system, plays an essential role in hearing by capturing and channeling sound waves into the middle ear. It can be anatomically divided into two primary components: the auricle and the external auditory canal.

11.2.1.1 Auricle

The auricle (Fig. 11.22), the pinna, is the most visible part of the ear and serves as the initial point of contact for sound waves entering the auditory pathway. The auricle primarily comprises a flexible framework of elastic cartilage covered with skin. The shape of the auricle is designed to capture sound waves and funnel them into the external auditory canal. This concentration of sound waves enhances the amplitude, making softer sounds more audible. The complex geometry of the auricle selectively amplifies certain frequencies while attenuating others. This property assists in identifying the elevation and distance of sound sources. The auricle's multiple curves, folds, and grooves provide essential cues for the brain to determine the direction of sound sources, particularly in the vertical plane.

11.2.1.2 External Acoustic Canal

The external acoustic canal (EAC) (Fig. 11.23) is the anatomical and functional bridge between the external and middle ear. The EAC is a tubular, S-shaped passage extending approximately 2.5–3 cm from the auricle to the tympanic membrane (eardrum) in adults. It is anatomically divided into two segments:

- Cartilaginous Portion: Constitutes about one-third of the EAC closer to the auricle and comprises elastic cartilage. This portion houses hair follicles and glands.

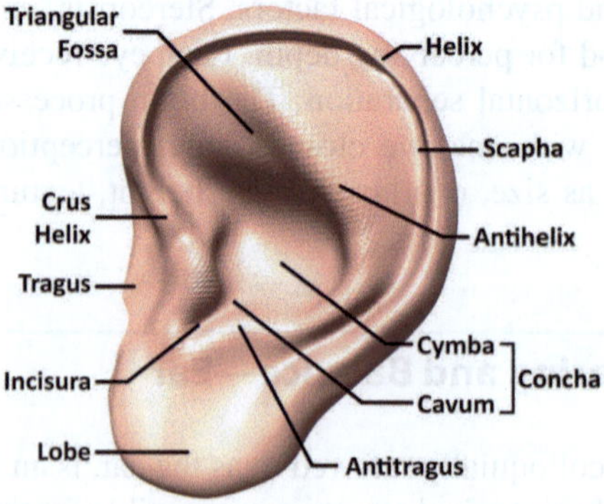

Fig. 11.22 The auricle

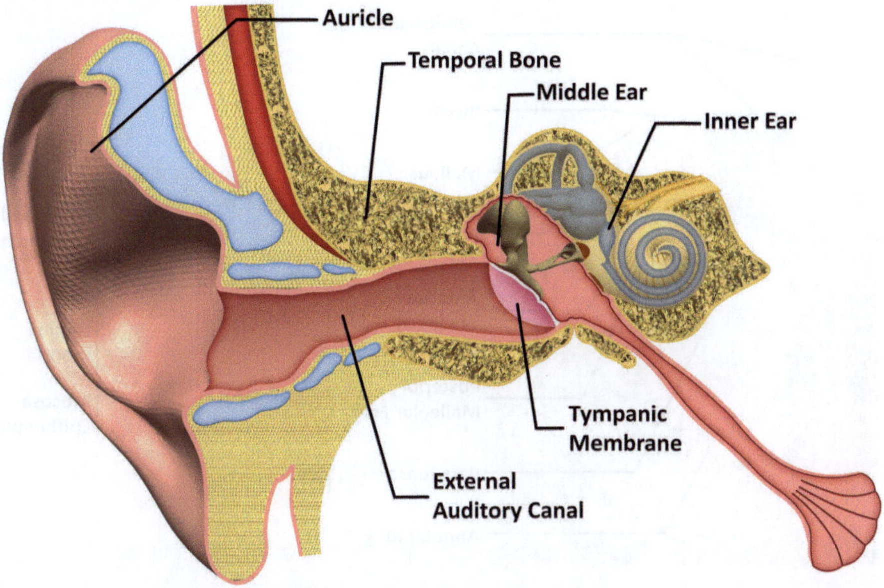

Fig. 11.23 The structure of the ear, showing the detailed structure of the external ear

- Osseous (Bony) Portion: Makes up the remaining two-thirds and is embedded in the temporal bone. It is devoid of hair and glands.

The EAC is lined with stratified squamous epithelium, which gradually thins as it approaches the tympanic membrane. The EAC guides sound waves collected by the auricle toward the tympanic membrane. Its tubular shape aids in the resonant amplification of specific frequency bands, notably those essential for human speech comprehension.

The EAC possesses several protective mechanisms: Ceruminous glands secrete cerumen (earwax), which has antimicrobial properties and traps foreign particles. Hair captures dust and small particles. The epithelium of the EAC moves laterally, effectively removing trapped debris and old cerumen.

11.2.1.3 Tympanic Membrane

The tympanic membrane (Fig. 11.24), known as the eardrum, is the interface between the external auditory canal (EAC) and the middle ear. The tympanic membrane is a thin, oval-shaped membrane at the EAC's terminal end. It measures approximately 9–10 mm in diameter and has an area of about 55 mm^2. The membrane can be anatomically divided into two main parts:

- Pars Tensa: This is the larger, more rigid portion of the tympanic membrane and consists of three layers—an outer epithelial layer, a fibrous middle layer, and an inner mucosal layer.

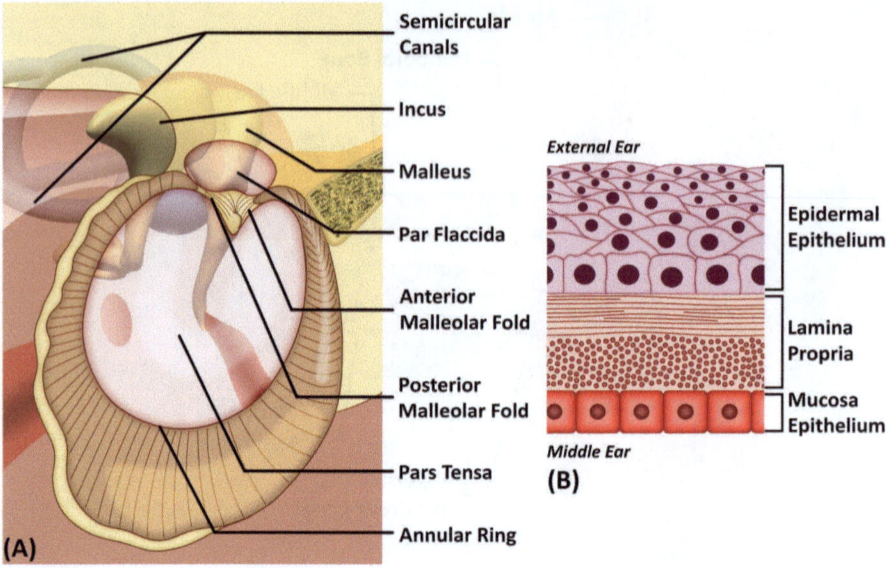

Fig. 11.24 Tympanic membrane: (**a**) Gross structure from outside. (**b**) The microscopic structure

- Pars Flaccida: Located in the upper region, this part lacks the fibrous middle layer, making it more flaccid than the pars tensa.

The tympanic membrane is oriented obliquely in the ear canal, forming an angle with the canal's floor. Important anatomical landmarks include:

- Umbo: The central point where the tympanic membrane is most indrawn.
- Cone of Light: A triangular reflection of light commonly used as a clinical indicator of tympanic membrane health.
- Manubrium of Malleus: This bone is partially embedded in the tympanic membrane and extends from the umbo.

The primary physiological function of the tympanic membrane is the conversion of sound waves into mechanical vibrations. These vibrations are transmitted to the ossicles (malleus, incus, and stapes), leading to further mechanical processes in the middle and inner ear. The tympanic membrane responds to a wide range of frequencies, typically from as low as 20 Hz to as high as 20,000 Hz in humans. The membrane's flexibility and tension can adapt to various sound intensities, allowing for a dynamic range of hearing.

11.2.2 Middle Ear

The middle ear is between the tympanic membrane (eardrum) and the inner ear's labyrinth. It is a complex mechanical amplifier designed to efficiently transfer sound waves arriving at the tympanic membrane to the fluid-filled labyrinth. This region is an air-filled cavity within the temporal bone and contains a series of connected structures.

Tympanic Cavity: The tympanic cavity (Fig. 11.25) is medial to the tympanic membrane and lateral to the inner ear. The tympanic cavity is enclosed within the

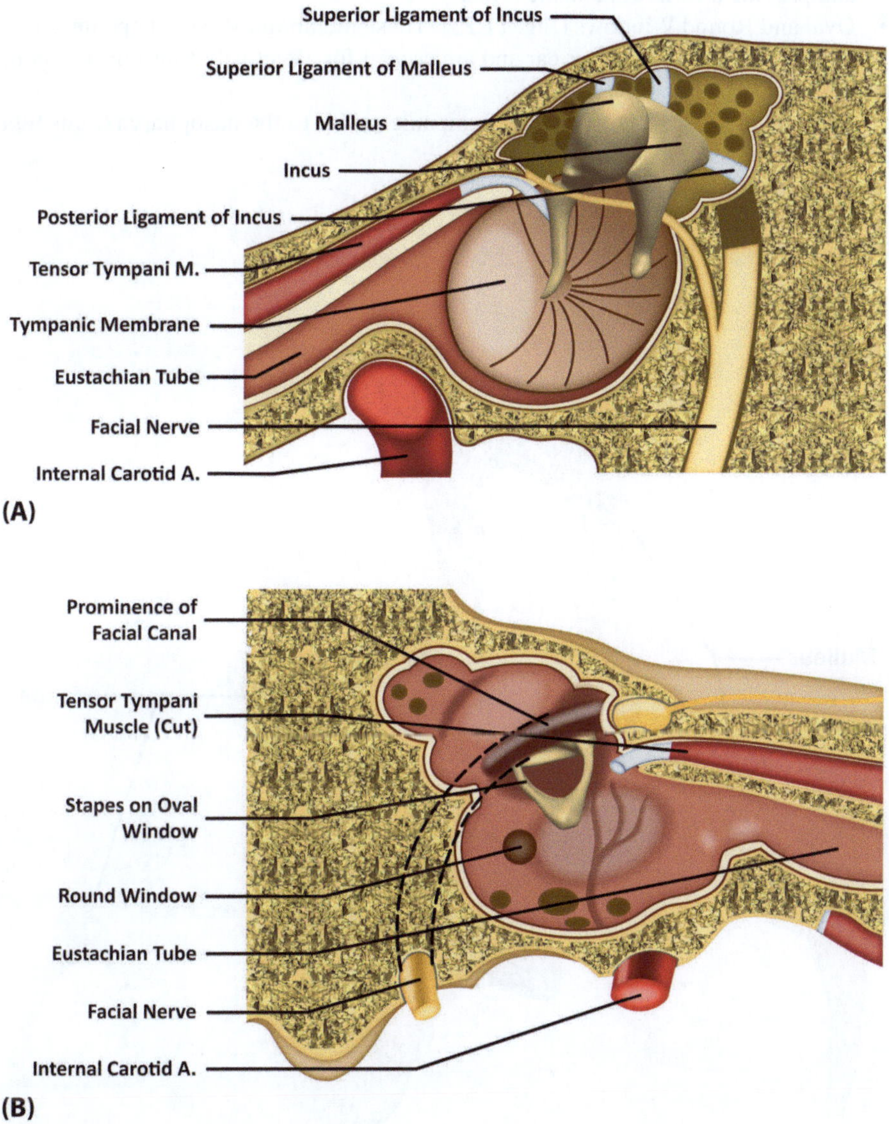

Fig. 11.25 Tympanic cavity: (**a**) Sagittal view showing the medial wall. (**b**) Sagittal view showing the lateral wall

petrous part of the temporal bone and can be described as having a biconcave, box-like shape. Components of the tympanic cavity include:

- Auditory Ossicles (Fig. 11.26): Consisting of the malleus, incus, and stapes, the smallest bones in the human body amplify and transmit vibrations from the tympanic membrane to the inner ear.
- Muscles: The tensor tympani and stapedius muscles modulate ossicle movement and provide a form of acoustic reflex.
- Oval and Round Windows (Fig. 11.25): These membrane-covered openings provide a gateway to the inner ear and are crucial for effectively transmitting sound vibrations.
- Eustachian Tube: Connecting the tympanic cavity to the nasopharynx, this tube helps in pressure equalization.

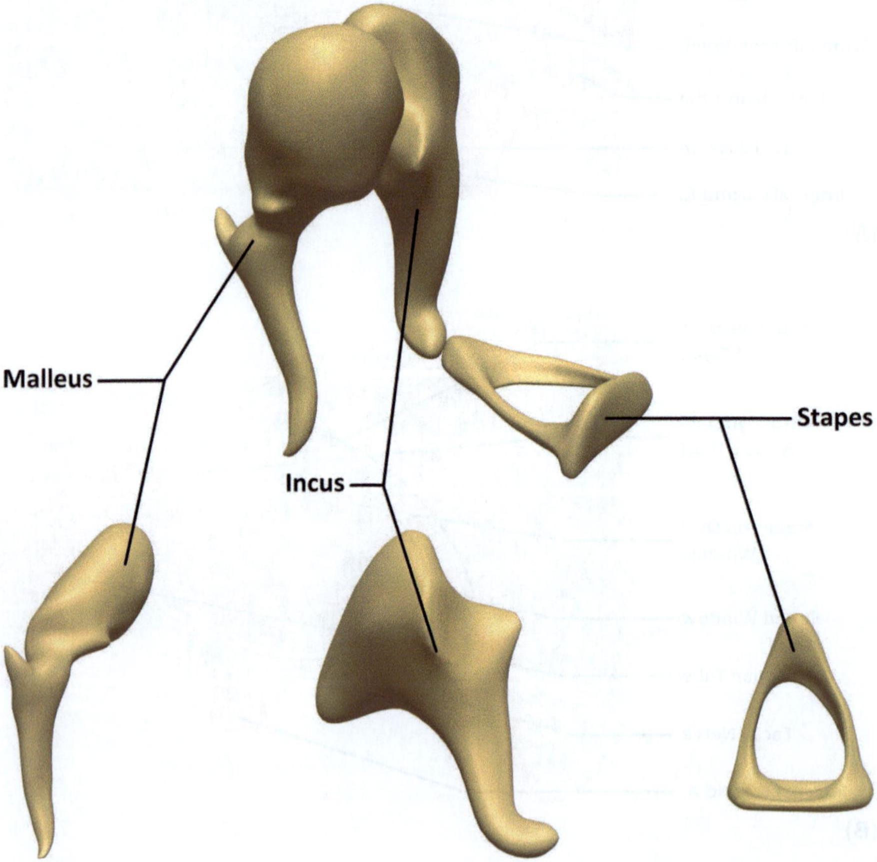

Fig. 11.26 The auditory ossicles

The tympanic cavity performs several critical functions. The ossicles act as levers, magnifying the vibrations received by the tympanic membrane before passing them to the oval window. The tympanic cavity helps to match the impedance between the air-filled external auditory canal and the fluid-filled cochlea, facilitating efficient energy transfer. The Eustachian tube ensures that pressure on both sides of the tympanic membrane remains balanced, optimizing the membrane's vibratory characteristics. The muscles within the tympanic cavity can contract reflexively to dampen the transmission of loud noises, serving a protective role.

11.2.3 Inner Ear

The inner ear, also known as the labyrinth, serves as the sensory organ for hearing and balance. The inner ear is a complex anatomical structure comprising two principal components: the bony labyrinth and the membranous labyrinth (Fig. 11.27).

11.2.3.1 Bony Labyrinth

The bony labyrinth resides within the petrous portion of the temporal bone. It serves as a rigid, bony casing that encloses the membranous labyrinth. Resembling a snail's shell, the cochlea consists of a spiral-shaped bony tube that makes 2.5–2.75 turns around a central bony pillar called the modiolus. The cochlear canal houses the scala vestibuli, scala tympani, and scala media of the membranous labyrinth. Situated between the cochlea and the semicircular canals, the vestibule is the central component of the bony labyrinth. It houses the utricle and saccule of the membranous labyrinth and serves as the intersection for auditory and vestibular pathways. Three ring-like structures, oriented in the three space planes, extend from the vestibule. Each canal encloses a corresponding semicircular duct of the membranous labyrinth. The bony labyrinth is filled with perilymph, a fluid similar in composition to cerebrospinal fluid, which surrounds the membranous labyrinth filled with endolymph. The bony labyrinth provides the necessary rigidity and protective encasement for the delicate structures of the membranous labyrinth. The perilymph within the bony labyrinth facilitates the transmission of sound vibrations and head movements to the sensory cells in the membranous labyrinth.

11.2.3.2 Membranous Labyrinth

The membranous labyrinth resides within the confines of the bony labyrinth. It comprises fluid-filled tubes and sacs and is the sensory organ for hearing and balance. Structurally, the membranous labyrinth consists of the cochlear duct, the semicircular ducts, and the utricle and saccule.

Utricle and Saccule

The utricle and saccule (Fig. 11.28), collectively called the otolith organs, detect linear acceleration and changes in head position relative to gravity. By doing so, these structures play an indispensable role in maintaining postural equilibrium and spatial orientation. The utricle is larger and more horizontally oriented, while the

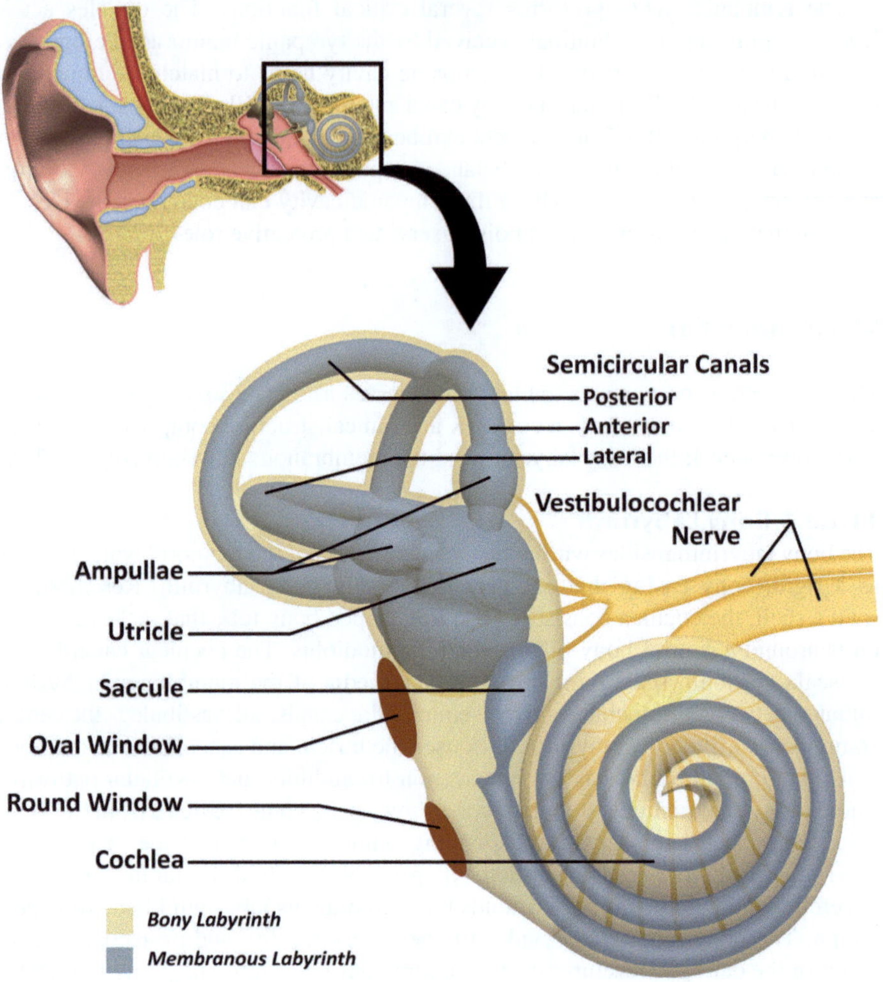

Fig. 11.27 The inner ear. The membranous labyrinth (gray) lies within the chambers of the bony labyrinth (gold)

saccule is smaller and vertically oriented. Each otolith organ consists of a membranous sac filled with endolymph. The sensory epithelium, the macula, is found in each otolith organ. The maculae are integral to our perception of linear acceleration and gravitational forces.

Comprising a bed of hair cells and supporting cells, these structures interact with an overlying gelatinous layer studded with otoliths to transduce mechanical signals into neural impulses (Fig. 11.29). The sensory receptors present are categorized into Type I and Type II hair cells. Protruding from the apical surface of each hair cell are multiple, closely packed, actin-filled protrusions known as stereocilia. They are organized in a bundle, often in a staircase-like arrangement in order of height, where

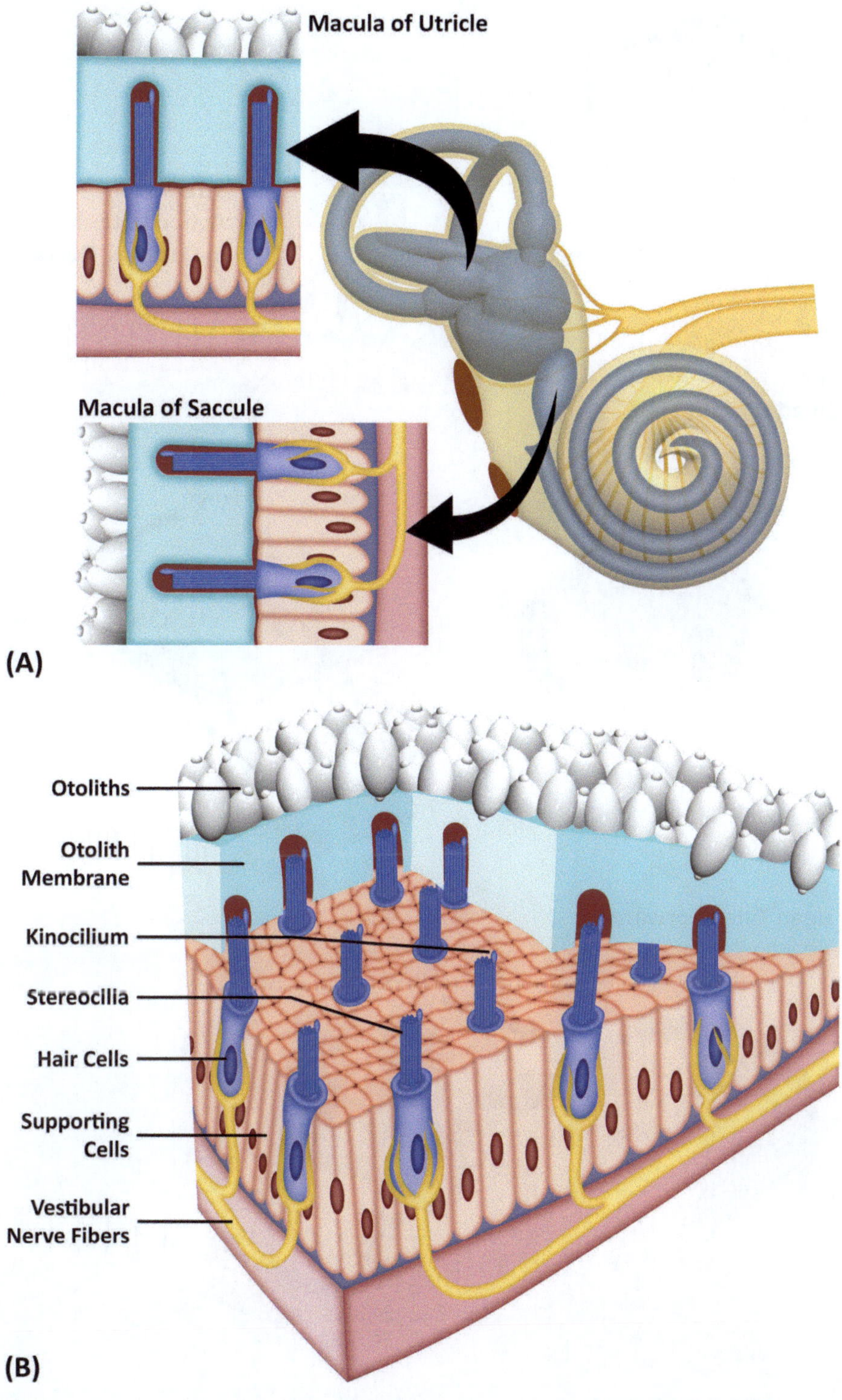

Fig. 11.28 Inner ear macula: (**a**) orientation of macula in utricle and saccule. (**b**) Detailed structure of macula

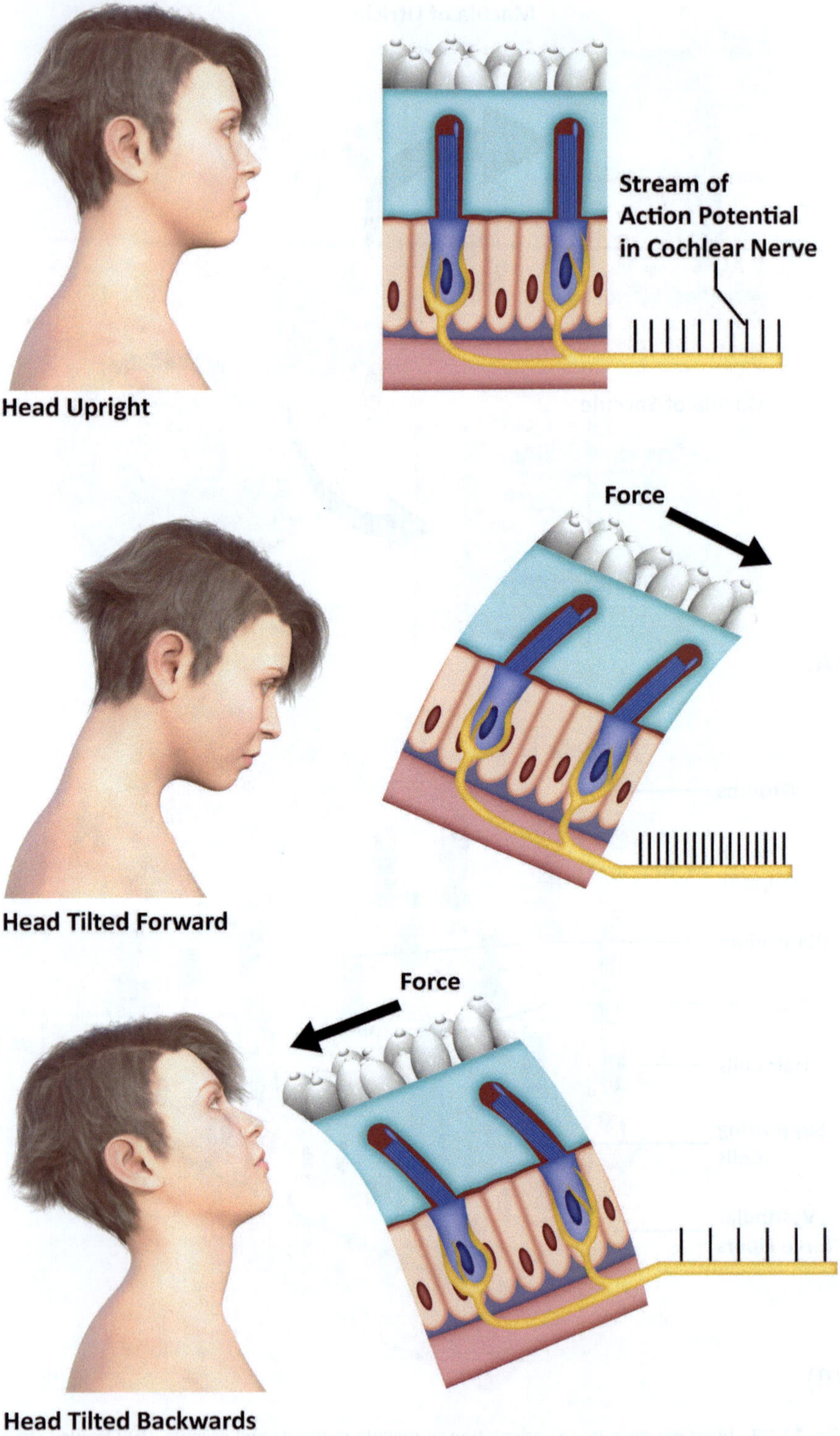

Fig. 11.29 The mechanism of detecting the head position by the macula of the utricle

the tallest is situated farthest from the kinocilium. In addition to stereocilia, a single, larger, true cilium known as the kinocilium is also on the apical surface. Unlike stereocilia, the kinocilium contains microtubules and is usually located at the edge of the stereocilia bundle. The stereocilia and kinocilium are embedded in a gelatinous otolithic membrane containing calcium carbonate crystals (otoliths). The tips of the stereocilia are often anchored to the otolithic membrane through specialized tip links, which are protein filaments. At the basal end of the hair cell, specialized ribbon synapses are formed with the afferent nerve fibers. These synapses are responsible for the release of neurotransmitters in response to mechanical stimulation.

When linear acceleration or gravitational forces act upon the head, the inertial mass of the otolith induces a shift in the otolithic membrane. This displacement bends the stereocilia protruding from the apical surface of the hair cells, resulting in the opening or closing of mechanically-gated ion channels. The resulting ion flux generates a receptor potential, which triggers the release of neurotransmitters at the synaptic junction with the afferent nerve fibers. These neurotransmitters initiate action potentials sent to the vestibular nuclei in the brainstem for processing.

Semicircular Canals

The semicircular canals (Fig. 11.27) detect angular movements and maintain dynamic equilibrium. Three semicircular canals—superior (anterior), posterior, and lateral (horizontal)—are arranged orthogonally to each other, enabling the detection of rotational movements in three-dimensional space. An expansion known as the ampulla is at one end of each membranous semicircular duct. The ampulla contains the crista ampullaris (Fig. 11.30), a sensory organ responsible for transducing rotational movements. Crista ampullaris comprises hair cells, supporting cells, and a gelatinous structure called the cupula. The hair cells are categorized into Type I and Type II. Hair cells in the cristae ampullares are crowned with stereocilia, which are actin-based protrusions. The stereocilia are organized in ascending order of height to form a bundle, somewhat akin to a staircase shape. Each hair cell also possesses a single, true cilium called a kinocilium. Unlike the otolith organs, the stereocilia and kinocilium are embedded in a gelatinous structure known as the cupula, which extends into the endolymph of the semicircular canal. The cupula has no otoliths but is nonetheless critical for the mechanotransduction process. The base of the hair cell contains specialized ribbon synapses for neurotransmitter release, connecting the hair cells to the afferent nerve fibers. Rotational movement causes a flow of endolymphatic fluid within the semicircular ducts. The endolymph's inertia lags behind the head's movement, causing deflection of the cupula. This deflects the stereocilia of the hair cells. The bending of the hair cells opens mechanically gated ion channels, resulting in depolarization or hyperpolarization, depending on the bending direction. This change in membrane potential triggers the release of neurotransmitters into the synaptic cleft, affecting the firing rate of the innervating sensory neurons.

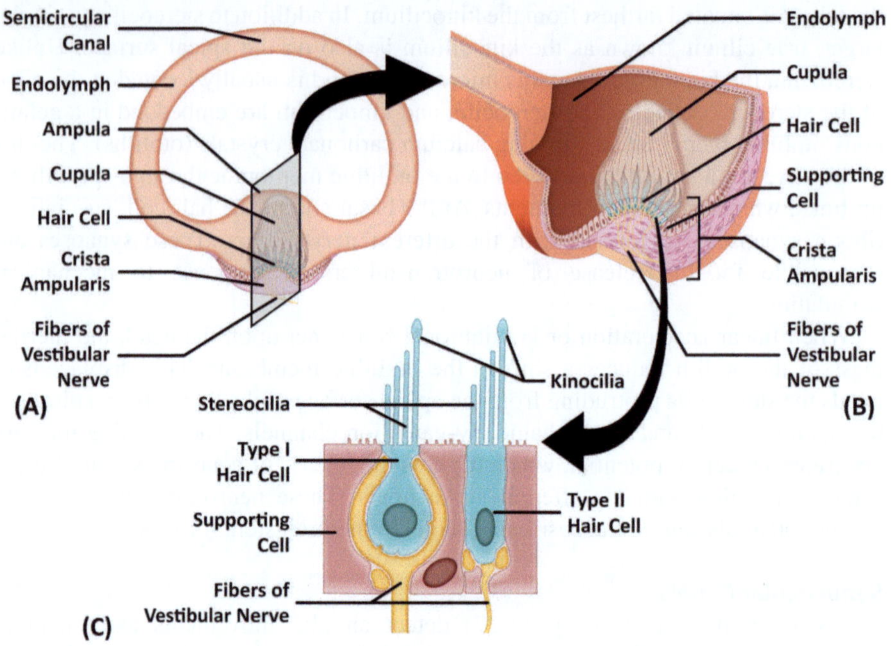

Fig. 11.30 Detailed structure of the crista ampullaris in the internal ear

Cochlear Duct

The cochlear duct, also known as the scala media (Fig. 11.31), is a structure within the bony cochlea. It is a fluid-filled membranous labyrinth situated within the bony cochlea. It spirals for approximately 2.5–2.75 turns around the modiolus, a bony core, and is sandwiched between two other perilymph-filled compartments: the scala vestibuli above and the scala tympani below.

- Reissner's Membrane: This separates the cochlear duct from the scala vestibuli and is composed of a thin, nonvascular layer of epithelial cells.
- Basilar Membrane: This forms the base of the cochlear duct and separates it from the scala tympani. It is a compound structure made of collagen fibers and ground substance, becoming increasingly broader and less stiff from the base to the apex of the cochlea.

The organ of Corti (Fig. 11.32) is situated on the cochlear duct's basilar membrane, the cochlea's middle compartment. It extends along the spiral length of the cochlea, from the basal end to the apical end, varying subtly in structure and function along this axis. Hair cells are specialized sensory cells within the organ of Corti and are responsible for transducing mechanical sound vibrations into electrical signals. Two types exist inner and outer hair cells. Inner hair cells (IHCs) are arranged in a row of about 3500 cells. They are flask-shaped and have fewer, more rigid

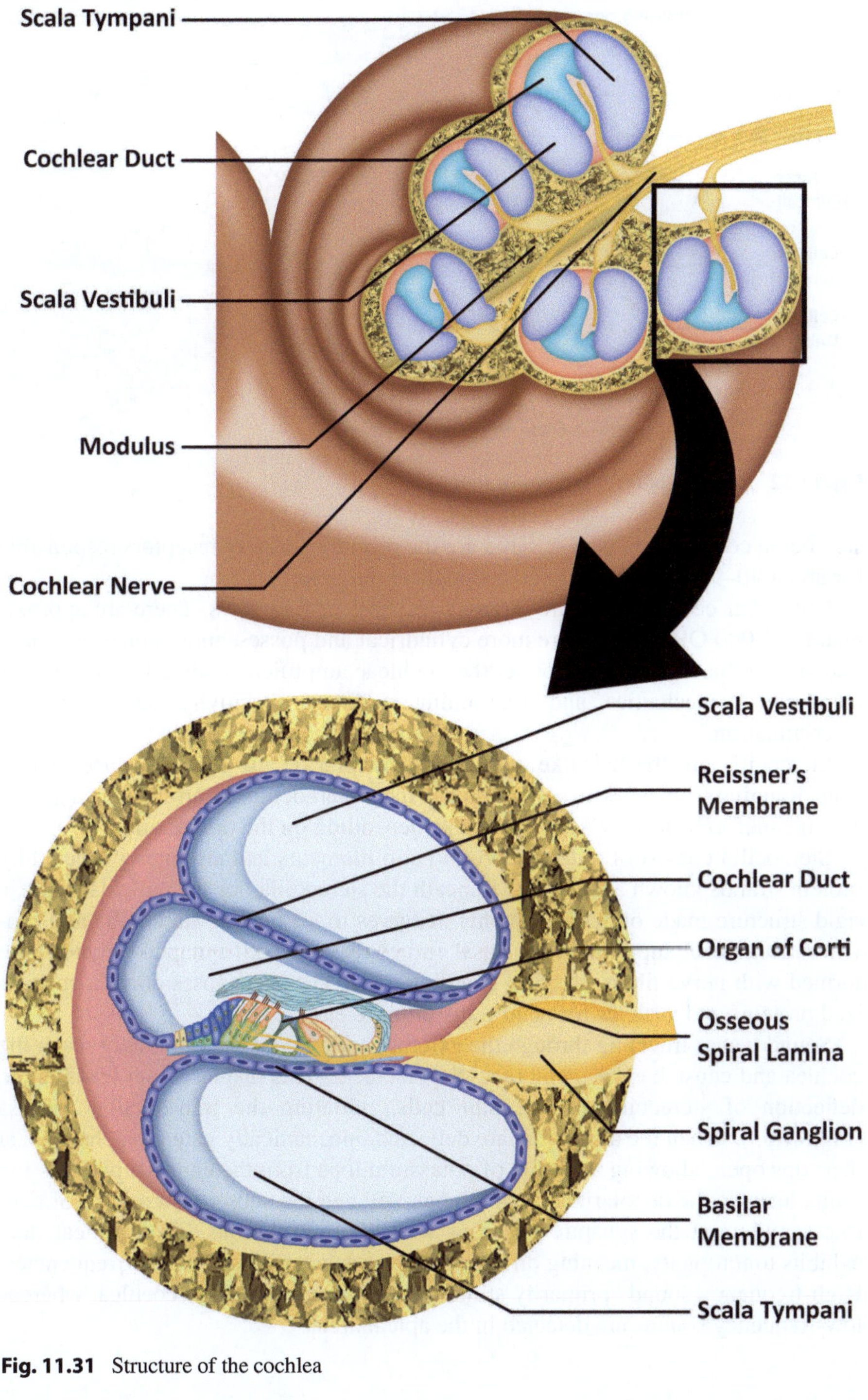

Fig. 11.31 Structure of the cochlea

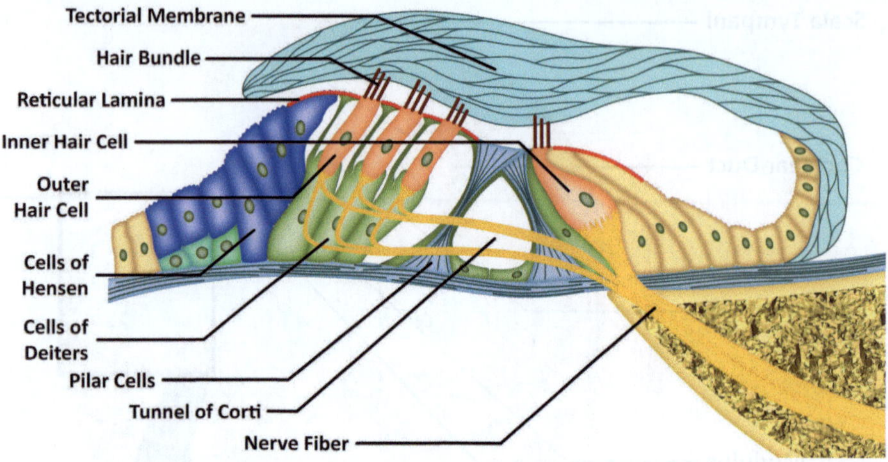

Fig. 11.32 Structure of the organ of Corti

stereocilia compared to OHCs. IHCs are the primary auditory receptors responsible for about 90–95% of afferent nerve signals to the brain.

Outer hair cells (OHCs) are arranged in three or four rows. There are approximately 12,000 OHCs. They are more cylindrical and possess more numerous, flexible stereocilia. OHCs function as the cochlear amplifier, modulating the basilar membrane's mechanics and fine-tuning auditory sensitivity and frequency discrimination.

Stereocilia are the hair-like projections situated on the apical surface of hair cells. Ranging from a few to several dozen, these stereocilia are arranged in a graduated manner to form a "V" or "W" shape, depending on the orientation.

Stereocilia consist of a dense core of actin filaments and are interconnected by protein strands known as tip links. Beneath the stereocilia lies the cuticular plate, a rigid structure made of actin filaments. It serves to anchor the stereocilia and provides mechanical support. At the basal end of the hair cell, synaptic junctions are formed with nerve fibers of spiral ganglion cells. These synapses contain specialized proteins and vesicles for neurotransmitter release.

Sound waves traveling through the external and middle ear eventually reach the cochlea and cause basilar membrane displacement. This displacement leads to the deflection of stereocilia on the hair cells, initiating the transduction process (Fig. 11.33). When the stereocilia are deflected, mechanically gated ion channels at their tips open, allowing an influx of potassium ions from the endolymph. This ion influx leads to the depolarization of the hair cell and the subsequent release of neurotransmitters at the synaptic junctions with afferent neurons. The cochlear duct exhibits tonotopicity, meaning different regions detect different sound frequencies. High-frequency sounds primarily stimulate the basal turn of the cochlea, whereas low-frequency sounds are detected in the apical areas.

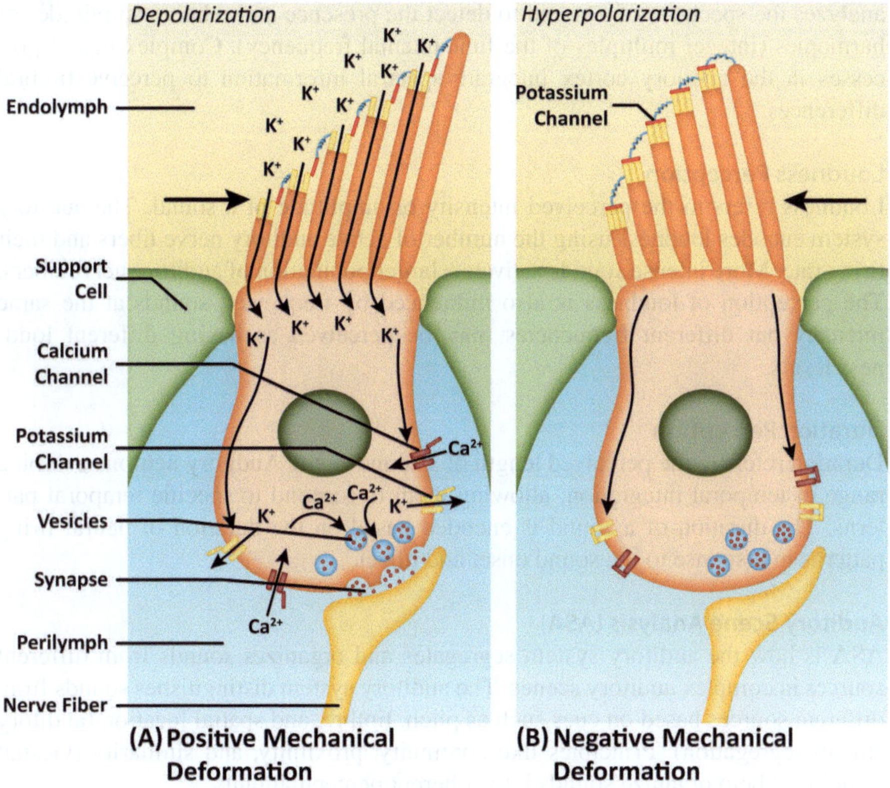

Fig. 11.33 Structure and the functional mechanism of the hair cells in the organ of Corti

11.2.3.3 Sound Characters Perception

Sound character perception in the auditory system involves a complex interplay of sensory processes and neural mechanisms that allow us to differentiate between various auditory qualities such as pitch, timbre, loudness, and duration.

Pitch Perception

Pitch refers to the perceived frequency of a sound wave. Different regions of the basilar membrane in the cochlea vibrate preferentially in response to specific frequencies. The frequency of action potentials in auditory nerve fibers corresponds to the frequency of the sound wave, providing information about pitch. Higher-frequency sounds are encoded by regions of the cochlea closer to the base, while lower-frequency sounds are encoded by areas closer to the apex.

Timbre Perception

Timbre encompasses a sound's unique quality or tone color, allowing us to distinguish between musical instruments or voices producing the same pitch. The harmonic content and envelope of a sound wave influence timbre. The auditory system analyzes the spectrum of a sound to detect the presence and relative amplitudes of harmonics (integer multiples of the fundamental frequency). Complex neural processes in the auditory cortex integrate spectral information to perceive timbral differences.

Loudness Perception

Loudness refers to the perceived intensity or amplitude of a sound. The auditory system encodes loudness using the number of active auditory nerve fibers and their firing rate. More intense sounds activate a larger population of auditory nerve fibers. The perception of loudness is also influenced by frequency; sounds at the same intensity but different frequencies may be perceived as having different loudness levels.

Duration Perception

Duration refers to the perceived length of a sound event. Auditory neurons exhibit a range of temporal integration, allowing them to respond to specific temporal patterns. The duration of a sound is encoded based on the duration of neural firing patterns in response to the sound onset and offset.

Auditory Scene Analysis (ASA)

ASA is how the auditory system segregates and organizes sounds from different sources in complex auditory scenes. The auditory system distinguishes sounds from different sources based on cues such as pitch, timbre, and spatial location (auditory stream segregation). Principles like continuity, proximity, and similarity (Gestalt principles) help organize sounds into coherent perceptual units.

Central Auditory Processing

The central auditory system, including the auditory cortex, is crucial in integrating and processing auditory information. Complex neural networks in the auditory cortex analyze sound characteristics further, allowing for higher-level auditory perception, such as sound localization and speech recognition.

11.3 Part 3: Chemical Sensing

Chemical sensing is a fundamental biological process that allows humans to interact with their environment. The olfactory (smelling) and gustatory (tasting) systems are the chemical sensing systems in humans.

11.3.1 Smell

The sense of smell, or olfaction, serves many functions, from identifying food and potential hazards to involving complex social and emotional experiences.

11.3.1.1 Olfactory Receptor Cells

The olfactory receptor cells are the primary sensory neurons that serve as the initial interface between external chemical stimuli and the neural processing that culminates in the perception of smell. These cells are located in the olfactory epithelium in the nasal cavity (Fig. 11.34). The olfactory epithelium is a pseudostratified columnar epithelium composed of several cell types, including olfactory receptor cells, supporting cells, and basal cells. It is situated on the cribriform plate of the ethmoid bone, extending along the superior nasal concha. Olfactory receptor cells (Fig. 11.35) are interspersed between supporting cells and are rooted in the basal layer of the epithelium. Olfactory receptor cells are bipolar neurons with a single dendrite and axon. The dendrite extends towards the apical surface of the olfactory epithelium, while the axon projects through the basal lamina towards the olfactory bulb in the brain. The dendrite of each olfactory receptor cell culminates in an enlargement known as the dendritic knob. From this knob emanate several long, thin cilia, which project into the mucus layer of the nasal cavity. The cilia are the primary sites of odorant reception and are endowed with an array of receptor proteins sensitive to different odor molecules. The sensory cilia extending from the dendritic knobs are particularly enriched in ion channels and signaling molecules, making them highly sensitive to chemical stimuli. The axon projects through the cribriform plate on the opposite pole of the cell body, penetrating the skull to synapse within the olfactory bulb. This axon is unmyelinated and is bundled with axons from other olfactory receptor cells to form nerve fascicles. Olfactory receptor cells have a relatively short lifespan compared to other neurons, averaging around 30–60 days. They are continually regenerated from basal cells in the olfactory epithelium, ensuring a functional population of receptor cells.

11.3.1.2 Olfactory Bulb

The olfactory bulb (Fig. 11.35), situated beneath the frontal lobe and nestled within the cribriform plate of the ethmoid bone, is a bilateral, oval-shaped structure. It is an extension of the olfactory tract, and its proximal end adjoins the olfactory epithelium via olfactory nerves. The olfactory bulb can be conceptually divided into

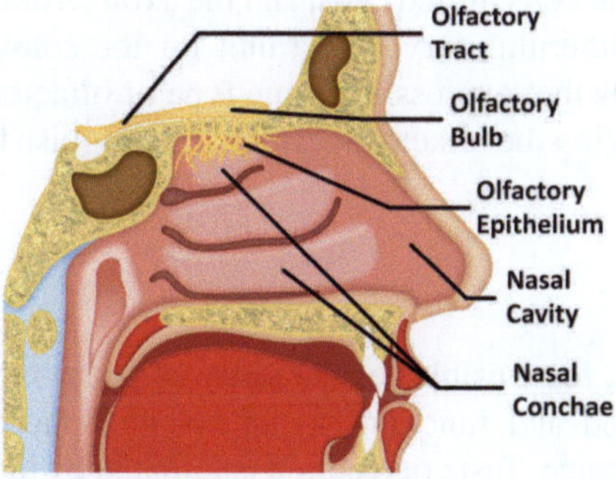

Fig. 11.34 Olfactory epithelium in the superior part of the nasal cavity

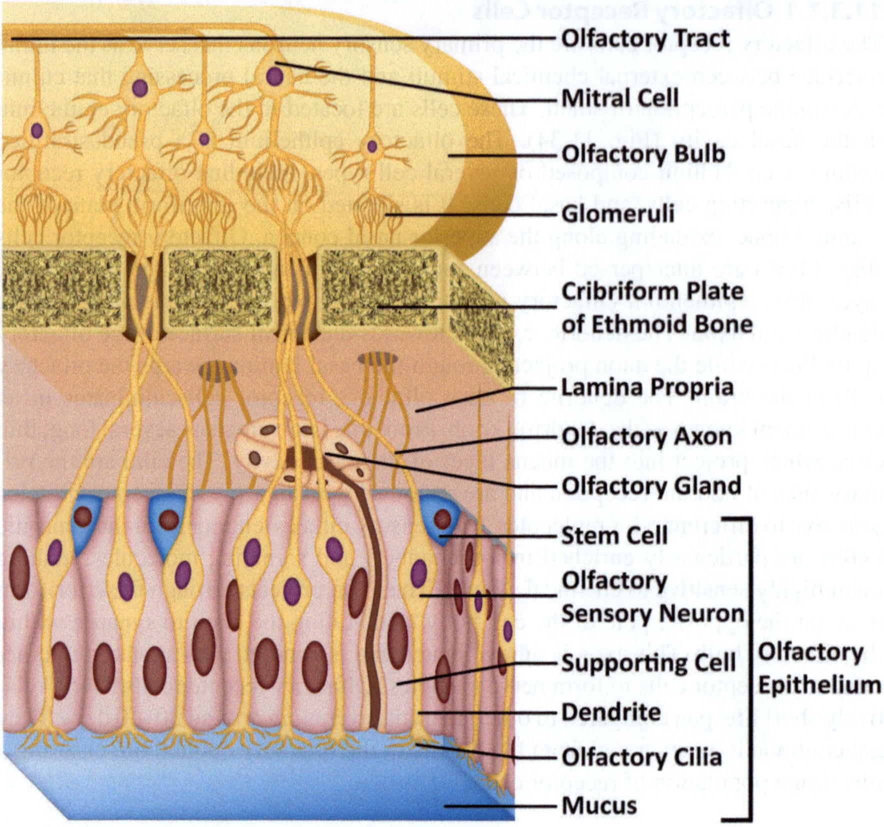

Fig. 11.35 Detailed structure of the olfactory epithelium and bulb

several layers, including the outermost olfactory nerve layer, the glomerular layer, the external plexiform layer, the mitral cell layer, the internal plexiform layer, and the granule cell layer. Olfactory glomeruli are specialized, spheroidal neuropil structures located in the olfactory bulb. Each glomerulus comprises a dense tangle of dendrites from mitral cells, tufted cells, and the axon terminals of olfactory receptor neurons. Each glomerulus serves as a unit for the convergence of input from olfactory receptor cells that express the same type of olfactory receptor. This leads to spatial coding, enabling the olfactory system to distinguish between various odors.

11.3.2 Taste

The gustatory system, responsible for the sense of taste, facilitates the perception and enjoyment of food and functions as an essential mechanism for nutritional intake and toxin avoidance. Taste perception is initiated by the interaction of soluble chemicals—often derived from food—with specialized receptor cells located primarily on the tongue but also in other regions of the oral cavity. These receptor cells are components of structures known as taste buds, which are embedded in the taste papillae's epithelial tissue.

11.3.2.1 Papillae

The papillae on the human tongue can be primarily classified into four categories (Fig. 11.36):

- Fungiform Papillae: These mushroom-shaped papillae are located primarily on the tip and sides of the tongue and house a modest number of taste buds. They detect various taste modalities, including sweet, salty, bitter, and sour.
- Circumvallate Papillae: These large papillae are surrounded by a trench-like moat and positioned at the back of the tongue, forming a V-shaped row. They are rich in taste buds, critical for detecting bitter tastes. These papillae have Von Ebner's glands that secrete fluids to flush the trench, aiding taste molecule diffusion.
- Foliate Papillae: These are characterized by a series of folds or leaf-like structures and located on the lateral borders of the tongue. They are primarily involved in the perception of sour and salty tastes.
- Filiform Papillae: These are conical or cylindrical and constitute the most abundant papillae covering the entire dorsal surface. They do not contain taste buds and are primarily mechanical, aiding in gripping and manipulating food.

11.3.2.2 Taste Buds

Taste buds are predominantly found in the tongue papillae, each containing varying numbers of taste buds. In addition to the tongue, taste buds are also present in other parts of the oral cavity, including the buccal surface, palate, pharynx, and epiglottis. A taste bud (Fig. 11.36) is an oval or elongated structure measuring 50–100 μm in diameter and typically containing between 50 and 100 cells. These cells are densely packed and enveloped by a capsule-like basal lamina that separates them from the surrounding epithelium. These cells include:

- Type I Cells: Also known as "supporting cells," these cells have glial-like properties and envelop other cell types within the taste bud. They are implicated in ionic balance and the removal of neurotransmitters.
- Type II Cells: These "receptor cells" are the primary cells detecting sweet, umami, and bitter tastes. They lack synapses but communicate via paracrine signaling.
- Type III Cells: Also known as "presynaptic cells," these cells have synapses with afferent nerve fibers and are mainly responsible for detecting sour and salty tastes. They serve as the principal output cells of the taste bud.
- Basal Cells: Basal cells, located at the base of the taste bud, act as progenitor cells that differentiate into mature taste cells, enabling the taste bud's regular turnover and regeneration.

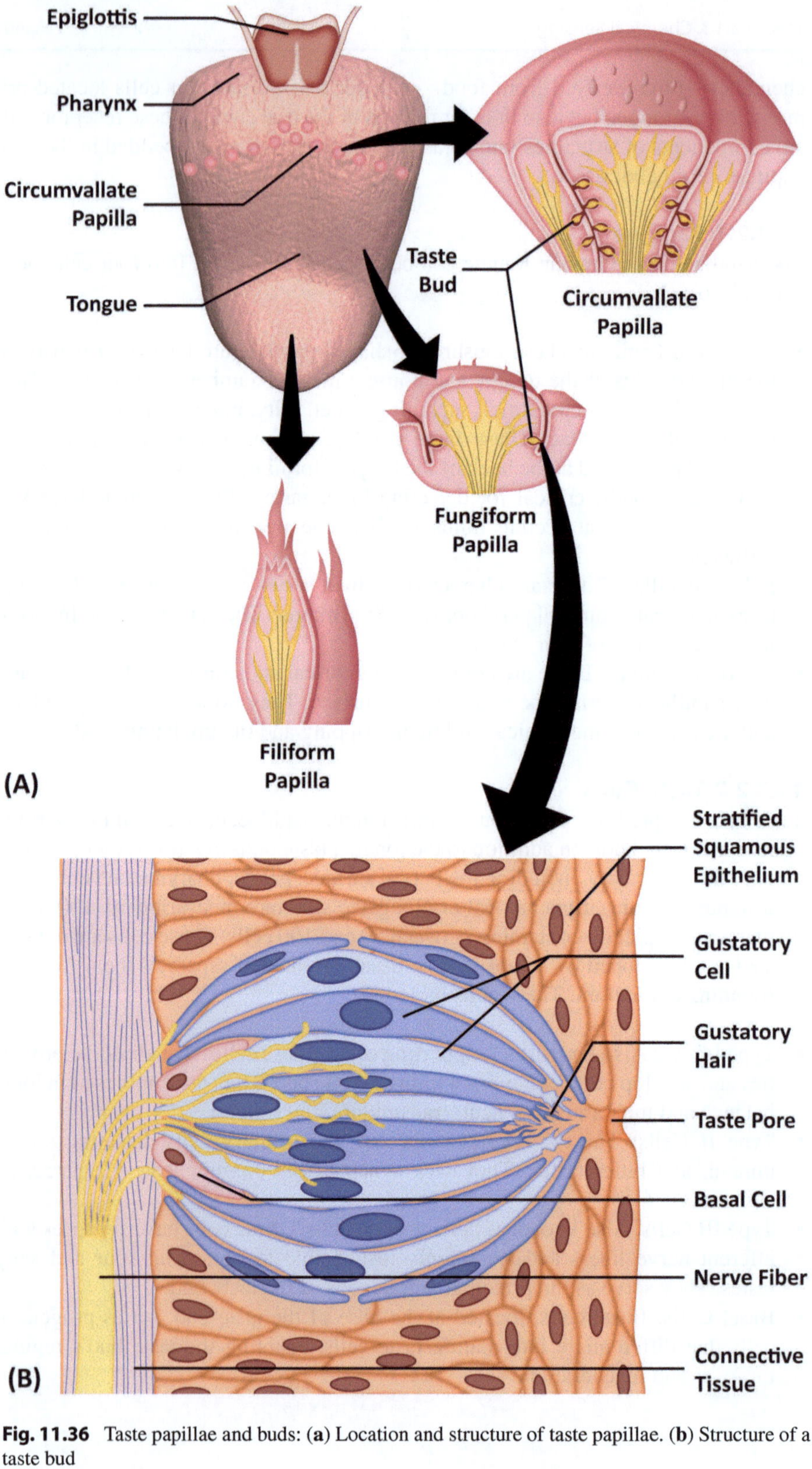

Fig. 11.36 Taste papillae and buds: (**a**) Location and structure of taste papillae. (**b**) Structure of a taste bud

Microvilli, tiny, hair-like structures on the apical surface of the Type II and III cells, protrude into the taste pore and are the primary sites for taste molecule interaction.

Various types of receptors are expressed on the membranes of taste cells:

- GPCRs: G-protein-coupled receptors, specifically the T1R and T2R families, are crucial for sweet, umami, and bitter detection.
- Ion Channels: Specific ion channels, such as ENaC for salt and PKD2L1 for sour, are integral for perceiving these tastes.

11.3.2.3 Tastes

There are five basic tastes, called classical quintet: sweet, salty, sour, bitter, and umami.

Sweet: Sweet taste is mediated by G-protein-coupled receptors (GPCRs) in the T1R family. Attraction to sweet taste likely evolved to direct humans toward energy-rich carbohydrates. Sweetness is often associated with pleasure and reward, manifesting in the widespread consumption of sweetened foods and beverages.

Salty: The epithelial sodium channel (ENaC) primarily mediates the sensation of saltiness. Salt is essential for physiological processes like nerve function and fluid balance. Salt has been a staple in human civilization for millennia as a preservative and flavor enhancer.

Sour: Primarily mediated by ion channels such as PKD2L1, sour taste is the perception of acidity. It helps identify spoiled or fermented foods, which could be either beneficial or harmful. Sour foods are often associated with caution or acquired taste, although they are integral to various cuisines.

Bitter: Bitter taste is the most structurally diverse, mediated by a wide range of GPCRs in the T2R family. Generally aversive, bitterness likely evolved as a defense mechanism against toxic substances. Despite its aversive nature, bitter foods and beverages like coffee and certain vegetables are culturally significant.

Umami: Detected by T1R1 + T1R3 GPCR heterodimers, umami has a savory taste associated primarily with amino acids like glutamate. Umami may guide humans toward protein-rich foods. While always present in diets, the concept of umami was scientifically recognized only in the twentieth century.

Emerging Taste Modalities

- Fatty Acids: Recent research suggests that humans may have a distinct taste for fatty acids, although this remains contentious.
- Metallic: Though not universally recognized, some consider metallic taste a separate category, often associated with blood or metal ingestion.

- Calcium: Some scientists propose a distinct taste for calcium, although this is not widely accepted. Three major cranial nerves transmit taste information from the tongue and other taste organs:
  - Facial Nerve (VII): Innervates the anterior two-thirds of the tongue.
  - Glossopharyngeal Nerve (IX): Innervates the posterior one-third of the tongue.
  - Vagus Nerve (X): Innervates the epiglottis and lower pharynx.

Suggested Reading

Advanced Physiology and Pathophysiology: Essentials for Clinical Practice, 2nd Ed. Nancy Tkacs, Linda Herrmann, Springer Publishing Company, 2024
Anatomy & Physiology, S. Meloni, et al. Independent Publisher, 2023
Anatomy & Physiology: The Unity of Form and Function, 10th Ed. Kenneth S. Saladin, McGraw Hill, 2023
Gray's Anatomy for Students, 5th Ed. Richard L. Drake, et al. Elsevier, 2020
Human Anatomy & Physiology, 11th Ed. Elaine N. Marieb and Katja Hoehn, Pearson, 2022
Moore's Clinically Oriented Anatomy, 9th Ed. Keith L. Moore, et al. Lippincott Williams & Wilkins, 2020
Netter Atlas of Human Anatomy: Classic Regional Approach, 8th Ed. Frank H. Netter, Elsevier, 2022
Principles of Anatomy and Physiology, 16th Ed. Gerard J. Tortora, Bryan H. Derrickson, Wiley, 2020

Endocrine System 12

In this chapter, you will learn:

- Basic organization of the endocrine system.
- Hormones and their mechanisms of action, including interaction with specific receptors.
- Detailed insights into signal transduction pathways, emphasizing G-protein-coupled receptors and second messengers like cyclic AMP.
- Understanding of negative and positive feedback mechanisms that regulate hormone levels and ensure homeostasis.
- Hormone half-life and its implications for hormonal activity and regulation.
- Duration of action of hormones and their impact on physiological functions.
- Central role of the hypothalamus in linking the nervous system to the endocrine system.
- Details on the hypothalamic releasing and inhibiting hormones and their roles.
- Detailed exploration of the pituitary gland, differentiating between the anterior and posterior pituitary and their hormones.
- Anatomy and hormone production of the thyroid and parathyroid glands, focusing on metabolic regulation and calcium homeostasis.
- Structure and function of the adrenal glands, including cortisol, aldosterone, and catecholamine production.
- Insights into the endocrine functions of the pancreas, focusing on the roles of insulin and glucagon in glucose metabolism.
- Hormonal regulation of reproductive functions, including hormones produced by the gonads.
- Transportation of hormones in plasma, highlighting the role of binding proteins.

The endocrine system (Fig. 12.1) stands out as an intricate and vital network that orchestrates various bodily functions through the subtle yet powerful mechanism of hormones. These biochemical messengers regulate myriad physiological processes,

H. Hosseinzadeh, *Fundamentals of Medicine for Biomedical Engineering*,
https://doi.org/10.1007/978-3-031-74085-5_12

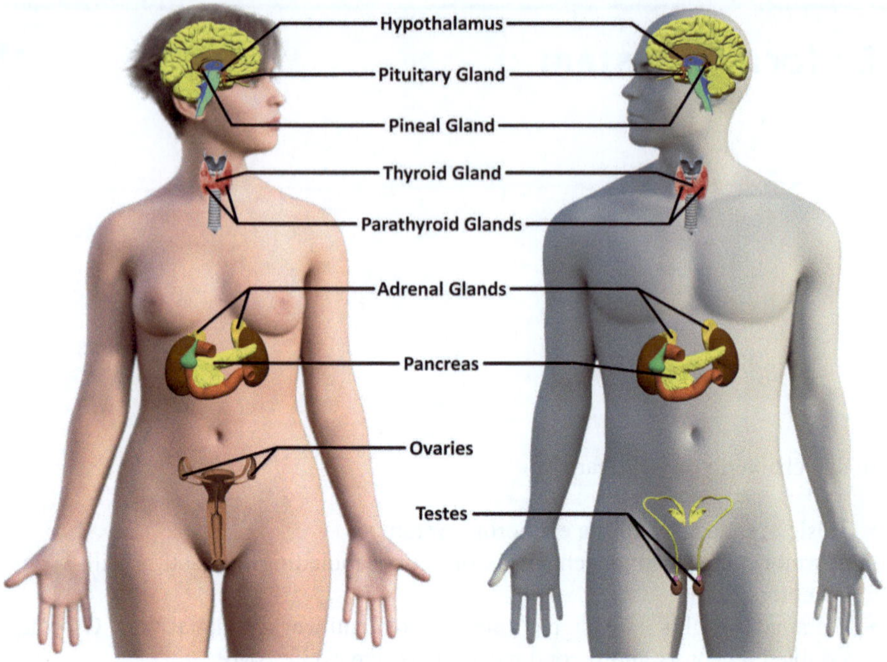

Fig. 12.1 Endocrine organs of the human body

ranging from the simplest cellular interactions to the complexities of growth, metabolism, and homeostasis.

12.1 Part 1: Hormones

Hormones are the body's chemical messengers. Specialized glands and cells secrete them and travel through the bloodstream to target tissues where they exert their effects. Hormones' structural diversity is quite broad, reflecting their many roles in the body.

12.1.1 Structure

Hormones can be categorized into three groups based on their chemical structures: amino acid derivatives, peptide and protein hormones, and steroid hormones.

Amino Acid Derivatives: Amino acid-derived hormones are synthesized from the amino acids tyrosine and tryptophan. Hormones in this group include thyroid hormones (thyroxine and triiodothyronine) and catecholamines (epinephrine and norepinephrine).

Peptide and Protein Hormones: Peptide and protein hormones comprise chains of amino acids of various lengths and complexities. Peptide and protein hormones are primarily synthesized as larger precursors within specific endocrine cells, often preprohormones or prohormones. These precursors undergo a series of enzymatic cleavages and modifications to yield the active hormone. These hormones include antidiuretic hormone (ADH), oxytocin, follicle-stimulating hormone (FSH), luteinizing hormone (LH), thyroid-stimulating hormone (TSH), insulin, glucagon, and growth hormone (GH).

Steroid Hormones: Steroid hormones are lipid-soluble molecules synthesized from cholesterol. Various functional groups can be attached to this basic skeleton, defining the specific steroid hormone classes, including glucocorticoids, mineralocorticoids, androgens, and estrogens. Steroid hormones are lipophilic, allowing them to cross cell membranes readily.

12.1.2 Receptors

Hormone receptors are specialized proteins that bind to specific hormones with high affinity and act as molecular switches that modulate various cellular functions. Hormone receptors can be broadly categorized into cell surface receptors and intracellular receptors.

Cell Surface Receptors: Cell surface receptors, predominantly associated with peptide and protein hormones, are embedded in the cellular membrane. They facilitate the transduction of extracellular signals into intracellular responses. These receptors have an intracellular second messenger (G-protein or tyrosine kinase) (Fig. 12.2).

Intracellular Receptors: Intracellular receptors, typically associated with steroid hormones, reside within the cell and migrate to the nucleus upon hormone binding. They directly influence gene expression and comprise DNA-binding and ligand-binding domains (Fig. 12.3).

12.1.3 Transportation of Hormones in Plasma

Hormones are produced by specialized glands or cells and released directly into the bloodstream. These hormones are then carried throughout the body, where they can exert their effects on distant target cells that possess the appropriate receptors. The classic endocrine glands include the pituitary, thyroid, adrenals, gonads (ovaries and testes), and pancreas (Fig. 12.1).

Once secreted by endocrine glands, hormones navigate the vascular pathways conveyed within the plasma to their respective target tissues and organs. The carrying of hormones in plasma encompasses various mechanisms. Binding proteins

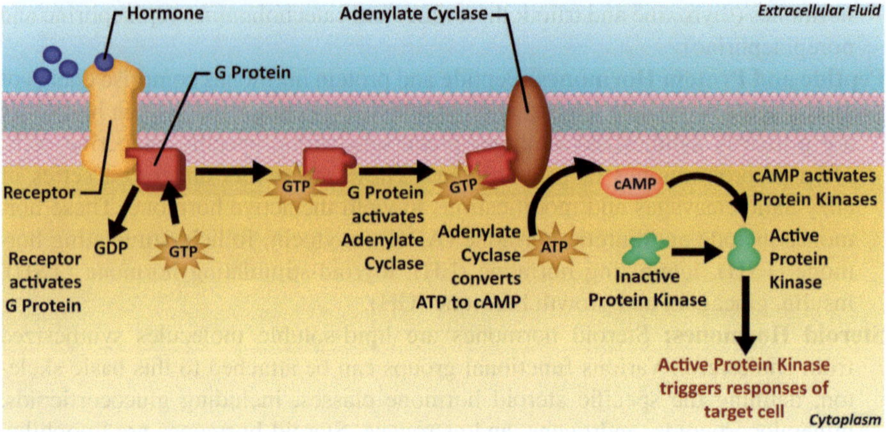

Fig. 12.2 Signal transduction pathway involving G-protein-coupled receptors (GPCRs): A hormone (ligand) binds to a specific receptor on the cell membrane. The receptor is a G-protein-coupled receptor (GPCR), which undergoes a conformational change upon hormone binding. The receptor activates an associated G protein by exchanging GDP for GTP on the alpha subunit of the G protein. The GTP-bound alpha subunit of the G protein dissociates from the beta and gamma subunits and activates adenylate cyclase, an enzyme embedded in the cell membrane. Adenylate cyclase converts ATP to cAMP, which serves as a second messenger in the cell. cAMP amplifies the signal by activating downstream targets. cAMP activates protein kinase A (PKA) by binding to its regulatory subunits, causing the release and activation of the catalytic subunits. PKA phosphorylates various target proteins, leading to changes in their activity. These responses can include changes in gene expression, enzyme activity, ion channel permeability, and other cellular processes

emerge as pivotal in the plasma transportation of hormones, primarily steroid and thyroid hormones, facilitating hormone solubilization, protection, and regulated delivery. Binding proteins exhibit specificity and affinity towards particular hormones. Due to their lipophilic nature, steroid hormones predominantly rely on binding proteins for effective plasma transportation. Sex hormone-binding globulin (SHBG) and corticosteroid-binding globulin (CBG) carry the major steroid hormones in plasma. Thyroid hormones engage with specific binding proteins (thyroxine-binding globulin [TBG], with high affinity for thyroxine (T4), and albumin and prealbumin, with less affinity). Other hormones are usually transported in plasma without carrier molecules as free molecules.

12.1.4 Hormone Half-Life

Hormone half-life refers to the time required for the concentration of a hormone to reduce to half of its initial value, embodying the rate of hormone metabolism and elimination. This chronological parameter is intrinsic to the modulation of hormonal effects, influence on target tissues, and overall contribution to endocrine homeostasis and physiological regulation. Peptide hormones possess shorter half-lives, ranging from minutes to hours, allowing for rapid regulation and adaptability.

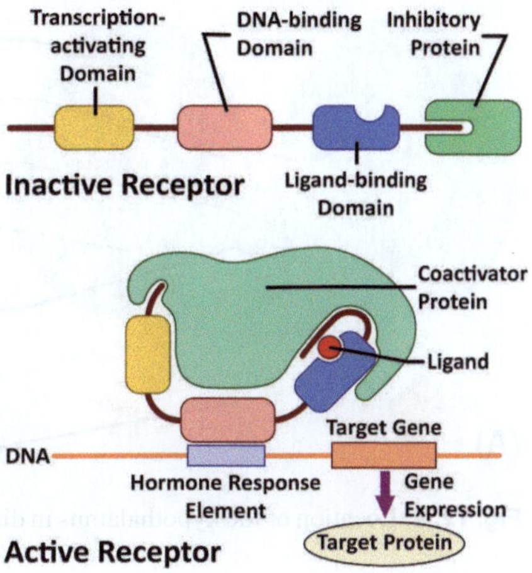

Fig. 12.3 Intracellular receptor

In comparison, steroid hormones have longer half-lives, ranging from several hours to days, supporting sustained actions and regulatory influences.

12.1.5 Hormone Duration of Action

The duration of action of hormones symbolizes the temporal spectrum over which hormones exert their physiological influences, reflecting the synchronization of their secretion, systemic circulation, receptor interaction, and eventual degradation and elimination. Peptide and protein hormones are characterized by shorter durations of action, allowing for rapid and transient physiological adaptations. Steroid hormones exhibit extended durations of action, facilitating sustained influences and regulatory roles in various physiological processes.

12.2 Part 2: Endocrine Gland

12.2.1 Hypothalamus

The hypothalamus is a maestro of hormonal harmonization called the "master of master gland." In this pivotal role, the hypothalamus integrates neural inputs with endocrine responses, navigating the confluence of mind and body and guiding the rhythms of homeostasis, adaptation, and biological synchronicities. Anatomically nestled within the brain's basal region (Section 10.2.2.3), the hypothalamus is strategically positioned to receive, interpret, and respond to diverse physiological signals, embodying a nexus of neuroendocrine integration (Fig. 12.4).

The hypothalamus employs intricate signaling mechanisms to communicate its regulatory influences. A specialized portal system facilitates the direct

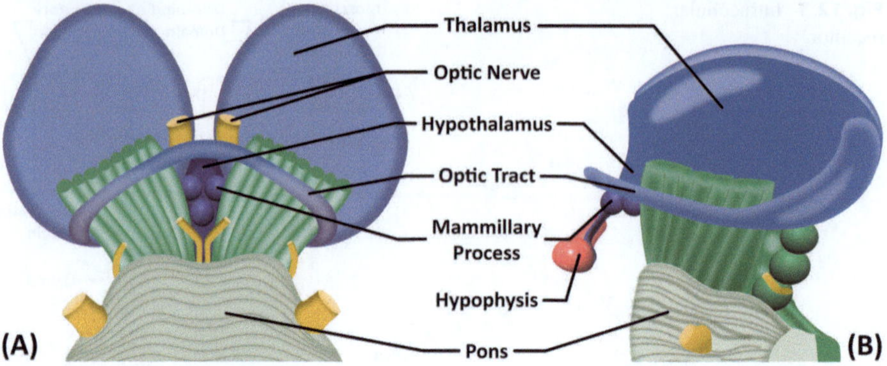

Fig. 12.4 Location of the hypothalamus in diencephalon: (**a**) Anteroinferior view. (**b**) Lateral view

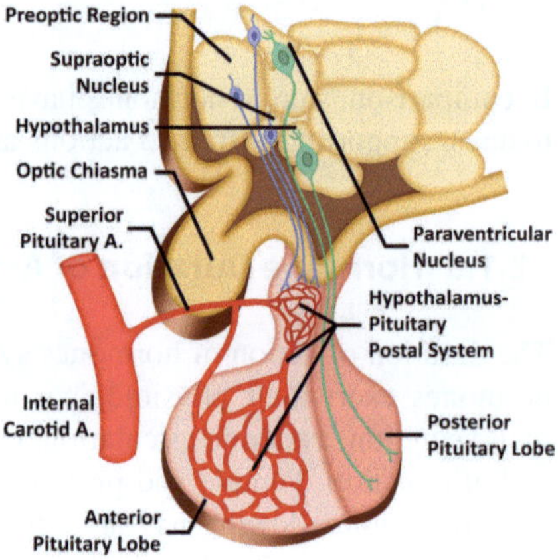

Fig. 12.5 The anatomy of the hypothalamus and pituitary glands and the connections (neural and vascular) between them

communication between the hypothalamus and the anterior pituitary, optimizing the efficiency of hormonal signals. Neurosecretory pathways enable the hypothalamus to convey hormonal messages directly and effectively to the posterior pituitary (Fig. 12.5). The hypothalamus houses a spectrum of hormones integral to the body's symphony of physiological regulation.

12.2.1.1 Hypothalamic Releasing and Inhibiting Hormones

The hypothalamus releases these hormones into the hypothalamic-pituitary portal system to reach the anterior pituitary directly.

- Gonadotropin-releasing hormone (GnRH): GnRH regulates the release of gonadotropins from the anterior pituitary, thereby modulating reproductive functions and hormonal cycles.
- Growth hormone-releasing hormone (GHRH) and somatostatin: These hormonal counterparts control the regulation of growth hormone (GH), with GHRH encouraging its release and somatostatin inhibiting it.

- Thyrotropin-releasing hormone (TRH): TRH stimulates the pituitary secretion of thyroid-stimulating hormone (TSH), controlling the rhythms of thyroid function.
- Corticotropin-releasing hormone (CRH): In stress responses, CRH directs the release of adrenocorticotropic hormone (ACTH), modulating adrenal activities and cortisol rhythms.

12.2.1.2 Oxytocin and Vasopressin

These hormones are produced in the cell body of some hypothalamic neurons and then transferred through their axons to the posterior pituitary to be released into the bloodstream.

- Oxytocin: A hormone with social and reproductive functions, oxytocin regulates bonding, childbirth, and lactation, creating connection and care.
- Vasopressin (antidiuretic hormone, ADH): Vasopressin conducts the body's water balance, modulating renal functions and regulating osmotic equilibrium.

12.2.2 Pituitary Gland

The pituitary gland, often called the "master gland" of the endocrine system, is a small yet intricately vital organ nestled within the sella turcica of the sphenoid bone at the base of the brain (Fig. 12.6). Its remarkable control over many physiological processes has earned it this distinguished title. The pituitary gland consists of two distinct lobes: the anterior pituitary (adenohypophysis) and the posterior pituitary (neurohypophysis) (Fig. 12.5). The anterior pituitary, comprising approximately 75% of the total pituitary mass, is characterized by its glandular nature. This part includes different secretory cells such as somatotrophs (producing growth hormone), lactotrophs (producing prolactin), corticotrophs (producing adrenocorticotropic hormone), thyrotrophs (producing thyroid-stimulating hormone), and gonadotrophs (producing follicle-stimulating hormone and luteinizing hormone). The posterior pituitary is neural in origin. It comprises nervous tissue extensions from the hypothalamus and does not produce hormones directly. Instead, it stores and releases hormones synthesized in the hypothalamus, including oxytocin and vasopressin (antidiuretic hormone).

12.2.2.1 Anterior Pituitary

The anterior pituitary, or adenohypophysis, comprises an array of hormone-secreting cell types, each responsible for producing specific hormones that govern essential bodily functions. These hormones, collectively called tropic hormones, target various endocrine organs throughout the body, thereby regulating their hormone production and maintaining systemic balance (Table 12.1).

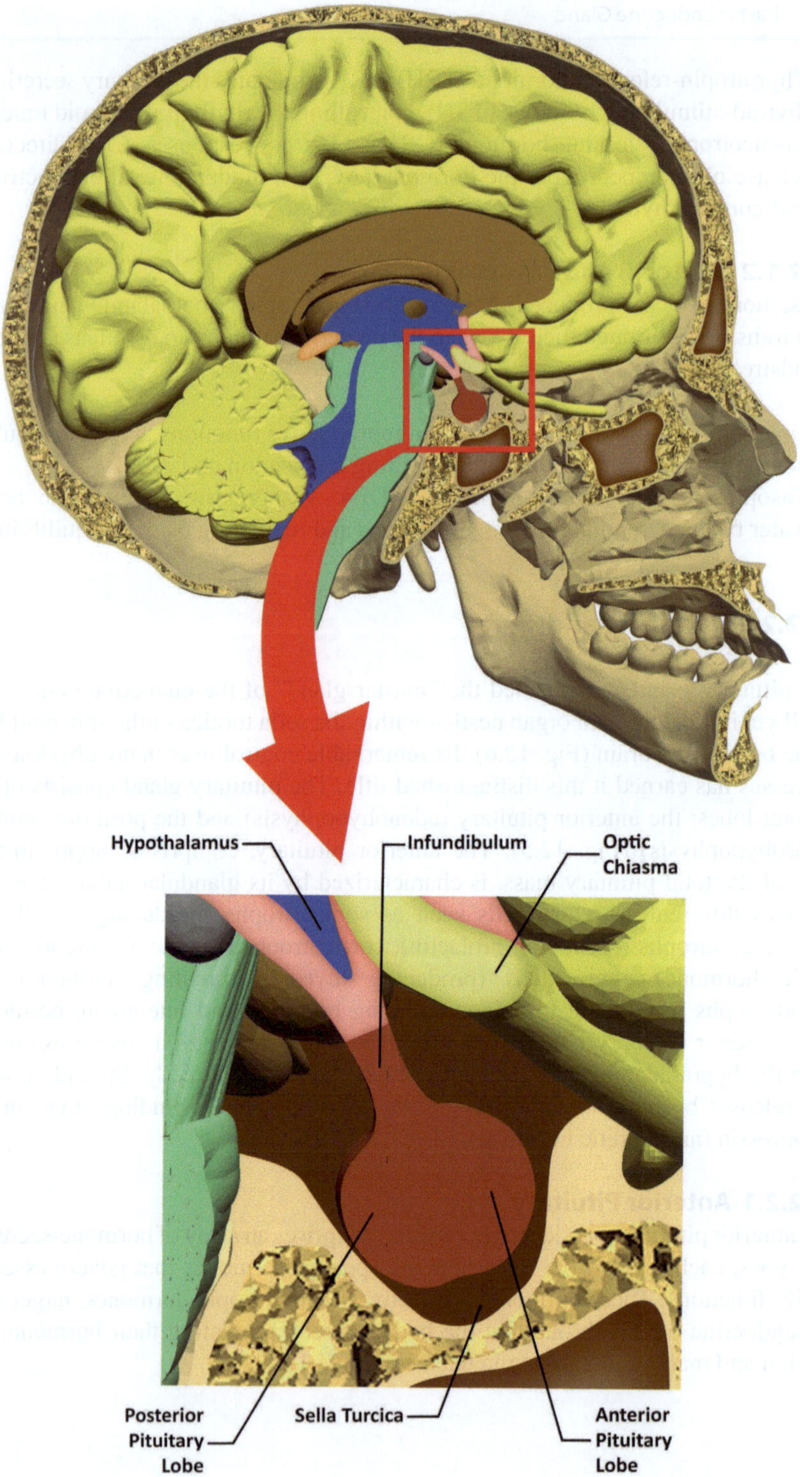

Fig. 12.6 Anatomical location of the pituitary gland in the brain (right) and inside the sella turcica (left)

Table 12.1 Anterior pituitary hormones: structure, function, stimulator, inhibitor, and feedback mechanism

Hormone name	Structure	Function	Stimulator(s)	Inhibitor(s)	Feedback mechanism
Growth hormone (GH)	Protein hormone	Stimulates growth, cell reproduction, and cell regeneration. Increases muscle mass and bone density.	GHRH (growth hormone-releasing hormone)	Somatostatin (GHIH), high GH levels	Negative feedback by IGF-1 and GH itself
Thyroid-stimulating hormone (TSH)	Glycoprotein hormone	Stimulates the synthesis and release of thyroid hormones from the thyroid gland.	TRH (thyrotropin-Releasing hormone)	Somatostatin, high thyroid hormones	Negative feedback by thyroid hormones
Adrenocorticotropic hormone (ACTH)	Peptide hormone	Stimulates the adrenal cortex to produce and secrete glucocorticoids.	CRH (corticotropin-releasing hormone)	High glucocorticoid levels	Negative feedback by glucocorticoids
Follicle-stimulating hormone (FSH)	Glycoprotein hormone	In females: Stimulates growth of ovarian follicles. In males: Stimulates spermatogenesis.	GnRH (gonadotropin-releasing hormone)	Inhibin, high sex steroid levels	Negative feedback by inhibin and sex steroids
Luteinizing hormone (LH)	Glycoprotein hormone	In females: Triggers ovulation and corpus luteum formation. In males: Stimulates testosterone production.	GnRH (gonadotropin-releasing hormone)	Inhibin, high sex steroid levels	Negative feedback by inhibin and sex steroids
Prolactin (PRL)	Protein hormone	Promotes milk production in females postpartum. Also has roles in metabolism, regulation of the immune system, and development.	Estrogen, TRH, nipple stimulation	Dopamine (prolactin-inhibiting hormone), high prolactin levels	Negative feedback by prolactin itself and dopamine
Melanocyte-stimulating hormone (MSH)	Peptide hormone	Influences pigmentation in some species, but its role in humans is less clear.	CRH, UV radiation	High cortisol levels	Not fully established

12.2.2.2 Posterior Pituitary

The posterior pituitary gland, often called the neurohypophysis, merely stores and releases hormones (oxytocin and vasopressin) produced by the hypothalamus, its neuroendocrine master (Table 12.2). These hormones are structurally peptides.

12.2.3 Thyroid Gland

The thyroid gland is significant due to its crucial involvement in maintaining metabolic homeostasis, growth, and development. It is located in the lower part of the neck, anterior to the trachea, and inferior to the thyroid cartilage of the larynx (Fig. 12.7).

12.2.3.1 Anatomy

The thyroid gland typically consists of two lobes, the right and left, connected by a thin bridge of thyroid tissue known as the isthmus. Its overall shape resembles a butterfly or a bowtie. These lobes are situated on either side of the trachea, with the isthmus spanning the front of the trachea. The thyroid gland generally weighs approximately 15–25 g in adults. The fundamental structural unit of the thyroid gland is the thyroid follicle. These spherical structures are the sites of thyroid hormone synthesis and storage.

Table 12.2 Posterior pituitary hormones: structure, function, stimulator, inhibitor, and feedback mechanism

Hormone name	Synthesis and secretion	Function	Stimulator(s)	Inhibitor(s)	Feedback mechanism
Oxytocin	Synthesized in the hypothalamus (paraventricular nucleus), stored and released by the posterior pituitary	Stimulates uterine contractions during childbirth and milk ejection during breastfeeding. Also involved in bonding, sexual activity, and social behavior.	Stretching of the cervix and uterus, suckling at the breast, sexual activity, sound of baby crying		Primarily positive feedback during childbirth
Vasopressin (ADH)	Synthesized in the hypothalamus (supraoptic nucleus), stored, and released by the posterior pituitary	Increases water retention in the kidneys, thus reducing urine volume. Also increases blood pressure.	High plasma osmolality, low blood volume, stress	Ethanol, high water intake	Negative feedback by blood osmolality and blood volume

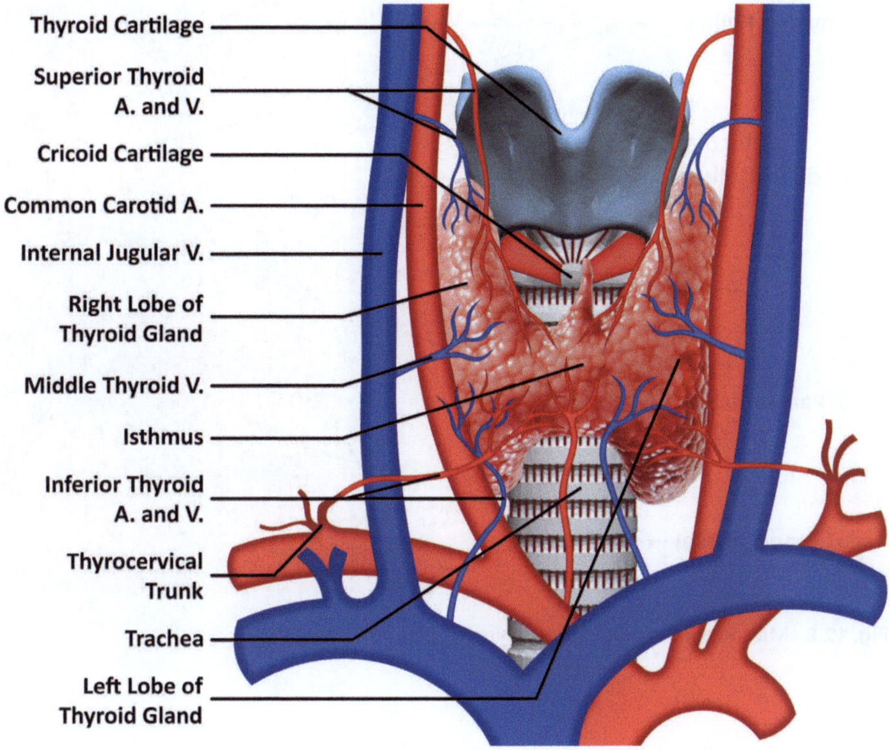

Fig. 12.7 Anatomical location of the thyroid gland

12.2.3.2 Follicles

A typical thyroid follicle (Fig. 12.8) is lined by a simple cuboidal or squamous epithelium, which changes in height depending on the gland's activity status, cuboidal during active hormone synthesis, and squamous during less active phases.

The interior of each follicle contains an eosinophilic, proteinaceous colloid rich in the glycoprotein thyroglobulin, which serves as a precursor for thyroid hormone synthesis. The colloid serves as a reservoir for these precursors, allowing the storage of large amounts of hormones that can be mobilized when needed. Thyroglobulin (Tg) is a large, dimeric glycoprotein, synthesized by the thyroid follicular cells, that serves as the scaffold for the synthesis of the thyroid hormones thyroxine (T4) and triiodothyronine (T3) (Table 12.3). Tg provides tyrosine residues as the substrate for iodination and hormone synthesis. The iodination and coupling reactions that produce T3 and T4 occur at specific tyrosyl residues within the Tg molecule. Once iodinated, Tg forms a colloid within the follicle, effectively storing large quantities of thyroid hormones in an inactive form that can be mobilized when needed.

Interspersed between the follicles are the parafollicular cells, also known as C cells. These cells are larger and contain a paler-staining cytoplasm than the

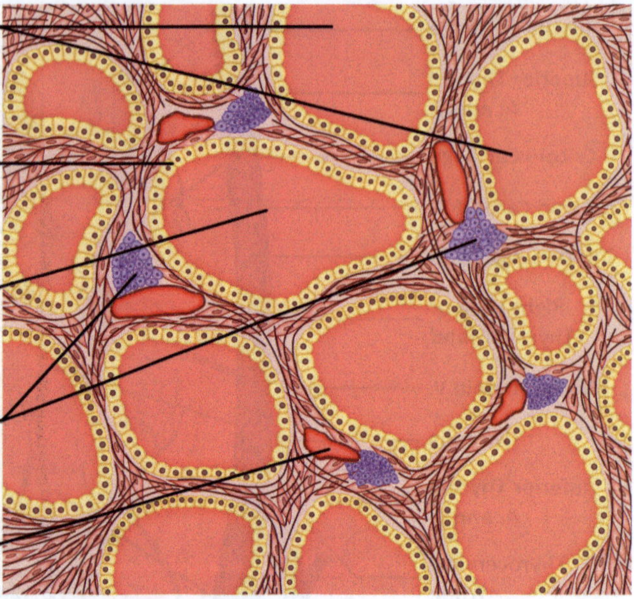

Fig. 12.8 Microscopic structure of the thyroid gland

follicular cells. Parafollicular cells are responsible for synthesizing and secreting calcitonin, a calcium homeostasis hormone (Table 12.3).

The follicles are supported by a delicate network of reticular fibers and a scant amount of interfollicular connective tissue, or stroma, which contains blood vessels, lymphatics, and nerves.

12.2.3.3 Thyroid Hormones Synthesis

The first step in thyroid hormone synthesis (Fig. 12.9) is the active uptake of iodide (I^-) from the bloodstream into the thyroid follicular cells. This process is mediated by a sodium-iodide symporter (NIS) located on the basolateral membrane of these cells. Once iodide is inside the thyroid follicular cells, it is oxidized to iodine (I_2) by an enzyme called thyroid peroxidase (TPO). This occurs at the thyroid follicular cells' apical (luminal) membrane. Iodine molecules (I_2) are attached to specific tyrosine residues on thyroglobulin to form monoiodotyrosine (MIT) and diiodotyrosine (DIT). TPO catalyzes these reactions. MIT and DIT residues on thyroglobulin combine to create thyroid hormones. Combining one MIT and one DIT molecule forms triiodothyronine (T3), while combining two DIT molecules forms thyroxine (T4).

Thyroxine is produced in larger quantities than triiodothyronine. The iodinated thyroglobulin is endocytosed into the follicular cells, which are contained in vesicles and are called colloid droplets. Proteolytic enzymes cleave the T3 and T4 molecules within these vesicles from the thyroglobulin backbone. Once cleaved from

Table 12.3 Thyroid gland hormones: structure, function, stimulator, inhibitor, feedback mechanism, and features

Hormone name	Structure	Function	Stimulator(s)	Inhibitor(s)	Feedback mechanism	Important features
Thyroxine (T4)	Iodine-containing amine	Increases the rate of metabolism and oxygen consumption; influences growth and development, heart rate, and body temperature.	TSH (thyroid-stimulating hormone)	High levels of thyroid hormones	Negative feedback on the hypothalamus and pituitary to decrease TSH and TRH secretion.	T4 is the primary hormone produced by the thyroid gland but is less active than T3; most of it is converted to T3 in the peripheral tissues.
Triiodothyronine (T3)	Iodine-containing amine	More potent than T4, it performs similar functions in regulating metabolism, growth, and development, and influencing heart rate and body temperature.	TSH (thyroid-stimulating hormone)	High levels of thyroid hormones	Negative feedback on the hypothalamus and pituitary to decrease TSH and TRH secretion.	T3 is more active than T4 but is in lower quantities in the blood.
Calcitonin	Peptide hormone	Lowers blood calcium levels by inhibiting bone osteoclast activity and increasing calcium excretion in the kidneys.	High blood calcium levels	Low blood calcium levels	Not primarily regulated by a feedback mechanism, but by calcium levels in the blood.	

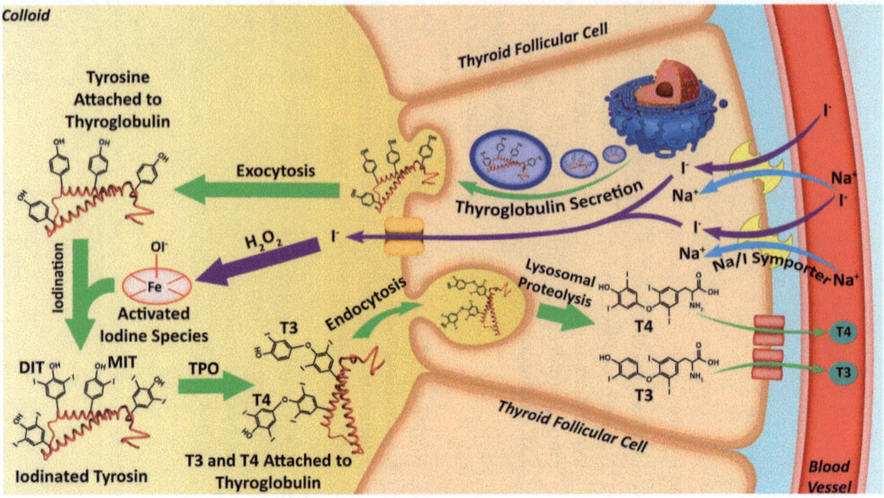

Fig. 12.9 Synthesis of thyroid hormones. (TPO Thyroid peroxidase, MIT Monoiodotyrosine, DIT Diiodotyrosine, T3 Triiodothyronine, T4 Thyroxin)

thyroglobulin, T3 and T4 are released into the bloodstream. Thyroid hormones are hydrophobic and bind to plasma proteins, primarily thyroxine-binding globulin (TBG), for transport in the blood to target tissues. Although T4 is the predominant thyroid hormone released by the thyroid gland, it is relatively inactive. Peripheral tissues, especially the liver and kidneys, convert T4 into the more active T3 by deiodination. T3 is the thyroid hormone that interacts with thyroid receptors in target cells to exert its effects.

12.2.4 Parathyroid Glands

The parathyroid glands are small yet formidable endocrine glands essential for fine-tuning the regulation of calcium homeostasis in the human body. Typically four in number, these diminutive structures are nestled on the posterior surface of the thyroid gland (Fig. 12.1) and secrete parathyroid hormone (PTH), the primary regulator of serum calcium and phosphate levels.

12.2.4.1 Structure

Each parathyroid gland is roughly the size of a grain of rice and is typically ovoid or bean-shaped. The microscopic anatomy of the parathyroid glands primarily consists of two cell types: chief cells are the predominant cell type in the parathyroid glands. They are responsible for synthesizing and secreting parathyroid hormone (PTH). Chief cells are rich in secretory granules containing PTH. Oxyphil cells are less numerous and have a granular, eosinophilic cytoplasm. Their exact function in the parathyroid gland remains somewhat enigmatic.

12.2.4.2 Parathyroid Hormone

PTH, also known as parathormone, is a peptide hormone composed of 84 amino acids and has a molecular weight of approximately 9400 daltons. It is synthesized as a preprohormone and undergoes posttranslational processing to produce the biologically active form, a single-chain peptide. The primary function of PTH is to regulate calcium homeostasis in the body. It achieves this by increasing renal calcium reabsorption, stimulating calcium release from bone, stimulating intestinal calcium absorption, and inhibiting renal phosphate reabsorption. PTH acts on the kidneys to increase calcium reabsorption in the distal renal tubules. This reduces calcium loss in the urine. PTH stimulates the release of calcium from bone tissue. It activates osteoclasts, which are bone-resorbing cells, leading to the breakdown of bone matrix and the release of calcium into the bloodstream. PTH enhances calcium absorption from the small intestine by promoting vitamin D activation, which is necessary for efficient calcium uptake in the gut. PTH decreases phosphate reabsorption in the renal tubules, increasing phosphate excretion in the urine. This helps maintain an appropriate calcium-to-phosphate ratio.

Several factors can stimulate the release of PTH, including low blood calcium levels, low 1,25-dihydroxyvitamin D levels, and high blood phosphate levels. Conversely, factors inhibiting PTH secretion include high blood calcium and low phosphate levels.

A negative feedback loop regulates the release of PTH. When blood calcium levels fall below the normal range, the parathyroid glands are stimulated to secrete PTH. Once calcium levels return to the normal range, the negative feedback mechanism inhibits further PTH secretion, helping to maintain calcium homeostasis.

12.2.5 Adrenal Glands

The adrenal glands are small, pyramidal-shaped. These glands are highly symmetrical in their placement.

12.2.5.1 Anatomy

The adrenal glands are situated immediately above each kidney, adhering to the superior aspect of the kidney's upper pole (Fig. 12.10). They receive their blood supply through a rich network of arteries, including the superior, middle, and inferior adrenal arteries, derived from branches of the aorta. The adrenal glands are innervated by sympathetic nerves originating from the celiac plexus and the splanchnic nerves. These nerves regulate adrenal medullary function and the release of catecholamines in response to stress.

12.2.5.2 Histologic Regions

The adrenal glands have two primary regions: the outer adrenal cortex and the inner adrenal medulla (Fig. 12.11). The adrenal cortex, the outermost layer of the adrenal glands, is a complex and highly organized structure responsible for synthesizing and secreting a variety of steroid hormones, classified into mineralocorticoids, glucocorticoids, and androgens, all steroid hormones (Table 12.4). The adrenal medulla,

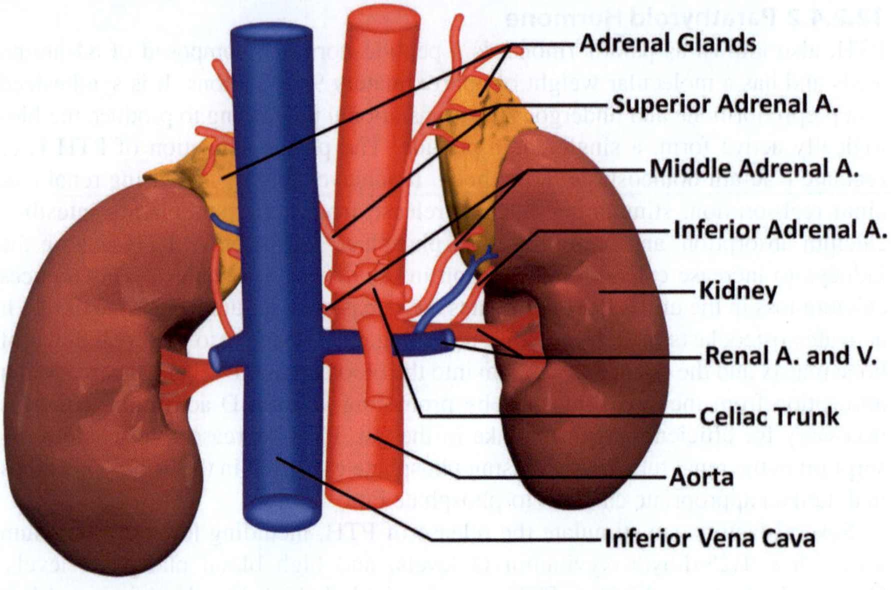

Fig. 12.10 The anatomical location of the adrenal glands

the innermost region of the adrenal glands, consisting primarily of specialized neuroendocrine cells known as chromaffin cells, is responsible for the synthesis and secretion of catecholamines, primarily epinephrine (adrenaline) and norepinephrine (noradrenaline) (Table 12.5). These hormones are central to the body's acute stress response and significantly affect heart rate, blood pressure, and various physiological processes.

12.2.6 Pancreas

The pancreas gland is a multifunctional organ located in the abdomen, nestled between the stomach and the spine (Fig. 12.12). The pancreas is a complex and multifaceted structure consisting of both exocrine and endocrine components. The exocrine portion of the pancreas constitutes the majority of its mass and is responsible for producing digestive enzymes (Section 16.3.6). The endocrine portion of the pancreas comprises pancreatic islets, also known as islets of Langerhans (Fig. 12.13). These islets are dispersed throughout the exocrine tissue and consist of several types of endocrine cells. Beta cells are the most abundant endocrine cells within the islets and are primarily responsible for producing insulin. Insulin is pivotal in regulating blood glucose levels by facilitating glucose uptake into cells. Alpha cells secrete glucagon, which has the opposite effect of insulin. Glucagon raises blood glucose levels by stimulating the liver to release stored glycogen and convert it into glucose. Delta cells produce somatostatin, a hormone that inhibits insulin and glucagon secretion.

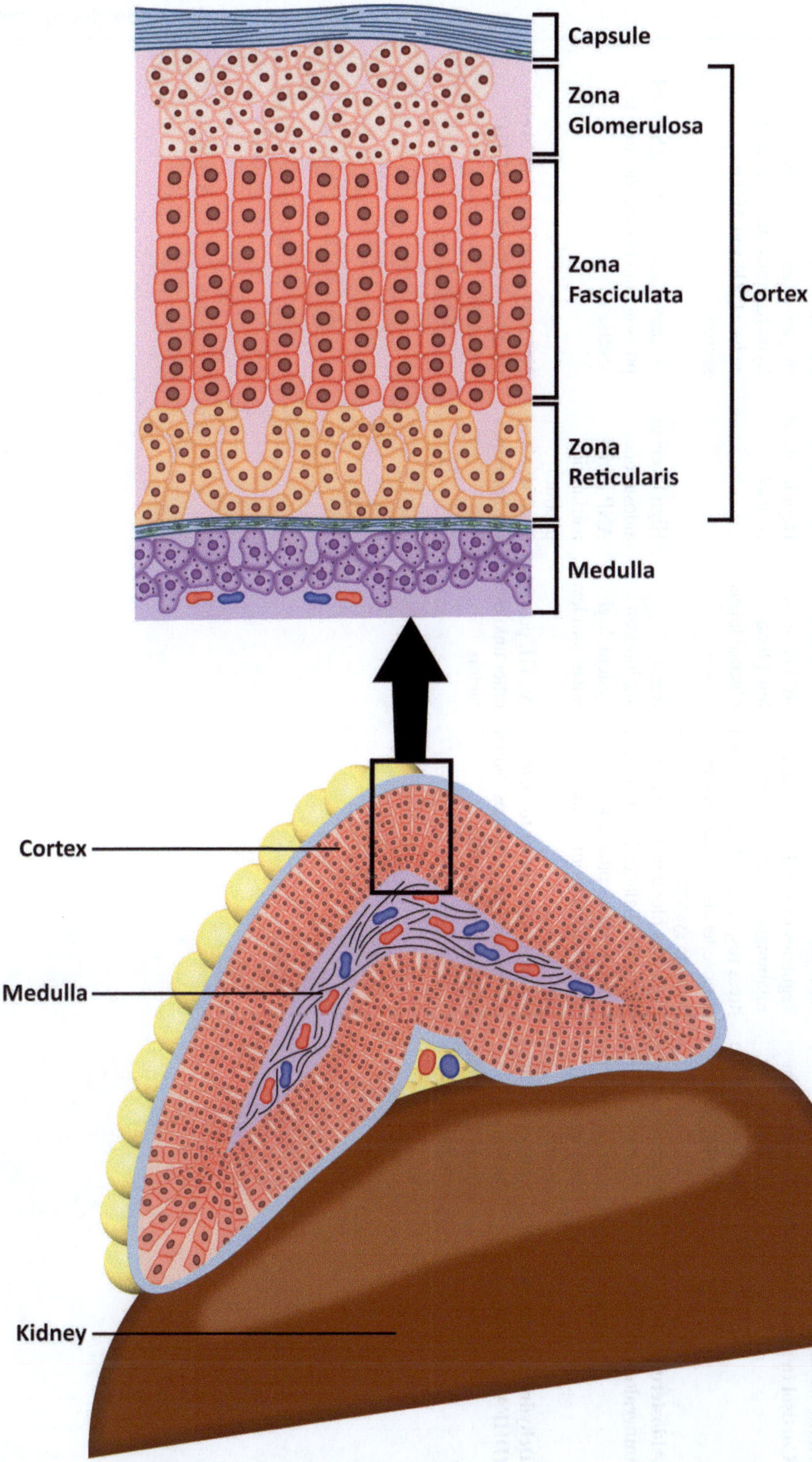

Fig. 12.11 Internal structure of the adrenal gland

Table 12.4 Hormones of adrenal gland cortex: origin, function, stimulator, inhibitor, and feedback mechanism

Hormone name	Origin	Function	Stimulator(s)	Inhibitor(s)	Feedback mechanism
Cortisol (a glucocorticoid)	Synthesized in the zona fasciculata of the adrenal cortex.	Regulates metabolism, reduces inflammation, and assists in stress response. Influences blood glucose levels and suppresses immune system.	ACTH, stress, low blood glucose levels	High levels of cortisol	Negative feedback to the hypothalamus and pituitary to reduce CRH and ACTH secretion.
Aldosterone (a mineralocorticoid)	Synthesized in the zona glomerulosa of the adrenal cortex.	Regulates blood pressure and electrolyte balance by increasing sodium reabsorption and potassium excretion in the kidneys.	Renin-angiotensin system, high potassium levels	High levels of aldosterone, ANP (atrial natriuretic peptide)	Negative feedback through blood pressure and plasma electrolyte levels.
Dehydroepiandrosterone (DHEA) (an androgen)	Synthesized in the zona reticularis of the adrenal cortex.	Serves as a precursor to male and female sex hormones. Minor direct androgenic effects.	ACTH, possibly other unknown factors	High levels of sex hormones	Not fully established but likely involves feedback on ACTH secretion.

Table 12.5 Hormones of adrenal gland medulla: structure, origin, function, stimulator, and features

Hormone name	Structure	Origin	Function	Stimulator(s)	Important features
Epinephrine (adrenaline)	Catecholamine	Synthesized from tyrosine in the chromaffin cells of the adrenal medulla. Majority of the medulla's output is epinephrine.	Increases heart rate and stroke volume, dilates bronchioles, and stimulates gluconeogenesis and lipolysis. Enhances blood flow to muscles and brain.	Sympathetic nervous system activation (stress, physical activity, low blood sugar)	Vital for the fight-or-flight response, increases alertness, and prepares the body for immediate physical action.
Norepinephrine (noradrenaline)	Catecholamine	Synthesized from dopamine in the chromaffin cells of the adrenal medulla. Comprises a smaller percentage of the medulla's output than epinephrine.	Increases heart rate, causes vasoconstriction leading to increased blood pressure, and aids in converting glycogen to glucose in the liver.	Sympathetic nervous system activation (stress, physical activity, low blood sugar)	Functions in concert with epinephrine in the fight-or-flight response also plays a more significant role in maintaining baseline blood pressure.

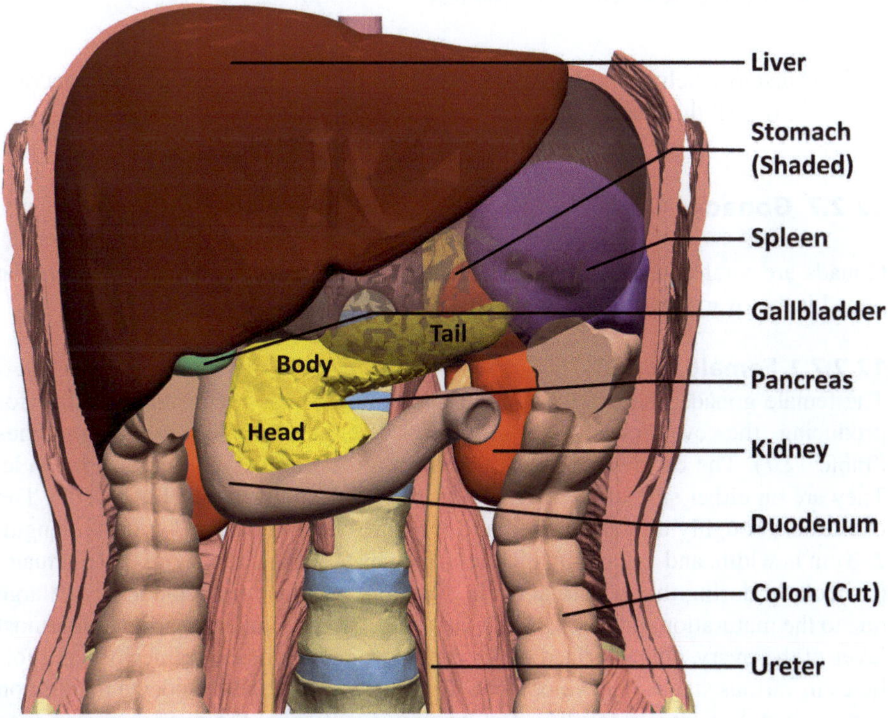

Fig. 12.12 Anatomical location of the pancreas

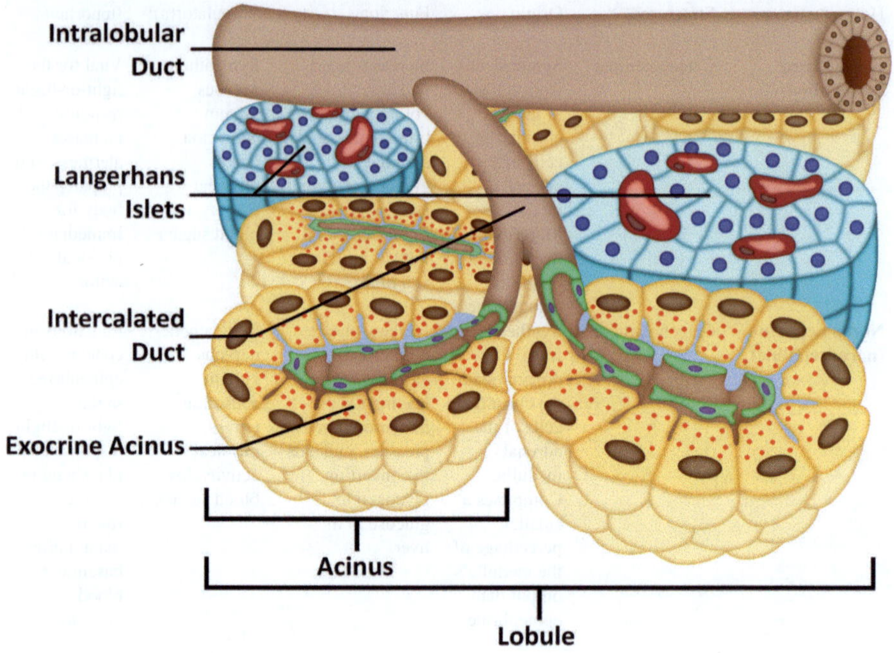

Fig. 12.13 Microscopic structure of the pancreas

Somatostatin helps regulate the overall endocrine function of the pancreas (Table 12.6). All the pancreatic hormones are peptides.

12.2.7 Gonads

Gonads are vital components of the human reproductive system. In females, the gonad is the ovary, and in males, it is the testis.

12.2.7.1 Female Gonads

The female gonads, the ovaries, are paired, almond-shaped organs responsible for producing the ova (eggs) and secreting essential female sexual hormones (Table 12.7). The ovaries are bilaterally within the true pelvis, one on each side. They are on either side of the uterus, near the lateral pelvic wall (Fig. 12.14). The ovaries are roughly almond-shaped and measure approximately 3–5 cm in length, 2–3 cm in width, and 1–2 cm in thickness. However, these dimensions can fluctuate, particularly during the menstrual cycle, when the ovaries may temporarily enlarge due to the maturation of follicles and ovulation. The ovarian cortex is the outermost layer of the ovary, characterized by its dense connective tissue framework and follicles in various stages of development. This region is critical for oocyte maturation and is pivotal in female fertility. The primary function of the ovarian cortex is to house follicles, which are the structural and functional units of the ovary. These

Table 12.6 Hormones of pancreas gland: function, stimulator, inhibitor, and feedback mechanism

Hormone name	Function	Stimulator(s)	Inhibitor(s)	Feedback mechanism
Insulin	Lowers blood glucose levels by facilitating cellular glucose uptake and promoting glycogen, fat, and protein synthesis.	High blood glucose levels, amino acids, GIP, GLP-1	Low blood glucose levels, somatostatin	Negative feedback by decreased blood glucose levels.
Glucagon	Increases blood glucose levels by stimulating glycogenolysis and gluconeogenesis in the liver.	Low blood glucose levels, amino acids	High blood glucose levels, insulin, somatostatin	Negative feedback by increased blood glucose levels.
Somatostatin	Inhibits the secretion of insulin and glucagon. In the gut, it reduces gastric secretion, motility, and absorption of nutrients.	High blood glucose levels, high amino acid levels	–	Feedback inhibition on its release in response to insulin and glucagon.

Table 12.7 Hormones of female gonads: origin, function, stimulator, inhibitor, and feedback mechanism

Hormone name	Origin	Function	Stimulator(s)	Inhibitor(s)	Feedback mechanism
Estrogen	Synthesized in the ovaries, primarily by the follicles' granulosa cells, and in other tissues like adipose tissue.	Promotes the development of female secondary sexual characteristics, regulates the menstrual cycle, supports pregnancy, and maintains bone density.	FSH (follicle-stimulating hormone) and LH (luteinizing hormone)	High levels of estrogen (in certain phases of the menstrual cycle)	Negative and positive feedback mechanisms with the hypothalamus and pituitary, depending on the menstrual cycle phase.
Progesterone	Synthesized in the ovaries by the corpus luteum after ovulation and also by the placenta during pregnancy.	Prepares the endometrium for embryo implantation, maintains pregnancy, and regulates the menstrual cycle.	LH	High levels of progesterone (in certain phases of the menstrual cycle)	Negative feedback on the hypothalamus and pituitary in the luteal phase of the menstrual cycle.
Inhibin	Synthesized and secreted by the granulosa cells in the ovaries.	Inhibits the secretion of FSH from the anterior pituitary, thus playing a role in regulating the menstrual cycle.	FSH		Negative feedback on the anterior pituitary to decrease FSH secretion.

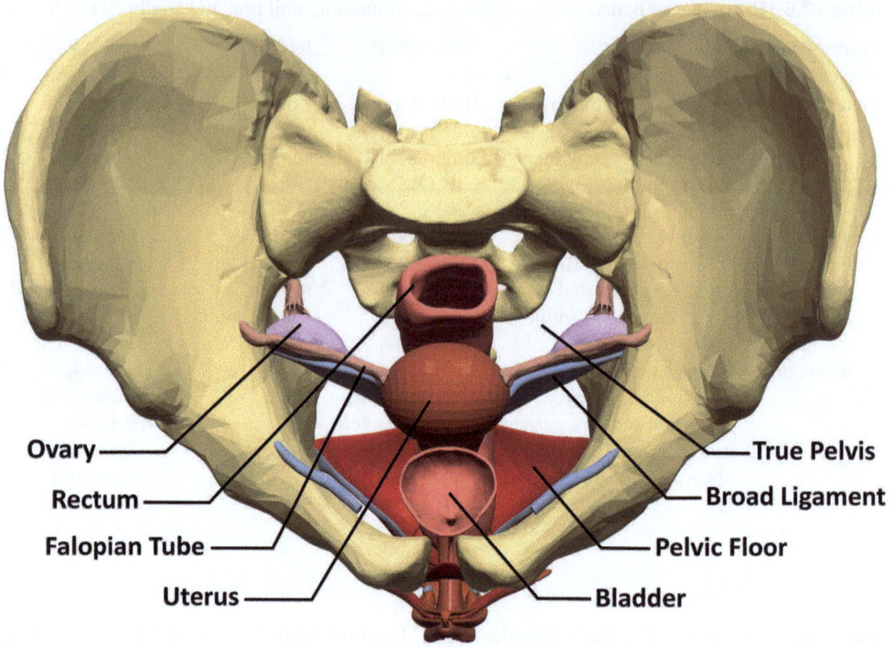

Fig. 12.14 The anatomical location of the female gonads

follicles consist of an oocyte (egg) surrounded by layers of granulosa cells. There are four main categories of follicles (Fig. 12.15):

- Primordial Follicles: These represent the earliest stage of follicular development. Each primordial follicle contains a dormant oocyte enveloped by a single layer of flattened granulosa cells.
- Primary Follicles: Primary follicles feature a larger oocyte surrounded by a single layer of cuboidal granulosa cells. As folliculogenesis progresses, these cells become more stratified.
- Secondary Follicles: In secondary follicles, the granulosa cell layer becomes multilayered, and the oocyte is now surrounded by a clear zone called the zona pellucida. Additionally, thecal cells begin to form around the follicle.
- Tertiary (Graafian) Follicles: Tertiary follicles, also known as Graafian follicles, represent the most mature stage of follicular development. They feature multiple layers of granulosa cells, a prominent antrum (fluid-filled cavity), and a well-defined theca with inner and outer layers.

Following ovulation, the mature follicle transforms into the corpus luteum, a temporary endocrine structure. If pregnancy does not occur, the corpus luteum regresses into the corpus albicans, characterized by dense fibrous tissue. Beneath

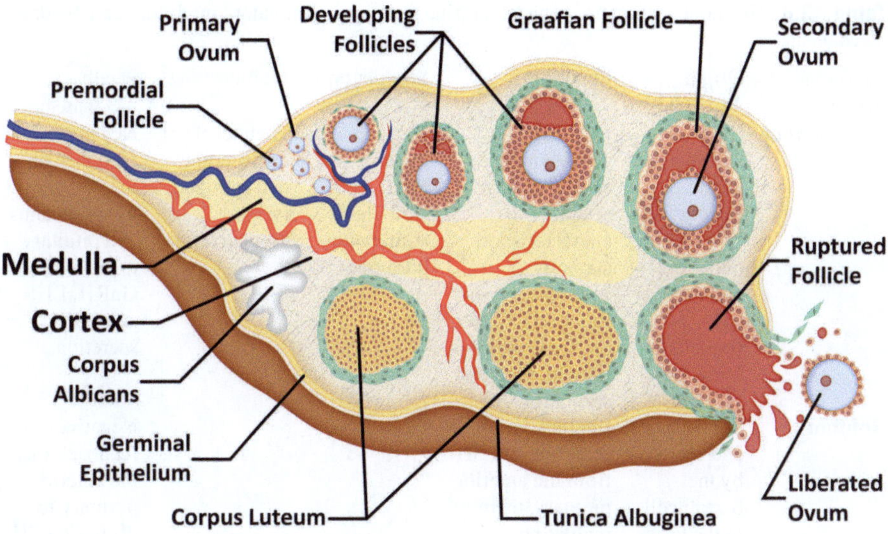

Fig. 12.15 Microscopic structure of the ovary

the cortex lies the ovarian medulla, a region characterized by loose connective tissue, blood vessels, lymphatics, and nerves. The medulla plays a vital role in maintaining the ovarian vasculature and innervation.

12.2.7.2 Male Gonads

The male gonads also called the testes, are paired organs that serve as the cornerstone of masculine vitality, governing reproduction and influencing many hormonal processes (Table 12.8). The testes are situated within the scrotum (Fig. 12.16), a specialized pouch of skin and connective tissue that hangs below the male pelvis. The scrotum's external position allows the testes to be maintained at approximately 2–4 °C lower than the core body temperature. This cooler environment is essential for optimal sperm production (spermatogenesis). The typical adult testis is an ovoid structure measuring about 4–5 cm in length, 2.5 cm in breadth, and 3 cm in height. The surface of each testis is smooth and white, enveloped by a tough, fibrous membrane called the tunica albuginea. This capsule extends inward to form septa, dividing the testis into 250–300 lobules. Within the testes, a complex network of tubules and cells, the seminiferous tubules, are the primary sites of sperm production within the testes (Fig. 12.17). Within the seminiferous tubules, Sertoli cells, also known as sustentacular cells, provide essential support and nourishment to developing sperm cells. The interstitial tissue between the seminiferous tubules contains Leydig cells, also known as interstitial cells, responsible for producing androgens, the male sex hormones. All sex hormones, male and female, are steroid hormones. Inhibin is a glycoprotein.

Table 12.8 Hormones of male gonads: origin, function, stimulator, inhibitor, and feedback mechanism

Hormone name	Origin	Function	Stimulator(s)	Inhibitor(s)	Feedback mechanism
Testosterone	Synthesized and secreted by the Leydig cells in the testes, under the control of LH (luteinizing hormone).	Promotes the development of male secondary sexual characteristics, supports spermatogenesis, influences libido, and contributes to muscle mass and strength.	LH, and to a lesser extent, FSH (follicle-stimulating hormone)	High levels of testosterone and estrogen (via aromatization)	Negative feedback on the hypothalamus and pituitary to decrease GnRH, LH, and FSH secretion.
Inhibin	Synthesized and secreted by the Sertoli cells in the testes, under the control of FSH.	Inhibits the secretion of FSH from the anterior pituitary, thereby regulating spermatogenesis.	FSH		Negative feedback on the anterior pituitary to decrease FSH secretion.

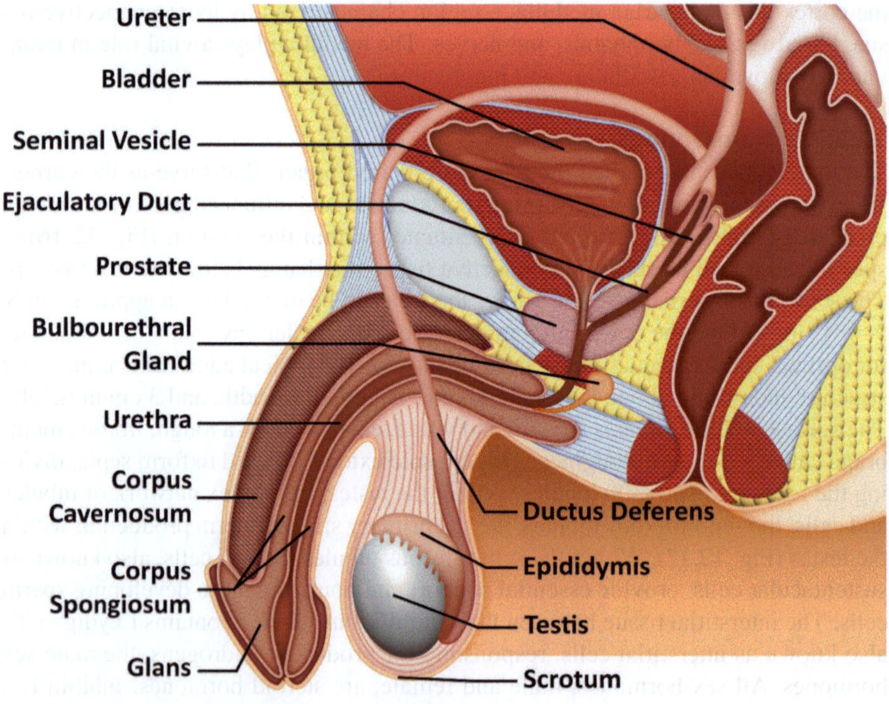

Fig. 12.16 The anatomical structure of the male gonads

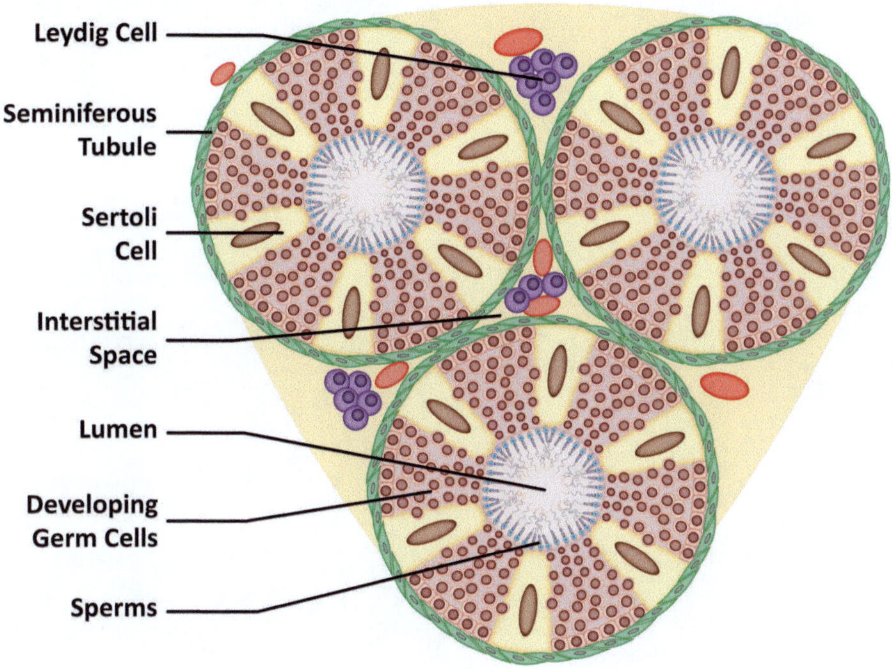

Fig. 12.17 Microscopic structure of the testis

Suggested Reading

Advanced Physiology and Pathophysiology: Essentials for Clinical Practice, 2nd Ed. Nancy Tkacs, Linda Herrmann, Springer Publishing Company, 2024

Anatomy & Physiology, S. Meloni, et al. Independent Publisher, 2023

Anatomy & Physiology: The Unity of Form and Function, 10th Ed. Kenneth S. Saladin, McGraw Hill, 2023

Gray's Anatomy for Students, 5th Ed. Richard L. Drake, et al. Elsevier, 2020

Human Anatomy & Physiology, 11th Ed. Elaine N. Marieb and Katja Hoehn, Pearson, 2022

Moore's Clinically Oriented Anatomy, 9th Ed. Keith L. Moore, et al. Lippincott Williams & Wilkins, 2020

Netter Atlas of Human Anatomy: Classic Regional Approach, 8th Ed. Frank H. Netter, Elsevier, 2022

Principles of Anatomy and Physiology, 16th Ed. Gerard J. Tortora, Bryan H. Derrickson, Wiley, 2020

Cardiovascular System 13

In this chapter, you will learn:

- Anatomy of the heart, including chambers, valves, and major blood vessels.
- Coverings of the heart, including the pericardium.
- Layers of the heart wall (epicardium, myocardium, endocardium).
- The conduction system of the heart, including the SA node and AV node.
- Electrical activity governing heart contractions.
- Details of the cardiac cycle, including phases like systole and diastole.
- Dynamics of blood flow through the heart, lungs, and systemic circulation.
- Stroke volume and factors affecting it.
- Ejection fraction as a measure of heart pumping efficiency.
- Cardiac output calculation and its clinical significance.
- Heart sounds and their origins in heart function.
- Mechanisms underlying blood pressure regulation.
- Blood supply to the heart muscle through coronary circulation.
- Use of electrocardiograms (ECG) in diagnosing heart conditions.
- Types and functions of blood vessels.
- Capillary types and their roles in tissue perfusion.
- Mechanisms and principles of blood flow.
- Exchange processes at the capillary level.
- Reflex mechanisms involving baroreceptors and chemoreceptors in cardiovascular regulation.
- Unique features of portal and nephron circulation.
- Major blood vessels and their branches.
- Composition of blood, including plasma and cellular components.
- Role of hemoglobin in oxygen transport.
- Coagulation process and the factors involved in blood clotting.
- Importance of blood typing and considerations during transfusions.

The cardiovascular system, often called the circulatory system, plays a fundamental role in maintaining homeostasis and ensuring the delivery of oxygen,

H. Hosseinzadeh, *Fundamentals of Medicine for Biomedical Engineering*,
https://doi.org/10.1007/978-3-031-74085-5_13

nutrients, and hormones to cells while removing waste products. This intricate network of blood vessels, the heart, and the blood constitutes a dynamic and highly organized system that functions tirelessly throughout an individual's lifespan.

13.1 Part 1: Heart

The heart, located within the thoracic cavity, is a muscular pump that tirelessly propels blood throughout the body. Its rhythmic contractions, regulated by a sophisticated electrical system, allow for continuous blood circulation, making it one of the most vital organs.

The heart is situated within the mediastinum, a central compartment of the thoracic cavity (Fig. 13.1). The mediastinum is the space between the lungs, extending from the superior thoracic aperture to the diaphragm. The heart occupies a relatively fixed position within this mediastinal region, nestled between the sternum (breastbone) anteriorly and the vertebral column posteriorly.

The heart's orientation within the thoracic cavity is such that it leans slightly to the left side, resulting in approximately two-thirds of its mass lying to the left of the midline. This deviation is due to the asymmetric positioning of the heart's chambers. Consequently, the apex of the heart, formed by the pointed tip of the left ventricle, is found at the fifth intercostal space, approximately 7–9 cm to the left of the midline. The heart's superior-inferior position typically spans from the second to

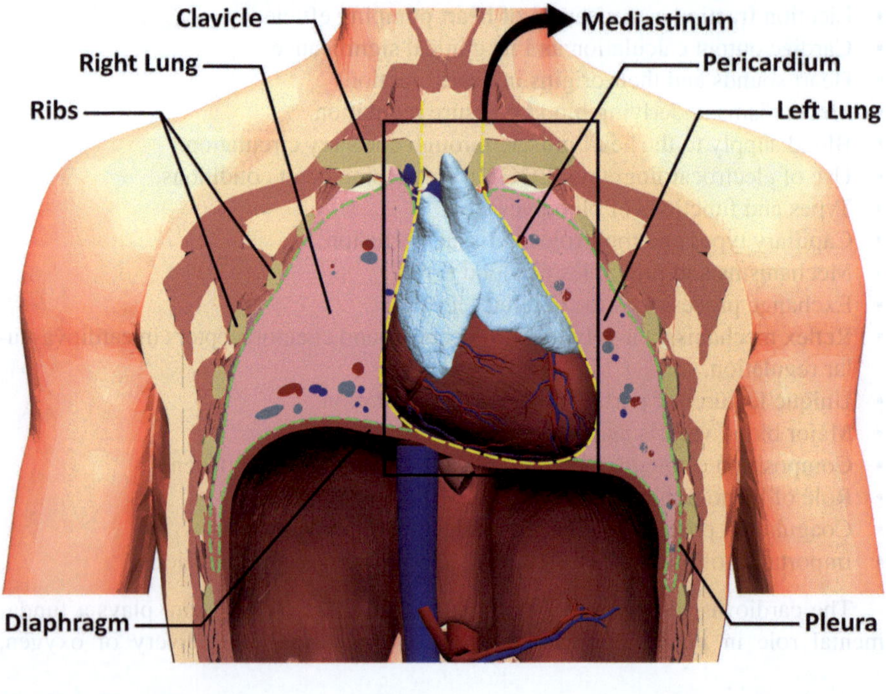

Fig. 13.1 The heart location in the mediastinum

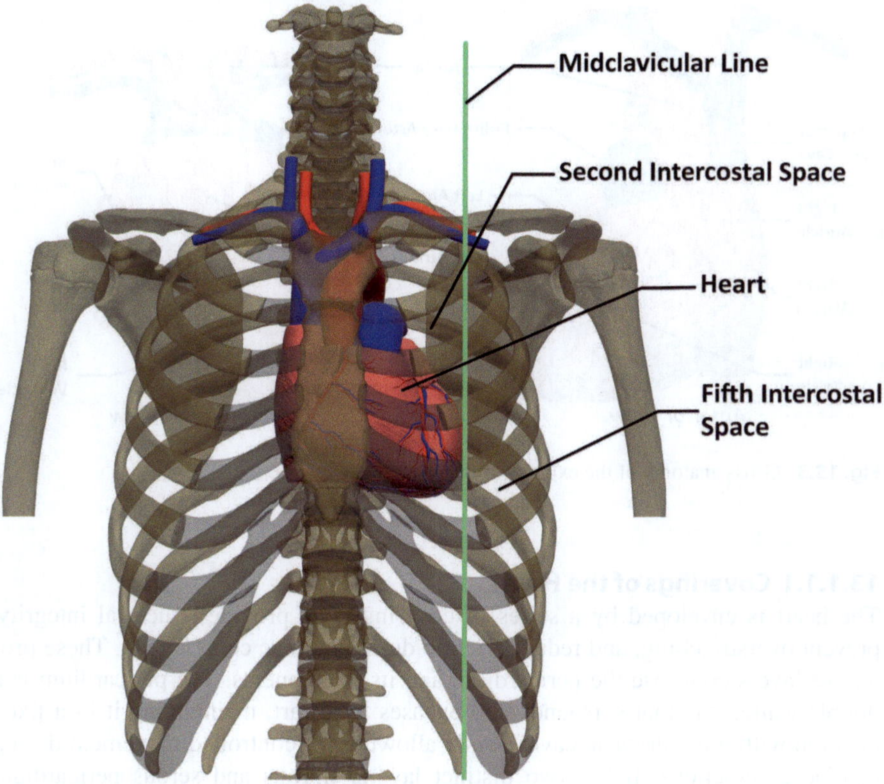

Fig. 13.2 Location of heart concerning the ribs

the fifth intercostal space along the midclavicular line (Fig. 13.2). The heart size can exhibit natural variations among individuals based on age, sex, body size, and overall health. The average dimensions of a healthy adult heart are approximately 12–14 cm in length, 8–9 cm in width, and 6–8 cm in thickness.

13.1.1 Structure

The heart's four-chambered design (Fig. 13.3), encompassing two atria and two ventricles, allows for the efficient separation of oxygenated and deoxygenated blood. Blood circulation occurs through two distinct but interconnected pathways: the pulmonary and systemic circuits. These circuits ensure the transport of oxygen and nutrients to tissues while facilitating the removal of waste products. The pulmonary circuit facilitates the exchange of gases between the blood and the lungs. The systemic circuit delivers oxygenated blood to all cells and tissues while collecting deoxygenated blood for subsequent oxygenation in the pulmonary circuit.

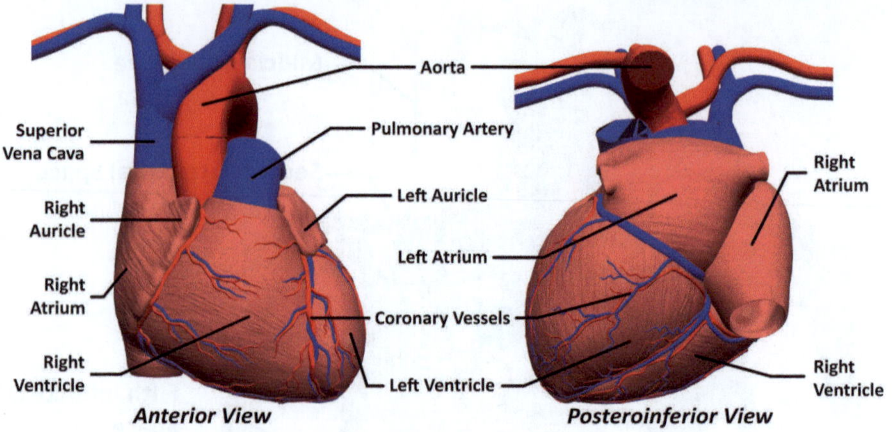

Fig. 13.3 Gross anatomy of the exterior of the heart

13.1.1.1 Coverings of the Heart

The heart is enveloped by a series of coverings that provide structural integrity, prevent overstretching, and reduce friction during cardiac contractions. These protective layers comprise the pericardium and its components. The pericardium is a double-walled sac that surrounds and encases the heart, maintaining it in a fixed position within the thoracic cavity while allowing for controlled movement during cardiac contractions. It has two distinct layers: fibrous and serous pericardium (Fig. 13.4).

Fibrous Pericardium

The outer layer, the fibrous pericardium, comprises dense connective tissue. It is tough and inelastic, serving as a protective shield against external trauma and overdistention of the heart. Additionally, the fibrous pericardium anchors the heart to surrounding structures, including the diaphragm and sternum, preventing excessive movement. The fibrous pericardium is continuous with the walls of the great vessels leaving the heart, such as the aorta and pulmonary trunk.

Serous Pericardium

The serous pericardium is a delicate, two-layered membrane beneath the fibrous pericardium. It is composed of two layers: the parietal layer, which lines the inner surface of the fibrous pericardium, and the visceral layer (also known as the epicardium), which adheres closely to the surface of the heart. A small potential space between these layers, the pericardial cavity, contains a thin layer of pericardial fluid. This fluid acts as a lubricant, reducing friction between the heart's surface and the surrounding pericardial layers during cardiac contractions.

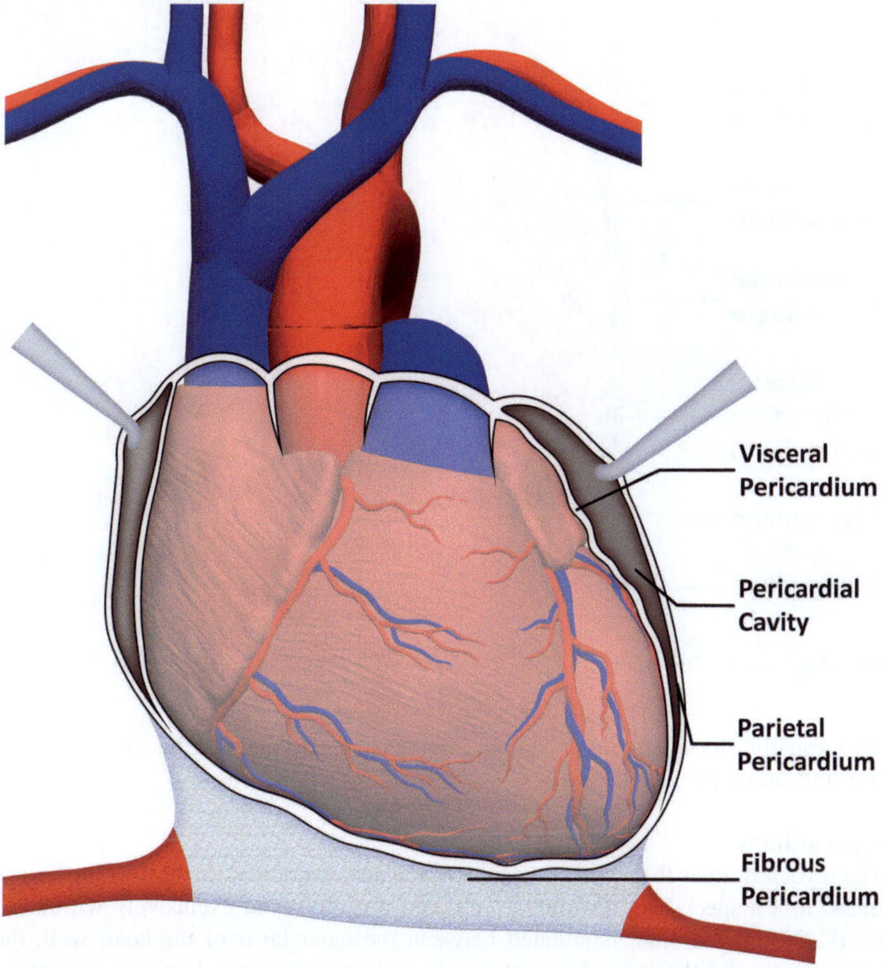

Fig. 13.4 Coverings of the heart

13.1.1.2 Layers of the Heart Wall

The heart is a muscular organ comprising multiple layers forming its intricate structure. These heart wall layers are crucial to its function, providing mechanical strength and the ability to contract rhythmically. The heart wall consists of three primary layers: epicardium, myocardium, and endocardium (Fig. 13.5).

Epicardium

The epicardium, also known as the visceral layer of the pericardium, is the outermost layer of the heart wall. It is the site of origin for the coronary arteries, which branch out and penetrate the myocardium to supply it with oxygen and nutrients.

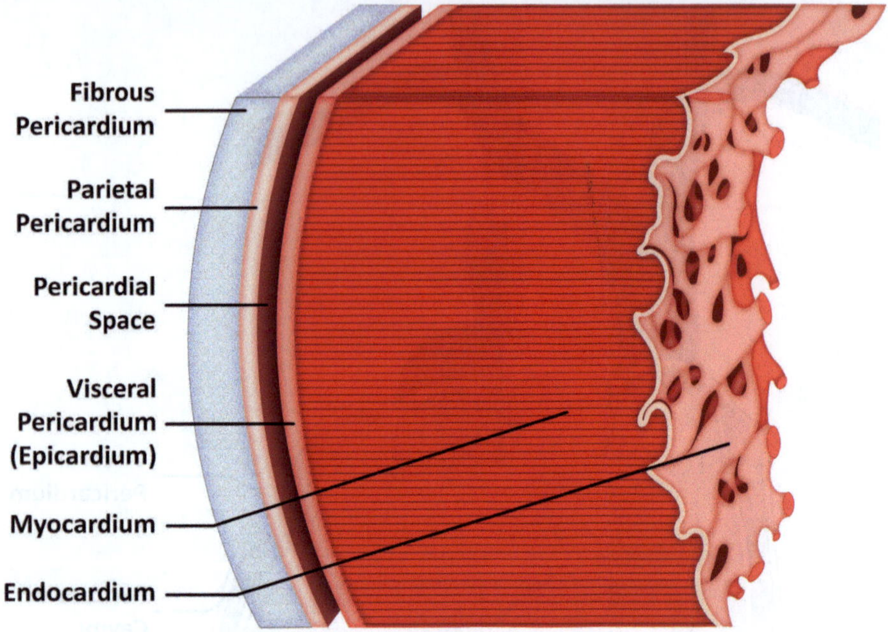

Fig. 13.5 Layers of the heart

Additionally, the epicardium contains nerves that regulate cardiac function and transmit sensory information.

Myocardium

The myocardium is the heart's muscular tissue and constitutes the bulk of the heart's mass. It is a specialized involuntary striated muscle found exclusively within the heart. The myocardium is situated between the outer layer of the heart wall, the epicardium, and the inner layer, the endocardium. Myocardial cells (cardiomyocytes) are the building blocks of the myocardium, and they exhibit several distinctive characteristics (Fig. 13.6):

- Striated Appearance: Cardiomyocytes are striated muscle cells containing alternating dark and light bands or striations when viewed under a microscope. These striations result from the arrangement of contractile proteins, primarily actin and myosin, within the myofibrils of cardiomyocytes, like skeletal muscle (Section 9.1.2.6).
- Intercalated Discs: Intercalated discs are specialized junctions found between adjacent cardiomyocytes. They serve two crucial functions: facilitating rapid electrical communication and synchronizing contractions between neighboring cells. Intercalated discs contain gap junctions (Page 53), which allow for the passage of electrical impulses, and desmosomes (Page 52), which provide mechanical stability and keep the cells connected during contractions.

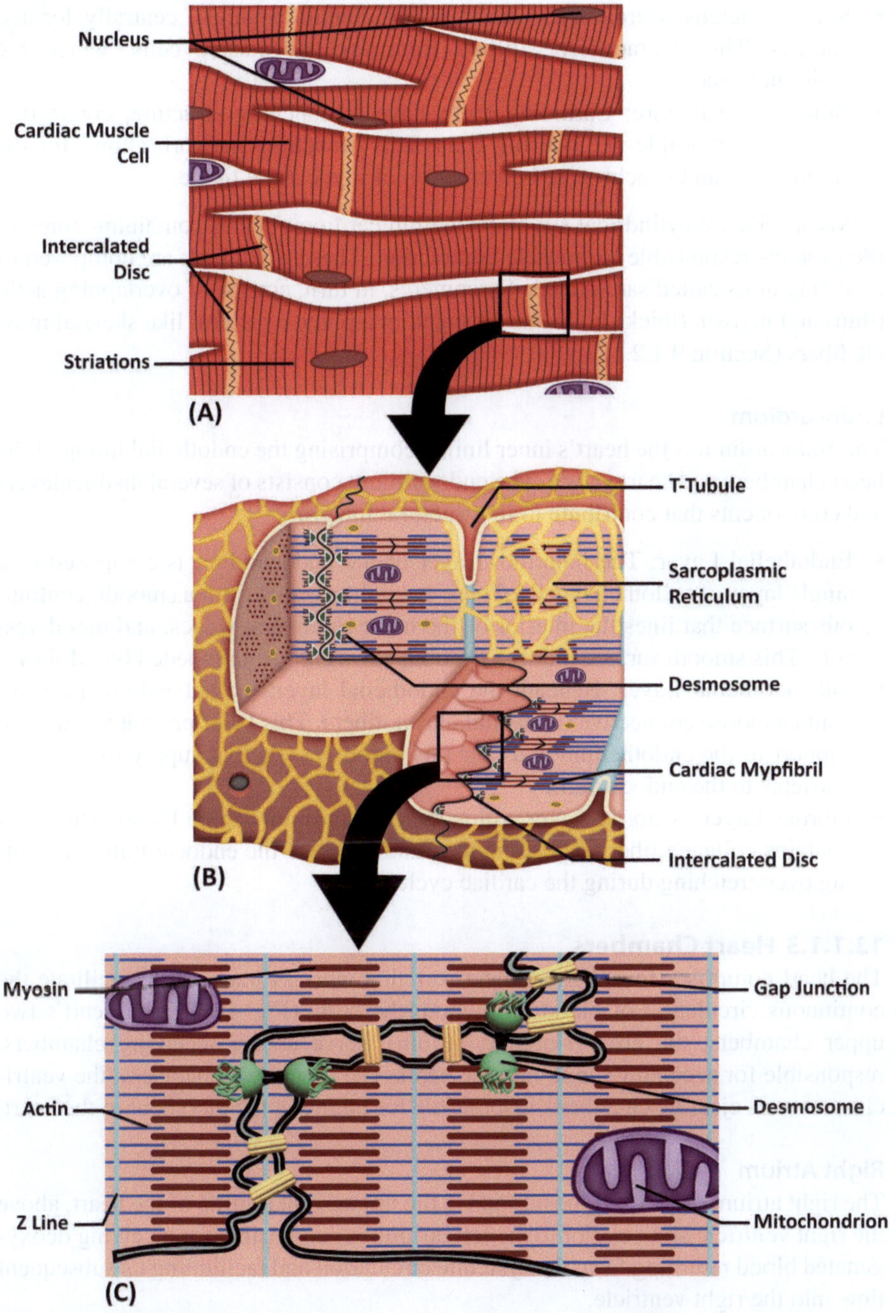

Fig. 13.6 Microscopic hierarchical structure of myocardium. (**a**) Arrangement of cardiac muscle fibers. (**b**) Internal structure of cardiac muscle fiber. (**c**) The structure of the intercalated disc

- Single Nucleus: Cardiomyocytes typically contain a single, centrally located nucleus. This characteristic differs from skeletal muscle cells, which are multinucleated.
- Branching Structure: Cardiomyocytes have a branching structure, connecting one cell to multiple neighboring cells. This branching network allows for the coordinated and synchronized contraction of myocardial tissue.

Myofibrils are cylindrical structures within cardiomyocytes containing contractile proteins responsible for muscle contraction. These myofibrils are composed of repeating units called sarcomeres. Sarcomeres, in turn, consist of overlapping actin (thin) and myosin (thick) filaments arranged in a precise pattern, like skeletal muscle fibers (Section 9.1.2.6).

Endocardium

The endocardium is the heart's inner lining, comprising the endothelial lining of the heart chambers and heart valves. The endocardium consists of several distinct layers and components that contribute to its structural integrity:

- Endothelial Layer: The innermost layer of the endocardium is composed of a single layer of endothelial cells. These specialized cells form a smooth, continuous surface that lines the interior of the heart chambers, valves, and blood vessels. This smooth surface reduces friction, allowing for unimpeded blood flow.
- Subendothelial Layer: Beneath the endothelial layer, the subendothelial layer contains loose connective tissue and elastic fibers. This layer provides structural support to the endothelium and contains blood vessels that supply oxygen and nutrients to the endocardium.
- Fibrous Layer: A fibrous layer within the endocardium can be found. This layer contains collagen fibers that strengthen and stabilize the endocardium, preventing overstretching during the cardiac cycle.

13.1.1.3 Heart Chambers

The heart comprises four distinct chambers that work in harmony to facilitate the continuous circulation of blood throughout the body (Fig. 13.7). The heart's two upper chambers, the atria (singular: atrium), serve as the receiving chambers, responsible for accepting blood into the heart. The lower two chambers, the ventricles, serve as ejecting chambers, responsible for pumping the blood out of the heart.

Right Atrium

The right atrium (Fig. 13.7), positioned in the upper right portion of the heart, above the right ventricle and anterior to the left atrium, is responsible for receiving deoxygenated blood returning from the systemic circulation and facilitating its subsequent flow into the right ventricle.

Left Atrium

The left atrium (Fig. 13.7) is positioned in the upper left portion of the heart, above the left ventricle and posterior to the right atrium. It is responsible for receiving oxygenated blood returning from the pulmonary circulation and facilitating its subsequent flow into the left ventricle.

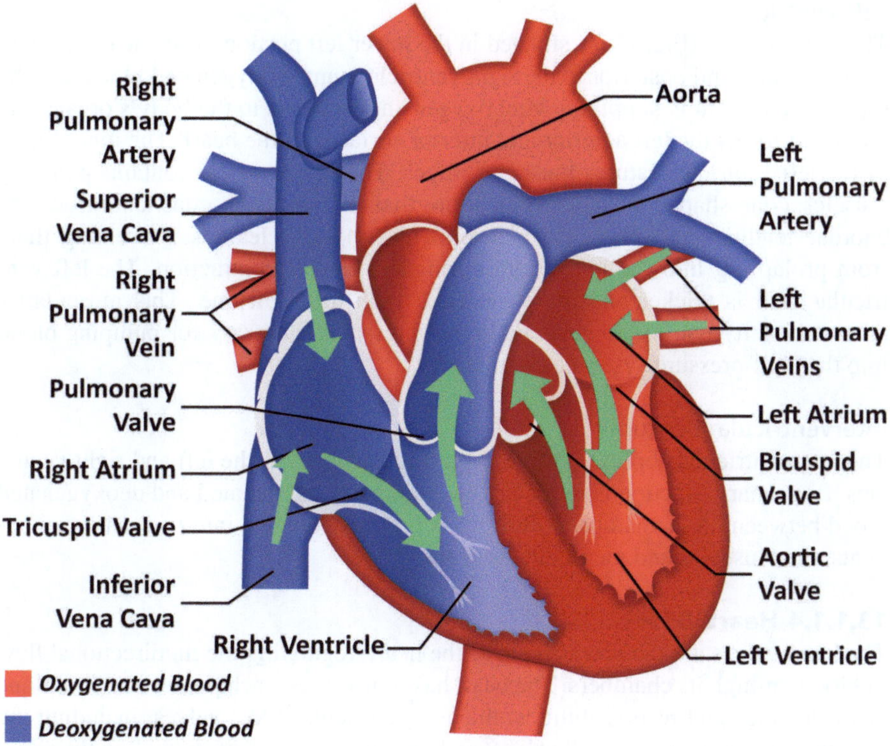

Fig. 13.7 Heart chambers: coronal section

Auricles

The heart auricles, also known as atrial appendages (Fig. 13.3), are small, ear-like structures located on the superior surface of the heart, adjacent to the main atrial chambers, and continuous with the atrial chambers. The auricles act as expandable reservoirs temporarily storing blood when the atria cannot accommodate the entire venous return during high-flow periods.

Right Ventricle

The right ventricle (Fig. 13.7), situated in the lower right portion of the heart, beneath the right atrium and anterior to the left ventricle, is responsible for receiving deoxygenated blood from the right atrium and pumping it into the pulmonary circulation for oxygenation in the lungs. It forms the anterior surface of the heart. The inner lining of the right ventricle features muscular ridges called trabeculae carneae. These structures contribute to the ventricle's muscular texture and help prevent the ventricular wall from collapsing during contraction. Papillary muscles, cone-shaped muscular projections, are present within the right ventricle. These muscles anchor the chordae tendineae, which attach to the tricuspid valve leaflets. The contraction of the papillary muscles prevents the tricuspid valve from prolapsing into the right atrium during ventricular contraction.

Left Ventricle

The left ventricle (Fig. 13.7), situated in the lower left portion of the heart, beneath the left atrium and posterior to the right ventricle, pumps oxygenated blood into the systemic circulation, supplying vital oxygen and nutrients to the body's organs and tissues. It forms the left anterior and inferior surfaces of the heart. The inner lining of the left ventricle features muscular ridges. The left ventricle contains papillary muscles, cone-shaped muscular projections that anchor the chordae tendineae. The chordae tendineae attach to the bicuspid (mitral) valve leaflets, preventing them from prolapsing into the left atrium during ventricular contraction. The left ventricular wall is thicker and more muscular than the right one. This muscularity allows the left ventricle to generate higher pressure, necessary for pumping blood into the high-pressure systemic circulation.

Interventricular Septum

The interventricular septum (IVS) is a partition separating the left and right ventricles. Its primary function is to prevent the mixing of oxygenated and deoxygenated blood between these chambers. The IVS is typically divided into two main components: the muscular and membranous septa.

13.1.1.4 Heart Valves

The heart valves are the gatekeepers of the heart, regulating the unidirectional flow of blood through its chambers. The heart has four valves, each possessing its unique characteristics and responsibilities: the atrioventricular (AV) valves, including the tricuspid and mitral (bicuspid) valves, and the semilunar valves, including the aortic and pulmonary valves (Fig. 13.7). Heart valve rings, also known as annuli fibrosi, serve as the structural foundations for the heart valves. These fibrous rings provide stability and support to the valves, anchoring them in place and facilitating their proper function in regulating blood flow.

Right Atrioventricular Valve (Tricuspid Valve)

The right atrioventricular valve, commonly known as the tricuspid valve (Fig. 13.8), is positioned between the right atrium and ventricle. The tricuspid valve consists of three leaflets, or cusps: the anterior, posterior, and septal cusps. Commissures serve as attachment points between the cusps or leaflets of heart valves, where two cusps meet. The tricuspid valve has three commissures. Chordae tendineae are tendinous cords that connect the free edges of the tricuspid valve cusps to the papillary muscles within the right ventricle. These tendinous attachments anchor the valve leaflets and prevent them from prolapsing into the right atrium during ventricular contraction.

Left Atrioventricular Valve (Mitral Valve)

The left atrioventricular valve, known as the mitral or bicuspid valve (Fig. 13.8), is positioned between the left atrium and the left ventricle. The mitral valve consists of two leaflets or cusps: the anterior and posterior cusps. These cusps create a bicuspid configuration, allowing for effective valve closure. Chordae tendineae connect the

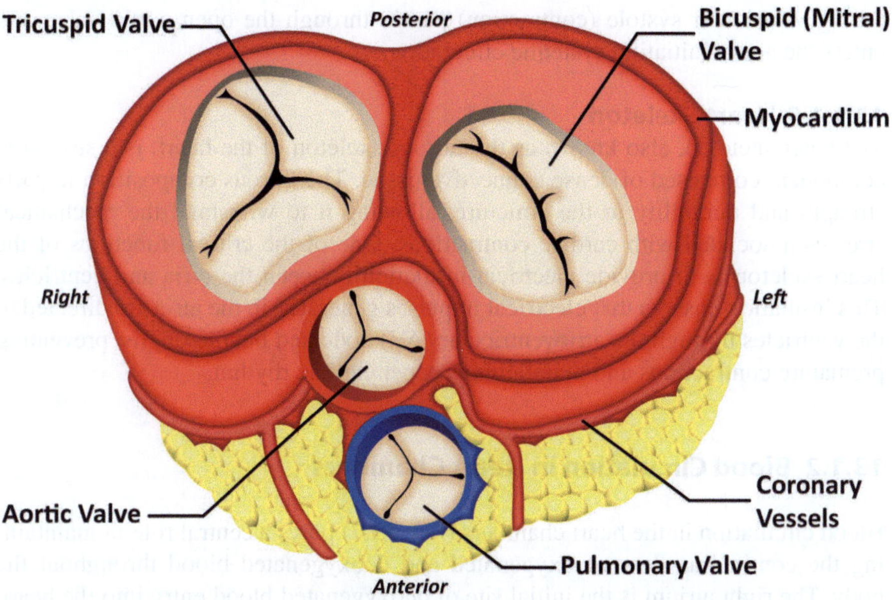

Fig. 13.8 Heart valves: superior view (atria removed)

free edges of the mitral valve cusps to the papillary muscles within the left ventricle and prevent them from prolapsing into the left atrium during ventricular contraction.

Right Semilunar Valve (Pulmonary Valve)

The right semilunar valve, also known as the pulmonary valve (Fig. 13.8), is positioned between the right ventricle exit and the pulmonary trunk entrance. The pulmonary valve comprises three semilunar cusps: the anterior, right, and left cusps. These cusps are shaped like half-moons and are situated within the pulmonary artery immediately distal to the right ventricle. The pulmonary valve cusps are associated with three pulmonary sinuses. These sinuses are dilated regions located just above the cusps in the pulmonary trunk and play an essential role in preventing blood flow turbulence. The pulmonary valve serves as the gateway through which deoxygenated blood is ejected from the right ventricle into the pulmonary artery for transport to the lungs.

Left Semilunar Valve (Aortic Valve)

The left semilunar valve, known as the aortic valve (Fig. 13.8), is positioned at the exit of the left ventricle and the entrance of the aorta. The aortic valve consists of three semilunar cusps: the right, left, and posterior cusps. These cusps are shaped like half-moons and are situated within the aortic root. The aortic valve cusps are associated with three dilated aortic sinuses, known as the sinuses of Valsalva. These sinuses prevent turbulence of blood flow. Blood ejected from the left ventricle during ventricular systole (contraction) passes through the open aortic valve and enters the aorta, initiating systemic circulation.

13.1.1.5 Heart Skeleton

The heart skeleton, also known as the fibrous skeleton of the heart, is a structural component composed of dense connective tissue. The fibrous composition imparts strength and durability to the structure, allowing it to withstand the mechanical stresses associated with cardiac contractions. One of the critical functions of the heart skeleton is to provide electrical insulation between the atria and ventricles. This insulation ensures that electrical impulses generated in the atria are directed to the ventricles through the atrioventricular (AV) node and bundle of His, preventing premature contractions and maintaining proper cardiac rhythm.

13.1.2 Blood Circulation in Heart Chambers

Blood circulation in the heart chambers (Fig. 13.7) plays a central role in maintaining the continuous flow of oxygenated and deoxygenated blood throughout the body. The right atrium is the initial site of deoxygenated blood entry into the heart. Deoxygenated blood from the superior and inferior vena cava enters the right atrium. As the right atrium contracts, it pushes blood through the tricuspid valve into the right ventricle. Once the blood enters the right ventricle, contraction of the right ventricle forces deoxygenated blood through the pulmonary valve into the pulmonary artery. The pulmonary valve prevents backflow of blood into the right ventricle. Blood is then directed into the pulmonary artery, bifurcating into the right and left pulmonary arteries, carrying deoxygenated blood to the lungs. Oxygenated blood from the lungs returns to the heart via the pulmonary veins and enters the left atrium. As the left atrium contracts, it pushes blood through the mitral (bicuspid) valve into the left ventricle. Once oxygenated blood enters the left ventricle, contraction of the left ventricle forces oxygenated blood through the aortic valve into the aorta. The aortic valve prevents backflow of blood into the left ventricle. Blood is then directed into the ascending aorta, which branches into various systemic arteries, supplying oxygenated blood to the body's tissues and organs.

Circulation in the heart chambers is tightly regulated to ensure that oxygenated and deoxygenated blood is directed to the appropriate locations. This coordination is achieved through the opening and closing of the heart valves, as well as the rhythmic contractions of the atria and ventricles. The atria contract first, followed by the ventricles, ensuring efficient blood flow and preventing regurgitation.

13.1.3 Coronary Circulation

The coronary vessels (Fig. 13.9) supply the heart with oxygen and nutrients and remove waste products, ensuring the continuous function of this essential organ. There are two main coronary arteries: the left coronary artery (LCA) and the right

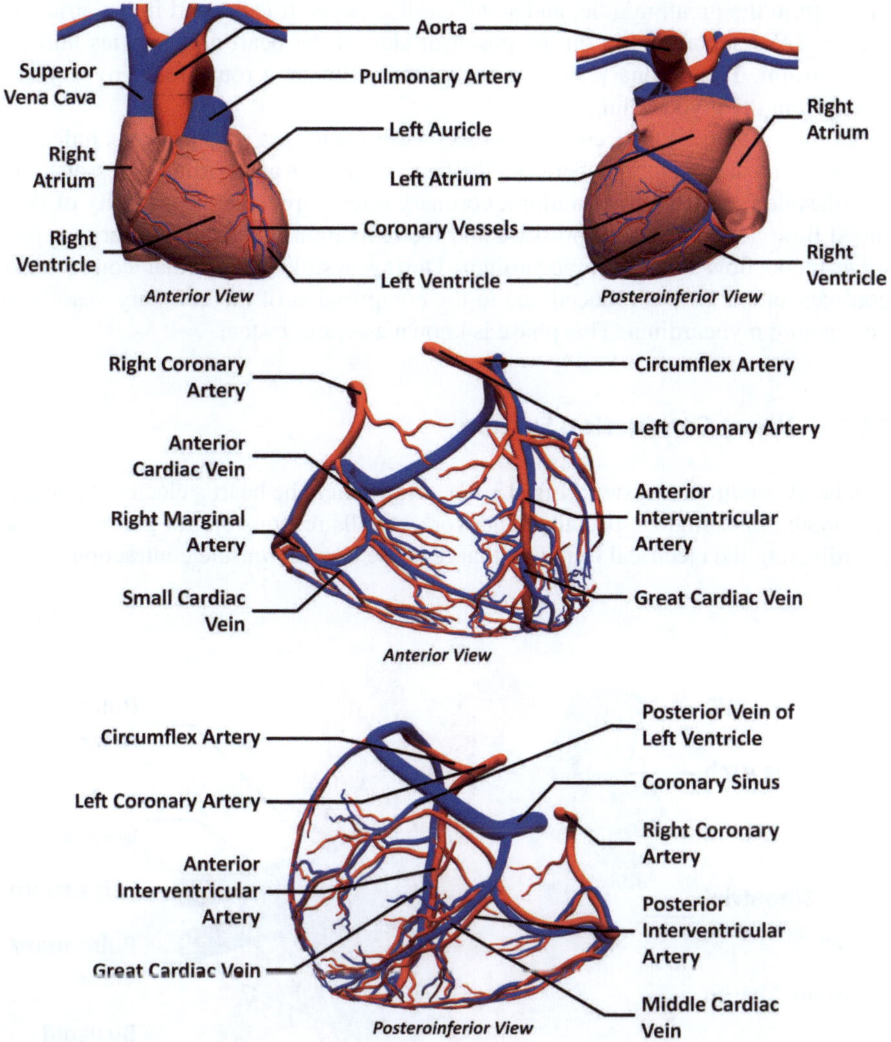

Fig. 13.9 Coronary vessels

coronary artery (RCA). The left coronary artery is a major branch of the ascending aorta, originating just above the aortic valve in the left aortic sinus. It is subdivided into two primary branches: the left anterior descending artery (LAD) and the left circumflex artery (LCx). The right coronary artery originates from the ascending aorta's right aortic sinus and supplies the heart's right side. It follows a variable course and gives rise to several branches: right marginal arteries and posterior descending artery (PDA).

The major coronary veins include the great, middle, and small cardiac veins and the coronary sinus. The coronary sinus is a larger venous structure that receives blood from the great, middle, and small cardiac veins. It is located in the atrioventricular (AV) septal region on the posterior side of the heart and empties into the right atrium. The coronary sinus is the primary drainage route for deoxygenated blood from the myocardium.

Coronary blood flow occurs in distinct phases during the cardiac cycle, reflecting the mechanical interactions between the heart chambers and coronary vessels: during diastole (ventricular relaxation), coronary arteries receive the majority of their blood flow. The high aortic pressure and the relaxation of coronary arterioles promote blood flow into the myocardium. During systole (ventricular contraction), coronary blood flow is reduced due to the compression of the coronary vessels by contracting myocardium. This phase is known as systolic steal.

13.1.4 Heart Conduction System

The heart conduction system (Fig. 13.10), often called the heart's electrical conduction system, is a highly specialized network of cells responsible for generating and coordinating the electrical impulses that regulate cardiac muscle contraction.

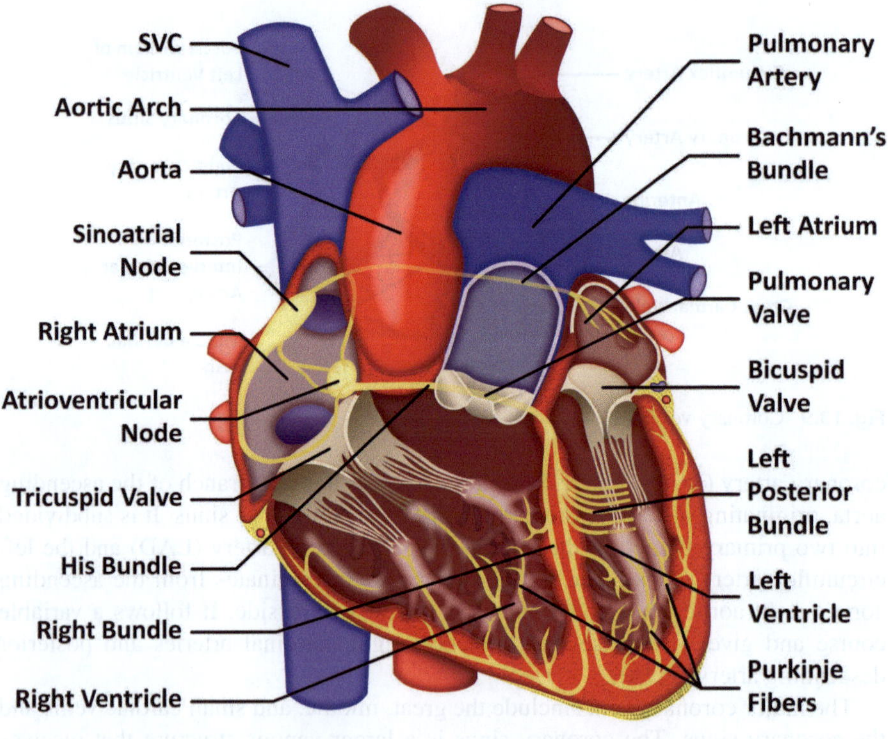

Fig. 13.10 Heart conduction system. (SVC Superior vena cava)

13.1.4.1 Components

The heart conduction system consists of several critical components:

- Sinoatrial (SA) Node: The SA node near the superior vena cava in the right atrium is the heart's primary pacemaker. It initiates electrical impulses and determines the heart's baseline rate and rhythm.
- Atria: Specialized conducting pathways distribute the electrical impulse rapidly across the atria, causing atrial contraction and facilitating the transfer of blood into the ventricles.
- Atrioventricular (AV) Node: The AV node is near the atrioventricular septum, between the atria and ventricles. It serves as a relay station, briefly delaying the transmission of electrical impulses from the atria to the ventricles. This delay allows the atria to contract and blood to fill the ventricles.
- Bundle of His: After passing through the AV node, the electrical impulse enters the bundle of His, a bundle of specialized conducting fibers located in the interventricular septum.
- Right and Left Bundle Branches: His bundle is divided into the right and left bundle branches. These branches carry the electrical impulse to the respective ventricles.
- Purkinje Fibers: The bundle branches further divide into Purkinje fibers, which extensively ramify throughout the ventricular walls. These fibers rapidly transmit the electrical impulse, ensuring synchronized ventricular contraction.

13.1.4.2 Impulse Generation

The SA node and AV node cells possess intrinsic automaticity, allowing them to spontaneously depolarize and generate action potentials. This property is essential for setting the heart's rhythm. Gap junctions allow for rapid electrical communication between cells, ensuring the synchronized contraction of the heart. Pacemaker cells, primarily found in the SA node, have less myofibrillar structure than typical cardiomyocytes, favoring electrical impulses' rapid initiation and conduction. The pacemaker potential, a vital feature of these cells, is responsible for initiating the cardiac cycle. The SA node holds the highest rank in the conduction hierarchy due to its fastest intrinsic rate of automaticity. It usually dictates the rhythm of the entire heart by initiating impulses at approximately 60–100 times per minute in a resting adult. The atrioventricular (AV) node, bundle of His, and Purkinje fibers possess pacemaker cells that can generate impulses, albeit at slower intrinsic rates (40–60 and 20–40 times per minute, respectively). The SA node, firing at the highest rate, continually depolarizes other potential pacemaker cells before they reach their threshold to initiate an impulse. This phenomenon, known as overdrive suppression, maintains the dominance of the SA node. The AV node serves as a checkpoint, slowing the conduction to ensure that ventricles have ample time to fill with blood.

13.1.4.3 Regulation

Control of heart rhythm involves a complex interplay between the cardiac conduction system, autonomic nervous system, endocrine influences, and the intrinsic properties of cardiac cells:

- Sympathetic Stimulation: The release of catecholamines (e.g., norepinephrine) increases heart rate and contractility by enhancing calcium influx into cells.

- Parasympathetic Control: Vagal nerve stimulation releases acetylcholine, which slows the heart rate by increasing potassium efflux and decreasing calcium influx during pacemaker potential.
- Endocrine Regulation
  - Adrenal Hormones: Epinephrine and norepinephrine, released from the adrenal medulla, increase heart rate and force of contraction in response to stress (the "fight-or-flight" response).
  - Thyroid Hormones: Thyroid hormones upregulate adrenergic receptors and enhance calcium ATPase activity, increasing heart rate and contractility.
- Intrinsic Cardiac Reflexes
  - Bainbridge Reflex: An increase in venous return stretches the atria, triggering a reflex increase in heart rate to maintain cardiac output.
  - Baroreceptor Reflex: Baroreceptors in the carotid sinus and aortic arch (Page 467) respond to changes in blood pressure by adjusting the heart rate accordingly.

13.1.5 Cardiac Cycle

The cardiac cycle is a repetitive sequence of events during one complete heartbeat. It involves the coordinated contraction and relaxation of the heart's chambers to pump blood efficiently through the circulatory system. The cardiac cycle is typically divided into two main phases: diastole and systole (Fig. 13.11).

13.1.5.1 Diastole

Diastole is the phase during which the heart muscle relaxes following a contraction (systole), allowing the heart chambers, particularly the ventricles, to fill with blood in preparation for the next contraction. This phase is crucial for maintaining an adequate cardiac output and ensuring efficient blood flow throughout the body. Although atrial contraction is not part of diastole per se, it is intimately related to ventricular filling. The atria contract in response to an electrical impulse from the sinoatrial (SA) node, propelling the remaining blood into the ventricles. This atrial "kick" contributes the final 20–30% of the blood volume the ventricles receive before the systole begins.

13.1.5.2 Systole

Systole represents a phase within the cardiac cycle during which the heart muscle contracts, leading to the ejection of blood into the arterial system. This phase is essential for propelling oxygenated blood from the left heart to the systemic circulation and deoxygenated blood from the right heart to the pulmonary circulation. The atrial systole occurs just before ventricular systole and is considered part of the diastole period. Ventricular systole initiates with a phase where the ventricles contract with no corresponding ejection of blood because all the heart valves are closed.

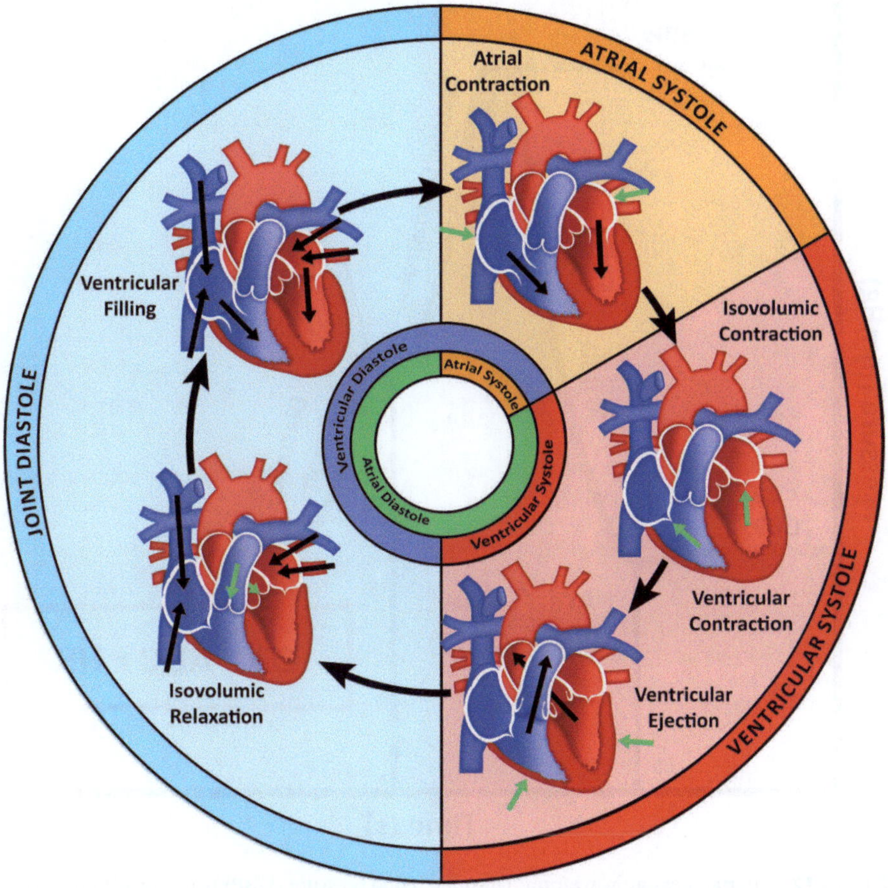

Fig. 13.11 Cardiac cycle: phases of heart activity during a single heartbeat

As the pressure within the ventricles exceeds the pressure in the aorta and pulmonary artery, the aortic and pulmonary valves open, allowing blood to be ejected.

13.1.5.3 Stroke Volume

Stroke volume (SV) represents the volume of blood ejected by the heart's left ventricle in a single contraction. In conjunction with heart rate (HR), SV determines cardiac output (CO), the volume of blood pumped by the heart per minute. Several methods are employed to measure stroke volume based on distinct principles: echocardiography, thermodilution, Fick method, and pulse contour analysis (Fig. 13.12). Echocardiography is currently the most commonly used technique. It utilizes high-frequency sound waves (ultrasound) to visualize the heart's structures and assess blood flow. In particular, the Doppler technique measures blood flow velocities to calculate stroke volume. SV can be accurately determined by examining the left ventricle dimensions and aortic flow velocities.

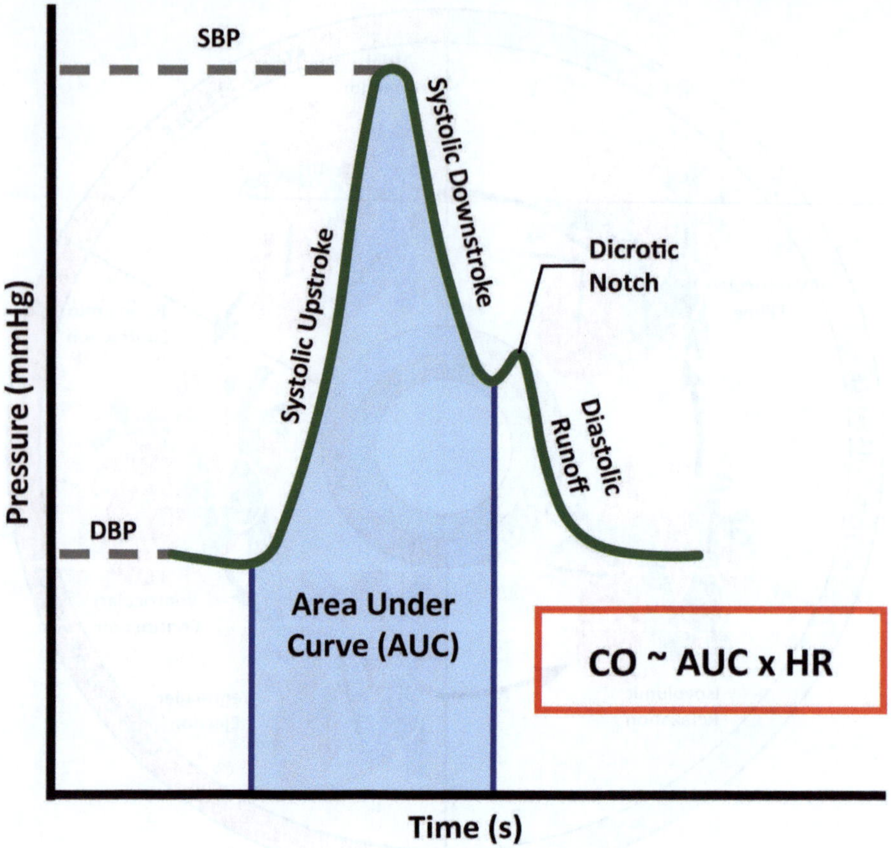

Fig. 13.12 Arterial pressure waveform: Diastolic blood pressure (DBP) is the lowest pressure in the arteries, occurring just before the heart contracts (end of diastole). Systolic blood pressure (SBP) is the highest pressure in the arteries, occurring during ventricular contraction (systole). The systolic upstroke is the initial rapid rise in pressure as the ventricles contract and blood is ejected into the aorta and arteries. It indicates the rate of pressure increase during systole, influenced by cardiac contractility and arterial compliance. The systolic downstroke is the decline in pressure after the peak systolic pressure is reached. It represents when the aortic valve closes, and the heart transitions to diastole. It reflects the decrease in ventricular pressure and the beginning of diastolic runoff. Diastolic runoff is the continuous decline in arterial pressure during diastole as blood flows into the peripheral circulation, reflecting the blood flow rate out of the arterial system and into the capillaries. It is influenced by peripheral resistance and arterial compliance. Dicrotic notch is a slight, brief increase in arterial pressure that occurs when the aortic valve closes and causes a slight backflow of blood. The area under curve (AUC) is the area under the systolic portion of the pressure waveform, representing the mean arterial pressure (MAP) during systole. The AUC is directly related to the stroke volume and overall cardiac output (CO). This relation can be expressed as CO ~ AUC x HR (heart rate)

13.1.5.4 Ejection Fraction

Ejection fraction (EF) assesses the heart's pumping efficiency, specifically the left ventricle. The ejection fraction is a percentage representing the proportion of blood ejected from the left ventricle during each heartbeat. Ejection fraction is the stroke volume ratio (the amount of blood ejected from the left ventricle during systole) to the end-diastolic volume (the blood volume in the left ventricle at the end of diastole).

$$\text{Ejection Fraction}\left(\text{EF}\right) = \frac{\text{Stroke Volume}}{\text{End Diastolic Volume}} \times 100$$

A normal ejection fraction typically ranges from 50% to 70%. This means that, under normal conditions, the left ventricle ejects 50–70% of the blood it contains during each contraction. Ejection fraction can be determined through various imaging techniques, including echocardiography (most commonly used), magnetic resonance imaging (MRI), and radionuclide ventriculography. Echocardiography is noninvasive and widely accessible, making it the preferred method for routine clinical assessment.

13.1.5.5 Cardiac Output

Cardiac output (CO) is the volume of blood ejected by the heart in 1 min. It is a critical determinant of tissue perfusion and oxygen delivery, reflecting the heart's ability to meet the body's metabolic demands. Cardiac output can be calculated using heart rate and stroke volume:

$$\text{Cardiac Output}\left(\text{CO}\right) = \text{Heart Rate}\left(\text{HR}\right) \times \text{Stroke Volume}\left(\text{SV}\right)$$

13.1.5.6 Heart Sounds

Normal heart sounds (Fig. 13.13), often referred to as "lub" and "dub," are produced by the closure of heart valves and the turbulent flow of blood within the heart chambers. The two primary heart sounds are designated as S1 (the first heart sound) and S2 (the second heart sound), with some subdivisions:

- S1 (First Heart Sound): S1 is the "lub" sound. It is generated by the closure of the atrioventricular (AV) valves— specifically, the tricuspid and mitral valves—at the onset of ventricular systole (contraction). S1 is a low-pitched sound with a "lub" quality. It is typically longer and more intense than S2. The sound is synchronous with the apical pulse.
- S2 (Second Heart Sound): S2 is the "dub" sound. It is produced by the closure of the semilunar valves—specifically, the aortic and pulmonary valves—at the end of ventricular systole, signifying the termination of ventricular ejection. S2 is typically split into two components: A2 and P2. A2 corresponds to the closure of the aortic valve, while P2 corresponds to the closure of the pulmonary valve. S2 is a high-pitched sound with a "dub" quality. It is shorter and less intense than S1. The splitting of S2 into A2 and P2 is normal during inspiration due to changes in intrathoracic pressure.

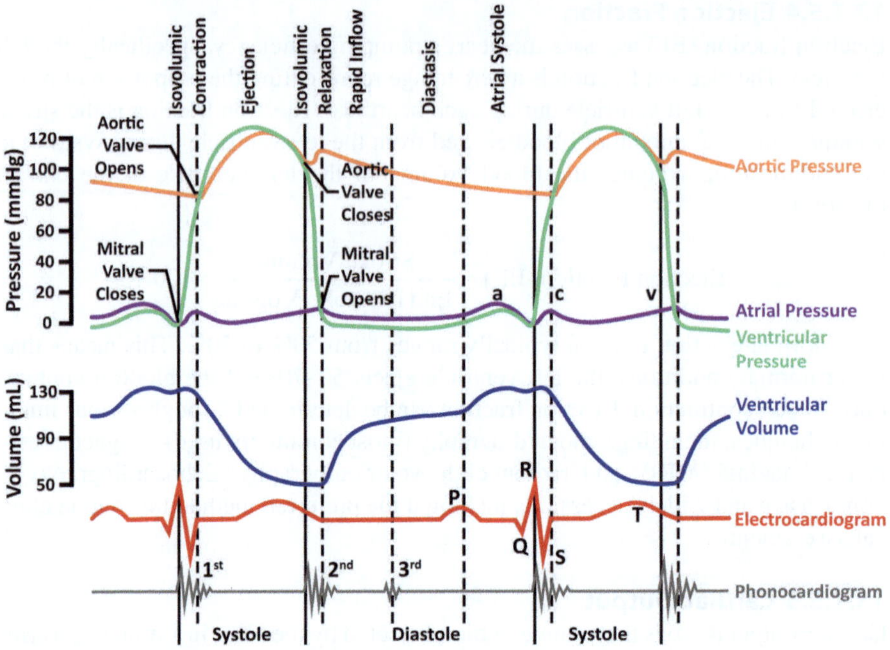

Fig. 13.13 Phases of the cardiac cycle corresponding to correlating pressures, volumes, electrocardiogram (ECG), and phonocardiogram recordings

13.1.6 Electrocardiography

Electrocardiography (ECG or EKG) is a noninvasive diagnostic tool that records the heart's electrical activity over time. It is a fundamental and widely used method in cardiology and clinical medicine for assessing cardiac function, diagnosing arrhythmias, and identifying various cardiac conditions. ECG records cardiac cells' electrical impulses as they initiate and propagate action potentials. ECG electrodes are placed on the skin's surface to detect electrical changes the heart generates. The standard ECG uses 12 leads, each consisting of a positive electrode (measuring electrical potential) and a reference electrode (providing a baseline) (Fig. 13.14). Bipolar leads measure voltage differences between two electrodes. The standard limb leads (I, II, and III) are bipolar leads. Unipolar leads measure voltage regarding a central point. The augmented limb leads (aVR, aVL, and aVF) and the precordial (chest) leads (V1–V6) are unipolar leads. ECG records the sum of electrical vectors created by cardiac cell depolarization. The direction and magnitude of these vectors provide information about the heart's electrical activity (Fig. 13.15).

The ECG waveform consists of P, QRS, and T waves, each representing specific electrical events (Fig. 13.16). The P wave is the first waveform of the ECG and represents atrial depolarization, which signifies the initiation of atrial contraction. The normal P wave duration is typically less than 0.12 s (120 milliseconds). The

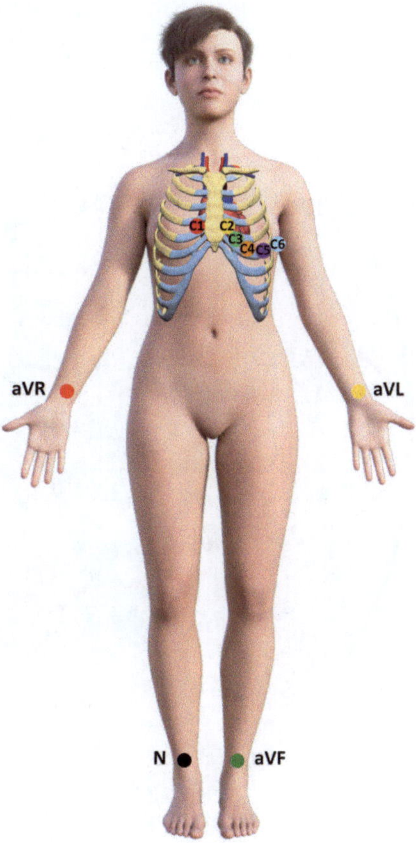

Fig. 13.14 Placement of electrodes for recording a 12-lead electrocardiogram (ECG)

amplitude of the P wave is generally less than 2.5 millimeters (0.25 millivolts) in the standard limb leads. The QRS complex is the second waveform of the ECG and represents ventricular depolarization, indicating the onset of ventricular contraction. The normal QRS complex duration is typically less than 100 milliseconds. The amplitude of the QRS complex varies with lead placement but is generally higher than that of the P wave. The T wave is the third waveform of the ECG and represents ventricular repolarization, indicating the end of ventricular contraction. The normal T wave duration is typically less than 200 milliseconds. The amplitude of the T wave is generally smaller than that of the QRS complex. The PR interval is the segment from the beginning of the P wave to the start of the QRS complex, representing the time it takes for the electrical impulse to travel from the atria to the ventricles via the atrioventricular (AV) node. The normal PR interval duration is typically between 120 and 200 milliseconds. The QT interval is the segment from the start of the QRS complex to the end of the T wave, representing the total duration of ventricular depolarization and repolarization. The normal QT interval duration varies with heart rate but is typically less than 440 milliseconds at a heart rate of 60–100 beats per minute. The ST segment is between the end of the QRS complex and the

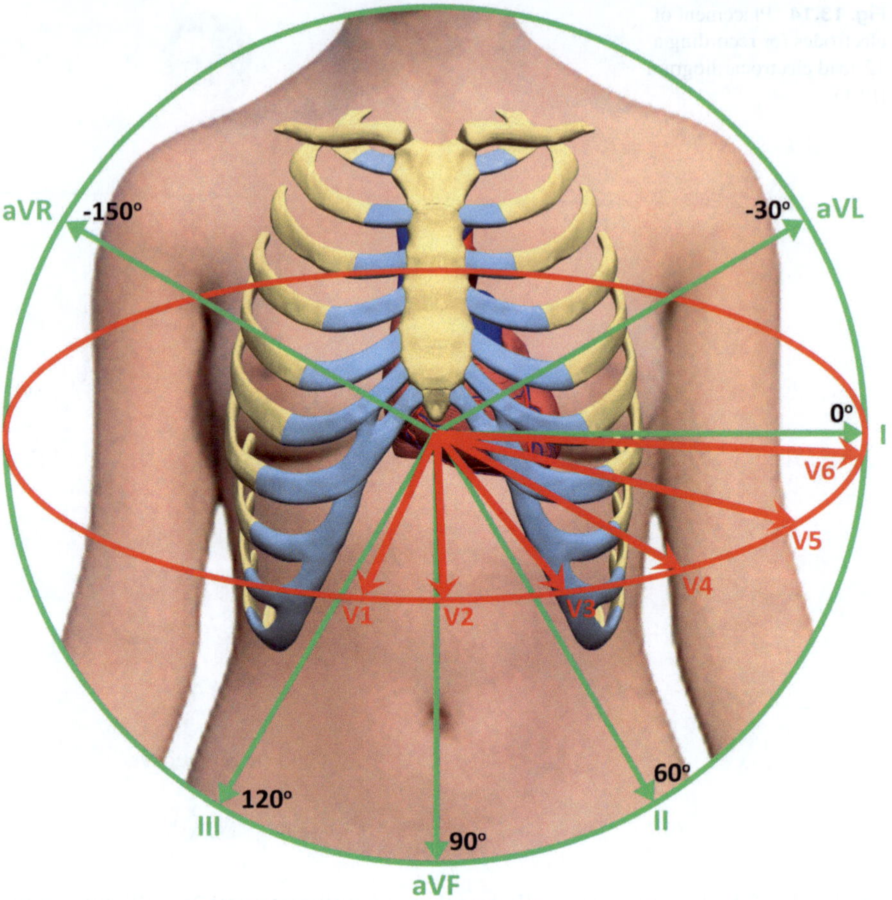

Fig. 13.15 Axes of the leads in a 12-lead electrocardiogram

start of the T wave, representing the early phase of ventricular repolarization. The normal ST segment is generally isoelectric (at the baseline), with any deviation suggesting myocardial ischemia or injury.

13.1.7 Echocardiography

Echocardiography is a noninvasive and widely used diagnostic imaging technique that provides real-time visualization of the heart's structure and function. It utilizes ultrasound waves to produce detailed images of the heart and its chambers, valves, walls, and blood flow patterns. Ultrasound waves are emitted from a transducer placed on the patient's chest or inside the esophagus. These sound waves travel through the tissues, bounce off cardiac structures, and return to the transducer. The transducer receives the reflected sound waves and converts them into electrical

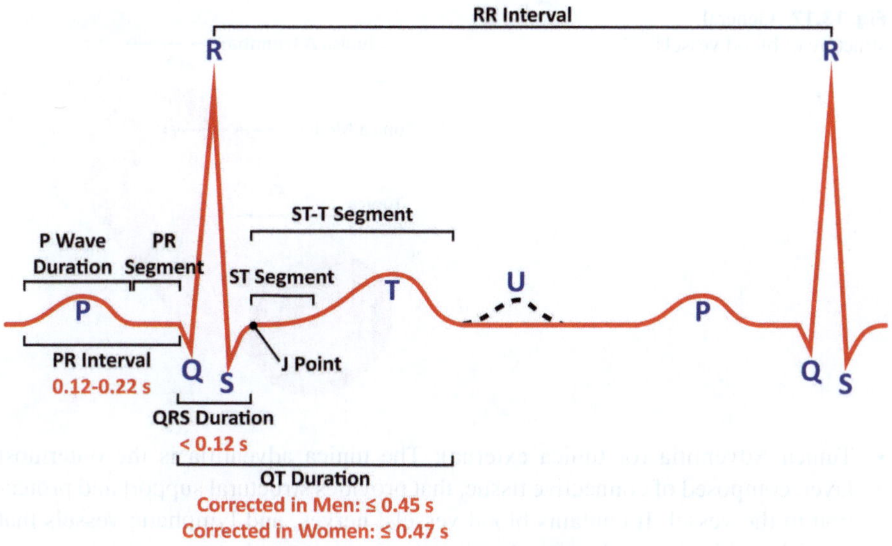

Fig. 13.16 A typical normal electrocardiogram waveform

signals. The electrical signals are processed and used to create real-time images of the heart on a monitor. Doppler echocardiography assesses blood flow by detecting changes in the frequency of sound waves reflected by moving blood cells. This allows for the assessment of blood velocity and direction.

13.2 Part 2: Blood Vessels

The circulatory system comprises an intricate network of blood vessels, serving as the conduits for the continuous blood flow throughout the human body. These vessels transport oxygen, nutrients, hormones, and waste products to and from various tissues and organs.

13.2.1 Structure and Types

Blood vessels are composed of three distinct layers (Fig. 13.17):

- Tunica Intima: It is the innermost layer of blood vessels and comprises a single layer of endothelial cells in direct contact with the blood. The functions of this layer include regulating blood flow, preventing clot formation, and facilitating the exchange of substances between the blood and surrounding tissues.
- Tunica Media: The middle layer primarily consists of smooth muscle cells. The tunica media is responsible for regulating the diameter of the vessel, thereby controlling blood pressure and flow. Elastic fibers in arteries allow them to stretch and recoil, maintaining steady blood flow.

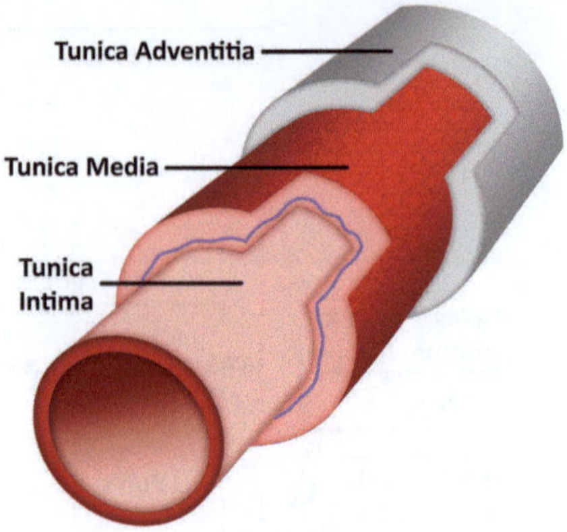

Fig. 13.17 General structure of blood vessels

- Tunica Adventitia (or tunica externa): The tunica adventitia is the outermost layer, composed of connective tissue, that provides structural support and protection to the vessel. It contains blood vessels, nerves, and lymphatic vessels that nourish and innervate the vessel wall.

Blood vessels are generally classified into three main types: arteries, veins, and capillaries.

13.2.1.1 Arteries

Arteries (Fig. 13.18) carry blood away from the heart to various tissues and organs throughout the body. Arteries are composed of the same three concentric layers; however, the structural components and properties of each layer in arteries are adapted for their high-pressure environment. Tunica intima has the same endothelium and subendothelial connective tissue layer. At the border with the tunica media, the internal elastic lamina is a condensed sheet of elastin that imparts resilience and elasticity, enabling the artery to absorb the energy of the systolic pulse wave and maintain blood pressure during diastole. The most unique characteristic of the arterial wall is its tunica media. The tunica media of arteries contains a substantial amount of smooth muscle cells arranged circularly. These muscle cells are responsible for vasoconstriction and vasodilation, thus regulating blood pressure and flow. In some arteries, particularly medium-sized muscular arteries, the external elastic lamina defines the boundary between the tunica media and the adventitia. In larger arteries, the tunica adventitia houses microvessels (vasa vasorum) that supply blood to the outer layers of the arterial wall and nerve fibers (nervi vascularis) that modulate smooth muscle contraction. The combination of smooth muscle, elastic fibers, and collagen in arterial walls provides the strength to withstand the high-pressure environment created by the heart's contractions.

The arterioles, often termed resistance vessels, are the smallest branches of the arterial tree before the capillary network. They are the primary regulators of microcirculatory blood flow and play a pivotal role in systemic vascular resistance and blood pressure regulation. The arteriolar wall typically comprises the same three concentric layers, albeit thinner and less distinct than larger arteries. Circumferentially

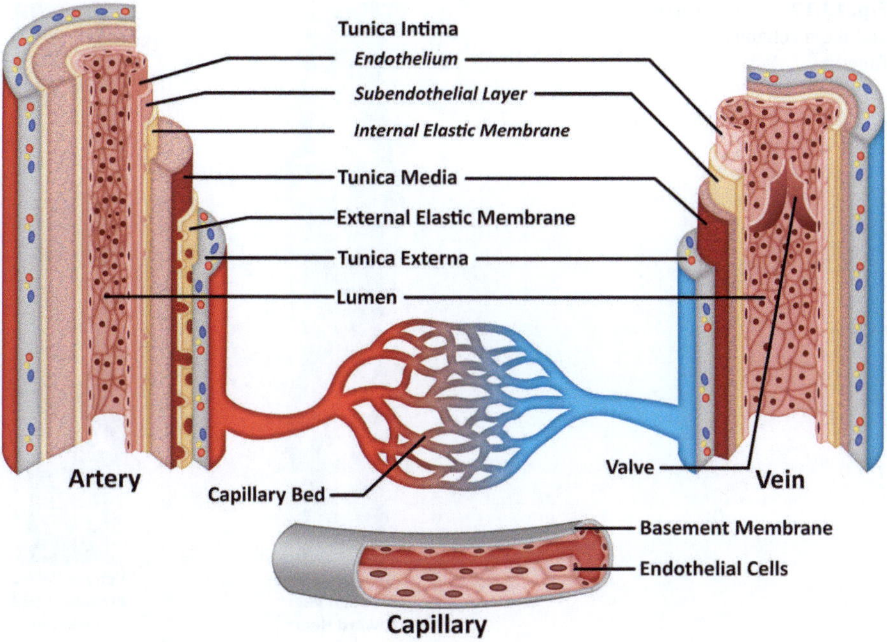

Fig. 13.18 Detailed arteries, veins, and capillary structure

arranged smooth muscle cells (SMC) are the dominant feature of arterioles. Depending on the arteriole's size and location, the number of SMC layers can range from one to five or six. Arterioles exhibit an intrinsic ability to regulate their diameter, a phenomenon known as autoregulation. This allows tissues to maintain a relatively constant blood flow despite changes in arterial blood pressure. The autonomic nervous system innervates arterioles, with sympathetic stimulation generally causing vasoconstriction. Local metabolic conditions, such as hypoxia, low pH, and high carbon dioxide levels, can induce vasodilation to increase blood flow and oxygen delivery. Arterioles are also responsive to mechanical stimuli, like pressure and flow.

13.2.1.2 Veins

Veins (Fig. 13.18) transport blood back to the heart. Their unique structural features enable them to efficiently transport blood against gravity and accommodate changes in blood volume. Veins are composed of the same three concentric layers; however, the structural components and properties of each layer in veins are adapted for their low-pressure environment. The tunica media is generally thinner in veins than arteries, reflecting the lower pressure and the lesser need for active caliber regulation in the venous system. It primarily comprises circularly arranged smooth muscle cells interspersed with collagen and elastic fibers. One of the most significant structural features of veins, especially those in the limbs, is the presence of semilunar valves (Fig. 13.19). These valves are formed by folds of the tunica intima that project into

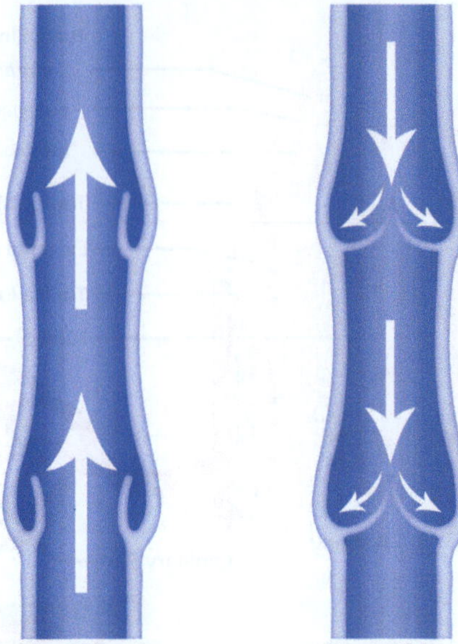

Fig. 13.19 Venous valve and the mechanism of function

the lumen of the vein. They prevent the backflow of blood, ensuring unidirectional flow towards the heart.

This is particularly crucial in the extremities, where blood must travel against gravity. Veins are highly compliant, meaning they can accommodate large volumes of blood with only modest increases in pressure. This is facilitated by the elasticity of the tunica adventitia and the relatively thin tunica media. Venous pressure is typically much lower than arterial pressure, with central venous pressure (CVP) in the supine position range of 2–8 mmHg.

Venules are conduits that collect blood from the capillary beds and channel it into larger veins. As the primary site of fluid and leukocyte exchange between the blood and surrounding tissues, venules play a crucial role in circulatory and immune system functions. Venules can be classified based on their diameter and function into postcapillary venules, which are the immediate recipients of blood from capillaries, ranging from 8 to 30 micrometers in diameter, and collecting venules, measuring between 30 and 50 micrometers in diameter and exhibiting a more developed wall structure with a few layers of smooth muscle cells, which transport blood to small veins.

13.2.1.3 Capillaries

Capillaries (Fig. 13.18) represent the circulatory system's smallest and most numerous blood vessels. These microscopic vessels significantly facilitate the exchange of oxygen, nutrients, and waste products between the bloodstream and surrounding tissues. Capillaries exhibit a unique structural design that allows for the efficient exchange of substances between blood and tissues: they consist of a single layer of endothelial cells. Endothelial cells are flattened, forming a thin barrier that facilitates the exchange of substances such as oxygen, nutrients, and waste products. Beneath the endothelial layer, capillaries often have a basement membrane composed of extracellular matrix proteins. The basement membrane provides structural support and helps regulate permeability. Unlike larger arteries and veins, capillaries lack tunica media (smooth muscle layer) and tunica adventitia (connective tissue layer). The vast network of capillaries provides an extensive surface area for exchange. This increased surface area allows for the efficient diffusion of substances. The thin endothelial layer of capillaries minimizes the distance over which substances must diffuse.

The capillaries are classified into continuous, fenestrated, and sinusoidal (discontinuous):

Continuous Capillaries

Continuous capillaries (Fig. 13.20) are the most common type of capillary found within the body's microvasculature. They are characterized by an uninterrupted endothelial lining without fenestrations, supported by a continuous basal lamina. These capillaries are found in most tissues, including muscle, skin, lungs, and the central nervous system. They form an extensive network of blood vessels that facilitate the exchange of gases, nutrients, and waste products between the bloodstream and tissues.

Fenestrated Capillaries

Fenestrated capillaries (Fig. 13.20) are a specialized subtype of capillaries that exhibit numerous pores or fenestrations within their endothelial lining. Each fenestration is a circular opening that spans 50–100 nanometers and is often covered by a thin diaphragm of glycoprotein. These capillaries are predominantly located in tissues with high rates of molecular exchange, such as endocrine glands, the intestinal mucosa, and the kidney glomeruli.

Sinusoid Capillaries

Sinusoid capillaries, also referred to as sinusoidal capillaries or sinusoids (Fig. 13.20), are specialized capillaries characterized by their wide, tortuous lumens and discontinuous nature, both in terms of their endothelial lining and often incomplete or absent basal lamina. These features facilitate the exchange of large molecules and cells between the blood and the surrounding tissue. Sinusoids are the least common type of capillary. Sinusoids can be five to ten times wider than typical capillaries, providing a large surface area for exchange. Their tortuous shape and irregular caliber slow down blood flow, which is crucial for their function in filtration and cellular exchange. Sinusoids are primarily located in the liver, spleen, bone marrow, lymph nodes, and some endocrine glands. Their presence correlates with the need for active filtration, phagocytosis, and cell migration.

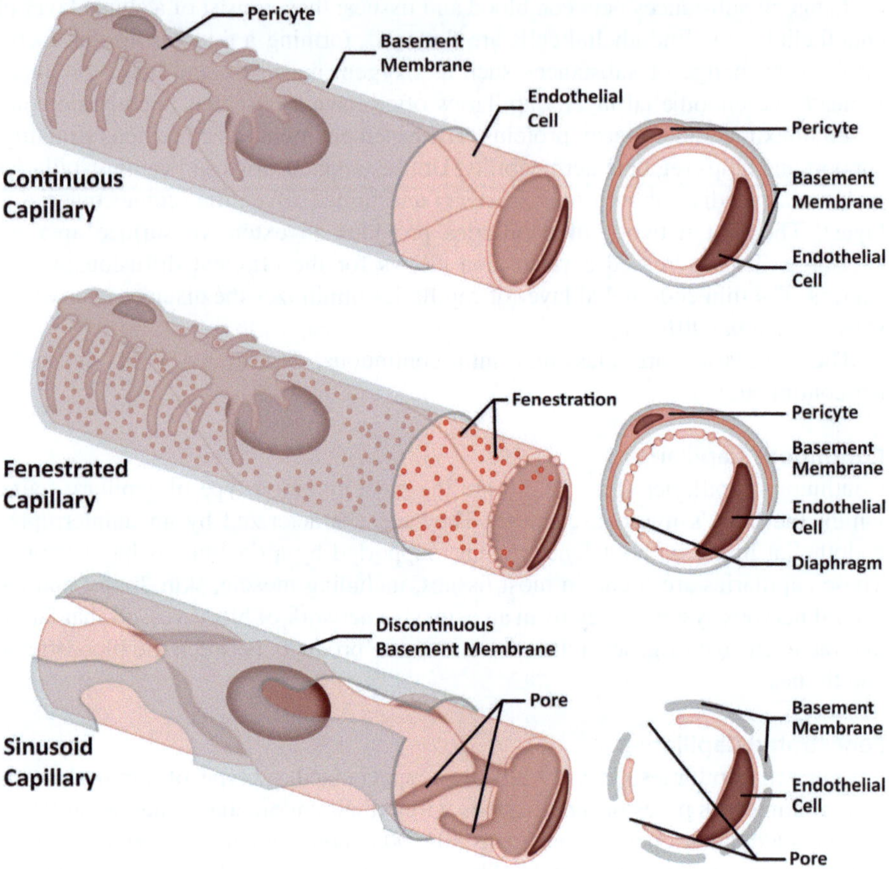

Fig. 13.20 Types of capillaries

Capillary Bed

The capillary bed (Fig. 13.21) represents the site within the circulatory system where blood perfusion reaches its finest and most intricate point. The bed is a branching network of capillaries functioning as the primary locale for exchanging gases, nutrients, and waste between the bloodstream and the tissues. The components of a capillary bed include (Fig. 13.22) terminal arterioles, serving as primary feeders into the capillary network and controlling blood entry through muscular sphincters, capillaries, the smallest blood vessels comprised of a single layer of endothelial cells where exchange with tissue cells occurs, postcapillary venules, which collect the blood from capillaries and start the return journey to the heart, precapillary sphincters, which are rings of smooth muscle at the origin of capillaries that regulate the entry of blood into each capillary, composed of a few layers of

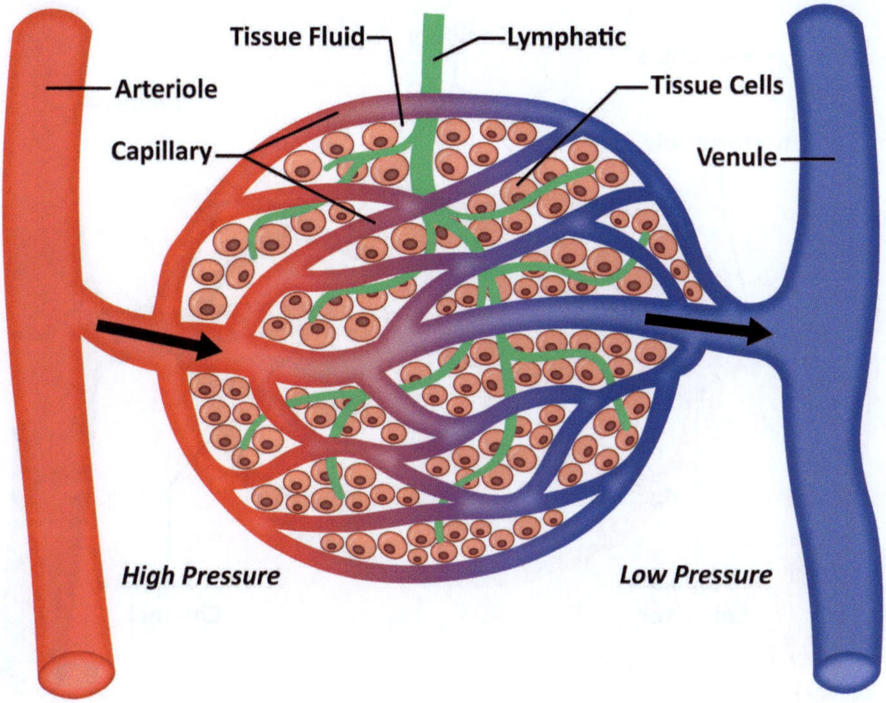

Fig. 13.21 Capillary bed

smooth muscle cells, capable of sustained contraction and relaxation, and metarterioles, which are small arterioles that form a transitional pathway between the arterial and venous ends of the microcirculation, functioning as a conduit for blood to bypass the capillary network via arteriovenous anastomoses during states of low perfusion demand and serve to control blood flow into the capillary network when demand increases. Typically, capillary pressures are estimated to be around 25–30 mmHg at the arteriolar end and 10–15 mmHg at the venular end.

The Starling equation describes the forces that govern fluid movement across the capillary wall, where blood pressure is a central component. Starling forces include:

- Capillary Hydrostatic Pressure (P_c): The physical force exerted by the blood pushes fluid out of the capillaries.
- Interstitial Hydrostatic Pressure (P_i): Often close to zero or negative, it slightly opposes the outward movement of fluid.
- Plasma Colloid Osmotic Pressure (π_c): Generated by plasma proteins, it draws fluid into the capillaries.
- Interstitial Colloid Osmotic Pressure (π_i): Exerted by interstitial proteins, it promotes fluid exit from the capillaries.

The balance of these forces leads to either filtration or reabsorption.

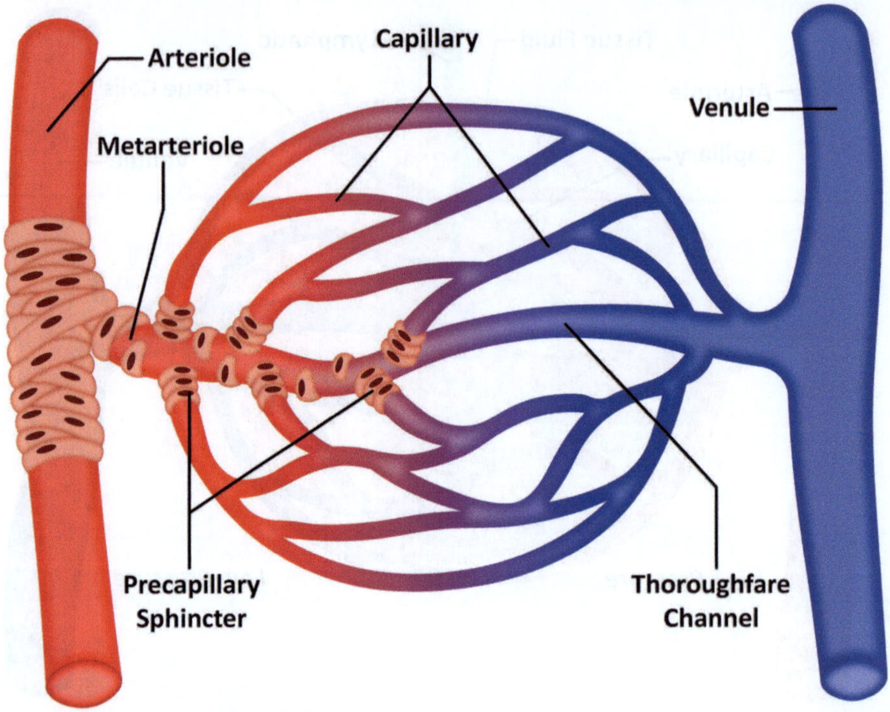

Fig. 13.22 Structural components of a capillary bed

13.2.2 Physiology of Circulation

The physiology of circulation encompasses the complex mechanisms that govern the movement of blood throughout the cardiovascular system. This intricate process involves the coordinated action of the heart, blood vessels, and various regulatory mechanisms to continuously deliver oxygen and nutrients and remove metabolic waste products to maintain tissue and organ function.

13.2.2.1 Regulation of Circulation

Central Regulation Cardiovascular center, located in the medulla oblongata, integrates sensory input and orchestrates autonomic responses to modulate heart rate and vascular tone. Baroreceptors are specialized sensory neurons that respond to changes in blood pressure (BP) by adjusting heart rate (HR) and vascular resistance. Baroreceptors (Fig. 13.23) are primarily located in the carotid sinuses and the aortic arch, with additional receptors found in the walls of nearly all large thoracic and cervical arteries. The carotid sinus, located at the bifurcation of the common carotid arteries, contains high-pressure baroreceptors. Aortic arch baroreceptors are positioned within the arch and major branches of the aorta. Baroreceptors consist of nerve endings that intertwine with the elastic fibers within the arterial walls, mainly

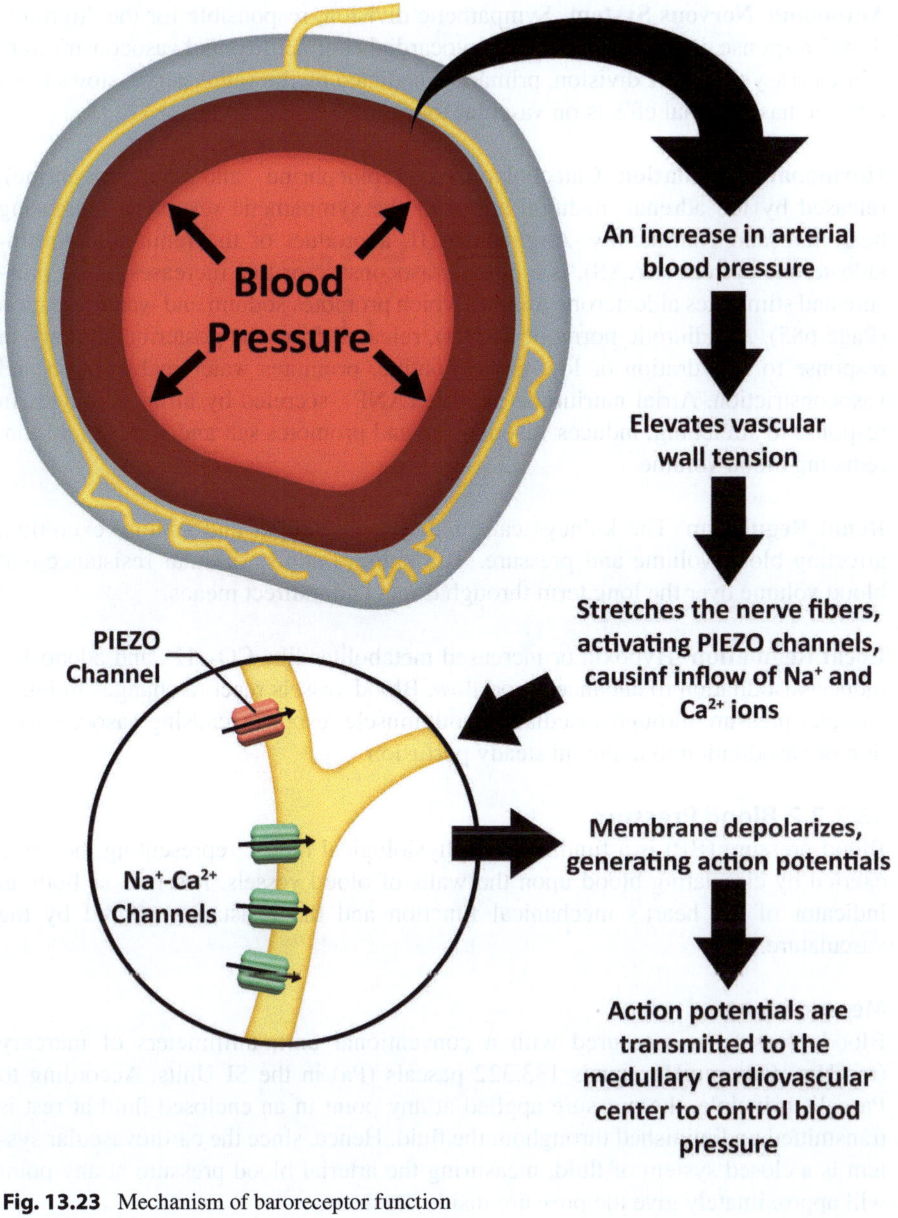

Fig. 13.23 Mechanism of baroreceptor function

concentrated in regions subjected to pulsatile stretch. An increase in blood pressure distorts the arterial wall, activating the baroreceptors. Carotid baroreceptors have a lower threshold for activation and are more sensitive to changes in blood pressure than aortic baroreceptors. Peripheral chemoreceptors, sensitive to blood oxygen, carbon dioxide, and pH levels, initiate reflexes to adjust cardiovascular parameters.

Autonomic Nervous System Sympathetic division, responsible for the "fight-or-flight" response, increases heart rate, myocardial contractility, and vasoconstriction, while parasympathetic division, primarily mediated by the vagus nerve, slows heart rate and has minimal effects on vascular tone.

Hormonal Regulation Catecholamines (epinephrine and norepinephrine), released by the adrenal medulla, augment the sympathetic response, increasing heart rate and contractility. Angiotensin II, a product of the renin-angiotensin-aldosterone system (RAAS), is a potent vasoconstrictor that increases blood pressure and stimulates aldosterone release, which promotes sodium and water retention (Page 685). Antidiuretic hormone (ADH), released from the posterior pituitary in response to dehydration or low blood volume, promotes water reabsorption and vasoconstriction. Atrial natriuretic peptide (ANP), secreted by atrial myocytes in response to stretching, induces vasodilation and promotes salt and water excretion, reducing blood volume.

Renal Regulation The kidneys can modify urine output and sodium excretion, affecting blood volume and pressure. The kidneys adjust vascular resistance and blood volume over the long term through direct and indirect means.

Local Regulation Hypoxia or increased metabolites like CO_2, H^+, and adenosine induce vasodilation to enhance blood flow. Blood vessels react to changes in intravascular pressure through vascular smooth muscle response, causing vasoconstriction or vasodilation to maintain steady perfusion.

13.2.2.2 Blood Pressure

Blood pressure (BP) is a fundamental physiological metric representing the force exerted by circulating blood upon the walls of blood vessels. It serves as both an indicator of the heart's mechanical function and the resistance offered by the vasculature.

Measurement

Blood pressure is measured with a conventional unit, millimeters of mercury (mmHg). One mmHg equals 133.322 pascals (Pa) in the SI Units. According to Pascal's principle, the pressure applied at any point in an enclosed fluid at rest is transmitted undiminished throughout the fluid. Hence, since the cardiovascular system is a closed system of fluid, measuring the arterial blood pressure at any point will approximately give the pressure inside the heart.

Systolic Blood Pressure (SBP): The maximum pressure exerted on the arteries during the heart muscle contraction.

Diastolic Blood Pressure (DBP): The minimum pressure within the arteries when the heart is relaxed and filling with blood.

Pulse Pressure: Pulse pressure is calculated as the difference between SBP and DBP. Increased pulse pressure increases cardiovascular risk, particularly in older adults.

Mean Arterial Pressure (MAP): MAP is the average pressure in the arteries during one cardiac cycle and is considered a better indicator of perfusion to vital organs than SBP or DBP alone. The formula can approximate MAP:

$$\text{MAP} = \text{DBP} + 1/3(\text{SBP} - \text{DBP}).$$

Regulation of Blood Pressure

Blood pressure regulation ensures the proper functioning of the cardiovascular system and maintains blood flow to various tissues and organs. It involves a complex interplay of mechanisms that control systemic blood pressure (the pressure in the arteries throughout the body) and local blood flow to specific tissues.

Baroreceptor Reflex Baroreceptors (Fig. 13.23) are specialized stretch receptors located in the walls of certain arteries, primarily in the carotid sinuses and the aortic arch. These receptors continuously monitor blood pressure by detecting changes in arterial wall stretch. When blood pressure increases (e.g., during physical activity or stress), the baroreceptors are stretched more, leading to increased firing of action potentials. This increased firing is transmitted to the cardiovascular control center in the brain, primarily the medulla oblongata. In response, the cardiovascular control center initiates a reflex that decreases sympathetic nervous system activity and increases parasympathetic nervous system activity, decreasing heart rate and relaxation of blood vessels (vasodilation), reducing blood pressure. Conversely, when blood pressure decreases, the opposite response occurs, raising blood pressure back to normal levels.

The carotid body (Fig. 13.24) is typically found bilaterally, situated at the point where the common carotid artery bifurcates into the internal and external carotid arteries. This location ensures that it has access to oxygenated blood directly from the heart and can quickly respond to changes in blood composition. The carotid body is a small, oval, or round-shaped structure, roughly measuring 3–5 mm in diameter. Its size may vary among individuals and can change in response to factors like chronic hypoxia. The carotid body is composed of several cell types. Type I or glomus cells are the primary sensory cells of the carotid body and are responsible for detecting changes in blood gas concentrations (Section 15.2.7). They are richly supplied with blood vessels and contain numerous mitochondria. Sustentacular (supporting) cells provide structural support and nourishment to the glomus cells. Baroreceptor nerve endings are specialized sensory nerve endings located within the carotid body, sensitive to mechanical deformation or stretch of the arterial wall. The carotid body is innervated by sensory nerve fibers, which transmit sensory information from the carotid body to the central nervous system. The carotid body has an extensive blood supply, which is critical for its function. The rich blood flow ensures that the carotid body can continuously monitor blood gas composition and blood pressure. The carotid body is innervated by sensory nerve fibers of the glossopharyngeal nerve (cranial nerve IX). These nerve fibers transmit sensory information from the carotid body to the central nervous system, primarily to the medulla oblongata.

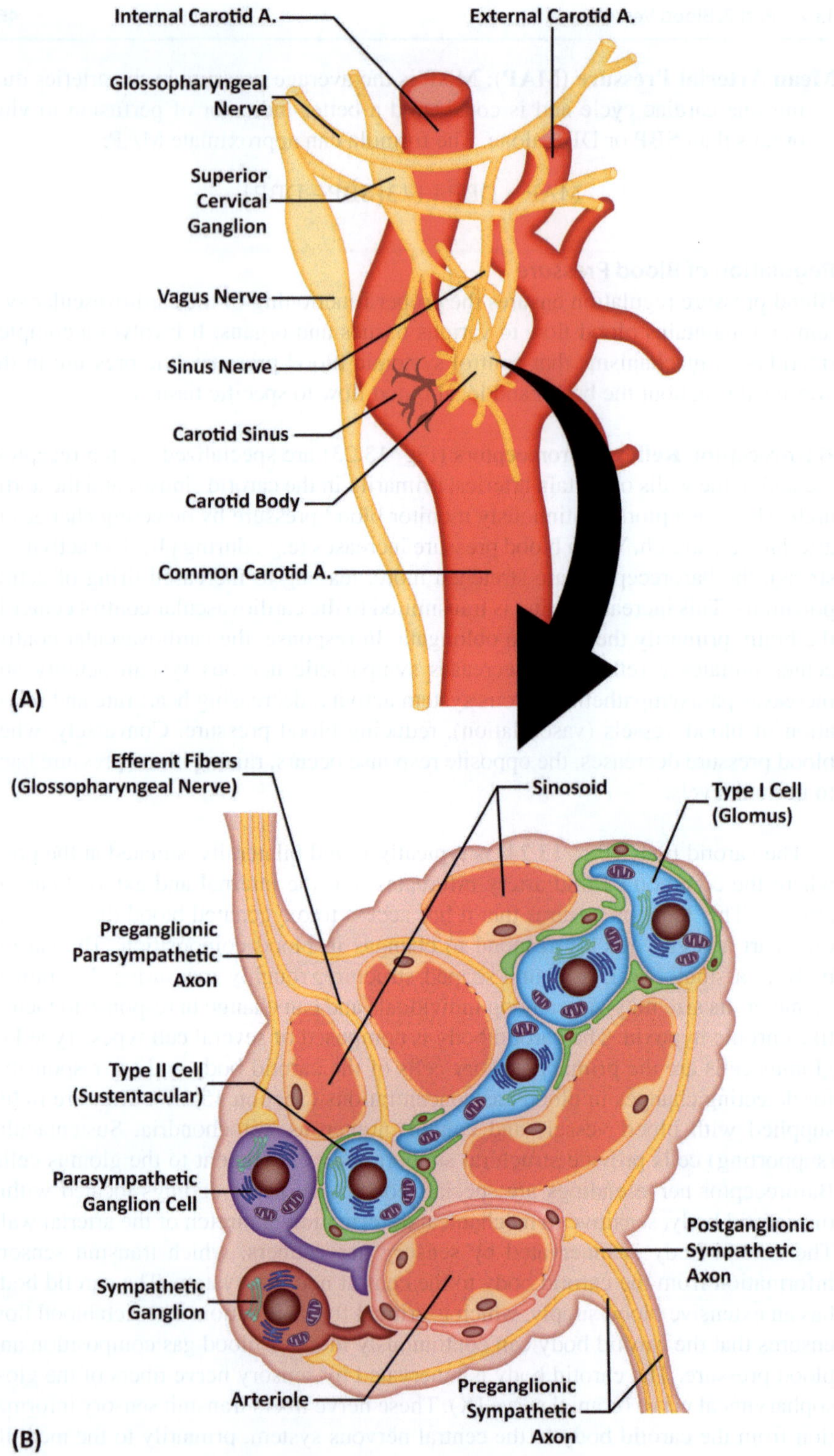

Fig. 13.24 Carotid body: (**a**) Gross anatomic structure. (**b**) Microscopic structure

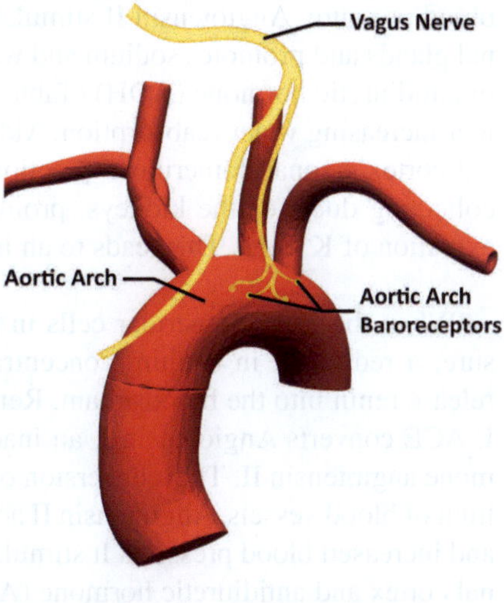

Fig. 13.25 Aortic arch baroreceptors

The aortic arch baroreceptors (Fig. 13.25) are situated within the wall of the aortic arch, which is the curved portion of the aorta as it exits the heart. These baroreceptors are nerve endings located within the walls of the aortic arch. These endings are specialized nerve fibers sensitive to mechanical stretch, which occurs when the arterial wall expands or contracts due to changes in blood pressure. The supportive connective tissue surrounding the baroreceptor endings and nerve fibers helps anchor and protect the sensory structures within the arterial wall. The sensory information from the baroreceptor endings is transmitted by afferent nerve fibers, part of the glossopharyngeal nerve (cranial nerve IX). The aortic arch baroreceptors function by detecting changes in arterial pressure caused by variations in blood volume and cardiac output. When blood pressure increases, the arterial walls stretch, leading to mechanical deformation of the baroreceptor endings. This mechanical deformation is detected by the sensory nerve endings, which are sensitive to changes in stretch or distension.

Renin-Angiotensin-Aldosterone System The renin-angiotensin-aldosterone system (RAAS) is a hormonal system that regulates blood pressure, electrolyte balance, and fluid balance. The RAAS involves a series of enzymatic reactions and the release of several essential hormones. Renin is an enzyme produced and released by juxtaglomerular cells (Section 17.2.3) in the kidneys. Renin is the initiating enzyme in the RAAS cascade. Angiotensinogen is an inactive precursor globulin protein synthesized by the liver and released into the bloodstream. It circulates in the blood until it is acted upon by renin. Renin cleaves angiotensinogen to produce angiotensin I, an inactive peptide. Angiotensin-converting enzyme (ACE) is predominantly found in the lungs and endothelial cells of blood vessels. ACE converts inactive angiotensin I into the active form, angiotensin II. Angiotensin II is a biologically active hormone that exerts numerous effects on the cardiovascular and renal systems. It is a potent vasoconstrictor, causing blood vessels to narrow and increase blood pressure. Angiotensin II stimulates the release of aldosterone from the adrenal glands and promotes sodium and water retention. It also stimulates the secretion of antidiuretic hormone (ADH) (Table 12.2) from the posterior pituitary gland, further increasing water reabsorption. Aldosterone is a hormone produced by the adrenal cortex's zona glomerulosa (Section 12.2.5.2). It acts on the distal tubules and collecting ducts of the kidneys, promoting the reabsorption of Na^+ ions and the excretion of K^+ ions. This leads to an increase in blood volume and blood pressure.

When the juxtaglomerular cells in the kidneys sense a decrease in blood pressure, a reduction in sodium concentration, or a decrease in blood volume, they release renin into the bloodstream. Renin cleaves angiotensinogen into angiotensin I. ACE converts Angiotensin I, an inactive peptide, to the biologically active hormone angiotensin II. This conversion occurs primarily in the lungs and the endothelium of blood vessels. Angiotensin II acts on blood vessels, causing vasoconstriction and increased blood pressure. It stimulates the release of aldosterone from the adrenal cortex and antidiuretic hormone (ADH) from the posterior pituitary gland.

Autoregulation Autoregulation is the ability of local tissues to regulate their blood flow. It ensures that organs and tissues receive a constant blood supply, regardless of changes in systemic blood pressure. It prevents excessive blood flow when systemic blood pressure is high and maintains sufficient blood flow when it is low.

Autoregulation works through two mechanisms: myogenic and metabolic mechanisms. In response to changes in blood pressure, vascular smooth muscle cells within arterioles can contract or relax to regulate blood flow (myogenic mechanism). When blood pressure increases, arteriolar smooth muscle cells contract to reduce blood flow; when blood pressure decreases, they relax to increase blood flow. Tissues release metabolic byproducts (e.g., adenosine, carbon dioxide) due to increased metabolic activity. These byproducts can cause vasodilation by relaxing smooth muscle cells in the arterioles, increasing blood flow to meet the tissue's demands (metabolic mechanism).

The brain is susceptible to changes in blood pressure because it requires a constant and reliable blood supply to function correctly. Cerebral autoregulation maintains cerebral blood flow within a relatively narrow range, allowing the brain to receive adequate oxygen and nutrients under varying systemic blood pressure conditions.

The kidneys are essential for regulating blood pressure and fluid balance. Renal autoregulation ensures the kidneys receive consistent blood flow for filtration and waste removal. The myogenic and tubuloglomerular feedback mechanisms play critical roles in renal autoregulation (Section 17.3.1.2).

The coronary arteries that supply blood to the heart muscle also exhibit autoregulation. When the heart's metabolic demands increase during exercise, coronary autoregulation ensures it receives adequate blood flow to meet its needs.

Hormonal Regulation Besides aldosterone and ADH, other hormones also play significant roles in controlling blood pressure. Epinephrine and norepinephrine (released by the adrenal medulla) can rapidly increase blood pressure during the "fight-or-flight" response. Atrial natriuretic peptide (ANP), released by the heart's atria in response to high blood pressure or volume, promotes vasodilation and increases sodium excretion, reducing blood volume and pressure.

Central Nervous System Regulation The central nervous system, particularly the medulla oblongata, integrates input from baroreceptors, chemoreceptors (detecting oxygen and carbon dioxide levels), and other sensory receptors to fine- tune blood pressure.

13.2.2.3 Blood Flow

Blood flow is the continuous blood circulation in the cardiovascular system, driven by the heart's pumping action. Rheology, the study of the flow of matter, is fundamental in understanding the behavior of blood as it circulates through the vascular system. Factors governing the blood flow, especially in the capillaries, include the Fahraeus-Lindqvist effect, which describes the reduction in apparent viscosity as blood flows through capillaries (this effect facilitates blood flow in microvessels despite the high hematocrit), and Poiseuille's law, also known as the Hagen-Poiseuille equation, which describes the flow of a viscous fluid in a long, cylindrical pipe. The derivation of Poiseuille's law begins with the assumption that the flow is steady, incompressible, and fully developed (meaning the velocity profile does not change along the length of the pipe). The flow must also be laminar, consisting of layers that do not mix turbulently, and the pipe must be circular. Under these conditions, the flow rate (Q) can be determined by the following equation:

$$Q = \frac{\pi r^4 \Delta P}{8\mu L}$$

where:

- Q is the volumetric flow rate
- r is the radius of the pipe
- ΔP is the pressure difference between the two ends of the pipe
- μ is the dynamic viscosity of the fluid
- L is the length of the pipe over which the pressure difference is measured

From the law, we can deduce that the flow rate is proportional to the fourth power of the radius. Hence, even small changes in the vessel's radius significantly impact the flow rate.

13.2.3 Circulatory Pathways

The human circulatory system can be categorized into two distinct but interrelated circulatory pathways: pulmonary and systemic (Fig. 13.26). Two specific subtypes of systemic circulation are portal circulation and nephron circulation.

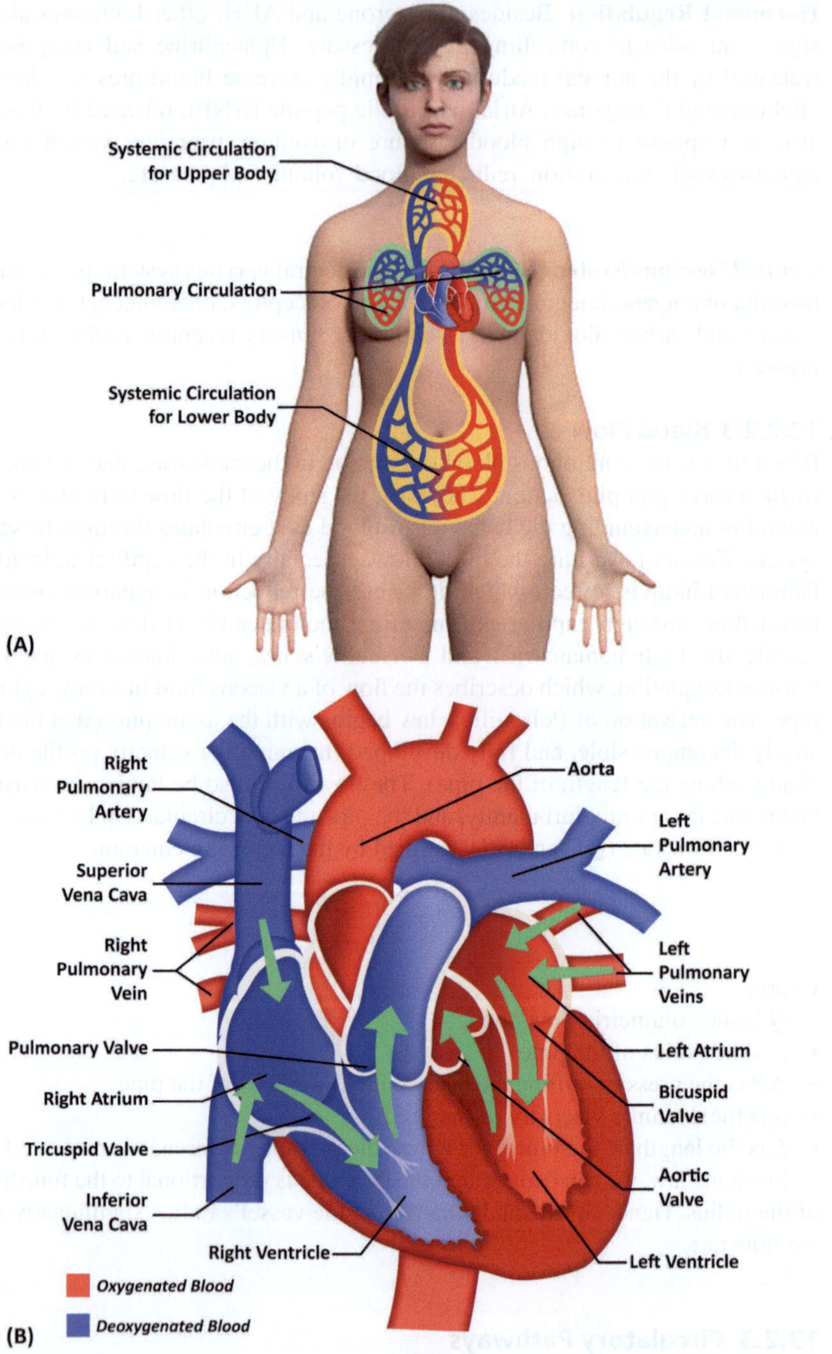

Fig. 13.26 (**a**) Pulmonary and systemic circulations. (**b**) Circulation in heart chambers

13.2.3.1 Pulmonary Circulation

The pulmonary circulatory pathway (Fig. 13.27) is specifically designed to transport deoxygenated blood from the heart to the lungs and return oxygenated blood to the heart. The journey begins in the right ventricle of the heart. As the right ventricular muscle contracts, it generates enough force to push the deoxygenated blood into the pulmonary artery through the pulmonary valve. The pulmonary artery is unique as it is one of the few arteries in the body that carries deoxygenated blood. It exits the heart and bifurcates into the right and left pulmonary arteries corresponding to the two lungs. As the right and left pulmonary arteries reach the lungs, they subdivide into smaller and smaller arteries and, eventually, arterioles. This branching network ensures blood is distributed to the entire lung tissue. The arterioles further branch into a dense network of capillaries that envelop the alveoli, which are tiny air sacs in the lungs. It is here that the critical exchange of gases takes place. The deoxygenated blood releases carbon dioxide into the alveoli and absorbs oxygen from the inhaled air.

The oxygen-rich blood is now collected into venules, which merge into larger pulmonary veins. Unlike systemic veins, pulmonary veins carry oxygenated blood. The oxygenated blood from the lungs is then returned to the heart via four pulmonary veins. Two veins from each lung drain into the left atrium. Once the left atrium fills with blood, it contracts, pushing the blood into the left ventricle through the

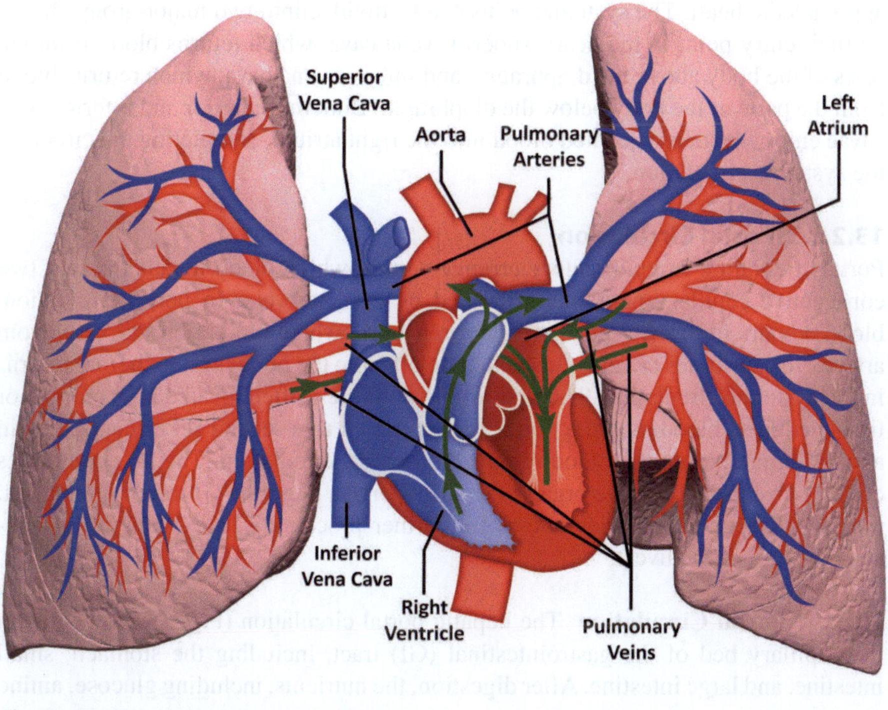

Fig. 13.27 Detailed pulmonary circulation

mitral valve. The pulmonary arteries exhibit a lower vascular resistance compared to systemic vessels, accommodating the entire cardiac output with minimal pressure elevation.

13.2.3.2 Systemic Circulation

Systemic circulation represents the segment of the cardiovascular system that carries oxygenated blood away from the heart to the body's tissues and returns deoxygenated blood to the heart. Systemic circulation originates from the left ventricle, which ejects oxygenated blood into the aorta during systole. The aorta arches upwards as the ascending aorta, then curves posteriorly and downward as the aortic arch, and finally extends down the thoracic and abdominal cavity as the descending aorta. The aorta continues as the thoracic aorta, which gives rise to numerous branches to the chest wall and organs within the thorax. After passing through the diaphragm, it is called the abdominal aorta. It continues into the abdomen, supplying abdominal organs and splitting into the common iliac arteries to serve the pelvis and lower limbs.

The arteries branch into smaller arterioles as they reach the tissues. The arterioles further branch into capillaries, the smallest blood vessels in the body. The walls of capillaries are thin enough to allow for the exchange of nutrients, gases, and waste products between the blood and the surrounding tissues. The deoxygenated blood then begins its return journey via venules, which are small veins that drain blood from the capillaries and then merge into larger veins. Veins grow larger as they approach the heart. The systemic veins can be divided into two major groups based on their entry point to the heart: superior vena cava, which returns blood from the parts of the body above the diaphragm, and inferior vena cava, which returns blood from the parts of the body below the diaphragm. Both the superior and inferior vena cavae empty the deoxygenated blood into the right atrium, completing the circuit of the systemic circulation.

13.2.3.3 Portal Circulation

Portal circulation is a circulatory arrangement in which blood travels through two consecutive capillary networks separated by a portal vein. In portal circulation, blood detours from the typical circulation pattern, where blood flows directly from arteries to capillaries to veins. Instead, it diverts into the portal system before returning to systemic circulation. The first set of capillaries, often located in one organ or tissue, collects blood, and the portal vein carries it to a second set of capillaries in another organ before the blood rejoins the general circulation. The first capillaries collect substances from the organ or tissue of origin (e.g., the GI tract). In contrast, the second set of capillaries allows for further processing or exchanging of substances (e.g., in the liver).

Hepatic Portal Circulation The hepatic portal circulation (Fig. 13.28) begins in the capillary bed of the gastrointestinal (GI) tract, including the stomach, small intestine, and large intestine. After digestion, the nutrients, including glucose, amino

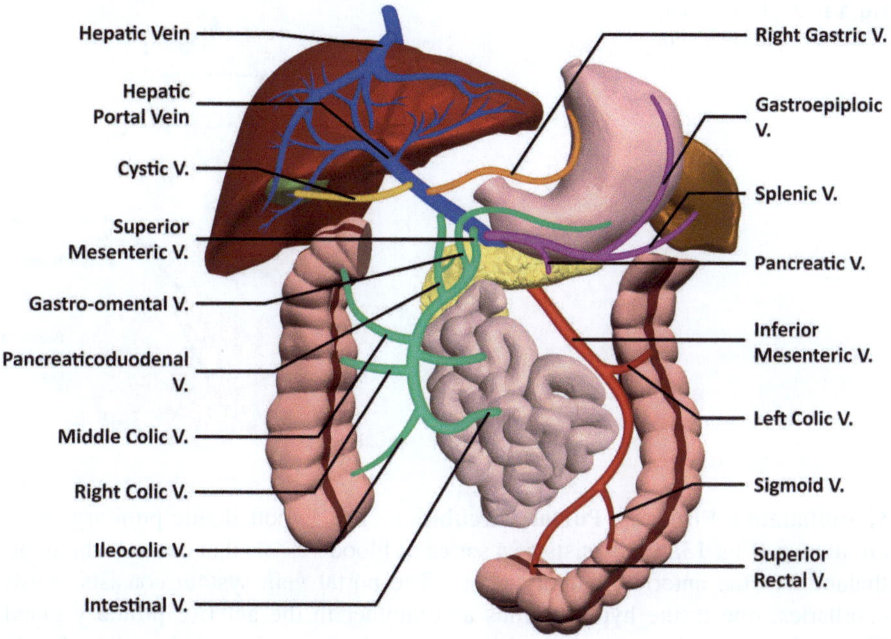

Fig. 13.28 Hepatic portal circulation

acids, and some other products of digestion, are absorbed into the bloodstream in the GI capillaries. The blood rich in nutrients and other substances from the GI tract flows into the hepatic portal vein. The hepatic portal vein is a large blood vessel that carries blood from the GI organs directly to the liver. It is formed by the convergence of several veins, including the superior mesenteric vein (draining the small intestine) and the splenic vein (draining the spleen and part of the stomach). The hepatic portal vein enters the liver and branches into a network of tiny blood vessels called hepatic sinusoids (Page 655) within the liver's functional units known as hepatic lobules. As blood flows through the hepatic sinusoids, hepatocytes metabolize and process the nutrients, detoxify harmful substances, and store excess nutrients.

After passing through the hepatic sinusoids and undergoing metabolic processing, the hepatic veins collect blood within the hepatic portal circulation. Hepatic veins converge into a single large vein called the hepatic vein. The hepatic vein eventually joins the inferior vena cava, a major blood vessel part of the systemic circulation.

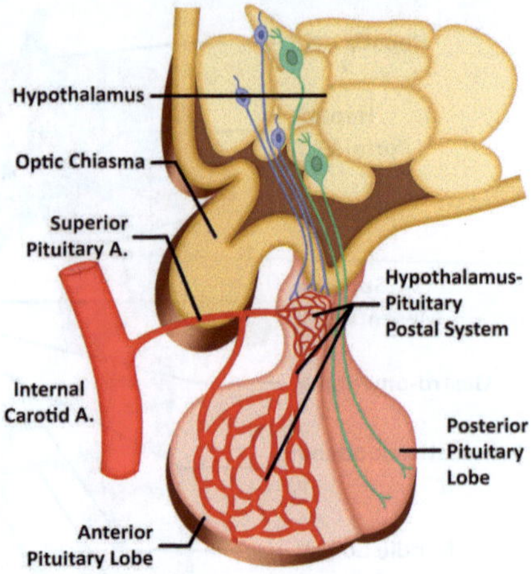

Fig. 13.29 Hypothalamic-pituitary portal circulation

Hypothalamic-Pituitary Portal Circulation The hypothalamic-pituitary portal circulation (Fig. 13.29) consists of a series of blood vessels that connect the hypothalamus to the anterior pituitary gland. The portal vein system consists of two capillaries: one in the hypothalamus and another in the anterior pituitary gland. These capillary beds are connected by a network of portal veins that allow for the direct exchange of hormones between the hypothalamus and the anterior pituitary.

13.2.3.4 Nephron Circulation

Nephron circulation (Fig. 13.30) is a unique circulatory bed that contains a capillary bed between two arterioles (instead of one arteriole and one vein) and two capillary beds connected by an arteriole. The components of nephron circulation are afferent arteriole, glomerulus, efferent arteriole, peritubular capillaries, and vasa recta (Section 17.2.1.2).

The afferent arteriole is a small renal artery branch that supplies blood to the nephron. It carries oxygenated blood containing waste products and nutrients to the nephron. The afferent arteriole enters the renal corpuscle, forming a unique network of tiny blood capillaries called the glomerulus. Blood exits the glomerulus through the efferent arteriole.

Positioned between two arterioles, capable of effectively changing their diameter, it provides a unique structural and functional configuration that can control the blood pressure inside the glomerulus. The efferent arteriole can either be a high-resistance arteriole (afferent and efferent arterioles have similar diameters) or a low-resistance arteriole (efferent arteriole has a smaller diameter than the afferent arteriole). After leaving the glomerulus, the efferent arteriole gives rise to a network of capillaries called the peritubular capillaries, surrounding the renal tubules (proximal convoluted tubule, loop of Henle, distal convoluted tubule, and collecting duct) throughout the nephron. In juxtamedullary nephrons (Section 17.2.1.2), peritubular capillaries have a unique structure called the vasa recta. Blood from the peritubular capillaries and vasa recta collects into venules, eventually merging to form larger veins and draining into the systemic circulation.

Fig. 13.30 Nephron circulation

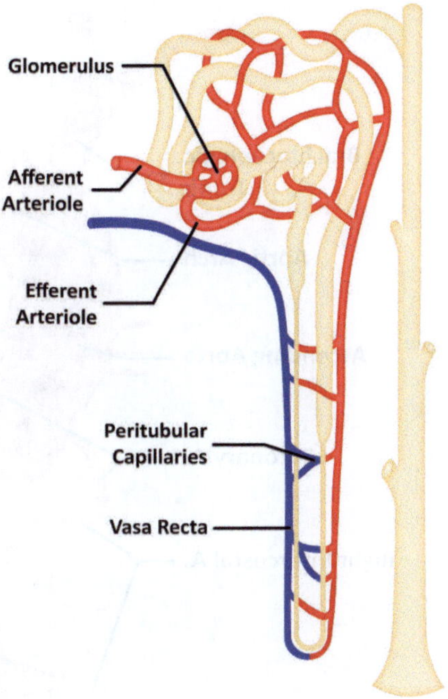

13.2.4 Gallery of Vessels (Tables 13.1, 13.2, 13.3, 13.4, 13.5, 13.6, 13.7, 13.8, 13.9, 13.10, 13.11, 13.12, 13.13 and 13.14)

13.3 Part 3: Blood

Blood, the "river of life," fulfills many crucial bodily functions. Its roles can be categorized broadly into transportation, regulation, and protection, each encompassing several intricate processes:

Transportation One of the primary functions of blood is to transport oxygen from the lungs to the tissues and organs. Hemoglobin, a protein in red blood cells, binds oxygen in the lungs and releases it in the tissues. Blood also carries nutrients, including glucose, amino acids, and fatty acids, absorbed from the digestive tract to various body parts, which are utilized for energy, growth, and repair. Blood plays a critical role in removing waste products of metabolism, like carbon dioxide and urea, transporting them to the lungs and kidneys for excretion. The blood carries hormones that the endocrine glands produce to target organs, facilitating communication between different body parts.

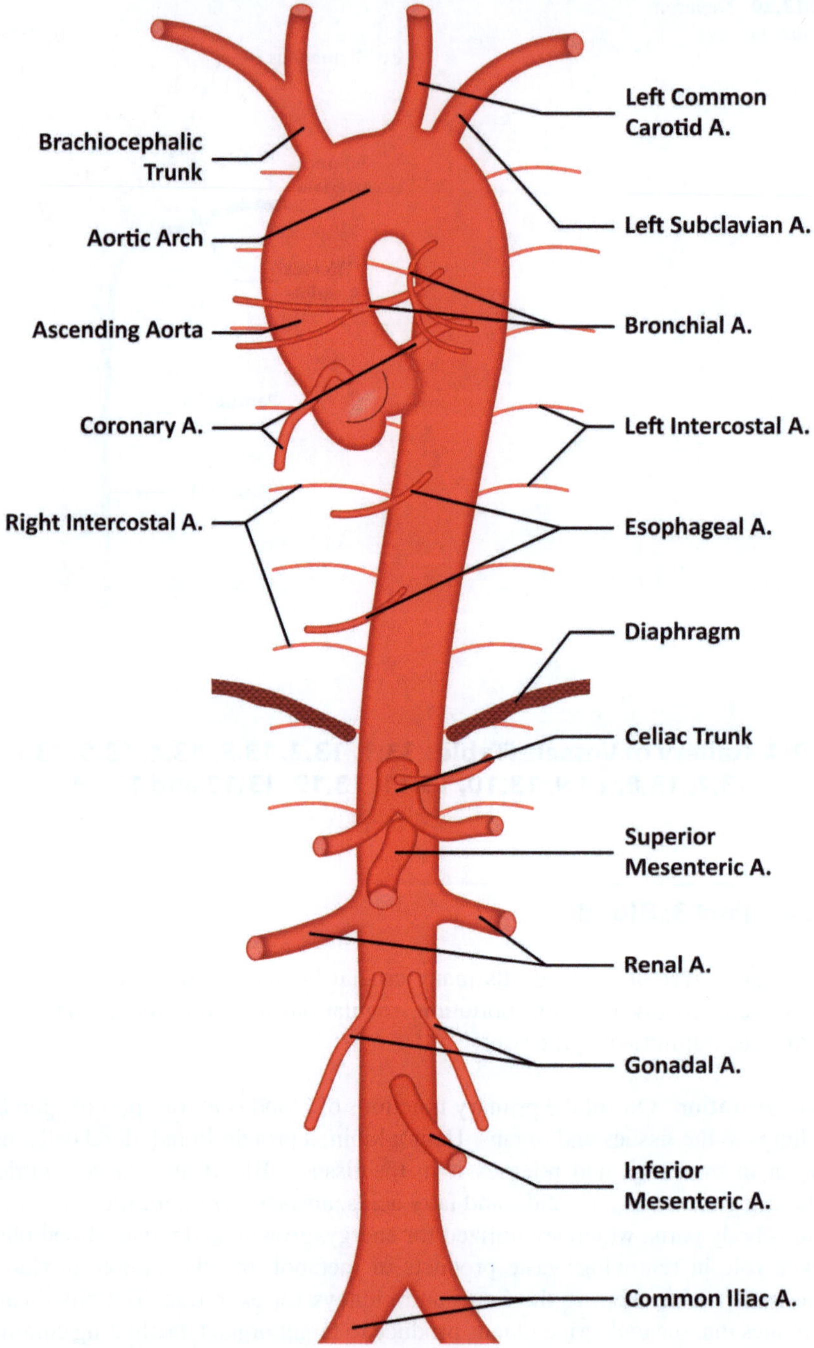

Fig. 13.31 Major branches of the aorta

Table 13.1 Branches of aorta: origin, course, and supply distribution (Fig. 13.31)

Name of branch	Origin	Course	Function
Coronary arteries	Ascending aorta	Arise near the start of the aorta from the aortic sinuses	Supply blood to the heart muscle
Brachiocephalic trunk	Aortic arch	Extends upward and to the right before bifurcating into the right subclavian and right common carotid arteries	Supplies blood to the right side of the head, neck, and right arm
Left common carotid artery	Aortic arch	Ascends next to the trachea to the neck, where it bifurcates into the internal and external carotid arteries	Supplies blood to the left side of the head and neck
Left subclavian artery	Aortic arch	Passes laterally toward the left shoulder, eventually becoming the axillary artery	Supplies blood to the left arm, with branches to the head and thorax
Bronchial arteries	Descending thoracic aorta	Small branches that directly enter the lung tissue from the posterior surface of the aorta	Provide oxygenated blood to the lung tissue for nourishment
Esophageal arteries	Descending thoracic aorta	Small branches that descend along the esophagus	Supply blood to the esophagus
Mediastinal arteries	Descending thoracic aorta	Numerous small branches supplying the mediastinum	Supply blood to the lymph nodes and other structures in the mediastinum
Celiac trunk	Abdominal aorta	A short, wide branch that trifurcates into the left gastric, splenic, and common hepatic arteries	Supplies blood to the stomach, liver, spleen, and other abdominal organs
Superior mesenteric artery	Abdominal aorta	Extends horizontally to supply the intestine	Supplies blood to the small intestine and part of the large intestine
Renal arteries	Abdominal aorta	Lateral branches that extend to the kidneys	Supply blood to the kidneys
Gonadal arteries (testicular or ovarian)	Abdominal aorta	Descend to the pelvis to supply the testes or ovaries	Supply blood to the male and female gonads
Inferior mesenteric artery	Abdominal aorta	Extends downward to supply the distal part of the large intestine	Supplies blood to the descending colon, sigmoid colon, and rectum
Common iliac arteries	Abdominal aorta (bifurcation)	At the level of the fourth lumbar vertebra, the aorta bifurcates into these arteries	Supply blood to the pelvis and lower limbs

Regulation Blood helps maintain the body's pH balance by buffering acids and bases. It also regulates body temperature by redistributing heat and maintaining water balance by transporting water and its dissolved substances, thus influencing the cellular environment.

Protection White blood cells, or leukocytes, in the blood, are crucial for the immune response. They identify and attack pathogens like bacteria, viruses, and

Table 13.2 Arteries supplying head and neck: origin, course, branches, and supply distribution (Fig. 13.32)

Name of artery	Origin	Course	Branches	Function
Internal carotid artery	Common carotid artery	Enters the skull through the carotid canal; does not branch in the neck	Ophthalmic artery, anterior cerebral artery, middle cerebral artery, posterior communicating artery	Supplies blood to the brain and eyes
External carotid artery	Common carotid artery	Ascends the neck, dividing into multiple branches	Superior thyroid artery, lingual artery, facial artery, occipital artery, posterior auricular artery, maxillary artery, superficial temporal artery	Supplies blood to the anterior (front) part of the scalp, face, and neck
Vertebral artery	Subclavian artery	Ascends through the transverse foramina of the cervical vertebrae, then through the foramen magnum into the cranium	Anterior spinal artery, posterior spinal artery, basilar artery	Supplies blood to the posterior part of the brain
Basilar artery	Union of the two vertebral arteries at the base of the skull	Travels along the ventral surface of the brainstem	Pontine arteries, labyrinthine artery, superior cerebellar artery, posterior cerebral artery	Supplies blood to the brainstem and cerebellum
Thyrocervical trunk	Subclavian artery	Short trunk that branches off the subclavian artery	Inferior thyroid artery, ascending cervical artery, suprascapular artery, transverse cervical artery	Supplies blood to thyroid gland, neck muscles, and upper back
Costocervical trunk	Subclavian artery	Branches near the neck	Deep cervical artery, supreme intercostal artery	Supplies blood to the deep muscles of the neck and upper intercostal spaces
Lingual artery	External carotid artery	Runs along the underside of the tongue	Dorsal lingual branches, deep lingual artery, sublingual artery	Supplies blood to the floor of the mouth and tongue
Facial artery	External carotid artery	Winds along the face to the medial angle of the eye	Superior and inferior labial arteries, lateral nasal branches, angular artery	Supplies blood to the facial structures, including muscles and skin
Maxillary artery	External carotid artery	Passes through the infratemporal fossa	Middle meningeal artery, inferior alveolar artery, and others	Supplies deep facial structures, teeth, maxilla, mandible, and dura mater
Superficial temporal artery	External carotid artery	Ascends in front of the ear	Transverse facial artery, anterior auricular branches, zygomatico-orbital artery	Supplies blood to the lateral aspect of the scalp

Table 13.3 Arteries supplying upper limb: origin, course, branches, and supply distribution (Fig. 13.33)

Name of artery	Origin	Course	Branches	Function
Axillary artery	Continuation of the subclavian artery after the first rib	Runs through the axilla	Superior thoracic artery, thoracoacromial artery, lateral thoracic artery, subscapular artery, anterior circumflex humeral artery, posterior circumflex humeral artery	Supplies the axilla, shoulder, and proximal part of the upper arm
Brachial artery	Continuation of the axillary artery at the lower margin of the teres major muscle	Runs down the anteromedial aspect of the upper arm to the elbow	Profunda brachii artery, nutrient arteries to the humerus, muscular branches, superior ulnar collateral artery, inferior ulnar collateral artery	Main blood supply to the upper arm; continues to the forearm
Radial artery	Terminal branch of the brachial artery, starting at the cubital fossa	Runs laterally down the forearm to the wrist	Radial recurrent artery, palmar carpal branch, superficial palmar branch, dorsal carpal branch, princeps pollicis artery, radialis indicis artery	Supplies the lateral aspect of the forearm and hand
Ulnar artery	Terminal branch of the brachial artery, starting at the cubital fossa	Runs medially down the forearm, anterior to the flexor carpi ulnaris muscle	Common interosseous artery, palmar carpal branch, dorsal carpal branch, superficial palmar branch, deep palmar branch	Supplies the medial aspect of the forearm and hand
Deep palmar arch	Mainly from the radial artery	Located deep in the palm, closer to the metacarpals	Palmar metacarpal arteries, perforating arteries, recurrent branches	Supplies the deep structures of the hand, the palmar metacarpal area
Superficial palmar arch	Mainly from the ulnar artery	Formed superficially in the palm	Common palmar digital arteries, proper palmar digital arteries	Supplies the palm and fingers

other foreign bodies. Blood contains platelets and clotting factors that prevent excessive bleeding by forming clots. The circulatory system also plays a role in surveillance, with specific cells in the blood identifying and eliminating cancerous cells.

13.3.1 Components

Blood is more than just a simple fluid; it is a complex mixture with numerous components. The blood components include plasma and the formed elements (Fig. 13.43).

Table 13.4 Arteries supplying thorax: origin, course, branches, and supply distribution (Fig. 13.33)

Name of artery	Origin	Course	Branches	Function
Internal thoracic (mammary) artery	Subclavian artery	Descends behind the ribs, about 1–2 cm lateral to the sternum	Pericardiophrenic artery, anterior intercostal arteries, musculophrenic artery, superior epigastric artery	Supplies the anterior chest wall and breasts
Anterior intercostal arteries	Internal thoracic artery	Run along the inferior margin of the ribs within the intercostal spaces		Supply blood to the upper intercostal spaces
Bronchial arteries	Thoracic aorta	Run along the bronchi within the lungs		Supply oxygenated blood to the lung tissue
Esophageal arteries	Thoracic aorta	Follow the course of the esophagus through the thorax		Supply blood to the esophagus
Posterior intercostal arteries	Thoracic aorta	Located along the posterior part of the intercostal spaces	Lateral and medial branches	Supply the thoracic wall, spinal cord, and associated structures
Superior phrenic arteries	Thoracic aorta	Descend anterior to the thoracic vertebrae to reach the superior surface of the diaphragm		Supply blood to the superior surface of the diaphragm

13.3.1.1 Plasma

Blood plasma is the straw-colored, liquid matrix in which blood cells and other components are suspended, constituting about 55% of the total blood volume. Plasma's composition is complex and includes:

Water: The primary constituent of plasma, water makes up about 90–92% of its volume. It serves as a solvent and medium for carrying other substances.

Proteins: Blood plasma proteins, comprising about 7–8% of the plasma's composition, play many essential roles in the body.

Albumin: It is the most abundant protein in blood plasma, constituting about 60% of the total protein content. Its primary function is maintaining oncotic pressure in the blood, which helps keep the fluid from leaking out of blood vessels.

Albumin is a transport protein for various substances, including fatty acids, hormones, bilirubin, and drugs. It is synthesized in the liver.

Globulins: These proteins transport lipids and some metal ions. They include high-density lipoproteins (HDL), which transport cholesterol. Beta globulins also transport lipids and vitamins. Transferrin, a beta globulin, is crucial for iron

Table 13.5 Arteries supplying spinal column: origin, course, branches, and supply distribution

Name of artery	Origin	Course	Branches	Function
Anterior spinal artery	Arises from branches of the vertebral arteries	Descends along the anterior median fissure of the spinal cord	None directly; gives rise to sulcal arteries that penetrate the spinal cord	Supplies the anterior two-thirds of the spinal cord, including the anterior horns and corticospinal tracts
Posterior spinal arteries	Arises from branches of the vertebral arteries or the posteroinferior cerebellar arteries	Descend on the dorsolateral surfaces of the spinal cord, one on each side	Dorsal branches that supply the posterior one-third of the spinal cord	Supplies the posterior one-third of the spinal cord, including the dorsal columns and dorsal horns
Segmental medullary arteries	Various, including vertebral, ascending cervical, deep cervical, intercostal, and lumbar arteries	Course varies; they pass through the intervertebral foramina and along the spinal nerves to reach the spinal cord	Radicular arteries supply nerve roots, and pial branches that supply the spinal cord	Provide additional blood supply to the spinal cord at different levels, including the cervical, thoracic, and lumbar regions
Sulcal arteries	Anterior spinal artery	Penetrate the spinal cord along the anterior median fissure	None; they directly supply the spinal cord tissue	Supply the central part of the spinal cord, including the anterior horns and the base of the dorsal horns
Radicular arteries	Segmental arteries (like intercostal arteries)	Follow the nerve roots into the spinal canal	Branch into dorsal and ventral radicular arteries that supply the spinal nerve roots and dura	Support the blood supply to the nerve roots and their surrounding structures

transport. Gamma globulins are primarily immunoglobulins (antibodies) produced by plasma cells. They are essential for immune responses, as they identify and neutralize foreign substances like bacteria and viruses.

Fibrinogen: Fibrinogen is a key factor in blood coagulation. During the clotting process, thrombin converts it into fibrin, which forms the meshwork that traps blood cells and forms a blood clot. Like albumin, fibrinogen is produced in the liver.

Complement Proteins: These proteins are part of the immune system and fight infections, promote inflammation, and dispose of dead cells and antibody complexes (Section 14.2.1.5). They are generally present in an inactive form and are activated in response to immune reactions.

Table 13.6 Arteries supplying abdomen: origin, course, branches, and supply distribution (Figs. 13.34 and 13.35)

Name of artery	Origin	Course	Branches	Function
Celiac artery	Abdominal aorta	Short trunk that arises from the aorta just below the diaphragm	Left gastric artery, splenic artery, common hepatic artery	Supplies the stomach, liver, spleen, upper duodenum, and pancreas
Superior mesenteric artery	Abdominal aorta	Extends inferiorly from the aorta, anterior to the third lumbar vertebra	Inferior pancreaticoduodenal, jejunal, ileal, ileocolic, right colic, and middle colic arteries	Supplies most of the small intestine and the beginning of the large intestine
Inferior mesenteric artery	Abdominal aorta	Arises from the aorta just above the bifurcation at the level of L3	Left colic artery, sigmoid arteries, superior rectal artery	Supplies the distal third of the transverse colon, descending colon, sigmoid colon, and rectum
Renal arteries	Abdominal aorta	Lateral branches from the aorta, extending to the kidneys	Segmental arteries, inferior suprarenal artery	Supply blood to the kidneys, adrenal glands, and surrounding structures
Gonadal arteries	Abdominal aorta	Descend along the psoas major muscle	No significant branches in the abdomen (divided into testicular or ovarian arteries in the pelvis)	Supply the testes in males or ovaries in females
Internal iliac artery	Common iliac artery	Branches within the pelvis	Numerous, including the umbilical, superior gluteal, inferior gluteal, and internal pudendal arteries	Supplies the pelvic floor, pelvic organs, gluteal region, and perineum
External iliac artery	Common iliac artery	Travels along the pelvic brim to the inguinal ligament	Inferior epigastric artery, deep circumflex iliac artery	Supplies the lower abdominal wall and perineum and becomes the femoral artery in the thigh
Inferior phrenic arteries	Abdominal aorta	Ascend to supply the diaphragm	No major branches noted	Supply the diaphragm
Middle suprarenal arteries	Abdominal aorta	Small branches that directly target the adrenal glands	No major branches noted	Supply the adrenal glands

Table 13.7 Arteries supplying pelvis: origin, course, branches, and supply distribution (Fig. 13.36)

Name of artery	Origin	Course	Branches	Function
Umbilical artery	Internal iliac artery	Initially courses anteromedially within the pelvis, part of it obliterates after birth, forming the medial umbilical ligament	Superior vesical arteries	Supplies the superior part of the bladder and, in the fetus, it is involved in the circulation of blood to the placenta
Superior vesical artery	Umbilical artery (patent part)	Runs toward the bladder	Branches to the bladder and ureter	Supplies the upper parts of the bladder and the ureter
Inferior vesical artery	Internal iliac artery (in males)	Directs toward the base of the bladder	Branches to the bladder, prostate, and seminal vesicles	Supplies the base of the bladder, prostate, seminal vesicles, and ureter
Middle rectal artery	Internal iliac artery	Courses toward the rectum	Branches to the rectum, prostate, or vagina	Supplies the rectum and contributes to the supply of the prostate or vagina
Internal pudendal artery	Internal iliac artery	Exits the pelvis through the greater sciatic foramen, then reenters via the lesser sciatic foramen	Inferior rectal artery, perineal artery, and others	Supplies the perineum, external genitalia, and lower part of the rectum
Obturator artery	Internal iliac artery	Passes through the obturator foramen to enter the medial thigh	Branches to the head of the femur and obturator externus muscle	Supplies the medial compartment of the thigh and pelvic muscles
Superior gluteal artery	Internal iliac artery	Exits the pelvis through the greater sciatic foramen above the piriformis muscle	Branches to gluteal muscles	Supplies the gluteal region and tensor fasciae latae
Inferior gluteal artery	Internal iliac artery	Leaves the pelvis via the greater sciatic foramen below the piriformis	Branches to the gluteal muscles, internal pudendal artery	Supplies the muscles and skin of the buttocks
Lateral sacral artery	Internal iliac artery	Courses medially towards the sacrum	Branches to sacral canal and sacral foramina	Supplies the sacrum, coccygeus muscle, and erector spinae muscles
Vaginal artery	Internal iliac artery (in females)	Directed toward the vagina and the base of the bladder	Branches to the vagina, bladder, and rectum	Supplies the vagina, bladder, and rectum
Uterine artery	Internal iliac artery (in females)	Travels medially to the uterus, crossing the ureter anteroinferiorly	Branches to the uterus, tubes, and ovaries	Supplies blood to the uterus and associated structures

Table 13.8 Arteries supplying lower limb: origin, course, branches, and supply distribution (Fig. 13.36)

Name of artery	Origin	Course	Branches	Function
Femoral artery	External iliac artery	Runs through the femoral triangle and down the anterior compartment of the thigh	Deep artery of the thigh, lateral circumflex femoral artery, medial circumflex femoral artery, descending genicular artery	Supplies the thigh, hip joint, and head of the femur
Deep artery of the thigh	Femoral artery	Extends posteriorly between the pectineus and adductor longus muscles	Perforating arteries	Supplies the muscles of the thigh and contributes to the knee joint
Popliteal artery	Continuation of the femoral artery at the adductor hiatus	Travels through the popliteal fossa, behind the knee	Anterior tibial artery, posterior tibial artery, sural arteries	Supplies the knee joint, knee muscles, and skin; continues to the lower leg and foot
Anterior tibial artery	Popliteal artery	Travels from the upper part of the popliteal fossa to the anterior compartment of the leg, becoming the dorsalis pedis artery on the foot	Recurrent tibial artery, anterior medial malleolar arteries, anterior lateral malleolar arteries, dorsalis pedis artery	Supplies the anterior compartment of the leg and dorsal aspect of the foot
Posterior tibial artery	Popliteal artery	Descends in the posterior compartment of the leg	Circumflex fibular branch, medial malleolar branches, lateral plantar artery, medial plantar artery	Supplies the posterior part of the leg and plantar surface of the foot
Fibular (peroneal) artery	Posterior tibial artery or popliteal artery	Runs along the medial side of the fibula	Perforating branches, calcaneal branches	Supplies the lateral compartment of the leg and the calcaneal region of the foot
Dorsalis pedis artery	Anterior tibial artery	Continuation of the anterior tibial artery on the dorsum of the foot	Tarsal arteries, metatarsal arteries, arcuate artery, first dorsal metatarsal artery	Supplies the dorsal aspect of the foot
Medial plantar artery	Posterior tibial artery	Runs on the medial side of the foot	Digital branches to the big toe and second toe	Supplies the medial aspect of the sole of the foot
Lateral plantar artery	Posterior tibial artery	Runs laterally in the sole and then curves medially to form the plantar arch	Digital branches to the toes, deep plantar arch	Supplies the lateral and plantar aspects of the foot

Table 13.9 Veins draining brain: origin, course, drainage field, drainage destination (Fig. 13.37)

Vein name	Origin	Course	Drainage field	Drainage to
Superior sagittal sinus	Along the attached margin of the falx cerebri	Runs anteroposteriorly in the superior margin of the falx cerebri	Drains blood from the superior surface of the brain	Transverse sinuses
Inferior sagittal sinus	Along the free edge of the falx cerebri	Runs within the posterior part of the free edge of the falx cerebri	Drains blood from the medial surfaces of the hemispheres	Straight sinus
Straight sinus	Confluence of the inferior sagittal sinus and the great cerebral vein	Extends from the deep center of the brain towards the back, ending in the confluence of sinuses	Drains blood from the deep midline structures of the brain	Confluence of sinuses, then to transverse sinuses
Transverse sinuses	Confluence of sinuses	Extends laterally from the internal occipital protuberance, following the attached margin of the tentorium cerebelli	Drains blood predominantly from the posterior parts of the brain	Sigmoid sinuses
Sigmoid sinuses	Continuation of the transverse sinuses	S-shaped sinus that courses along the posterior cranial fossa to the jugular foramen	Drains blood from the posterior fossa of the cranium	Internal jugular veins
Internal cerebral veins	Deep within the cerebral hemispheres	Run posteriorly in the deep midline, beneath the corpus callosum	Drain blood from deep structures of the cerebral hemispheres	Great cerebral vein, then to straight sinus
Great cerebral vein (of Galen)	Confluence of internal cerebral veins	Located in the midline just beneath the splenium of the corpus callosum	Drains blood from the internal cerebral veins and basal veins	Straight sinus
Basal veins (of Rosenthal)	Anterior perforated substance	Runs posteriorly around the cerebral peduncles and joins the great cerebral vein	Drains blood from the deep middle parts of the cerebral hemispheres	Great cerebral vein

Other Components of Plasma

Electrolytes: Blood plasma electrolytes are minerals with an electric charge in the blood. These electrolytes include sodium, potassium, calcium, magnesium, chloride, bicarbonate, and phosphate.

Nutrients: Plasma carries essential nutrients, such as glucose, amino acids, and fatty acids, absorbed from the digestive tract to various tissues for energy and growth.

Waste Products: Metabolic waste products like urea, creatinine, and bilirubin are transported by plasma to the kidneys and liver for excretion or processing.

Table 13.10 Veins draining head and neck: origin, course, drainage field, drainage destination (Fig. 13.38)

Vein name	Origin	Course	Drainage field	Drainage to
Facial vein	Angular vein (near the eye)	Runs downward and backward, parallel to the facial artery, and drains into the internal jugular vein	Drains the superficial areas of the face and oral cavity	Internal jugular vein
Retromandibular vein	Confluence of the superficial temporal vein and maxillary vein	Divides into anterior and posterior branches; the anterior joins with the facial vein, and the posterior with the external jugular vein	Drains blood from the temporal region, parotid gland, and face	External jugular vein (posterior branch)
Anterior jugular vein	Submental region	Runs downward in the midline of the neck and drains into the external jugular or subclavian vein	Drains the front of the neck	External jugular or subclavian vein
Internal jugular vein	Junction of the sigmoid sinus and inferior petrosal sinus in the jugular foramen	Descends vertically within the carotid sheath alongside the carotid artery	Drains blood from the brain, superficial parts of the face, and neck	Brachiocephalic vein
External jugular vein	Retromandibular vein and posterior auricular vein	Runs diagonally across the sternocleidomastoid muscle to the subclavian vein	Drains blood from the scalp and superficial parts of the face	Subclavian vein
Subclavian vein	Continuation of the axillary vein at the outer border of the first rib	Runs under the clavicle and joins with the internal jugular vein to form the brachiocephalic vein	Drains the superficial areas of the neck, and part of the thorax	Brachiocephalic vein
Vertebral vein	Posterior aspect of the suboccipital triangle	Descends within the transverse foramina of the cervical vertebrae and drains into the brachiocephalic vein	Drains the cervical spinal cord and posterior part of the skull	Brachiocephalic vein

Table 13.11 Veins draining upper limb: origin, course, drainage field, drainage destination (Fig. 13.39)

Vein name	Origin	Course	Drainage field	Drainage to
Cephalic vein	Radial aspect of the dorsal venous network of the hand	Ascends along the lateral aspect of the bicep and then travels through the deltopectoral groove to the shoulder	Drains the lateral aspect of the arm and forearm	Axillary vein
Basilic vein	Ulnar aspect of the dorsal venous network of the hand	Runs along the medial aspect of the forearm and arm and joins the brachial veins to form the axillary vein	Drains the medial aspect of the arm and forearm	Brachial veins, then axillary vein
Median cubital vein	Communication between the cephalic and basilic veins	Crosses the anterior aspect of the elbow (cubital fossa)	Connects the superficial and deep venous systems of the arm	Cephalic vein and basilic vein
Radial vein	Deep palmar venous arch	Ascends along the lateral side of the forearm	Drains the lateral aspect of the hand and forearm	Merges with the ulnar vein
Ulnar vein	Deep palmar venous arch	Runs up the medial aspect of the forearm	Drains the medial aspect of the hand and forearm	Merges with the radial vein
Brachial veins	Continuation of the radial and ulnar veins	Run along each side of the brachial artery	Drain the arm and deep structures of the forearm	Merge to form the axillary vein
Axillary vein	Continuation of the basilic vein	Extends from the lower border of the teres major muscle to the lateral border of the first rib	Drains the upper limb, shoulder, and lateral thoracic wall	Merges with the cephalic vein to form the subclavian vein
Subclavian vein	Continuation of the axillary vein	Extends from the lateral border of the first rib to the internal jugular vein, forming the brachiocephalic vein	Drains the upper limb and some of the thoracic wall	Brachiocephalic vein

Hormones and Enzymes: Plasma transports hormones secreted by endocrine glands. Various enzymes are also present in plasma, playing roles in different biochemical reactions.

Gases: Dissolved gases such as oxygen and carbon dioxide are present in plasma.

Table 13.12 Veins draining lower limb: origin, course, drainage field, drainage destination (Fig. 13.40)

Vein name	Origin	Course	Drainage field	Drainage to
Great saphenous vein	Dorsal venous arch of the foot	Ascends along the medial aspect of the leg and thigh and drains into the femoral vein in the groin	Drains the medial side of the leg, thigh, and foot	Femoral vein
Small saphenous vein	Lateral side of the dorsal venous arch of the foot	Runs up the back of the leg to drain into the popliteal vein behind the knee	Drains the lateral side of the lower leg and foot	Popliteal vein
Anterior tibial vein	Venous plexuses of the foot	Runs upward along the anterior aspect of the lower leg and joins with the posterior tibial vein	Drains the anterior compartment of the lower leg	Popliteal vein
Posterior tibial vein	Venous plexuses of the foot	Ascends along the posterior aspect of the lower leg and joins with the anterior tibial vein	Drains the posterior compartment of the lower leg	Popliteal vein
Peroneal vein	Lower leg	Runs parallel to the fibula and joins the posterior tibial vein before it empties into the popliteal vein	Drains the lateral compartment of the lower leg	Popliteal vein
Popliteal vein	Confluence of the anterior and posterior tibial veins	Runs through the popliteal fossa at the back of the knee	Drains the lower leg	Femoral vein
Femoral vein	Continuation of the popliteal vein	Runs through the thigh and becomes the external iliac vein as it enters the pelvis	Drains the thigh and lower leg	External iliac vein
External iliac vein	Continuation of the femoral vein	Runs in the pelvis and becomes the common iliac vein	Drains blood from the lower limb	Common iliac vein
Common iliac vein	Confluence of the external and internal iliac veins	Ascends to the lower abdomen and joins with its counterpart to form the inferior vena cava	Drains the pelvis and lower limbs	Inferior vena cava

13.3.1.2 Formed Elements

Blood is also composed of crucial cellular components collectively known as the formed elements. These formed elements comprise red blood cells, white blood cells, and platelets.

Red Blood Cells

Red blood cells (RBCs), also known as erythrocytes (Fig. 13.44), are the most abundant cells in human blood and a cornerstone of the circulatory system. Their primary function is transporting oxygen from the lungs to the body's tissues and

Table 13.13 Veins draining abdomen: origin, course, drainage field, drainage destination (Figs. 13.41 and 13.42)

Vein name	Origin	Course	Drainage field
Portal vein	Confluence of the superior mesenteric and splenic veins	Runs posterior to the neck of the pancreas and enters the liver	Carries nutrient-rich blood from the gastrointestinal tract and spleen to the liver
Superior mesenteric vein	Intestinal tract and pancreas	Ascends in the mesentery, to the right of the superior mesenteric artery, and joins the splenic vein to form the portal vein	Drains blood from the small intestine, cecum, ascending colon, and transverse colon
Splenic vein	Spleen	Runs along the posterior side of the pancreas, joining with the superior mesenteric vein	Drains blood from the spleen, pancreas, and part of the stomach
Renal veins	Kidneys	Emerge from the hilum of the kidneys and drain directly into the IVC	Drain deoxygenated blood from the kidneys
Hepatic veins	Liver	Drain the liver and empty into the IVC just below the diaphragm	Carry deoxygenated blood from the liver to the IVC
Gonadal veins (testicular/ ovarian)	Testes in males, ovaries in females	Left vein drains into the left renal vein, while the right drains directly into the IVC	Drain blood from the testes or ovaries
Lumbar veins	Posterior abdominal wall	Drain the posterior abdominal wall and vertebral column, emptying into the IVC	Drain blood from the muscles and skin of the back
Inferior vena cava (IVC)	Confluence of the left and right common iliac veins	Ascends along the posterior abdominal wall, to the right of the aorta, and enters the right atrium of the heart	Drains blood from the lower limbs, pelvis, and abdominal organs

returning carbon dioxide to the lungs for exhalation. RBCs are characterized by their distinctive biconcave disc shape, which optimizes their surface area-to-volume ratio, enhancing their gas exchange capabilities. This shape is maintained by a cytoskeletal structure underlying the cell membrane. The average diameter of an RBC is approximately 6–8 micrometers, and they have a thickness of 2–2.5 micrometers at the thickest point. The RBCs can deform, essential for navigating through the narrow capillaries.

Hemoglobin

Hemoglobin (Hb), the quintessential protein within red blood cells (RBCs), is the cornerstone of respiratory gas transport in the human body. Human adult hemoglobin, known as HbA (Fig. 13.45), predominantly consists of two alpha (α) chains and two beta (β) chains. Each α-chain comprises 141 amino acids, while each β-chain is longer, comprising 146 amino acids. The quaternary structure of hemoglobin is its assembly into a tetramer, with two αβ dimers joining to form a functional protein.

Table 13.14 Veins draining thorax: origin, course, drainage field, drainage destination (Fig. 13.39)

Vein name	Origin	Course	Drainage field
Intercostal veins	Intercostal spaces (between the ribs)	Run along the bottom of each rib and drain into the azygos and hemiazygos veins (or directly into the superior vena cava at higher levels)	Drain the intercostal muscles and surrounding tissues
Internal thoracic vein	Tissues along the inside of the anterior thoracic wall	Run alongside the internal thoracic artery and drain into the brachiocephalic veins	Drains the anterior chest wall and breasts
Hemiazygos vein	Ascends on the left side of the vertebral column	Crosses to the right side (usually around T9 level) to join the azygos vein	Drains the lower half of the left thoracic wall and abdominal wall
Azygos vein	Lumbar regions and lower thoracic veins	Ascends in the right side of the thoracic vertebral column and empties into the superior vena cava	Drains blood from the posterior walls of the thorax and abdomen
Brachiocephalic veins	Formed by the union of the internal jugular and subclavian veins	Run inferiorly and medially to join and form the superior vena cava	Drain the head, neck, and upper limbs
Superior vena cava (SVC)	Confluence of the left and right brachiocephalic veins	Runs vertically downward and enters the right atrium of the heart	Drains blood from the upper half of the body, excluding the heart

The heme group comprises a porphyrin ring with an iron (Fe^{2+}) at its center (Fig. 13.46). The iron atom is the site of oxygen binding.

Hemoglobin's primary function is to bind oxygen in the lungs and transport it to various body parts. Each hemoglobin molecule has four subunits, each containing an iron-bound heme group. Oxygen molecules bind to these iron atoms (Fig. 13.47). Each hemoglobin molecule can bind up to four oxygen molecules, one on each heme group. In the lungs, where oxygen concentration is high, oxygen molecules bind to the iron in the heme groups, forming oxyhemoglobin. In the body's tissues, where the oxygen concentration is lower, hemoglobin releases oxygen, diffusing into the cells.

Hemoglobin exhibits cooperative binding, meaning the binding of one oxygen molecule increases the affinity of the remaining sites for oxygen (Fig. 13.48). This property is crucial for efficient oxygen loading in the lungs and unloading in the tissues. The cooperative binding is facilitated by allosteric transitions in the hemoglobin structure, shifting between a low-affinity Tense (T) state and a high-affinity Relaxed (R) state. The Bohr effect describes how hemoglobin‘s oxygen affinity is inversely related to the concentration of carbon dioxide and the acidity (pH) of the blood. In areas with high CO_2 concentration and lower pH (like active tissues), hemoglobin releases more oxygen. This effect is essential for meeting the higher oxygen demands of metabolically active tissues.

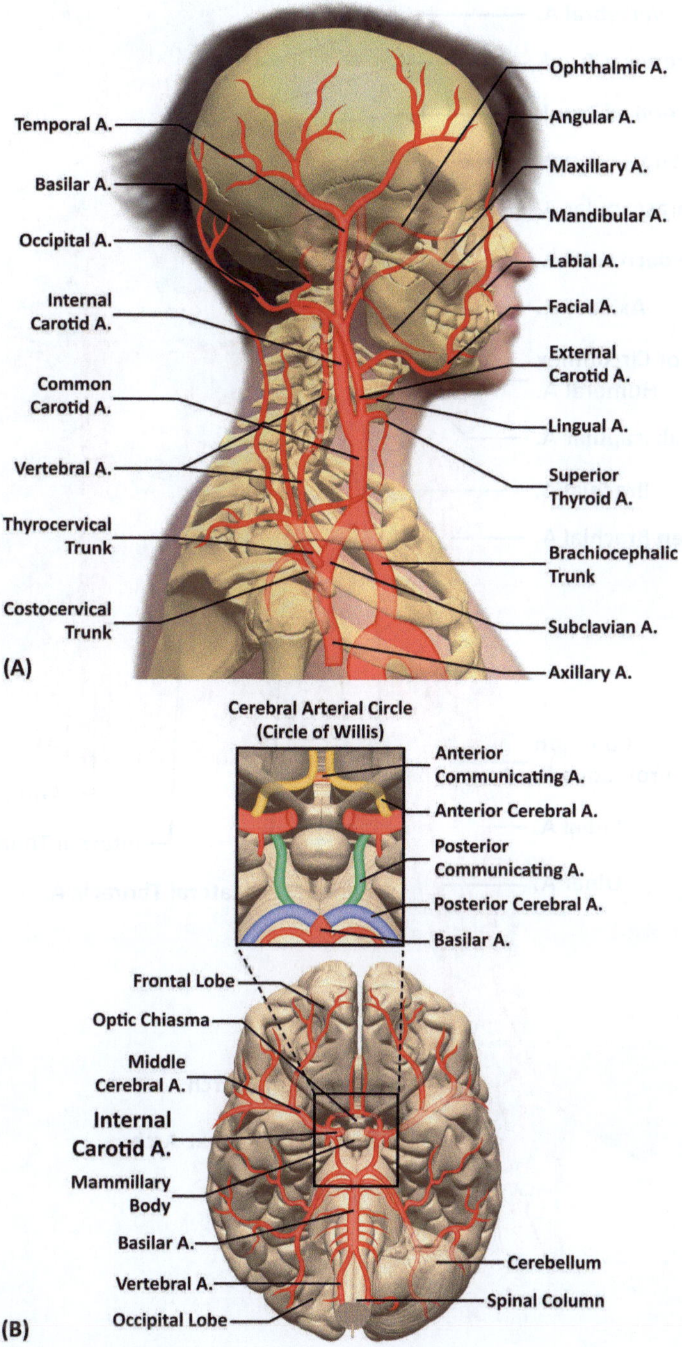

Fig. 13.32 Arteries of the head, neck, and brain: (**a**) Arteries of the head and neck, right lateral view. (**b**) Major arteries of the brain, inferior view

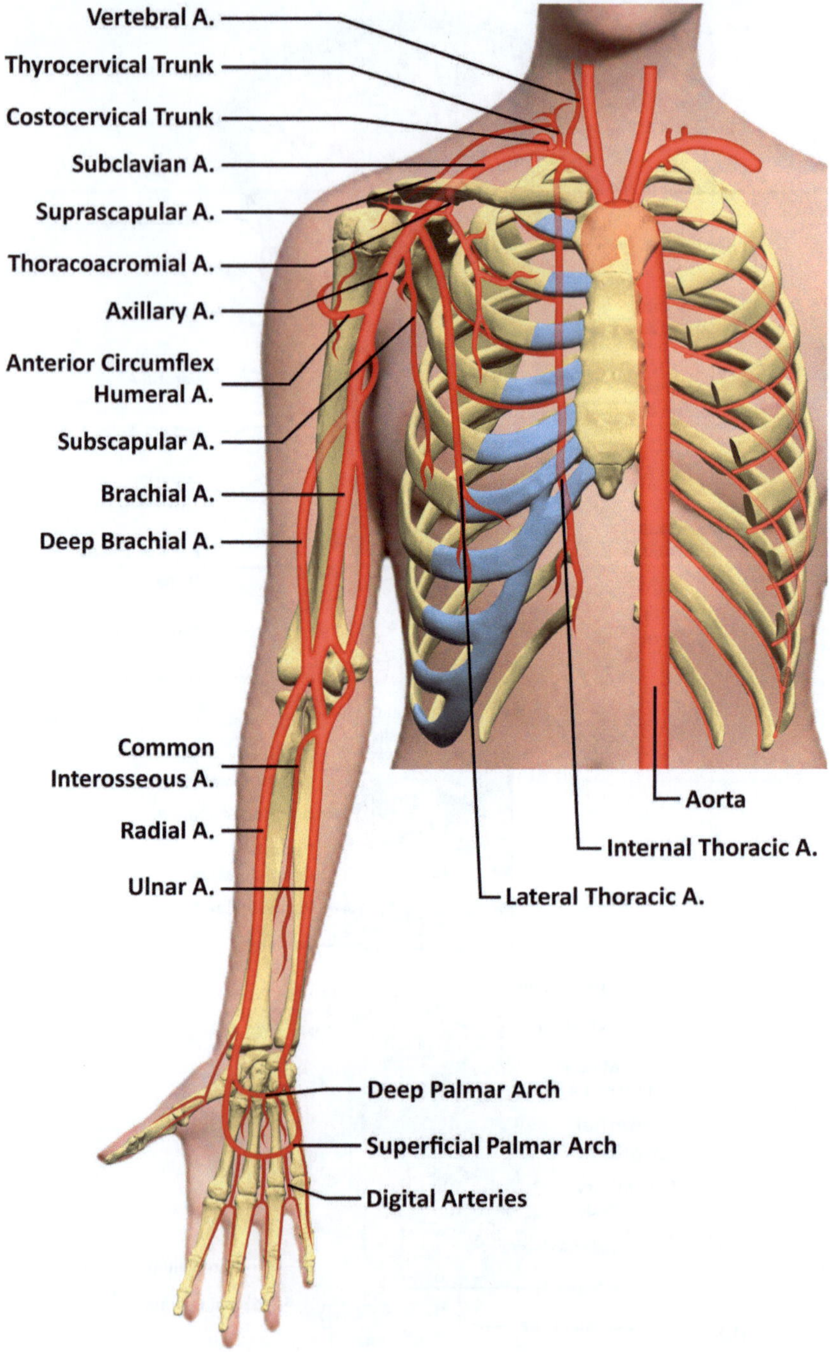

Fig. 13.33 Arteries of the thorax and right upper limb, anterior view

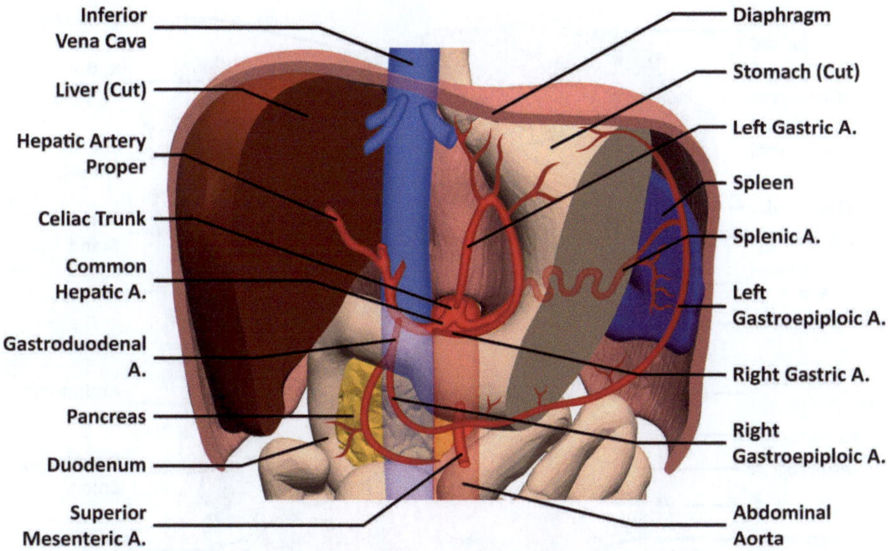

Fig. 13.34 Arteries of the upper abdomen, the territory of the celiac trunk, anterior view. (Parts of the liver and stomach are removed)

Hemoglobin also transports carbon dioxide from the body's tissues back to the lungs. About 20–25% of carbon dioxide is carried by hemoglobin. Carbon dioxide binds to the amino acids in hemoglobin, forming carbamino hemoglobin, which is then transported back to the lungs.

RBC Production

The production of red blood cells (RBCs), or erythropoiesis (Fig. 13.49), occurs primarily in the bone marrow. Erythropoiesis begins with hematopoietic stem cells (HSCs) in the bone marrow. These multipotent stem cells have the potential to differentiate into various blood cell types, including RBCs. Bone marrow provides a specialized microenvironment, or niche, that supports and regulates the differentiation of HSCs. The hormone erythropoietin (EPO), primarily produced by the kidneys, stimulates RBC production. RBC maturation involves extensive changes, including hemoglobin accumulation, size reduction, and loss of cellular organelles. The bone marrow ensures only fully functional RBCs are released into the bloodstream.

RBC Destiny

RBCs typically have a lifespan of about 120 days in circulation. This limited lifespan is due to the mechanical stress they endure while traversing the circulatory system and their lack of organelles to repair damage. They lose flexibility, their membranes become more fragile, and certain surface proteins are modified or lost. These changes are recognized by macrophages, particularly in the spleen, liver, and bone marrow. Macrophages phagocytize senescent RBCs. This process

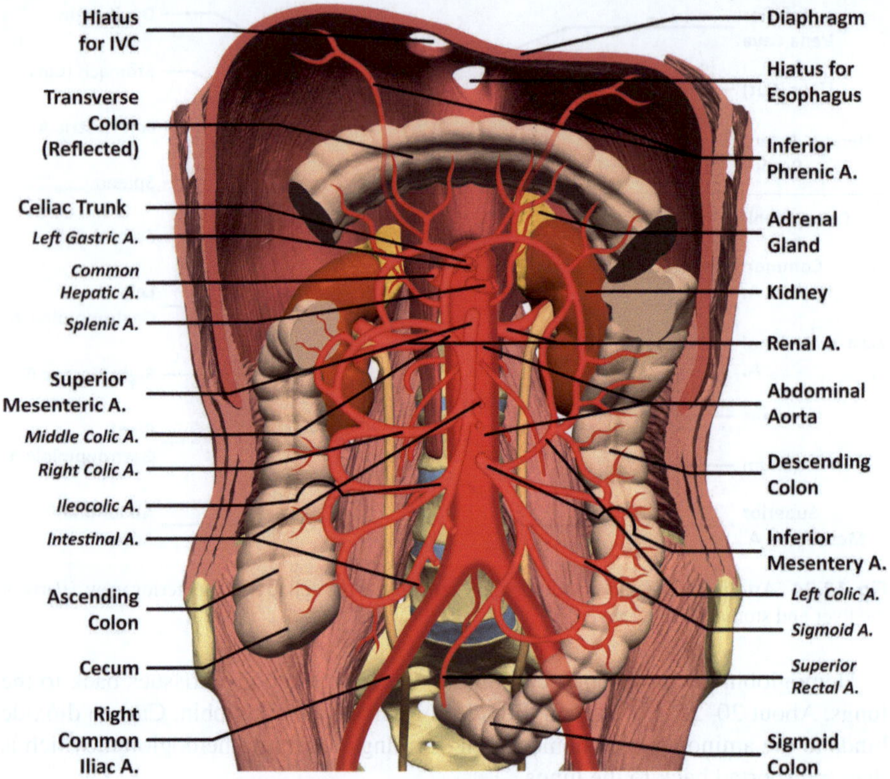

Fig. 13.35 Arteries of the middle and lower abdomen, the territory of the superior and inferior mesenteric arteries, anterior view. (All viscera except colon, kidneys, and adrenals are removed.) (IVC Inferior vena cava)

predominantly occurs in the spleen. Within macrophages, hemoglobin is broken down. The heme portion is separated from the globin chains. Globin chains are further broken down into amino acids, which are recycled. The heme is converted to biliverdin by the enzyme heme oxygenase. Biliverdin is subsequently converted to bilirubin, a yellow pigment. Bilirubin is transported to the liver bound to albumin. In the liver, bilirubin is conjugated (made water-soluble) and excreted into the bile (Fig. 13.50). The iron released from heme is salvaged and transported to the bone marrow for reuse in erythropoiesis. This recycling is crucial, as the body has limited mechanisms for excreting excess iron.

White Blood Cells

White blood cells (WBCs), or leukocytes, protect the body against infections and foreign invaders. They function in the body's defense mechanisms, both in innate and adaptive immunity. White blood cells are broadly categorized into two groups: granulocytes and agranulocytes.

Fig. 13.36 Arteries of the right pelvis and lower extremity, anterior view

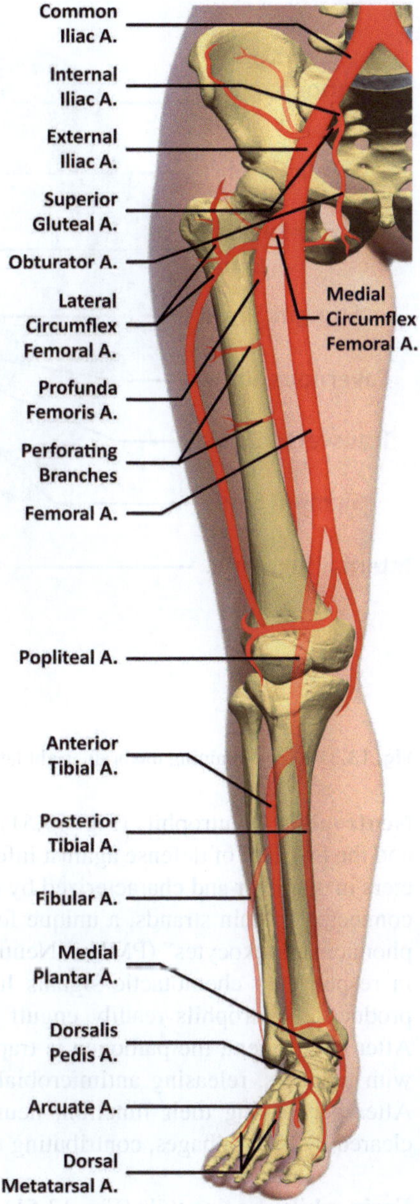

Granulocytes

Granulocytes (Fig. 13.51) are characterized by granules in their cytoplasm, visible under a light microscope. These granules contain enzymes and other substances necessary for the cell's function. Granulocytes are further subdivided into three types: neutrophils, eosinophils, and basophils.

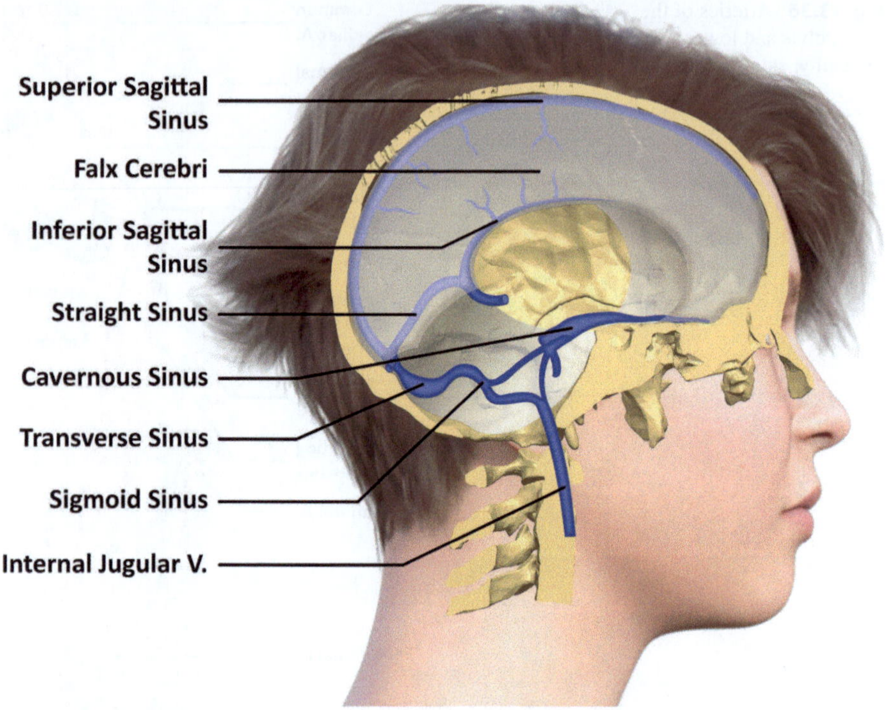

Fig. 13.37 Veins draining the brain, right lateral view, parasagittal section

Neutrophils Neutrophils (Fig. 13.51) are the most abundant type of granulocytes and the first line of defense against infection. Neutrophils are about 12–15 micrometers in diameter and characterized by a multilobed nucleus, usually with 2–5 lobes connected by thin strands, a unique feature that earned them the name "polymorphonuclear leukocytes" (PMNs). Neutrophils rapidly migrate toward infection sites in response to chemotactic signals like interleukins, leukotrienes, and bacterial products. Neutrophils readily engulf pathogens, particularly bacteria and fungi. After engulfment, the pathogen is trapped in a phagosome. The phagosome fuses with granules, releasing antimicrobial agents and killing the engulfed pathogen. After performing their function, neutrophils undergo apoptosis (death) and are cleared by macrophages, contributing to the resolution of inflammation.

Eosinophils Eosinophils (Fig. 13.51) play a significant role in the body's immune response, especially in allergic reactions and parasitic infections. Characteristically, the nucleus has two lobes connected by a thin strand, giving it a bilobed appearance. Cytoplasmic granules are large and uniform in size, and they stain bright red-orange with eosin dye. These granular contents are toxic to parasites. The eosinophils attach to the surface of large parasites that cannot be phagocytized and release granular contents, causing damage to the parasite.

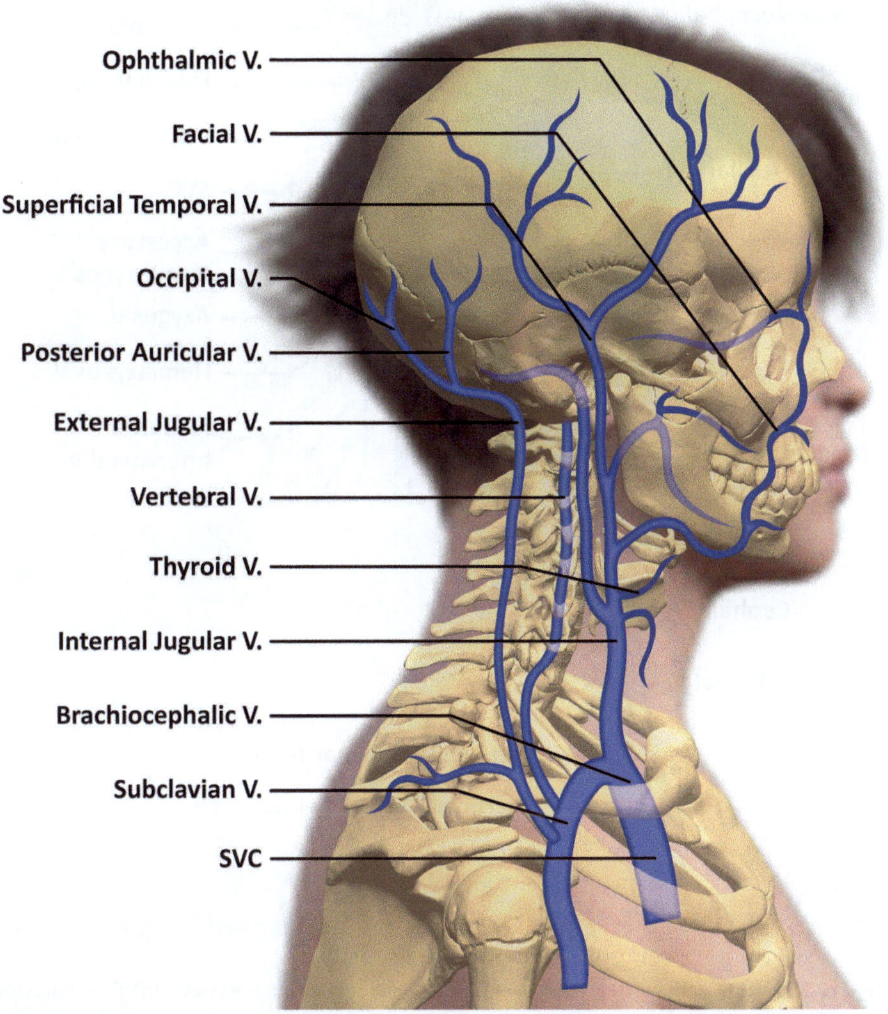

Fig. 13.38 Veins draining the head and neck, right lateral view. (SVC Superior vena cava)

Basophils Basophils (Fig. 13.51) are the least common type of white blood cell. Typically, the nucleus is bilobed or S- shaped but is often obscured by the granules in the cytoplasm. Cytoplasmic granules are characterized by large, dark blue-purple granules that contain histamine, heparin, and other mediators. These granules stain intensely with basic dyes. These substances are vital to the basophil's role in immune responses, particularly in allergic reactions.

Agranulocytes

Agranulocytes (Fig. 13.52) lack visible cytoplasmic granules and have a more uniform cytoplasm. This group includes lymphocytes and monocytes.

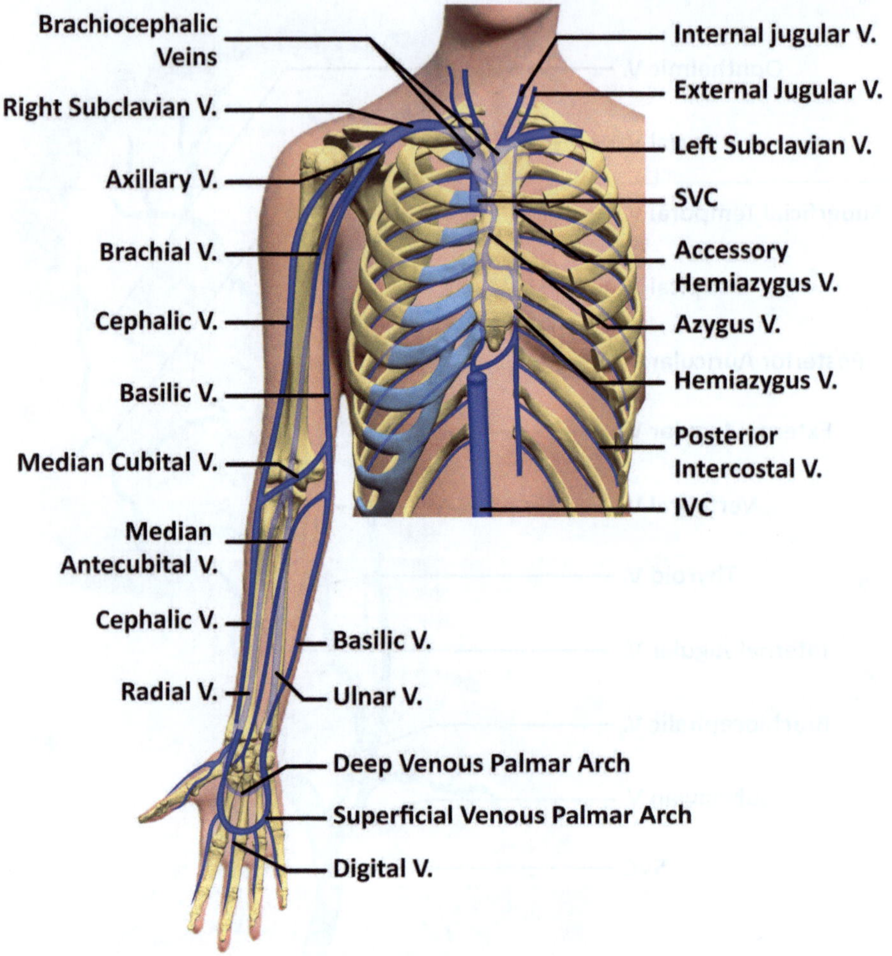

Fig. 13.39 Veins draining the thorax and right upper limb, anterior view. (SVC = Superior vena cava)

Lymphocytes Lymphocytes (Fig. 13.52) are key players in the adaptive immune system (Section 14.2.2). Lymphocytes are relatively small, typically ranging from 6 to 15 micrometers in diameter. They have a large, round nucleus that occupies most of the cell volume, with a thin cytoplasm rim containing few organelles. Lymphocytes are classified into three main groups: B lymphocytes (responsible for antibody production), T lymphocytes (involved in cell-mediated immunity), and natural killer (NK) cells (play a role in innate immunity by targeting virus-infected cells and tumors) (Section 14.2.1.3).

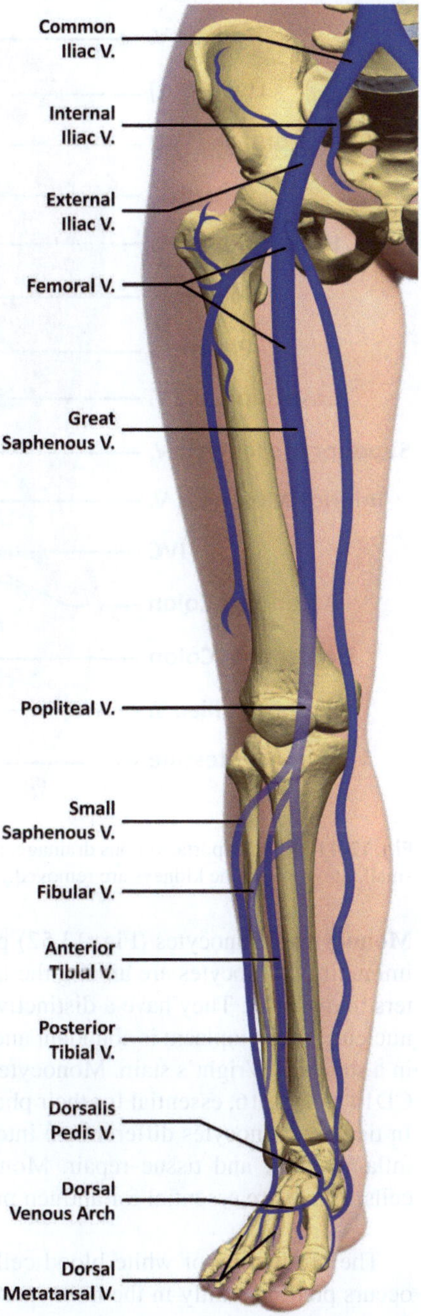

Fig. 13.40 Veins draining the right pelvis and lower limb, anterior view

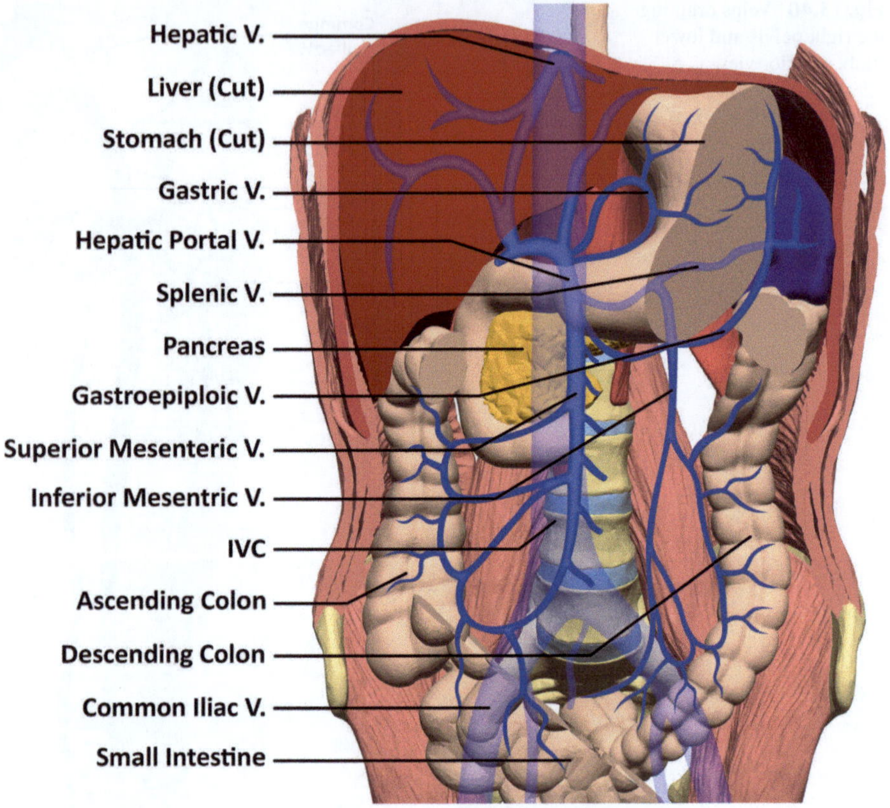

Fig. 13.41 Hepatic portal venous drainage, anterior view. (Parts of the liver, stomach, colon, and small intestine and the kidneys are removed.) (IVC Inferior vena cava)

Monocytes Monocytes (Fig. 13.52) play a significant role in innate and adaptive immunity. Monocytes are among the largest leukocytes, typically 12–20 micrometers in diameter. They have a distinctive, often kidney-shaped or horseshoe-shaped nucleus. The cytoplasm is abundant and may appear gray-blue with shaped granules in a standard Wright's stain. Monocytes express various surface markers, including CD14 and CD16, essential for their phagocytosis and antigen presentation function. In tissues, monocytes differentiate into macrophages, key players in phagocytosis, inflammation, and tissue repair. Monocytes can also differentiate into dendritic cells, which are essential for antigen presentation and activation of T cells.

The production of white blood cells, also known as leukopoiesis (Fig. 13.49), occurs predominantly in the bone marrow. Like other blood cells, WBCs originate from pluripotent hematopoietic stem cells in the bone marrow. These stem cells differentiate into two primary lineages, the myeloid and lymphoid lineages, which give rise to different types of WBCs. The myeloid lineage produces granulocytes (neutrophils, eosinophils, and basophils) and monocytes. The lymphoid lineage

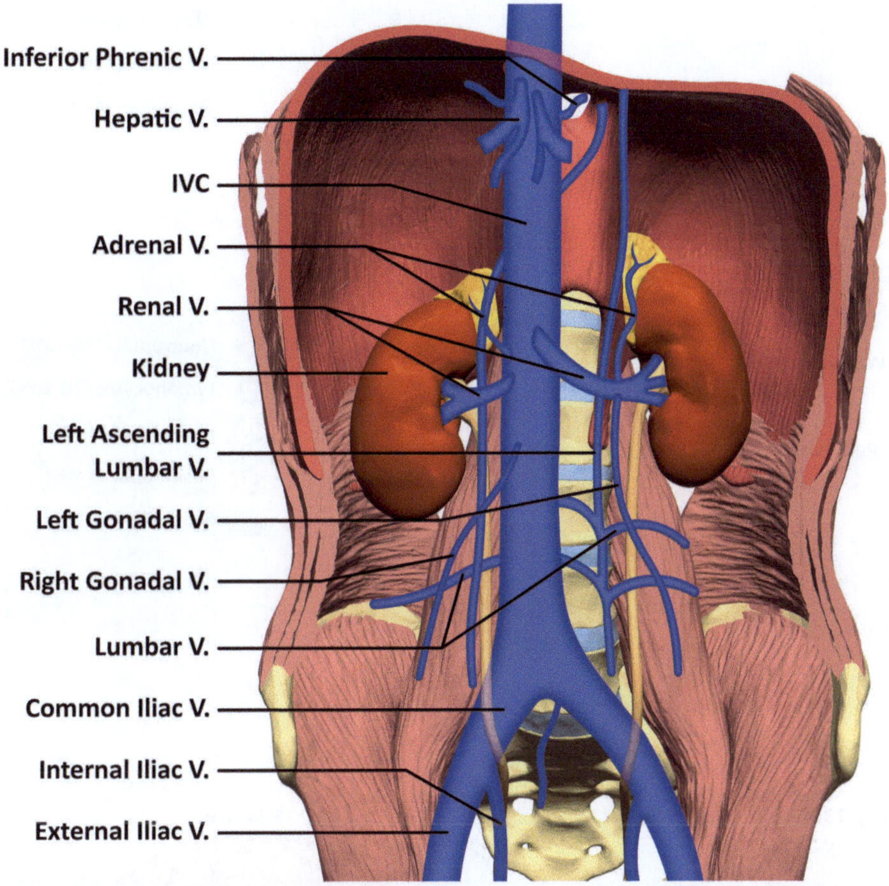

Fig. 13.42 Systemic venous drainage of the abdomen, anterior view. (All viscera except for kidneys and adrenals are removed.) (IVC Inferior vena cava)

produces lymphocytes, including B cells, T cells, and natural killer (NK) cells. Lymphocytes undergo different pathways for maturation. B cells mature in the bone marrow, while T cells mature in the thymus.

Platelets

Platelets, also known as thrombocytes (Fig. 13.53), are small, anucleate cell fragments crucial for blood clotting and wound healing. Platelets are small, with a diameter of 2–3 micrometers and a discoid shape when inactive. Upon activation, they change shape, extending filopodia and lamellipodia to aid in clot formation. Platelets have a lifespan of about 7–10 days in circulation. Aged or damaged platelets are removed from the circulation by the spleen and liver.

Platelets contain three types of granules:

Alpha Granules: Contain clotting factors, growth factors, and fibronectin.

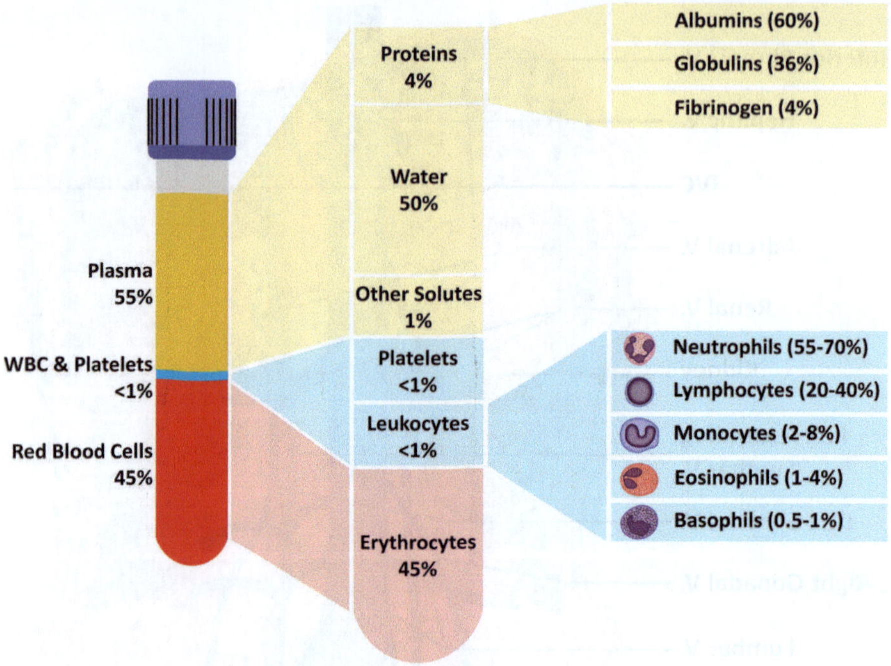

Fig. 13.43 The major components of whole blood

Fig. 13.44 Structure of an erythrocyte

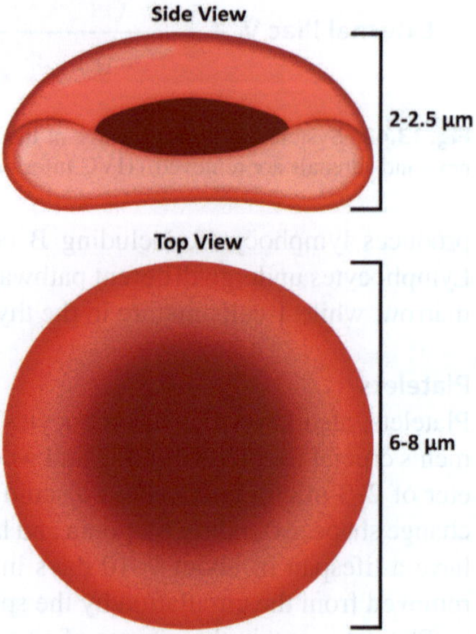

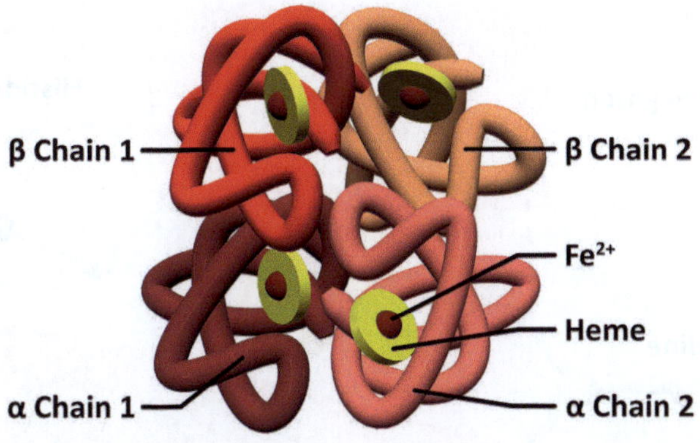

Fig. 13.45 Structure of hemoglobin A

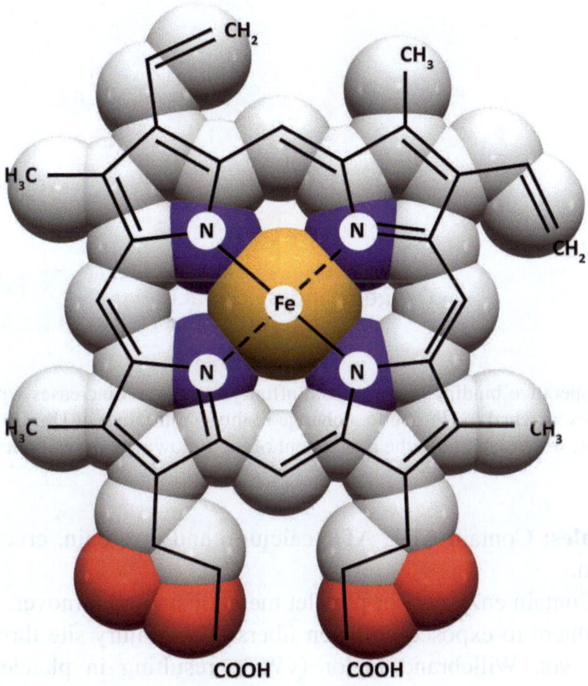

Fig. 13.46 Molecular structure of heme group

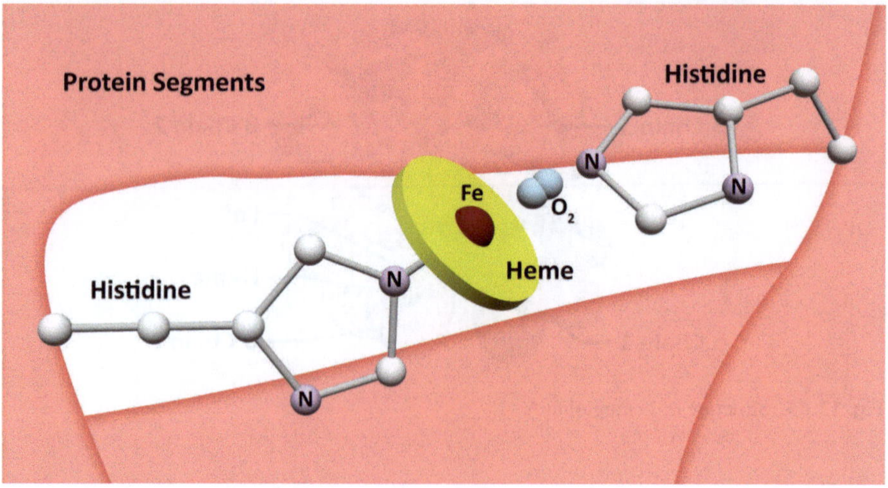

Fig. 13.47 Spatial structure of heme group binding to oxygen

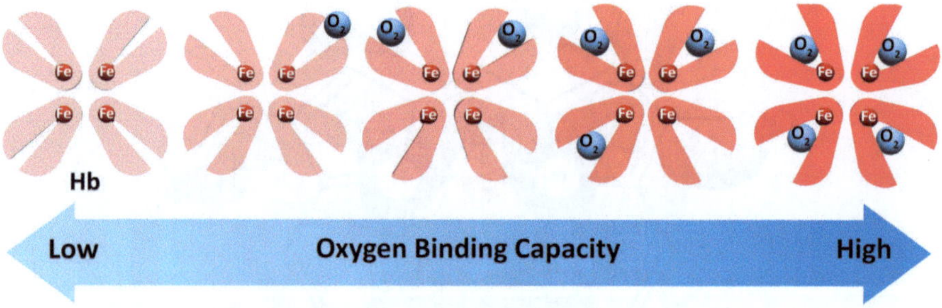

Fig. 13.48 Cooperative binding: Hemoglobin affinity for oxygen increases with the number of oxygen molecules attached to Hb due to a change in the 3D structure of Hb with binding of each oxygen molecule, which facilitates the attachment of the following oxygen molecule

Dense Granules: Contain ADP, ATP, calcium, and serotonin, crucial for platelet aggregation.

Lysosomes: Contain enzymes for platelet metabolism and turnover.

Platelets adhere to exposed collagen fibers at the injury site through a process facilitated by von Willebrand factor (vWF), resulting in platelet aggregation. Platelets are an essential player in the coagulation cascade. Activated platelets provide phospholipid surfaces to assemble coagulation factor complexes, which is necessary for the coagulation cascade.

Platelets support the conversion of prothrombin to thrombin, a key enzyme in fibrin clot formation. Thrombin mediates the conversion of fibrinogen to fibrin, forming a mesh stabilizing the platelet plug, resulting in a more robust clot.

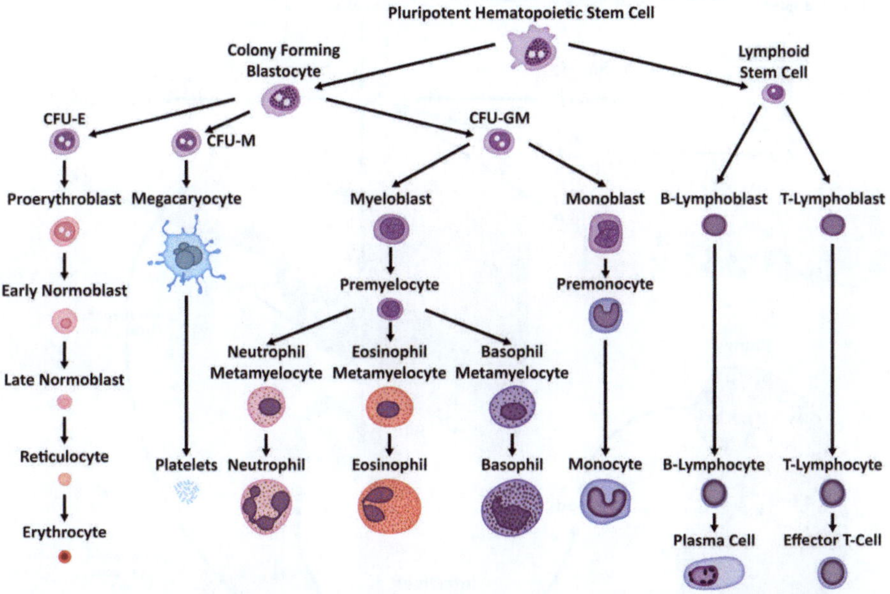

Fig. 13.49 Hematopoiesis, the process by which all blood cells are produced from pluripotent hematopoietic stem cells (HSCs) in the bone marrow. The HSC is the progenitor of all blood cells. It can self-renew and differentiate into various specialized blood cells. HSCs give rise to two main lineages: myeloid and lymphoid stem cells. **Myeloid lineage:** Colony-forming blastocyte (CFU-GEMM) gives rise to multiple myeloid cell types. **CFU-E (erythroid progenitor)** differentiates into proerythroblast, the earliest recognizable precursor of red blood cells. It differentiates into early normoblast and then late normoblast. In the late normoblast, the nucleus is expelled. It differentiates into reticulocytes, an immature red blood cell released into the bloodstream. Reticulocytes mature into erythrocytes, the mature red blood cells responsible for oxygen transport. **Megakaryocyte progenitor (CFU-M):** It is the precursor to this lineage and differentiates into megakaryoblast and then megakaryocyte. Megakaryocytes produce platelets (thrombocytes) through cytoplasmic fragmentation. **Granulocyte-monocyte progenitor (CFU-GM):** This multipotent stem cell is the granulocyte and monocyte lineage precursor. It differentiates into either myeloblast or monoblast. Myeloblasts differentiate into promyelocytes. Promyelocytes can differentiate into neutrophil, eosinophil, and basophil metamyelocytes. These metamyelocytes later differentiate into the corresponding granulocytes (neutrophil, eosinophil, and basophil). Monoblasts differentiate into promonocytes and then into monocytes, which circulate in the blood and differentiate into macrophages in tissues. **Lymphoid lineage:** The lymphoid stem cell gives rise to lymphoblasts (either B or T lineage). Lymphoblasts differentiate into corresponding lymphocytes

13.3.2 Hemostasis

Hemostasis is a physiological process that stops bleeding at the injury site while maintaining normal blood flow elsewhere in the circulation. It involves a complex interplay of mechanisms, including vascular constriction, platelet plug formation, and the blood coagulation cascade.

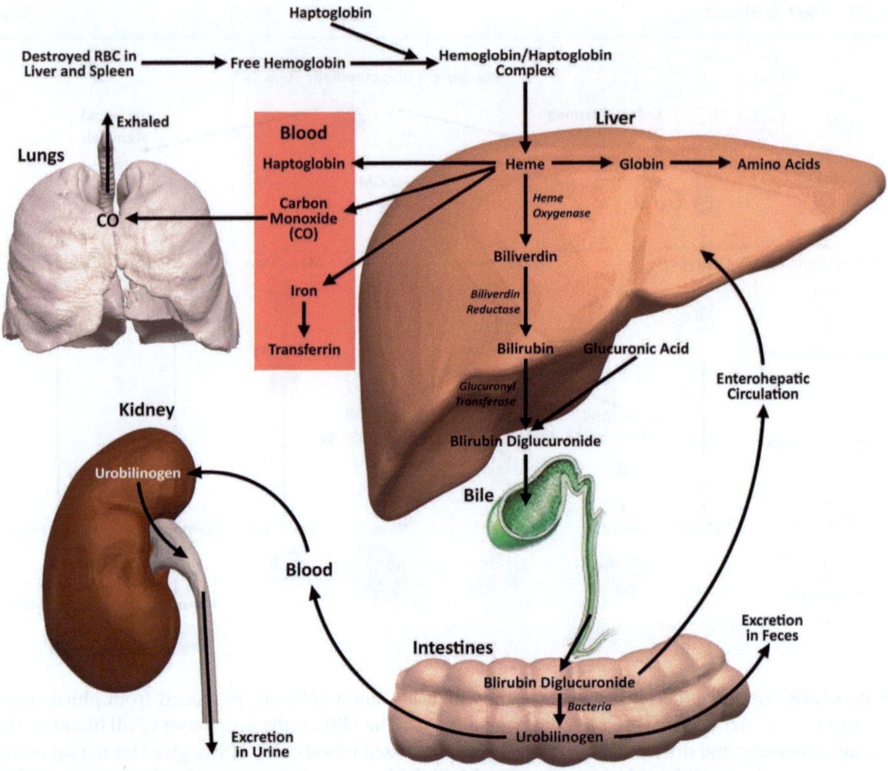

Fig. 13.50 Destiny of erythrocytes and the metabolism of heme

Fig. 13.51 Granulocytes

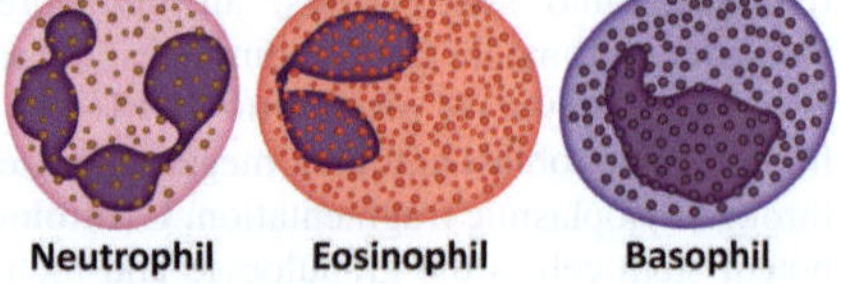

Fig. 13.52 Agranulocytes and their differentiated forms

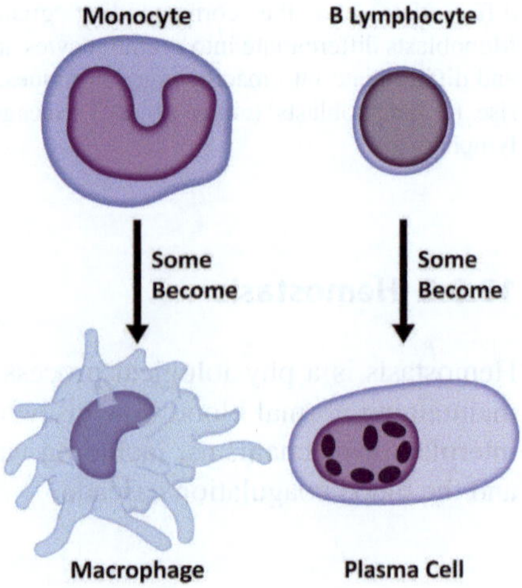

Fig. 13.53
Megakaryocyte and platelets

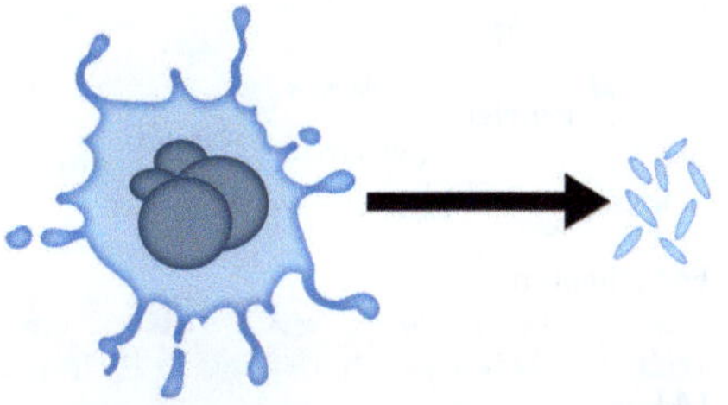

13.3.2.1 Vascular Spasm

Immediately following vessel injury, the vascular smooth muscle contracts, causing vasoconstriction. This response reduces blood flow to the injury site and limits blood loss.

13.3.2.2 Platelet Plug Formation

When a blood vessel is injured, platelets rapidly adhere to the exposed extracellular matrix, particularly to collagen, via specific receptors (like GPIa and GPIb-V-IX complex) (Fig. 13.54). The von Willebrand factor (vWF) bridges platelets and the damaged vessel wall, facilitating adhesion. Platelets change from disk-shaped to a more spherical form with extended pseudopodia upon activation. This increases their surface area and facilitates interaction with other platelets and the vessel wall. Activated platelets release contents of their granules, which include ADP, serotonin, and calcium, which are essential for further recruitment and activation of platelets. Fibrinogen binding to GPIIb/IIIa receptors on activated platelets mediates platelet aggregation, forming a platelet plug at the injury site. This platelet plug temporarily halts bleeding and is crucial for maintaining vascular integrity.

13.3.2.3 Coagulation Cascade

The coagulation cascade (Fig. 13.55) is a series of enzymatic reactions involving clotting factors, leading to thrombin formation. Thrombin converts fibrinogen into fibrin, which forms a mesh that traps blood cells, consolidating the platelet plug into a stable clot. The cascade has two pathways:

Intrinsic Pathway: The intrinsic pathway is initiated when blood contacts negatively charged surfaces, such as exposed collagen from a damaged vessel wall. This contact activates Factor XII, which in turn activates Factor XI. Activated Factor XI (XIa) activates Factor IX, which, with its cofactor Factor VIII, forms a complex that activates Factor X.

Extrinsic Pathway: The extrinsic pathway is triggered by blood exposure to tissue factor (TF) or thromboplastin, a protein found in tissues outside the blood vessels. TF binds and activates Factor VII. The TF-Factor VII complex activates Factor X. Factor X is where the intrinsic and extrinsic pathways converge.

Common Pathway: Activated Factor X (Xa), with its cofactor Factor V, forms the prothrombinase complex, which converts prothrombin (Factor II) into thrombin (Factor IIa). Thrombin then converts fibrinogen into fibrin. Fibrin strands form a mesh that traps blood cells, creating a stable clot. Thrombin also activates Factor XIII, which cross-links the fibrin strands, stabilizing the clot (Fig. 13.55).

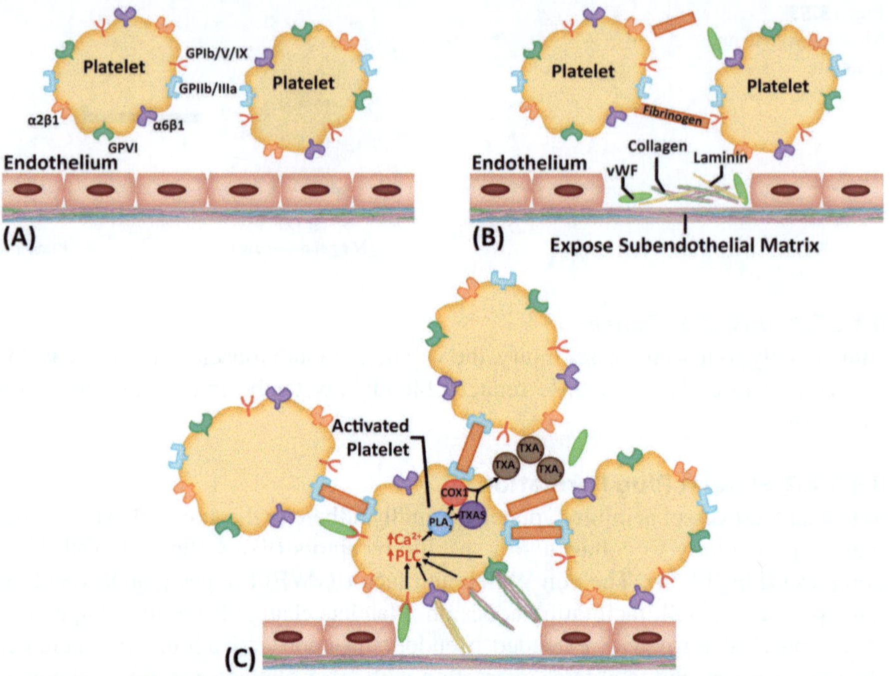

Fig. 13.54 Platelet activation and platelet plug formation: (**a**) The endothelium maintains a smooth surface that prevents platelet adhesion under normal conditions. Platelets circulate in an inactive form and do not adhere to the endothelium. (**b**) Vascular injury disrupts the endothelium, exposing the subendothelial matrix containing collagen, laminin, and von Willebrand factor (vWF). Platelets adhere to the exposed subendothelial matrix. (**c**) Upon adhesion to the subendothelial matrix, platelets become activated. Activation leads to shape change, the release of granule contents (including ADP, TXA2 (thromboxane A2), and serotonin, which further promote platelet activation and recruitment), and the expression of additional surface receptors (Increased expression of GPIIb/IIIa, which binds fibrinogen, forming bridges between platelets)

13.3.3 Transfusion

Blood transfusion is a medical procedure involving transferring blood or blood components from a donor to a recipient. It replaces lost blood components, commonly performed in severe anemia, trauma, surgery, and certain medical conditions. Due to its potential risks and the complexity of human blood types, blood transfusion is strictly regulated to ensure patient safety and the optimal use of blood products. Donated blood is tested for various infectious agents, such as HIV, hepatitis B and C, syphilis, and others. Blood is also typed based on the ABO blood group system and Rh factor. Before transfusion, compatibility between donor and recipient blood is confirmed through cross-matching tests.

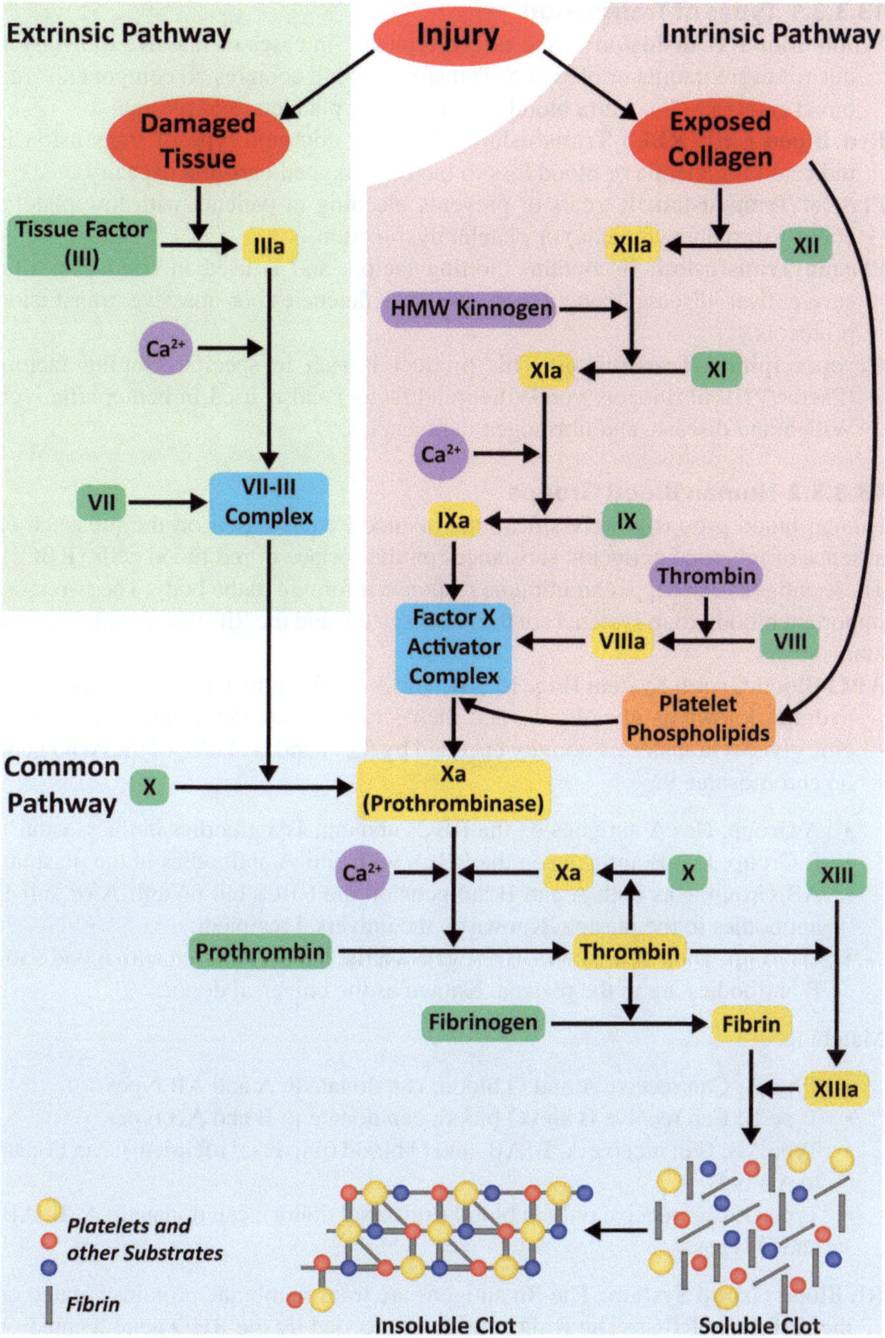

Fig. 13.55 Coagulation cascade

13.3.3.1 Types of Transfusion

Whole Blood Transfusion: Used rarely, typically in cases of massive transfusion due to severe trauma or bleeding. Transfused blood contains all components: red blood cells (RBCs), white blood cells (WBCs), platelets, and plasma.

Red Blood Cell (RBC) Transfusion: The most common type of transfusion is indicated for anemia or blood loss to increase oxygen-carrying capacity.

Platelet Transfusion: It treats or prevents bleeding in patients with low platelet counts (thrombocytopenia) or platelet dysfunction.

Plasma Transfusion: It contains clotting factors and is used in conditions like severe liver disease, coagulation factor deficiencies, or massive transfusion protocols.

Cryoprecipitate Transfusion: This product is rich in specific clotting factors (Factor VIII, fibrinogen, von Willebrand factor) and is used in hemophilia, von Willebrand disease, and fibrinogen deficiencies.

13.3.3.2 Human Blood Groups

Human blood groups are classifications of blood types based on the presence or absence of inherited antigenic substances on the surface of red blood cells (RBCs). These antigens can trigger an immune response if foreign to the body. The two most important blood group systems are the ABO system and the Rh system, but there are many others.

ABO Blood Group System (Fig. 13.56): The A and B antigens are complex carbohydrates known as glycolipids, and their structures are determined by the specific glycosyltransferase activity encoded by the A and B alleles of the ABO gene on chromosome 9:

- A Group: Has A antigens on the RBCs and anti-B antibodies in the plasma.
- B Group: Has B antigens on the RBCs with anti-A antibodies in the plasma.
- AB Group: Has both A and B antigens on the RBCs but no anti-A or anti-B antibodies in the plasma. Known as the universal recipient.
- O Group: There are no A or B antigens on the RBCs, but both anti-A and anti-B antibodies are in the plasma. Known as the universal donor.

Matching:

- Type A: Can receive A and O blood; can donate to A and AB types.
- Type B: Can receive B and O blood; can donate to B and AB types.
- Type AB: Can receive A, B, AB, and O blood (universal recipient); can donate to AB only.
- Type O: Can receive only O blood (universal donor); can donate to A, B, AB, and O types.

Rh Blood Group System: The Rh antigens are transmembrane proteins located on the surface of RBCs. The RhD antigen is encoded by the RHD gene located on chromosome 1. Individuals who possess the RhD antigen on their RBCs are classified as Rh-positive (e.g., A^+, B^+), while those lacking the antigen are categorized as Rh-negative (e.g., A^-, B^-). Rh-positive is more common than Rh-negative.

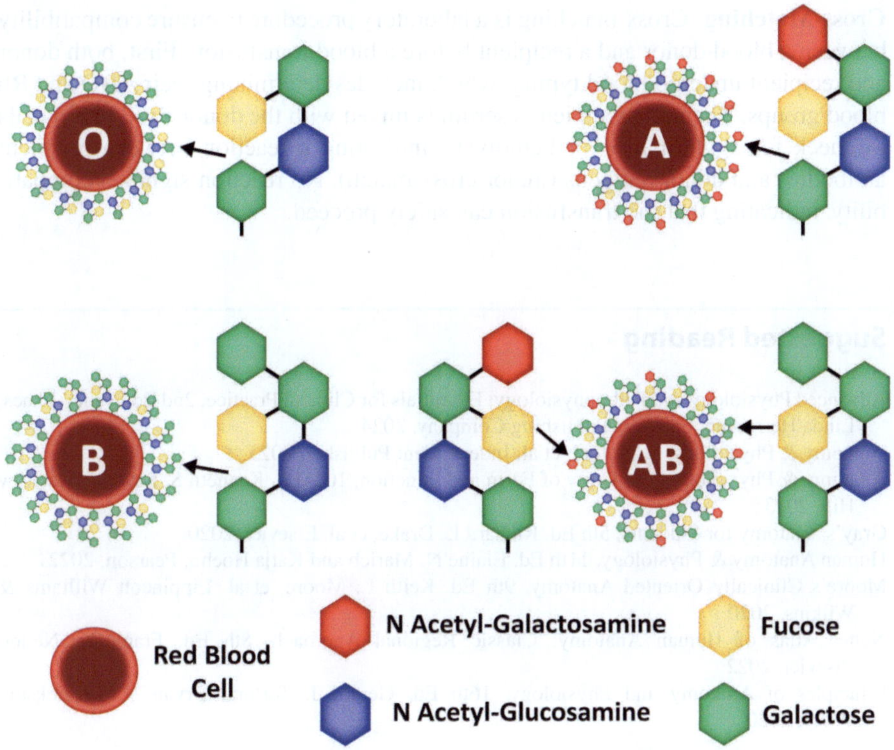

Fig. 13.56 ABO blood groups

Matching:

- Rh-Positive: Can receive Rh-positive and Rh-negative blood.
- Rh-Negative: Should ideally receive Rh-negative blood to avoid sensitization and potential hemolytic reactions.

13.3.3.3 Blood Typing

Blood typing is a laboratory procedure that determines an individual's blood group based on the presence or absence of specific antigens on the surface of red blood cells (RBCs). The most recognized blood group systems are the ABO system and the Rh system. ABO typing is done by mixing the blood with anti-A and anti-B antibodies and observing agglutination (clumping of RBCs). Rh typing involves mixing the blood with anti-Rh antibodies. Agglutination of RBCs indicates the presence of the corresponding antigen on the RBCs.

Cross-Matching Cross-matching is a laboratory procedure to ensure compatibility between a blood donor and a recipient before a blood transfusion. First, both donor and recipient undergo blood typing, which includes determining their ABO and Rh blood groups. Then, the recipient's serum is mixed with the donor's red blood cells to check for agglutination or hemolysis, indicating a reaction between recipient antibodies and donor antigens (major cross-match). No reaction signifies compatibility, indicating that the transfusion can safely proceed.

Suggested Reading

Advanced Physiology and Pathophysiology: Essentials for Clinical Practice, 2nd Ed. Nancy Tkacs, Linda Herrmann, Springer Publishing Company, 2024
Anatomy & Physiology, S. Meloni, et al. Independent Publisher, 2023
Anatomy & Physiology: The Unity of Form and Function, 10th Ed. Kenneth S. Saladin, McGraw Hill, 2023
Gray's Anatomy for Students, 5th Ed. Richard L. Drake, et al. Elsevier, 2020
Human Anatomy & Physiology, 11th Ed. Elaine N. Marieb and Katja Hoehn, Pearson, 2022
Moore's Clinically Oriented Anatomy, 9th Ed. Keith L. Moore, et al. Lippincott Williams & Wilkins, 2020
Netter Atlas of Human Anatomy: Classic Regional Approach, 8th Ed. Frank H. Netter, Elsevier, 2022
Principles of Anatomy and Physiology, 16th Ed. Gerard J. Tortora, Bryan H. Derrickson, Wiley, 2020

14 Lymphatic and Immune Systems

In this chapter, you will learn:

- Overview of the lymphatic system.
- Anatomy and physiology of lymphatic vessels.
- Mechanisms of lymph transport.
- Structure and function of primary and secondary lymphoid organs, such as the thymus, spleen, and lymph nodes.
- Lymphoid cells' roles in pathogen recognition, response, and defense.
- Innate immunity, detailing the components and role as the first line of defense against pathogens.
- Phagocytes' roles and types, including neutrophils and macrophages, and the process of phagocytosis.
- Natural killer (NK) cells, their development, functions, and role in recognizing and destroying infected or abnormal cells.
- The inflammation process, its triggers, mediators, and roles in pathogen clearance and tissue repair.
- Antimicrobial proteins, their types like defensins and lysozyme, and their functions in direct pathogen combat.
- Activation and function of B lymphocytes in humoral immunity, including antibody production.
- Structure and functions of different classes of antibodies (IgG, IgA, IgM, IgE, IgD).
- T lymphocytes and cellular immunity, covering activation and function of helper, cytotoxic, and regulatory T cells.
- Differences between primary and secondary immune responses and their significance.
- Differences between active and passive immunity with examples of each.
- The complement system, including its components, pathways, and functions in immune response enhancement.

H. Hosseinzadeh, *Fundamentals of Medicine for Biomedical Engineering*,
https://doi.org/10.1007/978-3-031-74085-5_14

- Cytokines, their types, functions in immune responses, roles in inflammation, and cell signaling.

14.1 Part 1: The Lymphatic System

The lymphatic system (Fig. 14.1) is a complex network of vessels, nodes, organs, and tissues. Its primary functions extend far beyond its role as a mere drainage system for excess fluids. The lymphatic system is a vital part of the immune system,

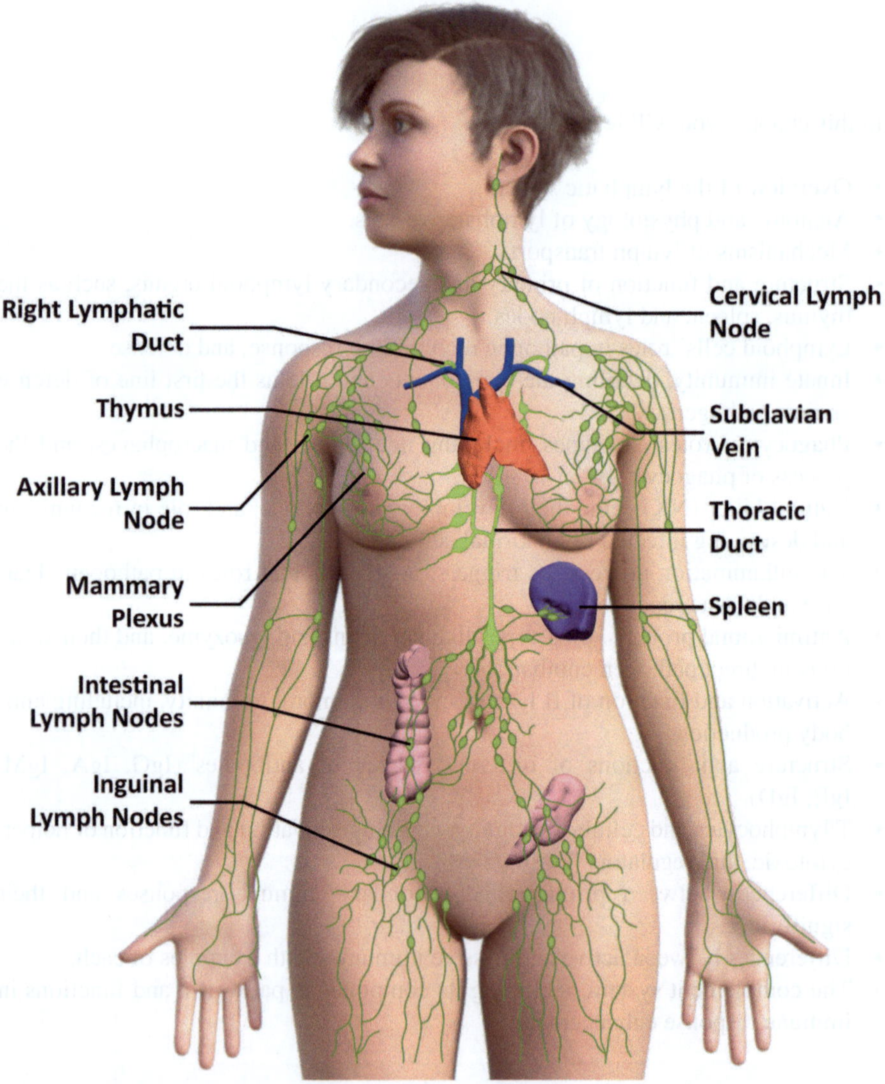

Fig. 14.1 The lymphatic system

facilitating immune responses, pathogen defense, and the surveillance of abnormal cells. Furthermore, it is an essential player in nutrient transport, lipid absorption, and the maintenance of tissue fluid balance.

14.1.1 Lymphatic Vessels

The lymphatic system comprises a vast network of lymphatic vessels, also known as lymphatics. These vessels form an intricate web that pervades nearly all tissues in the body. They are present alongside blood vessels. A few organs do not have lymphatic vessels: the central nervous system (CNS), bone marrow, teeth, the epidermis of skin, cartilage, cornea, and lens of the eye.

The initial lymphatic capillaries are at the smallest level of the lymphatic vascular system (Fig. 14.2). These delicate, blind-ended vessels serve as the entry points for lymph formation. The endothelial cells lining initial lymphatic capillaries possess several distinct structural features that distinguish them from blood capillaries (Fig. 14.3):

Overlapping Junctions: Unlike blood capillaries with continuous basement membranes, initial lymphatic capillaries lack a continuous basement membrane. Instead, their endothelial cells overlap, forming loose junctions that permit easy entry of interstitial fluid, solutes, and cellular components.

Flap-like Valves: Initial lymphatic capillaries are characterized by flap-like, anchoring filaments that attach the endothelial cells to surrounding tissues.

High Permeability: The absence of a continuous basement membrane and the presence of loose junctions make initial lymphatic capillaries highly permeable, allowing them to collect a wide range of substances from the surrounding interstitial space, including large proteins and antigens.

Initial lymphatic capillaries converge to form larger collecting lymphatic vessels with a more traditional tubular structure resembling blood vessels. The lymphatic vessels have three distinct layers: the endothelial lining, the smooth muscle layer, and the connective tissue layer.

Collecting lymphatic vessels progressively merge into larger trunks (Fig. 14.4), ultimately leading to the body's two primary lymphatic ducts: the thoracic and right lymphatic ducts.

The thoracic duct (Fig. 14.4) is the largest lymphatic vessel in the human body. It serves as the main conduit for lymph from the lower and left upper body parts back into the bloodstream. The thoracic duct originates in the abdomen from the cisterna chyli. (The cisterna chyli is located in the abdomen, anterior to the vertebral column, usually at the level of the first and second lumbar vertebrae. The cisterna chyli is typically a sac-like enlargement of the lymphatic vessels, receiving lymphatic drainage from three primary trunks: two lumbar lymphatic trunks and the intestinal trunk.) It then ascends through the diaphragm and runs through the posterior mediastinum, behind the esophagus, and in front of the vertebral column. It

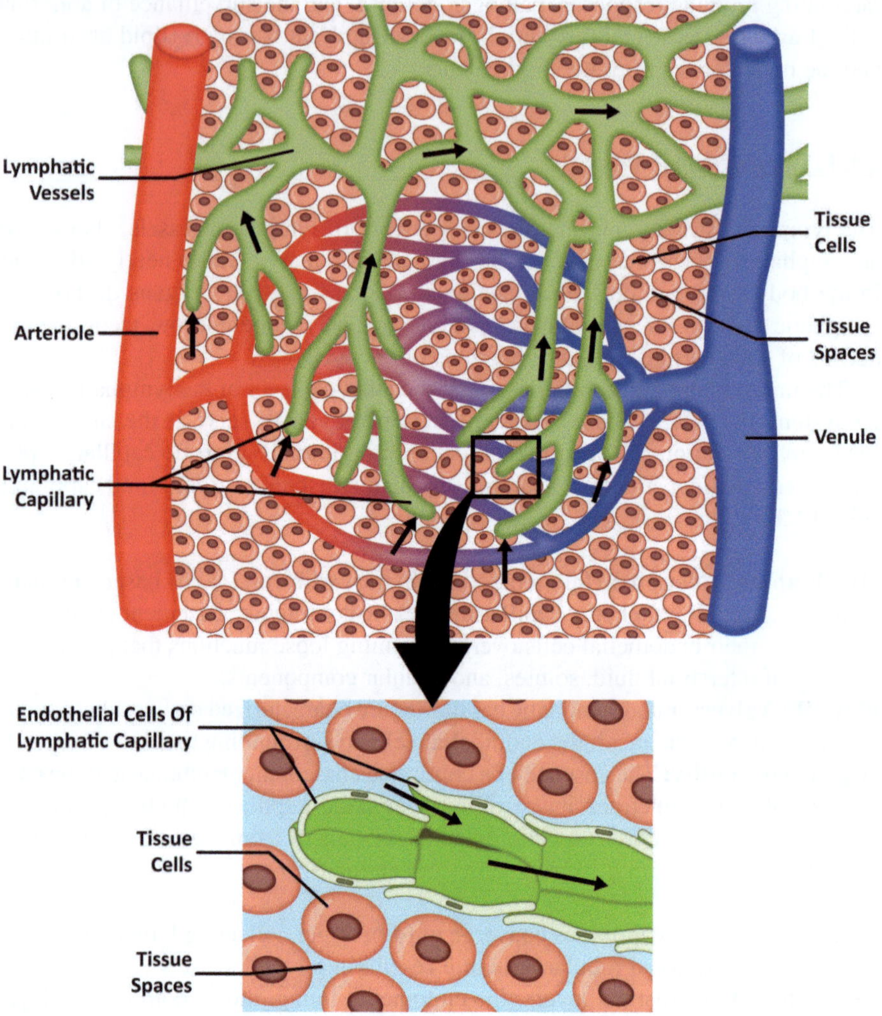

Fig. 14.2 Lymphatic capillaries: Relation of lymphatic capillaries with a capillary bed of the blood vascular system (left) and the microscopic structure of a single lymphatic vessel, showing adjacent endothelial cells overlap each other, forming unidirectional flaplike mini valves (right)

extends upward to the neck's left side, draining into the venous system at the junction of the left subclavian and internal jugular veins.

The right lymphatic duct (Fig. 14.4), smaller than the thoracic duct, is crucial for draining lymph from the right upper part of the human body. This duct is located at the root of the neck and is formed by merging the right jugular, right subclavian, and right bronchomediastinal lymphatic trunks. The right lymphatic duct empties its lymph into the venous system at the junction of the right subclavian and internal jugular veins.

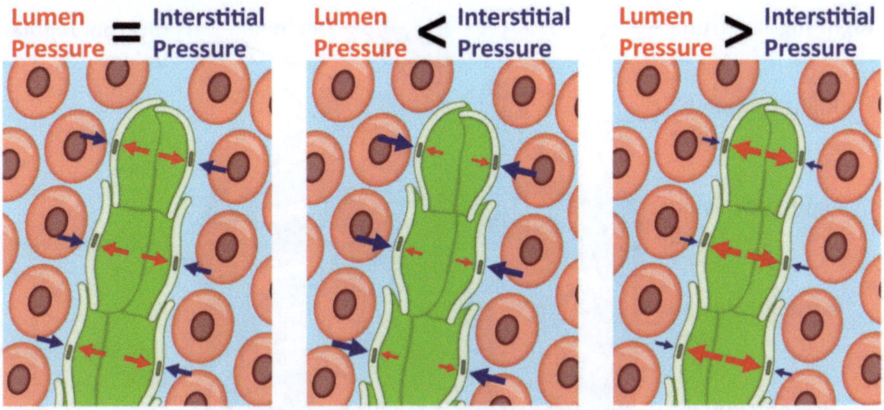

Fig. 14.3 Functional mechanism and the dynamics of fluid entering the lymphatic capillary

14.1.2 Lymph Production and Composition

Lymph, a clear, colorless fluid, plays a significant role in maintaining tissue fluid balance, transporting immune cells, and facilitating lipid absorption. Lymph formation begins with the filtration of blood plasma into the interstitial spaces, where it becomes tissue fluid. The initial lymphatic capillaries collect tissue fluid, dissolved substances, and cellular debris.

Flap-like mini-valves, overlapping endothelial cell junctions, and anchoring filaments ensure the entry of interstitial fluid and large molecules, such as proteins and antigens, into the lymphatic capillary lumen. The same factors that affect the filtration in blood capillaries play a role in lymph formation. The lymph's composition reflects its origin from tissue fluid and the additional elements acquired during its journey through the lymphatic system. Lymph typically contains water (making up the bulk of its volume), electrolytes (including ions such as sodium, potassium, calcium, and chloride), nutrients (such as glucose and amino acids), waste products (including carbon dioxide and metabolic byproducts), cellular components (lymphocytes, monocytes, and other immune cells are present, contributing to the lymph's immune function), and proteins (some plasma proteins and antibodies may enter lymphatic capillaries, enhancing the lymph's immune properties).

14.1.3 Lymph Transport

Lymph transport within the lymphatic vessels is under the influence of several factors as follows:

Contraction of Smooth Muscle: The walls of collecting lymphatic vessels contain smooth muscle that contracts rhythmically, generating peristaltic waves. These contractions are responsible for propelling lymph forward.

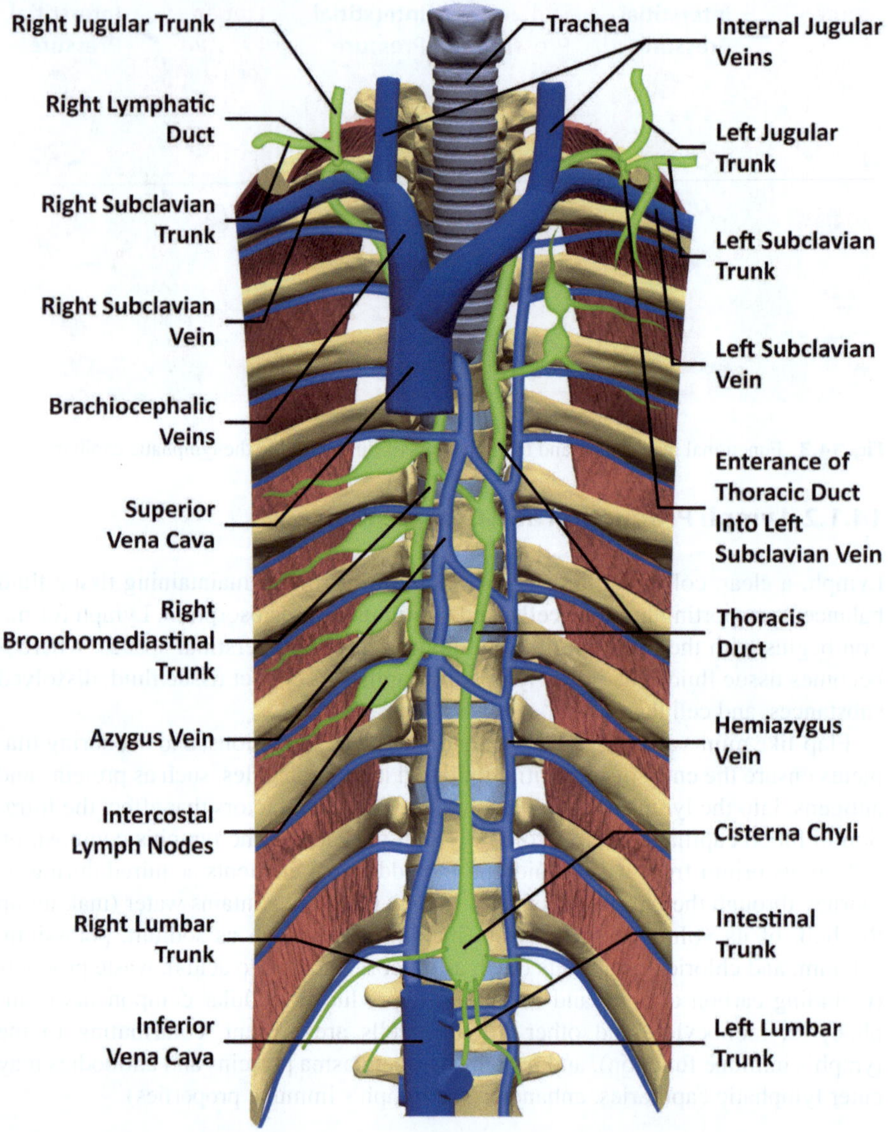

Fig. 14.4 Major lymphatic trunks and ducts. Anterior view of the posterior thoracic and abdominal wall

Valves (Fig. 14.5): Collecting lymphatic vessels are equipped with one-way valves that prevent the backflow of lymph. These valves ensure lymph flows in only one direction—toward larger lymphatic trunks and ducts.

External Factors: Physical activities such as muscle contractions during exercise and body movements aid in the compression of lymphatic vessels, enhancing lymph transport, specifically in the limbs.

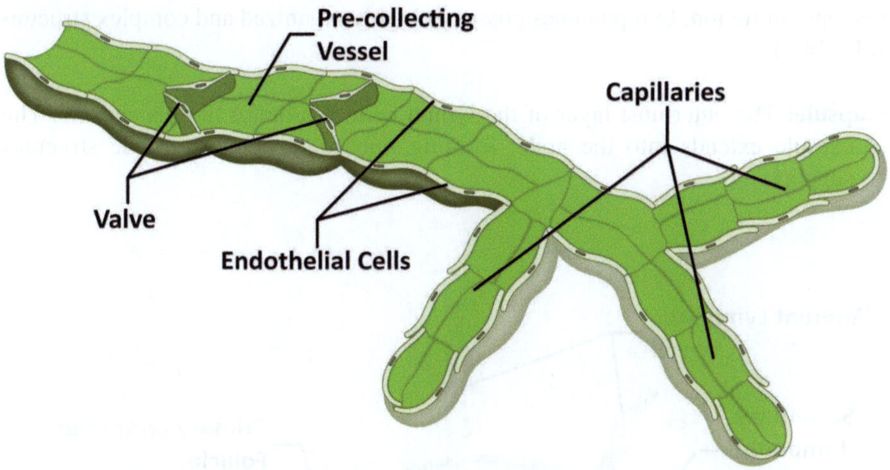

Fig. 14.5 Valves in lymphatic vessels ensure the unidirectional transport of absorbed fluid toward large central veins

14.1.4 Lymphoid Cells

Lymphoid cells are a diverse group of white blood cells responsible for recognizing, responding to, and defending the body against pathogens like bacteria, viruses, and fungi. Lymphoid cells can be broadly categorized into two main groups based on their function and characteristics: innate lymphoid cells (ILCs) and adaptive lymphoid cells (ALCs). Innate lymphoid cells are a diverse group of immune cells that lack antigen-specific receptors, distinguishing them from adaptive lymphoid cells. ILCs are primarily involved in the immediate defense against pathogens and maintaining tissue homeostasis. Adaptive lymphoid cells are characterized by their ability to express antigen-specific receptors, allowing them to recognize and respond to specific pathogens. ALCs include B, T, and natural killer (NK) cells (Section 14.2.1.3). Lymphoid cells are distributed throughout the body, with particular subsets residing in various tissues and organs, forming the lymphoid tissues.

14.1.5 Lymphoid Organs

Lymphoid organs are specialized structures within the body responsible for the production, maturation, and activation of lymphoid cells and initiating and coordinating immune responses.

14.1.5.1 Lymph Nodes

Lymph nodes are small, bean-shaped structures distributed throughout the body. They are vital hubs for immune surveillance, antigen filtration, and immune response initiation. Lymph nodes possess a highly organized and complex structure (Fig. 14.6):

Capsule: The outermost layer of the lymph node is a dense fibrous capsule. This capsule extends into the node, forming trabeculae, which provide structural support.

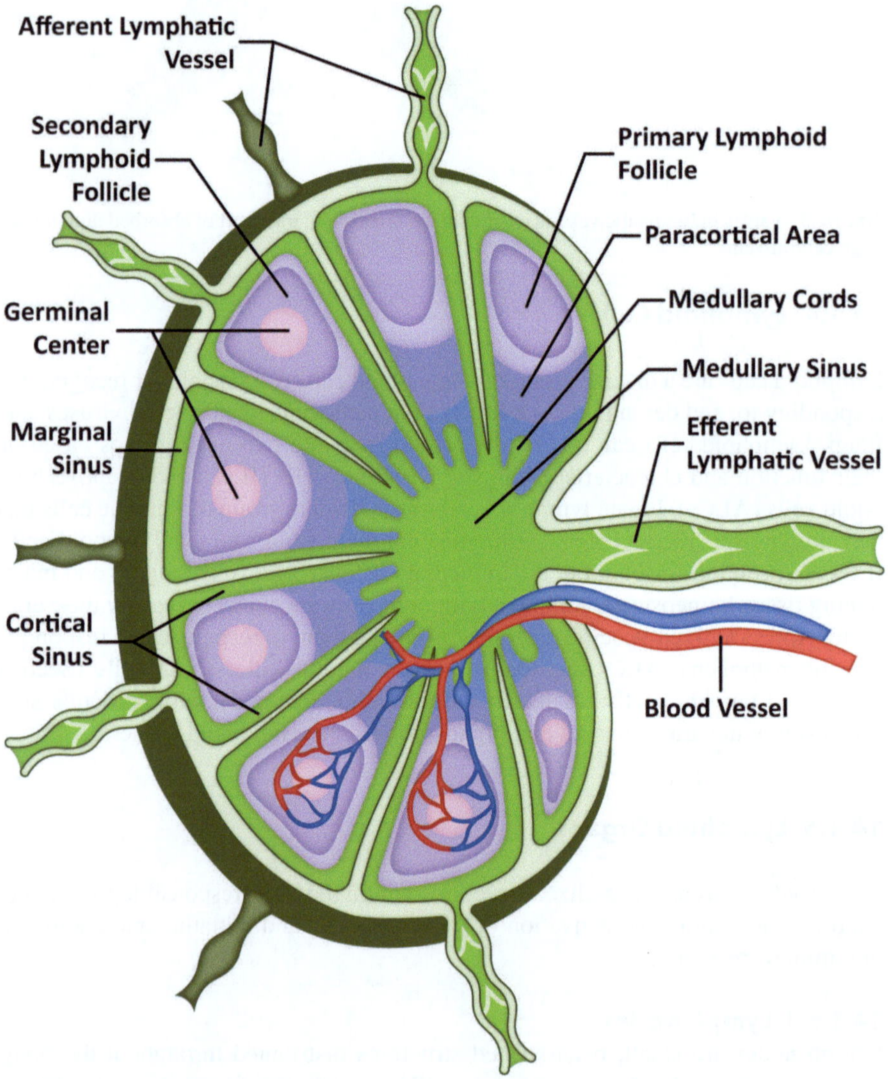

Fig. 14.6 Lymph node: Longitudinal section showing the internal structure of a lymph node and associated lymphatic vessels

Cortex: The outer part of the cortex contains spherical structures known as primary follicles, predominantly composed of B lymphocytes. Upon antigen stimulation, primary follicles can transform into secondary follicles, characterized by the presence of a germinal center. Germinal centers are active B cell proliferation, differentiation, and antibody production sites.

Paracortex: The paracortex lies between the cortex and the medulla. This area is rich in T cells and is the site of T cell activation.

Medulla: The innermost part of the lymph node consists of medullary cords, which are elongated cords of lymphoid tissue containing B cells, plasma cells, and macrophages. Medullary sinuses are channels interspersed between the medullary cords, lined by endothelial cells and macrophages. They carry lymph through the node and filter it.

Afferent Lymphatic Vessels: Lymph enters the lymph node through multiple afferent lymphatic vessels around the periphery.

Efferent Lymphatic Vessel: The lymph exits the node at the hilum through the efferent lymphatic vessel, which carries filtered lymph to the venous circulation.

Lymph nodes are strategically distributed throughout the body to optimize immune surveillance and response capabilities. The size and number of lymph nodes vary throughout the body. They are typically more numerous, where the body is more prone to injury or infection. Common locations include cervical, supraclavicular, axillary, mediastinal, abdominal, retroperitoneal, pelvic, inguinal, and popliteal lymph nodes.

Afferent lymphatic vessels bring lymph containing antigens, immune cells, and other substances into the lymph node. Lymph first accumulates in the subcapsular sinus, located just beneath the capsule. In this region, antigen-presenting cells (APCs) capture antigens from the lymph. Lymph then progresses deeper into the cortex and paracortex of the lymph node, where B and T cells are located, respectively. These regions are essential for immune responses. Lymph continues its journey through the medullary sinuses, undergoing further filtration, and interactions between lymphocytes and antigens occur. After traversing the lymph node and undergoing immune interactions, lymph exits through efferent lymphatic vessels. Efferent vessels carry filtered lymph, immune cells, and antigens away from the node, ultimately rejoining the lymphatic circulation and returning lymph to the bloodstream.

14.1.5.2 Spleen

The spleen is in the upper left abdomen, just beneath the ribcage (Fig. 14.7). The spleen is highly vascular and receives blood from the splenic artery, which branches into smaller arterioles that supply blood to the white and red pulp regions. A connective tissue layer encapsulates the spleen called the splenic capsule. This capsule gives the spleen its characteristic shape and structure. The internal structure of the spleen includes (Fig. 14.8):

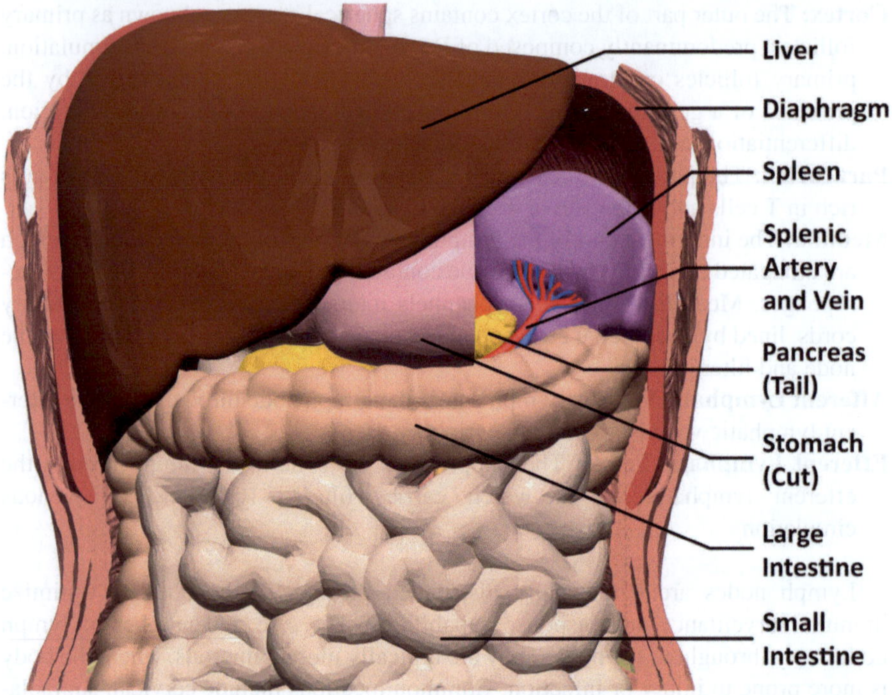

Fig. 14.7 Anatomical location of the spleen in the abdominal cavity, anterior view. (Part of the stomach has been removed)

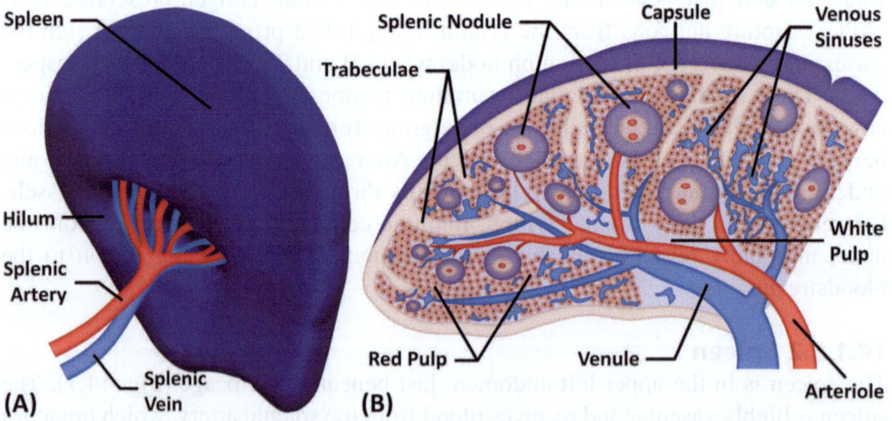

Fig. 14.8 The spleen: (**a**) Gross anatomy, anterior view. (**b**) Microscopic structure

White Pulp: The white pulp is composed of lymphoid tissue and is the site where the spleen's immune functions are carried out. The core of the white pulp is made up of lymphoid follicles, predominantly B lymphocytes. These follicles often contain germinal centers, sites of active B cell proliferation, differentiation, and antibody production. Surrounding the central arterioles, the periarteriolar lymphoid sheaths (PALS) predominantly comprise T lymphocytes. This structure facilitates the interaction between T cells, antigen-presenting cells, and antigens.

Red Pulp: The red pulp surrounds the white pulp and occupies most of the spleen's volume. It is characterized by a network of sinuses (vascular spaces) and splenic cords. Splenic sinuses are lined by endothelial cells and supported by a framework of reticular fibers. They filter blood and remove aged or damaged red blood cells. Splenic cords between the sinuses comprise a meshwork of reticular fibers, fibroblasts, and macrophages. The red pulp filters blood, removes defective blood cells, and recovers iron from hemoglobin.

Marginal Zone: The marginal zone is a transitional region between the white and red pulp. It contains specialized immune cells, such as marginal zone macrophages, dendritic cells, and B lymphocytes. The marginal zone plays a crucial role in capturing blood-borne antigens and presenting them to the lymphocytes in the white pulp.

The spleen is a crucial immune organ in innate and adaptive immunity. It is a primary site for removing blood-borne pathogens and initiating immune responses against these invaders. The spleen functions as a blood filter, removing old or damaged red blood cells and platelets from circulation. It also captures and removes foreign particles and pathogens from the bloodstream. During certain conditions, such as severe anemia or certain diseases, the spleen can serve as a site for hematopoiesis, producing red blood cells, white blood cells, and platelets to supplement the bone marrow's function.

14.1.5.3 Mucosa-Associated Lymphoid Tissues (MALT)

Mucosa-associated lymphoid tissues (MALT) are a diverse group of lymphoid structures in mucosal surfaces throughout the body. Key components of MALT include tonsils, Peyer's patches, gastrointestinal lymphoid follicles, and respiratory mucosal-associated lymphoid tissues (RMALT).

Tonsils

Tonsils (Fig. 14.9) are a group of lymphoid organs located in the throat and upper respiratory tract (Page 621). They are the body's initial defense against pathogens entering the oral and nasal passages. The bulk of the tonsillar tissue consists of lymphoid follicles. These follicles are spherical clusters of lymphocytes, primarily B cells. Within these follicles are germinal centers, sites for B cell proliferation, differentiation, and maturation. The germinal centers are more active during immune responses. Crypts are deep invaginations of the surface epithelium into the underlying lymphoid tissue, increasing the surface area of the tonsils and enhancing their ability to trap and expose antigens to the immune cells. The

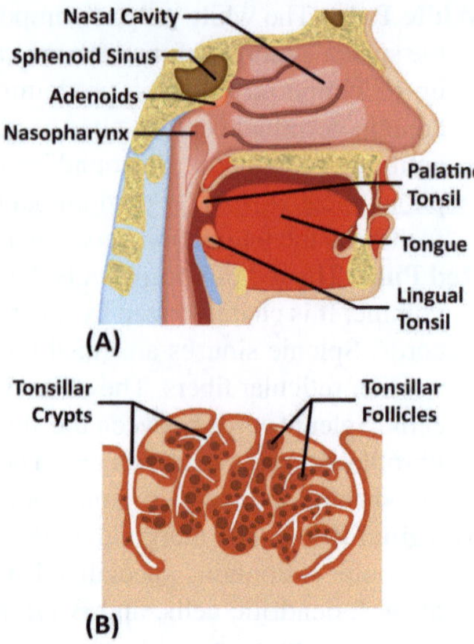

Fig. 14.9 Tonsils: (**a**) Anatomic location of tonsils, parasagittal lateral view of nasal and oral cavities. (**b**) Microscopic structure

tonsils are covered by stratified squamous epithelium, which is continuous with the mouth and throat lining.

Peyer's Patches

Peyer's patches (Fig. 14.10) are specialized lymphoid structures in the ileum. Peyer's patches appear microscopically as oval or round aggregates of lymphoid tissue visible in the mucosal layer. Overlying the Peyer's patches is a specialized epithelium known as follicle-associated epithelium (FAE). This epithelium is thinner than the surrounding intestinal epithelium and contains fewer goblet cells (mucus-secreting cells). Within the FAE are unique cells called M (microfold) cells. They facilitate the transport of antigens and microorganisms from the intestinal lumen to the lymphoid tissue of Peyer's patches. The core of Peyer's patches (Fig. 14.11) comprises multiple lymphoid follicles rich in B lymphocytes.

These follicles can contain germinal centers, especially during active immune responses, where B cell proliferation and differentiation occur. Between the follicles are areas rich in T lymphocytes and antigen-presenting cells, providing a site for T cell activation and interaction with other immune cells. Peyer's patches are crucial for sampling antigens from the intestinal lumen and initiating immune responses. Activated B cells in the germinal centers can differentiate into plasma cells, which produce antibodies, particularly IgA. This IgA is secreted into the intestinal lumen and plays a crucial role in mucosal immunity. Peyer's patches contribute to the development and maintenance of immunological memory. Memory B and T cells generated within these structures provide long-lasting protection against previously encountered pathogens. Peyer's patches also contribute to immune

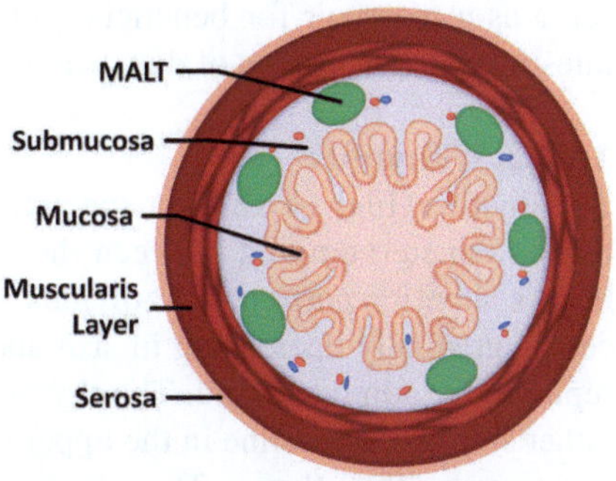

Fig. 14.10 General location of mucosa-associated lymphoid tissues (MALT)

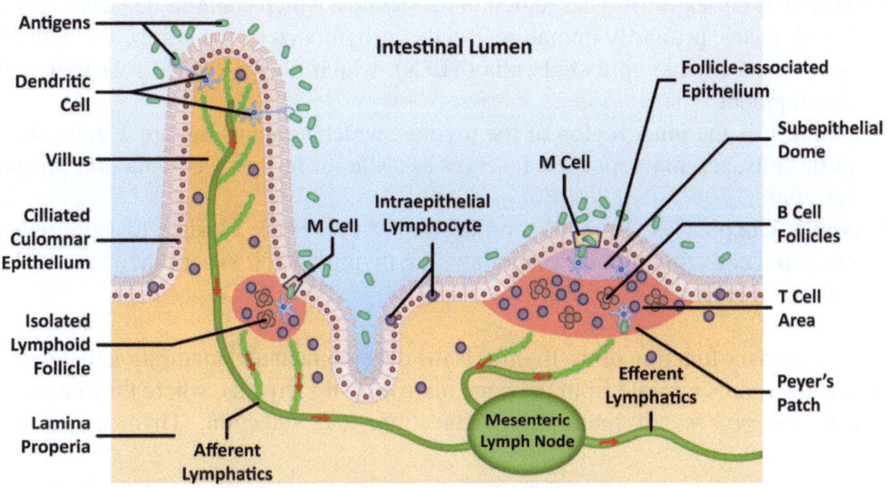

Fig. 14.11 Peyer's patches, microscopic structure. (Red arrows depict the direction of lymph flow)

tolerance in the gut. They help differentiate between harmless dietary antigens and potential threats, preventing unnecessary immune reactions to food.

14.1.5.4 Appendix

The appendix is a small, finger-like pouch attached to the cecum near the junction of the small and large intestines. The appendix is a tubular structure, typically around 3–4 inches long, with a diameter of approximately 8 mm. The wall of the appendix contains various layers, like other parts of the intestines, including mucosa, submucosa, muscularis, and serosa. The appendix contains lymphoid follicles similar to those found in other lymphoid organs. These lymphoid follicles are rich in immune cells, such as lymphocytes and antigen-presenting cells (APCs). The presence of these immune cells suggests that the appendix serves as a site for immune surveillance in the gastrointestinal tract. Recent research suggests the appendix may serve as a reservoir for beneficial gut bacteria, playing a role in repopulating the intestines after episodes of diarrhea or gut infections.

14.1.5.5 Thymus

The thymus (Fig. 14.12) is a specialized lymphoid organ in the upper chest, just behind the sternum and between the lungs. It extends from the neck's base to the heart's superior portion. The thymus is most active during childhood and adolescence, gradually decreasing in size and activity with age. In adults, it is gradually replaced by adipose tissue. The thymus is divided into two lobes, each located on either side of the midline in the upper chest. These lobes are connected by a central region called the isthmus. The internal structure of the thymus (Fig. 14.13) includes:

Cortex: The cortex is the outer region of the thymus, which contains densely packed lymphocytes, primarily immature T cells or thymocytes. The cortex also houses specialized thymic epithelial cells (TECs), which play a critical role in T cell development.

Medulla: it is the inner region of the thymus, which contains mature T cells, dendritic cells, and macrophages. It serves as a site for further T cell maturation and selection.

Hassall's Corpuscles: Hassall's corpuscles are concentric clusters of keratinized epithelial cells found in the medulla of the thymus. Their exact function is still a subject of research.

The primary function of the thymus is the development and maturation of T cells. Immature T cells migrate from the bone marrow to the thymus, where they undergo a complex process of maturation, education, and selection. Thymocytes that

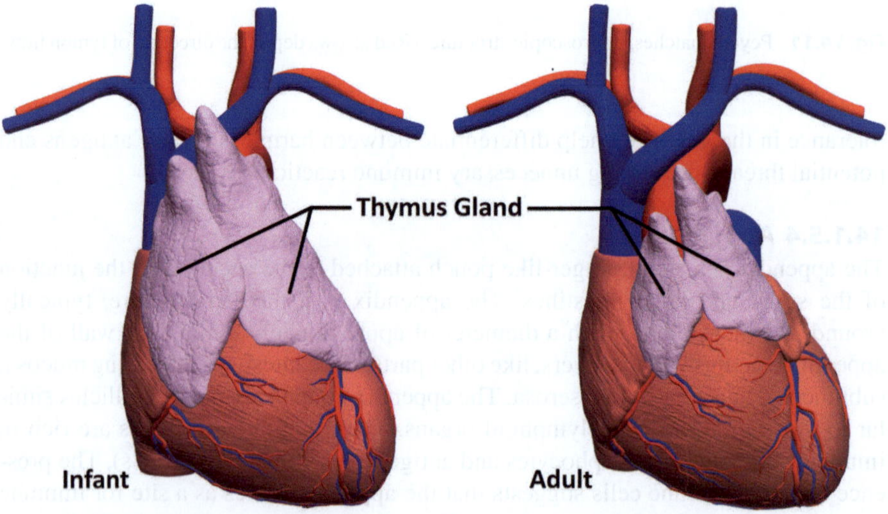

Fig. 14.12 Thymus: Anatomical location and size in (**a**) infants and (**b**) adults

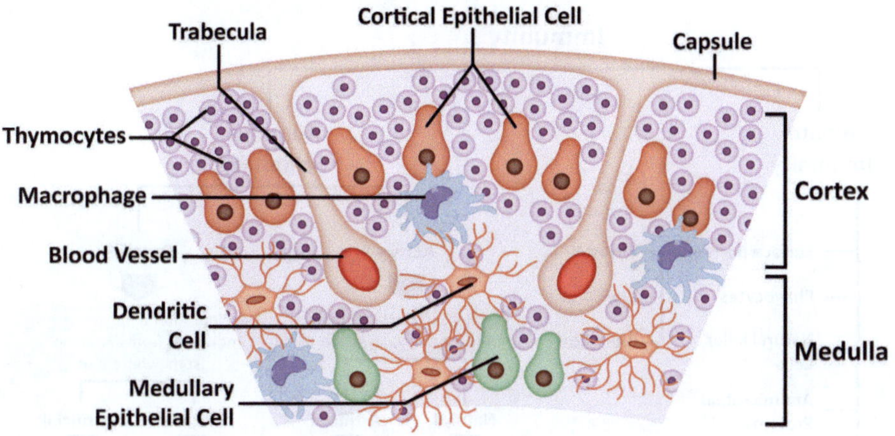

Fig. 14.13 Microscopic structure of thymus

successfully mature become functional T cells with specific antigen receptors. During T cell maturation, thymocytes are exposed to a wide range of self-antigens presented by thymic epithelial cells. T cells that recognize and react strongly to self-antigens are eliminated, preventing autoimmune reactions. Mature T cells generated in the thymus play a central role in cell-mediated immune responses (Section 14.2.2.5).

14.2 Part 2: The Immune System

The immune system is a complex and intricate network of cells, tissues, and molecules responsible for defending the human body against various pathogens, including bacteria, viruses, fungi, and parasites. The immune system comprises two major parts: innate and adaptive immunity (Fig. 14.14).

14.2.1 Innate Immunity

Innate immunity represents the first line of defense against invading pathogens (harmful agents) and is the body's immediate response to pathogens. Unlike adaptive immunity, characterized by specificity and memory, innate immunity is a non-specific defense system that provides immediate protection against microorganisms and other harmful agents.

14.2.1.1 Surface Barriers

The immune system's first line of defense against invading pathogens and environmental challenges is the establishment of surface barriers. These barriers include physical, chemical, and biological components that protect the body's external surfaces from infection and maintain homeostasis (Table 14.1).

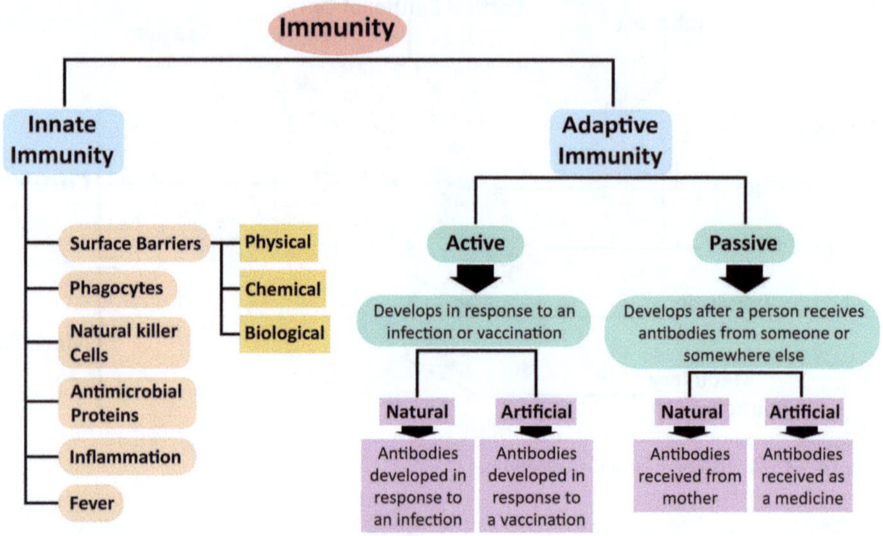

Fig. 14.14 Components of the immune system

14.2.1.2 Phagocytes

Phagocytes are specialized immune cells capable of engulfing and digesting invading microbes, cellular debris, and other foreign substances. The primary types of phagocytes include:

Neutrophils: Neutrophils are the most abundant white blood cells in the bloodstream and are often the first responders to infection or tissue injury.

Macrophages: Macrophages are tissue-resident phagocytes found in various organs and tissues, including the spleen, liver, lungs, and lymph nodes.

Dendritic Cells: Dendritic cells are another group of antigen-presenting cells that bridge the innate and adaptive immune systems. They capture antigens, process them, and present them to T cells, initiating adaptive immune responses.

Monocytes: They are circulating immune cells that can differentiate into macrophages or dendritic cells when they migrate to tissues in response to infection or inflammation.

Steps of Phagocytosis (Fig. 14.15): Phagocytes can identify pathogen-associated molecular patterns (PAMPs) in microbes. This recognition is essential for distinguishing between self- and nonself, triggering the phagocytic response. Once a phagocyte recognizes a pathogen, it extends pseudopodia to surround and engulf the microbe. The engulfed microbe is then encased in a phagosome, an intracellular vesicle. The newly formed phagosome undergoes a series of maturation steps, fusing with lysosomes to form a phagolysosome. Lysosomes contain enzymes, such as proteases and nucleases, and produce reactive oxygen species (ROS), which play a crucial role in pathogen degradation. Phagocytes also clear cellular debris and promote tissue repair after infection or injury. They release

Table 14.1 Surface barriers in innate immunity: location and protective mechanism(s)

Location		Protective mechanism
Physical surface barrier		
Skin		The outermost layer, the stratum corneum, is composed of tough keratinocytes that provide a strong, impermeable layer Desquamation (shedding skin cells) removes microbes adhered to skin surfaces.
Mucous membranes		Mucus, secreted by goblet cells, traps microbes and foreign particles Ciliary action in the respiratory tract moves mucus and trapped particles towards the throat for expulsion Tight junctions between epithelial cells prevent pathogen entry
Respiratory tract		Nasal hairs trap larger particles Mucus produced in the respiratory tract traps smaller particles and pathogens The ciliated epithelium propels the mucus towards the pharynx to be coughed out or swallowed
Gastrointestinal tract		Mucus lining the GI tract is a physical barrier, protecting the epithelium from damage and pathogen invasion Peristalsis helps move contents, including any trapped pathogens, through the digestive system for elimination
Urogenital tract		Urinary flow helps flush out pathogens Mucus in the genital tract traps pathogens
Eyes		Eyelashes trap dust and larger particles Blinking spreads tears across the eye surface, which helps wash away irritants and microbes
Ears		Earwax (cerumen) traps dirt and has antimicrobial properties Ear hairs filter larger particles, preventing them from entering deeper into the ear canal
Chemical surface barrier		
Sebum	Skin	Contains fatty acids and lactic acid that lower skin pH, inhibiting pathogen growth Antimicrobial properties prevent colonization of harmful bacteria and fungi
Sweat	Skin	Contains antimicrobial peptides (e.g., dermcidin) that kill bacteria and fungi
Gastric acid	Stomach	Highly acidic environment (pH 1.5–3.5) denatures proteins and kills most ingested bacteria and viruses

(continued)

Table 14.1 (continued)

Location		Protective mechanism
Lysozyme	Tears, saliva, nasal secretions, human milk	Destroys the peptidoglycan layer of bacterial cell walls, leading to cell lysis, which is particularly effective against gram-positive bacteria
Defensins	Mucosal surfaces (e.g., respiratory, gastrointestinal, urogenital tracts)	Insert into microbial membranes, forming pores that disrupt membrane integrity, leading to cell death
Mucus	Mucous membranes (e.g., respiratory, gastrointestinal, urogenital tracts)	Thick, viscous fluid physically traps pathogens; often contains immunoglobulins and antimicrobial peptides that neutralize or kill trapped microbes
Saliva	Mouth	Contains enzymes (e.g., amylase) and antimicrobial agents (e.g., lysozyme, lactoferrin) that inhibit microbial growth and degrade bacterial cell walls
Bile	Liver (stored in gallbladder, secreted into small intestine)	Bile salts disrupt microbial cell membranes and inhibit certain bacteria, contributing to intestinal homeostasis
Acidic pH	Skin, vagina	The acidic environment (pH ~4–5 in the vagina, slightly acidic on skin) inhibits the growth of many pathogenic microorganisms
Enzymes in pancreatic juice	Pancreas (secreted into small intestine)	Enzymes like trypsin have antimicrobial properties; they can degrade bacterial proteins and cell structures
Biological surface barrier		
Normal flora (microbiota)	Skin, gastrointestinal tract, urogenital tract, respiratory tract	Produces substances harmful to pathogens (e.g., bacteriocins) Lowers local pH (e.g., lactic acid production in the vagina), making the environment inhospitable to many pathogens Competes for binding sites on host tissues, preventing colonization by pathogens
Commensal bacteria in gut	Gastrointestinal tract	Stimulates the development and maturation of gut-associated lymphoid tissues (GALT) Influences the balance of T helper cells, playing a role in immune tolerance and defense.
Lactobacilli in vagina	Vagina	Produces lactic acid, maintaining an acidic pH that inhibits the growth of many harmful microbes Competes with pathogenic bacteria and yeast for adherence to the vaginal epithelium

(continued)

Table 14.1 (continued)

Location		Protective mechanism
Skin flora	Skin	Competes with potential pathogens for space and nutrients on the skin surface Some skin flora can produce antimicrobial substances that inhibit the growth of pathogenic bacteria
Oral and nasopharyngeal microbiota	Mouth	Competes with pathogenic microorganisms for binding sites and nutrients. Produces metabolic products (e.g., short-chain fatty acids) that inhibit pathogen growth

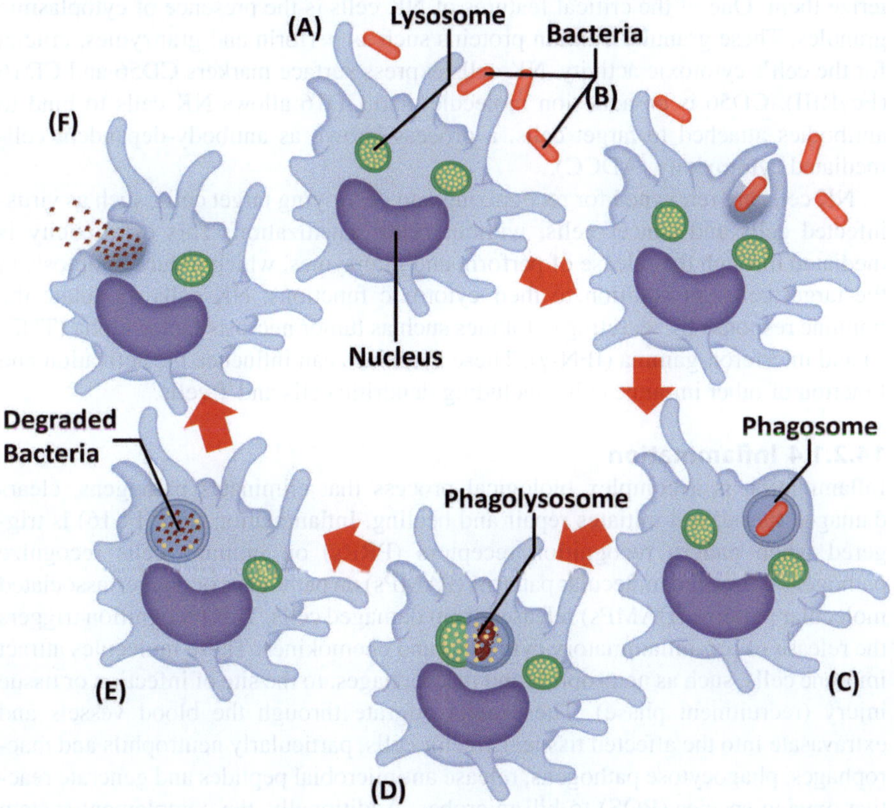

Fig. 14.15 Steps of phagocytosis: (**a**) Recognition and attachment: Phagocytes recognize and attach to the bacteria. (**b**) Engulfment: The phagocyte engulfs the bacteria by extending its pseudopodia around them, trying to form a phagosome. (**c**) Phagosome formation: The bacteria are fully enclosed in a phagosome within the cytoplasm of the phagocyte. (**d**) Phagolysosome formation: The phagosome fuses with a lysosome to form a phagolysosome. (**e**) Digestion: The digestive enzymes within the phagolysosome break down the bacterial components. (**f**) Exocytosis of debris: The phagocyte expels the indigestible debris from the degraded bacteria

growth factors and cytokines that aid in tissue regeneration. Macrophages and dendritic cells are major cells in antigen presentation, capturing and processing antigens (Section 14.2.2.5) to present them to T cells. This process initiates adaptive immune responses.

14.2.1.3 Natural Killer (NK) Cells

Natural killer (NK) cells are a unique subset of lymphocytes that originate from hematopoietic stem cells in the bone marrow. They undergo a distinct lineage commitment and development process, which differs from T and B lymphocytes. NK cells then enter the bloodstream, circulating and distributing them throughout various tissues, including the spleen, liver, and lungs. NK cells are larger than typical lymphocytes. A large, round nucleus and a noticeable amount of cytoplasm characterize them. One of the critical features of NK cells is the presence of cytoplasmic granules. These granules contain proteins such as perforin and granzymes, crucial for the cell's cytotoxic activity. NK cells express surface markers CD56 and CD16 (FcγRIII). CD56 is an adhesion molecule, and CD16 allows NK cells to bind to antibodies attached to target cells, a process known as antibody-dependent cell-mediated cytotoxicity (ADCC).

NK cells are renowned for recognizing and destroying target cells, such as virus-infected cells and cancer cells, without prior sensitization. This cytotoxicity is mediated through the release of perforin and granzymes, which induce apoptosis in the target cells. In addition to their cytotoxic functions, NK cells modulate the immune response by secreting cytokines such as tumor necrosis factor-alpha (TNF-α) and interferon-gamma (IFN-γ). These cytokines can influence the activation and function of other immune cells, including dendritic cells and T cells.

14.2.1.4 Inflammation

Inflammation is a complex biological process that eliminates pathogens, clears damaged tissue, and initiates repair and healing. Inflammation (Fig. 14.16) is triggered when pattern recognition receptors (PRRs) on immune cells recognize pathogen-associated molecular patterns (PAMPs) on pathogens or danger-associated molecular patterns (DAMPs) released from damaged cells. This recognition triggers the release of pro-inflammatory cytokines and chemokines. These molecules attract immune cells, such as neutrophils and macrophages, to the site of infection or tissue injury (recruitment phase). These cells migrate through the blood vessels and extravasate into the affected tissue. Effector cells, particularly neutrophils and macrophages, phagocytose pathogens, release antimicrobial peptides and generate reactive oxygen species (ROS) to kill microbes. Additionally, the complement system assists in pathogen clearance. Once the threat is eliminated, anti-inflammatory signals resolve the inflammation. Macrophages switch to an anti-inflammatory phenotype and promote tissue repair and healing. The inflammatory response can be broadly categorized into acute and chronic inflammation, each with distinct characteristics. Acute inflammation is the immediate and early response to injury or infection, typically lasting a few days. Chronic inflammation is a prolonged inflammatory response that may last months or even years. It often begins as acute inflammation but persists due to the non-resolution of the underlying cause.

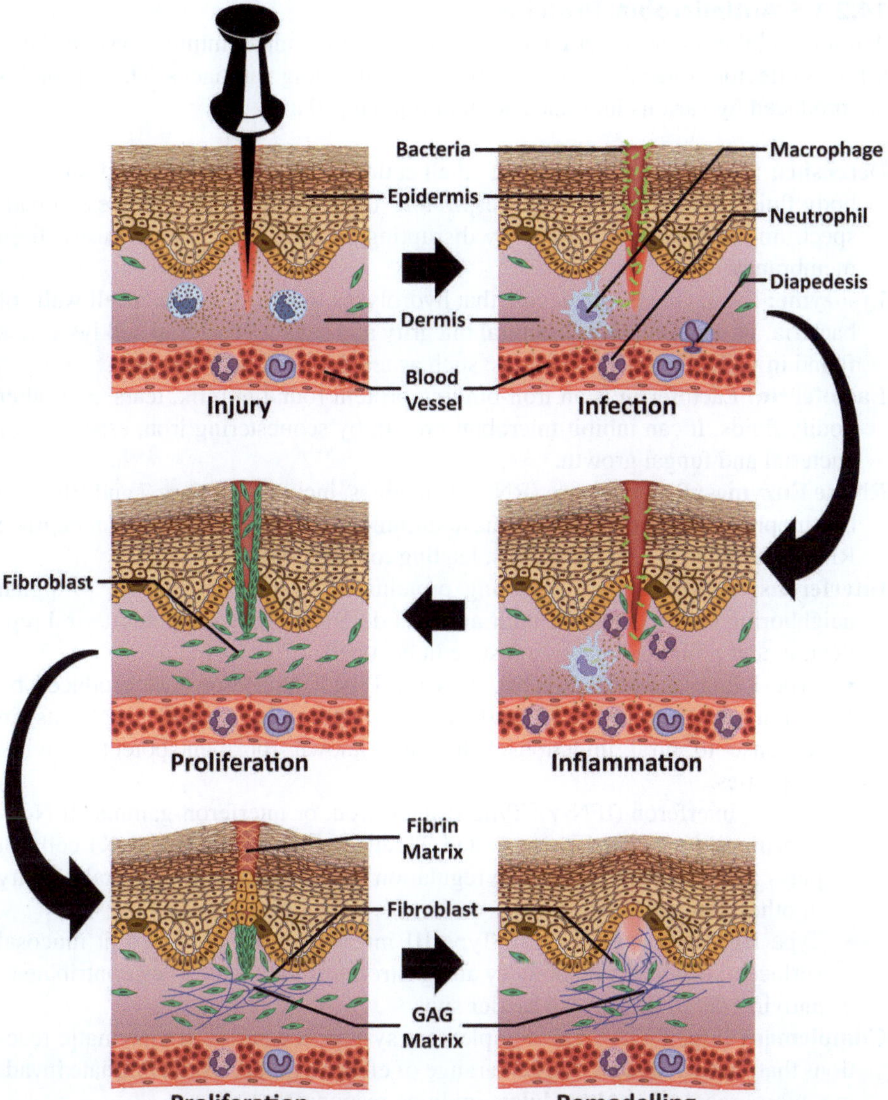

Fig. 14.16 Phases of inflammation: Following injury, bacteria grow in the injury site. The injury and the bacteria stimulate the basophils to release histamine and prostaglandins, which attract the macrophages and neutrophils to the injury site and dilate the vessels around it. These phagocytes ingest the bacteria. The dilated vessels distribute more blood to the region, causing swelling, warmth, and redness. Fibroblasts proliferate to repair the injured tissues after eradicating the bacteria and toxic chemicals. Finally, the local original tissues regenerate and remodel the injured tissues. (GAG Glucosaminoglycan)

14.2.1.5 Antimicrobial Proteins

Antimicrobial proteins are crucial components of the innate immune system. They serve as effector molecules that directly combat invading pathogens. These proteins are produced by various immune and nonimmune cells:

Defensins: Defensins are a family of small cationic peptides in various tissues and body fluids, including the skin, respiratory tract, and gut. They possess broad-spectrum antimicrobial activity by disrupting bacteria, fungi, and viruses' lipid membranes.

Lysozyme: Lysozyme is an enzyme that hydrolyzes the peptidoglycan cell walls of bacteria, weakening their structural integrity and causing bacterial cell lysis. It is found in various bodily secretions, such as tears, saliva, and mucus.

Lactoferrin: Lactoferrin is an iron-binding protein found in milk, tears, and other bodily fluids. It can inhibit microbial growth by sequestering iron, essential for bacterial and fungal growth.

RNase Enzymes: Ribonuclease (RNase) enzymes, including RNase 7 and RNase 3 (eosinophil-derived neurotoxin), have antimicrobial properties and can degrade RNA molecules in microbial cells, leading to their death.

Interferons: Interferons are signaling proteins infected cells produce to signal neighboring cells to increase their antiviral defenses. They can inhibit viral replication and promote an antiviral state in host cells (Fig. 14.17).

- Type I Interferons (IFN-α and IFN-β): Type I interferons are produced by various cells, particularly dendritic cells, macrophages, and fibroblasts, in response to viral infections. They are known for their potent antiviral properties.
- Type II Interferon (IFN-γ): Type II interferon, or interferon-gamma (IFN-γ), is primarily produced by activated T cells and natural killer (NK) cells. It plays a crucial role in immune regulation and enhances the antiviral activity of other immune cells.
- Type III Interferons (IFN-λ): Type III interferons are produced at mucosal surfaces, such as the respiratory and gastrointestinal tracts. They contribute to antiviral defenses at these barrier sites.

Complement (Fig. 14.18): The complement system is a series of enzymatic reactions that, once activated, trigger a range of effector functions to eliminate invading microorganisms and modulate immune responses.

The complement system includes soluble proteins (C1–9), cell surface receptors (CR 1–4), and regulatory molecules (factor H, factor I, CD55, and CD59 [Protectin]).

The complement cascade can be activated through three distinct pathways:

Classical Pathway: The classical pathway is initiated by antigen-antibody complexes, particularly IgM and IgG, which bind to C1q and activate C1. This leads to a cascade of proteolytic reactions, ultimately forming the membrane attack complex (MAC) and promoting opsonization.

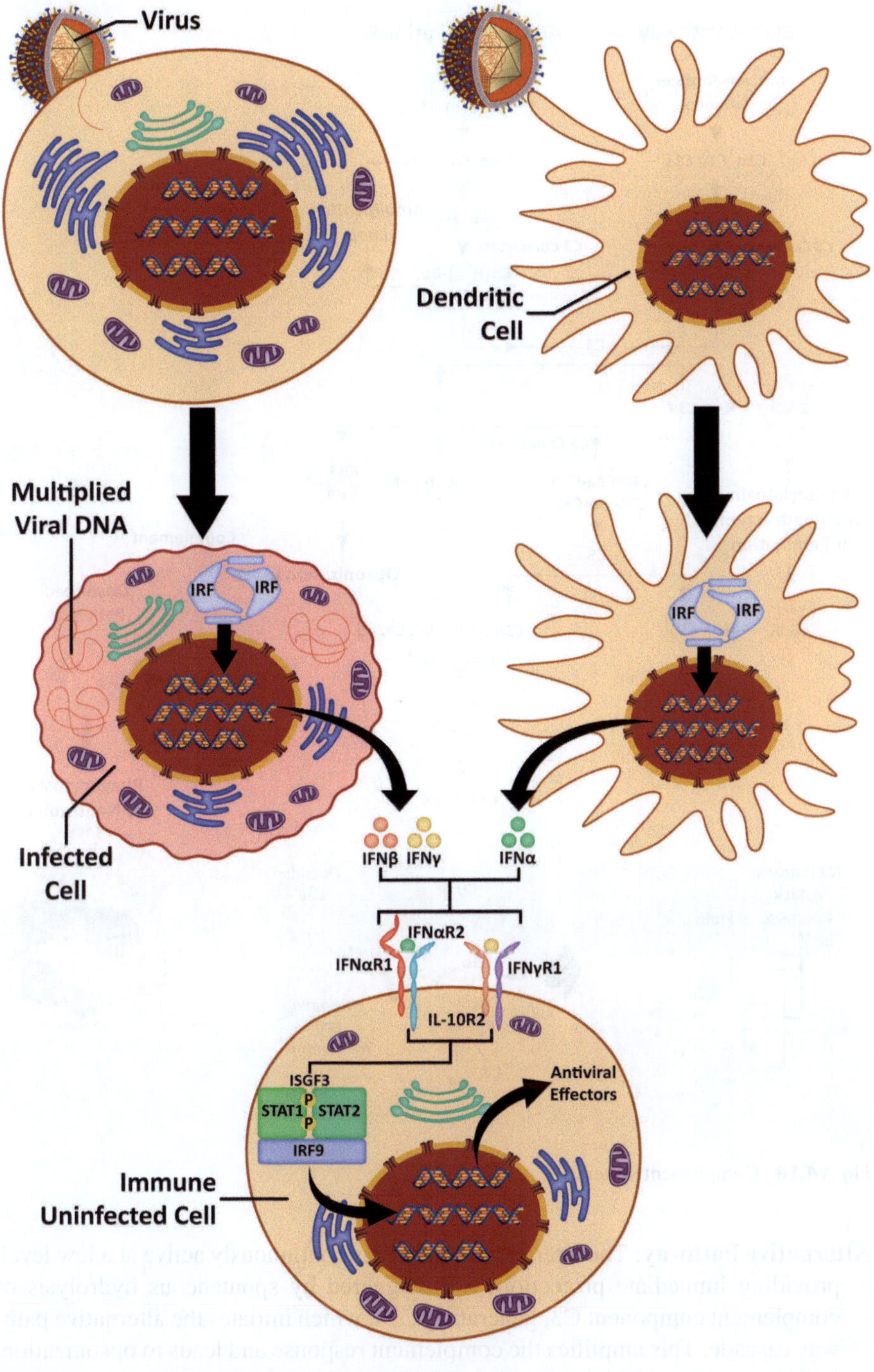

Fig. 14.17 Interferon mechanism against viral infections

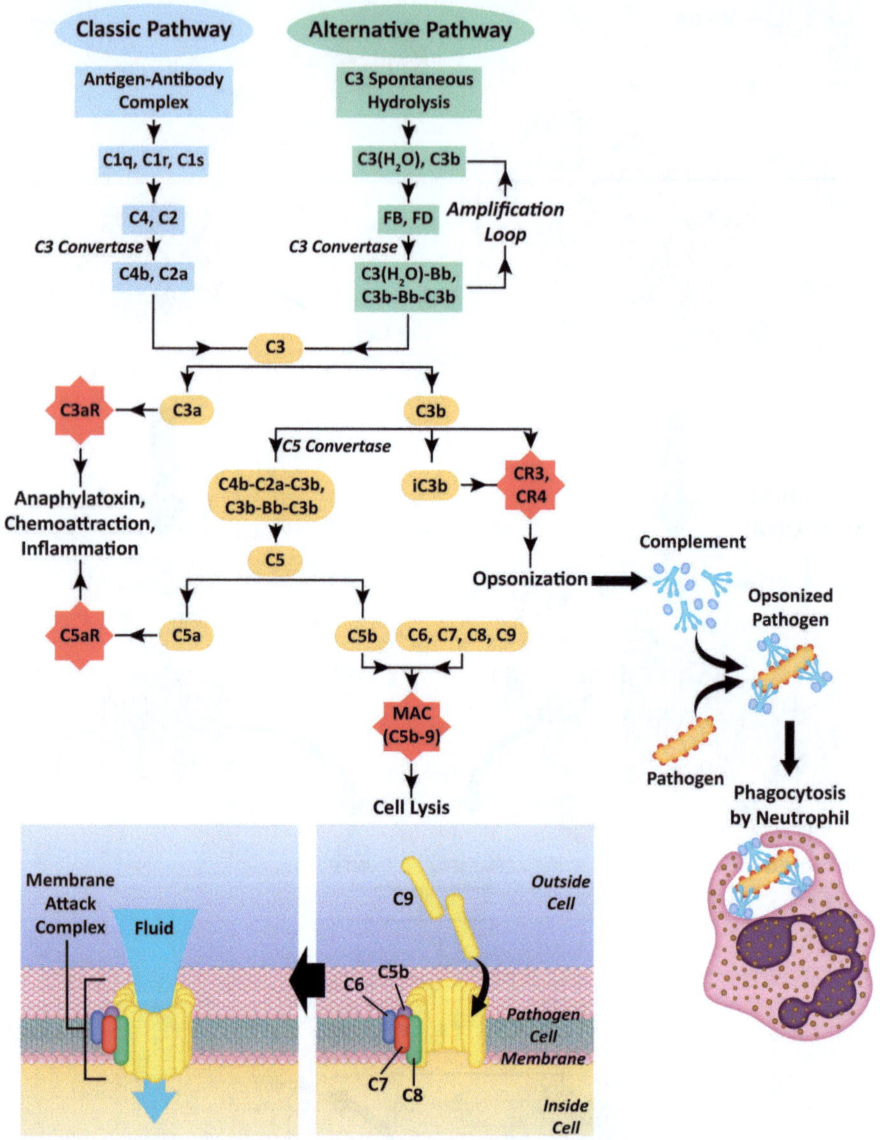

Fig. 14.18 Complement system

Alternative Pathway: The alternative pathway is continuously active at a low level, providing immediate protection. It is triggered by spontaneous hydrolysis of complement component C3, generating C3b, which initiates the alternative pathway cascade. This amplifies the complement response and leads to opsonization, inflammation, and MAC formation.

One of the primary functions of the complement system is opsonization, where complement proteins, particularly C3b, coat the surface of pathogens. Opsonized pathogens are more readily recognized and engulfed by phagocytic cells.

Complement activation generates C3a and C5a, which are potent inflammatory mediators. They induce vasodilation, increase vascular permeability, and attract immune cells to the site of infection, promoting an inflammatory response. Complement components, including C5a and C3b, bind to receptors on immune cells, such as neutrophils and mast cells, activating these cells and promoting their participation in the immune response. The complement system can lead to the formation of the membrane attack complex (MAC), a pore-forming complex composed of C6, C7, C8, and multiple C9 molecules. The MAC inserts into the lipid bilayer of target cells, causing cell lysis and destruction of the pathogen.

14.2.1.6 Fever

Fever is a body's physiological response to infection, inflammation, or other disturbances in homeostasis. It is a well- established innate immunity component and a critical defense mechanism against invading pathogens. Fever is an elevation in body temperature above the normal range (approximately 98.6 °F or 37 °C) in response to various stimuli, particularly infection. Macrophages and other immune cells recognize pyrogens through pattern recognition receptors (PRRs). Pyrogen recognition triggers the release of pro-inflammatory cytokines, particularly interleukin-1 (IL-1), interleukin-6 (IL-6), and tumor necrosis factor-alpha (TNF-α). Pro-inflammatory cytokines act on the hypothalamus, specifically on the thermoregulatory center in the preoptic area. This results in the generation of fever through various mechanisms:

- Vasoconstriction: Cytokines cause blood vessels in the skin to constrict, reducing heat loss.
- Shivering: Muscle contractions generate heat.
- Increased Metabolic Rate: Fever increases metabolic rate, producing additional heat.

The hypothalamus maintains the elevated body temperature until the infection or other pyrogenic stimulus is resolved. Fever confers several advantages in innate immunity: Enhanced immune function, inhibition of pathogen growth, increased antigen presentation, and improved tissue repair.

14.2.2 Adaptive Immunity

Adaptive immunity is one of the two primary branches of the immune system, alongside innate immunity. It is a highly specialized and dynamic defense system that allows the body to recognize and respond to specific pathogens, toxins, and other foreign substances.

14.2.2.1 Antigens

Antigens, short for "antibody generators," are molecules that can elicit an immune response when recognized by the immune system. Antigens can be classified into various categories based on their origin and properties:

Exogenous Antigens: Exogenous antigens originate from outside the body and can include:
- Pathogens: Bacteria, viruses, fungi, and parasites carry a variety of antigens.
- Toxins: Toxins produced by pathogens, such as bacterial exotoxins, can be antigens.
- Allergens: Substances like pollen, dust mites, or certain foods that trigger allergic responses.

Endogenous Antigens: Endogenous antigens are derived from within the body and may include:
- Cellular Components: Intracellular proteins and peptides can become antigens when released or presented on the surface of infected or damaged cells.
- Tumor Antigens: Abnormal proteins expressed by cancer cells can be recognized as antigens.

Autoantigens: Autoantigens are self-antigens (endogenous antigens) normally tolerated by the immune system. However, the immune system mistakenly recognizes self-antigens as foreign in autoimmune diseases and mounts an immune response against them.

Haptens: Haptens are small molecules that are not immunogenic but can become antigens when they bind to carrier molecules, forming a complex.

Superantigens: Superantigens are potent antigens that can activate many T cells by binding to the T cell receptor and MHC class II molecules (Section 14.2.2.5), leading to excessive immune responses.

14.2.2.2 Lymphocytes

Lymphocytes are highly specialized and form the basis of the body's ability to develop immunity against pathogens and respond to foreign substances. Two primary types of lymphocytes are involved in adaptive immunity: B lymphocytes and T lymphocytes.

B Lymphocytes

B lymphocytes, also known as B cells, are specialized in antibody-mediated immunity. B cells originate from hematopoietic stem cells (HSCs) in the bone marrow (Fig. 14.19). HSCs differentiate into pro-B cells, where they begin rearranging immunoglobulin (Ig) gene segments to generate diverse antigen receptors. Pro-B cells become pre-B cells, expressing a functional Ig heavy chain (IgH) and a surrogate light chain (SLC). Pre-B cells differentiate into immature B cells expressing IgH and Ig light chains (IgL). These immature B cells undergo a selection process to ensure that their antigen receptors are functional and not self-reactive. Successfully

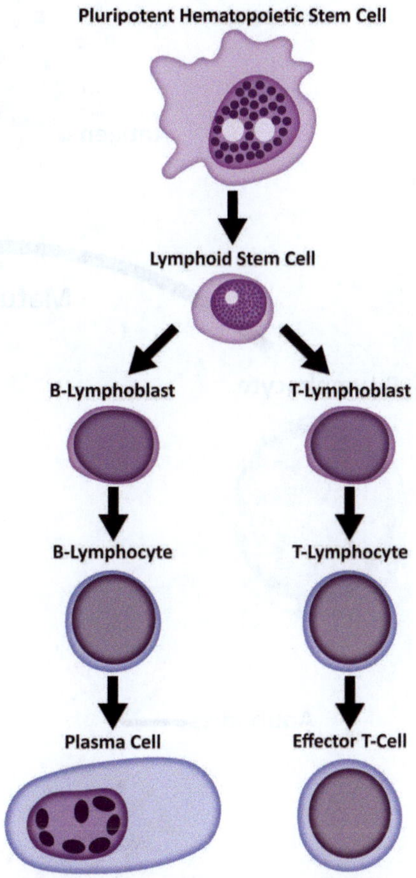

Fig. 14.19 Lymphopoiesis

screened immature B cells become mature B cells and are released into the bloodstream and secondary lymphoid organs.

B cells activate when their antigen receptors bind to specific antigens (Fig. 14.20). B cell antigen receptors (BCRs) are membrane-bound antibodies that recognize antigens with high specificity. When BCRs bind to antigens, B cells activate and internalize the antigen. Activated B cells require additional signals, typically provided by helper T cells, to complete their activation process. Activated B cells undergo clonal expansion, rapidly multiplying to form a population of two primary types of effector cells. Plasma cells (Fig. 14.20) are specialized antibody factories that produce large quantities of antibodies specific to the encountered antigen. Memory B cells are long-lived cells that "remember" the antigen and enable a quicker and more robust response upon reexposure to the same pathogen.

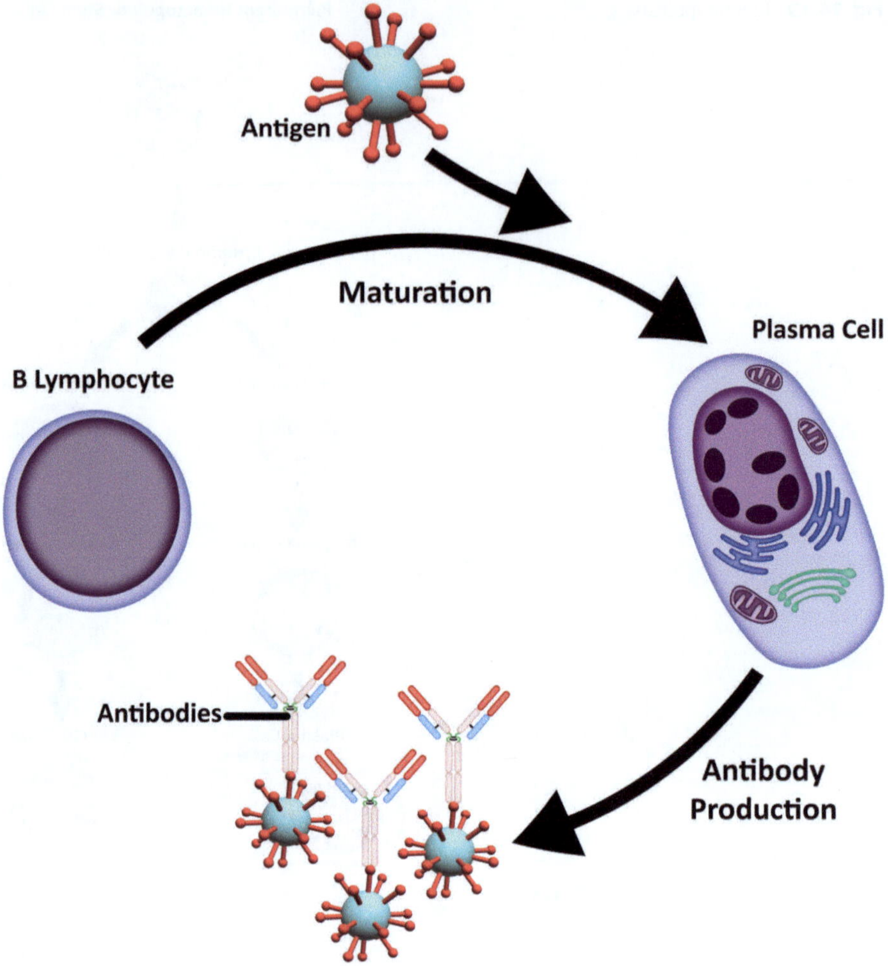

Fig. 14.20 Activation of B lymphocyte

T Lymphocytes

T lymphocytes, commonly called T cells, are responsible for cell-mediated immunity (Fig. 14.21). They are essential in recognizing and eliminating infected or abnormal cells and regulating immune responses. T lymphocytes express surface receptors called T cell receptors (TCRs). TCRs are membrane-bound protein complexes that specifically recognize and bind to antigens presented on the surface of antigen-presenting cells (APCs). T cells also express co-receptors, including CD4 and CD8, which help TCRs interact with MHC molecules. CD4 co-receptors are found on helper T cells ($CD4^+$), while CD8 co-receptors are found on cytotoxic T cells ($CD8^+$). Each T cell expresses a unique TCR with a distinct antigen-binding site, allowing the immune system to recognize various antigens. T cells recognize antigens presented in the context of major histocompatibility complex (MHC)

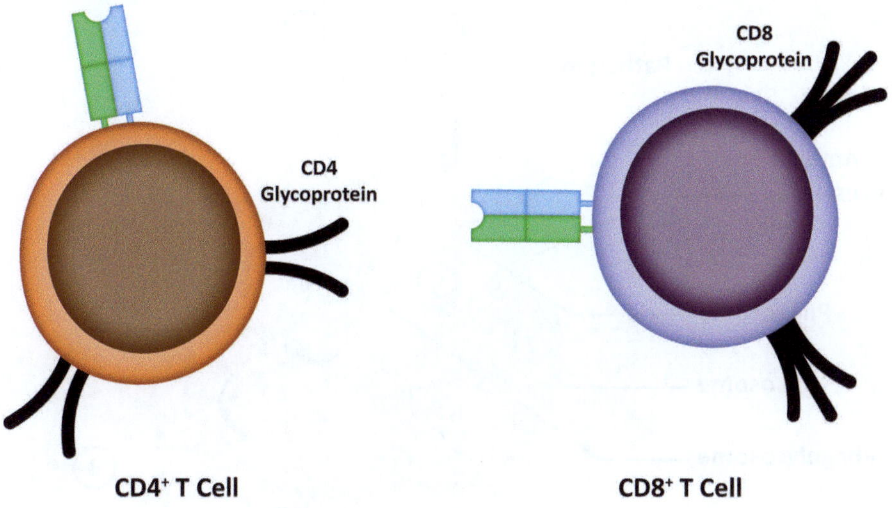

Fig. 14.21 T lymphocyte: CD4⁺ and CD8⁺

molecules on the surface of APCs. CD4⁺ T cells recognize antigens presented by MHC class II molecules, while CD8⁺ T cells recognize antigens presented by MHC class I molecules. Cytotoxic T cells are primarily responsible for recognizing and destroying infected or abnormal host cells.

They identify target cells displaying foreign antigens on their surface, leading to their elimination. Helper T cells play a central role in coordinating immune responses. They provide essential signals and cytokines that activate other immune cells, including B and cytotoxic T cells. Helper T cells are involved in both humoral and cell-mediated immunity.

Regulatory T cells are crucial for maintaining immune tolerance and preventing excessive or inappropriate immune responses. They suppress immune reactions to self-antigens and help control inflammation.

T lymphocytes undergo development and maturation in the thymus. Precursors migrate to the thymus, where they undergo positive and negative selection. Positive selection ensures that T cells recognize self-MHC molecules. Negative selection eliminates T cells that react too strongly to self-antigens, preventing autoimmunity. Successfully selected T cells mature into either CD4⁺ (helper) or CD8⁺ (cytotoxic) T cells based on the co-receptor they express.

14.2.2.3 Antigen-Presenting Cells

Antigen-presenting cells (APCs) (Fig. 14.22) are a group of immune cells whose primary function is to process and present antigens to T cells, thereby initiating a specific immune response. APCs are essential for the immune system to recognize and respond to foreign substances (antigens), including pathogens, and are also involved in tolerance to self- antigens, preventing autoimmune reactions.

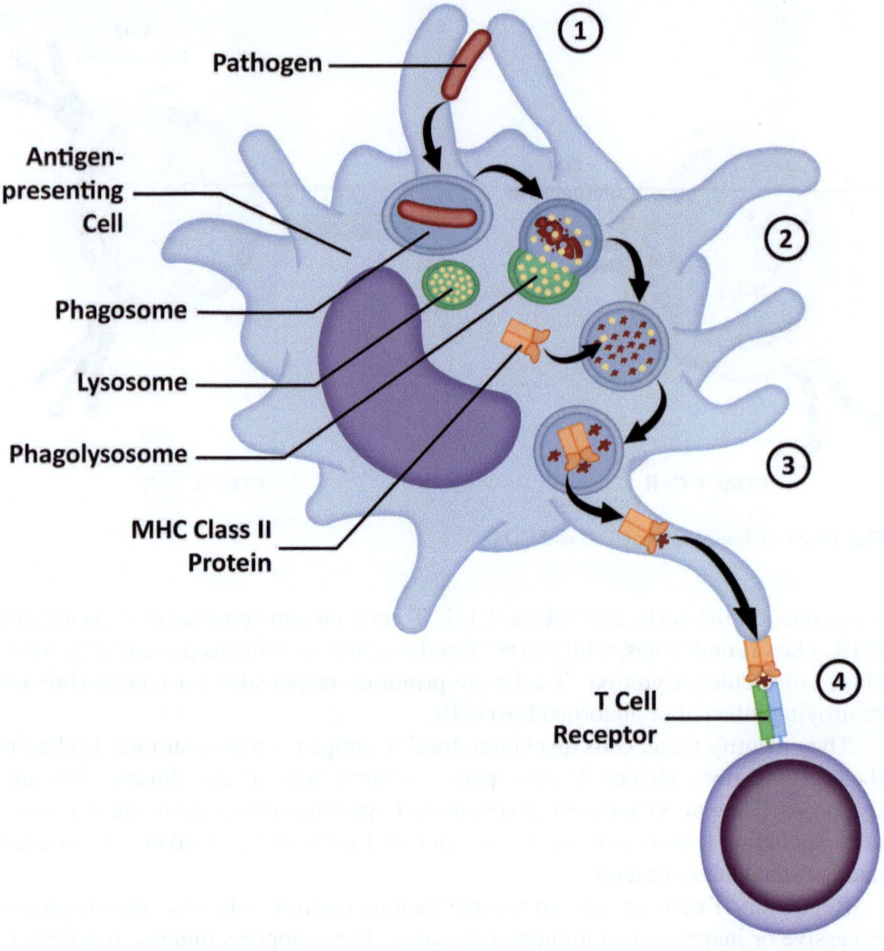

Fig. 14.22 Antigen processing and presentation mechanism: (1) An antigen-presenting cell, such as a dendritic cell or macrophage, engulfs a pathogen. (2) Its phagosome merges with a lysosome. (3) The pathogen is digested and fragments of the pathogen bind to MHC class II proteins and are presented on the cell surface. (4) The MHC class II protein can interact with helper T cells for evaluation and priming of the immune response

Types

Antigen-presenting cells include:

Dendritic Cells (DCs): These are the most potent and effective APCs in tissues exposed to the external environment, like skin (Langerhans cells) and mucous membranes. Upon encountering antigens, DCs mature and migrate to lymph nodes to present antigens to T cells.

Macrophages: In virtually all tissues, particularly at sites of infection, macrophages secrete various cytokines that modulate immune responses. They are essential in the initial, nonspecific defense and in presenting antigens to T cells.

B Lymphocytes: While primarily known for producing antibodies, B cells can also function as APCs. They are unique in presenting only the antigen their specific antibody (B cell receptor) recognizes.

Antigen Processing and Presentation

APCs can ingest antigens through various mechanisms, including phagocytosis (macrophages and dendritic cells) and receptor-mediated endocytosis (B cells) (Page 57). Inside the APC, the antigen is broken down into smaller peptides. These peptides are then loaded onto MHC molecules. MHC Class I molecules present antigens to CD8$^+$ T cells (cytotoxic T cells) and are expressed by almost all nucleated cells. MHC Class II molecules present antigens to CD4$^+$ T cells (helper T cells) and are expressed mainly by professional APCs (dendritic cells, macrophages, and B cells). The antigen-MHC complex is transported to the cell surface of the APC, where it can be recognized by T cells. By presenting antigens to T cells, APCs initiate specific immune responses, including the activation and proliferation of T cells.

14.2.2.4 Humoral Immunity

Humoral immunity provides defense against pathogens through the production of antibodies. B cells are responsible for antibody production in humoral immunity. Upon activation by antigens, B cells differentiate into plasma cells, which produce large quantities of antibodies with the same antigen specificity as the original BCRs. These antibodies are released into the bloodstream and body tissues. Plasma cells secrete antibodies continuously for several days to weeks. Antibodies can circulate in the blood, neutralize pathogens, and mark them for destruction by other immune cells.

Alongside plasma cells, activated B cells can differentiate into memory B cells. Memory B cells have a long lifespan and "remember" the antigen encountered during the initial infection. Memory B cells rapidly differentiate into plasma cells upon reexposure to the same antigen. This secondary immune response is faster and more potent than the primary response (Fig. 14.23).

Antibodies

Antibodies, known as immunoglobulins (Ig) (Fig. 14.24), are Y-shaped proteins B cells produce. They are the cornerstone of humoral immunity, recognizing and neutralizing pathogens, marking them for destruction, and regulating immune responses. Antibodies have a distinctive Y-shaped structure consisting of four polypeptide chains: two identical heavy chains (H) and two identical light chains (L). Each heavy chain consists of multiple domains, including a variable domain (VH) and three or four constant domains (CH1, CH2, CH3, and sometimes CH4). The heavy chain (VH) variable domain contains the antigen-binding site, the paratope. Each light chain has a variable domain (VL) and a constant domain (CL). The light chain (VL) variable domain also contributes to the antigen-binding site.

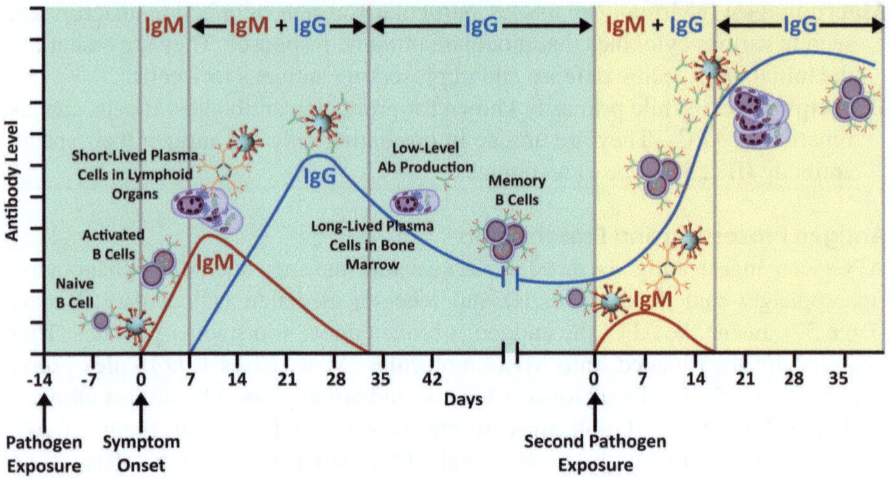

Fig. 14.23 Primary and secondary humoral responses. The primary response to antigens generates memory cells that give rise to an enhanced secondary response to antigens

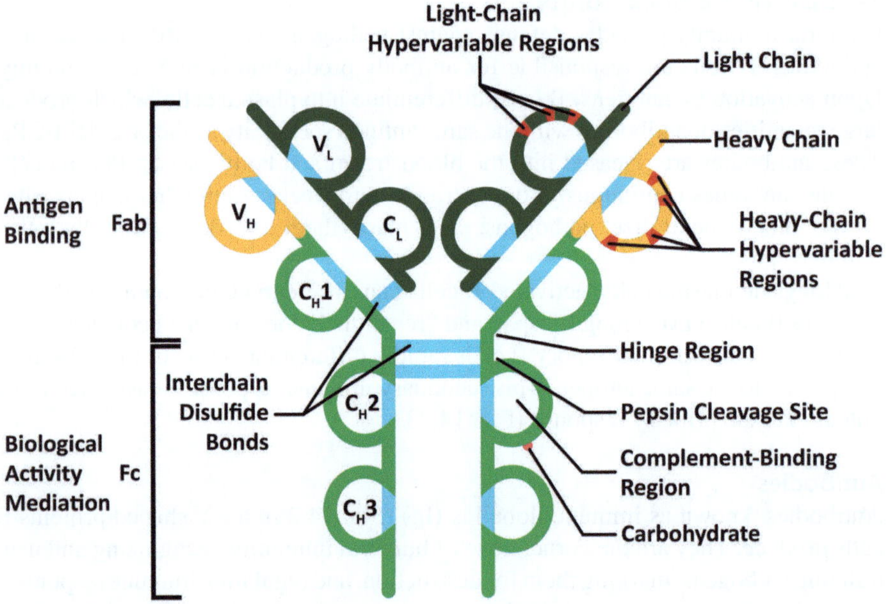

Fig. 14.24 Antibody structure: The basic structure of an antibody consists of four polypeptides—two short light chains and two long heavy chains joined together by covalent bonds between sulfur atoms called disulfide bonds (S-S). Each chain has a variable (V) region (which differs in antibodies made by different cells) and a constant (C) region (essentially identical in different antibodies of the same class). Together, the variable regions form the antigen-binding sites

The antigen-binding site of an antibody is responsible for binding to specific antigens and determining the antibody's specificity. The constant domains of both heavy and light chains form the crystallizable fragment (Fc) region. The Fc region determines the biological functions of the antibody, such as binding to immune cells and activating the complement system.

There are five major classes (isotypes) of antibodies, each with unique structural and functional characteristics. Each major class has different antibody subtypes or subclasses (Fig. 14.25) (Table 14.2). These subtypes have variations in their constant regions, leading to differences in their effector functions and antigen-binding properties.

Plasma cells synthesize antibodies based on the genetic information encoded in the B cell's rearranged immunoglobulin genes. The antibody components (heavy and light chains) are assembled and undergo posttranslational modifications, including glycosylation, in the endoplasmic reticulum and Golgi apparatus. The fully assembled antibodies are secreted into the bloodstream and tissues.

Primary and Secondary Responses in Humoral Immunity

The humoral immune response can be divided into primary and secondary, each with its timeline and characteristics (Fig. 14.23) (Table 14.3).

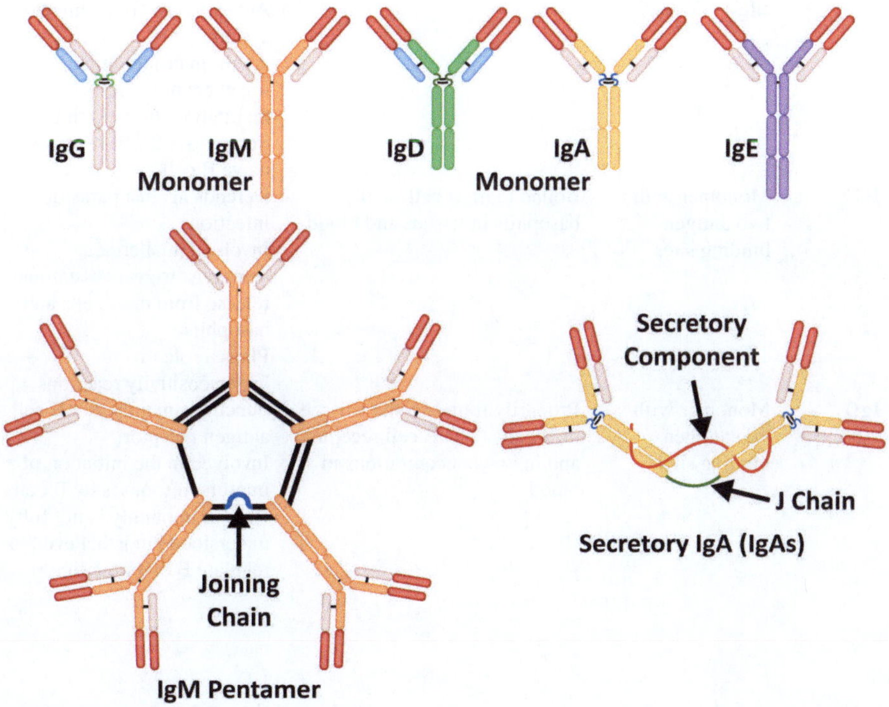

Fig. 14.25 Classes of antibodies

Table 14.2 Immunoglobulin classes: structure, location, and function

Antibody class	Structure	Location	Functions
IgG	Monomer with two antigen-binding sites	Found in blood and extracellular fluid and crosses the placenta	Provides the majority of antibody-based immunity against pathogens Opsonizes pathogens for phagocytosis Neutralizes toxins and viruses Activates the complement system Provides passive immunity to the fetus.
IgA	Exists as a monomer in serum and as a dimer in secretions	Predominantly found in mucosal areas (gut, respiratory tract, urogenital tract), saliva, tears, and breast milk	Protects mucosal barriers by neutralizing enzymes and toxins Prevents microbial adherence to epithelial cells Important in mucosal immunity and passive protection of infants through breast milk
IgM	Pentamer with five antigen-binding sites	Present in blood and lymphatic fluid	First antibody produced during an immune response Activates the complement system Agglutinates (clumps together) microbes, facilitating their elimination Serves as a B cell receptor in naïve B cells
IgE	Monomer with two antigen-binding sites	Bound to mast cells and basophils in tissues and blood	Defends against parasitic infections Involved in allergic reactions; triggers histamine release from mast cells and basophils Plays a role in hypersensitivity reactions
IgD	Monomer with two antigen-binding sites	Primarily found on the surface of B cells (as a B cell receptor) and in low concentrations in blood	Functions mainly as a B cell antigen receptor Involved in the initiation of immune responses by B cells Role in immunity is not fully understood but is believed to regulate B cell function

Table 14.3 Comparison of primary and secondary humoral immune responses

Feature	Primary humoral immune response	Secondary humoral immune response
Initiation	Triggered upon first exposure to a specific antigen	Triggered upon subsequent exposure(s) to the same antigen
Latency period	Longer latency period (days to weeks)	Shorter latency period (hours to a few days)
Magnitude of response	Lower magnitude of antibody production	Higher magnitude of antibody production
Duration of response	Relatively short-lived	Prolonged; can last for years or even a lifetime
Types of antibodies produced	Initially, more IgM is produced, followed by IgG	Predominantly IgG, IgA, or IgE, depending on the response site
Memory cells	Generation of memory B cells occurs	Memory B cells from the primary response are reactivated
Speed of antibody production	Slower; takes longer to reach peak antibody levels	Rapid; peak antibody levels are achieved quickly
Role in vaccination	Forms the basis for initial immunization	Basis for booster shots; ensures long-term immunity

The primary humoral immune response occurs when the immune system is first exposed to an antigen.

- Day 0 (Antigen Exposure): The process begins when an antigen enters the body for the first time. Antigen-presenting cells (APCs) process the antigen and present it to helper T cells.
- Days 1–5 (Initial Response): B cells that recognize the antigen, with the help of activated helper T cells, become activated. These B cells then differentiate into plasma cells and memory B cells.
- Days 6–10 (Antibody Production): Plasma cells produce antibodies specific to the antigen. The dominant antibody type produced initially is IgM. The concentration of antibodies in the blood gradually increases.
- Days 10–14 (Peak Antibody Response): The antibody levels peak, and IgG antibodies begin to replace IgM as the predominant type. IgG provides more effective and long-lasting immunity.
- Days 14+ (Decline and Memory Cell Formation): Antibody levels slowly decline after the peak. Memory B cells remain in the lymphoid tissues, ready to respond to future exposures.

The secondary humoral immune response is triggered upon subsequent exposure to the same antigen.

- Day 0 (Subsequent Antigen Exposure): The immune system quickly recognizes the antigen due to the presence of memory B cells from the primary response.
- Days 1–3 (Rapid Activation): Memory B cells are rapidly activated and differentiate into plasma cells. These cells produce antibodies more quickly and efficiently than during the primary response.

- Days 3–5 (High Antibody Levels): There is a rapid increase in antibody levels, mainly IgG, which is more effective than IgM. The antibodies produced are of higher affinity due to affinity maturation.
- Days 5+ (Sustained Response and Longevity): Compared to the primary response, antibody levels remain higher for longer. Memory B cells are replenished, ensuring long-lasting immunity.

Active and Passive Humoral Immunity

Active and passive humoral immunity are distinct ways individuals acquire immunity to specific antigens (Table 14.4).

Active Humoral Immunity

- Natural Active Immunity: Natural active immunity occurs when an individual is exposed to a pathogen in their environment. B cells recognize the pathogen's antigens, leading to the production of antibodies and memory B cells. After recovery, the individual is immune to future infections by the same pathogen.
- Artificial Active Immunity (Vaccination): Artificial active immunity is induced through vaccination, where individuals are exposed to a harmless form of the pathogen (antigen) or a related antigen. Vaccination stimulates the production of antibodies and memory B cells, resulting in long-lasting immunity without causing the disease.

Passive Humoral Immunity

- Natural Passive Immunity: Natural passive immunity occurs when antibodies are transferred from a mother to her infant through breast milk or across the placenta. These maternal antibodies temporarily protect the infant until its immune system matures.

Table 14.4 Comparison of active and passive humoral immunities

Feature	Active humoral immunity	Passive humoral immunity
Source of antibodies	Antibodies are produced by the individual's B cells after antigen exposure	Antibodies are obtained from another source, such as maternal antibodies, or through administration (e.g., immunoglobulins)
Duration of immunity	Long-lasting due to the production of memory cells. Immunity can last for years or a lifetime	Short-term; antibodies degrade over time, usually lasting a few weeks to months
Onset of immunity	Slow; takes days to weeks as the immune system must recognize the antigen and produce specific antibodies	Immediate; antibodies are already formed and start working as soon as they are introduced
Memory formation	Generates immune memory; subsequent exposures to the same antigen produce a rapid and robust response	Does not generate immune memory; no enhanced response upon subsequent exposures
Strength of response	Initially weaker and slower, it strengthens significantly with memory cell formation over time	Strong and immediate, but wanes as the transferred antibodies are metabolically degraded

- Artificial Passive Immunity (Antibody Administration): Artificial passive immunity involves the administration of preformed antibodies, such as monoclonal antibodies (mABs) (Monoclonal antibodies are created by cloning identical immune cells [usually B cells] to produce large quantities of a single type of antibody, hence the term "monoclonal." They are highly specific and designed to target a single antigenic site.) or immune globulins to individuals. These antibodies are derived from individuals with immunity or produced in a laboratory.

14.2.2.5 Cellular Immunity

Cellular immunity, also known as cell-mediated immunity, is a vital arm of the adaptive immune system responsible for defending the body against intracellular pathogens, abnormal cells, and foreign antigens. Unlike humoral immunity, which involves antibodies, cellular immunity relies on the activity of specific immune cells, particularly T lymphocytes (T cells). The components of cellular immunity include T lymphocytes (helper, cytotoxic, and regulatory) and antigen- presenting cells.

Major Histocompatibility Complex (MHC) Proteins and Antigen Presentation

The primary function of MHC proteins is to present antigens to T cells. MHC proteins help T cells distinguish between self- and nonself cells. T cells are educated in the thymus to recognize self-MHC molecules.

Intracellular pathogens, such as viruses, produce antigens within host cells. These antigens are degraded into peptides in the cytoplasm by the proteasome. Peptides are transported into the endoplasmic reticulum and loaded onto MHC class I molecules. Loaded MHC class I-antigen complexes are transported to the cell surface for recognition by $CD8^+$ cytotoxic T cells. APCs engulf extracellular pathogens and antigens. Within APCs, antigens are processed into peptides in lysosomes. MHC class II molecules in the ER bind to a protein called the Invariant Chain (I_i), preventing premature binding of self-antigens. In the endocytic pathway, I_i is degraded, and the antigenic peptides replace it in the MHC class II groove. MHC class II-antigen complexes are transported to the cell surface for recognition by $CD4^+$ helper T cells (Fig. 14.26).

Activation and Differentiation of T Cells

Activation and differentiation of T cells happen when they encounter specific antigens. Activation begins when a T cell encounters its cognate antigen presented on MHC molecules on the surface of APCs. The T cell receptor (TCR) recognizes the antigen-MHC complex. In addition to TCR engagement, co-stimulatory signals are required for full T cell activation. Co-stimulation is provided by interaction between co-stimulatory molecules on the T cell (e.g., CD_{28}) and ligands on the APC (e.g., $CD_{80/86}$). TCR engagement and co-stimulation trigger intracellular signaling cascades, producing IL-2 as an autocrine signal (self-signal) to promote T cell proliferation. Activated T cells undergo clonal expansion, resulting in the proliferation of antigen-specific T cell clones. Based on the local cytokine milieu, helper T cells

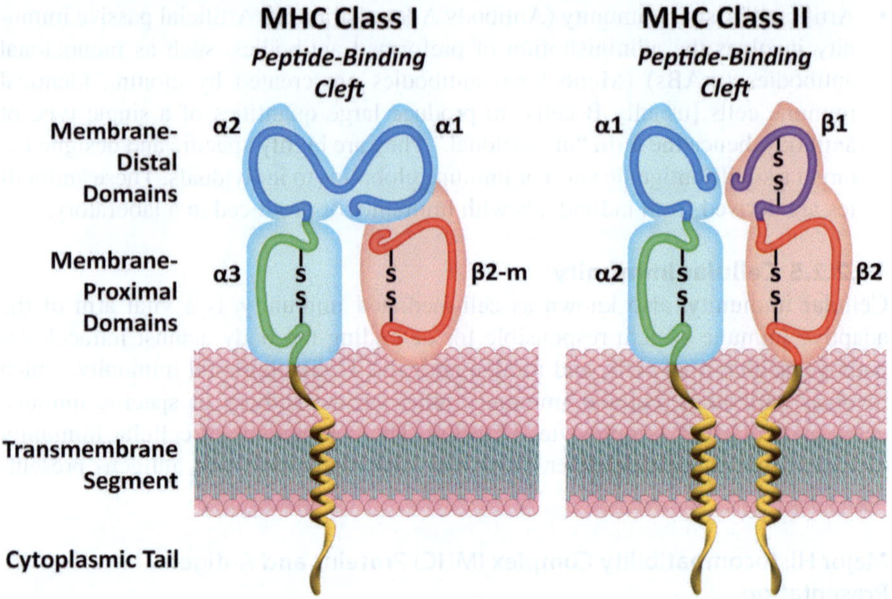

Fig. 14.26 Major histocompatibility complex (MHC) structure

differentiate into distinct subsets, including Th1, Th2, Th17, and Treg cells. Each subset plays specialized roles in immune responses (Fig. 14.27):

- Th_1 cells assist in cell-mediated immunity and activate macrophages.
- Th_2 cells aid humoral immunity and stimulate B cells.
- Th_{17} cells contribute to inflammation and host defense at mucosal surfaces.
- Treg cells regulate immune responses and maintain tolerance.

Cytotoxic T lymphocytes (CTLs) differentiate into effector cytotoxic T cells capable of directly targeting and killing infected or abnormal host cells. Perforin and granzymes are released to induce apoptosis in target cells (Fig. 14.28).

A subset of activated T cells becomes long-lived memory T cells. Memory T cells "remember" the antigen and respond more rapidly and effectively upon reexposure.

Cellular Immunological Memory

Immunological memory refers to the ability of the immune system to "remember" previous encounters with specific antigens, resulting in a rapid and enhanced immune response upon reexposure, leading to the secondary response.

Different timelines and efficacies characterize cellular immunity's primary and secondary responses (Table 14.5).

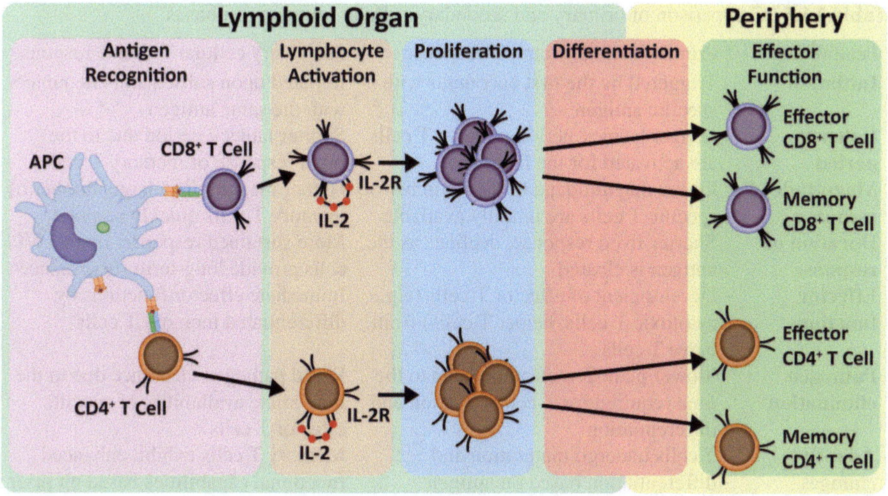

Fig. 14.27 T lymphocyte activation and differentiation.

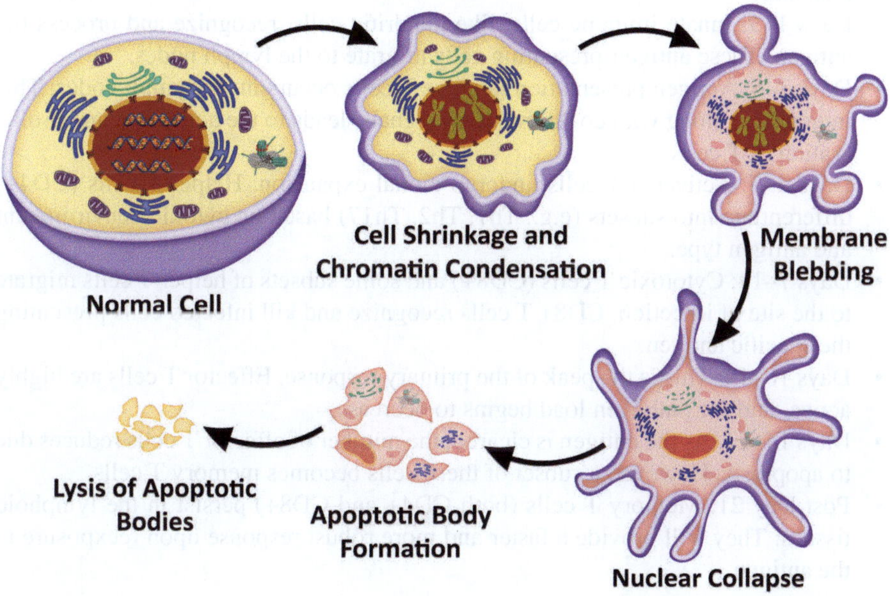

Fig. 14.28 Cell apoptosis

Table 14.5 Comparison of primary and secondary cellular immune responses

Feature	Primary cellular immune response	Secondary cellular immune response
Initiation	Triggered by the first encounter with a specific antigen	Initiated upon subsequent encounters with the same antigen
Latency period	Longer latency period as naïve T cells are activated for the first time	Shorter latency period due to the rapid response of memory T cells
Magnitude of response	Lower magnitude; a limited number of specific T cells are initially available	Higher magnitude; a large number of memory T cells quickly respond
Duration of response	Shorter-lived response; declines as the antigen is cleared	More sustained response; memory T cells provide long-term surveillance
Effector functions	Development of effector T cells (e.g., cytotoxic T cells, helper T cells) from naïve T cells	Immediate effector functions by differentiated memory T cells
Pathogen elimination	Slower pathogen clearance due to the time required for T cell activation and differentiation	Rapid pathogen clearance due to the immediate availability of specific effector T cells
Adaptive changes	T cells undergo maturation and differentiation based on antigen exposure	Memory T cells exhibit enhanced functional capabilities based on prior antigen exposure

Primary Immune Response

- Day 0: Exposure to a novel antigen occurs, initiating the primary immune response.
- Days 1–3: Innate immune cells, like dendritic cells, recognize and process the antigen. These antigen-presenting cells migrate to the lymph nodes.
- Days 3–5: Antigen presentation to naive T cells occurs in the lymph nodes. This interaction, along with co- stimulatory signals, leads to the activation and proliferation of T cells.
- Days 5–7: Activated T cells undergo clonal expansion. Helper T cells (CD4+) differentiate into subsets (e.g., Th1, Th2, Th17) based on cytokine environment and antigen type.
- Days 7–14: Cytotoxic T cells (CD8+) and some subsets of helper T cells migrate to the site of infection. CD8+ T cells recognize and kill infected cells presenting the specific antigen.
- Days 10–14: This is the peak of the primary response. Effector T cells are highly active, and the pathogen load begins to decrease.
- Days 14–21: As the antigen is cleared, the number of effector T cells reduces due to apoptosis. However, a subset of these cells becomes memory T cells.
- Post Day 21: Memory T cells (both CD4+ and CD8+) persist in the lymphoid tissues. They will provide a faster and more robust response upon reexposure to the antigen.

Secondary Immune Response

- Day 0: Reexposure to the same antigen.
- Hours to Days 1–3: Memory T cells rapidly recognize the antigen. These cells have a lower threshold for activation than naive T cells.

- Days 1–5: Accelerated clonal expansion of memory T cells. These cells differentiate into effector cells more swiftly than during the primary response.
- Days 3–7: Increased specific effector T cells lead to a more rapid and effective pathogen elimination.
- Days 5–10: Peak of the secondary response, typically more pronounced and quicker than the primary response. Enhanced cytotoxic activity and cytokine production contribute to rapid pathogen clearance.
- Days 10–14: As the antigen is eliminated, the effector cells undergo apoptosis. However, the pool of memory T cells is replenished and often expanded, providing even more efficient protection against future exposures.

14.2.3 Organ Transplants

Organ transplantation is a life-saving medical procedure involving transferring organs or tissues from one individual (the donor) to another (the recipient). While transplantation can significantly improve recipients' quality of life and survival, it poses a unique challenge to the immune system.

14.2.3.1 Immune Responses in Organ Transplantation

Innate Immune Responses: Innate immune cells, such as neutrophils and macrophages, play roles in the early stages of graft injury and inflammation.

Allograft Recognition: The immune system recognizes transplanted organs or tissues as foreign, primarily due to differences in MHC antigens between the donor and recipient. Allografts express donor MHC antigens that can activate recipient T cells.

Cell-Mediated Immune Responses: $CD8^+$ cytotoxic T cells recognize and attack donor cells expressing foreign MHC antigens. This cellular response leads to acute rejection and tissue damage.

Humoral Immune Responses: B cells produce antibodies against donor antigens, leading to antibody-mediated rejection (AMR). AMR can result in vascular damage and organ dysfunction.

14.2.3.2 Mechanisms of Rejection

Hyperacute Rejection: Occurs within minutes to hours after transplantation. Preexisting antibodies in the recipient react with donor antigens, leading to immediate graft destruction.

Acute Cellular Rejection: Develops within days to weeks after transplantation and involves activated T cells recognizing donor antigens and causing cellular damage.

Chronic Rejection: Develops over months to years and is characterized by fibrosis and narrowing of blood vessels, leading to gradual organ dysfunction. It involves both cellular and humoral responses.

14.2.3.3 Advances in Organ Transplantation

Desensitization Protocols: Strategies to reduce preexisting antibodies in sensitized recipients, allowing for transplantation in cases of high immunological risk.

Organ Preservation Techniques: Improved organ preservation and transportation methods extend donor organs' viability. HLA Matching: Advances in human leukocyte antigen (HLA) typing and matching help reduce the risk of rejection.

Tolerance Induction: Research into immune tolerance aims to achieve long-term graft acceptance without lifelong immunosuppression.

14.2.4 Cytokines

Cytokines are a broad and diverse group of small proteins crucial in controlling the growth, differentiation, and activation of immune system cells and blood cells. They are secreted primarily by immune cells like lymphocytes (T cells, B cells), macrophages, dendritic cells, and various other cell types, including endothelial cells and fibroblasts. Cytokines play key roles in the body's immune response, inflammation, tissue repair, and even in developing diseases like autoimmune disorders and cancer.

14.2.4.1 Characteristics of Cytokines

Cytokines are typically small peptides or glycoproteins. They are often produced transiently and locally, acting in a paracrine or autocrine fashion, although some can act in an endocrine manner (traveling through the bloodstream). A single cytokine can have multiple effects on different types of cells (pleiotropic effects). Different cytokines can have similar functions. Cytokines often work synergistically or antagonistically with each other.

14.2.4.2 Types of Cytokines (Table 14.6)

- Interleukins (ILs): These are primarily produced by leukocytes and act on other leukocytes. They have a wide range of functions in immunity and inflammation.
- Tumor Necrosis Factors (TNFs): They are involved in systemic inflammation and are among the cytokines that can cause apoptosis (cell death).
- Interferons (IFNs): These cytokines are involved in responding to viral infections and modulating the immune response.
- Growth Factors: Transforming growth factor-beta (TGF-β) and vascular endothelial growth factor (VEGF) play roles in cell growth and wound healing.
- Chemokines: They are specialized in attracting leukocytes to sites of infection or inflammation.

14.2.4.3 Functions of Cytokines

- Immune Response: Cytokines are vital in both innate and adaptive immunity. They help activate and regulate immune responses.
- Inflammation: Pro-inflammatory cytokines are critical in mediating the inflammatory response and essential for combating infections and wound healing.

Table 14.6 Types of cytokines: structure, source, and functions

Cytokine type	Structure	Source	Functions
Interleukins (ILs)	Typically, glycoproteins	Produced by various cells, predominantly by helper T cells, macrophages, and dendritic cells	Mediate communication between white blood cells Involved in inflammation, immune response regulation, and hematopoiesis Examples include IL-1 (pro-inflammatory), IL-2 (T cell growth factor), IL-4 (B cell differentiation), and IL-10 (anti-inflammatory)
Interferons (IFNs)	Proteins or glycoproteins	Type I IFNs (IFN-α, IFN-β) are produced by virus-infected cells. Type II IFN (IFN-γ) is produced by T cells and NK cells	IFN-α and IFN-β are involved in antiviral responses, increasing resistance of cells to viral infection IFN-γ activates macrophages and enhances antigen presentation, which is pivotal in cellular immunity
Tumor necrosis factors (TNFs)	Proteins	Mainly produced by macrophages, also by T cells and NK cells	TNF-α is a major mediator of inflammation and fever Involved in the regulation of immune cells Can induce apoptosis in tumor cells
Growth factors	Proteins or glycoproteins	Produced by various cells, including macrophages, T cells, and fibroblasts	Stimulate cell growth, proliferation, and differentiation Important in wound healing and tissue repair Examples include erythropoietin (EPO) and granulocyte colony-stimulating factor (G-CSF)
Chemokines	Small proteins	Produced by various cell types, including immune cells, fibroblasts, and endothelial cells	Induce chemotaxis, directing the movement of immune cells towards infection sites Coordinate migration of immune cells during inflammation and immune surveillance
Transforming growth factor-beta (TGF-β)	Protein	Produced by many cell types, including T cells, macrophages, and endothelial cells	Involved in the regulation of immune response and inflammation Plays a role in tissue regeneration and repair Can suppress immune function and promote tolerance

- Cell Proliferation and Differentiation: Certain cytokines stimulate or inhibit the proliferation and differentiation of various blood cells and immune cells.
- Regulation of Metabolic Processes: Some cytokines influence cell metabolic activities, like lipid metabolism.

Suggested Reading

Advanced Physiology and Pathophysiology: Essentials for Clinical Practice, 2nd Ed. Nancy Tkacs, Linda Herrmann, Springer Publishing Company, 2024

Anatomy & Physiology, S. Meloni, et al. Independent Publisher, 2023

Anatomy & Physiology: The Unity of Form and Function, 10th Ed. Kenneth S. Saladin, McGraw Hill, 2023

Gray's Anatomy for Students, 5th Ed. Richard L. Drake, et al. Elsevier, 2020

Human Anatomy & Physiology, 11th Ed. Elaine N. Marieb and Katja Hoehn, Pearson, 2022

Immunology: Clinical Case Studies and Disease Pathophysiology, Warren Strober, Susan R. Gottesman, Wiley-Blackwell, 2014

Moore's Clinically Oriented Anatomy, 9th Ed. Keith L. Moore, et al. Lippincott Williams & Wilkins, 2020

Netter Atlas of Human Anatomy: Classic Regional Approach, 8th Ed. Frank H. Netter, Elsevier, 2022

Principles of Anatomy and Physiology, 16th Ed. Gerard J. Tortora, Bryan H. Derrickson, Wiley, 2020

Respiratory System 15

In this chapter, you will learn:

- General structure of the respiratory system
- Detailed anatomy and function of the larynx, including its role in sound production and airway protection
- Description of the nasal cavity and its functions, including air filtration and olfaction
- Structure and function of the paranasal sinuses
- Anatomy and role of the pharynx in the respiratory and digestive systems
- Role of the epiglottis in protecting the airway during swallowing
- Detailed discussion on the structure of the lower respiratory tract
- Function of the trachea in air conduction
- Anatomy of the bronchi and their branching into the bronchial tree
- Description of the lungs, including lobe arrangement and lung surfaces
- Structure and function of the alveoli and the process of gas exchange
- Overview of pulmonary and systemic circulation and their roles in gas exchange
- Role of respiratory muscles in breathing
- Structure and function of the pleura and pleural cavity
- Mechanics of pulmonary ventilation, including inspiration and expiration
- Definition and significance of various lung volumes and capacities
- Concept of lung dead space and its types: anatomical and physiological
- Mechanisms of oxygen transport in the blood
- Methods of carbon dioxide transport in the blood
- Physical principles governing gas behaviors, including Dalton's and Henry's laws
- Neural and chemical control of respiration
- Role of the respiratory system in maintaining acid-base balance
- Ventilation-perfusion coupling and its importance in efficient gas exchange

The respiratory system (Fig. 15.1) is intricately designed to fulfill the vital function of gas exchange. This system allows for oxygen inhalation and carbon dioxide

H. Hosseinzadeh, *Fundamentals of Medicine for Biomedical Engineering*,
https://doi.org/10.1007/978-3-031-74085-5_15

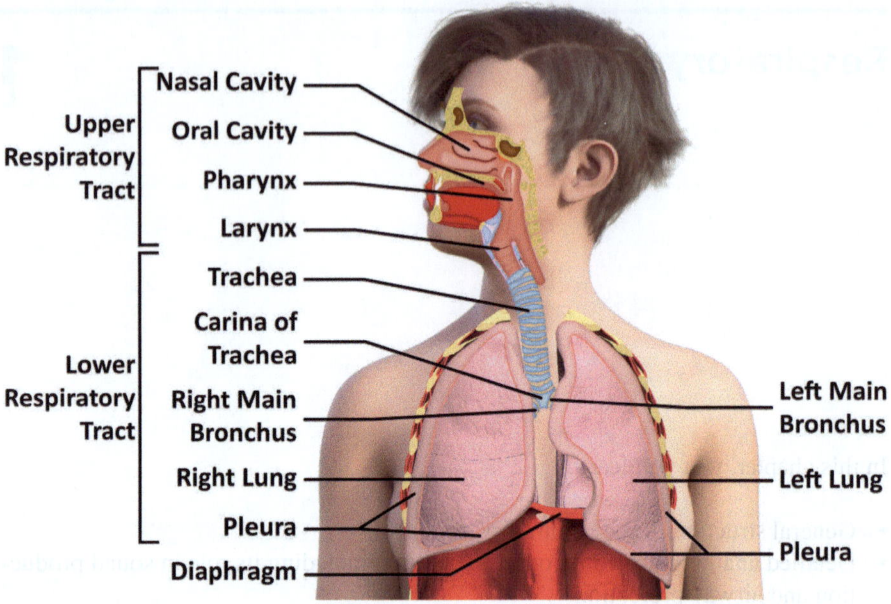

Fig. 15.1 The respiratory system organs

exhalation, two processes essential for sustaining life. The primary organs involved include the nose, pharynx, larynx, trachea, bronchi, and lungs. The respiratory system also plays a significant role in regulating the body's pH balance. It is intricately linked with the cardiovascular system to ensure the efficient transport of gases to and from body tissues. This system's importance extends beyond mere gas exchange; it filters incoming air, produces vocal sounds, and participates in the body's immune defense.

15.1 Part 1: Structure

The respiratory system comprises a series of interconnected organs and tissues that enable oxygen inhalation and carbon dioxide exhalation. It can be divided into the upper and lower respiratory tracts (Fig. 15.1).

15.1.1 Upper Respiratory Tract

The upper respiratory tract constitutes the initial segment of the respiratory system, serving as the gateway for air to enter the body. Comprising a series of specialized structures, this region plays a pivotal role in conditioning and preparing inhaled air for further transport into the lower respiratory tract.

15.1.1.1 The Nose and Paranasal Sinuses

The nose, located at the center of the face, serves as the primary entry point for external air into the respiratory system. It comprises two main sections: the external nose, visible on the face, and the internal nasal cavity.

External Nose

The external nose (Fig. 15.2) is a prominent facial feature positioned in the center of the face. The external nose protrudes above the oral cavity and provides an anterior opening to the nasal cavity. The upper portion of the nose is supported by the nasal bones, which are connected to the skull's frontal bone. These bones form the

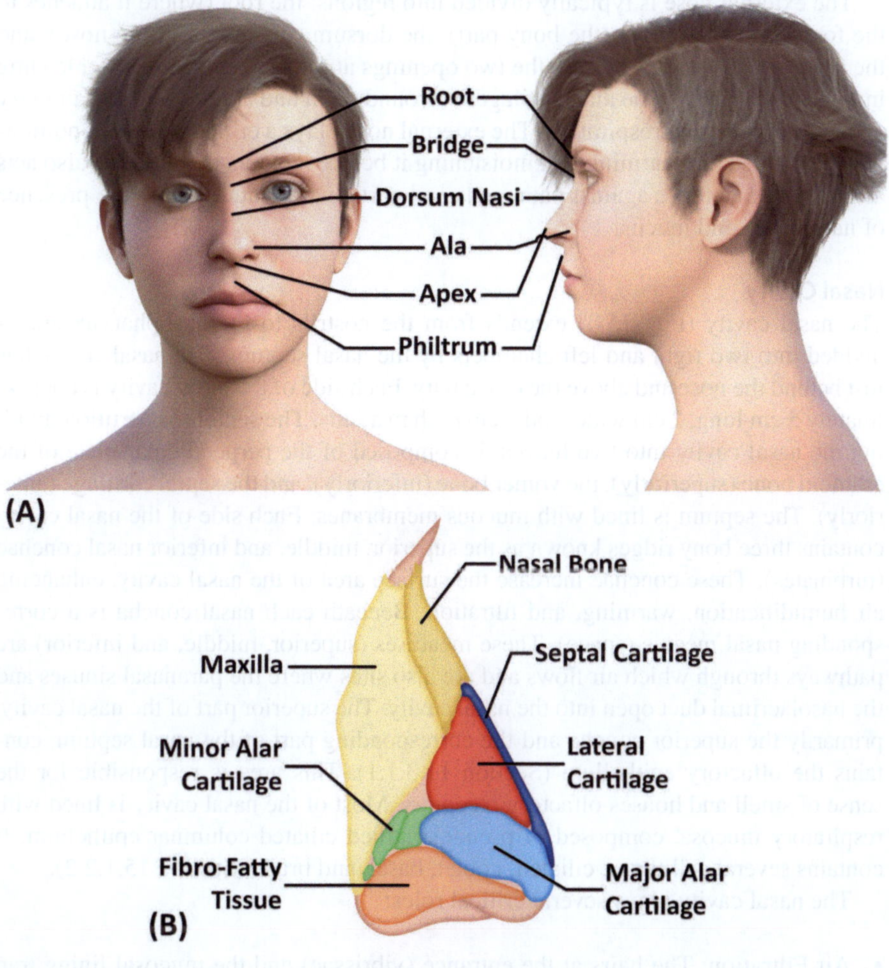

Fig. 15.2 The external nose. (**a**) The external anatomy. (**b**) The skeletal framework

bridge of the nose and contribute to its shape and strength. The lower portion, more flexible and mobile, is supported by several pieces of cartilage:

- Major Alar Cartilages (Lower Lateral Cartilages): These are two large pieces of cartilage that shape the tip of the nose and the nostrils. They are responsible for the width and projection of the nasal tip.
- Minor Alar Cartilages: Smaller fragments of cartilage near the major alar cartilages contribute to the nostrils' shape.
- Septal Cartilage: This cartilage forms part of the nasal septum, which divides the nasal cavity into two chambers. It extends from the nasal bones in the upper part of the nose to the tip.

The external nose is typically divided into regions: the root (where it attaches to the forehead), the bridge (the bony part), the dorsum (the length of the nose), and the apex or tip. The nostrils are the two openings at the bottom of the nose, leading into the nasal cavity. The alar cartilages surround them and are the primary air entry and exit route during respiration. The external nose plays a critical role in conditioning the inhaled air, warming and moistening it before it enters the lungs. It also acts as a first-line defense against pathogens and particulate matter due to the presence of nasal hairs and mucus.

Nasal Cavity

The nasal cavity (Fig. 15.3) extends from the nostrils to the nasopharynx and is divided into two right and left chambers by the nasal septum. The nasal cavity lies just behind the nose and above the oral cavity. Each side of the nasal cavity is approximately 5 cm long, 2 cm wide, and 7 cm high in adults. The septum, a partition dividing the nasal cavity into two halves, is composed of the perpendicular plate of the ethmoid bone (superiorly), the vomer bone (inferiorly), and the septal cartilage (anteriorly). The septum is lined with mucous membranes. Each side of the nasal cavity contains three bony ridges known as the superior, middle, and inferior nasal conchae (turbinates). These conchae increase the surface area of the nasal cavity, enhancing air humidification, warming, and filtration. Beneath each nasal concha is a corresponding nasal meatus (space). These meatuses (superior, middle, and inferior) are pathways through which air flows and are also sites where the paranasal sinuses and the nasolacrimal duct open into the nasal cavity. The superior part of the nasal cavity, primarily the superior concha and the corresponding part of the nasal septum, contains the olfactory epithelium (Section 11.3.1.1). This area is responsible for the sense of smell and houses olfactory receptors. Most of the nasal cavity is lined with respiratory mucosa, composed of pseudostratified ciliated columnar epithelium. It contains several cell types: ciliated, goblet, basal, and brush (Section 15.1.2.2).

The nasal cavity plays several critical roles:

- Air Filtration: The hairs at the entrance (vibrissae) and the mucosal lining trap dust, pollen, and other particulates.
- Humidification and Warming of Air: The extensive vascular network warms the air, while the mucus provides humidity.

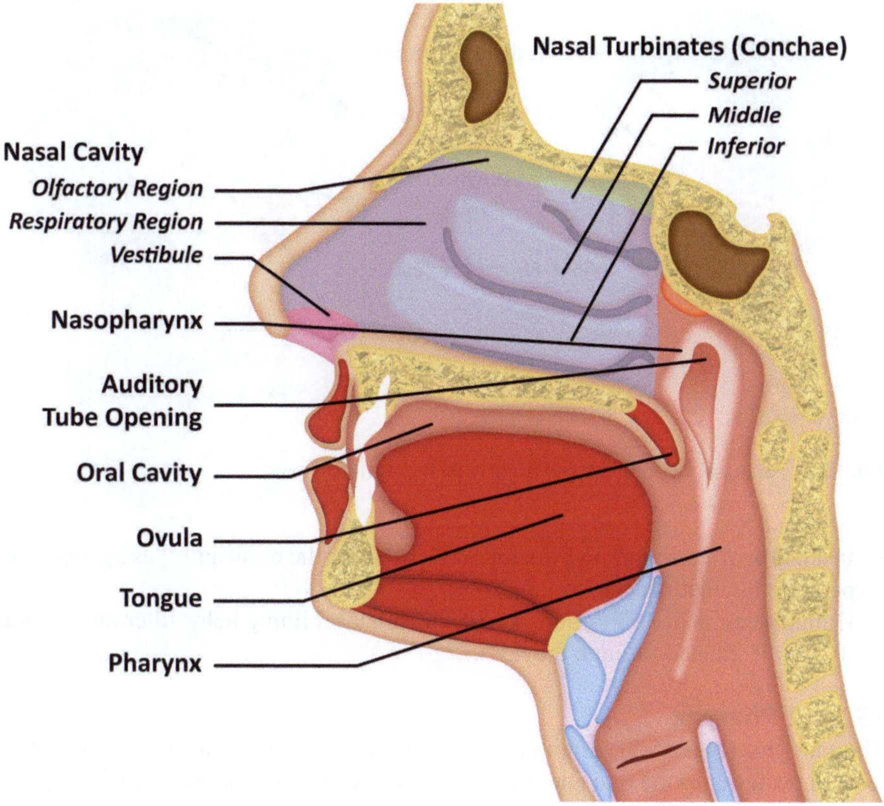

Fig. 15.3 The nasal cavity, parasagittal section

- Olfaction: The olfactory epithelium detects odorants dissolved in the nasal mucus.
- Resonating Chamber: The cavity affects the quality and tone of voice.
- Immune Defense: The mucus contains antibodies and enzymes that help neutralize pathogens.

Paranasal Sinuses

The paranasal sinuses are a group of paired air-filled spaces surrounding the nasal cavity, each named after the bone in which it is located: the maxillary, frontal, ethmoid, and sphenoid sinuses (Fig. 15.4). All sinuses are lined with a ciliated mucous membrane continuous with the lining of the nasal cavity. The cilia beat toward the sinus openings into the nasal cavity, helping to drain the mucus produced by the glands in the lining. The paranasal sinuses play several roles:

- Resonating Chambers for Voice: They contribute to the quality of the voice by acting as resonators.
- Decreasing Skull Weight: Air-filled spaces reduce the weight of the frontal part of the skull.

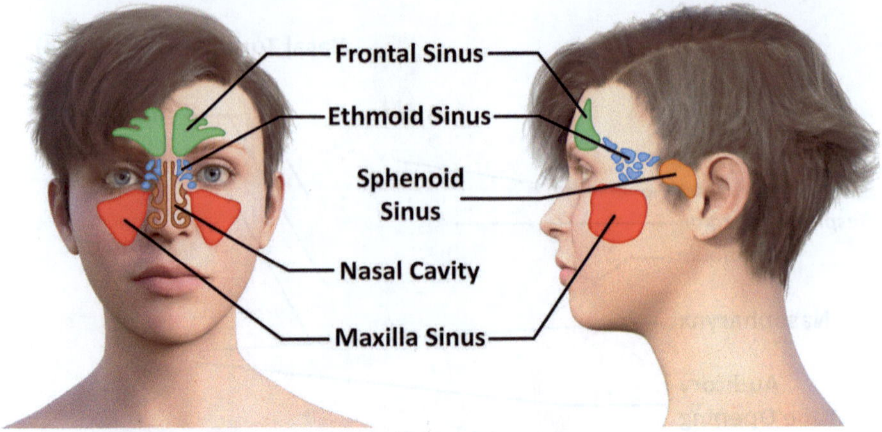

Fig. 15.4 Paranasal sinuses, anterior (right) and lateral (left) views

- Insulation of Surrounding Structures: They buffer the brain and eyes against temperature fluctuations and trauma.
- Humidification and Filtration of Air: The mucosal lining helps filter and humidify the air inhaled through the nose.

15.1.1.2 Pharynx

The pharynx (Fig. 15.5), often called the throat, is a muscular, tubular structure located posterior to the oral and nasal cavities. It is divided into three main regions: nasopharynx, oropharynx, and laryngopharynx.

Nasopharynx

The nasopharynx (Fig. 15.5) is the uppermost segment of the pharynx, situated behind the nasal cavity and above the oropharynx. It extends from the skull's base to the soft palate's upper edge. The nasopharynx communicates anteriorly with the nasal cavity through the posterior choanae (openings in the nasal cavity) and posteriorly with the oropharynx via the pharyngeal isthmus. Within the nasopharynx, several structures are notable, including the adenoids (pharyngeal tonsil) and the openings of the auditory tubes (Eustachian tubes) (Section 11.2.2). The adenoids are a collection of lymphoid tissue, part of the immune system's defense against pathogens (Page 622). The lining of the nasopharynx is composed of a respiratory epithelium (pseudostratified ciliated columnar epithelium) (Fig. 15.13) featuring tall columnar cells with cilia projecting from their apical surfaces. These cilia move in coordinated waves, aiding in the movement of mucus and trapped particles toward the pharyngeal isthmus.

The nasopharynx serves several vital functions, including passing air, defending against immune attack, and regulating the function of the auditory tube.

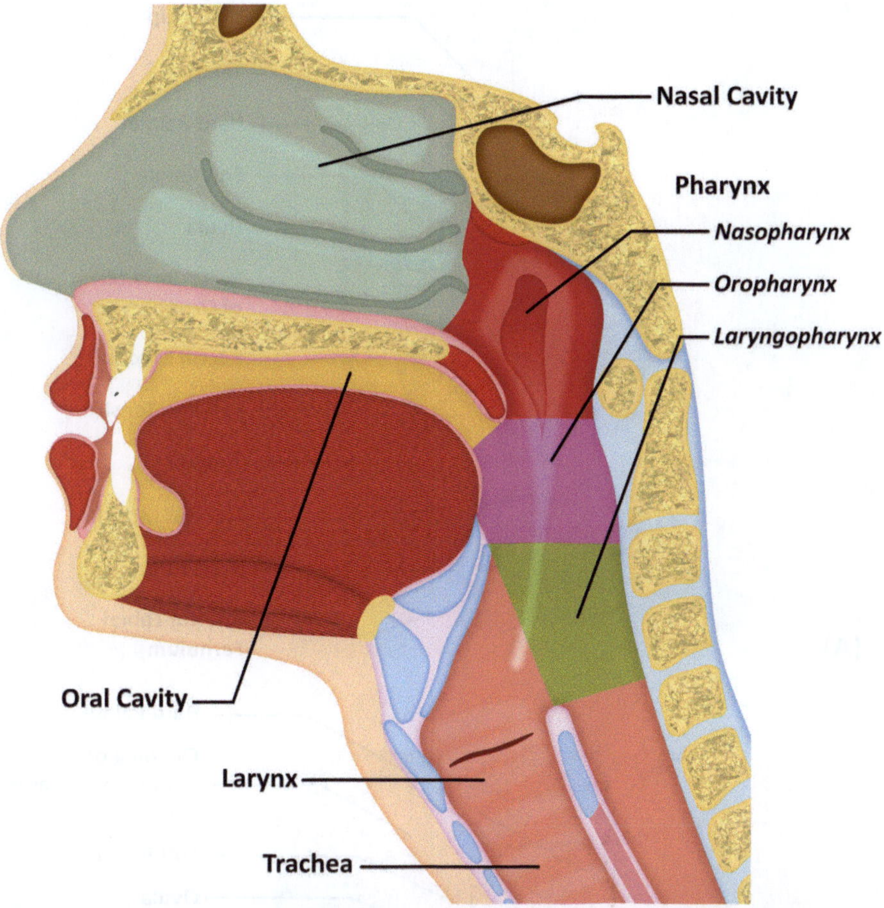

Fig. 15.5 The pharynx, parasagittal section

Oropharynx

The oropharynx (Fig. 15.5) is located behind the oral cavity, extending from the soft palate to the base of the tongue. It marks the transition between the nasopharynx and the laryngopharynx. Several structures are notable within the oropharynx, including palatine tonsils, lingual tonsils (Page 622), and the posterior pharyngeal wall. The lining of the oropharynx is composed of stratified squamous epithelium. This epithelial type is well-suited to withstand mechanical stresses associated with swallowing and speech production. The palatine tonsils are located on either side of the oropharynx, in the area of the tonsillar fossa. The lingual tonsils are situated on the posterior surface of the tongue.

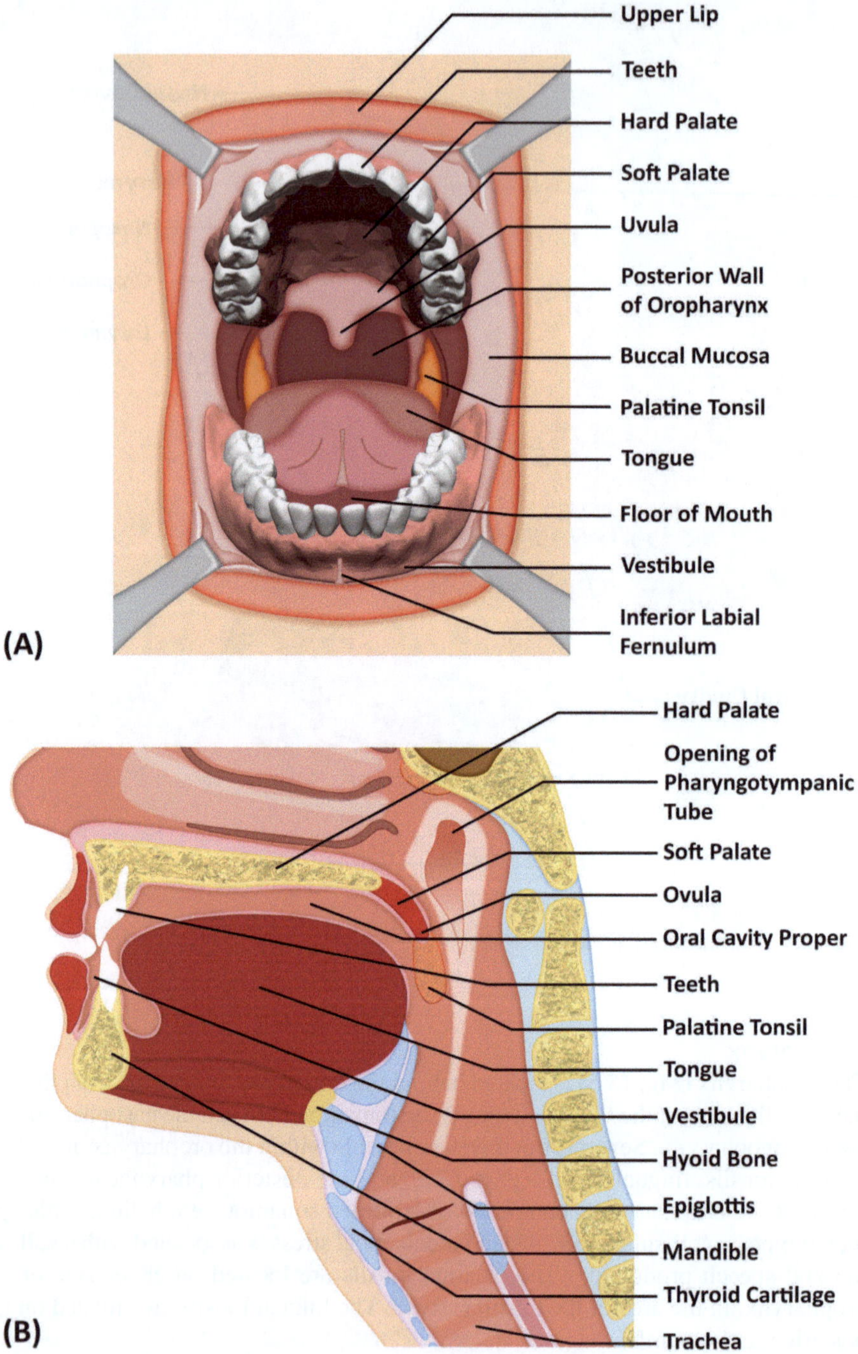

Fig. 15.6 Oral cavity and soft palate. (**a**) Anterior view. (**b**) Parasagittal view

The soft palate (Fig. 15.6) is a flexible and muscular extension of the hard palate, forming the posterior portion of the roof of the mouth. This structure is crucial in speech and swallowing. The soft palate marks the division between the oropharynx and nasopharynx at the back of the oral cavity. The soft palate comprises muscle and connective tissues, making it flexible and movable. The uvula is a conical projection at the back of the soft palate, a small, fleshy mass of tissue that hangs above the throat. It aids in speech and swallowing.

Laryngopharynx

The laryngopharynx (Fig. 15.5), the lowest segment of the pharynx, marks the division between the upper respiratory and digestive tracts. It plays a significant role in the passage of both air and food, ensuring proper routing and preventing aspiration. It is situated between the oropharynx and the esophagus. The laryngopharynx is characterized by the epiglottis, a flap-like cartilaginous structure, and the entrance to the larynx and esophagus. The lining of the laryngopharynx is composed of non-keratinized stratified squamous epithelium. This type of epithelium protects against mechanical stresses associated with swallowing and the passage of food.

The laryngopharynx serves several functions:

- Passage of Air and Food: The laryngopharynx is a common pathway for air and food. It ensures inhaled air continues to the lower respiratory tract while directing food into the esophagus.
- Protection of the Lower Airway: The epiglottis within the laryngopharynx prevents food and liquids from entering the larynx and trachea during swallowing, protecting the lower airway from aspiration.
- Speech Production: The laryngopharynx, in conjunction with other structures such as the vocal cords, contributes to speech production by modulating the resonance of vocal sounds.

Epiglottis

The epiglottis (Fig. 15.7) is an anatomical structure that functions as a switch between the trachea and the esophagus. The epiglottis is situated just above the larynx and behind the root of the tongue. It stands at the entrance of the glottis, the opening between the vocal cords. It is a leaf-shaped flap of elastic cartilage covered with a mucous membrane. The epiglottis is relatively small, measuring about 2–3 cm long. It is supported by the hyoid bone and connected to the thyroid cartilage by the thyroepiglottic ligament. Its primary function is to seal off the windpipe (trachea) during swallowing, preventing food or liquid from entering the lungs. When not swallowing, it stands upright to allow air to enter the trachea, facilitating breathing. The epiglottis is flexible and can bend up and down. Its movement is primarily involuntary. During swallowing, the larynx moves up, releasing the epiglottis and letting it fold back to cover the glottis. This action is part of the reflex mechanism that prevents food aspiration (inhalation).

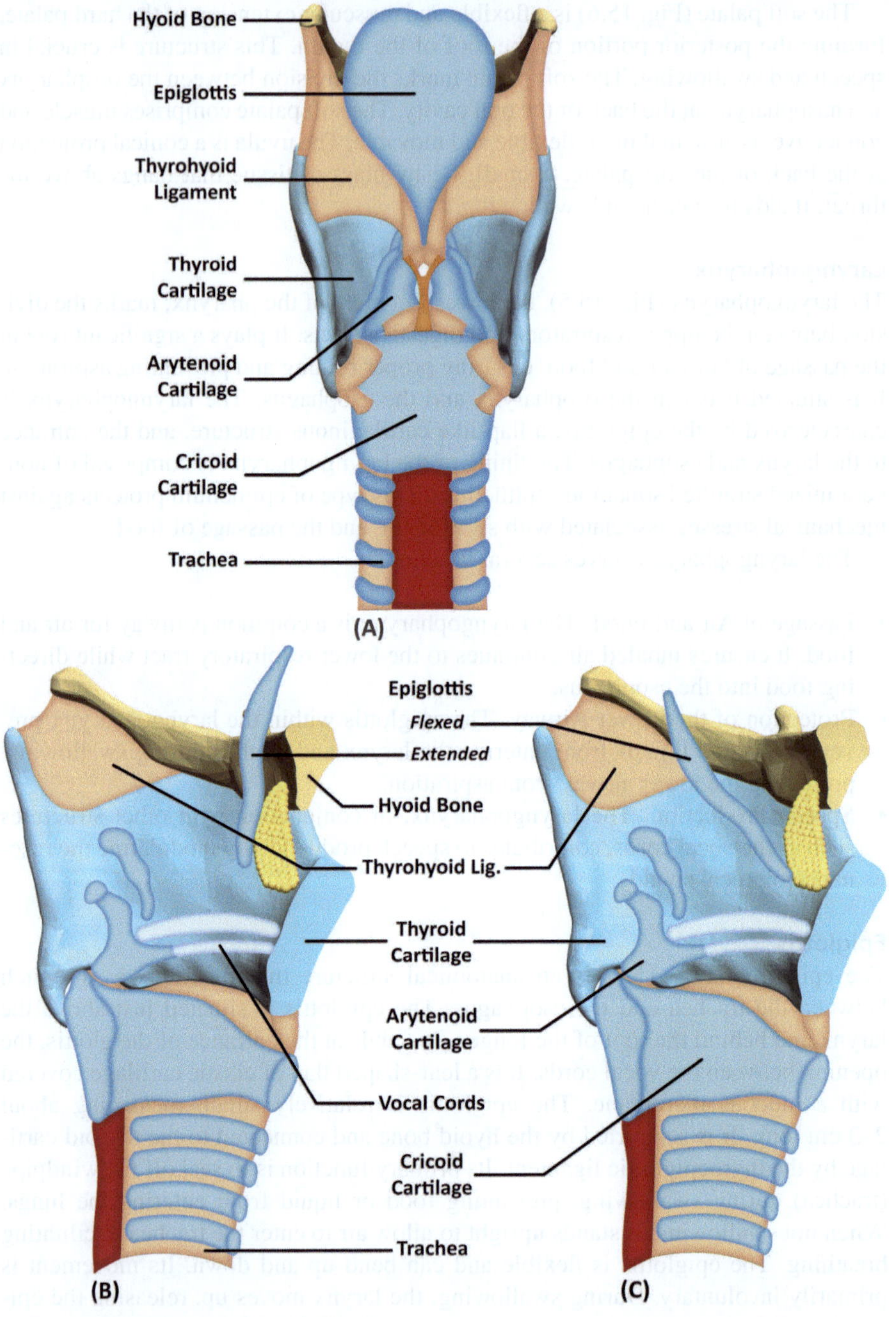

Fig. 15.7 The epiglottis. (**a**) Posterior view. (**b**) Sagittal section, with epiglottis extended. (**c**) Sagittal section, with epiglottis flexed

15.1.2 Lower Respiratory Tract

The lower respiratory tract extends from the larynx down to the alveoli in the lungs. This part of the respiratory tract is primarily responsible for delivering filtered, humid, and warm air to the pulmonary alveoli and exchanging oxygen with carbon dioxide in the blood.

15.1.2.1 Larynx

The larynx (Fig. 15.8), commonly known as the voice box, is a complex anatomical structure located in the neck region of the human body. The larynx is situated in the anterior portion of the neck, connecting the pharynx to the trachea. It lies between the tongue's base and the trachea's top. The larynx primarily comprises cartilage, including the thyroid cartilage, cricoid cartilage, and several smaller cartilages known as arytenoid, corniculate, and cuneiform cartilages. Within the larynx are the vocal cords (vocal folds), which consist of two pairs of fold-like structures. The true vocal cords are involved in sound production, while the false vocal cords play a role in sealing off the airway during swallowing, preventing aspiration. The lining of the larynx is composed of respiratory epithelium featuring ciliated pseudostratified columnar epithelial cells. The larynx has extrinsic and intrinsic muscles (Table 15.1) (Fig. 15.9). Intrinsic muscles control the tension and length of the vocal cords, affecting voice pitch and quality. Extrinsic muscles attach to the larynx and are responsible for its movement and positioning within the neck.

The true vocal cords (Fig. 15.10) are intricate structures within the larynx responsible for sound production, known as phonation. The true vocal cords are located within the larynx, specifically in the glottis region. The true vocal cords are paired structures, with the right and left cords situated side by side in the glottis. When viewed from above, they have a wedge-shaped appearance, tapering toward the front of the larynx. They comprise several layers, including the epithelium, vocal ligament, and muscle. The outermost layer of the vocal cords is lined with stratified squamous epithelium, which protects against mechanical stress and vibration. They act as sphincters, controlling airflow through the glottis for phonation (sound production). They play a crucial role in airway protection during activities such as swallowing and coughing. By precisely controlling glottal closure, cord tension, and length, individuals can modulate their voices to convey emotions, emphasize words, and create a wide range of sounds and pitches. This intricate control is fundamental to speech, singing, and communication.

The tension and length of the vocal cords control sound pitch. Higher tension and shorter length result in higher pitch. Volume is regulated by the force of the air expelled from the lungs—greater force results in louder sound. Once the sound is produced at the vocal cords, it resonates in the throat, mouth, and nasal cavities, which act as amplifiers and modify the quality of the sound. The shape and size of these cavities, along with the position of the tongue and lips, affect the resonance and are crucial for articulation. The final step in voice production is articulation, where the resonated sound is shaped into speech by the movements of the

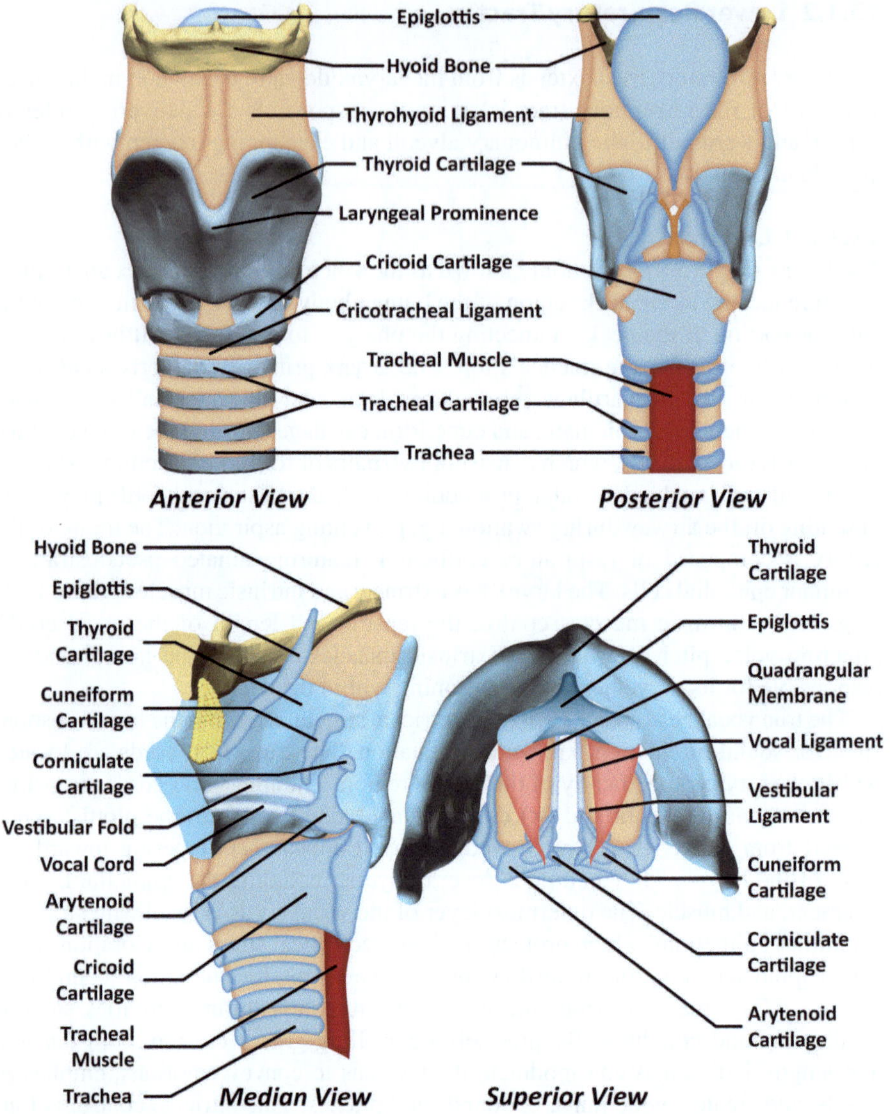

Fig. 15.8 The larynx skeleton

articulators—the tongue, lips, teeth, and palate. Different positions and movements of these structures produce different speech sounds.

15.1.2.2 Trachea

The trachea (Fig. 15.11), often called the windpipe, is a tubular structure that serves as a conduit for air. The trachea is a cylindrical tube located anteriorly in the neck and upper chest. The trachea is reinforced by a series of C-shaped rings made of

Table 15.1 Muscles of the larynx: origin, insertion, and function

Muscle	Origin	Insertion	Function
Intrinsic muscles			
Cricothyroid muscle	Anterior surface of the cricoid cartilage	Inferior horn of the thyroid cartilage	Tenses and elongates the vocal cords
Thyroarytenoid muscle (vocalis)	Inner surface of the thyroid cartilage	Inserts into the arytenoid cartilages	Relaxes and shortens the vocal cords
Lateral cricoarytenoid muscle	Lateral surface of the cricoid cartilage	Muscular process of the arytenoid cartilage	Adducts (closes) the vocal cords
Posterior cricoarytenoid muscle	Posterior surface of the cricoid cartilage	Muscular process of the arytenoid cartilage	Abducts (opens) the vocal cords
Aryepiglottic muscle	Arytenoid cartilage	Epiglottis	Closes the laryngeal inlet during swallowing
Oblique arytenoid muscle	Arytenoid cartilage	Arytenoid cartilage	Adducts the arytenoid cartilages
Transverse arytenoid muscle	Arytenoid cartilage	Arytenoid cartilage	Adducts the arytenoid cartilages
Extrinsic muscles			
Suprahyoid muscles			
Digastric muscle (anterior belly)	Digastric fossa of the mandible	Intermediate tendon that attaches to the body of the hyoid bone	Elevates the hyoid bone
Digastric muscle (posterior belly)	Mastoid notch of the temporal bone	Intermediate tendon that attaches to the body of the hyoid bone	Elevates the hyoid bone
Stylohyoid muscle	Styloid process of the temporal bone	Body of the hyoid bone	Elevates and retracts the hyoid bone
Mylohyoid muscle	Mylohyoid line of the mandible	Midline raphe and body of the hyoid bone	Elevates the hyoid bone, assists in tongue movement
Geniohyoid muscle	Mental spine of the mandible	Body of the hyoid bone	Pulls the hyoid bone superiorly and anteriorly
Hyoglossus muscle	Greater cornu of the hyoid bone	Lateral aspect of the tongue	Depresses and retracts the tongue
Genioglossus muscle	Mental spine of the mandible	Tongue	Protrudes and shapes the tongue
Infrahyoid muscles (strap muscles)			
Sternohyoid muscle	Posterior surface of the manubrium and clavicle	Body of the hyoid bone	Depresses the hyoid bone
Sternothyroid muscle	Posterior surface of the manubrium and first costal cartilage	Oblique line on the thyroid cartilage	Depresses the thyroid cartilage
Omohyoid muscle	Superior border of the scapula	Body of the hyoid bone	Depresses and retracts the hyoid bone

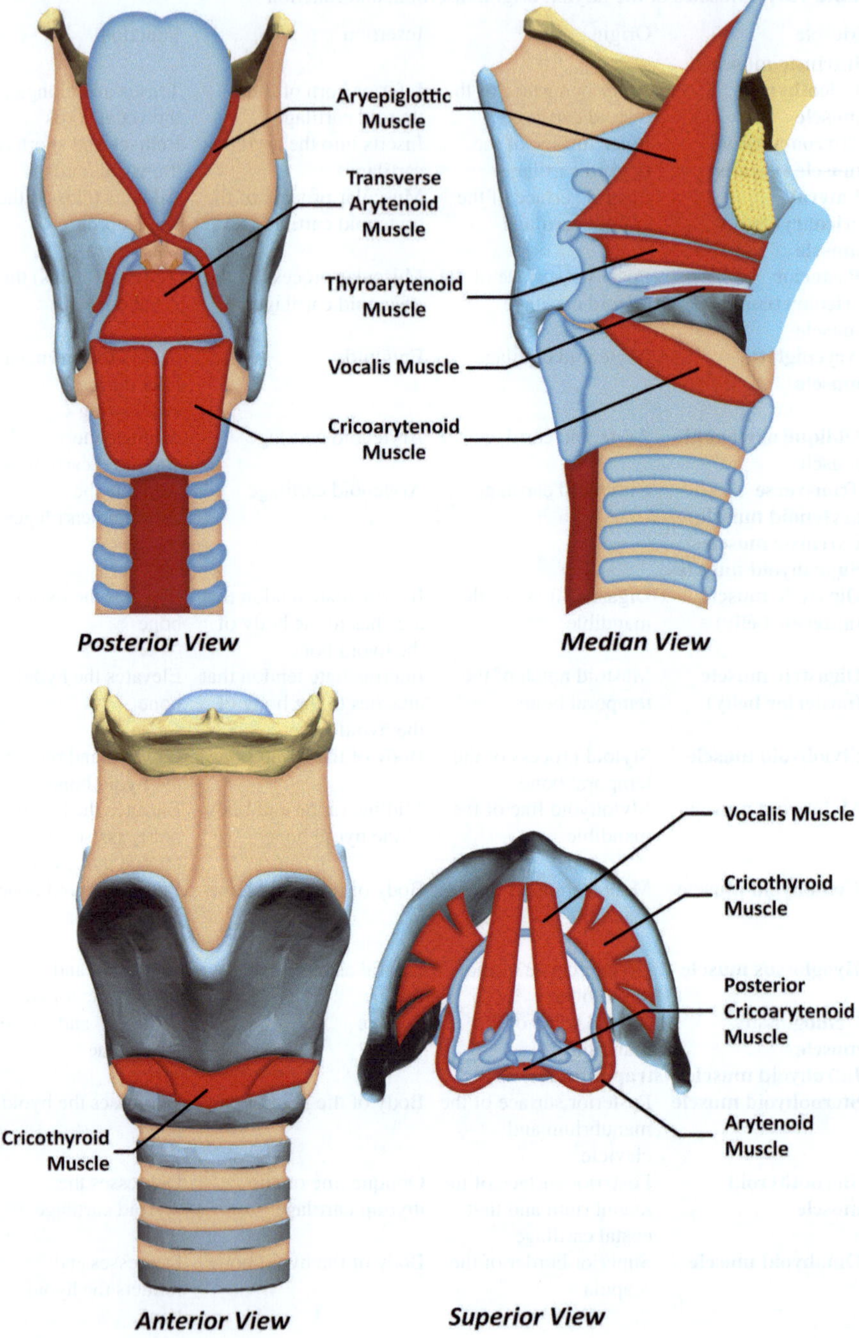

Fig. 15.9 Muscles of the larynx

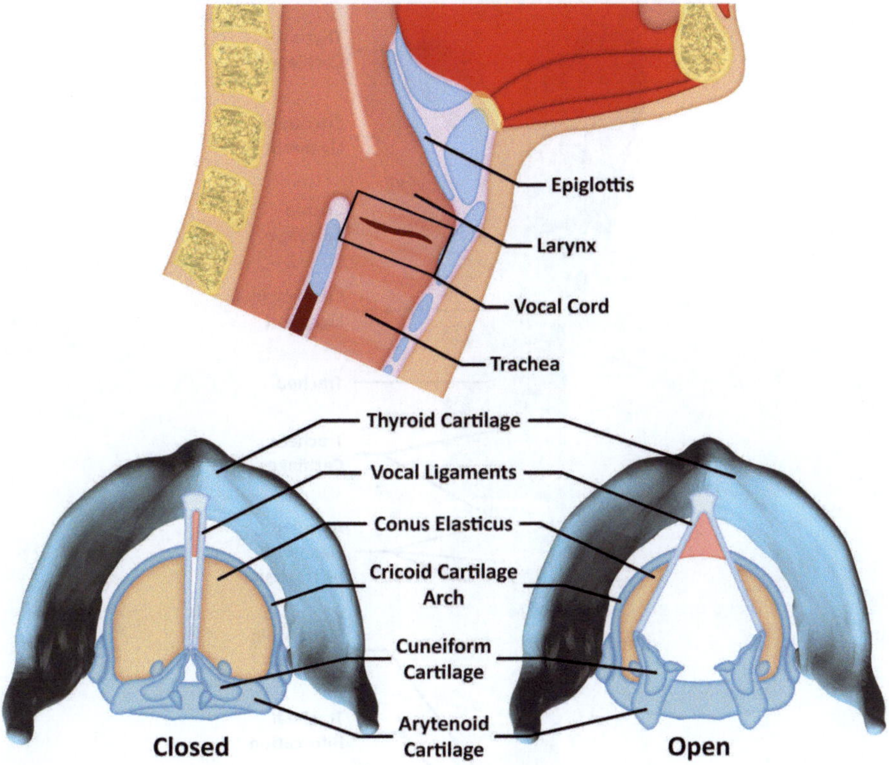

Fig. 15.10 Vocal cords

hyaline cartilage (Fig. 15.12). These rings provide structural support and prevent the trachea from collapsing during inhalation. The posterior aspect of the tracheal ring is open, allowing the esophagus, which lies directly behind the trachea, to expand during swallowing. The inner lining of the trachea is called the tracheal mucosa (Fig. 15.13). The innermost layer of the tracheal mucosa is lined with pseudostratified ciliated columnar epithelium. This specialized epithelium is equipped with cilia and goblet cells. Goblet cells secrete mucus. The cilia move in coordinated waves to propel mucus and trapped particles toward the pharynx, where they can be swallowed or expectorated.

Beneath the epithelium lies the lamina propria, a connective tissue layer that contains blood vessels, nerves, and lymphatic vessels. Deeper still, the submucosa contains seromucous glands that secrete mucus into the trachea, further aiding in trapping and removing foreign particles and pathogens. The posterior aspect of the trachea is not reinforced by cartilage rings but contains a smooth muscle known as the trachealis muscle. This muscle allows for flexibility and adjustment of the trachea's diameter, particularly during swallowing and speaking (Fig. 15.12). The trachea's primary role is to conduct air from the upper respiratory tract to the lower respiratory tract.

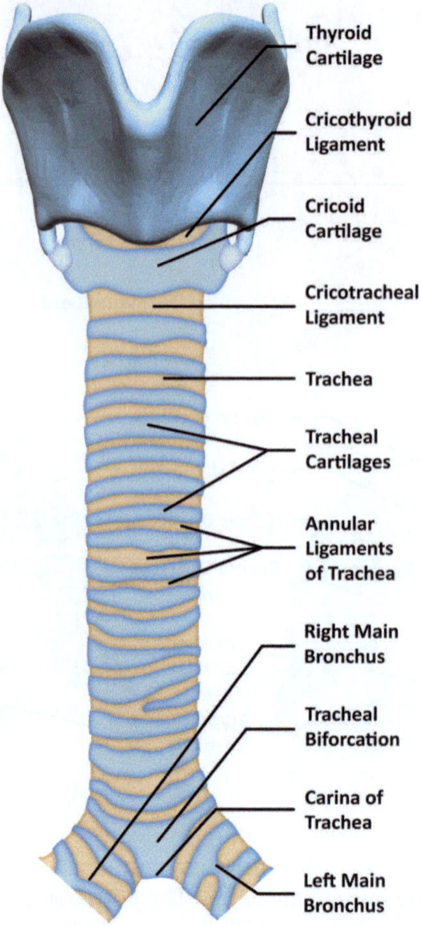

Fig. 15.11 Trachea: anterior view

Fig. 15.12 Cross section of trachea

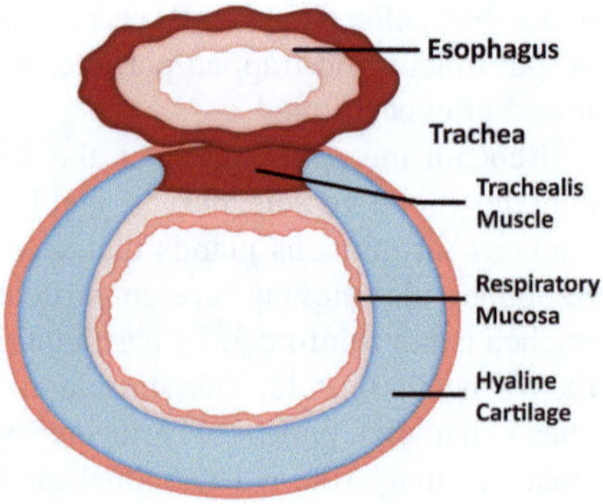

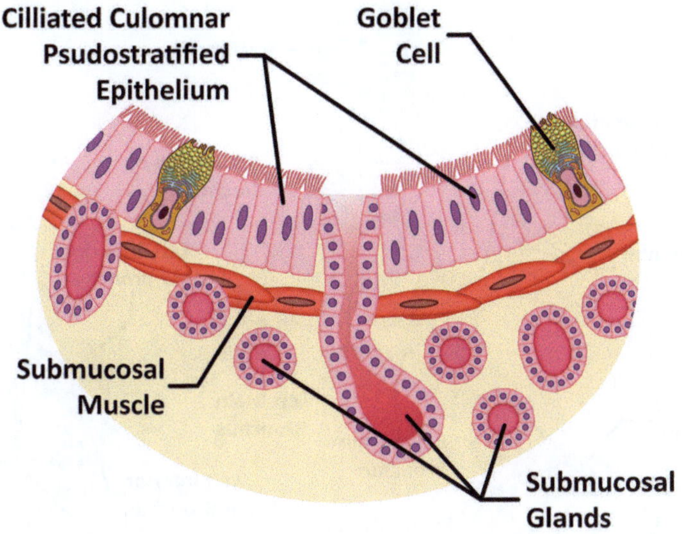

Fig. 15.13 The respiratory epithelium

15.1.2.3 Bronchi and Subdivisions

The bronchi (Fig. 15.14) serve as conduits that carry air from the trachea into the lungs. The bronchi (singular: bronchus) are a pair of tubular structures that branch off the trachea, with each bronchus leading to one lung. The branching pattern of the bronchi, from the trachea to the terminal bronchioles, resembles a tree-like structure collectively referred to as the bronchial tree. The structure of the bronchial walls is similar to the trachea except for the cartilage content. In large bronchi, the cartilages are ring-shaped, but they become more irregular and plate-like in the smaller bronchi (Fig. 15.15). The cartilage provides structural support and keeps the airway open. As the bronchi branch into smaller bronchi, the structural components change: the cartilage plates become sparser, and the relative amount of smooth muscle increases. Bronchioles are narrower than bronchi, with diameters less than 1 mm (Fig. 15.16). Their walls lack cartilage and glands, distinguishing them from the larger bronchi.

- Main Bronchi: At the carina, the trachea bifurcates into two main bronchi, the right and left main bronchi (Fig. 15.14). Each main bronchus enters its respective lung and further divides into lobar bronchi.
- Lobar Bronchi: Within each lung, the main bronchus divides into lobar bronchi (Fig. 15.14), also known as secondary bronchi. The right lung typically has three lobar bronchi, corresponding to its three lobes (upper, middle, and lower), while the left has two lobar bronchi, corresponding to its two lobes (upper and lower).
- Segmental Bronchi: Lobar bronchi continue to divide into smaller segmental bronchi (Fig. 15.15), supplying air to specific lung segments. The number of segmental bronchi varies between individuals but can range from eight to ten in the right lung and nine to ten in the left lung.

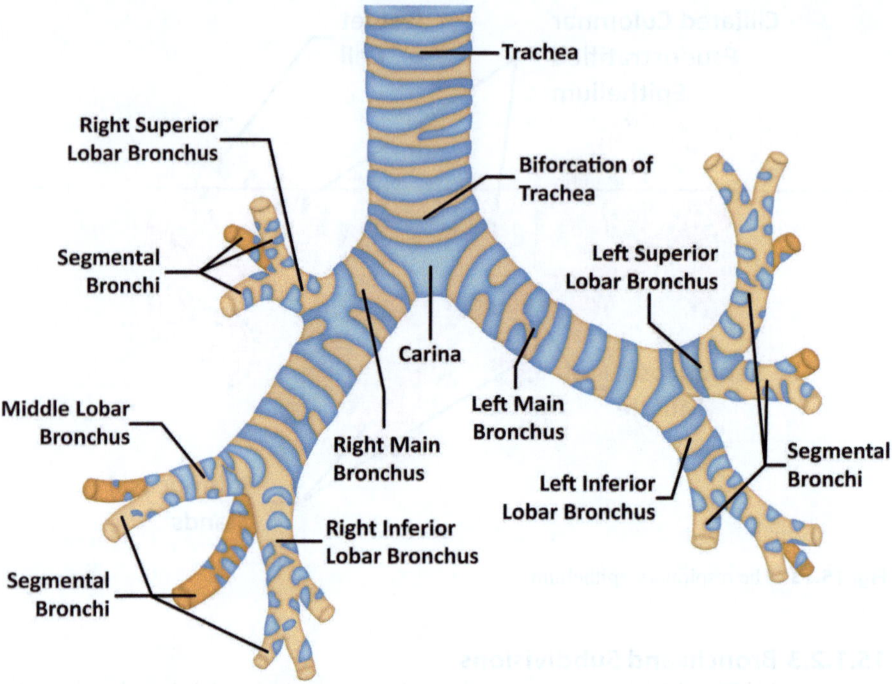

Fig. 15.14 Bronchi, anterior view

- Bronchopulmonary Segments: The segmental bronchi are divided into bronchopulmonary segments (Fig. 15.15), anatomically and functionally independent lung units. Specific bronchial and vascular arrangements define these segments, which vary in number and location throughout the lungs.
- Subsegmental Bronchi: Within each bronchopulmonary segment, the subdivision continues with smaller subsegmental bronchi (Fig. 15.15), also called terminal bronchi.
- Conducting Bronchioles: Subsegmental (terminal) bronchi eventually transition into conducting bronchioles (Fig. 15.15), characterized by a gradual decrease in cartilage and smooth muscle in their walls.
- Terminal Bronchioles: The conducting bronchioles further divide into terminal bronchioles (Fig. 15.15), which mark the end of the conducting portion of the respiratory system. These bronchioles do not participate in gas exchange but serve as airflow conduits.

The respiratory zone (Fig. 15.15), the structures past the terminal bronchioles, is where oxygen is delivered to the bloodstream and carbon dioxide is removed. The respiratory zone begins with respiratory bronchioles. These airways are lined with thin walls. Respiratory bronchioles lead into alveolar ducts, thin-walled passages of smooth muscle lined with alveoli. These ducts serve as conduits for air and lead to clusters of alveoli. Alveolar ducts terminate in alveolar sacs, which are grape-like

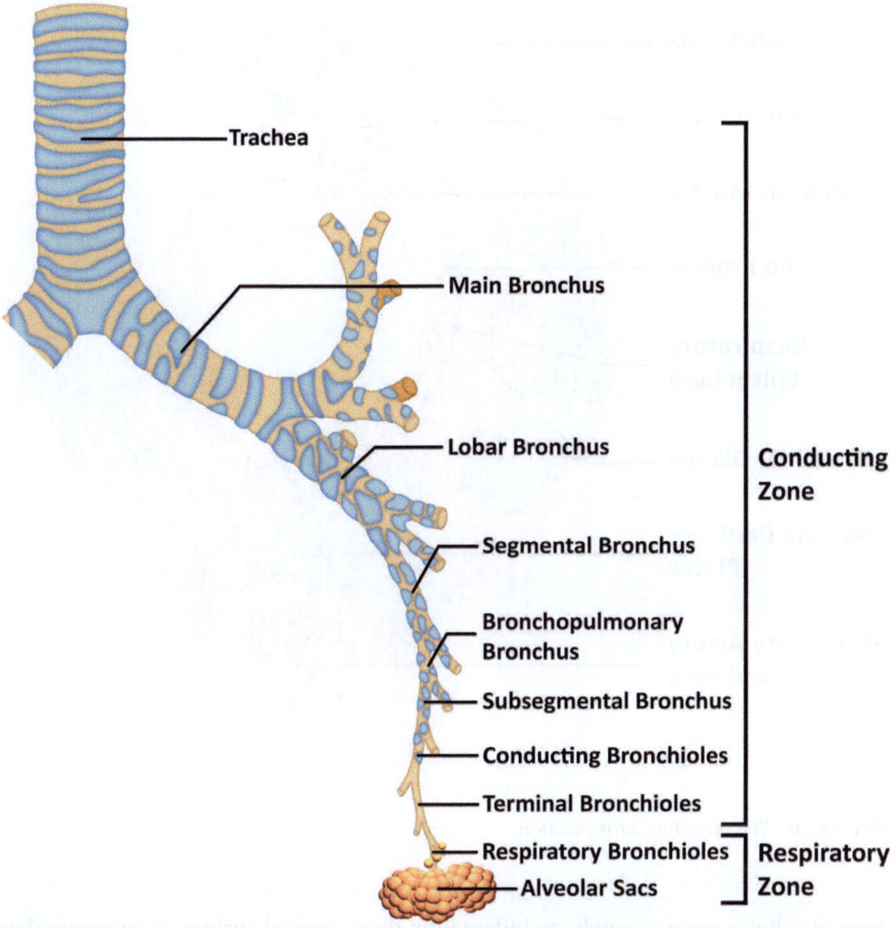

Fig. 15.15 The bronchial tree

clusters of individual alveoli. Each alveolus is a tiny, thin-walled sac with a rich network of capillaries surrounding it. It is within these alveoli that gas exchange occurs. Alveoli are the functional units of the respiratory zone, composed of a single layer of squamous epithelial cells, and are well-suited for efficient gas exchange.

15.1.2.4 Lungs

The lungs are a pair of spongy, air-filled organs in the chest (thoracic cavity), primarily responsible for respiration—the exchange of oxygen and carbon dioxide between the bloodstream and the environment. The lungs are pinkish gray, with the left lung slightly smaller than the right, accommodating space for the heart. They have a soft, spongy texture and are highly elastic. The right lung (Fig. 15.17) is divided into three lobes—the superior, middle, and inferior lobes—separated by two fissures: the horizontal fissure and the oblique fissure. The left lung (Fig. 15.17) has two superior and inferior lobes, separated by a single oblique fissure. The left

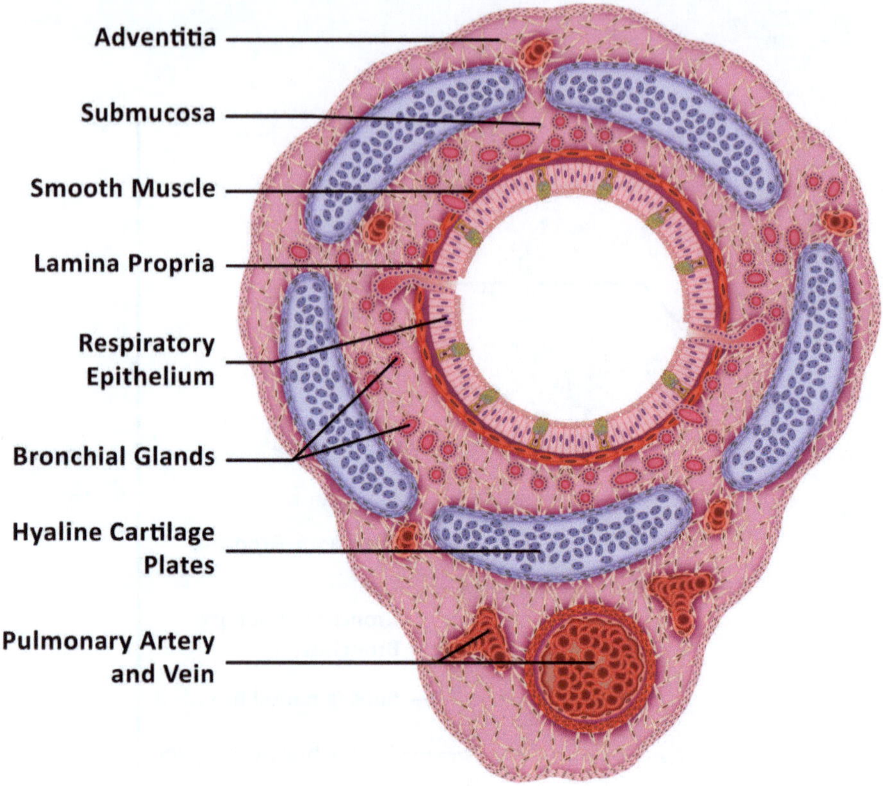

Fig. 15.16 The bronchus, cross section

lung also has a cardiac notch, an indentation on its medial surface to accommodate the heart. Each lung extends from just above the clavicle (collarbone) to the diaphragm at the bottom of the thoracic cavity. Each lung has three surfaces—costal (against the ribs), mediastinal (toward the midline of the chest), and diaphragmatic (against the diaphragm)—and three borders: anterior, posterior, and inferior (Fig. 15.18). The hilum is a region on the mediastinal surface of each lung, through which the bronchi, blood vessels, lymph vessels, and nerves enter and exit. The root of the lung is a collection of these structures at the hilum, connecting the lung to the mediastinum.

Each lung receives air through its primary bronchus (right and left main bronchi), which enters the hilum. Inside the lungs, these bronchi branch into secondary (lobar) bronchi, one for each lobe, and then into smaller tertiary (segmental) bronchi. The bronchial tree continues to branch into smaller bronchioles, eventually leading to respiratory bronchioles. These connect the terminal bronchioles to the alveolar ducts. Respiratory bronchioles have alveoli budding from their walls, marking the beginning of the gas exchange region (Fig. 15.19). Alveolar ducts are

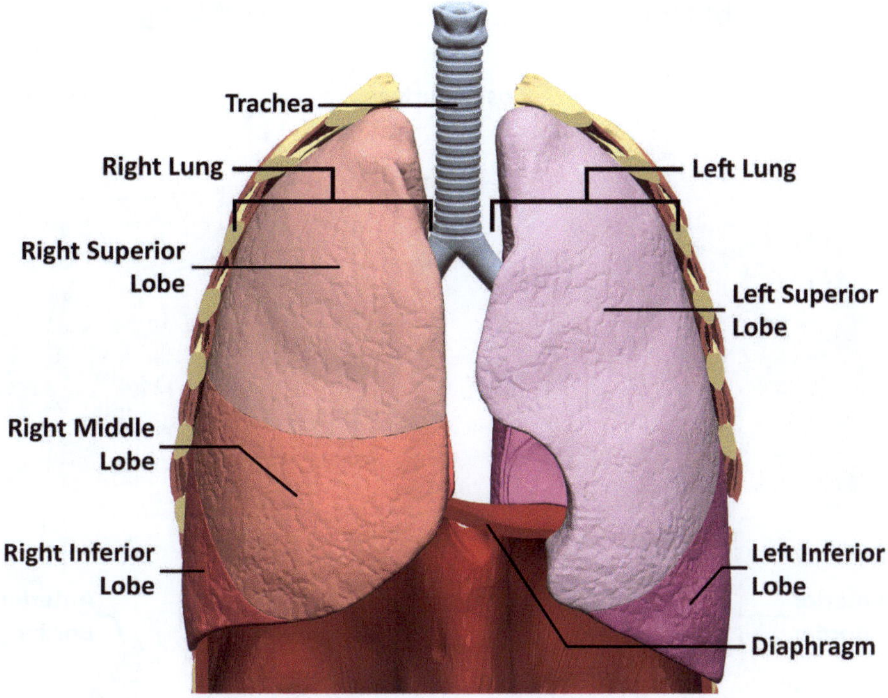

Fig. 15.17 Lung lobes and fissures

elongated airways that have walls composed almost entirely of alveoli. Alveoli (Fig. 15.20) are tiny, balloon-like structures and are the primary sites of gas exchange. Each lung contains about 300–600 million alveoli. Each alveolus is spherical, with a diameter ranging from 200 to 500 micrometers. The wall of each alveolus, also known as the alveolar septum, is extremely thin to facilitate gas exchange. It comprises a single layer of epithelial cells, a thin interstitial space, and a capillary endothelium (Fig. 15.21). The walls of the alveoli contain two types of cells:

- Type I Alveolar Cells: These are thin, flat cells covering about 95% of the alveolar surface area. They are involved in gas exchange.
- Type II Alveolar Cells: These are cuboidal and fewer in number, covering about 5% of the alveolar surface. They are responsible for producing and secreting pulmonary surfactant, which reduces surface tension within the alveoli, preventing collapse during exhalation.

Surfactant is a complex mixture of phospholipids and proteins. It lines the alveoli and reduces surface tension, facilitating lung expansion and preventing alveolar

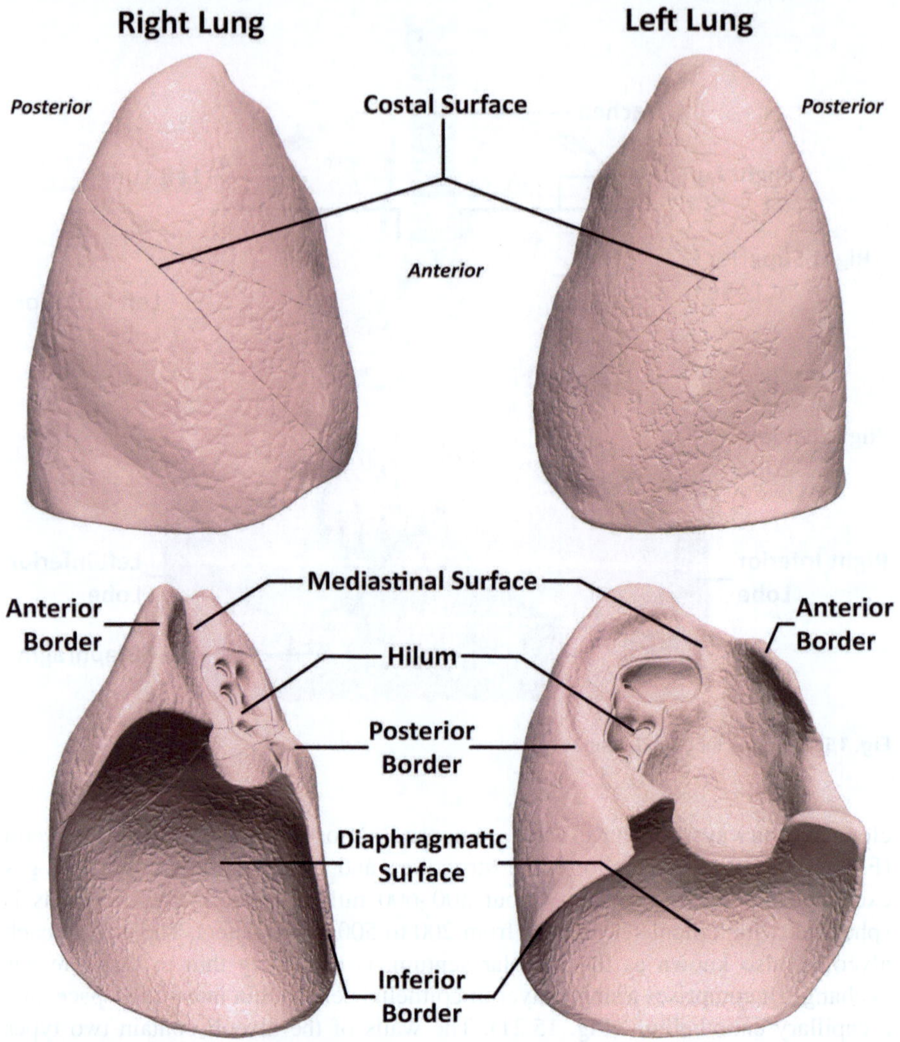

Fig. 15.18 Gross anatomy of the lungs

collapse. Small openings, called alveolar pores (pores of Kohn), in the alveolar walls allow air between alveoli, providing collateral ventilation and equalizing air pressure within the lungs. Alveoli also contain macrophages patrolling the alveolar surface, engulfing and removing particulate matter, bacteria, and other foreign substances.

15.1.2.5 Lung Blood Supply

Lungs have a dual blood supply. They receive blood from both the pulmonary and systemic circulations.

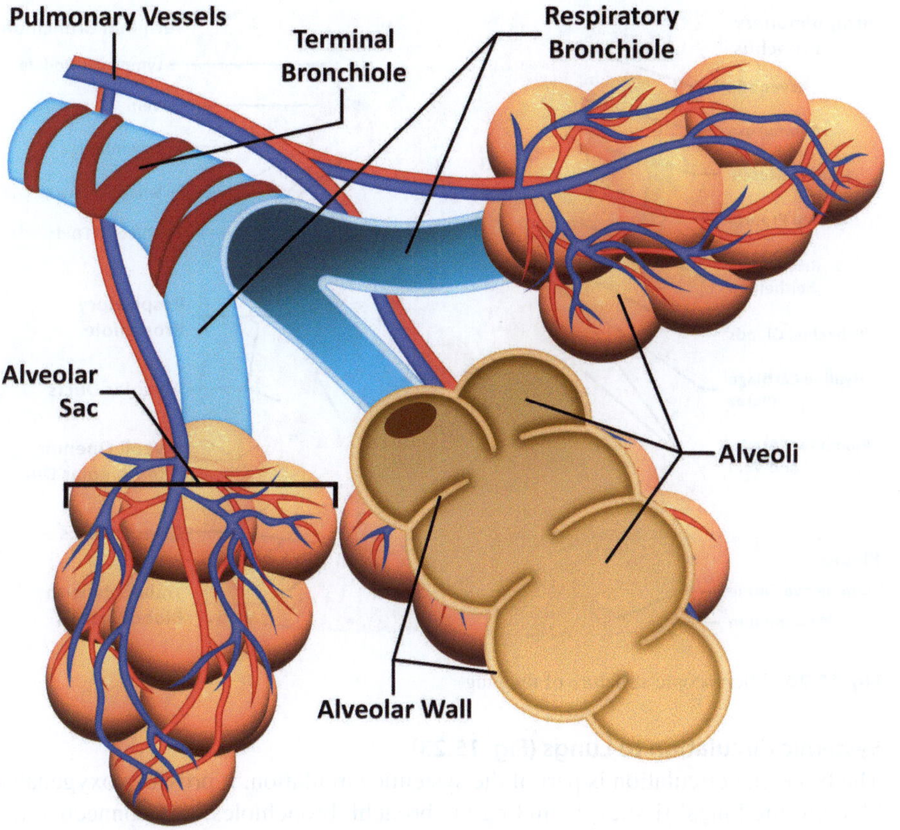

Fig. 15.19 Respiratory zone structure: the respiratory bronchiole and alveolar sac

Pulmonary Circulation to Lungs (Fig. 15.22)
The pulmonary artery is the only one in the body carrying deoxygenated blood. It exits the right ventricle and bifurcates into the left and right pulmonary arteries corresponding to each lung. These arteries branch repeatedly as they enter the lungs, becoming progressively smaller. The smallest branches of the pulmonary arteries are the pulmonary capillaries. These capillaries form extensive networks around the alveoli. The thin walls of the capillaries and alveoli facilitate the exchange of gases. Oxygen diffuses from the air in the alveoli into the blood in the capillaries, while carbon dioxide diffuses into the alveoli. Once the blood is oxygenated, it leaves the capillaries and enters small venules, which merge into larger pulmonary veins. Unlike systemic veins, the pulmonary veins carry oxygenated blood. Four pulmonary veins (two from each lung) typically transport the oxygenated blood back to the heart.

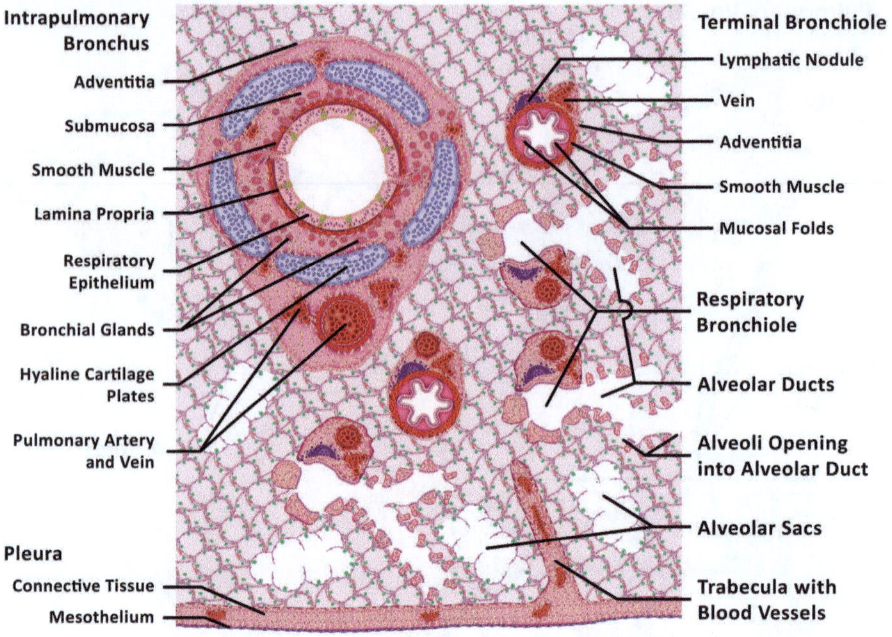

Fig. 15.20 Microscopic structure of the lung

Systemic Circulation to Lungs (Fig. 15.23)

The bronchial circulation is part of the systemic circulation. It provides oxygenated blood to the lungs' tissues, including the bronchi, bronchioles, and connective tissue. The bronchial arteries, which supply the bronchial circulation, typically arise from the descending aorta. There are usually two left bronchial arteries and one right bronchial artery. The bronchial arteries follow the branching pattern of the bronchial tree, providing oxygenated blood to the lung tissue. The bronchial veins drain the venous blood from the bronchial circulation. The bronchial veins drain into the pulmonary veins (carrying a small amount of deoxygenated blood back to the left atrium) and into the azygos and hemiazygos veins (part of the systemic venous system).

15.1.2.6 Innervation

The innervation of the lungs regulates and coordinates various processes, such as breathing, coughing, and the reflexes that protect the airway. The innervation of the lungs involves both autonomic (involuntary) and somatic (voluntary) nervous systems.

- Sympathetic Nervous System: Sympathetic fibers originate from the thoracic and upper lumbar spinal cord segments and enter the lungs via sympathetic ganglia. Activation of the sympathetic nervous system leads to bronchodilation

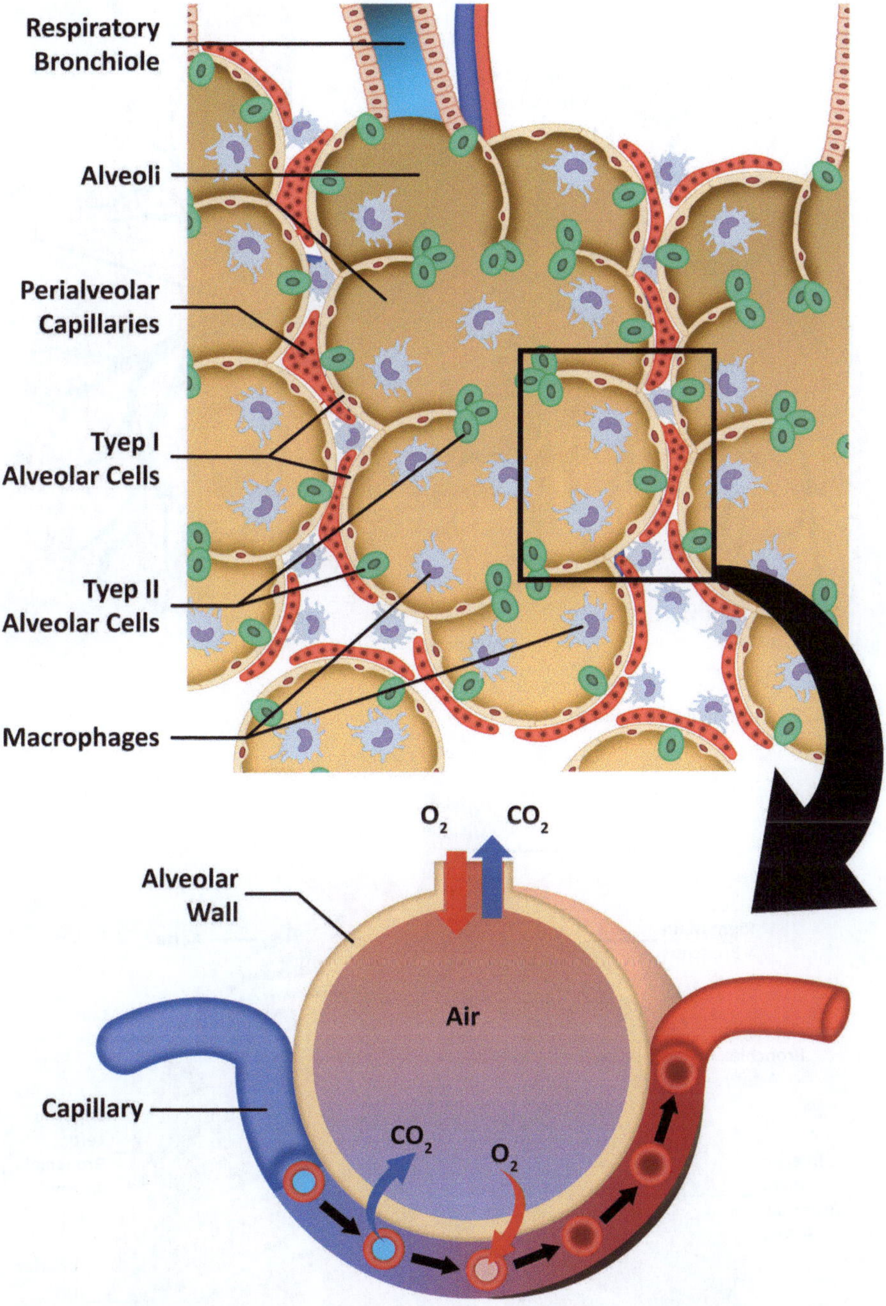

Fig. 15.21 Alveoli and the respiratory membrane

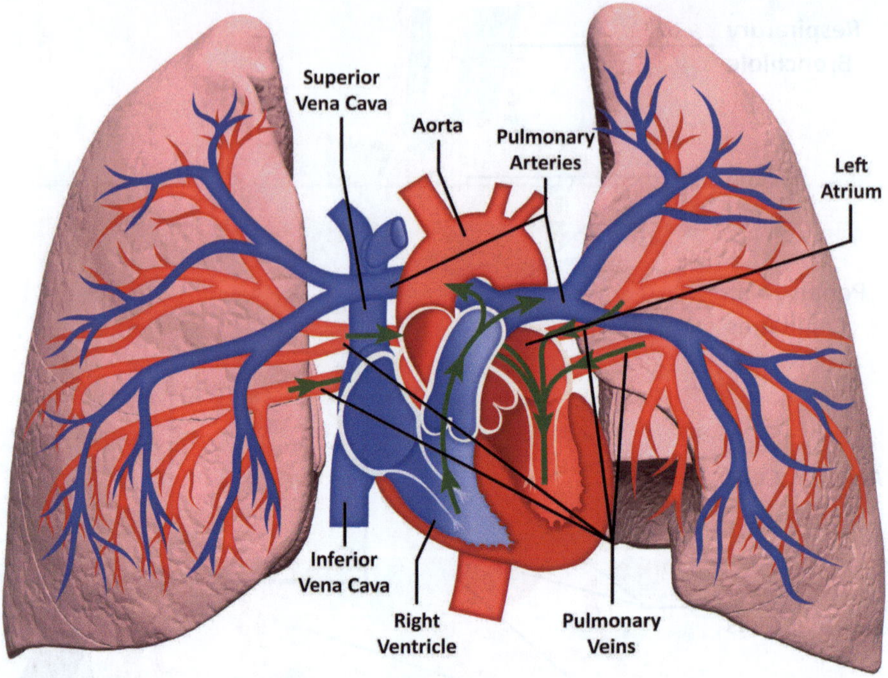

Fig. 15.22 Pulmonary circulation

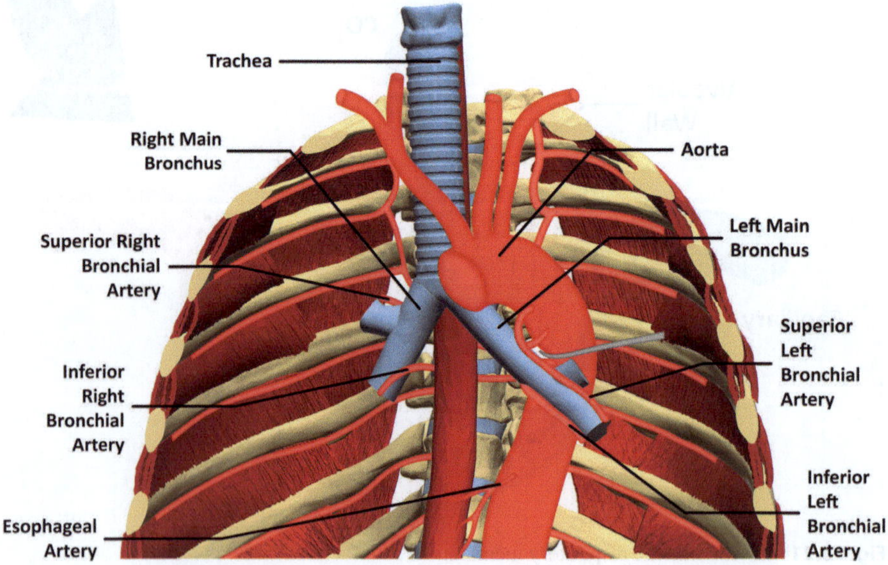

Fig. 15.23 Systemic circulation (bronchial arteries) of the lungs

(expansion of airways) and decreased mucus production, facilitating increased airflow.

- Parasympathetic Nervous System: Parasympathetic fibers originate from the vagus nerve (cranial nerve X) and primarily lead to bronchoconstriction (narrowing of airways) and increased mucus production, which can restrict airflow.
- Somatic Nervous System: The somatic nervous system provides voluntary control over specific respiratory actions, such as breathing and coughing. Motor neurons in the spinal cord send signals to the respiratory muscles, including the diaphragm and intercostal muscles, to initiate and control inhalation and exhalation.
- Sensory Nerves: Sensory nerves, known as afferent nerves, transmit information from the lungs to the central nervous system. They play a vital role in reflexes such as coughing and sneezing triggered by irritants in the airway.

15.1.3 Pleura

The pleura (Fig. 15.24) is a double-layered serous membrane that lines the thoracic cavity and envelops the lungs. It is crucial in facilitating smooth lung movements during respiration while ensuring airtight separation between the lungs and the chest wall. The pleura comprises two visceral and parietal layers, forming a closed sac around each lung. The parietal pleura lines the inner surface of the thoracic cavity. The visceral pleura is the innermost layer of the pleura and adheres closely to and directly covers the surface of the lung. It closely follows the contours of the lung, extending into the fissures and enveloping each lobe. The visceral pleura extends inferiorly to form pulmonary ligaments, which connect the inferior aspect of the lung to the diaphragm and anchor it within the thoracic cavity. Both layers of the pleura are composed of a single layer of mesothelial cells, which are simple squamous epithelial cells. These cells are flat and form a smooth, slippery surface that minimizes friction during lung movement. These cells secrete pleural fluid, a lubricating serous fluid that fills the space between the visceral and parietal pleura layers (pleural cavity). The parietal pleura is richly innervated and sensitive to pain, receiving innervation from the intercostal nerves (costal pleura) and the phrenic nerve (diaphragmatic and mediastinal pleura). The visceral pleura, in contrast, lacks sensory innervation.

The pleural space (Fig. 15.24) is a potential space, meaning under normal circumstances, it is a virtual space due to the close apposition of the two pleural layers. The pleural space contains a small amount of pleural fluid. Pleural fluid is a clear, serous fluid resembling plasma with fewer cells and proteins. In a healthy adult, the volume of pleural fluid is small, typically ranging from 10 to 20 milliliters per pleural cavity. The production process is influenced by hydrostatic and oncotic pressures in the capillaries, as described by Starling's forces. Hydrostatic pressure tends to push fluid out of the capillaries, while oncotic pressure draws it in. Pleural fluid circulates within the pleural space, driven by respiratory movements and gravity.

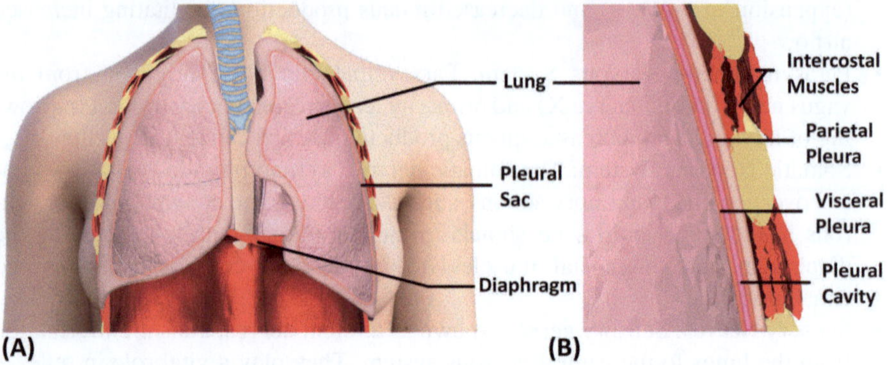

Fig. 15.24 The pleura. (**a**) Anterior view of the pleural sac covering the lungs (the anterior part of the pleura is removed). (**b**) The cross section of the pleura shows its parts and their relation to the surrounding structures

The fluid tends to move from the costal surface of the parietal pleura downward and medially toward the diaphragm, where absorption primarily occurs. Absorption of pleural fluid is facilitated mainly by the lymphatic system, particularly the lymphatic openings in the parietal pleura, predominantly located in the diaphragmatic and mediastinal regions. The balance between the production and absorption of pleural fluid is crucial for maintaining a stable volume and pressure within the pleural space. The fluid acts as a lubricant, reducing friction between the lungs and the chest wall during breathing.

15.1.4 Respiratory Muscles

Respiratory muscles are responsible for the mechanical act of ventilation, which involves air movement into and out of the lungs. The muscles involved in respiration are categorized into two main groups: the primary and accessory respiratory muscles (Table 15.2).

15.2 Part 2: Physiology of Respiration

Respiration is a vital physiological process that involves the exchange of gases, primarily oxygen (O_2) and carbon dioxide (CO_2), between the external environment and the body's cells. Respiration consists of two main processes: external respiration and internal respiration.

Table 15.2 Muscles of respiration: origin, insertion, innervation, function(s), and features

Muscle name	Origin	Insertion	Innervation	Function	Important features
Diaphragm	Xiphoid process, lower ribs, lumbar vertebrae	Central tendon	Phrenic nerve (C3–C5)	Principal muscle of inspiration; separates thoracic and abdominal cavities	Dome-shaped; has openings for the aorta, esophagus, and vena cava
External intercostal	Inferior border of the rib above	Superior border of the rib below	Intercostal nerves (T1–T11)	Elevate ribs during inspiration	Run inferoanteriorly from rib to rib
Internal intercostal	Superior border of the rib below	Inferior border of the rib above	Intercostal nerves (T1–T11)	Depress ribs during forced expiration	Run inferoposteriorly from rib to rib
Innermost intercostal	Inner aspect of ribs	Inner aspect of adjacent ribs	Intercostal nerves (T1–T11)	Similar to internal intercostals in function	Deeper layer of intercostal muscles
Serratus posterior superior	Ligamentum nuchae, spinous processes of C7–T3	Ribs 2–5	Intercostal nerves (T2–T5)	Elevate ribs during forced inspiration	
Serratus posterior inferior	Spinous processes of T11–L2	Ribs 9–12	Intercostal nerves (T9–T12)	Depress ribs during forced expiration	
Scalene (anterior, middle, posterior)	Transverse processes of cervical vertebrae	First and second ribs	Cervical spinal nerves (C3–C8)	Elevate first and second ribs during forced inspiration	Group of muscles in the lateral neck

External respiration refers to exchanging gases between the external environment and the lungs. It occurs in the alveoli of the lungs and involves the following steps:

- Pulmonary Ventilation (Breathing): It is the process of inhaling (inspiration) and exhaling (expiration) air into and out of the lungs. During inspiration, the diaphragm and external intercostal muscles contract, increasing the thoracic cavity volume and lowering pressure, causing air to enter the lungs. During expiration, these muscles relax, allowing the thoracic cavity to contract and expel air.
- Gas Diffusion: Oxygen in inhaled air diffuses from the alveoli into the bloodstream, while carbon dioxide diffuses into the alveoli for exhalation.

Internal respiration is exchanging gases between the bloodstream and the body's cells. It takes place in the systemic capillaries and involves the following steps:

- Oxygen Delivery: The heart pumps oxygen-rich blood from the pulmonary circulation to systemic capillaries. Here, oxygen diffuses into the cells used for cellular respiration.
- Carbon Dioxide Removal: Carbon dioxide produced by cellular metabolism diffuses from the cells into the bloodstream. It is then transported back to the lungs for elimination during exhalation.

15.2.1 Thoracic Cavity Pressure

Thoracic cavity pressure, often called intrathoracic pressure, is typically measured relative to atmospheric pressure (0 mmHg). The measurement of thoracic cavity pressure involves the use of a variety of techniques, including:

- Intrapleural Pressure (Pleural Pressure): Intrapleural pressure represents the pressure within the pleural cavity, the space between the visceral and parietal pleurae. It is measured using a pleural manometer, a thin catheter inserted into the pleural space. Under normal conditions, intrapleural pressure is subatmospheric, typically around −4 to −8 mmHg during quiet breathing.
- Intra-alveolar Pressure: Intra-alveolar pressure, the pressure within the lungs' alveoli, is measured using a spirometer or a pressure transducer. It fluctuates during the respiratory cycle, rising during inspiration and falling during expiration. At the end of a normal expiration, it is typically around 0 mmHg.
- Intra-esophageal Pressure: Intra-esophageal pressure is measured using a pressure-sensitive catheter inserted into the esophagus. It provides an indirect estimate of intrapleural pressure and helps assess changes in pleural pressure during breathing.

The mechanism of negative intrapleural pressure arises from the balance between the opposing forces acting on the pleural membranes (Fig. 15.25). The natural elastic recoil of the lungs tends to pull them inward, collapsing the alveoli and increasing intrapleural pressure. The chest wall has its elastic recoil, which tends to expand outward, decreasing intrapleural pressure. The surface tension created by the pleural fluid tends to draw the visceral pleura inward, increasing intrapleural pressure. The combination of these forces results in a delicate balance. Normally, the chest wall's tendency to expand slightly outweighs the lung's tendency to collapse, resulting in a negative intrapleural pressure, typically ranging from −4 to −8 mmHg during quiet breathing.

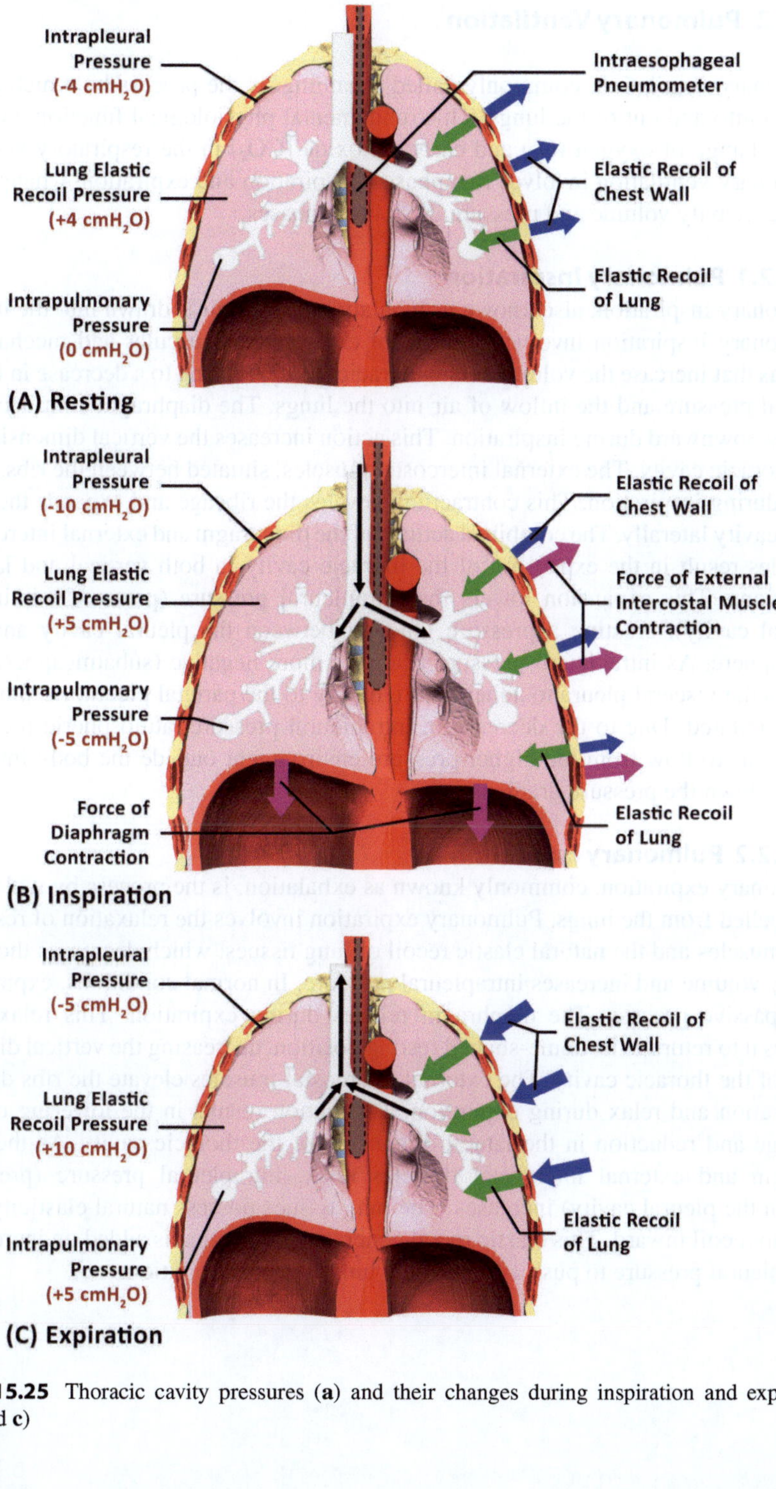

Fig. 15.25 Thoracic cavity pressures (**a**) and their changes during inspiration and expiration (**b** and **c**)

15.2.2 Pulmonary Ventilation

Pulmonary ventilation, commonly called breathing, is the process by which air is moved into and out of the lungs. This fundamental physiological function ensures the exchange of oxygen (O_2) and carbon dioxide (CO_2) in the respiratory system. Pulmonary ventilation involves two phases: inspiration and expiration. Changes in thoracic cavity volume and pressure drive these phases.

15.2.2.1 Pulmonary Inspiration

Pulmonary inspiration, also known as inhalation, is how air is drawn into the lungs. Pulmonary inspiration involves a series of coordinated muscular and mechanical actions that increase the volume of the thoracic cavity, leading to a decrease in intrapleural pressure and the inflow of air into the lungs. The diaphragm contracts and moves downward during inspiration. This action increases the vertical dimension of the thoracic cavity. The external intercostal muscles, situated between the ribs, contract during inspiration. This contraction elevates the ribcage and expands the thoracic cavity laterally. The combined actions of the diaphragm and external intercostal muscles result in the expansion of the thoracic cavity in both vertical and lateral directions. This expansion lowers the intrapleural pressure (pressure within the pleural cavity), creating a pressure gradient between the pleural cavity and the atmosphere. As intrapleural pressure becomes more negative (subatmospheric), it causes the visceral pleura to adhere more tightly to the parietal pleura, keeping the lungs inflated. Due to the decrease in intrapleural pressure, atmospheric pressure forces air to flow from the higher-pressure environment outside the body into the lungs, down the pressure gradient.

15.2.2.2 Pulmonary Expiration

Pulmonary expiration, commonly known as exhalation, is the process by which air is expelled from the lungs. Pulmonary expiration involves the relaxation of respiratory muscles and the natural elastic recoil of lung tissues, which decreases thoracic cavity volume and increases intrapleural pressure. In normal conditions, expiration is a passive process. The diaphragm relaxes during expiration. This relaxation allows it to return to its dome-shaped resting position, decreasing the vertical dimension of the thoracic cavity. The external intercostal muscles elevate the ribs during inspiration and relax during expiration. This action results in the lowering of the ribcage and reduction in the lateral dimension of the thoracic cavity. As the diaphragm and external intercostal muscles relax, intrapleural pressure (pressure within the pleural cavity) increases. The lung tissues possess natural elasticity and tend to recoil inward. This elastic recoil generates a force that is added on increased intrapleural pressure to push air out of the lungs during expiration.

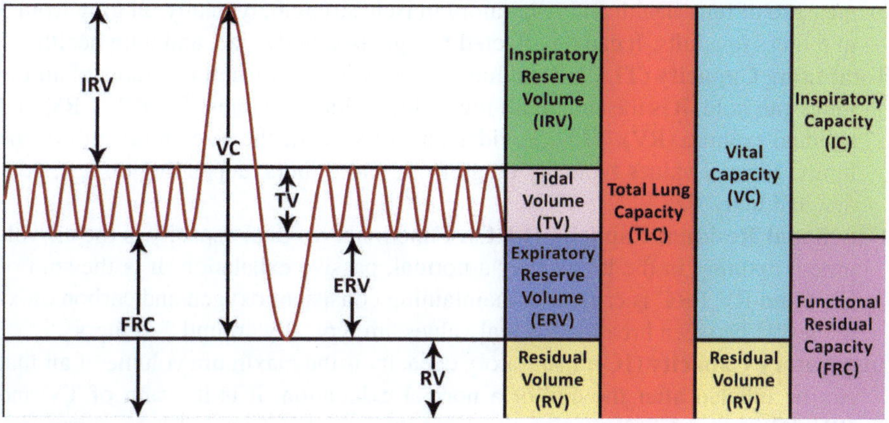

Fig. 15.26 Respiratory volumes and capacities

15.2.3 Lung Volumes

Lung volumes (Fig. 15.26) refer to the different volumes of air that the lungs can hold during the respiratory cycle phases:

Tidal Volume (TV): Tidal volume is the air inhaled or exhaled during normal, relaxed breathing. On average, a healthy adult's tidal volume is about 500 milliliters (mL).

Inspiratory Reserve Volume (IRV): The inspiratory reserve volume is the additional air that can be inhaled with maximum effort after a normal, quiet inspiration ends. IRV typically ranges between 1900 and 3100 mL, depending on age, sex, and physical condition.

Expiratory Reserve Volume (ERV): The ERV is the extra air that can be actively exhaled after a normal, quiet exhalation. In healthy adults, it is generally about 700–1200 mL.

Residual Volume (RV): Residual volume is the air that remains in the lungs after a maximal exhalation. It cannot be voluntarily expelled. RV is crucial for keeping the alveoli open and maintaining gas exchange between breaths. It averages around 1200–1500 mL.

15.2.4 Respiratory Capacities

Lung capacities (Fig. 15.26) represent combinations of different lung volumes and provide a more comprehensive assessment of lung function than individual volumes:

Vital Capacity (VC): Vital capacity is the air a person can expel from the lungs after a maximum inhalation. It is the sum of tidal volume (TV), inspiratory reserve volume (IRV), and expiratory reserve volume (ERV). VC is an important measure of lung health and respiratory muscle strength, typically ranging from 3 to 5 liters in adults. It can be affected by age, sex, body size, and lung health.

Total Lung Capacity (TLC): Total lung capacity is the maximum volume of air the lungs can hold. It is the sum of all four primary lung volumes: TV, IRV, ERV, and residual volume (RV). TLC provides an overview of the size and health of the lungs. Normal values typically range from 4 to 6 liters, depending on age, gender, and body size.

Functional Residual Capacity (FRC): Functional residual capacity is the air volume remaining in the lungs after a normal, passive exhalation. It is the sum of ERV and RV. FRC is critical in maintaining consistent oxygen and carbon dioxide levels between breaths. Normal values are typically around 2–3 liters.

Inspiratory Capacity (IC): Inspiratory capacity is the maximum volume of air that can be inhaled after the end of a normal exhalation. It is the sum of TV and IRV. IC is useful in assessing the capacity for inhalation and lung expansion, with normal values generally ranging from 2 to 4 liters.

15.2.5 Alveolar Ventilation

Alveolar ventilation describes the volume of fresh air that reaches the alveoli per unit of time. It is a crucial determinant of effective gas exchange within the lungs, influencing the concentration of oxygen (O_2) and carbon dioxide (CO_2) in the alveoli and arterial blood. Alveolar ventilation (VA) can be calculated using the following formula:

$$\mathrm{VA} = (\mathrm{VT} - \mathrm{VD}) \times \mathrm{RR}$$

where:

- VA: alveolar ventilation (mL/min)
- VT: tidal volume (mL)
- VD: dead space volume (mL)
- RR: respiratory rate (breaths/min)

15.2.6 Dead Space

Lung dead space (Fig. 15.27) refers to areas within the respiratory system where air is present, but gas exchange does not occur. Dead space can be categorized into two types: anatomical dead space and physiological dead space.

Anatomical Dead Space: Anatomical dead space encompasses the volume of the conducting airways where gas exchange does not occur. This includes the nose, pharynx, larynx, trachea, bronchi, and bronchioles. The air in these regions is referred to as "dead" because it doesn't participate in gas exchange with blood in

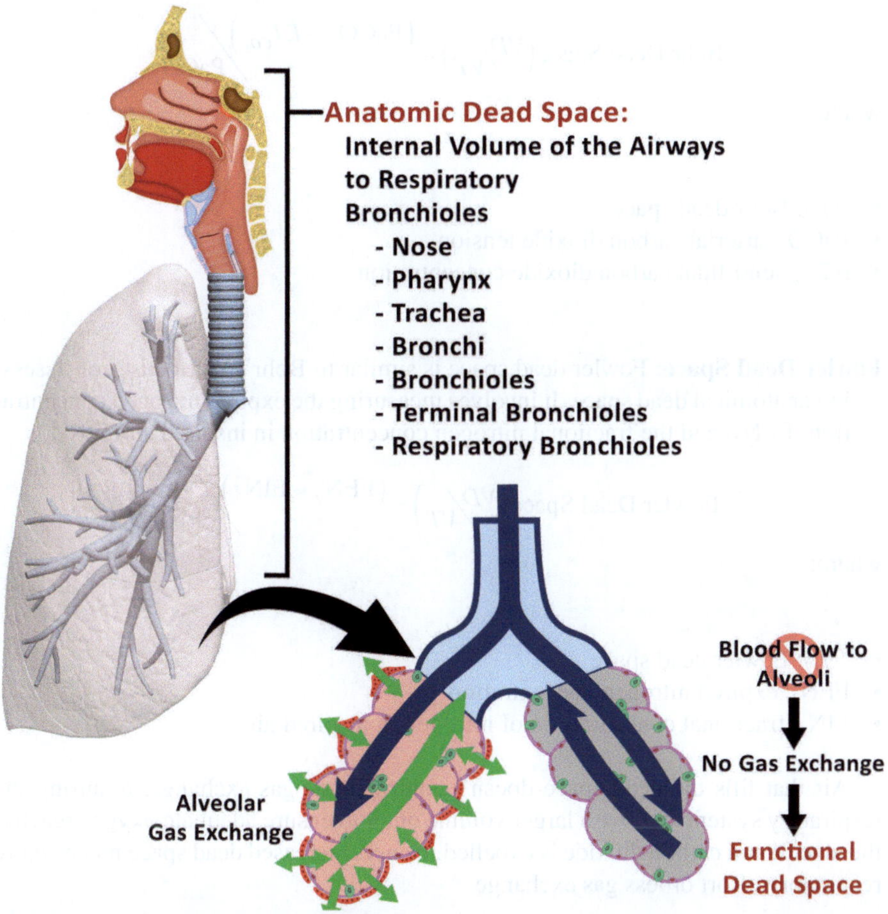

Fig. 15.27 Lung dead space

the alveoli. It primarily serves to conduct air to the lower respiratory tract. A healthy adult's anatomical dead space volume is roughly equivalent to about 1 mL per pound of ideal body weight, typically around 150 mL.

Physiological Dead Space: Physiological dead space is a broad concept that includes any alveolar volumes where ventilation happens but gas exchange is impaired or non-existent. This can be due to poor blood flow, damaged alveolar walls, or obstruction in pulmonary capillaries. It is a more clinical measure, reflecting the efficiency of the lungs in oxygenating blood and removing carbon dioxide.

Respiratory dead space can be measured using various techniques:

Bohr Dead Space: The Bohr dead space is calculated using the Bohr equation, which measures the physiological dead space.

$$\text{Bohr Dead Space}\left({VD}/{VT}\right) = \left(\text{PaCO}_2 - ET_{CO_2}\right)/\text{PaCO}_2$$

where:

- ${}^{VD}/_{VT}$: Bohr dead space
- $PaCO_2$: arterial carbon dioxide tension
- ET_{CO2}: end-tidal carbon dioxide concentration

Fowler Dead Space: Fowler dead space is similar to Bohr's but focuses on assessing anatomical dead space. It involves measuring the expired nitrogen concentration (FEN_2) and the fractional nitrogen concentration in inspired air (FIN_2).

$$\text{Fowler Dead Space}\left({VD}/{VT}\right) = \left(\text{FEN}_2 - \text{FIN}_2\right)/\text{FEN}_2$$

where:

- ${}^{VD}/_{VT}$: Fowler dead space
- FEN_2: expired nitrogen concentration
- FIN_2: fractional concentration of nitrogen in inspired air

Air that fills the dead space doesn't contribute to gas exchange, requiring the respiratory system to move a larger volume of air to ensure adequate oxygen reaches the alveoli and carbon dioxide is expelled. Hence, increased dead space means more respiratory effort or less gas exchange.

15.2.7 Gas Exchange

Gas exchange occurs in the lungs' alveoli, which are essential for maintaining cellular function and overall homeostasis. The primary mechanism of gas exchange is diffusion, where gases move from higher-concentration areas to lower-concentration regions. Oxygen diffuses from the alveoli into the pulmonary capillaries, while carbon dioxide diffuses into the alveoli. The respiratory membrane (Fig. 15.21), comprising the alveolar wall, capillary wall, and shared basement membrane, facilitates gas exchange. It is thin, allowing efficient diffusion of O_2 and CO_2. Differences in partial pressures drive gas exchange. The partial pressure of O_2 (PAO_2) in the alveoli is higher than in the pulmonary capillaries ($PACO_2$), promoting O_2 uptake by pulmonary capillaries. Conversely, the partial pressure of CO_2 ($PACO_2$) in the capillaries is higher than in the alveoli (PAO_2), facilitating CO_2 removal from pulmonary capillaries. Oxygen is transported in the blood bound to hemoglobin (Page 491). Each hemoglobin molecule can carry four O_2 molecules.

15.2.7.1 Physical Principles

Dalton's law of partial pressures describes the behavior of gas mixtures and individual gases' contributions to the mixture's total pressure. It states that the total pressure exerted by a mixture of gases equals the sum of the partial pressures of each gas in the mix.

$$P_{\text{Total}} = P1 + P2 + P3 + \cdots + Pn$$

where:

- P_{Total} is the total pressure of the gas mixture.
- P_1, P_2, P_3, ..., P_n are the partial pressures of each gas in the mixture.

Each partial pressure represents the pressure a single gas component would exert if it occupied the container's entire volume under the same temperature and volume conditions.

Henry's law states that the concentration of a gas dissolved in a liquid is directly proportional to the partial pressure of that gas above the liquid when the system is in equilibrium.

$$C = kH \times P$$

where:

- C is the gas concentration in the liquid (usually in mol/L or molarity).
- kH is Henry's law constant, specific to the gas-liquid system at a particular temperature.
- P is the partial pressure of the gas above the liquid (usually in atmospheres or pascals).

15.2.7.2 Alveolar Air Composition

The composition of alveolar gases refers to the specific mixture of gases in the lungs' alveoli. This composition differs from that of atmospheric air:

- Oxygen (O_2): The partial pressure of oxygen (PO_2) in the alveoli is lower than in atmospheric air. At sea level, the PO_2 in the alveoli is approximately 105 mmHg, whereas atmospheric PO_2 is about 160 mmHg. This difference is due to the mixing of inspired air with the residual air in the respiratory tract (anatomical dead space) and the exchange of oxygen with blood in the pulmonary capillaries.
- Carbon Dioxide (CO_2): The partial pressure of carbon dioxide (PCO_2) in the alveoli is higher than that in atmospheric air. It typically ranges from 35 to 45 mmHg. This elevated PCO_2 results from removing CO_2 from the blood as it passes through systemic capillaries and into the alveoli for exhalation.
- Nitrogen (N_2): Nitrogen is the most abundant gas in both atmospheric air and alveolar gases. The alveoli's partial pressure (PN_2) is similar to that in the atmosphere, around 570 mmHg.

- Water Vapor (H_2O): The respiratory system humidifies inspired air as it moves through the airways. Consequently, the alveoli contain water vapor, contributing to the total gas pressure. The partial pressure of water vapor (PH_2O) in the alveoli depends on humidity and body temperature, typically around 47 mmHg.
- Trace Gases: Alveolar gases also contain trace amounts of other gases, such as argon, neon, and helium, and trace amounts of volatile organic compounds. These gases are in minute concentrations and do not significantly impact gas exchange.

Oxygen is relatively insoluble in aqueous solutions, including blood plasma. However, the presence of hemoglobin in red blood cells dramatically enhances O_2 carrying capacity, as each hemoglobin molecule can bind to four O_2 molecules.

Carbon dioxide, on the other hand, is highly soluble in water and blood plasma. A substantial portion of CO_2 is carried in the blood as bicarbonate ions (HCO_3^-) following its conversion by the enzyme carbonic anhydrase (Section 15.2.9).

15.2.7.3 Ventilation-Perfusion Coupling

Ventilation-perfusion coupling (Fig. 15.28) refers to the matching of ventilation (airflow) and perfusion (blood flow) in the lungs. This coupling ensures efficient gas exchange by optimizing the distribution of ventilation and perfusion to different regions of the lungs. The mechanisms controlling ventilation and perfusion coupling involve:

- Gravity: Perfusion is favored at the base of the lungs, where alveoli are more distensible and blood vessels are under higher pressure. Ventilation is favored at the top of the lungs, where alveoli are less distensible and vascular pressure is lower.
- Regional Changes: During inhalation, as the chest cavity expands, alveoli at the base of the lungs are subject to increased ventilation, matching the increased perfusion in this region. During exhalation, the reverse occurs, with alveoli at the lung apex receiving better ventilation when perfusion is lower.
- Hypoxic Vasoconstriction: In response to low oxygen levels (hypoxia), pulmonary arterioles in poorly ventilated regions constrict. This diverts blood flow away from areas with inadequate ventilation, optimizing perfusion-ventilation matching.
- Bronchoconstriction and Bronchodilation: Local factors, such as irritants or inflammatory mediators, can induce bronchoconstriction or bronchodilation. These changes in airway diameter can influence airflow distribution and match it with local perfusion.
- Neural Control: The autonomic nervous system can regulate bronchomotor and vascular tone in response to changing metabolic demands or stress, further optimizing ventilation-perfusion matching.

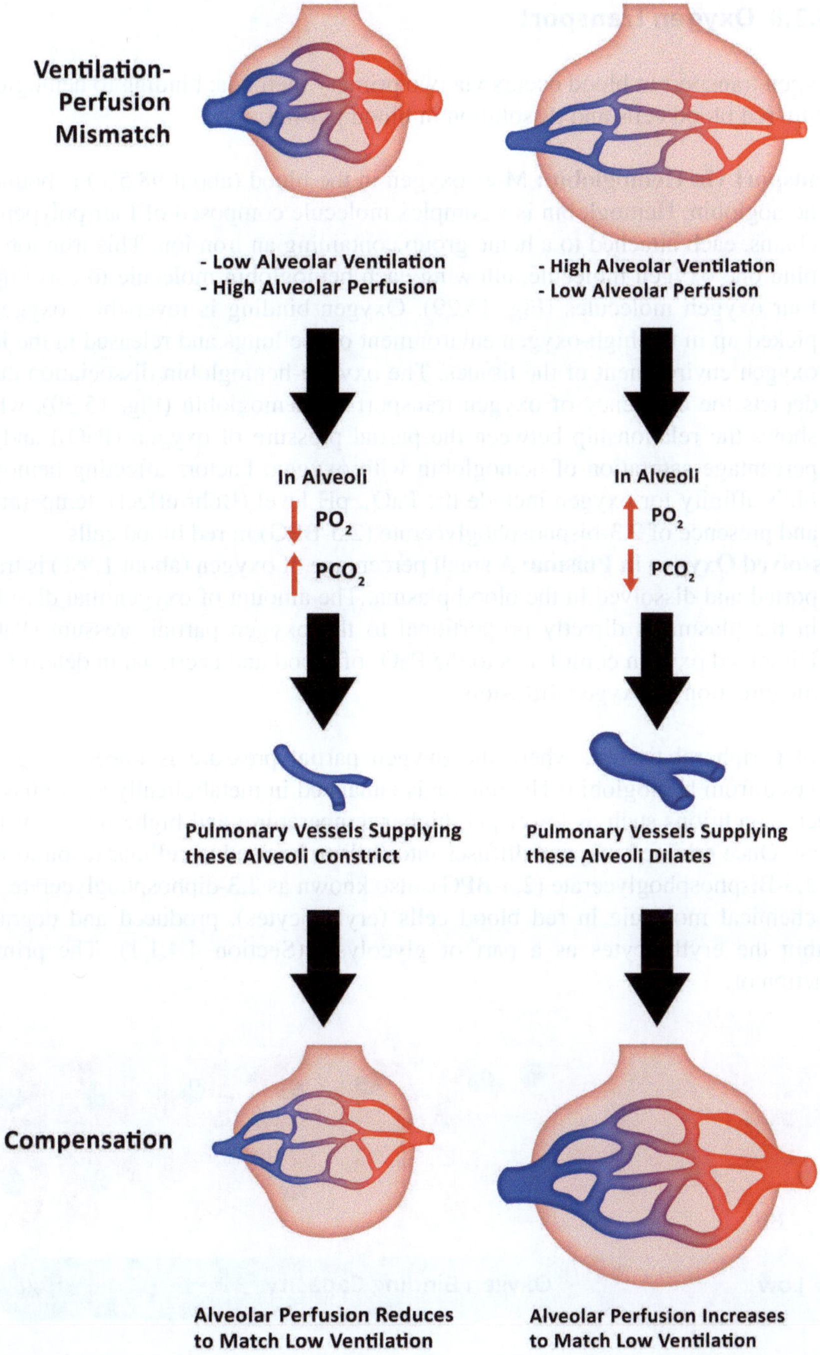

Fig. 15.28 Ventilation-perfusion coupling

15.2.8 Oxygen Transport

Oxygen transport in blood occurs via two primary methods: binding to hemoglobin within red blood cells and dissolution in blood plasma.

Transport via Hemoglobin: Most oxygen in the blood (about 98.5%) is bound to hemoglobin. Hemoglobin is a complex molecule composed of four polypeptide chains, each attached to a heme group containing an iron ion. This iron ion can bind one oxygen molecule, allowing each hemoglobin molecule to carry up to four oxygen molecules (Fig. 15.29). Oxygen binding is reversible; oxygen is picked up in the high-oxygen environment of the lungs and released in the low-oxygen environment of the tissues. The oxygen-hemoglobin dissociation curve depicts the efficiency of oxygen transport by hemoglobin (Fig. 15.30), which shows the relationship between the partial pressure of oxygen (PaO_2) and the percentage saturation of hemoglobin with oxygen. Factors affecting hemoglobin's affinity for oxygen include the PaO_2, pH level (Bohr effect), temperature, and presence of 2,3-bisphosphoglycerate (2,3-BPG) in red blood cells.

Dissolved Oxygen in Plasma: A small percentage of oxygen (about 1.5%) is transported and dissolved in the blood plasma. The amount of oxygen that dissolves in the plasma is directly proportional to the oxygen partial pressure (PaO_2). Dissolved oxygen contributes to the PaO_2 of blood and is crucial in determining the direction of oxygen diffusion.

In peripheral tissues, where the oxygen partial pressure is lower, oxygen is released from hemoglobin. The release is enhanced in metabolically active tissues, where conditions such as lower pH, higher temperature, and higher levels of CO_2 occur. Once released, oxygen diffuses into cells to be used in cellular respiration.

2,3-Bisphosphoglycerate (2,3-BPG), also known as 2,3-diphosphoglycerate, is a biochemical molecule in red blood cells (erythrocytes), produced and degraded within the erythrocytes as a part of glycolysis (Section 4.4.1.1). The primary function of.

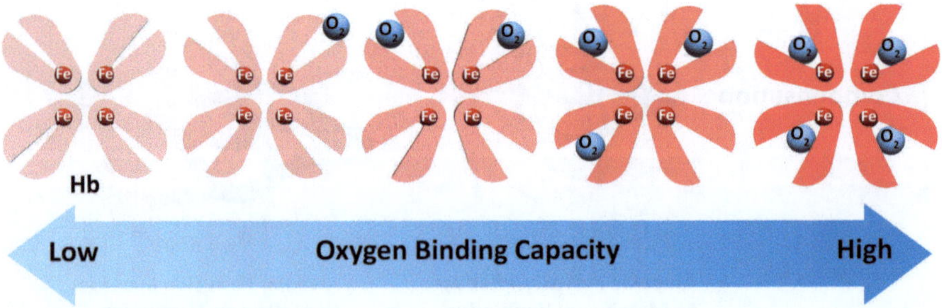

Fig. 15.29 Cooperative binding: hemoglobin affinity for oxygen increases with the number of oxygen molecules attached to Hb due to a change in the 3D structure of Hb with binding of each oxygen molecule, which facilitates the attachment of the following oxygen molecule

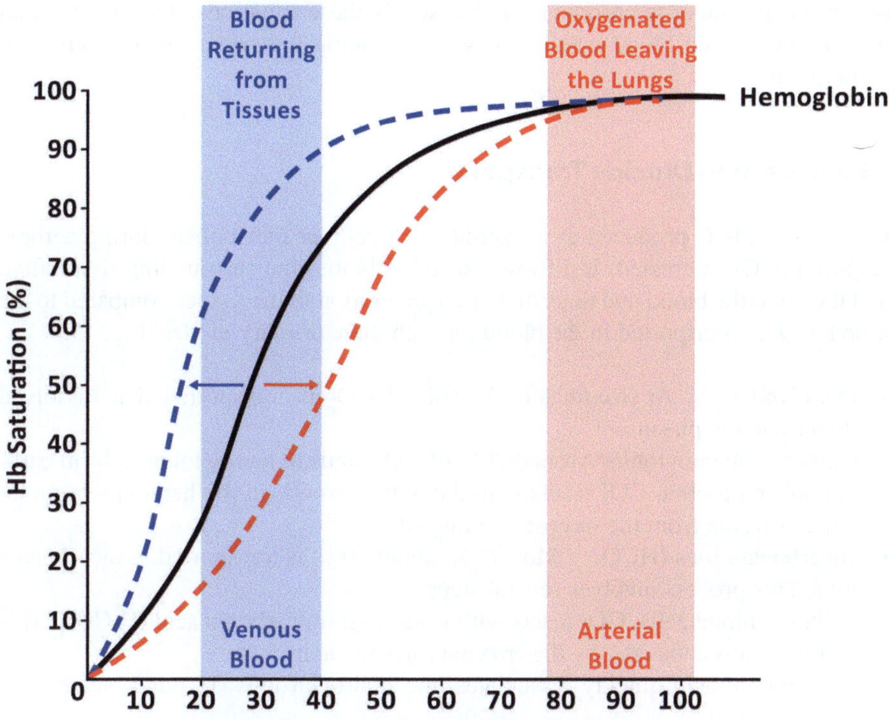

Left Shift (↓O_2 Unloading to Tissues)	↓H^+ (↑pH)
	↓PCO_2
	↓2,3-BPG
	↓Temperature
Right Shift (↑O_2 Unloading to Tissues)	↑H^+ (↓pH)
	↑PCO_2
	↑2,3-BPG
	↑Temperature
	Exercise
	High Altitude

Fig. 15.30 Oxygen-hemoglobin dissociation curve (upper) and the factors affecting Hb's affinity for oxygen (lower)

2,3-BPG is to regulate oxygen binding to hemoglobin. It binds with deoxyhemoglobin (hemoglobin not bound to oxygen) and stabilizes it, making it less likely to bind oxygen. By doing so, 2,3-BPG reduces hemoglobin's affinity for oxygen, facilitating the release of oxygen into the tissues. 2,3-BPG's role becomes crucial when increased oxygen delivery to tissues is needed, such as at high altitudes, during physical exertion, or in chronic lung diseases. In these situations, the concentration of 2,3-BPG in red blood cells increases, enhancing the release of oxygen from hemoglobin.

15.2.9 Carbon Dioxide Transport

Carbon dioxide is produced as a byproduct of cellular metabolism during aerobic respiration. Once created, it diffuses out of cells into the surrounding tissue fluid and then into the blood due to its higher concentration in the tissues compared to the blood. CO_2 is transported in the blood through three primary modes (Fig. 15.31):

- Dissolved CO_2: Approximately 5–10% of CO_2 is transported that dissolved directly in the plasma.
- Carbaminohemoglobin: Around 20% of CO_2 binds to hemoglobin to form carbaminohemoglobin. CO_2 reacts with the amino groups in the hemoglobin molecule, differing from the oxygen-binding site.
- Bicarbonate Ions (HCO_3^-): Most CO_2 (about 70%) is transported as bicarbonate ions. This process involves several steps:
  - In red blood cells, CO_2 reacts with water to form carbonic acid (H_2CO_3). This reaction is catalyzed by the enzyme carbonic anhydrase.
  - Carbonic acid quickly dissociates into hydrogen ions (H^+) and bicarbonate ions (HCO_3^-).
  - Bicarbonate ions are then transported out of the red blood cells into the plasma in exchange for chloride ions, a process known as the chloride shift or Hamburger phenomenon.
  - In the lungs, these processes are reversed. Bicarbonate ions return to the red blood cells and combine with hydrogen ions to form carbonic acid. Carbonic acid is then converted back to CO_2 and water by carbonic anhydrase. The CO_2 diffuses from the red blood cells into the alveoli and is expelled from the body during exhalation.

The carbon dioxide transport system also plays a crucial role in maintaining the acid-base balance in the body. The reversible conversion of CO_2 into bicarbonate ions acts as a buffer system for the blood. Changes in respiratory rate can alter blood pH by increasing or decreasing CO_2 elimination, a mechanism utilized in the body's response to acidosis or alkalosis (Section 18.4.2).

15.2.10 Control of Respiration

The control of respiration involves the coordination of neural and chemical mechanisms to ensure efficient gas exchange and to meet the body's metabolic demands.

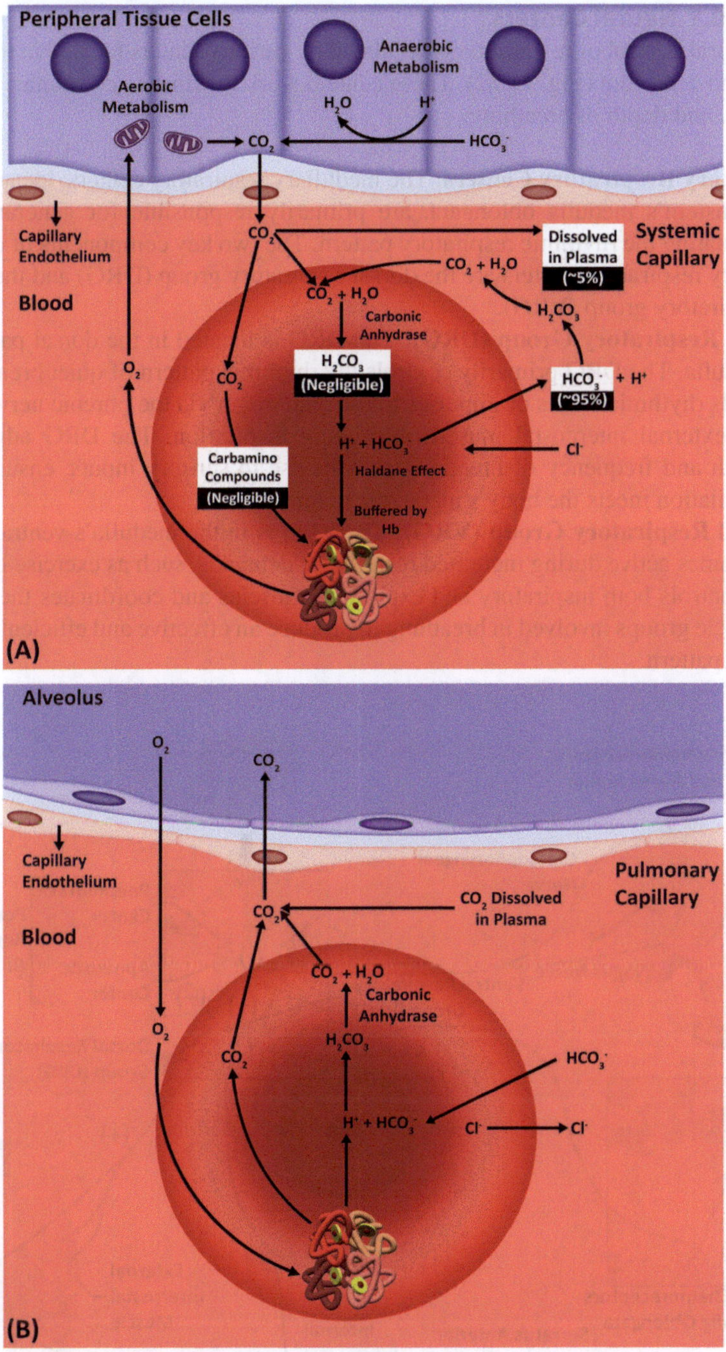

Fig. 15.31 Carbon dioxide transport: (**a**) in peripheral tissues and (**b**) in alveoli

15.2.10.1 Neural Centers

The neural control of respiration is regulated by various centers located in the brainstem and cerebrum (Fig. 15.32). These centers work together to coordinate the rate, rhythm, and depth of breathing.

Medullary Respiratory Centers: The medullary respiratory centers, located in the brainstem's medulla oblongata, are primarily responsible for generating and regulating the rhythmic respiratory pattern. The two key components of the medullary respiratory centers are the dorsal respiratory group (DRG) and the ventral respiratory group (VRG).

Dorsal Respiratory Group (DRG): The DRG is located in the dorsal part of the medulla. The DRG primarily controls the rhythmic pattern of quiet breathing. It sends rhythmic bursts of impulses to the diaphragm via the phrenic nerve and to the external intercostal muscles, initiating inspiration. The DRG adjusts the depth and frequency of breathing in response to sensory inputs, ensuring that ventilation meets the body's metabolic demands.

Ventral Respiratory Group (VRG): The VRG is in the medulla's ventral part. It becomes active during increased respiratory demands, such as exercise or stress. It controls both inspiratory and expiratory muscles and coordinates the various muscle groups involved in breathing to produce an effective and efficient respiratory pattern.

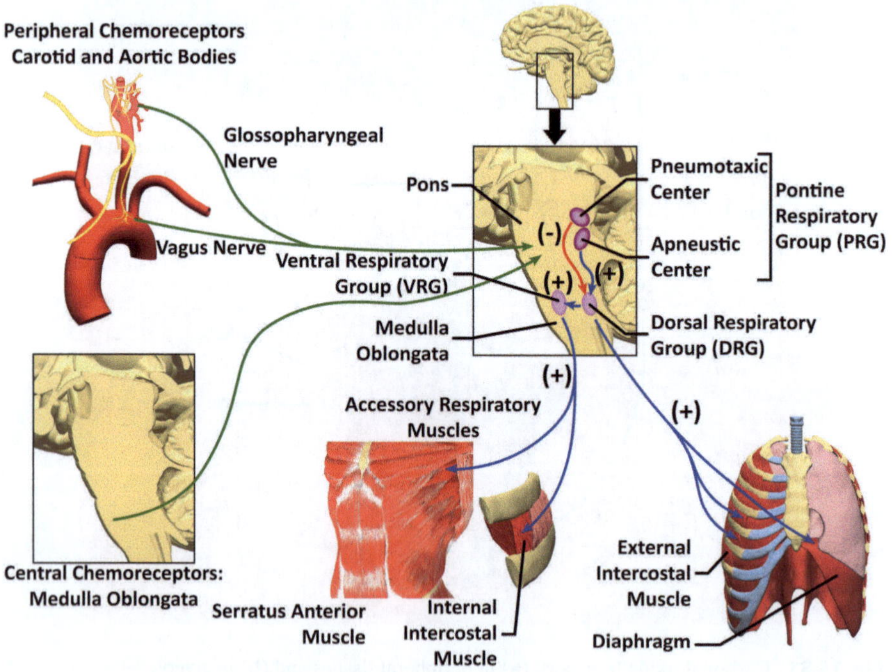

Fig. 15.32 Neural and chemical respiratory control centers

The medullary respiratory centers work with the pontine respiratory centers (the pneumotaxic and apneustic centers) to regulate the overall pattern and rhythm of breathing. The pontine centers modulate the activity of the DRG and VRG, fine-tuning the respiratory rate and depth, especially during activities like speaking, singing, or sleeping. The pneumotaxic center regulates the switch-off point of the inspiratory neurons in the medulla oblongata, thereby influencing the duration of inhalation and the respiratory rate. It effectively controls the depth and rate of breathing. The apneustic center provides a constant stimulatory signal to the DRG, promoting prolonged and deep inspiratory movements. This center is involved in fine-tuning the depth of breathing.

The higher brain centers in the cerebral cortex and limbic system add voluntary and emotional control to the primarily automatic process managed by the lower brain centers.

15.2.10.2 Chemical Factors

The chemical factors primarily include carbon dioxide (CO_2), oxygen (O_2), and the blood's and cerebrospinal fluid's pH. The blood's partial pressure of carbon dioxide ($PaCO_2$) is the most potent chemical factor affecting respiratory rate and depth. The respiratory centers in the brainstem are highly sensitive to changes in CO_2 levels. $PaCO_2$ (hypercapnia) increases during hypoventilation or increased metabolic activity. This leads to a decrease in blood pH (respiratory acidosis), stimulating the central chemoreceptors in the medulla oblongata and increasing the rate and depth of breathing to expel more CO_2. Conversely, a decrease in $PaCO_2$ (hypocapnia) typically results from hyperventilation and causes an increase in blood pH (respiratory alkalosis). This condition can inhibit respiratory drive, reducing breathing rate and depth.

Oxygen levels in the blood, specifically the partial pressure of oxygen (PaO_2), are monitored by peripheral chemoreceptors in the carotid and aortic bodies (Page 467). A significant drop in PaO_2 (hypoxemia) stimulates these chemoreceptors, particularly when PaO_2 falls below 60 mmHg (the normal range is 75–100 mmHg). This results in increased respiratory activity to enhance oxygen intake. However, the body's response to hypoxemia is generally less sensitive than hypercapnia.

The respiratory system helps regulate blood pH by controlling the exhalation of CO_2. CO_2 reacts with water in the blood to form carbonic acid, which dissociates into hydrogen ions (H^+) and bicarbonate ions (HCO_3^-), influencing pH. Acidosis (low pH) due to non-respiratory causes (metabolic acidosis) can also stimulate increased ventilation to remove more CO_2, thus raising blood pH. Conversely, alkalosis (high pH) can depress the respiratory system, reducing the breathing rate and depth.

Carotid Bodies: The carotid bodies (Fig. 15.33) are small, oval-shaped organs located bilaterally at the bifurcation of the common carotid arteries into the internal and external carotid arteries near the carotid sinus. They comprise clusters of chemoreceptor cells, known as glomus cells or type I cells, surrounded by a network of blood vessels and supporting sustentacular (type II) cells. The glomus

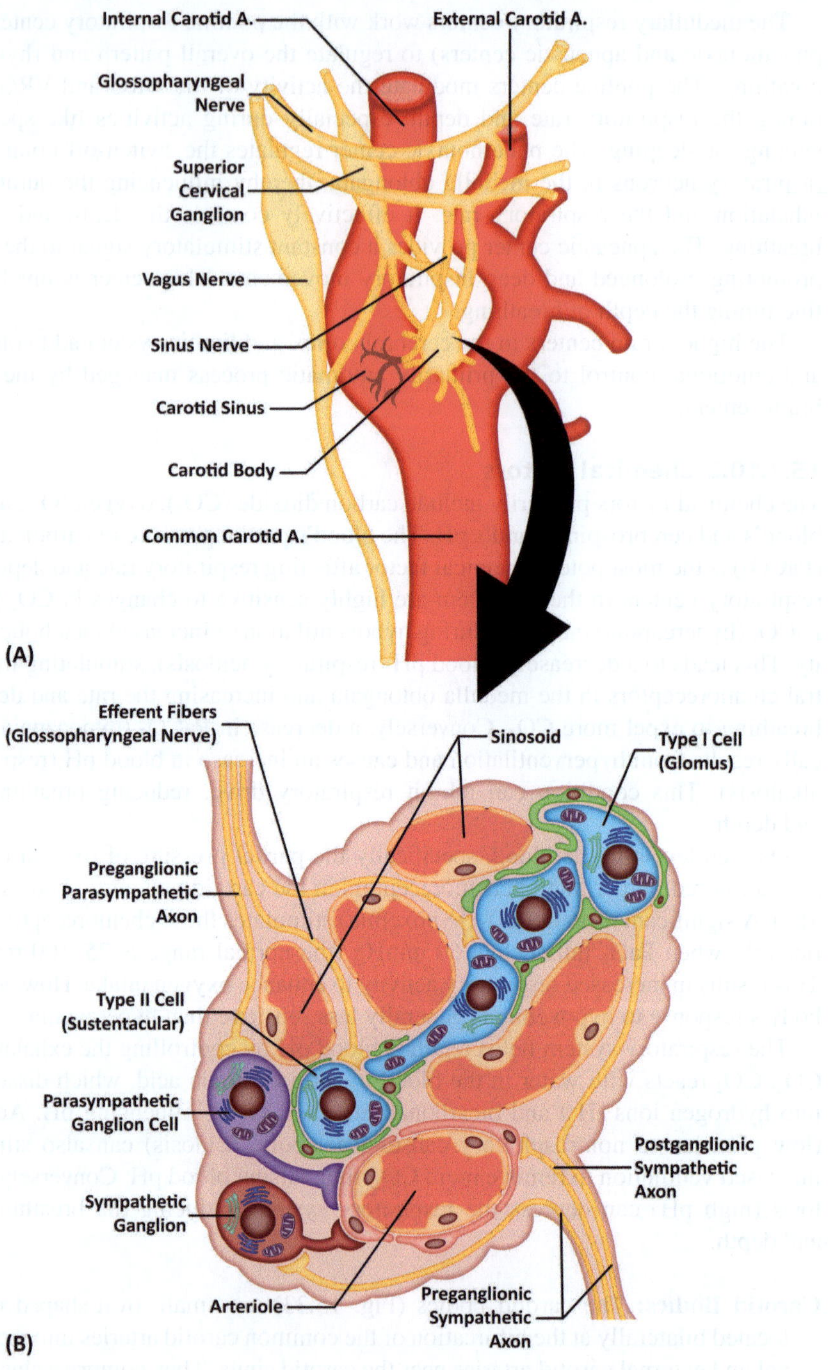

Fig. 15.33 Carotid body. (**a**) Gross anatomical structure. (**b**) Microscopic structure

cells are the primary chemosensitive elements, containing numerous neurotransmitters, including dopamine, acetylcholine, and adenosine. The carotid bodies have a rich blood supply, allowing them to constantly monitor the blood's chemical composition. They are innervated by the carotid sinus nerve, a branch of the glossopharyngeal nerve (cranial nerve IX), which carries sensory information from the carotid bodies to the brainstem. These chemoreceptors are particularly sensitive to changes in the partial pressure of arterial oxygen (PaO_2). They are also responsive to changes in $PaCO_2$ and blood pH.

Aortic Arch Chemoreceptors: The aortic bodies (Fig. 15.34), or aortic arch chemoreceptors, are located on the aortic arch and are associated with the vagus nerve (cranial nerve X). Like the carotid bodies, the aortic bodies contain chemosensitive cells that respond to changes in blood gases and pH. These structures are smaller and less numerous than the carotid bodies. They are primarily innervated by the aortic nerve, a branch of the vagus nerve, which conveys sensory information to the medulla oblongata. Their response to hypoxemia is generally less pronounced, and they primarily respond to changes in $PaCO_2$ and pH levels in the blood.

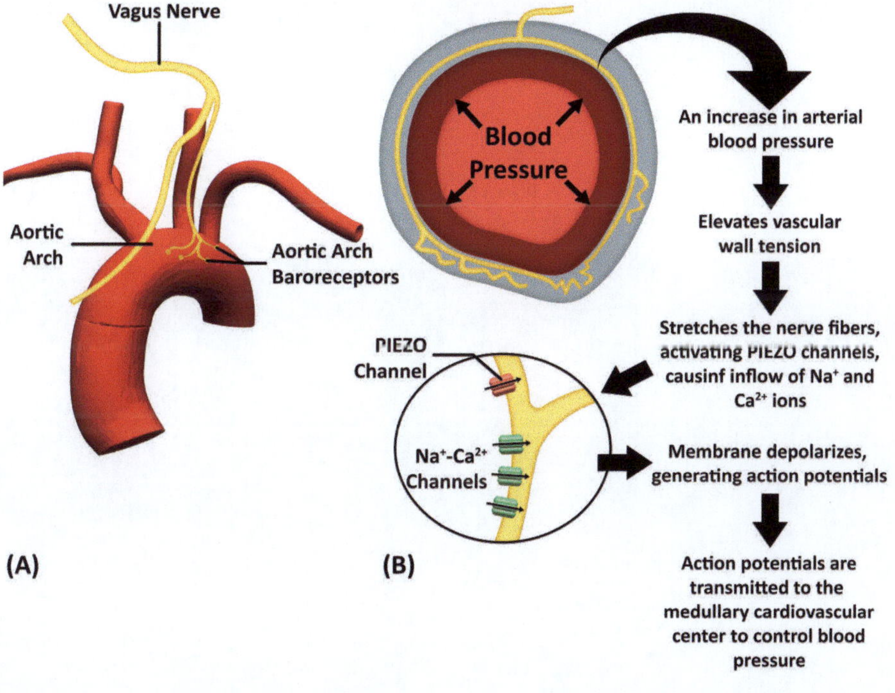

Fig. 15.34 Aortic arch baroreceptors. (**a**) Gross anatomical structure. (**b**) Mechanism of action

Suggested Reading

Advanced Physiology and Pathophysiology: Essentials for Clinical Practice, 2nd Ed. Nancy Tkacs, Linda Herrmann, Springer Publishing Company, 2024

Anatomy & Physiology, S. Meloni, et al. Independent Publisher, 2023

Anatomy & Physiology: The Unity of Form and Function, 10th Ed. Kenneth S. Saladin, McGraw Hill, 2023

Gray's Anatomy for Students, 5th Ed. Richard L. Drake, et al. Elsevier, 2020

Human Anatomy & Physiology, 11th Ed. Elaine N. Marieb and Katja Hoehn, Pearson, 2022

Moore's Clinically Oriented Anatomy, 9th Ed. Keith L. Moore, et al. Lippincott Williams & Wilkins, 2020

Netter Atlas of Human Anatomy: Classic Regional Approach, 8th Ed. Frank H. Netter, Elsevier, 2022

Principles of Anatomy and Physiology, 16th Ed. Gerard J. Tortora, Bryan H. Derrickson, Wiley, 2020

Digestive System

16

In this chapter, you will learn:

- General structure and function of the digestive system
- Structure and function of the mouth and pharynx in digestion
- Digestive functions of the teeth and process of chewing
- Structure of the esophagus and its function in transporting food
- Anatomy and functions of the stomach, including motility and digestion processes
- Detailed structure of the small intestine and its function in nutrient absorption
- Structure and function of the colon, including its role in water and electrolyte absorption
- The process of oral digestion, including mastication and the role of saliva
- The mechanics and phases of swallowing
- Composition and roles of gastric secretions
- Specific enzymatic actions in the small intestine for final digestion
- The physiology of the defecation process
- Role of the intestinal microflora in digestion and health
- Mechanisms of peristalsis and segmentation in the digestive tract
- Role of the large intestine in water absorption and feces formation
- Role of the rectum and anus in the storage and expulsion of feces
- Detailed anatomy and function of the salivary glands
- Anatomy and functions of the liver in metabolism and detoxification
- Bile secretion, storage in the gallbladder, and its role in digestion
- Pancreatic structure and functions and the role of pancreatic juices in digestion
- Various digestive enzymes, their sources, and functions
- Mechanisms of carbohydrate, protein, and lipid digestion in the intestines
- Nutrient absorption mechanisms in the digestive system

H. Hosseinzadeh, *Fundamentals of Medicine for Biomedical Engineering*,
https://doi.org/10.1007/978-3-031-74085-5_16

- Process of lipid emulsification and absorption
- Importance of water and electrolyte balance and their absorption processes
- Absorption of vitamins and minerals and the mechanisms involved
- Hormones involved in digestion and their regulatory functions

The digestive system (Fig. 16.1) is a complex assembly of organs and structures whose primary function is the breakdown of food into nutrients, which are then absorbed into the body to provide energy, growth, and cellular repair. This system encompasses a series of interconnected organs, including the mouth, esophagus, stomach, small intestine, large intestine, rectum, and anus, each performing distinct yet interrelated functions.

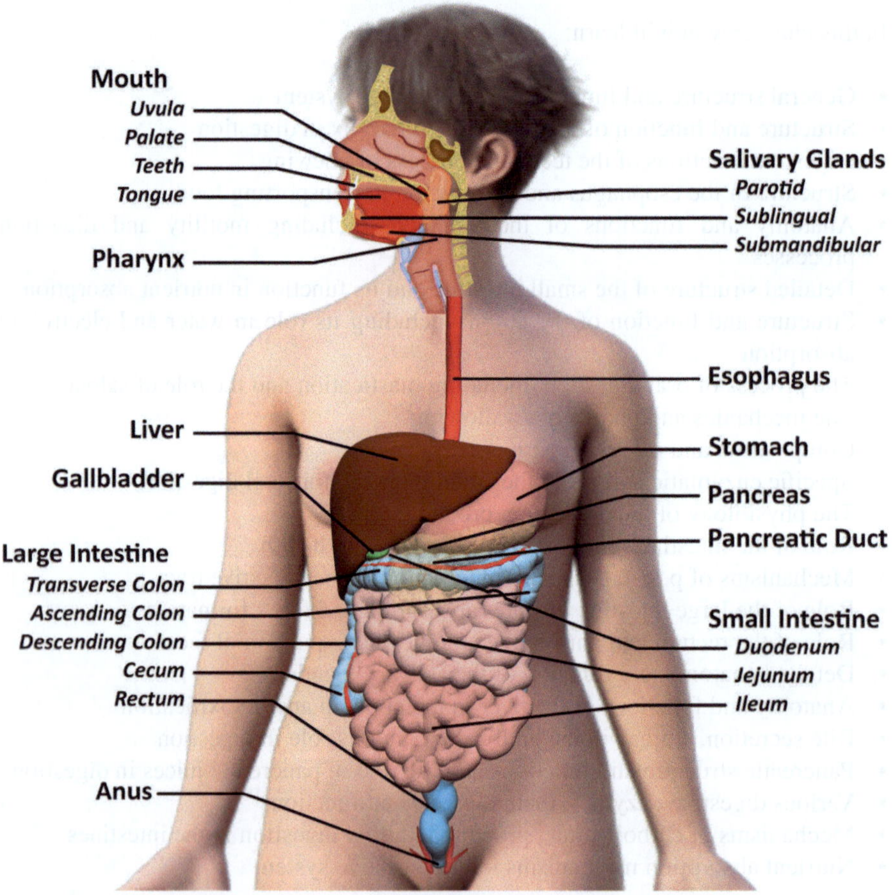

Fig. 16.1 The gastrointestinal system and the accessory digestive organs

16.1 Part 1: General Structure

The digestive system's microscopic structure reveals a highly specialized and organized arrangement of tissues and cells designed to optimize digestion and nutrient absorption. This microscopic architecture varies throughout the different parts of the digestive system, reflecting the specific functions of each organ; however, at its most fundamental level, the walls of the digestive tract share a common basic structure, which includes four layers: mucosa, submucosa, muscularis externa, and serosa or adventitia (Fig. 16.2).

16.1.1 Mucosa

The mucosa is the innermost layer of the digestive tract, pivotal in digestion and absorption. Its structure is highly specialized and varies throughout the digestive system to accommodate the specific functions of each segment. Despite these variations, the mucosa consists of three primary components: the epithelium, the lamina propria, and the muscularis mucosae (Fig. 16.2).

Epithelium Epithelium is a layer of cells that lines the luminal surface of the digestive tract. The type of epithelial cells varies depending on the location within the tract. For instance, the esophagus is lined with stratified squamous epithelium to protect against abrasion, while the stomach and intestines are lined with simple columnar epithelium to facilitate secretion and absorption. This layer serves multiple functions, including protection, secretion, and absorption. In the stomach, the

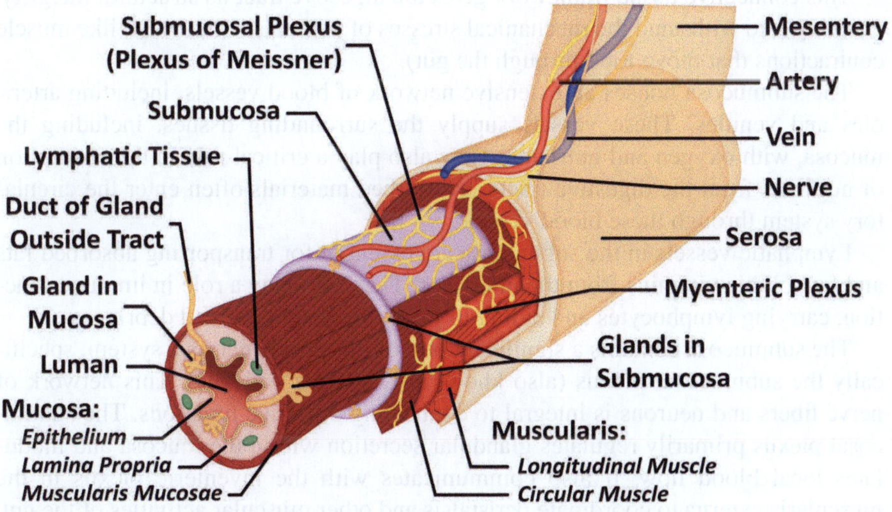

Fig. 16.2 General structure of the gastrointestinal tract

epithelium contains cells that secrete gastric juices, while in the intestines, it comprises cells specialized for nutrient absorption.

Lamina Propria Directly beneath the epithelium lies the lamina propria, a layer of loose connective tissue. It is rich in blood, lymphatic vessels, and various cells, including fibroblasts, immune cells (like lymphocytes and macrophages), and occasional glands. The lamina propria supports the epithelium and plays a crucial role in nutrient absorption by providing a rich vascular network to transport absorbed nutrients. It also houses components of the immune system, which defends against pathogens.

Muscularis Mucosae The muscularis mucosae is a thin layer of smooth muscle fibers beneath the lamina propria. It is typically organized into two layers: an inner circular layer and an outer longitudinal layer. This muscle layer enables local movement of the mucosa, enhancing its interaction with the contents of the digestive tract. Movements of the muscularis mucosae can help mix food with digestive enzymes and expose nutrients to the absorptive surfaces.

16.1.2 Submucosa

The submucosa (Fig. 16.2) is a supportive layer in the digestive tract located immediately outside the mucosa. This layer provides the mucosa's structural support, nourishment, and neural connections. The submucosa primarily consists of dense, irregular connective tissue. This tissue is rich in collagen fibers, providing strength and elasticity. It also contains variable amounts of elastic fibers, imparting flexibility and allowing the digestive tract to expand and contract.

This connective tissue framework gives the digestive tract its structural integrity, enabling it to withstand the mechanical stresses of peristalsis (the wave-like muscle contractions that move food through the gut).

The submucosa houses an extensive network of blood vessels, including arterioles and venules. These vessels supply the surrounding tissues, including the mucosa, with oxygen and nutrients. They also play a critical role in the absorption of nutrients from the digestive tract, as absorbed materials often enter the circulatory system through these blood vessels.

Lymphatic vessels in the submucosa are essential for transporting absorbed fats and fat-soluble vitamins from the intestines. They also play a role in immune function, carrying lymphocytes and aiding in removing pathogens and debris.

The submucosa contains a significant part of the enteric nervous system, specifically the submucosal plexus (also known as Meissner's plexus). This network of nerve fibers and neurons is integral to controlling digestive functions. The submucosal plexus primarily regulates glandular secretion within the mucosa and modulates local blood flow. It also communicates with the myenteric plexus in the muscularis externa to coordinate peristalsis and other muscular activities of the gut.

16.1.3 Muscularis Externa

The muscularis externa (Fig. 16.2) is primarily responsible for food movement through the system via peristalsis. The muscularis externa typically consists of two main layers of smooth muscle:

- Inner Circular Layer: This layer of muscle fibers encircles the lumen of the digestive tract. When these fibers contract, they constrict the lumen, helping to propel the food forward.
- Outer Longitudinal Layer: Running along the length of the digestive tract, this layer's contraction shortens the tract, aiding food movement.

Regional variations:

- Esophagus: Here, the upper part of the muscularis externa may contain skeletal muscle instead of smooth muscle, reflecting the voluntary control of swallowing.
- Stomach: The stomach has an additional oblique muscle layer inside the circular layer, which aids in the mechanical breakdown of food and mixing it with gastric juices.
- Small and Large Intestines: These regions have standard dual layers, which are essential for the peristalsis that moves food and waste products along the digestive tract.

The myenteric plexus (Auerbach's plexus), located between the inner circular and outer longitudinal layers, is a network of neurons and nerve fibers that forms part of the enteric nervous system. This plexus plays a crucial role in regulating the motor activity of the gastrointestinal tract, including the coordination of peristaltic movements and the control of gastrointestinal secretions.

The primary function of the muscularis externa is to facilitate the movement of food, chyme, and waste material through the digestive tract. This movement is achieved through coordinated contractions known as peristalsis. In certain areas, like the sphincters (e.g., the pyloric sphincter at the stomach exit), the circular muscle layer is thickened to control the passage of material from one part of the digestive tract to another.

16.1.4 Serosa and Adventitia

The serosa and adventitia (Fig. 16.2) are the outermost layers of the digestive tract, crucially providing mechanical support and protection. These layers differ based on their location within the digestive tract and the organs' relative position within the abdominal cavity.

Serosa Serosa is present in the intraperitoneal organs of the digestive tract, such as the stomach, small intestine, and parts of the large intestine. It is a serous membrane consisting of a thin layer of simple squamous epithelial cells known as mesothelium, underlain by a thin layer of connective tissue. The serosa produces a lubricating serous fluid that reduces friction between the digestive organs and the surrounding structures as they move during peristalsis and other bodily movements. In several regions, the serosa forms part of the mesentery and omentum, extensions of the peritoneum that connect the intestines to the posterior abdominal wall. These structures allow blood vessels, nerves, and lymphatics to reach the intestines.

Adventitia Adventitia is found in the retroperitoneal organs of the digestive tract, including the esophagus, parts of the duodenum, ascending and descending colons, and the rectum. It consists of dense connective tissue that blends into the surrounding tissue, lacking the mesothelial covering seen in the serosa. Unlike the serosa, the adventitia does not produce a lubricating fluid. Its primary function is to anchor the digestive organs to nearby structures, ensuring their stable positioning within the body.

16.1.5 Peritoneum

The peritoneum (Fig. 16.3) is a significant and complex serous membrane in the abdominal cavity. It supports abdominal organs, facilitates their movements, and houses blood vessels, nerves, and lymphatics.

The peritoneum consists of two main layers:

- Parietal Peritoneum: Lines the internal surface of the abdominal and pelvic cavities.
- Visceral Peritoneum: Envelopes the abdominal organs, covering their surfaces. This layer is essentially the serosa of the intraperitoneal organs in the digestive tract.

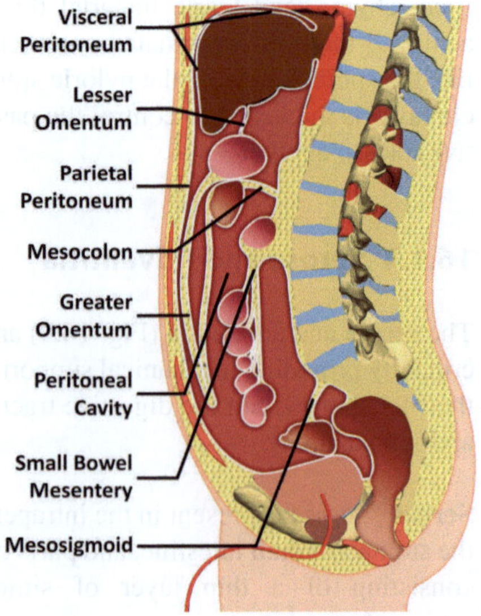

Fig. 16.3 Peritoneum and the major mesenteries of the digestive organs

The peritoneal cavity lies between the parietal and visceral layers, a potential space containing a small amount of serous fluid. This fluid, secreted by the cells of the peritoneum, acts as a lubricant, allowing the organs to glide smoothly against each other during movements like digestion.

Mesentery: A double layer of peritoneum that extends from the posterior abdominal wall to envelop the intestines. It provides a pathway for blood vessels, nerves, and lymphatics to reach the intestines and holds the organs in place.

Greater Omentum: A large apron-like fold of visceral peritoneum hanging from the stomach and covering the intestines. It contains fat deposits and immune cells, providing protection and insulation.

Lesser Omentum: Extends from the liver to the lesser curvature of the stomach and the beginning of the duodenum. It also contains vessels and nerves supplying these organs.

16.1.6 Blood Supply

The digestive tract's blood supply is extensive and complex, reflecting the high demand for oxygen and nutrients required for its functions. It involves a rich network of arteries, capillaries, and veins that provide blood to the various organs of the digestive system, each adapted to the specific needs of these organs.

Arterial Supply (Fig. 16.4): Except for the mouth and esophagus, which receive their blood supply from different sources, the arterial blood supply to the digestive tract is primarily derived from three major branches of the abdominal aorta: the celiac trunk, the superior mesenteric artery, and the inferior mesenteric artery.

- Celiac Trunk: This trunk supplies blood to the stomach, liver, spleen, and the first part of the duodenum. It branches into the left gastric artery, the splenic artery, and the common hepatic artery.
- Superior Mesenteric Artery (SMA): This artery supplies blood to the remainder of the duodenum, jejunum, ileum, cecum, ascending colon, and part of the transverse colon. It gives off branches like the jejunal, ileal, ileocolic, right colic, and middle colic arteries.
- Inferior Mesenteric Artery (IMA): This artery provides blood to the distal part of the transverse colon, descending colon, sigmoid colon, and rectum. Its branches include the left colic, sigmoid, and superior rectal arteries.

Venous Drainage: Venous blood from the digestive tract is collected by a series of veins corresponding to the arteries. These veins ultimately drain into the portal vein, which carries nutrient-rich blood to the liver for processing. The portal vein, formed by the convergence of the superior mesenteric and splenic veins, also receives blood from the inferior mesenteric vein. The portal vein then enters the liver, processing nutrients and toxins.

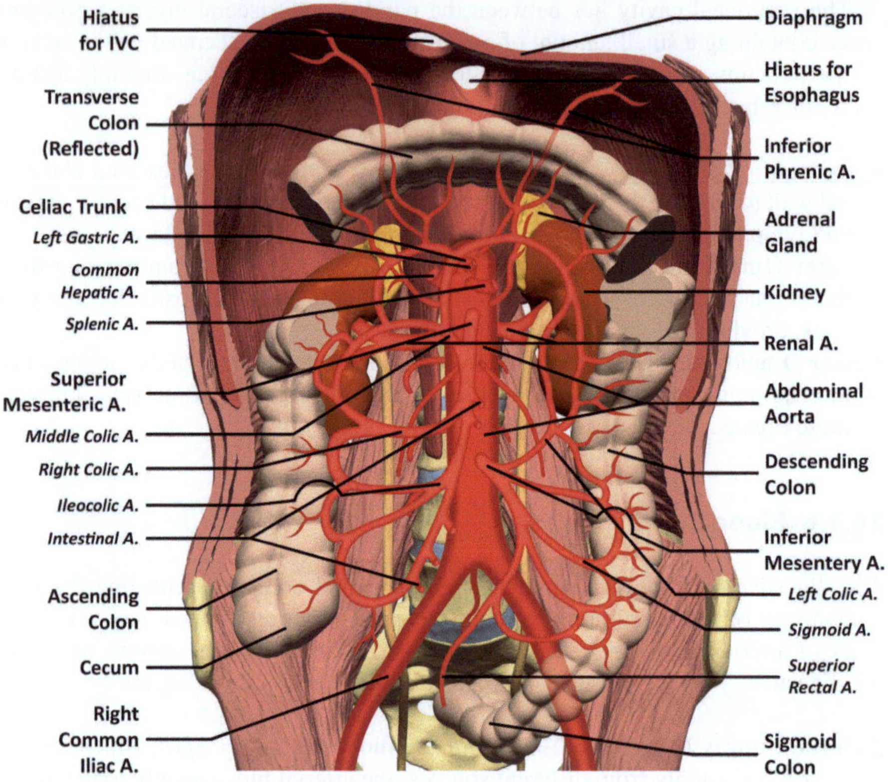

Fig. 16.4 Major arteries supplying the digestive system

Lymphatic Drainage: Alongside the blood vessels, there is a comprehensive system of lymphatic vessels crucial for transporting absorbed fats and fat-soluble vitamins from the intestines.

16.1.7 Nerve Supply

The nerve supply of the digestive tract is a sophisticated and intricate system that regulates various digestive functions, including motility, secretion, and blood flow. This system consists of two major components: the enteric nervous system (ENS) and the extrinsic nervous system, including autonomic and sensory nerve fibers.

Enteric Nervous System The enteric nervous system (ENS) is intrinsic to the gastrointestinal (GI) tract. Often referred to as the “second brain,” the ENS is an extensive network of neurons and nerve fibers that govern the digestive process with a high degree of autonomy. It spans the entire length of the GI tract, from the esophagus to the rectum. The ENS neurons are organized into two primary plexuses.

- Myenteric Plexus (Auerbach's Plexus): It is located between the layers of the muscularis externa (the inner circular and outer longitudinal muscle layers) and extends along the full length of the GI tract. It is primarily responsible for regulating the motor activity of the GI tract, including peristalsis and muscular contractions.
- Submucosal Plexus (Meissner's Plexus): This plexus is found in the submucosa, situated just beneath the mucosa. It mainly controls glandular secretions, absorption, and local blood flow and is more prominent in the small and large intestines.

16.2 Part 2: Functional Anatomy

The gastrointestinal (GI) system is intricately designed to perform the critical functions of digestion, absorption, and elimination. Encompassing a series of hollow organs joined in a long, twisting tube from the mouth to the anus, the GI system is a marvel of functional anatomy, each part specializing in a distinct aspect of the digestive process. This elaborate system consists of two compartments: main and accessory organs. The main organs include the mouth, esophagus, stomach, small intestine, large intestine, rectum, and anus. The accessory organs are the salivary glands, liver, gallbladder, and pancreas.

16.2.1 Main Organs

The main organs of the GI system, forming a continuous tubular system, starting at the mouth and ending at the anus, work in a coordinated manner to ensure the efficient breakdown of food into nutrients and the removal of waste products.

16.2.1.1 Mouth

The mouth, or oral cavity (Fig. 16.5), is the entry point of the gastrointestinal (GI) system and the starting point of digestion. It is not just a mere cavity for food intake but a complex structure equipped with various components, each contributing to mechanical breakdown, chemical digestion, and the initiation of swallowing.

The mouth is bounded at the front by the lips, at the sides by the cheeks, above by the hard and soft palates, below by the floor of the mouth and the tongue, and at the rear by the oropharynx. It is divided into two main parts: the vestibule (the area between the cheeks, lips, and teeth) and the oral cavity proper, which lies inside the teeth and gums.

Lips

The lips, forming the anterior part of the oral cavity, play significant roles in speech, facial expression, and the initial phase of digestion. The outermost layer of the lips is composed of thin, stratified squamous epithelium, similar to the skin on the face but with fewer layers, making it more delicate and translucent. At the vermilion border, where the skin transitions to the mucosal tissue, there is a notable change in the type of epithelial cells. The skin-like stratified squamous epithelium becomes more

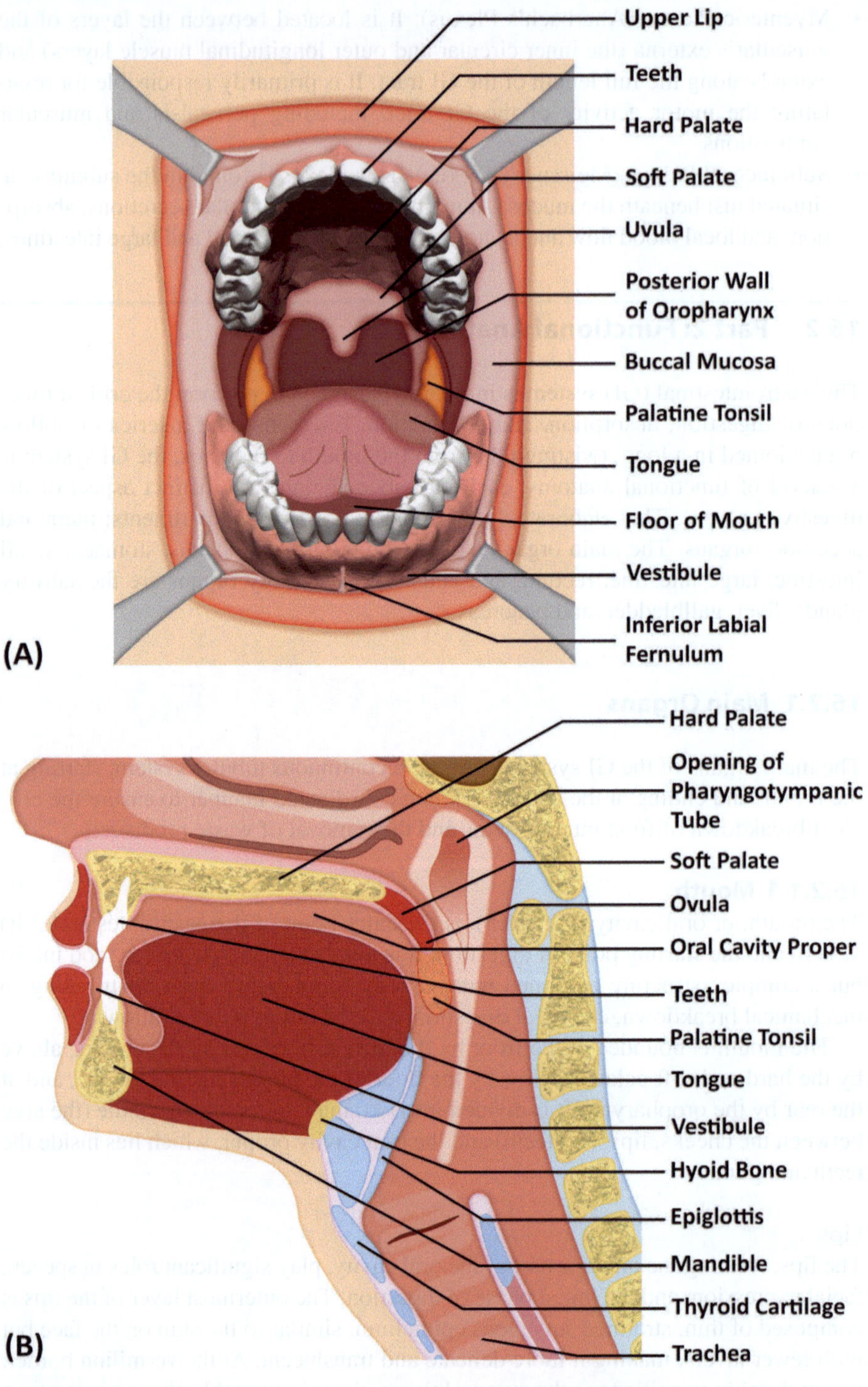

Fig. 16.5 The oral cavity (mouth). (**a**) Anterior view. (**b**) Sagittal section

mucosal-like, lacking hair follicles, sweat glands, or sebaceous glands, making it prone to dryness. The color is due to a high density of capillaries just beneath the thin epithelial layer. A mucous membrane, a moist, non-keratinized stratified squamous epithelium, lines the inner part of the lip. The muscles of the lips are skeletal muscles, controlled voluntarily and innervated by the facial nerve (cranial nerve VII). The primary muscle of the lips is the orbicularis oris (Fig. 16.6). It encircles the mouth opening and acts like a sphincter. It is not a single, continuous muscle but a complex of several muscle fibers, some originating from other facial muscles and inserted into the lips. The orbicularis oris controls the movements necessary for closing the lips, puckering (as in kissing), and holding food within the mouth.

Cheeks

The cheeks, forming the lateral walls of the oral cavity, are multifunctional structures active in facial expression, speech, and mastication (chewing) and even in maintaining oral hygiene. The buccinator muscle (Fig. 16.6) forms the principal muscular component of the cheek. It lies deep within the cheek's structure, stretching horizontally across the cheek, from the maxilla to the mandible. It originates from the alveolar processes of the maxilla and mandible. Its fibers converge toward the lips, inserting into the orbicularis oris muscle at the corners of the mouth. The buccinator is innervated by the buccal branch of the facial nerve (cranial nerve VII). The buccinator maintains the tension of the cheeks during chewing, preventing them from bulging outward when the jaws are closed. It also aids in food manipulation within the mouth and is involved in actions such as blowing and sucking.

Part of the parotid gland is located near the cheek, just in front of the ear. The duct of the parotid gland (Stensen's duct) opens into the oral cavity near the upper second molar, secreting saliva into the mouth.

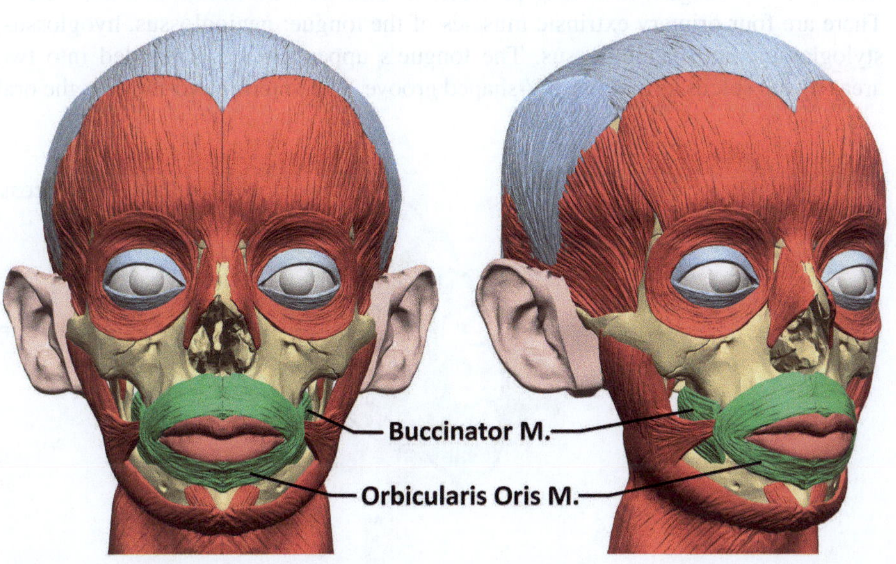

Fig. 16.6 Orbicularis oris and buccinator muscles

Palate

The palate forms the roof of the oral cavity, separating the oral and nasal cavities. It consists of two distinct parts: the hard and soft palates. The hard palate is the anterior, bony portion of the palate. It comprises two bones: the palatine processes of the maxillae (front) and the horizontal plates of the palatine bones (back) (Fig. 16.7). A thick mucosal lining covers the surface. This lining is slightly corrugated to aid food movement and has numerous small glands that secrete mucus. The hard palate provides structural support and a firm surface against which the tongue can press food during chewing. It also plays a role in speech production. The soft palate is a flexible, muscular extension of the hard palate hanging at the back of the mouth. Unlike the hard palate, it does not contain bone and is composed of muscle fibers, connective tissue, and a mucosal covering. The uvula is a conical projection from the posterior edge of the soft palate. It helps close the nasopharynx during swallowing and contributes to the articulation of certain sounds. The soft palate seals off the nasal passages during swallowing and speech. Its movement is crucial for proper speech production.

Tongue

The tongue is a highly versatile and essential organ that plays a vital role in various functions, including taste, mastication (chewing), swallowing, and speech. The tongue is primarily composed of a group of muscles. These muscles are unique because they do not attach to the bone at both ends like most skeletal muscles. The muscular structure of the tongue is categorized into intrinsic and extrinsic muscles. Intrinsic muscles are confined entirely within the tongue and are responsible for altering its shape, which is crucial for speech, mastication (chewing), and swallowing. Extrinsic muscles of the tongue are essential in the gross movement and positioning of the tongue, including protrusion, retraction, elevation, and depression. There are four primary extrinsic muscles of the tongue: genioglossus, hyoglossus, styloglossus, and palatoglossus. The tongue's upper surface is divided into two areas by the sulcus terminalis, a V-shaped groove. The anterior two-thirds is the oral

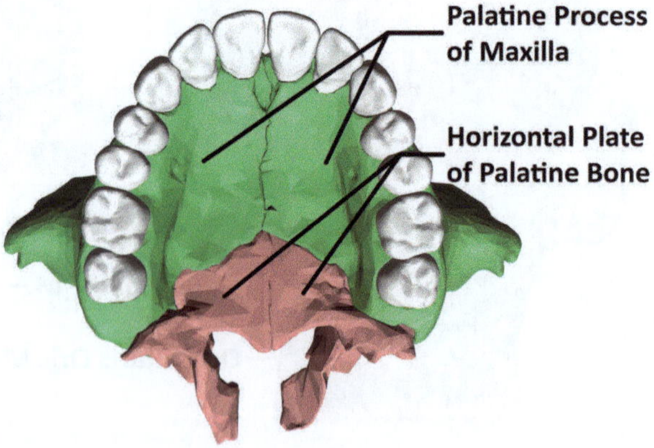

Fig. 16.7 Hard palate structure, inferior view

part, and the posterior one-third is the pharyngeal part. The oral part features numerous small bumps known as papillae, some containing taste buds. The pharyngeal part is smoother, with less distinct papillae, and contains the lingual tonsils. Taste buds are located mainly on the tongue's papillae, specifically the circumvallate, fungiform, and foliate papillae (Section 11.3.2.1). They are responsible for the sensation of taste (Section 11.3.2.3). Each taste bud contains receptor cells that detect sweet, sour, salty, bitter, and umami flavors. The anterior two-thirds of the tongue receives general sensation (touch, pain, temperature) from the lingual nerve (a branch of the trigeminal nerve) and taste sensation from the chorda tympani (a branch of the facial nerve). The posterior third receives both general and taste sensations from the glossopharyngeal nerve. All tongue muscles, except the palatoglossus (innervated by the vagus nerve), are innervated by the hypoglossal nerve.

The tongue is integral to several essential functions:

- Taste: The tongue is the primary organ of taste. It is covered with taste buds that contain gustatory receptor cells, capable of detecting the five fundamental tastes: sweet, sour, salty, bitter, and umami.
- Mastication: The tongue helps manipulate food within the mouth and works with the teeth and saliva to break down food into smaller, more manageable pieces. This action facilitates easier swallowing and ensures that food is adequately mixed with saliva, starting the process of chemical digestion.
- Swallowing: During swallowing (deglutition), the tongue is essential in moving the food bolus from the oral cavity to the pharynx. It pushes the food to the back of the mouth to initiate the swallowing reflex and then helps in propelling the food into the esophagus.
- Speech: The tongue movements produce different sounds by altering the shape and volume of the oral cavity and positioning against the teeth, palate, and other oral structures. They are particularly important in forming consonants and vowels, contributing to the clarity and distinctiveness of speech.
- Oral Hygiene: The tongue assists in cleaning the oral cavity. Its movements help in dislodging food particles from the teeth and gums. The lingual papillae and saliva also help cleanse the mouth after eating.
- Protective Functions: The tongue's sensory capabilities are crucial for detecting harmful substances, overly hot foods, or foreign objects, prompting reflex actions like spitting. Its agility and strength also help prevent choking by manipulating food away from the windpipe.
- Other Functions: The tongue plays a role in the sucking action, especially in infants. It is involved in specific reflex actions like coughing and gagging. The tongue contributes to facial expressions of emotions, such as sticking out the tongue in disgust or playfulness.

Teeth

Human teeth (dentition) are divided into two arches—the maxillary (upper) and mandibular (lower)—which interact through a process known as occlusion. This occlusal relationship is vital for effective mastication and articulation.

Human dentition phases encompass two distinct stages: the primary (deciduous) dentition phase and the permanent (secondary) dentition phase. The primary dentition phase typically begins around 6 months of age, although this can vary. The first teeth to erupt are usually the lower central incisors, followed by the upper central incisors. The complete set of primary teeth is usually in place by the age of 3 years. The shedding process typically starts with the lower central incisors and follows a predictable pattern over the next several years. The permanent (secondary) dentition phase commences around the age of 6 with the eruption of the first permanent molars (often called the "6-year molars"). The third molars, or wisdom teeth, are the last to emerge, usually appearing between 17 and 25 (Fig. 16.8).

A tooth is generally divided into two main parts (Fig. 16.9):

- Crown: The visible part of the tooth above the gum line
- Root: The part embedded in the jawbone, typically not visible

The periodontal ligament (PDL) supports and maintains the teeth within the alveolar processes. The PDL is located between the cementum, which covers the tooth's root, and the alveolar bone of the tooth socket (alveolus). It is composed of dense fibrous connective tissue. The primary components at this microscopic level include enamel, dentin, cementum, and dental pulp (Fig. 16.9). Enamel is the most

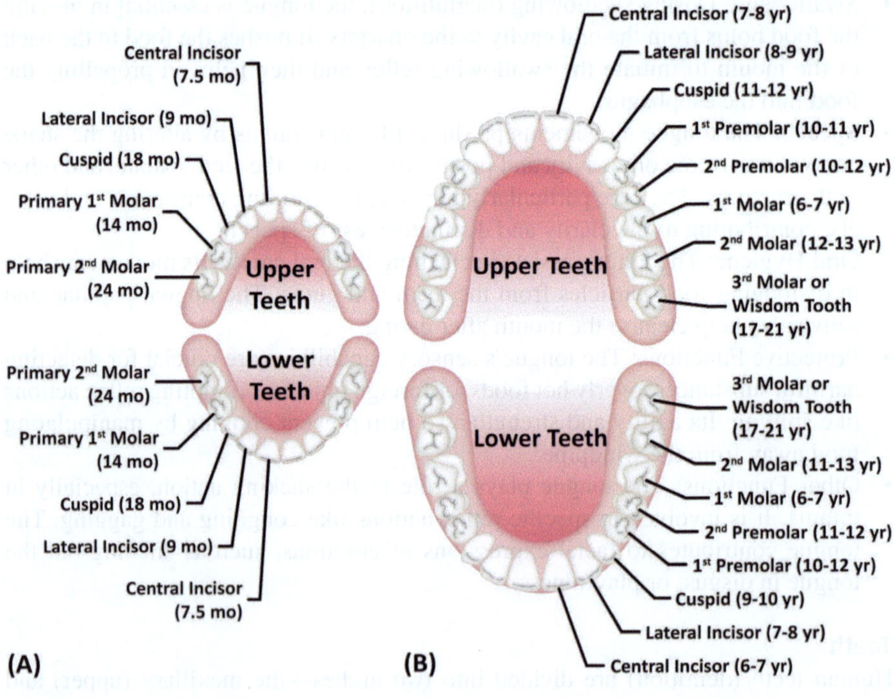

Fig. 16.8 Human dentition: the deciduous (**a**) and permanent (**b**) sets, with the approximate age at which the tooth erupts

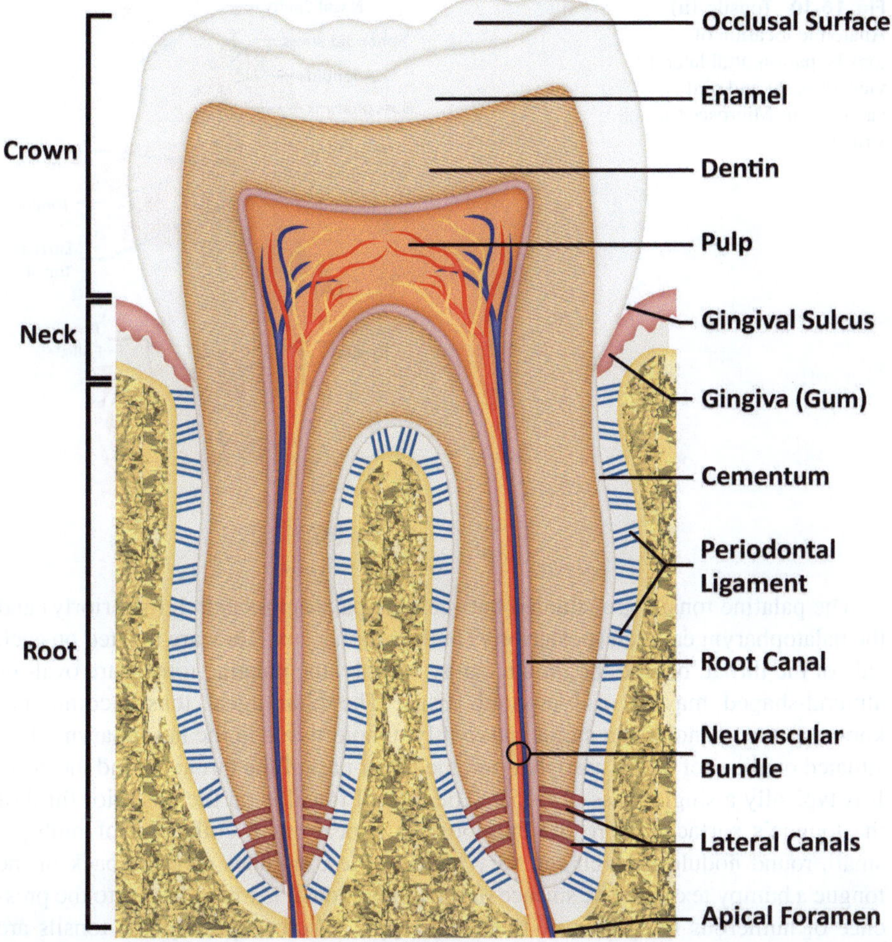

Fig. 16.9 The structure of a typical tooth, longitudinal section

rigid tissue in the human body, primarily composed of hydroxyapatite, a crystalline calcium phosphate. It consists of tightly packed enamel rods (or prisms). Cementum is a calcified tissue that covers the roots of teeth. Similar in composition to bone but less dense, it contains collagen fibers and hydroxyapatite.

Dentin lies beneath the enamel and cementum, forming the bulk of the tooth. It is a porous, yellowish tissue of hydroxyapatite, collagen fibers, and water. The pulp is the central part of the tooth, filled with soft connective tissue, blood vessels, and nerves. It is housed within the pulp chamber and root canals.

Tonsils

The tonsils (Fig. 16.10) are a group of lymphoid tissues located in the throat and nasopharynx. They form part of the body's immune system.

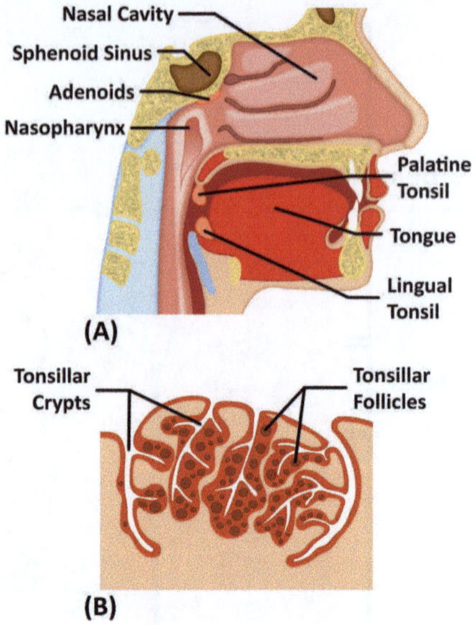

Fig. 16.10 Tonsils. (**a**) Anatomic location of tonsils, parasagittal lateral view of nasal and oral cavities. (**b**) Microscopic structure

The palatine tonsils are situated between the palatoglossal arch (anteriorly) and the palatopharyngeal arch (posteriorly) in the oropharynx. They are located on each side of the throat, behind the mouth cavity. Visibly, the palatine tonsils are oval- or almond-shaped masses of lymphoid tissue. The pharyngeal tonsil, commonly known as the adenoid, is a single mass of lymphoid tissue in the nasopharynx. It is situated on the roof of the nasopharynx, the upper part of the throat behind the nose. It is typically a single mass of tissue. The lingual tonsil is on the posterior third of the tongue's surface before the epiglottis. It appears as a collection of multiple small, round nodules, which can vary in size. These nodules give the back of the tongue a bumpy texture. The surface of the lingual tonsil is irregular due to the presence of numerous lymphatic nodules and crypts. Microscopically, the tonsils are made up of lymphoid tissue that contains many lymphocytes. The tissue is organized into follicles, with germinal centers rich in B lymphocytes. The tonsils have numerous crypts that extend deeply into their interior. These crypts in palatine and lingual tonsils are lined with stratified squamous epithelium, which is continuous with the oropharynx. However, the surface of the pharyngeal tonsil is lined with a pseudostratified columnar ciliated epithelium, which is continuous with that of the nasal passages. The ciliated epithelium helps move mucus and trap particles down the throat. The tonsils are part of the body's first line of defense in the immune system, especially against inhaled or ingested pathogens.

16.2.1.2 Pharynx

The pharynx (Fig. 16.11) is an anatomical structure located behind the nose and mouth, connecting these to the esophagus and trachea, serving as a pathway for air and food. The pharynx is a muscular, funnel-shaped tube that extends from the base of the skull to the level of the sixth cervical vertebra, where it continues as the

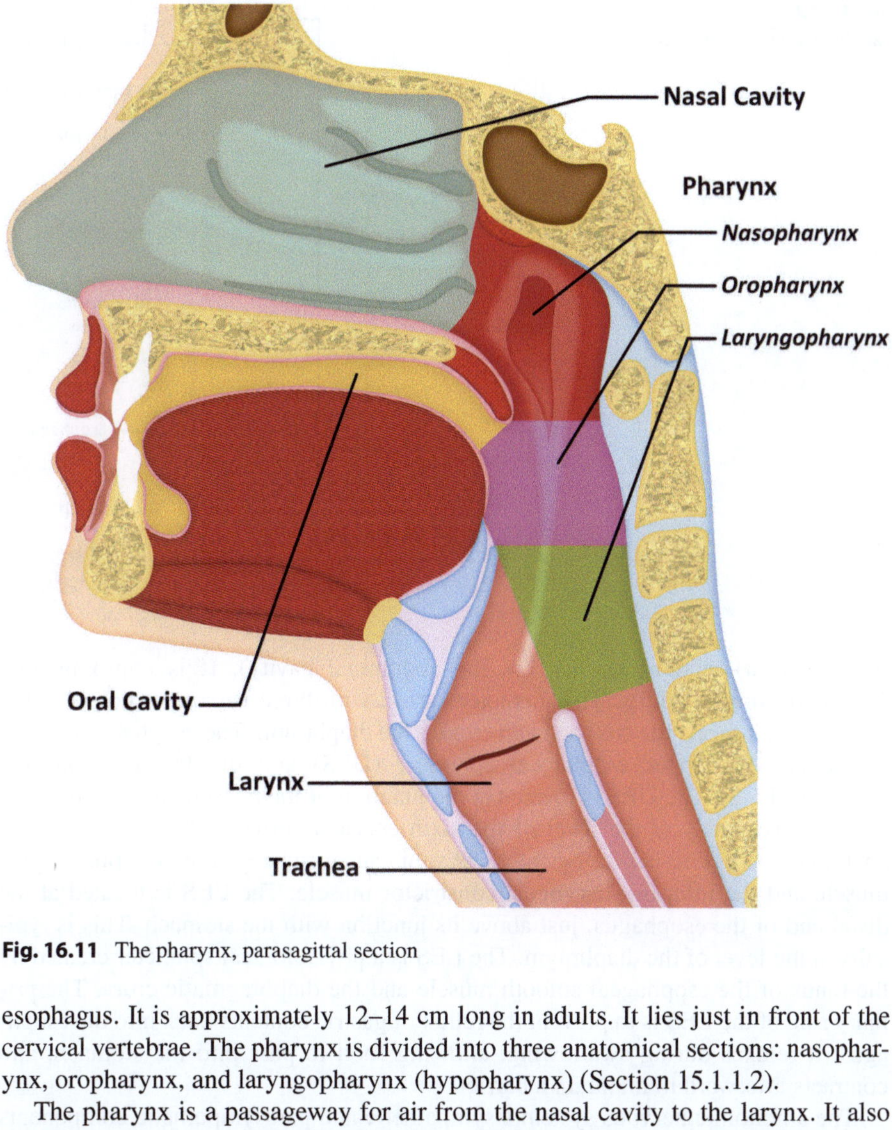

Fig. 16.11 The pharynx, parasagittal section

esophagus. It is approximately 12–14 cm long in adults. It lies just in front of the cervical vertebrae. The pharynx is divided into three anatomical sections: nasopharynx, oropharynx, and laryngopharynx (hypopharynx) (Section 15.1.1.2).

The pharynx is a passageway for air from the nasal cavity to the larynx. It also functions as a channel for food and liquids, carrying them from the oral cavity to the esophagus. The nasopharynx contributes to the resonance of speech and sound production.

16.2.1.3 Esophagus

The esophagus is a muscular tube that extends from the pharynx to the stomach.

Anatomy

The esophagus begins at the level of the sixth cervical vertebra, just behind the cricoid cartilage of the larynx, and descends the neck and thorax to enter the stomach (Fig. 16.12). The esophagus traverses three regions: cervical (in the neck), thoracic

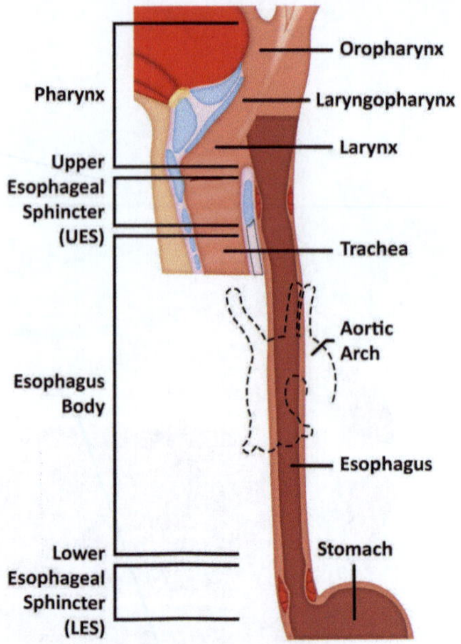

Fig. 16.12 The esophagus, gross structure

(in the chest), and abdominal (in the abdominal cavity). It is approximately 25–30 cm long in adults. The diameter varies, with three narrowest points at the start, where it meets the stomach and crosses the diaphragm. The esophagus has two sphincters—the upper esophageal sphincter (UES) and the lower esophageal sphincter (LES). The UES is located at the junction of the pharynx and the esophagus, approximately at the level of the sixth cervical vertebra. It is not a distinct anatomical structure but a physiological sphincter formed by the cricopharyngeus muscle and the inferior pharyngeal constrictor muscle. The LES is located at the distal end of the esophagus, just above its junction with the stomach. This is typically at the level of the diaphragm. The LES is a physiological sphincter created by the tonus of the esophageal smooth muscle and the diaphragmatic crura. The primary role of the LES is to prevent the reflux of gastric contents back into the esophagus. It relaxes during swallowing, allowing food to pass into the stomach, and contracts afterward to maintain a barrier.

The autonomic nervous system's sympathetic and parasympathetic components innervate the esophagus. The esophagus also contains elements of the myenteric (Auerbach's) plexus and the submucosal (Meissner's) plexus between the muscle layers and in the submucosa. These plexuses are crucial for the local, intrinsic regulation of esophageal motility and secretions.

Microscopic Structure

Like other parts of the GI tract, the esophagus has four anatomical layers (Fig. 16.13): mucosa, submucosa, muscularis externa, and adventitia. The innermost layer is lined with non-keratinized stratified squamous epithelium. This type of epithelium

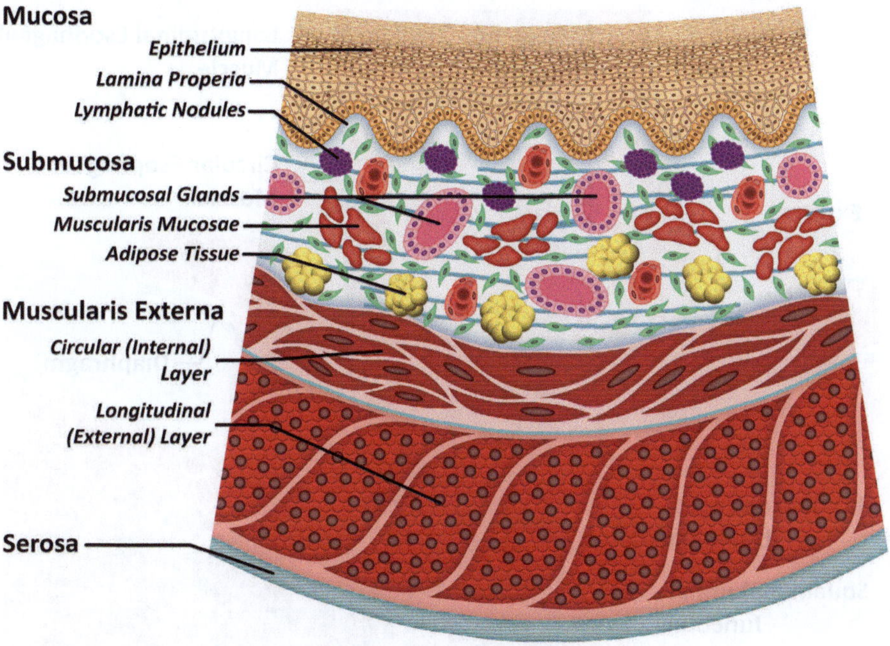

Fig. 16.13 Microscopic structure of the esophageal wall

protects against abrasion from food particles. The muscularis externa has two main layers: an inner circular layer and an outer longitudinal layer. The upper third of the esophagus contains striated muscle fibers, the middle third contains a mixture of striated and smooth muscle, and the lower third contains primarily smooth muscle fibers. During peristalsis, the circular muscles contract sequentially, creating a wave-like motion that propels the food bolus downward. In the lower esophagus, this layer is essential at the lower esophageal sphincter (LES) (Fig. 16.14), where it maintains a tonic contraction to prevent gastroesophageal reflux. The outer muscle layer is arranged longitudinally along the esophagus. Its contraction shortens and widens the esophagus, facilitating the passage of the food bolus. The outermost layer is the adventitia, consisting of loose fibrous connective tissue. Unlike most parts of the gastrointestinal tract, the esophagus has adventitia instead of a serosa.

Peristalsis

Peristalsis is a series of wave-like muscle contractions that move food through the digestive tract. In the esophagus, this involves the sequential contraction and relaxation of the muscularis layer. Primary peristalsis continues the peristaltic wave initiated in the pharynx (Fig. 16.15). Secondary peristalsis is triggered if the primary peristalsis fails to move all the food into the stomach, caused by the distention of the esophagus.

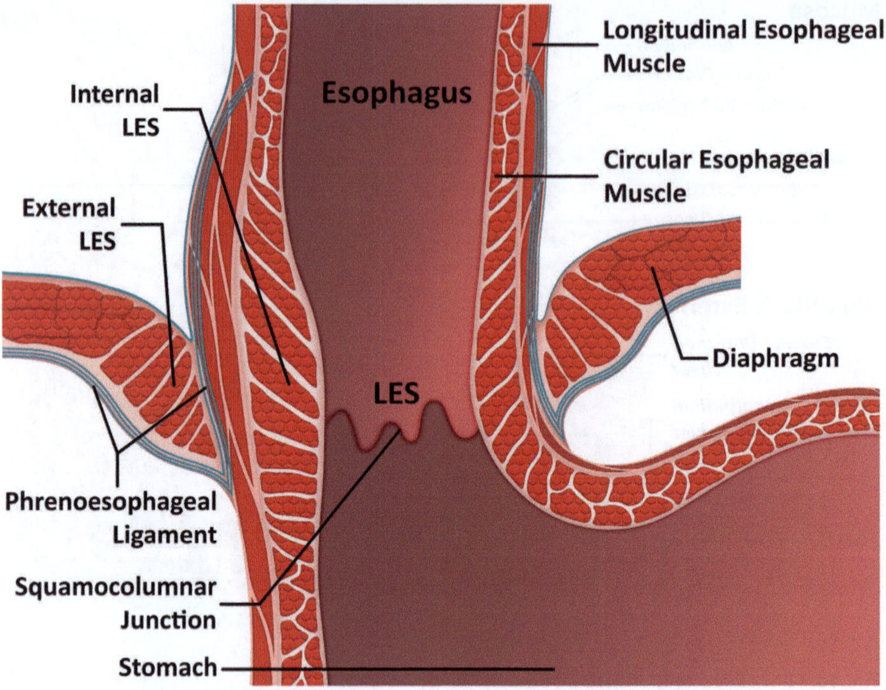

Fig. 16.14 Lower esophageal sphincter

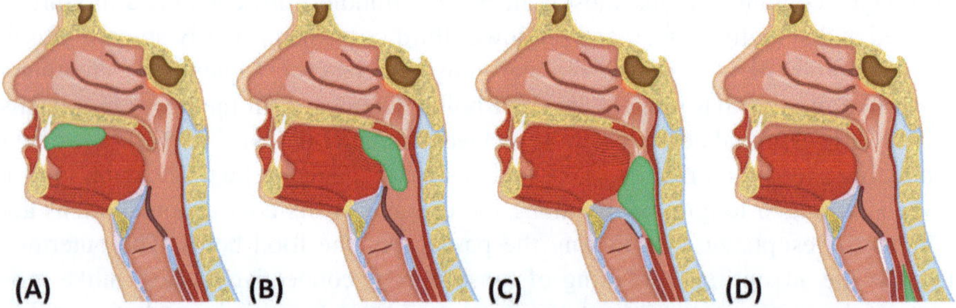

Fig. 16.15 Phases of swallowing (deglutition). **(a and b) Oral phase:** This is the voluntary phase where food is chewed and mixed with saliva to form a bolus (oral preparatory phase or buccal phase). Then, the tongue elevates to push the bolus to the back of the mouth and oropharynx, initiating the swallowing reflex. The soft palate rises to close off the nasopharynx, preventing food from entering the nasal cavity. **(c) Pharyngeal phase:** This involuntary phase involves the movement of the bolus through the pharynx and into the esophagus. Muscles of the pharynx contract to propel the bolus downward. The epiglottis is lowered to cover the laryngeal inlet, preventing food from entering the trachea. The nasopharynx is still closed. The upper esophageal sphincter relaxes to allow the bolus to enter the esophagus. **(d) Esophageal phase:** The bolus is transported through the esophagus to the stomach by peristaltic movements, which are coordinated, rhythmic contractions of the esophageal muscles

Swallowing

Swallowing (Fig. 16.15), also known as deglutition, is a complex process that involves the coordinated action of the mouth, pharynx, esophagus, and associated neural structures. This process can be divided into three distinct phases: the oral phase, the pharyngeal phase, and the esophageal phase. In the oral phase, after chewing food and mixing it with saliva, the tongue manipulates the cohesive bolus, pushing it against the hard palate and forming it into a bolus. Then, the tongue moves the bolus toward the oropharynx. Once the bolus reaches the back of the mouth, sensory receptors trigger the swallowing reflex. The soft palate and uvula lift to close off the nasopharynx, preventing food from entering the nasal cavity. The larynx elevates and moves anteriorly under the base of the tongue. The epiglottis folds backward to cover the larynx entrance, directing the bolus into the esophagus and protecting the airway. Sequential contraction of pharyngeal constrictor muscles moves the bolus into the esophagus. This action is involuntary. The upper esophageal sphincter (Fig. 16.16) relaxes, allowing the bolus to enter the esophagus. Once in the esophagus, coordinated peristaltic contractions of the esophageal muscles propel the bolus toward the stomach. The LES relaxes as the bolus approaches, allowing entry into the stomach.

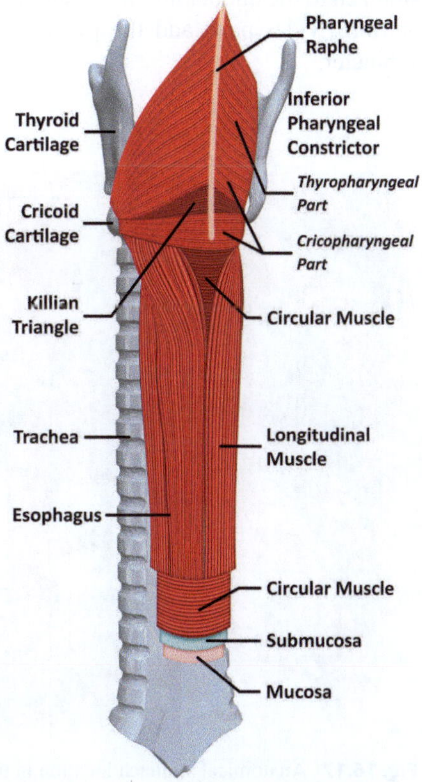

Fig. 16.16 Upper esophageal sphincter, anterior view

16.2.1.4 Stomach

The stomach is a key organ in the digestive system, known for its role in food digestion and storage.

Anatomy

It is a muscular, J-shaped sac in the left upper abdomen, just below the diaphragm (Fig. 16.17). It is situated between the esophagus and the small intestine. The stomach's orientation can vary depending on body habitus and the amount of content within it. It typically has a capacity of about 1–2 l. The stomach is divided into five distinct regions (Fig. 16.18), each with specific anatomical features and functions. The cardia surrounds the gastroesophageal junction, where the esophagus connects to the stomach. This relatively small region acts as a transitional zone between the esophagus and the stomach. The fundus is the uppermost portion of the stomach, lying above the level of the cardiac orifice. It forms a dome-shaped area that contacts the diaphragm. The body is the largest and central part of the stomach, extending from the fundus to the beginning of the pyloric region. The interior surface of the body displays rugae (Fig. 16.19), which are folds in the gastric mucosa that allow for expansion as the stomach fills. The antrum is the lower portion of the stomach, located between the body and the pylorus. It narrows as it approaches the pyloric canal. The pylorus is the distal segment of the stomach, connecting the stomach to the duodenum of the small intestine. It can be subdivided into the pyloric antrum (wider part) and the pyloric canal (narrower part) leading to the pyloric sphincter.

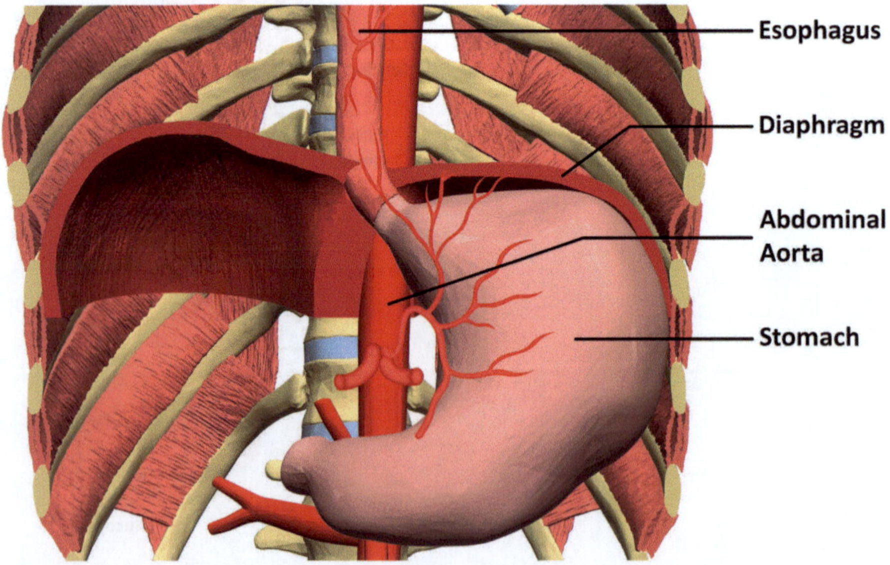

Fig. 16.17 Anatomical stomach location in the left upper quadrant of the abdomen

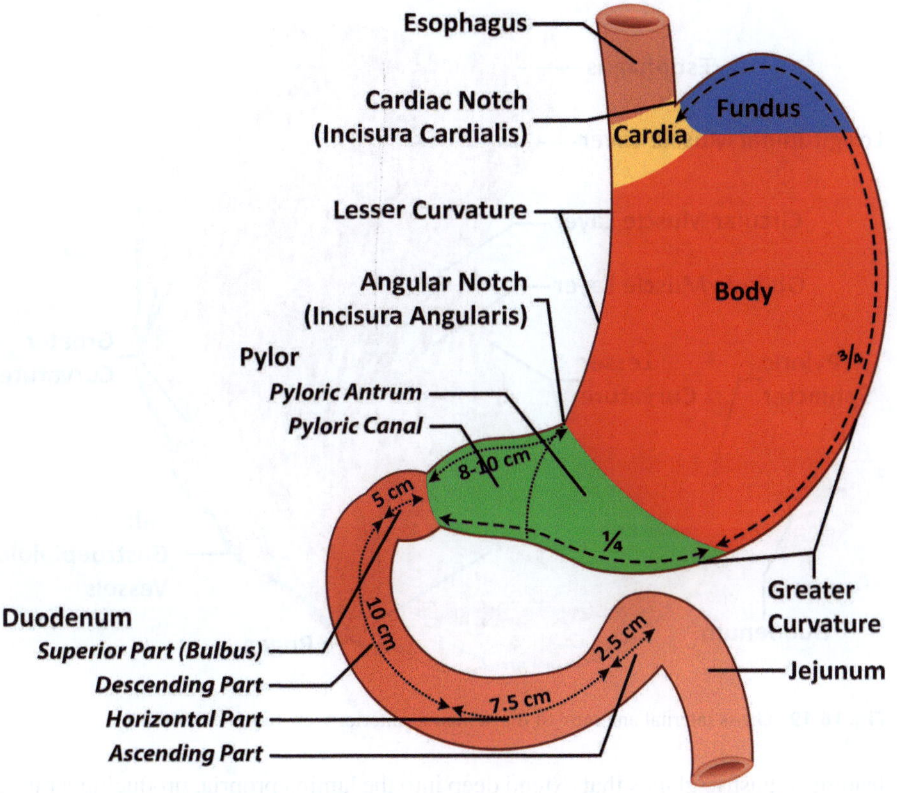

Fig. 16.18 Regions of the stomach and duodenum

The stomach has two curvatures (Fig. 16.18): the greater curvature extends from the junction of the esophagus and stomach (cardia) around the fundus and down the left side of the stomach. It's a long, convex border. The lesser curvature runs along the right side of the stomach from the cardia to the pylorus. It's shorter and concave.

There are two peritoneal folds (omenta) around the stomach (Fig. 16.20). The greater omentum is a large apron-like fold of visceral peritoneum that hangs down from the greater curvature of the stomach. It contains fat deposits and lymph nodes. Lesser omentum extends from the liver to the lesser curvature of the stomach and the beginning of the duodenum.

Microscopic Structure

The stomach's ability to perform its role in digestion is mainly due to its specialized microscopic structure. It comprises the same general four layers, with some specifications (Fig. 16.21). The innermost layer of the stomach lining is a simple columnar epithelium. It secretes mucus, which forms a protective barrier against the acidic gastric environment. The surface of the mucosa contains numerous gastric pits,

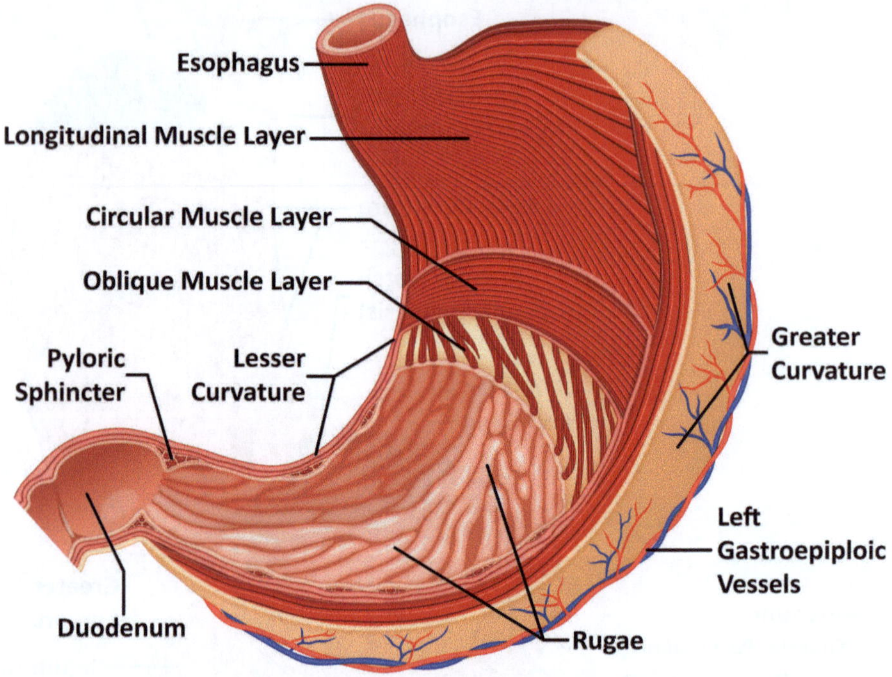

Fig. 16.19 Gross internal anatomy of the stomach, anterior view (coronal section)

leading to gastric glands that extend deep into the lamina propria, producing a range of secretions, including gastric acid, enzymes, and hormones. The gastric glands in the cardia region secrete mucus and bicarbonate, which protect the esophagus from the stomach's acidic environment. The glands located in the fundus and body of the stomach are the dominant type of gastric glands and have a complex structure with various kinds of cells (Fig. 16.22):

- Mucous Neck Cells: Produce mucus to protect the stomach lining.
- Parietal (Oxyntic) Cells: Secrete hydrochloric acid (HCl) and intrinsic factor (essential for vitamin B12 absorption).
- Chief (Peptic) Cells: Produce pepsinogen, the precursor to the enzyme pepsin that digests proteins.
- Enteroendocrine Cells: Secrete various hormones like gastrin.

Muscularis externa comprises three layers of smooth muscle: the inner oblique, middle circular, and outer longitudinal layers. These muscle layers are responsible for the peristaltic movements that mix and propel gastric contents. The innermost oblique layer is unique to the stomach. The oblique fibers run diagonally and are most evident in the upper part of the stomach, particularly in the region of the fundus and the body. They contribute to the churning and mixing actions of the stomach.

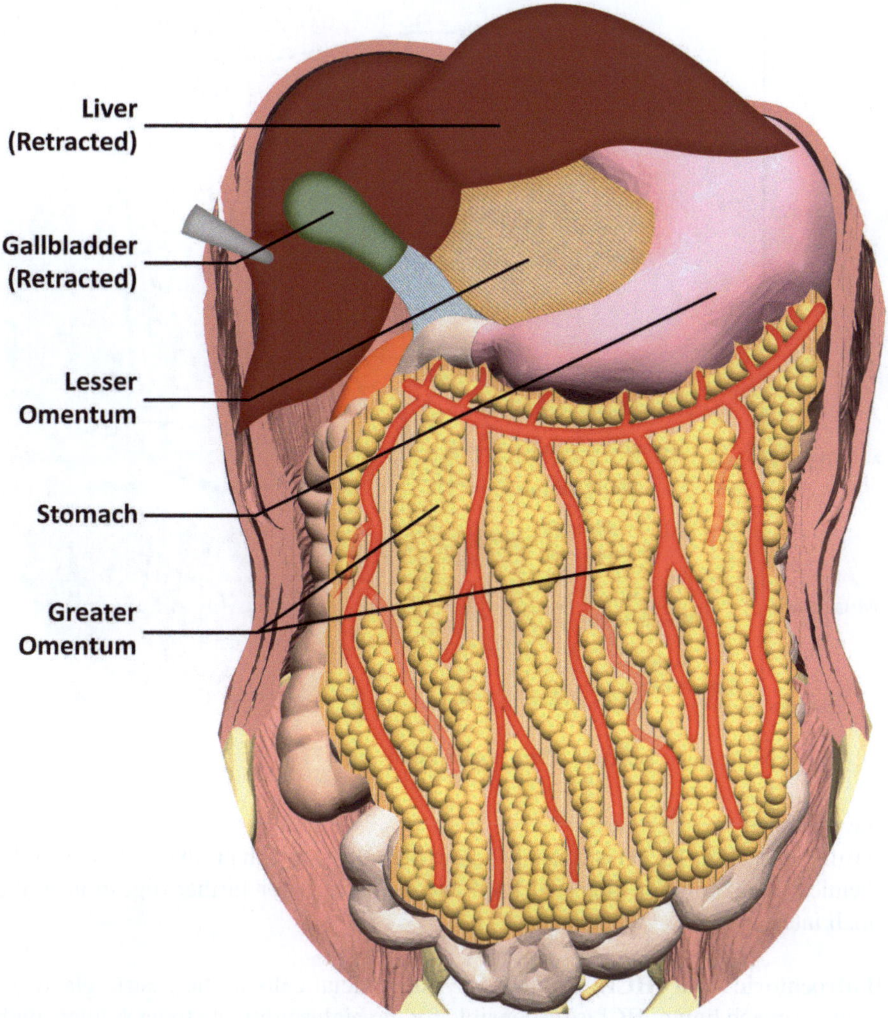

Fig. 16.20 Peritoneal folds around the stomach, anterior view

Pyloric Sphincter

The pyloric sphincter (Fig. 16.23), also known as the pylorus or pyloric valve, is situated at the distal end of the stomach, marking the junction between the stomach and the duodenum. It is located just distal to the pyloric antrum of the stomach and proximal to the duodenal bulb. The pyloric sphincter is a thickened band of circular smooth muscle. It's typically 2–3 cm long and has a diameter that varies, allowing the passage of digested material based on its degree of contraction. It also prevents the backflow of duodenal contents into the stomach.

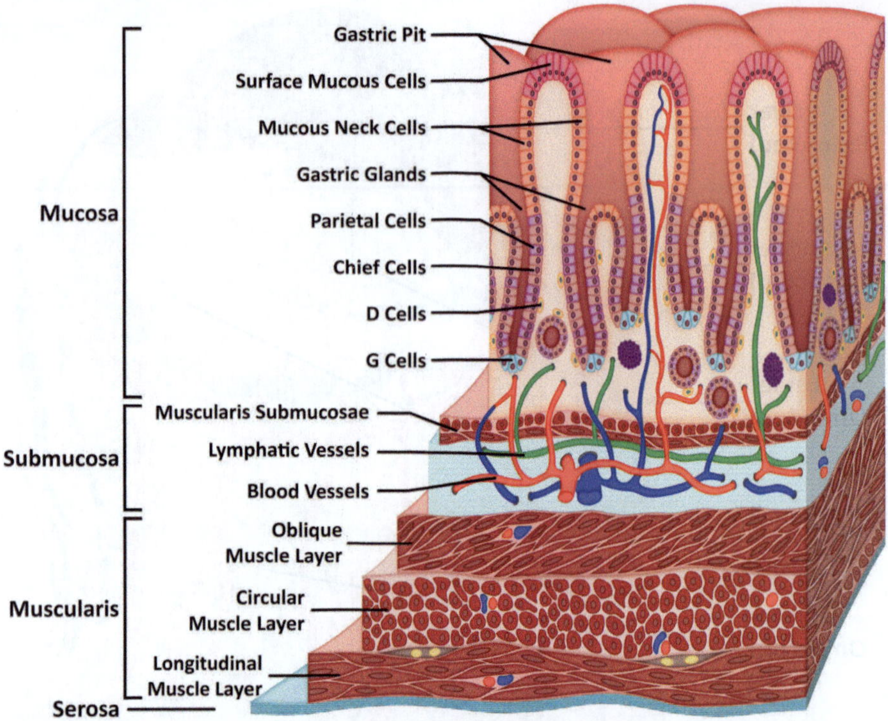

Fig. 16.21 Microscopic structure of the stomach wall

Gastric Juice

Gastric juice (Fig. 16.24) is a highly acidic fluid the stomach produces. It aids in the chemical breakdown of ingested food and prepares it for further digestion in the small intestine.

Hydrochloric Acid (HCl): Produced by the parietal cells in the gastric glands of the stomach lining, HCl is responsible for the high acidity of stomach juice, with a pH ranging from 1.5 to 3.5. This acidity serves several purposes: it denatures proteins in food, activates the enzyme pepsin, and helps kill bacteria and other pathogens ingested with food.

Pepsinogen: Secreted by the chief cells in the gastric glands, pepsinogen is an inactive enzyme precursor. When it comes into contact with HCl, it is converted into its active form, pepsin. Pepsin then begins the process of protein digestion by breaking down the peptide bonds in proteins.

Intrinsic Factor: Produced by the parietal cells, the same cells that secrete HCl, the intrinsic factor is essential for absorbing vitamin B12 in the small intestine. It binds to vitamin B12, allowing it to be absorbed in the ileum. This is crucial for preventing vitamin B12 deficiency and associated conditions like pernicious anemia.

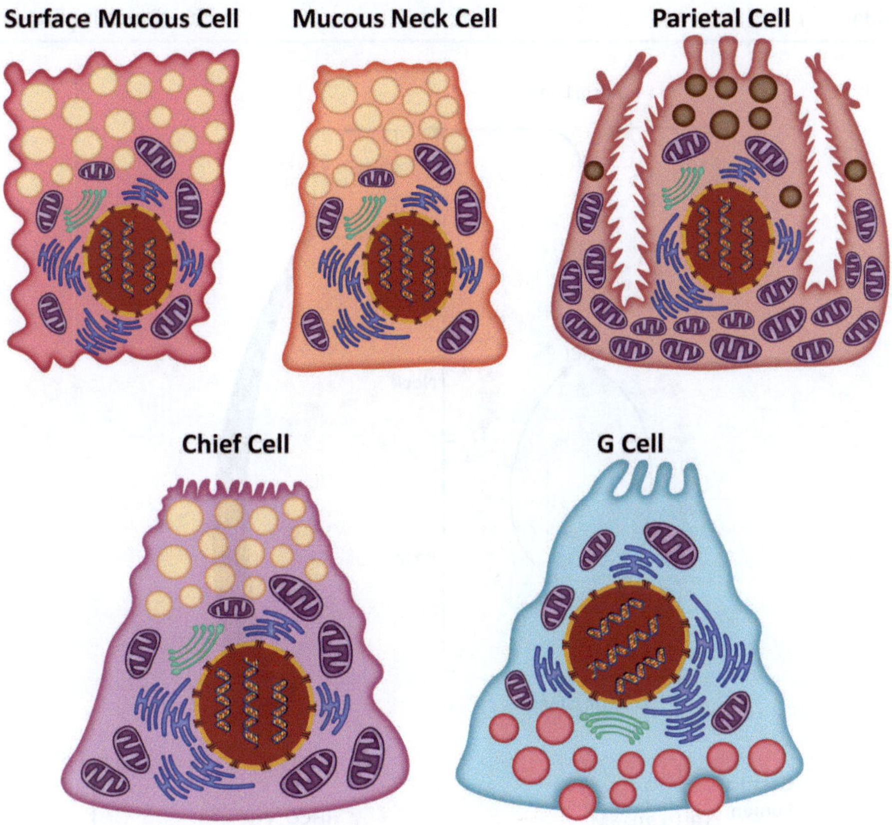

Fig. 16.22 Cells of the gastric mucosa

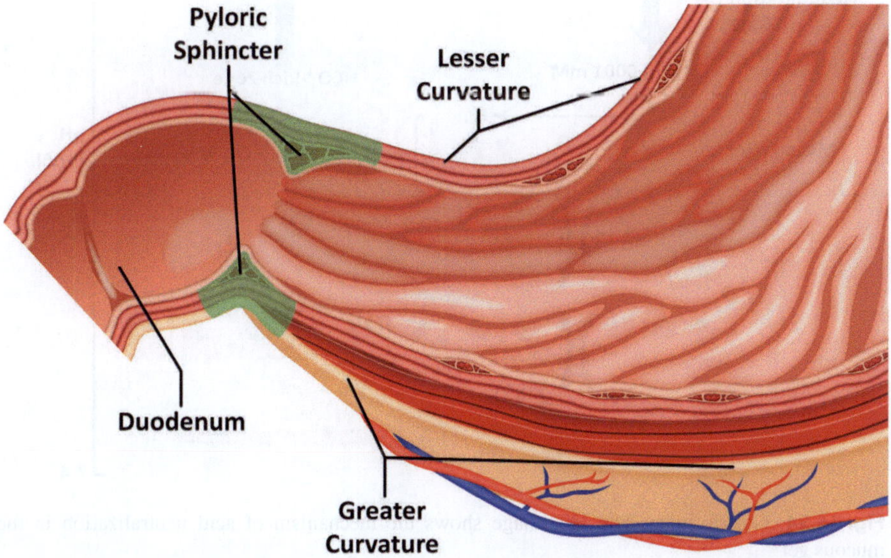

Fig. 16.23 Pyloric sphincter

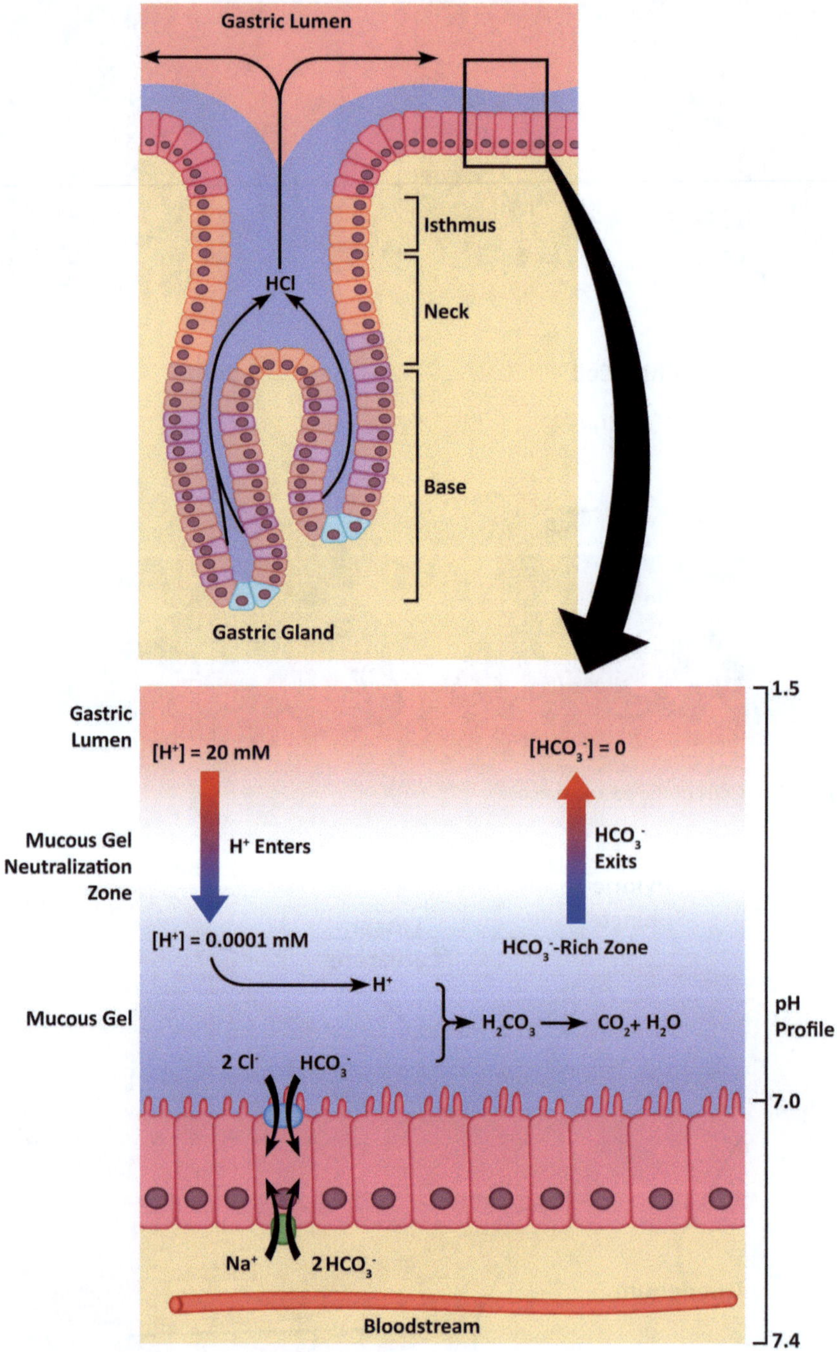

Fig. 16.24 Gastric juice. The left image shows the mechanism of acid neutralization in the mucous gel

Mucus: Secreted by the mucous cells and goblet cells in the stomach lining, mucus forms a protective layer that coats the stomach lining. This layer protects the stomach tissue from damage from the acidic environment and enzymatic activity within the stomach.

Water and Electrolytes: Stomach juice contains a significant amount of water and electrolytes such as potassium, chloride, and sodium.

Bicarbonate: Secreted by the stomach's epithelial cells, bicarbonate is secreted in the mucus layer. It neutralizes some of the acid near the stomach lining, providing a pH gradient that protects the stomach tissue from acid damage.

The stomach produces several hormones and hormone-like substances that regulate various aspects of digestion, appetite, and the coordination of digestive processes (Section 16.5).

16.2.1.5 Small Intestine

The small intestine is a long, tubular organ connecting the stomach to the large intestine. It extends from the pylorus of the stomach to the ileocecal valve, where it joins the large intestine (Fig. 16.25).

Anatomy

It is approximately 6 m (20 feet) long in an adult human, though this can vary considerably. The small intestine is situated in the abdomen, occupying the central and lower part of the abdominal cavity. The small intestine has three distinct sections: the duodenum is the first and shortest segment of the small intestine. It receives partially digested food (chyme) from the stomach and initiates further digestion with the help of enzymes from the pancreas and bile from the liver. It is predominantly retroperitoneal, located behind the peritoneum (the abdominal cavity lining). The duodenum is roughly 25–30 cm (10–12 inches) long. The jejunum is the middle section of the small intestine, located between the duodenum and the ileum. The jejunum begins at the duodenojejunal flexure, which transitions from the duodenum. It makes up approximately the upper two-fifths of the remaining small intestine, with the ileum constituting the lower three-fifths. Unlike the duodenum, the jejunum is intraperitoneal, encased in and suspended by the mesentery. This allows greater mobility. The jejunum is typically about 2.5 m (8 feet) in length. It has a larger diameter, thicker walls, and more extensive vascularization than the ileum. The ileum is the final and longest segment of the small intestine, following the jejunum and leading into the large intestine. It occupies the lower three-fifths of the small intestine, following the jejunum. It typically measures about 3.5 m (11 feet) in length, though this can vary. Like the jejunum, the ileum is intraperitoneal and is attached to the abdominal wall by the mesentery. It contains more Peyer's patches (aggregated lymphoid follicles), especially in the distal part, which are important for immune surveillance and response.

The mesentery of the small intestine is a fold of the peritoneum (Fig. 16.3). It is a fan-shaped fold of tissue that attaches the small intestine to the posterior abdominal wall. It comprises two layers of peritoneum that enclose a layer of connective

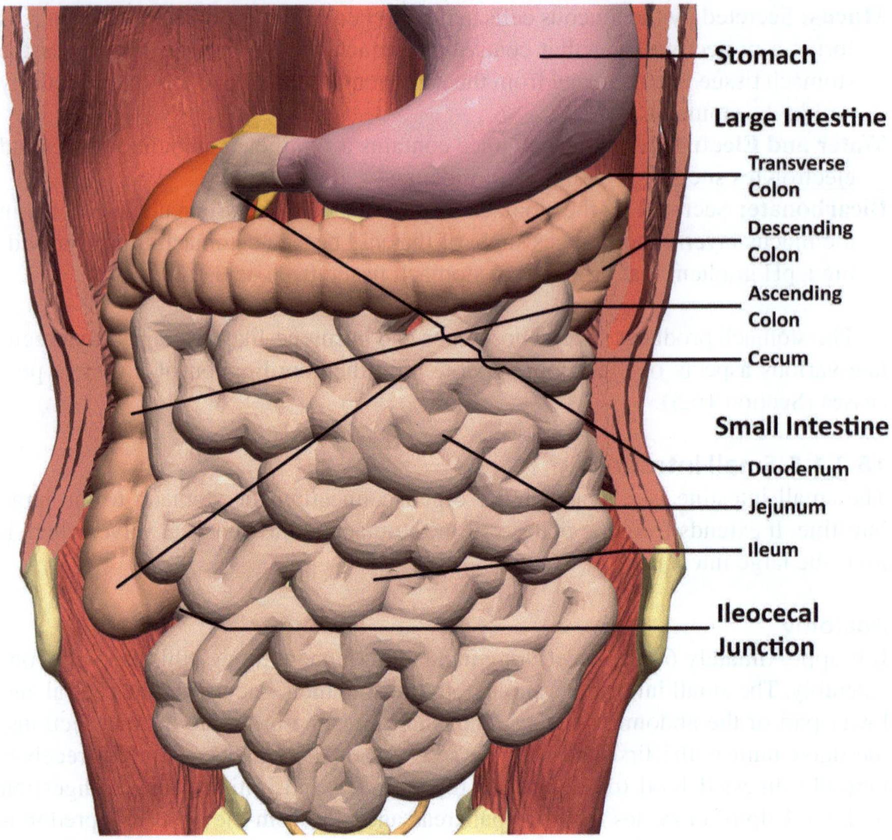

Fig. 16.25 Small intestine gross anatomy in the abdomen

tissue. The connective tissue layer contains blood vessels, lymphatics, nerves, and fat.

Microscopic Structure

The microscopic anatomy of the small intestine (Fig. 16.26) reveals a highly specialized structure adapted for efficient digestion and nutrient absorption. The innermost layer of the small intestine, which comes into direct contact with the luminal contents, is a single layer of columnar epithelial cells, including enterocytes (absorptive cells), goblet cells (secreting mucus), and enteroendocrine cells (secreting various hormones). The epithelial cells are tightly bound together by tight junctions, which control the passage of substances between the cells and maintain the barrier's integrity against pathogens. The epithelial cells of the small intestine are constantly renewed. Stem cells located in the crypts divide and give rise to new cells, which differentiate into the various cell types as they migrate up the villi. The inside surface of the small intestine is covered by multiple projections, villi, and microvilli

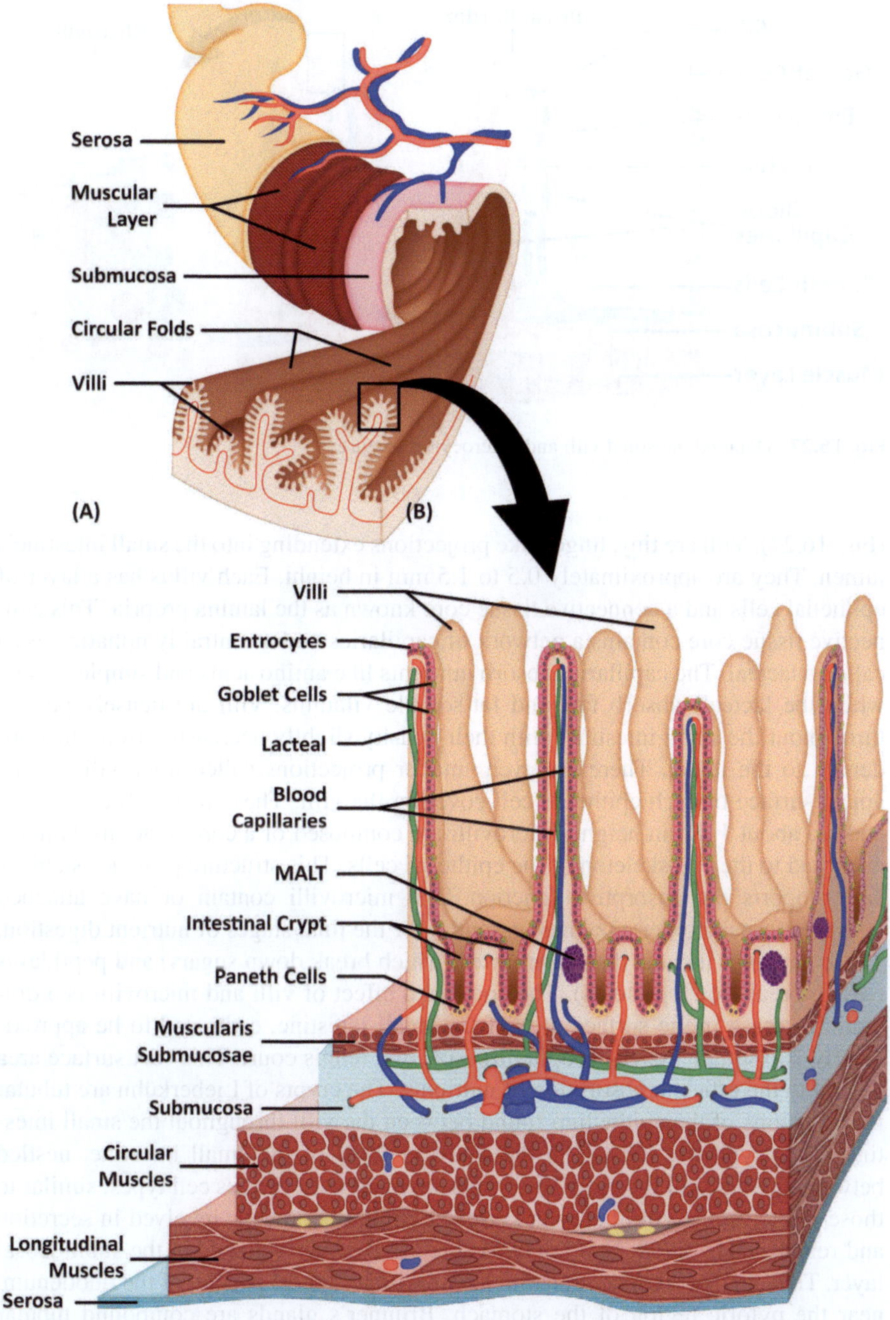

Fig. 16.26 Small intestine structure. (**a**) Macroscopic structure of the small intestine wall and the hierarchical configuration of intestinal mucosa to increase the surface area for absorption. (**b**) The microscopic structure of the small intestine wall

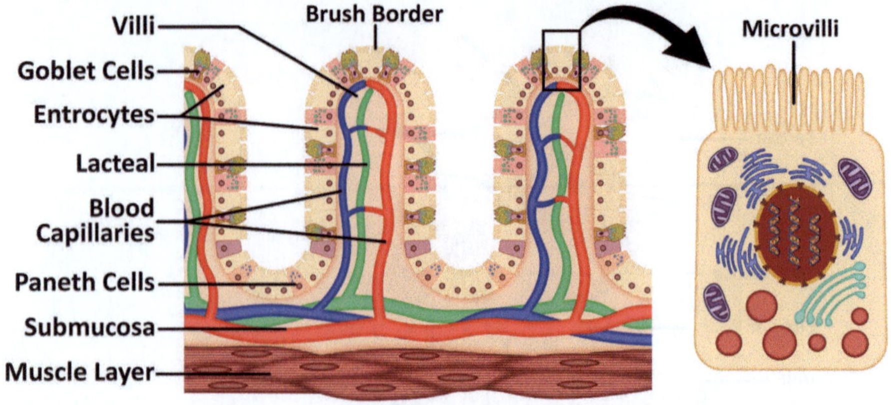

Fig. 16.27 Detailed intestinal villi and enterocyte structure

(Fig. 16.27). Villi are tiny, finger-like projections extending into the small intestine's lumen. They are approximately 0.5 to 1.5 mm in height. Each villus has a layer of epithelial cells and a connective tissue core known as the lamina propria. This connective tissue core contains a network of capillaries and a central lymphatic vessel called a lacteal. The capillaries absorb nutrients like amino acids and simple sugars, while the lacteals absorb fats and fat-soluble vitamins. Villi are densely packed throughout the small intestine, with their density slightly decreasing from the duodenum to the ileum. There are even smaller projections called microvilli on the apical surface of each epithelial cell covering the villi. These form a brush border and are about 1 μm in height. Microvilli are composed of a core of actin filaments anchored to the cytoskeleton of the epithelial cells. This structure provides stability and supports the absorptive function. The microvilli contain or have attached enzymes (brush border enzymes) essential for the final stages of nutrient digestion. These enzymes include disaccharidases (which break down sugars) and peptidases (which break down proteins). The combined effect of villi and microvilli is a dramatic increase in the surface area of the small intestine, estimated to be approximately 200 square meters, roughly the size of a tennis court. This vast surface area is vital to the efficient absorption of nutrients. The crypts of Lieberkühn are tubular invaginations of the epithelium found between the villi throughout the small intestine. The crypts are located in the mucosal layer of the small intestine, nestled between the bases of the villi. The crypts are lined with various cell types, similar to those found on the villi, but with a greater emphasis on cells involved in secretion and renewal. The duodenum contains specific Brunner's glands in the submucosal layer. They are most concentrated in the proximal (upper) portion of the duodenum, near the pyloric region of the stomach. Brunner's glands are compound tubular glands with a branched or complex structure. The secretory cells of Brunner's glands include goblet cells that produce mucus and columnar cells that produce bicarbonate-rich alkaline secretions. Brunner's glands secrete mucus and alkaline fluids into the duodenum. The mucus helps lubricate the lining of the duodenum, protecting it from the acidic chyme from the stomach. The alkaline secretions contain bicarbonate ions, which neutralize the acidic chyme, creating a more favorable pH environment for the digestion of nutrients by pancreatic enzymes.

Among the various cell types in the small intestine, I, S, K, and L cells are specialized enteroendocrine cells that secrete hormones that regulate multiple aspects of digestion and metabolism. These enteroendocrine cells are specialized secretory cells with apical microvilli that increase their surface area for nutrient sensing. I cells are primarily found in the epithelial lining of the duodenum and jejunum. I cells secrete cholecystokinin (CCK) (Section 16.5.2) in response to fatty acids and peptides in the small intestine. S cells are primarily located in the duodenum and jejunum, similar to I cells. They secrete secretin in response to acidic chyme entering the duodenum from the stomach. K cells are distributed throughout the small intestine, including the duodenum, jejunum, and ileum. They secrete glucose-dependent insulinotropic peptide (GIP), also known as gastric inhibitory peptide. GIP is released in response to glucose and fatty acids in the small intestine. L cells are more abundant in the ileum but are also found throughout the small intestine. They secrete glucagon-like peptide-1 (GLP-1) and peptide YY (PYY) in response to nutrients, especially glucose and fatty acids.

Peristalsis

Peristalsis is primarily responsible for moving chyme toward the large intestine in the small intestine. Peristalsis is typically initiated by the presence of chyme in the small intestine. The process of peristalsis involves a wave-like sequence of contractions. The circular muscles contract behind the bolus of chyme, constricting the lumen and pushing the contents forward, while the longitudinal muscles ahead of the bolus contract to shorten that segment of the intestine, easing the passage of chyme. Peristalsis in the small intestine is coordinated with other digestive processes, including gastric emptying from the stomach and the passage of chyme into the large intestine. This coordination ensures efficient digestion and nutrient absorption. Another form of contraction in the small intestine is segmental contraction. The rhythmic contraction and relaxation of the circular muscles in the small intestine characterize segmental contractions.

These contractions divide the intestine into segments, momentarily constricting the lumen at regular intervals and creating a "segmented" appearance. The primary function of segmental contractions is to mix the chyme with digestive enzymes and bring it into contact with the mucosal surface for nutrient absorption. Unlike peristaltic contractions, segmental contractions are non-propulsive, meaning they do not significantly move the contents along the length of the intestine.

Intestinal Juice

Small intestine juice is a clear to slightly yellowish fluid secreted by the glands in the mucosa of the small intestine. The volume of intestinal juice secreted daily varies, but it can range from 1 to 2 l daily, depending on dietary intake and physiological needs. The pH of intestinal juice is typically slightly alkaline, ranging from 7.5

to 8.0. The primary component of intestinal juice is water, which provides a medium for dissolving and transporting nutrients. Intestinal juice contains a variety of enzymes essential for the final stages of digestion, including disaccharidases (like sucrase, maltase, and lactase) that break down sugars into monosaccharides and peptidases (like aminopeptidase and dipeptidase) that continue the breakdown of proteins into amino acids. The juice contains several electrolytes, such as sodium, potassium, bicarbonate, and chloride. Bicarbonate is particularly important for neutralizing the acidic chyme received from the stomach.

Function

The small intestine performs many crucial functions necessary for the digestion, absorption, and transport of nutrients, as well as other associated roles. The duodenum is the primary site of chemical digestion. The duodenum receives chyme from the stomach, digestive enzymes from the pancreas (e.g., lipases, proteases, amylases), and bile from the liver and gallbladder. These secretions help break down fats, proteins, and carbohydrates. Digestive enzymes, both from pancreatic secretions and those in the intestinal juice (e.g., disaccharidases, peptidases), continue breaking down food into absorbable molecules. The internal surface of the small intestine is highly adapted for absorption, significantly increased by the presence of villi and microvilli. The primary function of the jejunum and ileum is the absorption of nutrients.

16.2.1.6 Large Intestine

The large intestine (Fig. 16.28), also known as the colon, is approximately 1.5 m (5 feet) in length and is divided into several distinct regions: the cecum, ascending colon, transverse colon, descending colon, sigmoid colon, rectum, and anal canal. The large intestine measures approximately 6–7 cm in diameter. Its diameter gradually decreases as it progresses toward the rectum. It encircles the small intestine, extending from the ileocecal valve to the anus. It occupies parts of the abdominal and pelvic cavities. The colon has several sections: cecum; ascending, transverse, descending, and sigmoid colons; rectum; and anal canal.

Cecum

The cecum (Fig. 16.28) is the initial segment of the large intestine and is situated in the lower right quadrant of the abdomen, at the junction where the small and large intestines meet. It lies below the ileocecal valve, the demarcation point between the small intestine's terminal ileum and the large intestine. The cecum is pouch-like, typically about 6–8 cm long. The appendix is a small, tube-like structure attached to the cecum, a part of the large intestine. The appendix is located in the lower right quadrant of the abdomen, attached to the posteromedial aspect of the cecum, just below the ileocecal valve. The ileocecal valve is situated in the lower right quadrant of the abdomen, at the junction of the ileum and the cecum. It is positioned slightly below and lateral to the umbilicus in most individuals. The valve comprises two distinct lips or folds of tissue that protrude into the lumen of the cecum. The invagination of the ileum forms these lips into the cecum. The muscle tone of the muscular

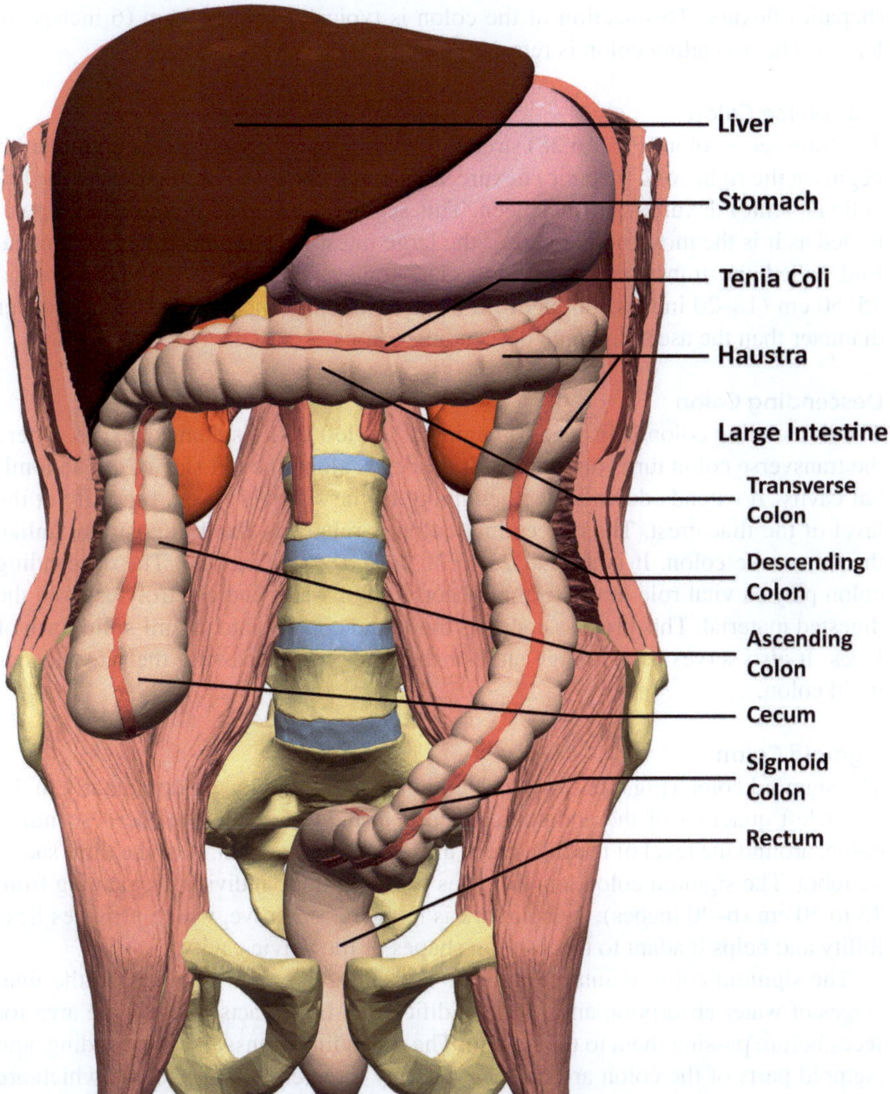

Fig. 16.28 Anatomy of the colon in the abdominal cavity, anterior view

layers contributes to the valve's ability to open and close, regulating the passage of intestinal contents.

Ascending Colon

The ascending colon (Fig. 16.28) is located on the right side of the abdominal cavity. It begins at the cecum, just distal to the ileocecal valve, and extends upward to the inferior surface of the liver, where it turns abruptly to form the right colic (hepatic) flexure. This section of the colon is typically about 15 cm (6 inches) in length. The ascending colon is retroperitoneal.

Transverse Colon

The transverse colon (Fig. 16.28) stretches from right to left across the abdomen. It begins at the right colic (hepatic) flexure, adjacent to the liver, and extends to the left colic (splenic) flexure near the spleen. This segment of the colon is uniquely positioned as it is the most mobile part of the large intestine, suspended by a peritoneal fold called the transverse mesocolon. The transverse colon is typically around 45–50 cm (18–20 inches) in length. It has a distinctive appearance with a broader diameter than the ascending and descending colons.

Descending Colon

The descending colon (Fig. 16.28) begins at the left colic (splenic) flexure, where the transverse colon turns downward. It descends along the left side of the abdominal cavity. It extends down to the beginning of the sigmoid colon, typically at the level of the iliac crest. The descending colon is relatively fixed in position, unlike the transverse colon. It is around 25 to 30 cm (10 to 12 inches). The descending colon plays a vital role in the further absorption of water and electrolytes from the digested material. This process converts the liquid content into a semi-solid form of feces. It also serves as a storage site for the feces before passing them to the sigmoid colon.

Sigmoid Colon

The sigmoid colon (Fig. 16.28) is located in the pelvic cavity, predominantly in the lower left quadrant of the abdomen. It starts at the termination of the descending colon, around the level of the iliac crest, and ends at the rectum, near the third sacral vertebra. The sigmoid colon length varies greatly among individuals, ranging from 15 to 50 cm (6–20 inches). Its hallmark is its S-shaped curve, which provides flexibility and helps it adapt to the varying shapes of the pelvic cavity.

The sigmoid colon is intraperitoneal. The sigmoid colon is critical in the final stages of water absorption and fecal solidification. It also acts as a storage area for feces before passing them to the rectum. The ascending, transverse, descending, and sigmoid parts of the colon are characterized by the presence of haustra, which are pouch-like sacs caused by the segmentation of the colon, and teniae coli, which are three bands of longitudinal smooth muscle running along the surface of the colon. Epiploic appendages, small fat-filled pouches of visceral peritoneum, are also present along their length.

Rectum

The rectum (Fig. 16.28) begins at the end of the sigmoid colon, at the level of the third sacral vertebra, and continues down to the anal canal. It is located in the pelvis, anterior to the sacrum and coccyx. The rectum is typically about 12–15 cm (4.7–5.9 inches) in length. Unlike other parts of the colon, the rectum is straight, lacking the haustra and teniae coli. The rectal walls are capable of considerable distension to store fecal matter.

Anal Canal

The anorectal junction (Fig. 16.29) is where the rectum joins the anal canal. This junction is marked internally by the anorectal line and externally by the anal verge. The anorectal junction is located at the level of the pelvic diaphragm. It is situated approximately 4–5 cm from the anal opening. The junction is characterized by a change in the lining epithelium from the columnar epithelium of the rectum to the stratified squamous epithelium of the anal canal. This transition zone is demarcated by the dentate (pectinate) line, which is visible as a series of small, tooth-like projections in the anal canal. Above this line is the rectum, and below is the anal canal. A key feature of the anorectal junction is the anorectal angle, formed between the axis of the rectum and the axis of the anal canal. This angle is approximately 90 degrees in a resting state and becomes more acute during defecation. The anal canal extends from the rectum to the anus, marking the exit point of the digestive tract. It is located at the level of the pelvic floor muscles, primarily within the pelvis and extending to the perianal skin. The anal canal is relatively short, measuring approximately 3–5 cm in length in adults. Its walls are lined with a series of longitudinal folds known as anal columns (or columns of Morgagni), which contain a network of blood vessels. The anal canal (below the dentate line) is lined by skin and is derived from the ectoderm. Due to its somatic innervation, this area is sensitive to pain, temperature, and touch. Two main sphincter muscles regulate the control of defecation: the internal anal sphincter (IAS) is a thick, involuntary ring of smooth muscle that forms the continuation of the circular muscle layer of the rectum. It is approximately 2–3 cm in length. It is situated at the upper part of the anal canal, encircling it just below the anorectal junction. As an involuntary muscle, the IAS is always contracted, ensuring the anal canal is closed at rest. This muscle relaxes during defecation. The external anal sphincter (EAS) is a voluntary sphincter composed of striated muscle fibers. It is broader and more superficial than the IAS and is divided

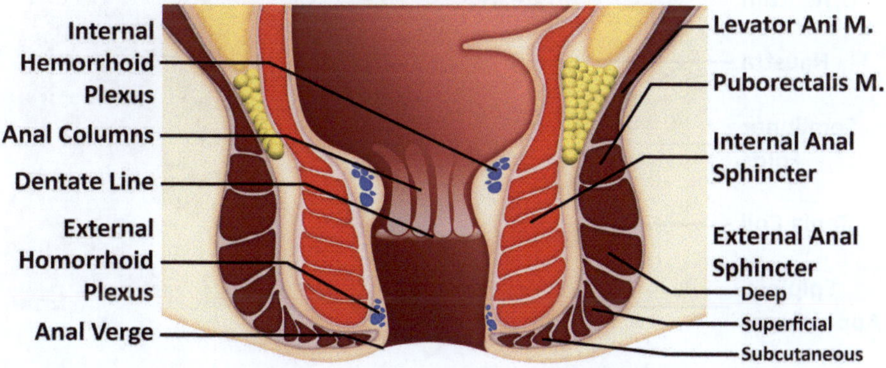

Fig. 16.29 Anal canal

into three parts: subcutaneous, superficial, and deep. The EAS encircles the outside of the internal sphincter and extends beyond it in both the superior and inferior directions. Being a voluntary muscle, it allows conscious control over defecation.

Tenia Coli

The teniae coli (Figs. 16.28 and 16.30) are distinctive anatomical features of the large intestine. The teniae coli are three separate longitudinal bands of smooth muscle on the surface of the colon. Each band extends from the base of the appendix to the rectosigmoid junction. Microscopically, the teniae coli consist of smooth muscle fibers. These fibers are oriented longitudinally along the length of the colon. At the rectosigmoid junction, they spread out and merge into the circular muscle layer of the rectum. The teniae coli are absent in the rectum, which is why the rectum lacks the haustral pattern seen in the rest of the colon. The primary function of the teniae coli is to exert constant tonic contraction. This contraction is less intense than typical peristalsis but is significant in maintaining a luminal tone. The tonic contraction of teniae coli creates the characteristic pouches or bulges in the colon known as haustra (Figs. 16.28 and 16.30).

Haustra

Haustra are the small pouches or sacculations along the colon, giving it a segmented appearance. There are small folds or creases in the colon wall between each haustrum, known as semilunar folds. Unlike the fixed folds of the small intestine (plicae circulares), haustra are dynamic structures. They can change their position along the colon as the muscle of the teniae coli contracts or relaxes.

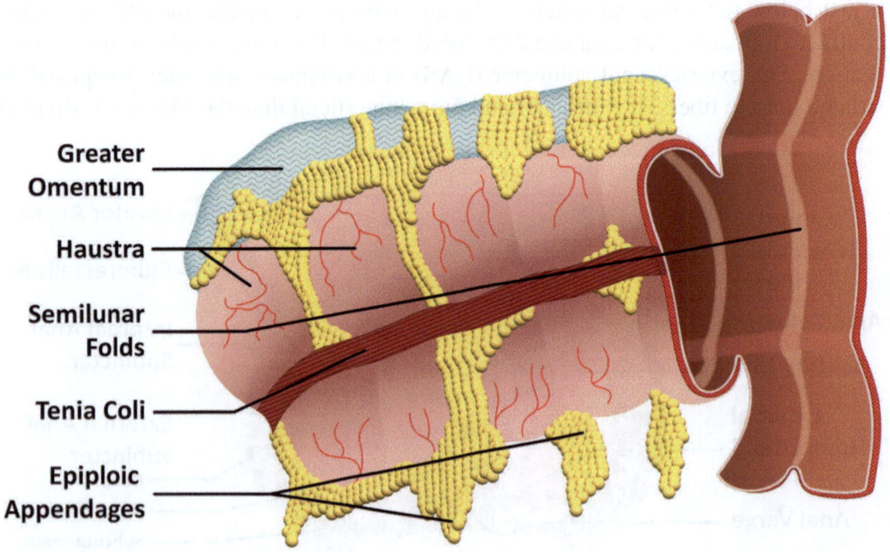

Fig. 16.30 Gross structure of the colon

Epiploic Appendages

Epiploic appendages, also known as appendices epiploicae (Fig. 16.30), are small, pouch-like structures attached to the outer surface of the large intestine. Epiploic appendages are found along almost the entire colon length, from the cecum to the rectosigmoid junction. They are most commonly associated with the transverse and sigmoid parts of the colon. Typically, they are absent in the rectum. These structures are small, fat-filled pouches or sacs, oval or elongated. They are usually 0.5–5 cm in length and contain adipose tissue.

Microscopic Structure of the Colon

The microscopic structure of the large intestine (Fig. 16.31) is intricately designed to fulfill its functions in water and electrolyte absorption, fecal formation, and immune defense. The mucosal surface is lined by a simple columnar epithelium, primarily composed of absorptive and goblet cells. The crypts of Lieberkühn are more numerous and densely packed in the large intestine than those in the small intestine, and they are relatively shallower. The crypts are lined with a simple

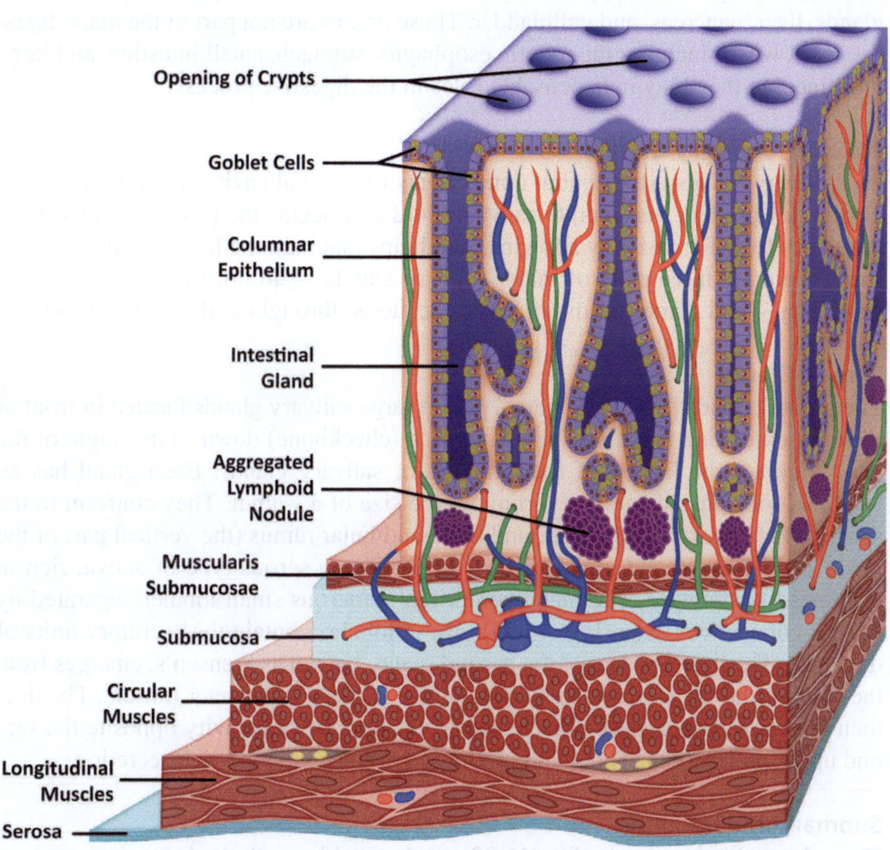

Fig. 16.31 Microscopic structure of the colon wall

columnar epithelium predominantly composed of absorptive, goblet, and stem cells. Apart from mucus production by goblet cells, the crypts are also involved in bicarbonate and water secretion. Unlike the small intestine, the mucosa of the large intestine does not have villi, reflecting its primary role in water and electrolyte absorption rather than nutrient absorption. The inner circular layer of the smooth muscle is continuous throughout the length of the large intestine. This layer is responsible for segmental contractions, essential for the mixing and slow propelling of intestinal contents, aiding in water absorption and fecal formation. In the rectum, this circular muscle layer thickens to form the internal anal sphincter, which is important for maintaining fecal continence. Instead of being continuous around the circumference, as in most of the gastrointestinal tract, the outer longitudinal muscle layer is organized into three distinct bands called teniae coli.

16.2.2 Accessory Organs

The digestive system is supported by several essential accessory organs: the salivary glands, liver, pancreas, and gallbladder. These organs are not part of the main digestive tract, which includes the mouth, esophagus, stomach, small intestine, and large intestine, but they play indispensable roles in the digestive process.

16.2.2.1 Salivary Glands

The salivary glands are essential components of the oral cavity. They are exocrine glands that produce and secrete saliva, a fluid that begins the process of digestion, lubricates food for easier swallowing, and helps maintain the health of the oral tissues. There are three pairs of major salivary glands—parotid, submandibular, and sublingual—and numerous minor glands scattered throughout the oral mucosa.

Parotid Glands

The parotid glands (Fig. 16.32) are a pair of large salivary glands located in front of the ears, extending from the zygomatic arch (cheekbone) down to the angle of the jaw. They are the largest of the three major salivary glands. Each gland has an irregular, lobulated shape and is roughly the size of a walnut. They conform to the space in the cheek, wrapping around the mandibular ramus (the vertical part of the lower jaw). The parotid glands primarily produce a serous type of saliva, rich in enzymes like amylase. The gland comprises numerous small lobules separated by fibrous connective tissue (Fig. 16.33). These lobules contain the secretory units of the gland. The primary duct of the parotid gland, known as Stensen's, emerges from the front of the gland, traversing horizontally across the masseter muscle. The duct then pierces the buccinator muscle and opens into the oral cavity opposite the second upper molar tooth. The gland produces a watery, enzyme-rich secretion.

Submandibular Glands

The submandibular glands (Fig. 16.32) are located beneath the lower jaw or mandible, more specifically in the submandibular triangle of the neck. They are situated

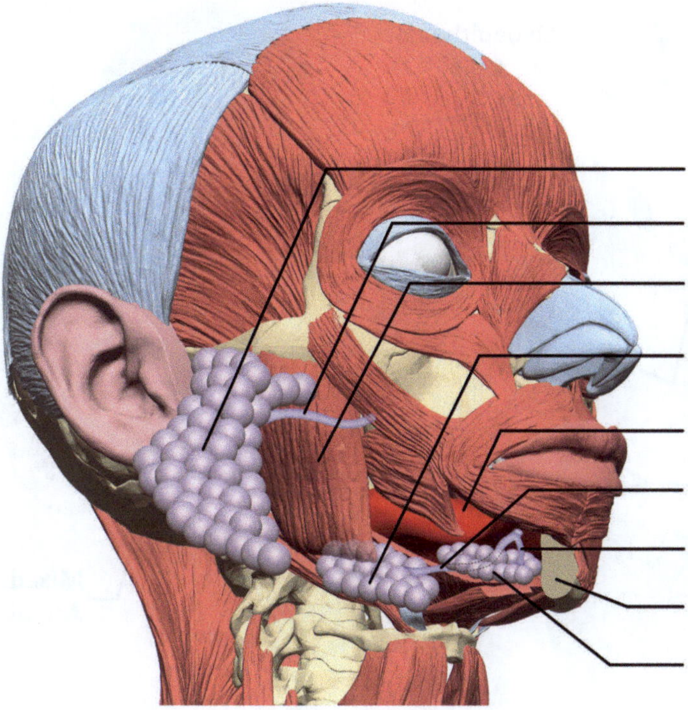

Fig. 16.32 Anatomical location of the salivary glands. (Part of the mandible and mandibular muscles are removed)

against the inner surface of the body of the mandible, near its angle. These glands are smaller than the parotid glands. Each gland is almond-shaped, with a size roughly comparable to that of a walnut. The submandibular glands produce approximately 60–70% of the saliva in the unstimulated mouth. Their secretion is a mixture of serous and mucous fluids, with the predominant serous component. The main duct of each submandibular gland, known as Wharton's duct, is relatively long and opens into the floor of the mouth at the sublingual caruncle, near the base of the lingual frenulum.

Sublingual Glands

The sublingual glands (Fig. 16.32) are located beneath the mouth's floor, under the oral cavity's mucous membrane. They are situated below the mucosa of the oral cavity. These glands are the smallest among the three major salivary glands.

Each gland is almond-shaped or elongated and relatively flattened. The sublingual glands predominantly produce mucous saliva, thicker than the serous saliva produced by the parotid glands. The main excretory duct of the sublingual gland is known as Bartholin's duct, which usually opens into the floor of the mouth via the sublingual papilla near the frenulum of the tongue. Additionally, numerous small ducts (Rivinus's ducts) open along the sublingual fold directly into the floor of the mouth.

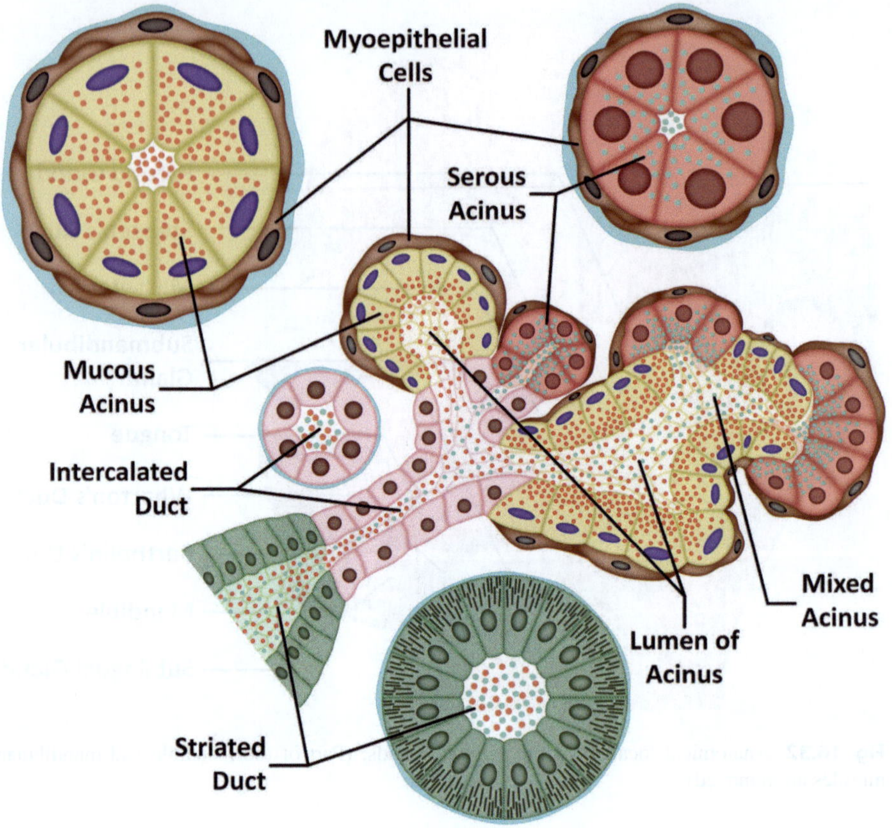

Fig. 16.33 Microscopic structure of the salivary glands

Minor Salivary Glands

The minor salivary glands (Fig. 16.34), while less prominent than the major salivary glands, are small, numerous, and widely distributed throughout the mucosa of the mouth. The minor salivary glands are found in the mucous membranes of the lips, cheeks, tongue, floor of the mouth, hard and soft palate, and pharynx. Each gland comprises a tiny lobular structure with its duct system that opens directly onto the surface of the mucous membrane. Most minor salivary glands are mixed type, secreting both serous and mucous fluids, though the proportion varies. Some glands are predominantly mucous (e.g., those in the lips and buccal mucosa), while others are predominantly serous (e.g., glands at the back of the tongue). The minor salivary glands continuously secrete small amounts of saliva, which keeps the oral mucosa moist.

Certain minor salivary glands, such as those at the back of the tongue (von Ebner's glands), have specialized functions, like secreting saliva rich in lingual lipase for fat digestion.

Fig. 16.34 Accessory salivary glands in the roof of the mouth. (The mucosa of the palate is removed)

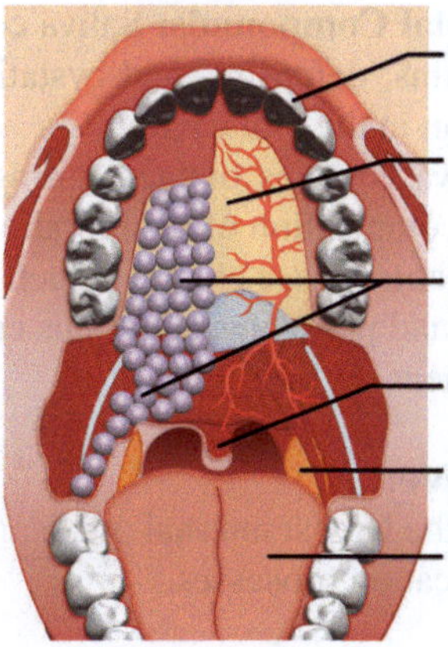

Saliva

Saliva is a complex and multifunctional fluid secreted by the salivary glands in the mouth. The composition of saliva is a finely balanced mixture of various components.

Water: Saliva is predominantly composed of water, accounting for about 95–99% of its total volume. This high water content is essential for its functions in lubrication and solvation.

Electrolytes: Saliva contains several electrolytes, including sodium, potassium, calcium, magnesium, bicarbonate, and chloride ions. Sodium and chloride maintain the osmotic balance. Potassium and bicarbonate are critical for neutralizing acids in the mouth, thus protecting teeth from decay. Calcium and phosphate ions are crucial in remineralizing the teeth, helping maintain tooth enamel, and preventing demineralization.

Enzymes:

- Amylase (ptyalin): Initiates the digestion of starches into simpler sugars.
- Lipase: Begins the digestion of lipids, although its role in oral lipid digestion is less significant.

Mucins (glycoproteins): Contribute to saliva's lubricating and viscoelastic properties.

Lysozyme: Has antibacterial properties and contributes to oral hygiene by lysing bacterial cell walls.

Lactoferrin: Binds to iron, inhibiting bacterial growth.

Immunoglobulins (mainly IgA): Play a role in immune defense by neutralizing pathogens.

Growth Factors: Saliva contains various growth factors, such as epidermal growth factor (EGF), which aid in the maintenance and repair of oral tissues.

Waste Products: Saliva also contains waste products such as urea and uric acid, which are excreted from the body through saliva.

Antimicrobial Compounds: Saliva contains various antimicrobial agents, including histatins, defensins, and cystatins, which help in controlling the microbial population in the oral cavity.

Gases: Saliva contains oxygen, nitrogen, and carbon dioxide, which reflects the gaseous composition of the blood.

Trace Elements: Trace elements such as zinc, copper, and iron are present in small amounts and play roles in various enzymatic and biological processes in the oral environment.

16.2.2.2 Liver

The liver is the largest internal organ and gland and plays a central role in metabolic and detoxification processes.

Anatomy

The liver is located in the upper right quadrant of the abdominal cavity (Fig. 16.28). It is positioned predominantly under the ribcage, which offers it some degree of protection from physical trauma. The liver is immediately below the diaphragm, with its largest lobe on the right side. In an average adult, it measures about 15–18.5 cm (6–7 inches) in vertical span, 21–22.5 cm (8–9 inches) in horizontal spread, and 10–12.5 cm (4–5 inches) in depth. Its weight typically ranges from 1.4 to 1.6 kg (3–3.5 pounds) in adults. This can represent about 2–3% of the body weight in a healthy adult. It has a significant reserve capacity and can regenerate, which allows it to maintain functionality even when partially damaged or surgically reduced.

The liver is typically smooth and has a soft yet firm consistency. In a healthy state, the liver has a reddish-brown color, owing to its rich blood supply and the presence of hepatocytes. The traditional description of the liver includes two main lobes, the right and left lobes, but a more detailed examination reveals two additional smaller lobes, the caudate and quadrate lobes (Fig. 16.35). The right lobe is the largest lobe of the liver. It extends from the right side of the body to the left, crossing the midline. The falciform ligament and the middle hepatic vein separate it from the left lobe. The right lobe is in contact with the diaphragm and the abdominal wall. It also houses the gallbladder in a small depression. Smaller than the right lobe, the left lobe of the liver extends across the midline toward the left side of the body. The left lobe is thinner and flatter than the right and contains the ligamentum teres, a remnant of the fetal umbilical vein. The caudate lobe is located on the posterior aspect of the liver, between the inferior vena cava and the left lobe. Situated on the underside of the liver, the quadrate lobe is bordered by the gallbladder and the round ligament (ligamentum teres). It lies adjacent to the stomach and is generally small and quadrilateral in shape.

Ligaments

The liver is held in place within the abdominal cavity by several ligaments and attachments (Fig. 16.36). The falciform ligament is a broad and thin peritoneal fold that attaches the liver to the anterior abdominal wall and diaphragm, separating the

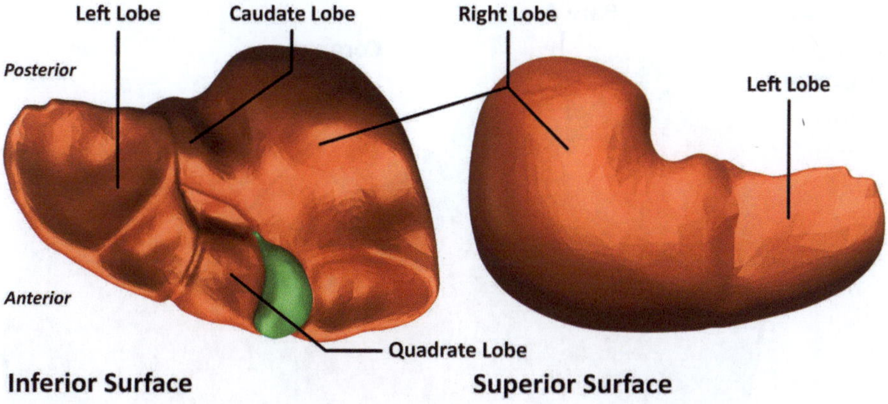

Fig. 16.35 Gross anatomy of the liver surface

right and left lobes of the liver. The coronary ligaments are reflections of the peritoneum that attach the liver to the diaphragm. Other ligaments include triangular, lesser omentum, round ligament (ligamentum teres hepatis), and ligamentum venosum.

Blood Supply

The liver's blood supply is unique and complex, receiving blood from two primary sources—the hepatic artery and the portal vein—making its blood supply dual (Fig. 16.37). The hepatic artery, a branch of the celiac trunk, supplies oxygen-rich arterial blood to the liver. The common hepatic artery branches into the proper hepatic artery, which then divides into the right and left hepatic arteries, supplying the corresponding liver lobes. The portal vein provides most (about 75%) of the liver's blood supply. It is formed by the confluence of the splenic vein and the superior mesenteric vein, bringing nutrient-rich blood from the gastrointestinal tract, spleen, and pancreas. This venous blood is rich in nutrients and contains toxins and other metabolites absorbed from the intestines, which the liver processes. Inside the liver, the portal vein divides into right and left branches, supplying the respective lobes. After passing through the sinusoids, the blood is collected in the central veins of each lobule, which then drain into larger hepatic veins. The hepatic veins eventually empty into the inferior vena cava, carrying deoxygenated blood from the liver to the heart.

Extrahepatic Biliary System

The extrahepatic biliary system (Fig. 16.38) is responsible for transporting bile from the liver and gallbladder to the duodenum and from the liver to the gallbladder. The liver produces bile, which initially drains into tiny canals within the liver called bile canaliculi and eventually converges into larger ducts. The right and left hepatic ducts emerge from the right and left lobes of the liver, respectively. The right and left hepatic ducts join outside the liver to form the common hepatic duct. The cystic

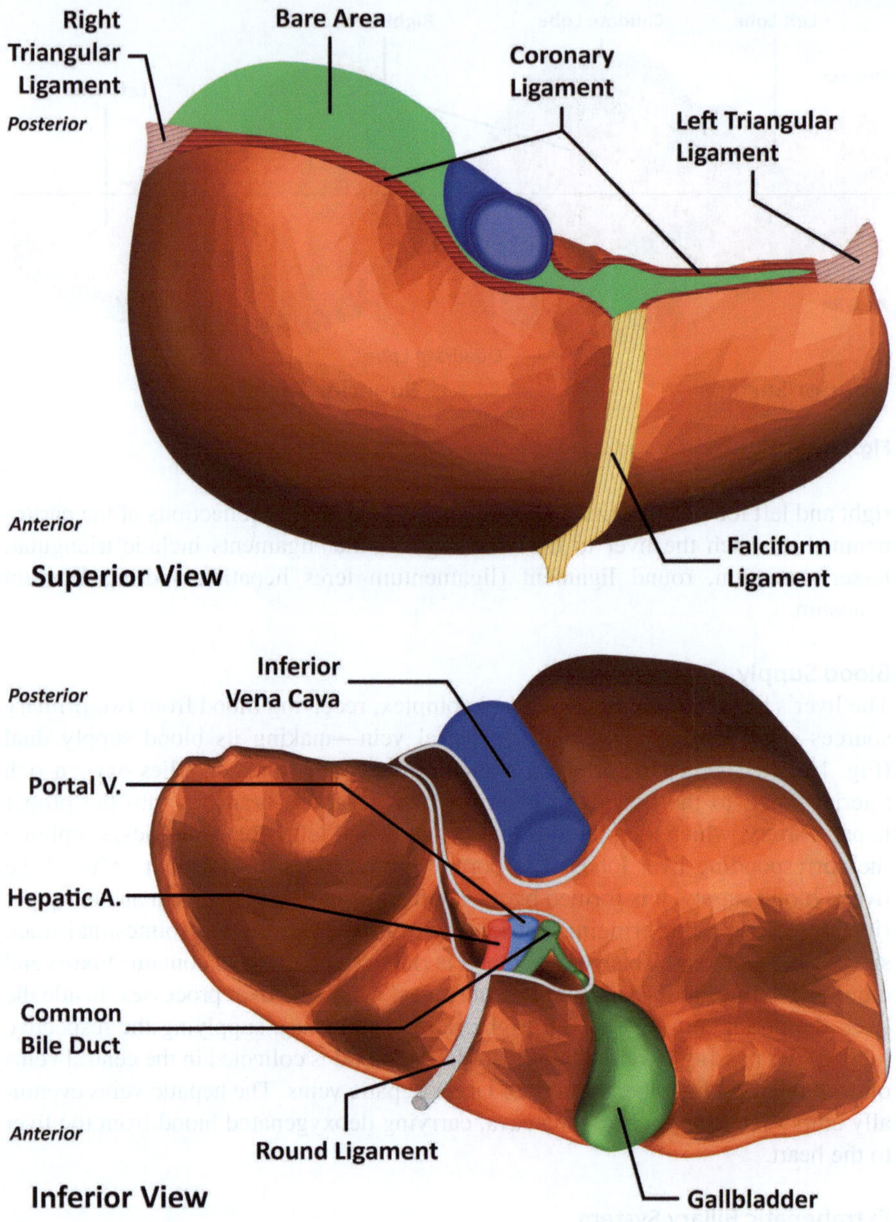

Fig. 16.36 Ligaments of the liver

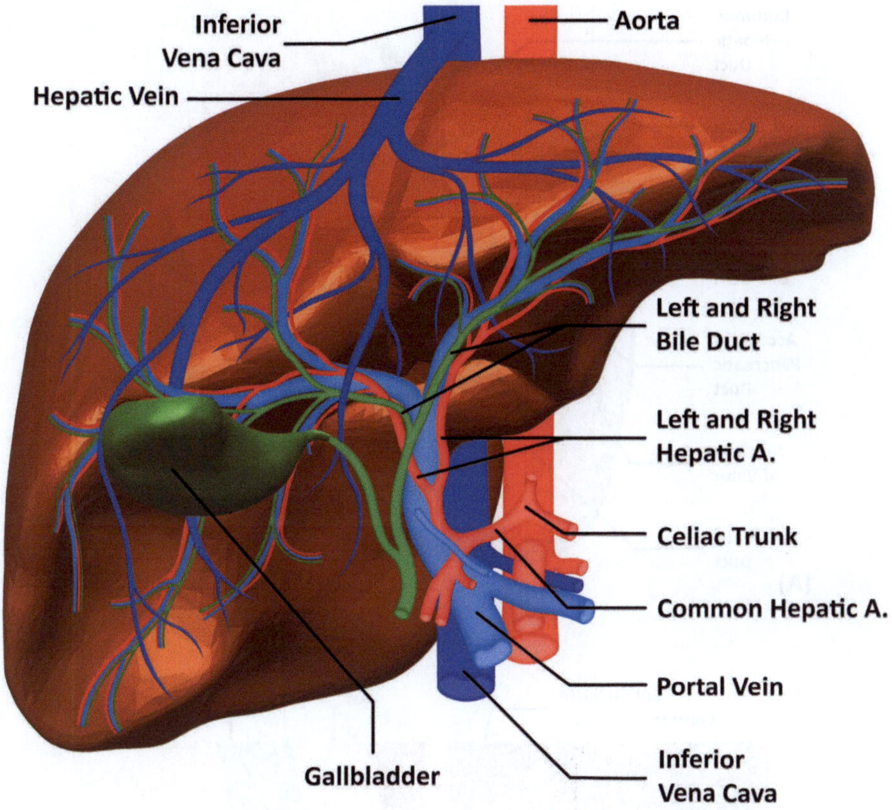

Fig. 16.37 Hepatic circulation. (The portal system is shown in light blue)

duct connects the gallbladder to the common hepatic duct. Bile can flow directly into the common bile duct or be diverted into the gallbladder for storage via the cystic duct. The common bile duct is formed by the union of the common hepatic duct and the cystic duct. This duct descends toward the duodenum, often running posterior to the first part of the duodenum and posterior to the pancreas. The sphincter of Oddi (Fig. 16.38) is a muscular valve that controls the flow of bile (and pancreatic juice) into the duodenum.

Microscopic Structure

The microscopic anatomy of the liver is characterized by its functional units called lobules, hepatocytes (liver cells), sinusoidal system, and the biliary system. Hepatocytes are the principal cells of the liver. They are large, polygonal cells, approximately 20–30 mm in diameter. They typically have a distinct, rounded shape with a well-defined cell boundary. Most hepatocytes contain a single, centrally located nucleus, but some may have two nuclei. The cytoplasm of hepatocytes is eosinophilic (stains pinkish-red with eosin dye) and abundant, containing various

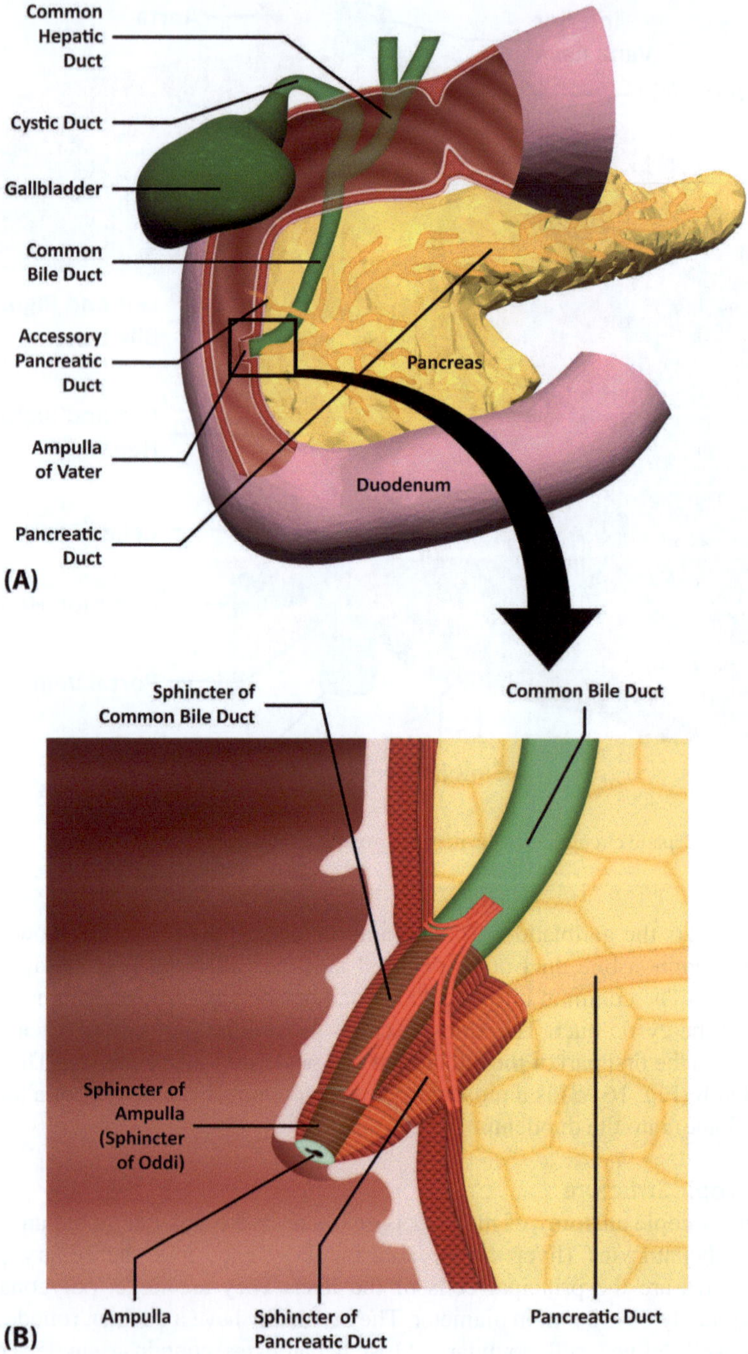

Fig. 16.38 Extrahepatic bile duct. (**a**) Gross anatomy. (**b**) Anatomy of the ampulla and sphincter of Oddi

organelles. Hepatocytes have numerous mitochondria, reflecting their high energy demands and metabolic activity.

The hepatic lobule is the functional unit of the liver (Fig. 16.39). Each lobule is roughly hexagonal, measuring about 1–2 mm in diameter. Portal triads mark the lobule's periphery, while its center contains the central vein. At each of the six corners of the hexagon, there is a portal triad consisting of a branch of the hepatic artery (supplying oxygenated blood), a branch of the portal vein (supplying nutrient-rich blood), and a bile duct for bile excretion. The triads are crucial for blood supply to the lobules and bile drainage from the hepatocytes. The central vein, located at the center of each lobule, collects blood from the sinusoids and drains it into larger veins, ultimately leading to the hepatic veins and the inferior vena cava. Blood flow within the lobule is from the periphery (portal triads) to the center (central vein). Hepatocytes are organized into plates or cords that radiate from the central vein toward the periphery of the lobule. These plates are one cell thick. Between the plates of hepatocytes are sinusoidal capillaries (sinusoids) (Fig. 16.40). These are unique blood vessels with a fenestrated endothelium, allowing the free exchange of molecules between the blood and hepatocytes. The sinusoids are lined with two main types of cells: endothelial cells and Kupffer cells (specialized macrophages).

Endothelial cells form the lining. Kupffer cells are specialized macrophages located in the liver. They are found attached to the endothelial lining of the hepatic sinusoids. They are strategically positioned to sample blood passing through the liver. These cells are characterized by an elongated, irregular shape that allows them to extend across the sinusoidal lumen. Kupffer cells are derived from monocytes in

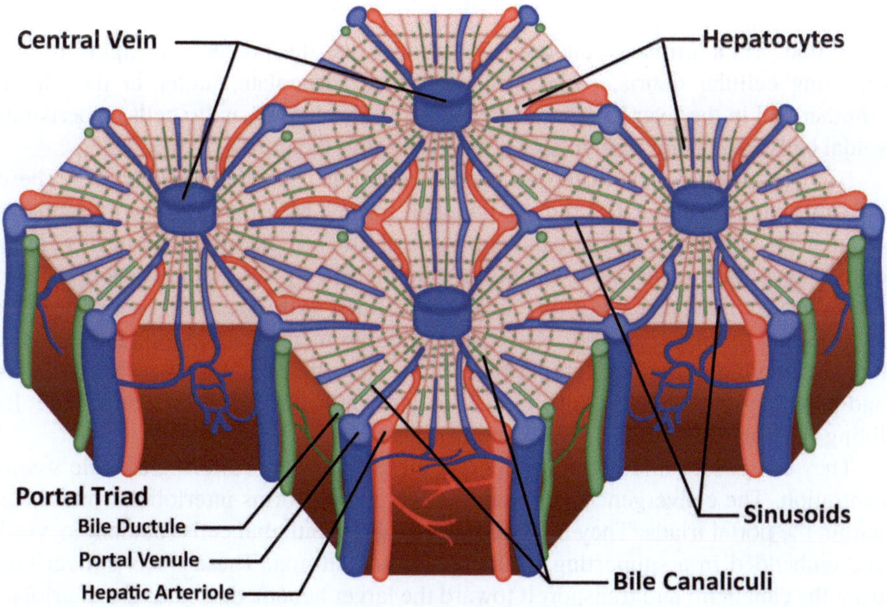

Fig. 16.39 Microscopic structure of liver lobules

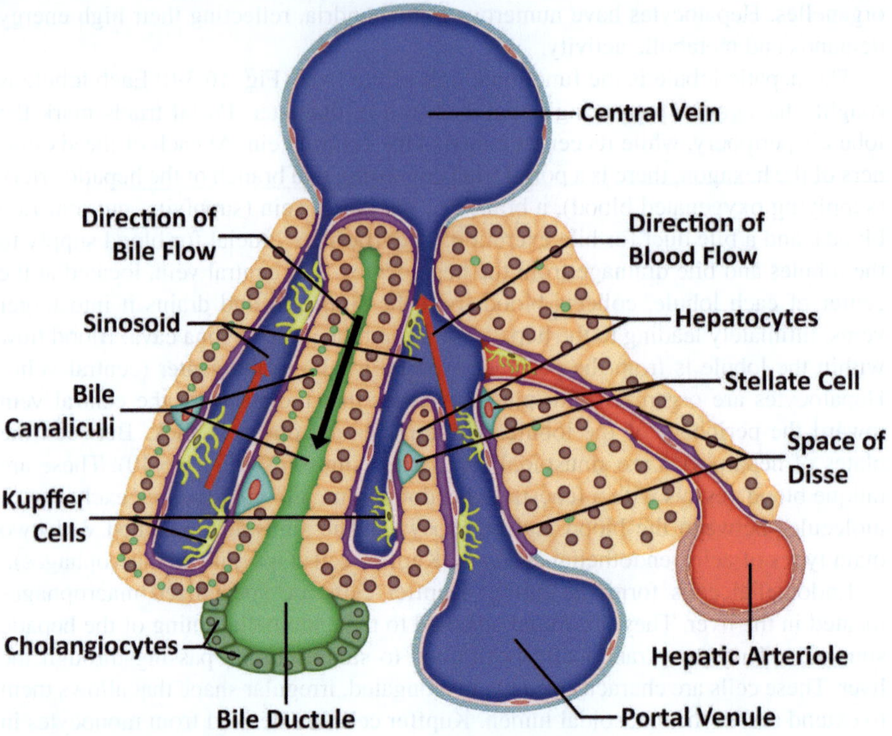

Fig. 16.40 Portal triad

the blood. Their primary function is phagocytosis, the process of engulfing and digesting cellular debris, pathogens, and other particulate matter in the blood. Another cell in the liver is hepatic stellate cells, also known as Ito cells or perisinusoidal cells, which are particularly active in storing vitamin A.

Bile produced by hepatocytes is secreted into bile canaliculi (Fig. 16.41). These are tiny channels, about 1–2 mm in diameter, formed by grooves in the membranes of adjacent hepatocytes. Bile canaliculi collect bile secreted by hepatocytes. They do not have their walls; instead, they are delineated by the hepatocytes that form them. The luminal surface of the canaliculi is lined with microvilli to increase the surface area for bile secretion. These short ductules mark the transition between the bile canaliculi and the interlobular bile ducts. They are lined partly by hepatocytes and partly by cholangiocytes (bile duct cells). Cholangiocytes, the epithelial cells lining the bile ducts, modify bile composition.

They can absorb and secrete water and electrolytes, thus altering the bile's concentration. The convergence of several Hering canals forms interlobular bile ducts within the portal triads. They are lined by cuboidal epithelial cells (cholangiocytes) and embedded in a supporting connective tissue stroma. These ducts collect bile from the canaliculi and transport it toward the larger hepatic ducts. As the interlobular bile ducts converge, they form larger ducts. These larger ducts eventually converge to create the right and left hepatic ducts, which exit the liver.

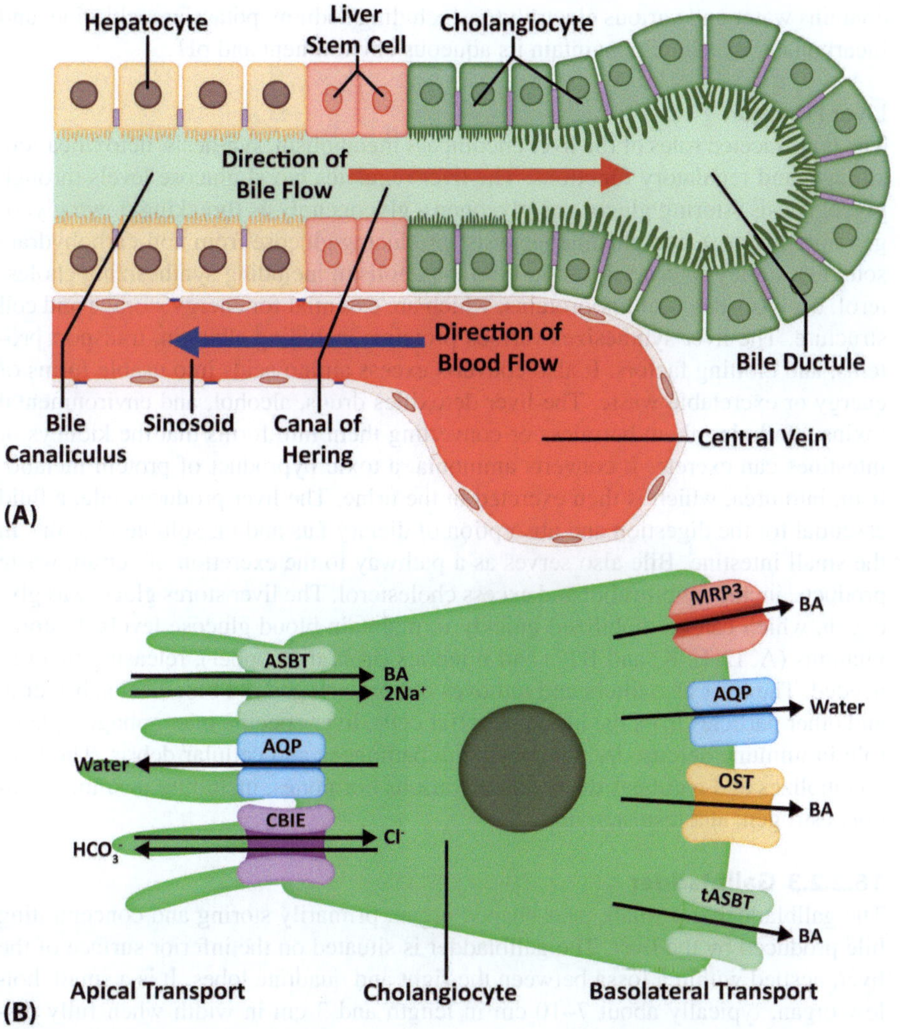

Fig. 16.41 Bile canaliculi. (**a**) Microscopic structure of bile canaliculus and ductule. (**b**) A cholangiocyte and the mechanism of bile formation.

Bile

Bile is synthesized by the liver and drained by the biliary system into the duodenum. The primary components of bile are bile acids and their conjugated forms, bile salts. Bile acids, synthesized from cholesterol in hepatocytes, are conjugated with amino acids (primarily glycine or taurine) to form bile salts, which are more effective in emulsifying fats. Bile also contains cholesterol, which is excreted from the body via bile. A breakdown product of hemoglobin, bilirubin gives bile its characteristic yellow-green color. Conjugated bilirubin is water-soluble and excreted in bile. Phospholipids, primarily lecithin, are essential for emulsifying dietary lipids. Bile contains water and various electrolytes, including sodium, potassium, chloride, and bicarbonate, which help maintain its aqueous environment and pH.

Liver Function

The multifaceted roles of the liver encompass metabolism, synthesis, detoxification, storage, and regulatory functions. The liver regulates blood glucose levels through glycogenesis (storing glucose as glycogen), glycogenolysis (breaking down glycogen into glucose), and gluconeogenesis (producing glucose from non-carbohydrate sources). It plays a crucial role in lipid metabolism, including synthesizing cholesterol, triglycerides, and lipoproteins, which are essential for energy storage and cell structure. The liver synthesizes various proteins, including albumin, transport proteins, and clotting factors. It also converts excess amino acids into usable forms of energy or excretable waste. The liver detoxifies drugs, alcohol, and environmental toxins, rendering them harmless or converting them into forms that the kidneys or intestines can excrete. It converts ammonia, a toxic byproduct of protein metabolism, into urea, which is then excreted in the urine. The liver produces bile, a fluid essential for the digestion and absorption of dietary fats and fat-soluble vitamins in the small intestine. Bile also serves as a pathway to the excretion of certain waste products, including bilirubin and excess cholesterol. The liver stores glucose as glycogen, which can be mobilized quickly to maintain blood glucose levels. It stores vitamins (A, D, E, K, and B12) and minerals (iron and copper), releasing them as needed. The liver also filters and removes old or damaged red blood cells, bacteria, and other particles from the blood. Kupffer cells, liver-specific macrophages, play a role in immune defense by phagocytizing pathogens and cellular debris. The liver metabolizes and regulates the levels of various hormones, including insulin, glucagon, estrogen, and testosterone.

16.2.2.3 Gallbladder

The gallbladder is a small, pear-shaped organ, primarily storing and concentrating bile produced by the liver. The gallbladder is situated on the inferior surface of the liver, nestled within a fossa between the right and quadrate lobes. It is a small, hollow organ, typically about 7–10 cm in length and 3 cm in width when fully distended. The gallbladder has a capacity of about 30–50 mL. The gallbladder is connected to the common bile duct via the cystic duct. The gallbladder's primary function is to store bile, which is crucial for fat digestion. It can store about 30–50 milliliters of bile, produced continuously by the liver but not always immediately needed for digestion. The gallbladder mucosa absorbs water and electrolytes from bile, concentrating it up to five to ten times its original potency. This concentration process enhances the efficiency of bile in the digestion of fats.

16.2.2.4 Pancreas

The pancreas is both an endocrine and exocrine organ, playing roles in both the digestive and endocrine systems.

Anatomy

The pancreas lies in the upper abdomen, extending horizontally across the posterior abdominal wall (Fig. 16.42). It is nestled between the stomach and the spine, situated retroperitoneally (behind the peritoneum). The pancreas has a soft, lobulated texture, and its color is typically a pale, creamy pink. It consists of several parts: the head of the pancreas is the broadest part, nestled in the C-shaped curve of the duodenum. A unique feature of the pancreatic head is the uncinate process. This hook-like extension projects from the lower part of the head and extends posteriorly and medially behind the superior mesenteric vessels. The common bile duct typically passes posteriorly through a groove in the pancreatic head or is embedded within its substance before joining the main pancreatic duct. The neck is the constricted pancreas segment that connects the head to the body. The body of the pancreas extends laterally from the neck, crossing the body's midline. The tail is the narrowest and most distal part of the pancreas, extending toward the left side of the body and reaching the hilum of the spleen.

Microscopic Structure

The fundamental units of the exocrine pancreas are the acinar cells (Fig. 16.43). These pyramidal cells are organized into clusters known as acini (singular: acinus).

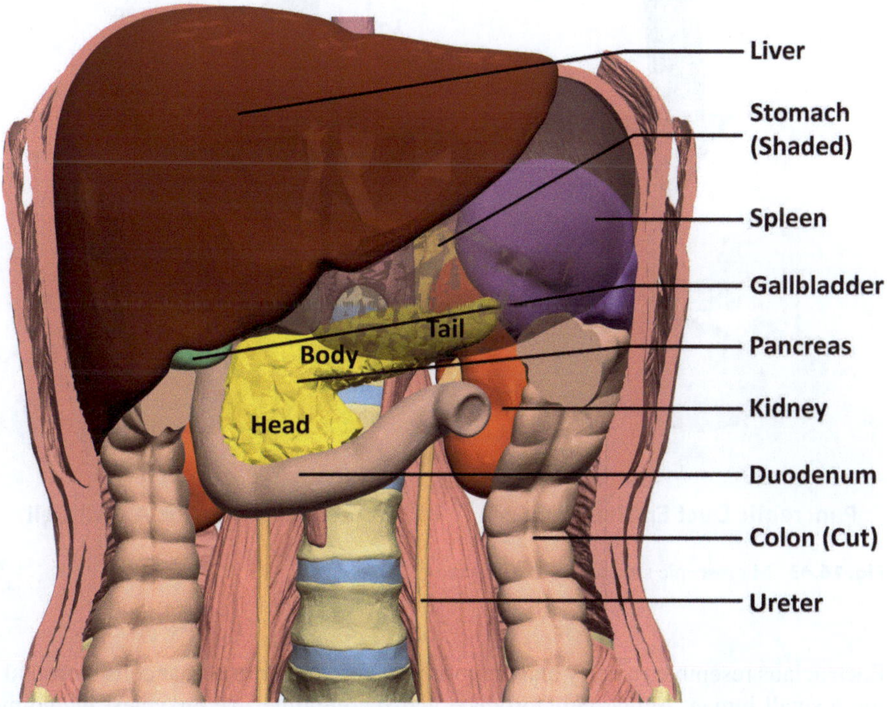

Fig. 16.42 Anatomy of the pancreas in the abdominal cavity, anterior view. (The stomach is shaded, and the small intestine and transverse colon are removed)

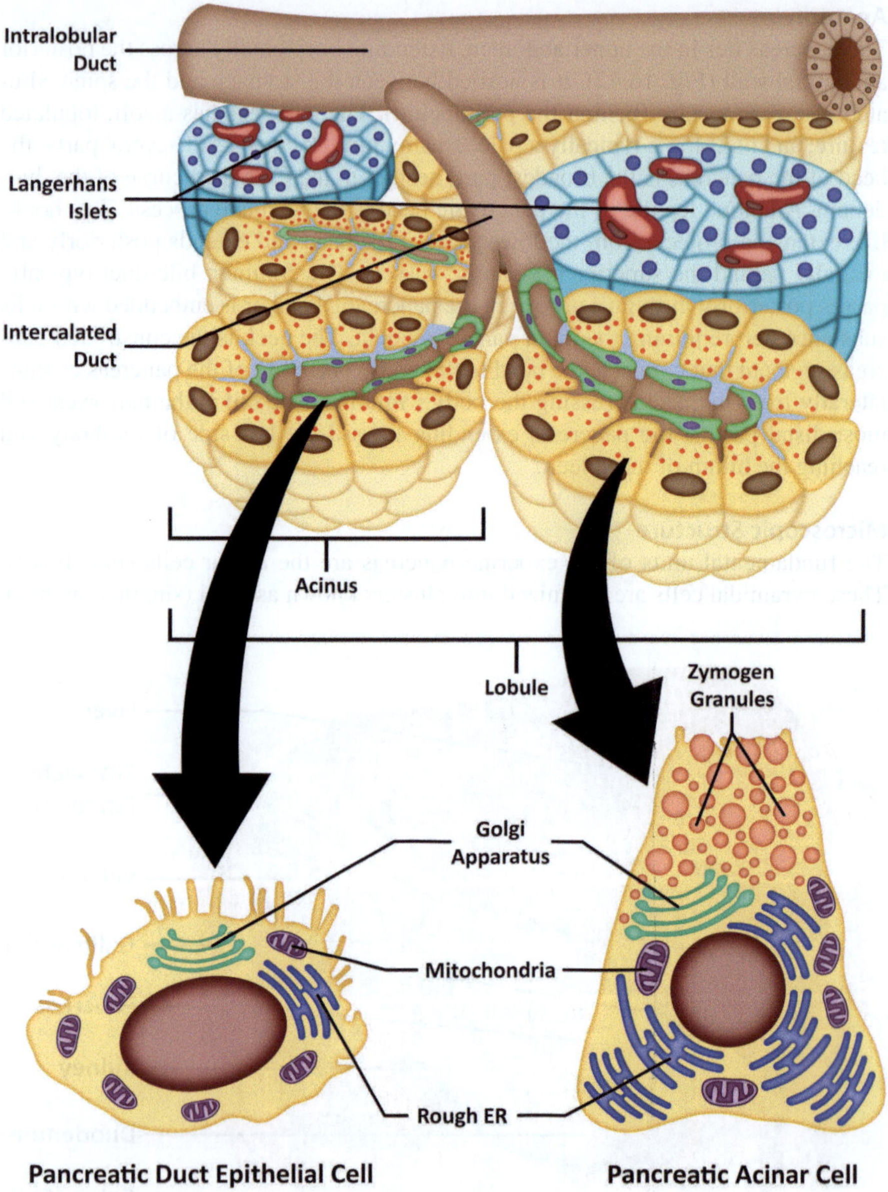

Fig. 16.43 Microscopic structure of the pancreas

Each acinus resembles a berry and comprises a single layer of acinar cells surrounding a small lumen. Acinar cells produce and secrete digestive enzymes, including amylase, lipase, and proteases like trypsinogen and chymotrypsinogen. These enzymes are initially produced in an inactive form and are activated in the small intestine. The cytoplasm of acinar cells contains a large amount of rough endoplasmic reticulum and zymogen granules.

Ductal System

The ductal system of the pancreas transports digestive enzymes from the pancreas to the duodenum (Fig. 16.44). Within each lobule of the pancreas, tiny intralobular ducts collect the exocrine secretions and merge to form larger interlobular ducts. Interlobular ducts collect secretions from the intralobular ducts and convey them to the main duct. The main pancreatic duct (duct of Wirsung) runs longitudinally through the entire pancreas. It begins in the tail, where it collects secretions from the pancreatic lobules, and runs toward the head of the pancreas. In the head of the pancreas, the main pancreatic duct typically joins with the common bile duct to form a dilated chamber known as the hepatopancreatic ampulla (ampulla of Vater). The ampulla opens into the duodenum at the major duodenal papilla, a slight elevation in the duodenum's lining, through the sphincter of Oddi, a muscular valve regulating the flow of bile and pancreatic juice into the intestine (Fig. 16.38). This muscular sphincter surrounds the end of the main pancreatic duct (and common bile duct) at the ampulla of Vater.

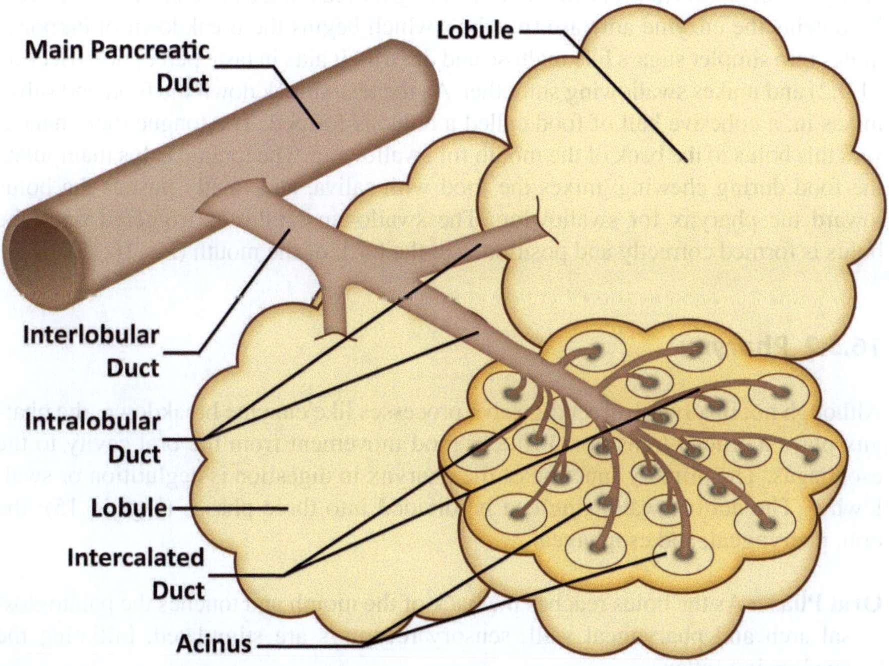

Fig. 16.44 Intrapancreatic ductal system

16.3 Part 3: Digestion Physiology

Digestion physiology concerns how our bodies convert food into energy and building blocks necessary for growth, repair, and sustenance. This complex process involves a series of mechanical and chemical transformations, where food is broken down into simpler forms that can be easily assimilated into the bloodstream and utilized by the body's cells.

16.3.1 Mouth

The process of digestion in the mouth, known as oral digestion, combines mechanical and chemical mechanisms to begin the breakdown of food, making it ready for further digestion in the stomach and intestines. Mechanical digestion begins with mastication, or chewing, where food is broken down into smaller, more manageable pieces. The chewing is facilitated by the teeth, which are specialized for different roles—incisors for cutting, canines for tearing, and molars for grinding. This mechanical breakdown increases the surface area of the food, making it more accessible for enzymatic action. Simultaneously, with mechanical digestion, chemical digestion starts in the mouth. The primary agent of chemical digestion in the oral cavity is saliva. Saliva moistens food, forming a bolus that can be easily swallowed. It contains the enzyme amylase (ptyalin), which begins the breakdown of carbohydrates into simpler sugars like maltose and dextrin. It aids in taste perception (Section 11.3.2) and makes swallowing smoother. As the teeth break down the food and saliva mixes in, a cohesive ball of food called a bolus is formed. The tongue then maneuvers this bolus to the back of the mouth for swallowing. The tongue helps manipulate the food during chewing, mixes the food with saliva, and finally pushes the bolus toward the pharynx for swallowing. The swallowing reflex is triggered once the bolus is formed correctly and positioned at the back of the mouth (Fig 16.15).

16.3.2 Pharynx

Although not involving direct digestive processes like enzyme breakdown, the pharynx plays a crucial transitional role in food movement from the oral cavity to the esophagus. The primary function of the pharynx in digestion is deglutition or swallowing. The act of swallowing can be divided into three phases (Fig. 16.15): the oral, pharyngeal, and esophageal.

Oral Phase: As the bolus reaches the back of the mouth and touches the palatoglossal arch and pharyngeal wall, sensory receptors are stimulated, initiating the swallowing reflex.

Pharyngeal Phase: The pharyngeal phase is a rapid, involuntary reflex. The soft palate elevates to close off the nasal passages, preventing food from entering the nasal cavity. The larynx, which houses the vocal cords and the opening to the

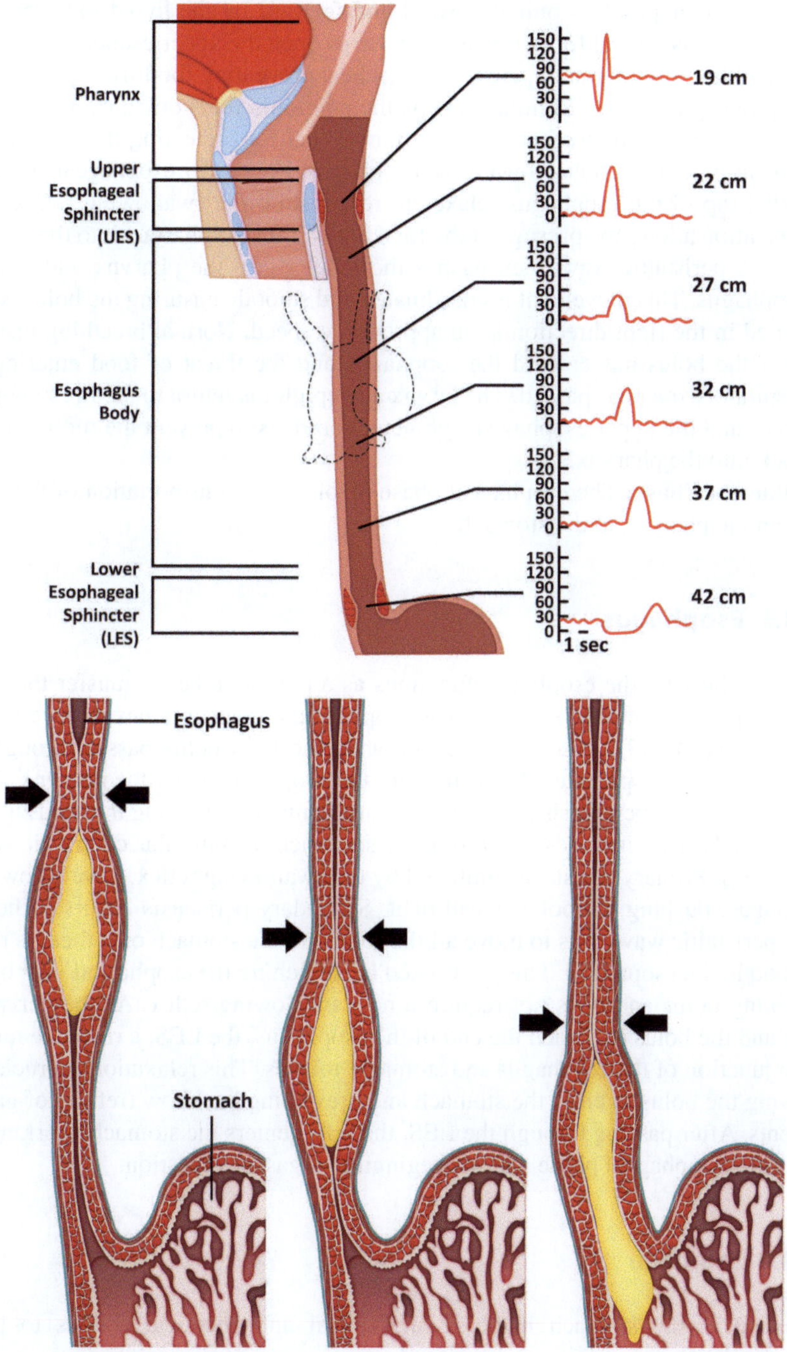

Fig. 16.45 Esophageal phase of swallowing. (**Upper**) Pressure generation in different parts of the esophagus. (**Lower**) Movement of the bolus toward the stomach by peristaltic wave

trachea (windpipe), is pulled upward and forward by the hyoid muscles. This movement is crucial for two reasons: it helps open the upper esophageal sphincter and, more importantly, closes the trachea, preventing food from entering the respiratory passages. Simultaneously, the epiglottis folds down over the glottis (the opening to the trachea). This action is vital for protecting the airways and directing the food bolus toward the esophagus. The upper esophageal sphincter at the top of the esophagus relaxes in response to the swallowing reflex. This relaxation allows the passage of the food bolus from the pharynx into the esophagus. A peristaltic wave then pushes the bolus down the pharynx and into the esophagus. This movement is coordinated and smooth, ensuring the bolus is propelled in the right direction at an appropriate speed. Normal breathing resumes after the bolus has entered the esophagus and the threat of food entering the respiratory tract has passed. The larynx and epiglottis return to their resting positions, and the upper esophageal sphincter contracts to prevent the reflux of food back into the pharynx.

Esophageal Phase: The esophageal phase involves the transportation of the bolus from the pharynx to the stomach.

16.3.3 Esophagus

Like the pharynx, the esophagus functions as a passing tube to transfer the bolus from the pharynx into the stomach. The esophageal phase continues the swallowing process (Fig. 16.45). This phase begins when the food bolus passes through the upper esophageal sphincter (UES) to enter the esophagus from the pharynx. Once in the esophagus, peristalsis is the primary mechanism for moving the bolus toward the stomach. This involves a coordinated sequence of muscular contractions and relaxations. Primary peristalsis, initiated by the swallowing reflex, travels down the esophagus, pushing the bolus ahead of it. Secondary peristalsis occurs if the primary peristaltic wave fails to move all the food into the stomach or if there is residual food in the esophagus. This is initiated by stretching the esophageal wall by the remaining bolus and does not require a new swallowing reflex. As the peristaltic wave and the bolus approach the end of the esophagus, the LES, a ring-like muscle at the junction of the esophagus and stomach, relaxes. This relaxation is crucial for allowing the bolus to enter the stomach and preventing backflow (reflux) of gastric contents. After passing through the LES, the bolus enters the stomach, marking the end of the esophageal phase and the beginning of gastric digestion.

16.3.4 Stomach

Digestion in the stomach involves mechanical and chemical actions to break down food.

16.3.4.1 Mechanical Digestion

The unique muscular structure and movements of the stomach primarily drive mechanical digestion. Once food enters the stomach, the muscular contractions of the stomach wall create a churning motion. This churning action mixes the food with gastric juices, transforming the solid food into a semi-liquid mixture called chyme. Rhythmic peristaltic movements are wave-like contractions that move along the stomach wall. Peristalsis in the stomach is not as regular as in the esophagus but occurs more vigorously. The strongest contractions occur in the antrum, which is the lower portion of the stomach. These intense contractions are crucial for grinding the food, especially more substantial and harder-to-digest items. Mechanical digestion in the stomach reduces the size of food particles, making them more manageable for chemical digestion and absorption in the small intestine. The churning action thoroughly mixes food with gastric juices, aiding in the chemical breakdown of proteins and other macronutrients.

16.3.4.2 Chemical Digestion

Chemical digestion in the stomach involves the breakdown of food using gastric juices. Pepsin breaks down proteins into smaller peptides by cleaving the peptide bonds. It targets specific amino acid sequences, particularly those involving aromatic or acidic amino acids. Pepsin is most effective in the acidic pH of the stomach and selectively breaks down specific protein structures, setting the stage for further protein digestion in the small intestine. While most carbohydrate digestion occurs in the mouth and small intestine, the stomach's acidic environment can continue the breakdown of carbohydrates begun by salivary amylase. The stomach does little in terms of fat digestion.

16.3.4.3 Gastric Secretion

Gastric secretions are a complex mixture of various components produced by different cells in the stomach lining. These secretions contribute to both the chemical breakdown of food and the protection of the stomach from its acidic environment.

Hydrochloric Acid (HCl): HCl is secreted by parietal (or oxyntic) cells in the gastric glands of the stomach lining. It provides an acidic environment for enzymatic activity, particularly for activating pepsin from pepsinogen, and helps break down food. It also acts as a defense mechanism by killing most bacteria and other pathogens ingested with food.

Pepsinogen and Pepsin: Chief cells, found in the deeper parts of the gastric glands, secrete pepsinogen, an inactive precursor of pepsin. In the stomach's acidic environment, pepsinogen is converted to pepsin, an active protease enzyme. Pepsin initiates the digestion of proteins by breaking down their peptide bonds.

Mucus: Mucous cells, which line the stomach's surface and necks of the gastric pits, secrete mucus. This viscous and alkaline layer coats the stomach lining and protects it from mechanical damage, the corrosive nature of HCl, and enzymatic digestion by pepsin.

Intrinsic Factor (IF) (Fig. 16.46): IF is also secreted by the parietal cells. It is essential to absorb vitamin B12 (cobalamin) in the small intestine.

Bicarbonate: Bicarbonate is secreted by the stomach's epithelial cells and is also present in mucus. It helps to neutralize the acid near the mucosal surface, providing a pH gradient in the mucus layer (Fig. 16.24) and protecting the stomach lining from acid-induced damage.

Gastric Lipase: Chief cells also secrete gastric lipase. It plays a minor role in digesting lipids in the stomach.

Water and Electrolytes: They are secreted along with other components to form gastric juice and help create the right consistency and volume for effective digestion.

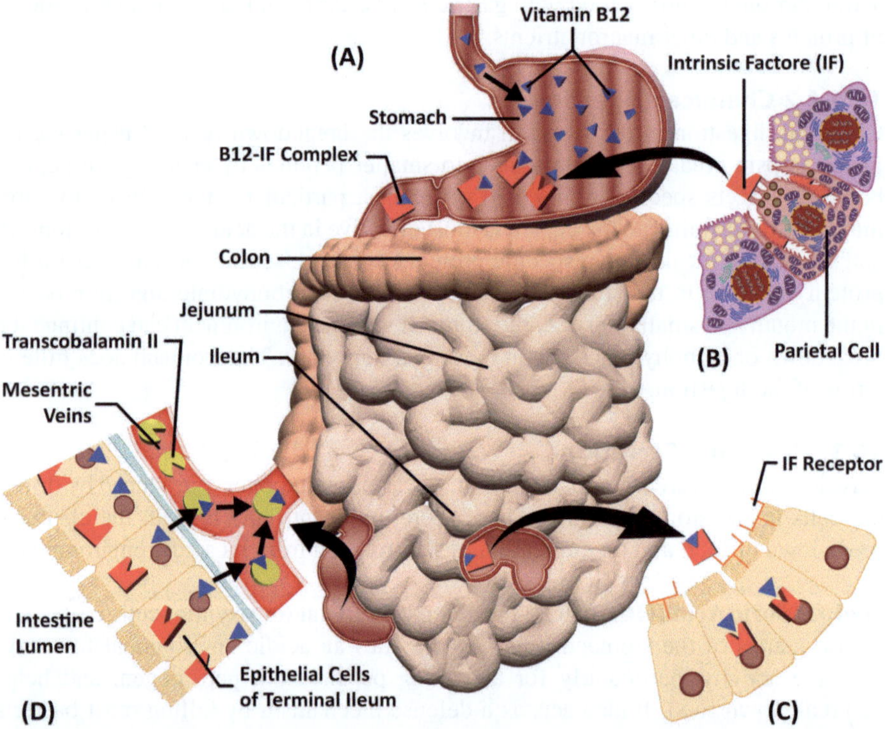

Fig. 16.46 Vitamin B12 transport and absorption in the gastrointestinal tract. (**a**) Vitamin B12 ingested through dietary sources such as meat, dairy, and fortified foods enters the stomach. (**b**) Intrinsic factor (IF), a glycoprotein produced by the parietal cells of the stomach lining, binds to vitamin B12 in the stomach's acidic environment to form the B12-IF complex. (**c**) The B12-IF complex travels through the small intestine unchanged because the intrinsic factor protects vitamin B12 from digestive enzymes. In the terminal part of the ileum, the epithelial cells of the intestinal mucosa have receptors for IF. The B12-IF complex attaches to the IF receptor through the IF factor and enters the epithelial cell. (**d**) Once inside the epithelial cells, vitamin B12 is released from the intrinsic factor, transported across the epithelial cells, and released into the bloodstream. In blood, B12 binds to transcobalamin II (TC II), forming the B12-TC II complex, and is transported to the liver and other tissues where vitamin B12 is utilized or stored

16.3.4.4 Gastric Emptying

Gastric emptying is transferring the stomach contents into the small intestine. The distal part of the stomach, the antrum, functions as a powerful pump. Strong antral contractions propel chyme toward the pylorus (the opening to the small intestine). The pyloric sphincter maintains a tonic contraction most of the time, which prevents the free passage of contents. During gastric emptying, the sphincter relaxes periodically, allowing small amounts of chyme to pass into the duodenum. This relaxation is coordinated with the antral contractions of the stomach. By regulating the amount and rate of chyme entering the duodenum, the pyloric sphincter ensures that the small intestine can effectively process and absorb nutrients.

16.3.5 Liver and Gallbladder: Bile

Bile, produced by the liver and stored in the gallbladder, primarily emulsifies and helps the small intestine absorb lipids.

16.3.5.1 Formation

Bile formation begins within the liver's hepatocytes, where various metabolic processes occur. Hepatocytes produce bile salts, bile pigments, cholesterol, phospholipids, and electrolytes, among other substances. Bilirubin is conjugated with glucuronic acid in the liver, converting it into water-soluble bilirubin diglucuronide. This conjugation is necessary for the excretion of bilirubin in bile. Hepatocytes synthesize primary bile salts from cholesterol, primarily cholic acid and chenodeoxycholic acid. Conjugated bile salts are actively transported into the bile canaliculi. Once synthesized, these bile components are transported to the bile canaliculi.

16.3.5.2 Composition

Bile composition includes bile salts, cholesterol, bilirubin, electrolytes, and water.

Bile Salts

Bile salts break down large fat globules into smaller droplets in a process known as emulsification. This increases the surface area available for the action of digestive enzymes like pancreatic lipase. By emulsifying fats, bile salts facilitate the digestion and absorption of dietary fats and fat-soluble vitamins (A, D, E, and K) in the small intestine. Bile salts also help solubilize cholesterol, aiding its absorption and preventing the formation of cholesterol-based gallstones. Bile salts possess antimicrobial properties, helping to control the population of microbes in the small intestine.

Bilirubin

Bilirubin is a yellowish pigment found in bile. It is the byproduct of the breakdown of hemoglobin, the oxygen-carrying component of red blood cells. As red blood cells age or become damaged, they are broken down by spleen, liver, and bone marrow macrophages. The heme portion of hemoglobin is converted into biliverdin and then reduced to unconjugated bilirubin. This form of bilirubin is not water-soluble and is bound to albumin for transport in the bloodstream. Once unconjugated bilirubin reaches the liver, it undergoes a process called conjugation. It is combined with glucuronic acid, catalyzed by the enzyme UDP-glucuronosyltransferase. This process transforms unconjugated bilirubin into conjugated bilirubin, which is water-soluble. Conjugated bilirubin is then secreted by liver cells into bile. Bilirubin gives bile its characteristic yellow-green color. In the intestine, conjugated bilirubin is metabolized by intestinal bacteria into urobilinogens. Some of the urobilinogens are absorbed back into the bloodstream and eventually excreted by the kidneys in urine, giving urine its characteristic yellow color. The rest is excreted in feces, contributing to its brown color.

Cholesterol

Cholesterol is a lipid molecule essential for the body's normal functioning. The liver synthesizes cholesterol, which becomes a constituent of bile. It's crucial for the formation of micelles (Fig. 16.47). Cholesterol is insoluble in water. In bile, cholesterol solubility is maintained by forming micelles, where bile salts and

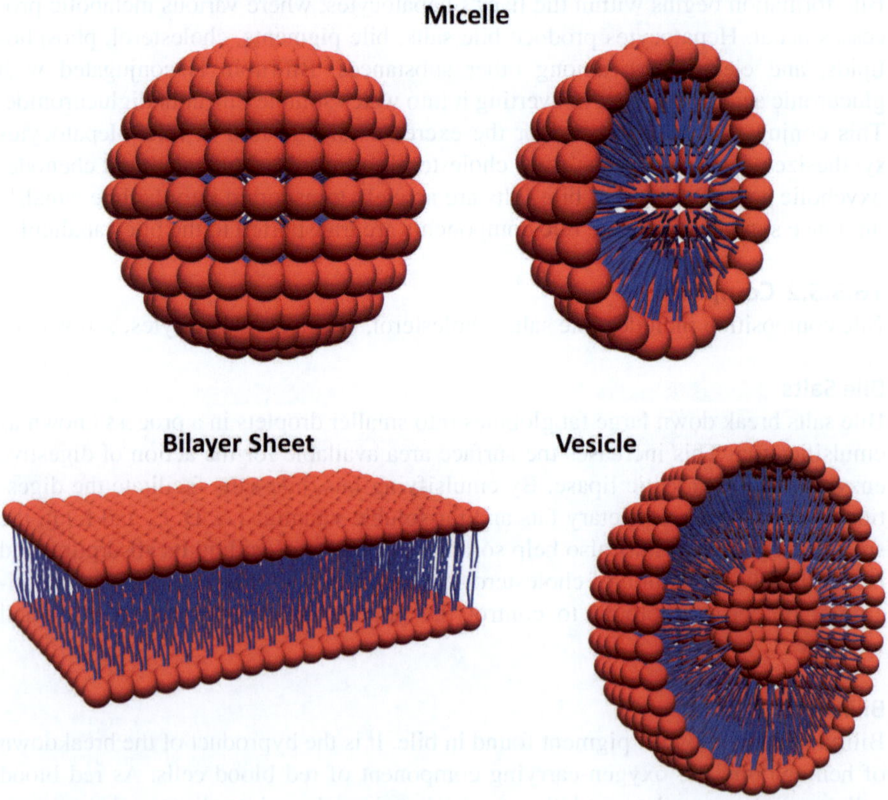

Fig. 16.47 Micelle and its comparison to lipid bilayer sheet (membranes) and vesicles

phospholipids, mainly lecithin, play a crucial role. These micelles keep cholesterol in a soluble form, preventing its precipitation. While cholesterol itself is not directly involved in the breakdown of dietary fats, the role of cholesterol in maintaining the integrity and function of bile ensures efficient fat digestion and absorption. Cholesterol in bile is also a means for the body to excrete excess cholesterol, thus playing a role in regulating overall cholesterol levels in the body.

Phospholipids
Phospholipids in bile, predominantly in the form of lecithin, are essential for various physiological functions, particularly in the digestion and absorption of fats. Phospholipids and bile salts help emulsify dietary fats in the small intestine. They break down large fat globules into smaller micelles, increasing the surface area for digestive enzymes to act upon. The emulsification of fats also aids in absorbing fat-soluble vitamins (A, D, E, and K) from the diet.

Electrolytes
The main electrolytes present in bile are sodium (Na^+), potassium (K^+), chloride (Cl^-), bicarbonate (HCO_3^-), and calcium (Ca^{2+}). Sodium and potassium ions help maintain the osmotic balance and pH of bile. Chloride ions are crucial for maintaining the ionic balance and the acid-base status of bile. Bicarbonate ions in bile are important for neutralizing the acid entering the small intestine from the stomach, thus providing an optimal pH for enzymatic activity in the intestine, particularly for pancreatic enzymes.

16.3.5.3 Enterohepatic Circulation

Enterohepatic circulation is a physiological process involving the recycling of bile acids between the liver, small intestine, and gallbladder. Bile acids are synthesized in the liver from cholesterol. The liver cells (hepatocytes) convert cholesterol into primary bile acids, primarily cholic acid and chenodeoxycholic acid. The liver secretes bile into small ducts that converge into the common hepatic duct. This bile flows directly into the small intestine via the common bile duct or is stored and concentrated in the gallbladder. After fat digestion, most bile acids are reabsorbed in the distal ileum, with about 95% of them being transported back to the liver through the portal vein. Once back in the liver, bile acids are taken up by hepatocytes and resecreted into bile. This recycling process occurs multiple times during a meal.

16.3.6 Pancreatic Juice

The exocrine part of the pancreas secretes pancreatic juice. This clear, alkaline juice is composed of a unique blend of enzymes, electrolytes, and water. The enzymes in pancreatic juice include amylase for carbohydrate digestion; lipase for fat digestion; proteases like trypsin, chymotrypsin, and carboxypeptidase for protein digestion; and nucleases like ribonuclease (RNase) and deoxyribonuclease

(DNase). These enzymes are initially secreted in an inactive form and activated in the small intestine to prevent damage to the pancreas. Trypsin is a serine protease that breaks down proteins by hydrolyzing peptide bonds. Specifically, it cleaves peptide bonds at the carboxyl side of the amino acids lysine and arginine. Trypsin also activates other pancreatic proenzymes, such as chymotrypsinogen (to chymotrypsin) and procarboxypeptidases (to carboxypeptidases). Chymotrypsin is also a serine protease that cleaves peptide bonds in proteins. It preferentially targets peptide bonds next to aromatic amino acids like phenylalanine, tyrosine, and tryptophan. Carboxypeptidases are exopeptidases, meaning they cleave one amino acid at a time from the carboxyl end of a peptide chain. This process helps in the final steps of protein digestion, breaking peptides into individual amino acids. Trypsin, chymotrypsin, and carboxypeptidase work together in a coordinated manner to efficiently break down dietary proteins.

RNase catalyzes the hydrolysis of RNA. This enzyme targets the phosphodiester bonds between nucleotides in RNA. It breaks down ribonucleic acid (RNA) into smaller components, specifically into nucleotides and nucleosides. DNase catalyzes the hydrolysis of DNA. DNase cleaves the phosphodiester bonds in the DNA backbone, resulting in the production of smaller DNA fragments.

In addition to its enzymatic components, pancreatic juice contains bicarbonate ions, which are critical in neutralizing the acidic chyme received from the stomach. This neutralization creates an optimal pH environment in the small intestine for the functioning of digestive enzymes.

16.3.7 Small Intestine

The small intestine plays a significant role in digestion and nutrient absorption. Stretching approximately 20 feet in length in adults, it is the site where most of the digestion and absorption of food occur.

Small intestine juice, also known as succus entericus, is a complex mixture of enzymes, mucus, bicarbonate, and electrolytes. These are essential for the final stages of digestion and the protection of the intestinal lining. Intestinal juice is secreted by the enterocytes (intestinal lining cells) and the cells of the intestinal glands (crypts of Lieberkühn).

16.3.7.1 Small Intestine Juice

Small intestine juice is composed of:

- Water: Constituting the bulk of intestinal juice, water serves as the solvent for all its other components, facilitating their interaction with food particles for adequate digestion.
- Enzymes: A variety of digestive enzymes are present in the intestinal juice, each targeting specific substrates:

  - Disaccharidases: These include enzymes like lactase, sucrase, and maltase, which break down lactose, sucrose, and maltose into their constituent monosaccharides (glucose, fructose, and galactose).
  - Peptidases (Dipeptidases and Aminopeptidases): These enzymes further break down small peptide chains into individual amino acids.
  - Enterokinase: This enzyme converts trypsinogen (from pancreatic juice) into trypsin, which then activates other pancreatic proteases.
- Mucus: Secreted by the goblet cells and Brunner's glands in the duodenum, mucus lubricates the intestinal contents, aiding in their smooth passage, and protects the intestinal lining from acidic chyme and mechanical damage.
- Bicarbonate and Alkaline Salts: These components help neutralize the acidic gastric juice from the stomach, creating an alkaline environment (pH around 7.5 to 8.0) necessary for the optimal function of intestinal and pancreatic enzymes.
- Electrolytes: The juice contains various electrolytes such as sodium, potassium, chloride, and calcium, which are crucial for maintaining the electrolyte balance and fluidity of the digestive contents.

The enzymes in the intestinal juice are responsible for the final breakdown of nutrients, particularly carbohydrates and proteins, into their absorbable forms. The bicarbonate and alkaline salts in the juice neutralize the acid from the stomach, protecting the intestine and providing a suitable pH for enzymatic activity. The mucus component lubricates the intestinal contents and protects the intestinal mucosa from enzymatic and mechanical damage.

16.3.7.2 Digestive Process in Duodenum

The duodenum, the first and shortest segment of the small intestine, receives partially digested food, known as chyme, from the stomach. This chyme is acidic and contains partially digested proteins and carbohydrates. Bicarbonate ions from the pancreas (in pancreatic juice) and Brunner's glands in the duodenal wall neutralize the stomach acid, raising the pH to create a suitable environment for digestive enzymes. Pancreatic juice contains vital enzymes: proteases (such as trypsin and chymotrypsin) for protein digestion, amylase for continuing carbohydrate digestion, and lipase, along with co-lipase, for fat digestion. Bile, released from the gallbladder in response to cholecystokinin (CCK), emulsifies fats. This process increases the surface area of fats, making them more accessible to pancreatic lipase.

16.3.7.3 Digestive Process in Jejunum

After the initial breakdown of food in the stomach and further digestion in the duodenum, the jejunum is responsible for most nutrient absorption. The jejunum's lining is highly specialized for absorption. It features finger-like projections, called villi, and tinier projections called microvilli, known as the brush border. This significantly increases the surface area for absorption. The epithelial cells on the villi are equipped with various transport mechanisms to facilitate the uptake of nutrients. The final breakdown of disaccharides (like sucrose, lactose, and maltose) into monosaccharides (glucose, fructose, and galactose) occurs at the brush border.

These monosaccharides are then absorbed by the epithelial cells and enter the bloodstream. Proteins, broken down into small peptides and amino acids by pancreatic enzymes in the duodenum, are further digested by peptidases at the brush border. Amino acids and dipeptides are absorbed by the epithelial cells and transported into the bloodstream. Fats, emulsified by bile in the duodenum and broken down by pancreatic lipase into fatty acids and monoglycerides, form micelles in the jejunum. These micelles facilitate the absorption of fats across the epithelial cell membrane. Inside the cells, fats are reassembled into triglycerides and packaged into chylomicrons before entering the lymphatic system. Jejunum absorbs most water-soluble vitamins and many minerals. Fat-soluble vitamins (A, D, E, K) are absorbed along with dietary fats. Water and electrolytes (sodium, potassium, chloride, bicarbonate) are absorbed through osmosis and active transport mechanisms, maintaining fluid and electrolyte balance.

16.3.7.4 Digestive Process in Ileum

The ileum is the final and longest segment of the small intestine. Its primary function is to complete the absorption of nutrients and other substances that begin in the duodenum and jejunum and to prepare the remaining undigested food for entry into the large intestine. The ileum has a highly folded mucosal surface lined with villi and microvilli, which increases the surface area for absorption. The ileum continues to absorb nutrients not fully absorbed in the jejunum, particularly certain vitamins, minerals, and remaining carbohydrates and amino acids. It is specifically essential for the absorption of vitamin B12 and bile salts. Vitamin B12, bound to intrinsic factor (secreted by the stomach), is absorbed explicitly in the ileum. Fatty acids and fat-soluble vitamins are absorbed and converted into chylomicrons within the enterocytes and enter the lymphatic system via lacteals in the villi. The ileum absorbs significant amounts of water and electrolytes, maintaining fluid and electrolyte balance in the body.

16.3.8 Large Intestine

While the earlier sections of the gastrointestinal tract are primarily involved in the digestion and absorption of nutrients, the colon's functions are centered around water and electrolyte absorption, fermentation of indigestible food matter, and formation and excretion of feces.

16.3.8.1 Microbiota

The bacterial microbiota in the colon, an extraordinarily diverse and dynamic population of microorganisms, is an ecosystem predominantly composed of bacteria (Firmicutes, Bacteroidetes, Actinobacteria, and Proteobacteria), along with archaea, fungi, and viruses. It is crucial for various physiological processes, including digestion, immune function, and mood regulation. The bacterial density in the colon is estimated to range from 10^{11} to 10^{12} cells per gram of intestinal content. Colonic bacteria ferment non-digestible carbohydrates, such as dietary fibers, producing short-chain fatty acids (SCFAs) like acetate, propionate, and butyrate. These SCFAs are critical energy sources for colonocytes (colon cells) and have systemic anti-inflammatory effects. The microbiota synthesizes certain vitamins, particularly vitamin K and some B vitamins, including folic acid and biotin. The microbiota plays a key role in the development and function of the immune system, helping to protect against pathogenic bacteria and modulating systemic immune responses. Emerging research suggests a significant interaction between the gut microbiota and the central nervous system, known as the gut-brain axis, impacting mood and behavior. As the chyme enters the colon, it is relatively liquid. The colon absorbs a significant amount of water from this chyme, gradually turning it into semi-solid feces. Water absorption in the colon occurs primarily through osmosis. The absorption of solutes, particularly sodium, and nutrients, creates an osmotic gradient that drives water absorption across the colonic mucosa. The colon is estimated to absorb about 1.5–2 l of water per day, depending on the volume and composition of the intestinal contents.

16.3.8.2 Feces Formation

Feces comprise water, indigestible food matter (such as fiber), sloughed-off intestinal cells, bacteria (both live and dead), and substances released from the liver, including bile pigments. The ascending and transverse sections of the colon primarily function in the absorption process and gradual formation of feces. The descending and sigmoid portions of the colon act as storage areas for the formed feces. Slow segmenting movements, known as haustral contractions, help mix fecal matter and its gradual propulsion toward the rectum. These contractions facilitate further water absorption and compaction of feces. Mass movements are strong peristaltic waves that occur a few times a day (often after meals) and move the fecal matter from the colon into the rectum. The rectum serves as a temporary storage site. As feces accumulate in the rectum, the walls stretch, triggering the defecation reflex. The odor of feces, often intense and unpleasant, is mainly due to the byproducts of bacterial fermentation and breakdown of food substances in the large intestine, including hydrogen sulfide (H_2S); methanethiol; dimethyl sulfide; mercaptans; indole; skatole; short-chain fatty acids such as acetate, propionate, and butyrate; amines such as putrescine and cadaverine; and phenols.

16.3.8.3 Motility in Large Intestine

Unlike the rapid mixing movements in the small intestine, colonic motility is generally slower and more segmented, optimized for its unique functions. Haustral contractions (Fig. 16.48) are distinctive to the colon and crucial for the absorption of water and electrolytes and for the gradual movement and processing of fecal matter. Unlike peristalsis, which is a propulsive movement, haustral contractions are segmental. This means they occur in one section (or haustrum) of the colon at a time without necessarily pushing the contents forward. These contractions facilitate thoroughly mixing the colonic contents with intestinal secretions, aiding in water absorption and electrolytes. Mass movements in the colon involve long segments characterized by a wave of strong muscular contraction that propels fecal matter over a considerable distance within the colon (Fig. 16.49). These movements typically occur a few times daily, often in response to certain stimuli. Mass movements

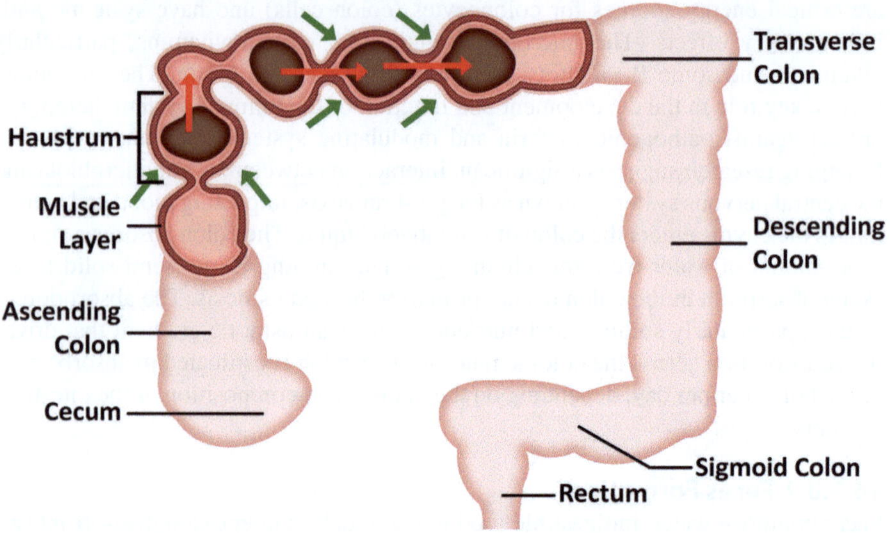

Fig. 16.48 Haustral contraction in the colon

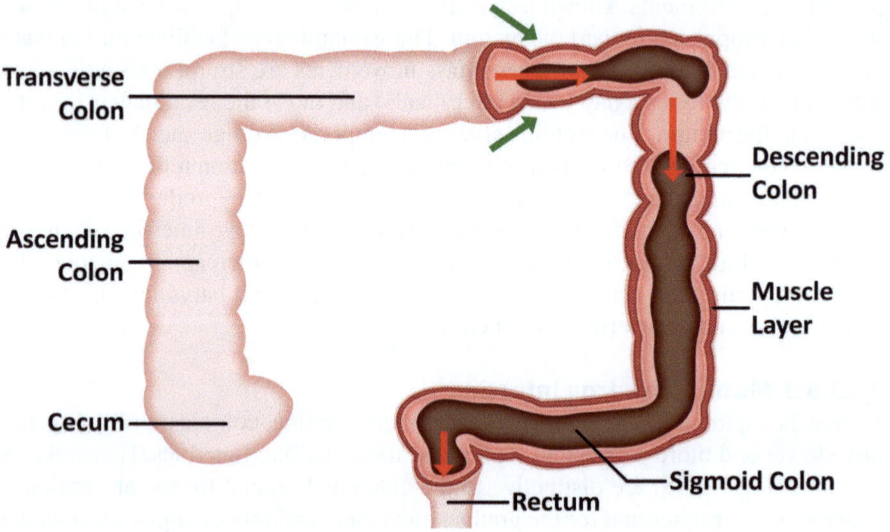

Fig. 16.49 Mass movement in the colon

are often triggered by the gastrocolic reflex, which is activated by food in the stomach or the duodenum. They are more likely to occur after meals, especially following breakfast.

16.3.8.4 Defecation

Defecation is the final step in the human digestive system, involving expulsing feces from the body. As the colon completes the absorption of water and electrolytes, fecal matter accumulates and moves into the rectum, causing its walls to stretch. The defecation reflex is initiated when fecal matter accumulates in the rectum, causing it to stretch. Neural signals from the activated stretch receptors are sent to the sacral spinal cord. From the spinal cord, motor nerves carry signals back to the colon and rectum, leading to a series of involuntary actions:

- The rectal muscles contract, increasing the pressure inside the rectum.
- The internal anal sphincter, which is an involuntary smooth muscle, relaxes automatically.

Unlike the internal sphincter, the external anal sphincter is under voluntary control. It is innervated by somatic nerves that allow conscious control over its relaxation. When a person decides to defecate, the brain sends signals to relax the external anal sphincter. Additionally, voluntary contractions of the abdominal and pelvic floor muscles increase intra-abdominal pressure, aiding in the expulsion of feces. Gravity and body position (such as squatting) can facilitate defecation.

16.4 Part 4: Absorption Physiology

Absorption is the process by which nutrients and other substances from our food are taken into the body. This process is essential for life, as it allows the body to obtain the necessary components for energy, growth, and cellular repair from diet. While some absorption occurs in the stomach and colon, most occur in the small intestine, particularly in the duodenum and jejunum. The large surface area of intestinal villi and microvilli facilitates efficient nutrient uptake.

Absorption involves several mechanisms, including passive diffusion, facilitated diffusion, active transport, and endocytosis, allowing for the uptake of various substances, ranging from simple ions to complex macromolecules.

16.4.1 Carbohydrates

Carbohydrates are predominantly consumed as starches, sugars, and fibers. The digestion of carbohydrates transforms complex carbohydrates into simple sugars that can be absorbed and utilized by the body. Carbohydrate digestion begins in the mouth, where salivary amylase breaks down starch into maltose and dextrin. In the small intestine, pancreatic amylase hydrolyzes starch into maltose and additional dextrins. Brush border enzymes, including maltase, sucrase, and lactase, located on the microvilli of enterocytes, break down disaccharides into monosaccharides (glucose, fructose, and galactose).

The small intestine, particularly the duodenum and jejunum, is the primary site for monosaccharide absorption. The intestinal epithelium's microvilli provide a large surface area, facilitating efficient absorption. The absorptive cells of the small intestine, known as enterocytes, line the mucosal surface and are directly involved in the uptake of monosaccharides.

Glucose and Galactose Absorption (Fig. 16.50) Glucose and galactose are absorbed via active transport through the sodium-glucose linked transporter 1 (SGLT1), located on the apical membrane (luminal side) of enterocytes, particularly in duodenum and jejunum. SGLT1 is primarily responsible for the secondary active transport of glucose and galactose into enterocytes. It does this by coupling the transport of these monosaccharides with sodium ions, utilizing the electrochemical gradient of sodium across the cell membrane. The transport process is driven by the sodium-potassium ATPase pump on the basolateral membrane of enterocytes, which maintains a low intracellular sodium concentration by pumping sodium out of the cell. This creates a favorable gradient for sodium to enter the cell via SGLT1,

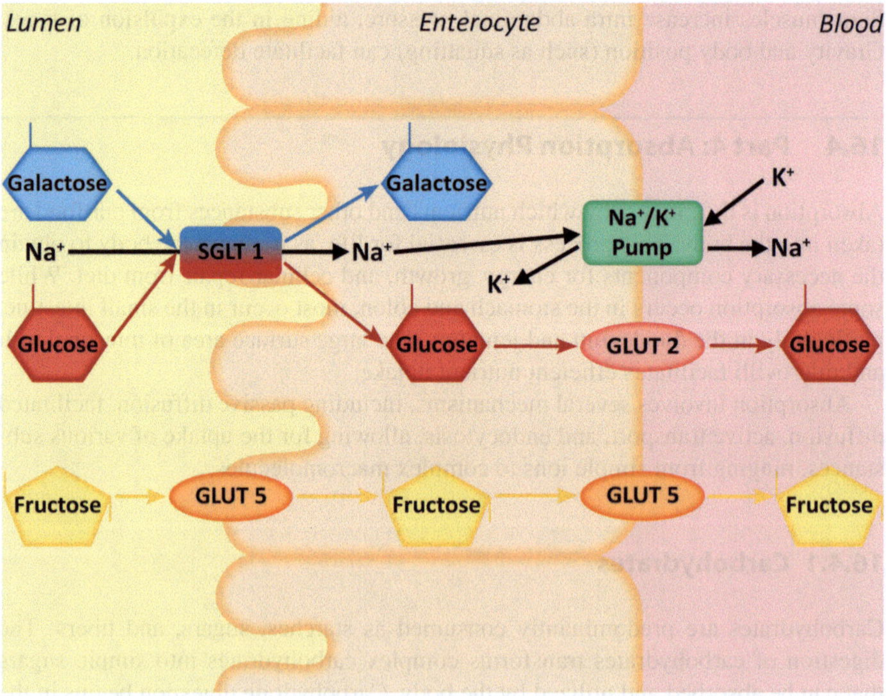

Fig. 16.50 Absorption of carbohydrates in the intestinal epithelial cells

bringing glucose or galactose molecules along with it against their concentration gradient.

Fructose Absorption (Fig. 16.50) Fructose is absorbed primarily by facilitated diffusion through the glucose transporter 5 (GLUT5), situated on the apical membrane of enterocytes in the small intestine, particularly in the jejunum. GLUT5 facilitates fructose uptake from the intestinal lumen into the enterocytes by facilitated diffusion along the fructose concentration gradient.

Transport into Bloodstream After absorption, monosaccharides are transported from the enterocytes into the bloodstream via the GLUT2 transporter on the basolateral membrane (Fig. 16.66). Sugars are transported across the basolateral membrane via GLUT2 by facilitated diffusion, relying on the concentration gradient of the monosaccharides between the enterocytes and the blood.

16.4.2 Proteins

The protein molecules are broken down into constituent amino acids and small peptides to be absorbed in the small intestine. Protein digestion begins in the stomach, where the acidic environment denatures proteins, unfolding them for easier enzymatic access. Pepsin, an enzyme secreted in the stomach, initiates the breakdown of proteins into smaller polypeptides. In the small intestine, pancreatic enzymes such as trypsin, chymotrypsin, and carboxypeptidase further break down polypeptides into shorter peptides and free amino acids.

Amino acids (Fig. 16.51) are absorbed across the enterocyte membrane via secondary active transport in the small intestine, particularly the jejunum and ileum. This process involves various amino acid transporters, each specific for certain types of amino acids (neutral, acidic, or basic). The transport is often coupled with sodium, utilizing the sodium gradient established by the Na^+/K^+-ATPase pumps to facilitate amino acid uptake into the cells.

Types of amino acid transporters include:

- System A (e.g., SLC38 family): Transports small, neutral amino acids
- System ASC (e.g., ASCT1): Transports neutral amino acids, including alanine, serine, and cysteine
- System B0$^+$ (e.g., SLC6A19): Transports neutral and basic amino acids
- System X-AG (e.g., SLC1A1): Transports acidic amino acids

Small peptides (Fig. 16.51) are transported into enterocytes via a different mechanism, primarily through the peptide transporter PEPT1. PEPT1 is predominantly expressed on the apical membrane of the enterocytes in the small intestine, particularly in the duodenum and jejunum. It specializes in transporting dipeptides and tripeptides, which are generated by the enzymatic breakdown of dietary proteins. PEPT1 operates via a proton-coupled symport mechanism. It transports peptides

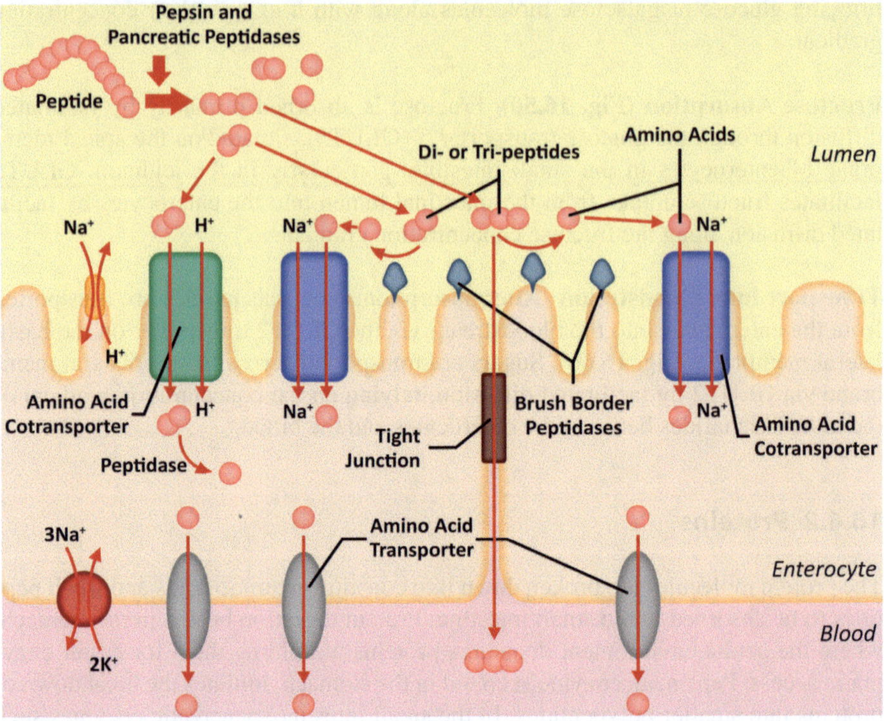

Fig. 16.51 Absorption of amino acids and short peptides in the intestinal epithelial cells

along with hydrogen ions (protons) into the enterocytes. This mechanism utilizes the proton gradient across the enterocyte membrane, effectively allowing the transport of peptides against their concentration gradient. Once inside the cell, these peptides are further broken down into free amino acids by intracellular peptidases.

Once inside the enterocytes, amino acids are transported to the opposite side of the cell (the basolateral side) to be released into the blood. The basolateral membrane contains various types of amino acid transporters, each specific for different groups of amino acids (e.g., neutral, basic, and acidic amino acids). These transporters work to move amino acids from the intracellular environment of the enterocytes into the interstitial fluid surrounding the intestinal cells. Once in the interstitial fluid, amino acids quickly diffuse into the capillaries due to the concentration gradient. The capillaries in the intestinal villi are part of the hepatic portal circulation, which transports absorbed nutrients directly to the liver.

16.4.3 Lipids

Before absorption, the lipids (fats) must be digested into their building blocks (diglycerides and free fatty acids). Lipid digestion begins in the stomach with mechanical churning and the action of gastric lipase, which starts the breakdown of triglycerides into diglycerides and free fatty acids. In the duodenum, bile salts secreted by the liver and stored in the gallbladder emulsify fats, breaking them into smaller droplets. This increases the surface area for enzymatic action.

Pancreatic lipase, aided by a cofactor called colipase, further breaks down triglycerides into monoglycerides and free fatty acids.

Micelle Formation The products of lipid digestion (monoglycerides and free fatty acids) and cholesterol and fat-soluble vitamins combine with bile salts to form micelles (Fig. 16.47). In a micelle, bile salts and phospholipids arrange themselves into a spherical structure with their hydrophilic (water-loving) ends facing outward and their hydrophobic (water-fearing) tails facing inward. This arrangement creates a hydrophobic core where lipid-soluble substances like monoglycerides and fatty acids are solubilized. Micelles are pretty small, typically around 4–7 nanometers in diameter, which allows them to stay in suspension in the intestinal chyme and easily interact with the brush border of the enterocytes. The primary function of micelles is to solubilize the products of lipid digestion and facilitate the transport of lipids to the surface of enterocytes. Once at the enterocyte surface, the lipid components of micelles are released and absorbed into the cells. After releasing their lipid cargo, bile salts in the micelles are left in the intestinal lumen. These are primarily reabsorbed in the ileum and recycled via enterohepatic circulation.

The small intestine's enterocytes absorb micelles' lipid components through passive diffusion across the brush border membrane. Lipid absorption primarily occurs via simple diffusion. The lipid bilayer of the enterocyte plasma membrane is naturally fluid, allowing the lipid-soluble monoglycerides and fatty acids to dissolve into it and pass through into the cell interior. After monoglycerides and free fatty acids are absorbed into enterocytes, these lipids are transported to the smooth endoplasmic reticulum. Here, they are re-esterified to form triglycerides. The newly formed triglycerides, cholesterol, phospholipids, and fat-soluble vitamins are assembled into chylomicrons. Apolipoproteins, especially ApoB-48, are vital components of the chylomicron surface, providing structural stability and mediating interactions with cell receptors (Fig. 16.52). Due to their high lipid content, chylomicrons are the largest and least dense class of lipoproteins (Fig. 16.53) (Table 16.1). They consist of a core of triglycerides and esterified cholesterol surrounded by a monolayer of phospholipids, free cholesterol, and apolipoproteins. Chylomicrons are released from the enterocytes by exocytosis and enter the lymphatic system through lacteals, bypassing the hepatic portal vein.

The primary site for lipid absorption is the small intestine, particularly the duodenum and jejunum, where the lipid digestion enzymes are most active, and the bile salts are in high concentrations. Hormones like cholecystokinin (CCK) regulate the secretion of bile and pancreatic enzymes, facilitating lipid digestion and absorption.

16.4.4 Nucleic Acids

Nucleic acids, primarily obtained from dietary sources like meat, fish, and plant-based foods, are broken down into constituent nucleotides and nucleosides, which are then absorbed in the small intestine.

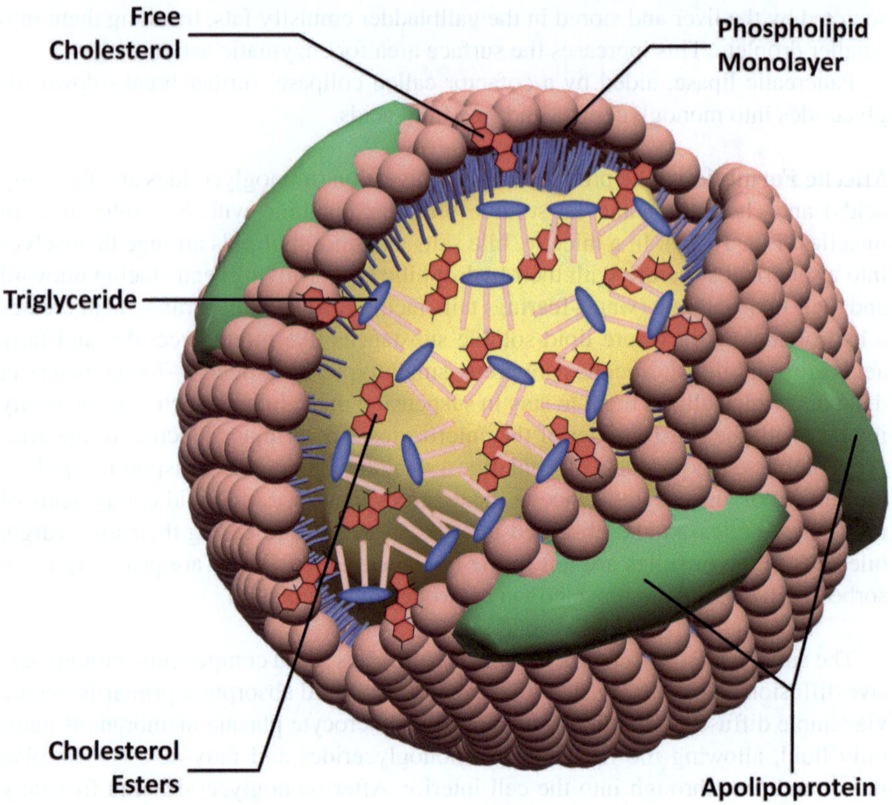

Fig. 16.52 General structure of lipoproteins

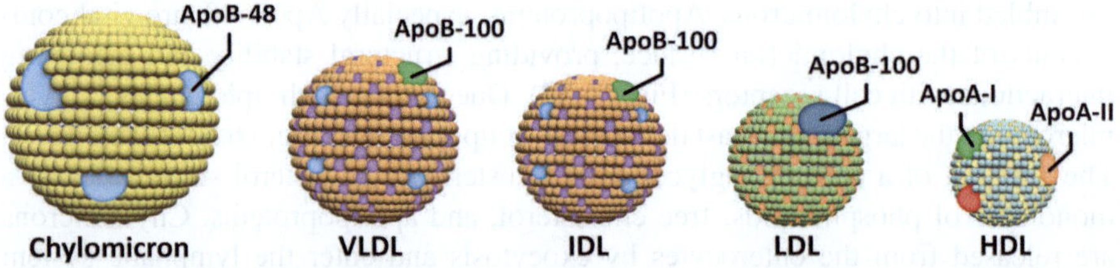

Fig. 16.53 Structure of different lipoproteins

Nucleic acids in food are initially broken down by gastric juices in the stomach and then further digested in the small intestine. Pancreatic enzymes, such as ribonuclease and deoxyribonuclease, cleave nucleic acids into smaller nucleotide chains. In the small intestine, enzymes located on the brush border of enterocytes, such as nucleotidases and phosphatases, further break down nucleotides into nucleosides (a nitrogenous base plus a sugar) and free phosphate ions. Nucleosides are

Table 16.1 Lipoproteins: types, structure, components, and function

Type of lipoprotein	Structure	Components	Function
Chylomicrons	Largest and least dense; composed of a single layer of phospholipids and proteins surrounding a core	Core: triglycerides and a small amount of cholesterol esters. Surface: Apolipoproteins (ApoB-48, ApoA-I, ApoA-II, ApoA-IV, ApoC, and ApoE), phospholipids, and free cholesterol	Transport dietary triglycerides and cholesterol from the intestines to other body parts, primarily the liver and adipose tissue
Very-low-density lipoprotein (VLDL)	Smaller than chylomicrons, with a similar structure	Core: Mainly triglycerides, with a small amount of cholesterol esters. Surface: apolipoproteins (mainly ApoB-100, ApoE, and ApoC), phospholipids, and free cholesterol	Transport endogenously synthesized triglycerides and cholesterol from the liver to various tissues, particularly muscle and adipose tissue
Intermediate-density lipoprotein (IDL)	Formed from the degradation of VLDL, transitional between VLDL and LDL	Core: triglycerides and cholesterol esters. Surface: apolipoproteins (mainly ApoB-100), phospholipids, and free cholesterol	Function as a transitional lipoprotein in the metabolism of VLDL to LDL. The liver takes some up, and others are further processed into LDL
Low-density lipoprotein (LDL)	Smaller and denser than VLDL and IDL, with a higher proportion of cholesterol	Core: mainly cholesterol esters with a small amount of triglycerides. Surface: apolipoproteins (mainly ApoB-100), phospholipids, and free cholesterol	Transport cholesterol from the liver to cells throughout the body, where it is used in cell membranes, hormone synthesis, and other functions
High-density lipoprotein (HDL)	Smallest and most dense; spherical particles	Core: cholesterol esters and triglycerides. Surface: apolipoproteins (mainly ApoA-I and ApoA-II), phospholipids, and free cholesterol	Involved in reverse cholesterol transport, carrying cholesterol away from cells and back to the liver for excretion or reutilization

absorbed into enterocytes via specific nucleoside transporters, which work through facilitated diffusion or active transport mechanisms. These transporters can be selective for purine or pyrimidine nucleosides. The free phosphate ions resulting from nucleotide breakdown are absorbed via sodium-dependent cotransporters in the small intestine. Once inside the enterocytes, nucleosides can be metabolized into nucleotides and incorporated into the nucleotide pool of the body for various biosynthetic processes. Alternatively, they can be catabolized to produce energy or excreted. Absorbed nucleosides and phosphate ions are transported from enterocytes into the bloodstream for distribution to various tissues.

16.4.5 Vitamins

Vitamins are absorbed predominantly in the small intestine, with each vitamin having specific mechanisms and sites of absorption. Based on their solubility, vitamins are classified into two groups: fat-soluble vitamins (A, D, E, K) and water-soluble vitamins (C and the B-complex vitamins).

16.4.5.1 Absorption of Fat-Soluble Vitamins (A, D, E, K)

Fat-soluble vitamins are primarily absorbed in the upper part of the small intestine, especially the duodenum and jejunum. These vitamins are absorbed via the same pathway as dietary fats. They are incorporated into micelles formed from bile salts and fatty acids. After being released from micelles, they diffuse into the small intestine's enterocytes.

Within enterocytes, these vitamins are incorporated into chylomicrons for transport via the lymphatic system to the general circulation.

16.4.5.2 Absorption of Water-Soluble Vitamins

Water-soluble vitamins are absorbed along the entire length of the small intestine, but the jejunum is particularly important.

Vitamin C (Ascorbic Acid): Vitamin C absorption primarily occurs in the small intestine, predominantly in the jejunum. Vitamin C is absorbed by both active transport and passive diffusion mechanisms. The active transport of ascorbic acid is primarily facilitated by the sodium-dependent vitamin C transporters (SVCTs), specifically SVCT1 and SVCT2. At higher concentrations, passive diffusion becomes more significant in the absorption process. After being absorbed through the enterocytes, vitamin C is transported into the bloodstream for systemic distribution. The absorption efficiency of vitamin C decreases as dietary intake increases. The body efficiently absorbs small amounts, but absorption rates diminish as intake exceeds about 1 gram per day. Certain minerals like iron are better absorbed in the presence of vitamin C.

Vitamin B1 (Thiamine): Primarily absorbed in the jejunum through active transport at low concentrations and passive diffusion at higher concentrations.

Vitamin B2 (Riboflavin): Absorbed in the upper part of the small intestine, mainly the duodenum and jejunum, primarily by an active transport mechanism.

Vitamin B3 (Niacin): Absorbed along the entire length of the small intestine by simple diffusion and, to some extent, by an active transport mechanism.

Vitamin B5 (Pantothenic Acid): Widely absorbed in the small intestine by a passive diffusion process.

Vitamin B6 (Pyridoxine): Absorbed in the jejunum by both passive diffusion and an active transport mechanism. **Vitamin B7 (Biotin):** Primarily absorbed in the small intestine via an active transport system.

Vitamin B9 (Folate): Folate is mostly absorbed in the proximal part of the small intestine, particularly the duodenum and jejunum. It is absorbed as methyltetrahydrofolate after being reduced from dietary folates. The absorption process involves active transport through specific folate transporters.

Vitamin B12 (Cobalamin): The freed vitamin B12 initially binds to R proteins (haptocorrins), which are present in saliva and gastric juice. In the alkaline environment of the duodenum, pancreatic enzymes degrade R proteins, releasing vitamin B12 to bind with intrinsic factors secreted by the stomach. The primary site for vitamin B12 absorption is the ileum, the latter part of the small intestine. Vitamin B12 remains bound to intrinsic factors until it reaches the ileum. Here, the intrinsic factor-vitamin B12 complex binds to specific enterocyte receptors, known as cubilin receptors.

Enterocytes take up the intrinsic factor-vitamin B12 complex by receptor-mediated endocytosis. Once inside the enterocytes, vitamin B12 is released from the intrinsic factor and binds to another protein, transcobalamin II. This new complex, transcobalamin II-vitamin B12, enters the bloodstream for transport to various tissues.

16.4.6 Electrolytes

Electrolytes, such as sodium, potassium, chloride, calcium, magnesium, phosphate, and bicarbonate, are absorbed in the small and large intestines through specific transport mechanisms.

16.4.6.1 Sodium (Na^+)

Cotransport Mechanisms in Duodenum and Jejunum

Sodium-Glucose Cotransport (SGLT) (Fig. 16.50) Sodium ions are cotransported with glucose (or galactose) into the enterocytes (intestinal cells), utilizing the sodium gradient as the driving force. SGLT1 is the primary cotransporter.

SGLT is primarily active in the proximal part of the small intestine (duodenum and jejunum). SGLT proteins span the apical membrane of enterocytes and have specific binding sites for sodium and glucose. SGLT proteins bind sodium and glucose simultaneously, enhancing the transporter's affinity for glucose. The process utilizes the electrochemical gradient of sodium, established by the sodium-potassium ATPase pump located on the basolateral membrane of enterocytes.

Sodium-Amino Acid Cotransport (Fig. 16.51) Like glucose, various amino acids are cotransported with sodium into the enterocytes. It facilitates the absorption of amino acids from the intestinal lumen into enterocytes using the sodium gradient. This cotransport predominantly occurs in the proximal segments of the small intestine (duodenum and jejunum). Multiple cotransporters are involved, each specific to different groups of amino acids (e.g., neutral, acidic, basic). These cotransporters are integral membrane proteins on enterocytes' apical (luminal) side. Sodium and amino acids bind to their respective sites on the cotransporter. The binding of sodium increases the transporter's affinity for amino acids. The sodium electrochemical gradient established by the sodium-potassium ATPase pump on the basolateral membrane provides the energy to transport amino acids against their concentration gradient.

Electroneutral Sodium Absorption in Ileum

Sodium-Hydrogen Exchanger (NHE3) The sodium-hydrogen exchanger (NHE) involves the exchange of extracellular sodium ions (Na^+) for intracellular hydrogen ions (H^+), helping to maintain acid-base balance and fluid-electrolyte homeostasis. They help remove excess intracellular H^+, thereby regulating intracellular pH while absorbing Na^+ from the lumen. NHEs are integral membrane proteins primarily located on epithelial cells' apical (luminal) side. NHEs are found throughout the GI tract, including the stomach, small intestine, and colon. NHEs transport Na^+ into the cell in exchange for H^+, which is expelled into the intestinal lumen or gastric juice. The process is not directly ATP-dependent but relies on the electrochemical gradient of Na^+, maintained by the basolateral Na^+/K^+-ATPase pump. There are several isoforms of NHE (e.g., NHE1 to NHE5). NHE1 is primarily involved in the regulation of intracellular pH and cell volume. NHE2 and NHE3 are essential in Na^+ and fluid absorption in the intestine. NHE3, in particular, is a significant player in the ileum and colon.

Absorption via Paracellular Pathway Paracellular transport involves the passage of sodium across the intestinal barrier through the intercellular spaces without entering the cells. Electrochemical and osmotic gradients drive it. Tight junctions between adjacent enterocytes regulate the paracellular pathway, controlling the passage of ions and molecules based on size and charge. Sodium concentration is typically higher in the intestinal lumen than in the intercellular space, creating a concentration gradient that favors the passive diffusion of sodium through tight junctions. The electrical potential difference across the epithelium also influences sodium movement. As water is absorbed transcellularly, it creates an osmotic gradient that facilitates the movement of sodium paracellularly.

Fine-Tuning Sodium Absorption in Colon

Electroneutral Sodium Absorption Electroneutral sodium absorption in the colon is essential for regulating sodium balance and water absorption in the body. It involves sodium-hydrogen exchange (NHE) and sodium-potassium exchange. NHEs are located on colonic epithelial cells' apical (luminal) membrane. NHEs transport sodium into the cell in exchange for hydrogen ions, which are expelled into the colonic lumen. This process is driven by the electrochemical gradient of sodium, maintained by the basolateral Na^+/K^+-ATPase pump. Sodium-potassium exchange involves potassium channels on the apical membrane, bringing sodium into the epithelial cells in exchange for potassium, which is secreted into the lumen.

Aldosterone Role in Sodium Absorption

Aldosterone enhances sodium reabsorption (and potassium secretion) in the colon. In the colon, aldosterone acts explicitly on the distal colon and rectum epithelial cells. It binds to mineralocorticoid receptors (MR) in the cytoplasm of colonic epithelial cells. The aldosterone-receptor complex then translocates to the nucleus, binding to specific response elements on the DNA and initiating the transcription of particular genes. The genes activated by aldosterone lead to the synthesis of various proteins, including the epithelial sodium channel (ENaC), sodium-potassium ATPase (Na^+/K^+-ATPase), and serum and glucocorticoid-regulated kinase (SGK1). It increases the number and activity of ENaC channels on the apical (luminal) membrane of the colonic epithelial cells, enhancing the entry of sodium from the colonic lumen into the cells. Aldosterone also stimulates the activity of Na^+/K^+-ATPase on the basolateral membrane, which pumps sodium out of the cell into the bloodstream in exchange for potassium.

16.4.6.2 Potassium (K^+)

Unlike sodium, potassium absorption is less tightly regulated and occurs along the entire length of the GI tract, including the small intestine and the colon. Most potassium absorption occurs passively, driven by concentration gradients and electrochemical gradients. Paracellular transport is a significant route, especially in the small intestine.

Duodenum and Jejunum The duodenum and jejunum, with their high luminal potassium concentration post-meal, facilitate the passive diffusion of potassium into the blood. During periods of high fluid absorption, potassium can be carried transcellularly or paracellularly via solvent drag.

Ileum Potassium is absorbed through active transport mechanisms, although these are less prominent than passive processes. The luminal flow rate can influence the ileum's absorption rate, with increased absorption at lower flow rates.

Colon The colon can both absorb and secrete potassium. The balance between these processes depends on dietary intake and body needs. Active potassium absorption in the colon is regulated by various factors, including aldosterone, which can enhance potassium secretion. Potassium secretion occurs in response to high blood potassium concentrations or high aldosterone levels, serving as a mechanism to prevent hyperkalemia.

16.4.6.3 Chloride (Cl^-)

Chloride absorption in the gastrointestinal (GI) tract is essential for maintaining electrolyte balance and contributing to gastric acid production, fluid balance, and overall acid-base homeostasis. Chloride is absorbed throughout the GI tract but predominantly in the small intestine and colon. Chloride often follows the electrochemical gradient established by sodium absorption.

Duodenum and Jejunum Chloride is absorbed alongside sodium, with water following osmotically. Chloride is also absorbed through various cotransporters, such as the sodium-chloride cotransporter (NCC) and chloride-bicarbonate exchangers.

Ileum The chloride-bicarbonate exchanger allows for the exchange of chloride ions with bicarbonate, aiding in acid-base balance. Direct chloride channels facilitate chloride absorption, especially in high luminal chloride concentration conditions.

Colon Chloride absorption in the colon occurs primarily through electroneutral mechanisms, often coupled with sodium absorption. Aldosterone enhances sodium reabsorption, which indirectly promotes chloride absorption. Like the small intestine, chloride-bicarbonate exchangers in the colon help maintain acid-base balance.

16.4.6.4 Calcium (Ca^{2+})

Calcium absorption in the gastrointestinal (GI) tract is a highly regulated process. It is essential for maintaining the body's calcium homeostasis and crucial for various physiological functions, including bone health, muscle function, and nerve signaling. This process predominantly occurs in the small intestine.

Duodenum and Jejunum Calcium absorption in the duodenum and jejunum is an active and vitamin D-dependent mechanism. Calcitriol, the active form of vitamin D, enhances calcium absorption by increasing the synthesis of calcium-binding proteins in enterocytes. Calcium enters the enterocytes through the apical membrane via the channels of TRPV6 (Transient Receptor Potential Vanilloid 6). Inside the enterocyte, calbindin (a calcium-binding protein) transports calcium across the cell, minimizing intracellular calcium concentration near the apical surface and maintaining the gradient for continued calcium entry. Calcium is extruded into the bloodstream via calcium-ATPase pumps (PMCA1b) and sodium-calcium exchangers (NCX1) at the basolateral membrane.

Ileum and Beyond When dietary calcium is high, absorption occurs passively via paracellular pathways. This process depends on the concentration gradient and is less regulated than active transport.

Regulatory Mechanisms Calcitriol (active vitamin D) is the primary hormone regulating calcium absorption. Parathyroid hormone (PTH) stimulates the production of calcitriol in response to low blood calcium levels. Dietary calcium, phosphate, and fat can influence absorption. For instance, high levels of dietary fat can form insoluble calcium soaps, reducing absorption.

16.4.6.5 Magnesium (Mg^{2+})

The body's magnesium balance is essential for numerous physiological functions, including enzymatic reactions, nerve function, and muscle contraction. Magnesium absorption occurs primarily in the small intestine and, to a lesser extent, in the large intestine. Active transport is the predominant absorption mechanism in the ileum under low or normal dietary magnesium conditions. Magnesium enters enterocytes primarily through TRPM6 (transient receptor potential melastatin 6) channels, a process upregulated by active vitamin D (calcitriol). Once inside the cell, magnesium binds to intracellular proteins or gets buffered in cellular organelles. Magnesium is then extruded into the bloodstream through various transporters, including magnesium-ATPases and sodium-magnesium exchangers. Passive diffusion occurs along the entire length of the intestine, more prominent when dietary magnesium is high.

16.4.6.6 Phosphate (PO_4^{3-})

Phosphate absorption in the GI tract is essential for various physiological functions such as bone mineralization, energy storage and transfer, and the formation of cell membranes.

Duodenum and Jejunum Phosphate is actively absorbed via sodium-phosphate cotransporters (primarily Npt2b) in the brush border membrane of enterocytes. Calcitriol, the active form of vitamin D, enhances phosphate absorption by upregulating the expression of sodium-phosphate cotransporters. Once inside the enterocytes, phosphate is transported across the cell and released into the bloodstream through the basolateral membrane.

Ileum The ileum also participates in phosphate absorption, though to a lesser extent than the duodenum and jejunum.

Colon The colon can absorb phosphate passively, particularly in high luminal phosphate concentrations.

16.4.6.7 Bicarbonate (HCO_3^-)

Bicarbonate absorption in the GI tract is essential for maintaining the body's acid-base balance. It is absorbed and secreted at various points in the GI tract, mainly in the small intestine and colon.

Bicarbonate transport is often coupled with sodium and chloride transport and is influenced by carbon dioxide levels. The pancreas secretes a high bicarbonate concentration into the duodenum to neutralize stomach acid. As bicarbonate neutralizes gastric acid, it's converted to carbon dioxide and water, which are absorbed and then reconverted to bicarbonate in the blood.

Duodenum and Jejunum Chloride-bicarbonate antiporter system exchanges intraluminal chloride for bicarbonate, facilitating bicarbonate absorption into the bloodstream. Carbon dioxide produced by neutralizing gastric acid diffuses into enterocytes and is hydrated to form bicarbonate.

Ileum Bicarbonate is absorbed along with sodium, which helps maintain an electrochemical gradient.

Colon Similar to the duodenum and jejunum, the colon absorbs bicarbonate in exchange for chloride. Colonic absorption of bicarbonate is crucial in maintaining systemic acid-base balance, especially in conditions like chronic renal failure.

16.4.7 Water

Water absorption in the human digestive system is vital for maintaining the body's fluid balance. It occurs predominantly in the small and large intestines through osmosis. Most water absorption occurs in the small intestine, particularly in the jejunum and ileum. The extensive surface area the intestinal villi and microvilli provide facilitates efficient water uptake. The colon also plays a significant role in water absorption. As the remaining digestive contents enter the colon, additional water is absorbed, contributing to the formation of solid feces.

Water absorption primarily occurs through osmosis, driven by the osmotic gradient established by the absorption of solutes such as electrolytes (sodium, potassium, chloride) and nutrients (like glucose and amino acids). As these solutes are actively absorbed from the intestinal lumen into the enterocytes, water follows passively to maintain osmotic equilibrium. Water can also be absorbed alongside solutes through a process known as solvent drag, where water is "dragged" along with the absorption of solutes, particularly electrolytes.

The amount of water and the osmolarity of foods and fluids ingested can influence the rate and volume of water absorption. The rate at which contents move through the digestive tract can affect water absorption. Rapid transit (as seen in diarrhea) can decrease water absorption, leading to fluid loss.

16.5 Part 5: Digestive System Hormones

The digestive tract secretes various hormones that play crucial roles in regulating multiple aspects of digestion, from gastric motility to the secretion of digestive enzymes and bile.

16.5.1 Gastrin

Gastrin is a peptide hormone. The primary active forms are gastrin-17 (G-17) and gastrin-34 (G-34), with numbers indicating the number of amino acids. G-17 is the most abundant form in the stomach. Gastrin is synthesized and secreted by G cells, primarily located in the antral region of the stomach. It is initially synthesized as preprogastrin, cleaved to progastrin, and finally to its active forms. Gastrin stimulates parietal cells in the stomach to secrete hydrochloric acid (HCl). The gastrin/CCK-B receptor achieves this by increasing intracellular calcium levels and activating the H^+/K^+ ATPase pump. Gastrin has a trophic effect on the gastric mucosa, promoting the growth of gastric epithelium and mucosal cells. It also enhances gastric motility and the mixing of stomach contents.

Gastrin release is stimulated by peptides and amino acids in the stomach, distention of the stomach, and vagal stimulation, mediated by the neurotransmitter gastrin-releasing peptide (GRP). Gastrin secretion is inhibited by low pH in the stomach (<2.5) (providing a negative feedback mechanism) and the hormone somatostatin (secreted by D cells in the stomach in response to high gastric acid levels).

16.5.2 Cholecystokinin (CCK)

CCK is a peptide hormone with several isoforms varying in amino acid chain length. The most biologically active forms are CCK-33, CCK-39, and, particularly, CCK-8 (the numbers denote the number of amino acids). CCK is synthesized and secreted by I cells (Page 639) located mainly in the mucosal lining of the duodenum and jejunum. It is initially synthesized as preprocholecystokinin and then processed to its active forms.

CCK triggers the contraction of the gallbladder, leading to the release of bile into the duodenum, aiding in fat digestion. It also stimulates the exocrine pancreas to secrete digestive enzymes, which are crucial for breaking down proteins, fats, and carbohydrates. Besides, CCK slows gastric emptying, allowing more time for digestive processes in the stomach and better mixing with bile and pancreatic juices in the duodenum. It has a role in appetite suppression, signaling satiety to the brain. In the small intestine, CCK release is stimulated by partially digested proteins (peptides) and fats (fatty acids). Negative feedback mechanisms, including the effects of bile acids and pancreatic enzymes on the small intestine, can inhibit further CCK release.

16.5.3 Secretin

Secretin is a peptide hormone consisting of 27 amino acids in its mature form. It has structural similarities to glucagon and glucagon-like peptides. Secretin is synthesized and secreted by S cells located predominantly in the mucosal lining of the duodenum. The primary stimulus for secretin release is the presence of acidic gastric contents in the duodenum.

Secretin stimulates the pancreas to secrete a bicarbonate-rich fluid, which helps neutralize the acidic chyme entering the duodenum from the stomach. It also inhibits gastric acid secretion from the stomach, providing a negative feedback mechanism. Secretin mildly stimulates bile secretion from the liver. Secretin may also have a trophic effect on the exocrine pancreas.

16.5.4 Ghrelin

Ghrelin is a peptide hormone composed of 28 amino acids. It is unique among peptide hormones because it requires posttranslational modification, specifically octanoylation (attachment of an octanoic acid molecule), for its biological activity.

This modification occurs at the serine residue in the third position of the peptide. Ghrelin is predominantly produced in the fundus region of the stomach. It is also synthesized in the small intestine, pancreas, and hypothalamus. Ghrelin is often termed the "hunger hormone" as it stimulates appetite. It acts on the hypothalamus to increase food intake. It plays a role in energy homeostasis by promoting the storage of fat. Ghrelin enhances gastrointestinal motility and gastric emptying. It stimulates the release of growth hormone from the pituitary gland. Ghrelin levels increase with fasting and decrease after meals, which is closely linked to the nutritional and energy status of the body. Ghrelin secretion is influenced by circadian rhythms, typically peaking before meals.

16.5.5 Motilin

Motilin is a peptide hormone consisting of 22 amino acids. The N-terminal portion of the motilin molecule is crucial for its biological activity. Motilin is synthesized and secreted by endocrine M cells, predominantly in the duodenum and jejunum. Its secretion follows a cyclical pattern, primarily associated with the fasting state.

The primary function of motilin is to increase GI motility. It initiates the migrating motor complex (MMC), a pattern of electromechanical activity observed in the GI tract during fasting. Motilin can enhance gastric emptying by increasing the amplitude of antral contractions. It also stimulates intestinal motility, aiding in the clearance of the intestines between meals. Motilin levels peak during fasting, particularly between digestive phases, to initiate the MMC. The presence of fats and acidic chyme in the small intestine can inhibit motilin secretion.

The MMC is primarily responsible for sweeping undigested residue through the digestive system, predominantly during periods of fasting. Phases of the MMC include:

Phase I: This phase is the period of inactivity, with minimal contractile activity. It lasts for about 40–60% of the total MMC cycle.

Phase II: This phase is characterized by intermittent contractions, which increase in frequency and intensity as the phase progresses. Phase II accounts for about 20–30% of the MMC cycle.

Phase III: Known as the "housekeeper wave," this phase is marked by a short period (about 5–15 min) of intense, regular contractions. These contractions sweep through the stomach and small intestine, moving undigested substances and debris forward. It constitutes about 5–10% of the MMC cycle.

Phase IV: This phase is a short transition period between Phase III's intense contractions and Phase I's quiescence.

The MMC helps prevent bacterial overgrowth in the small intestine by moving indigestible substances and secretions. It also helps to clear the stomach and intestines in preparation for the next meal. Regular MMC cycles are essential for maintaining overall gut health and function.

16.5.6 Somatostatin

Somatostatin is a peptide hormone. Somatostatin-14 is the predominant form in the central nervous system, while somatostatin-28 is more common in the gastrointestinal tract. Somatostatin is produced in various locations, including the hypothalamus, pancreatic delta cells, gastrointestinal mucosa, and adrenal glands. In addition to its endocrine actions, somatostatin acts in multiple tissues in a paracrine (local) and autocrine (self-regulating) manner. Somatostatin decreases gastric secretion and gastric emptying and reduces the motility of the stomach, small intestine, and gallbladder.

Somatostatin inhibits the secretion of insulin, glucagon, and pancreatic polypeptide. It inhibits the release of many other hormones, such as gastrin, cholecystokinin (CCK), and secretin. Nutrient levels (especially high glucose and amino acids), hormones, and neural signals can stimulate somatostatin secretion. Somatostatin secretion is regulated by feedback mechanisms, often inhibiting its release.

16.5.7 Glucagon-Like Peptide-1

Glucagon-like peptide-1 (GLP-1) is a peptide hormone that belongs to the glucagon-secretin family. It is derived from the proglucagon molecule. GLP-1 is primarily produced and secreted by the L cells in the ileum and colon of the small intestine. GLP-1 enhances insulin secretion from the pancreas in a glucose-dependent manner, stimulating insulin secretion only when blood glucose levels are high. It suppresses glucagon secretion, which helps in reducing hepatic glucose production. GLP-1 slows gastric emptying, which aids in regulating postprandial glucose levels. It acts on the brain to promote satiety and reduce food intake, contributing to weight loss in some individuals. Recent research suggests potential cardioprotective and neuroprotective roles for GLP-1. The presence of nutrients in the lumen of the small intestine, especially the ileum and colon, is the primary stimulus for GLP-1 secretion. Neural reflexes and various hormones also modulate the secretion of GLP-1.

16.5.8 Peptide YY

Peptide YY (PYY) is a 36-amino acid peptide, part of the pancreatic polypeptide (PP) family, which also includes neuropeptide Y (NPY) and pancreatic polypeptide (PP). PYY is produced and secreted by the L cells in the ileum, colon, and rectum. PYY(3-36) reduces appetite and food intake by acting on the hypothalamus, inhibiting neuropeptide Y, a potent appetite stimulant. It slows gastric emptying and intestinal motility, which contributes to the sensation of fullness and prolongs nutrient absorption time. PYY can inhibit pancreatic secretion, contributing to its role in digestive regulation. The primary stimulator of PYY secretion is the presence of nutrients, particularly fats and carbohydrates, in the lumen of the ileum and colon. PYY levels are influenced by circadian rhythms, showing fluctuations throughout the day.

16.5.9 Gastroinhibitory Peptide

Gastroinhibitory peptide (GIP) is a single-chain 42-amino acid peptide hormone. GIP is produced and secreted by K cells located in the duodenum and proximal jejunum of the small intestine. The primary stimuli for GIP secretion are fats, carbohydrates, and proteins in the chyme entering the small intestine. GIP decreases gastric acid secretion, slowing down gastric emptying and contributing to the inhibition of gastric motility. As an incretin hormone, GIP plays a critical role in glucose homeostasis by stimulating insulin secretion from the beta cells of the pancreas in a glucose-dependent manner. It also influences lipid metabolism by promoting lipid storage in adipose tissue.

16.5.10 Pancreatic Polypeptide

Pancreatic polypeptide (PP) is a 36-amino acid peptide hormone. It belongs to the family of peptides, including neuropeptide Y (NPY) and peptide YY (PYY), all sharing a similar structure. PP is produced and secreted by the PP cells, located in the islets of Langerhans in the pancreas, with a higher concentration in the head of the pancreas. PP modulates the exocrine and endocrine functions of the pancreas. It inhibits pancreatic bicarbonate and enzyme secretion. It influences gastrointestinal motility by slowing gastric emptying and reducing intestinal motility, thus regulating the rate of nutrient absorption. PP may regulate appetite, although this function is less understood than other gut hormones like ghrelin and PYY. Secretion of PP is stimulated by the presence of food in the stomach and the proximal small intestine, particularly proteins and fats. Vagal stimulation increases PP secretion, while vagotomy decreases it, indicating neural control over its release.

16.5.11 Vasoactive Intestinal Peptide (VIP)

Vasoactive intestinal peptide (VIP) is a peptide hormone composed of 28 amino acids. It belongs to the secretin-glucagon family of peptides. VIP is produced and secreted by neurons in the central and peripheral nervous system and endocrine cells in the gastrointestinal tract, particularly in the pancreas, small intestine, and salivary glands. Its distribution in the nervous system is widespread, including the brain, spinal cord, and enteric nervous system. One of the primary functions of VIP is inducing potent vasodilation, increasing blood flow to various tissues, including the GI tract. VIP stimulates the secretion of water and electrolytes in the intestines and pancreas. It relaxes smooth muscles in the gastrointestinal tract, airways, and blood vessels. VIP has immunomodulatory effects, influencing the activity of various immune cells.

16.5.12 Enteroglucagon

Enteroglucagon peptides are derived from the cleavage of proglucagon, a larger precursor molecule. The most significant peptides of enteroglucagon include GLP-1 and GLP-2, oxyntomodulin, and glicentin. Enteroglucagon peptides are produced and secreted by L cells, most numerous in the ileum and colon. Their secretion is stimulated by the presence of nutrients, particularly carbohydrates and fats, in the lumen of the small intestine. GLP-1 plays a crucial role in glucose metabolism by enhancing insulin secretion, inhibiting glucagon release, and slowing gastric emptying. GLP-2 is involved in intestinal growth and repair, enhancing intestinal absorptive capacity and reducing intestinal permeability.

Oxyntomodulin and glicentin have less well-defined roles but are believed to contribute to the regulation of glucose metabolism and may affect appetite.

Suggested Reading

Advanced Physiology and Pathophysiology: Essentials for Clinical Practice, 2nd Ed. Nancy Tkacs, Linda Herrmann, Springer Publishing Company, 2024

Anatomy & Physiology, S. Meloni, et al. Independent Publisher, 2023

Anatomy & Physiology: The Unity of Form and Function, 10th Ed. Kenneth S. Saladin, McGraw Hill, 2023

Clinical Nutrition, 2nd Ed. Rebecca J. Stratton, et al. Wiley-Blackwell, 2013

Gray's Anatomy for Students, 5th Ed. Richard L. Drake, et al. Elsevier, 2020

Human Anatomy & Physiology, 11th Ed. Elaine N. Marieb and Katja Hoehn, Pearson, 2022

Introduction to Human Nutrition, 3rd Ed. Hester H. Vorster, et al. Wiley-Blackwell, 2019

Moore's Clinically Oriented Anatomy, 9th Ed. Keith L. Moore, et al. Lippincott Williams & Wilkins, 2020

Netter Atlas of Human Anatomy: Classic Regional Approach, 8th Ed. Frank H. Netter, Elsevier, 2022

Principles of Anatomy and Physiology, 16th Ed. Gerard J. Tortora, Bryan H. Derrickson, Wiley, 2020

Urinary System 17

In this chapter, you will learn:

- Anatomy of the kidneys, including the renal cortex, medulla, and functional organization
- Structural and functional units of the kidney: nephrons and their components
- Structural features of glomeruli that facilitate selective filtration
- Blood supply to the kidneys, emphasizing the roles of renal arteries and veins
- The microscopic structure of nephrons, including renal corpuscles and tubular components
- The process of glomerular filtration and factors affecting filtration rates
- Mechanisms of tubular reabsorption and secretion along different parts of the nephron
- The countercurrent multiplication system's role in urine concentration
- Regulation of blood pH through renal processes, including bicarbonate handling
- Electrolyte balance maintained by kidney functions, especially for sodium, potassium, and calcium
- Juxtaglomerular apparatus and its importance in blood pressure regulation and renin secretion
- Collecting duct's role in final urine concentration, regulated by ADH and aldosterone
- The concept and significance of renal clearance and glomerular filtration rate (GFR) as diagnostic tools
- The importance of aquaporins in water reabsorption in collecting ducts
- Renal response to acid-base imbalances through bicarbonate reabsorption and hydrogen ion secretion

The urinary system, consisting primarily of the kidneys, ureters, bladder, and urethra (Fig. 17.1), is a sophisticated filtration mechanism and waste management. Its primary function is to remove excess water, salts, and metabolic waste products from the bloodstream, thus purifying the blood and maintaining a balance of fluids and electrolytes. The urinary system also plays a critical role in regulating blood pressure, red blood cell production, and the pH balance of the blood.

H. Hosseinzadeh, *Fundamentals of Medicine for Biomedical Engineering*,
https://doi.org/10.1007/978-3-031-74085-5_17

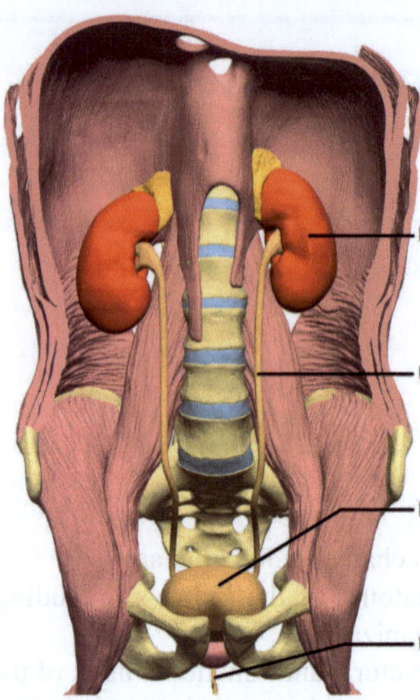

Fig. 17.1 The urinary system

17.1 Part 1: Kidney Anatomy

The kidneys are located on either side of the spine, just below the rib cage, in the retroperitoneal space of the abdominal cavity (Fig. 17.1).

17.1.1 Gross Anatomy

Each kidney, being a bean-shaped organ with a concave medial border known as the hilum, typically weighs about 150 grams and measures approximately 11–12 cm (4.3–4.7 inches) in length, 5–6 cm (2–2.4 inches) in width, and 2–3 cm (0.8–1.2 inches) in thickness. Kidneys are retroperitoneal organs. This positioning provides them with a degree of protection and stability. They are nestled against the posterior wall of the abdomen at about the T12 to L3 vertebrae level. The right kidney is generally positioned slightly lower than the left kidney. Each kidney has a smooth surface, characterized by a convex lateral side and a concave medial side.

The concave medial border features a notch known as the renal hilum (Fig. 17.2). This is the entry and exit point for structures such as the renal artery (which brings blood to the kidney), the renal vein (which carries filtered blood away), and the ureter (which carries urine from the kidney to the bladder).

The kidney consists of two sections: the renal cortex and the medulla. The outer region of the kidney is the renal cortex. It has a granular appearance and is the site where most filtration occurs. The inner area is the renal medulla, which consists of 10–18 renal pyramids. These pyramids appear striated due to parallel arrays of urine-collecting tubules and blood vessels. Between the pyramids are the renal

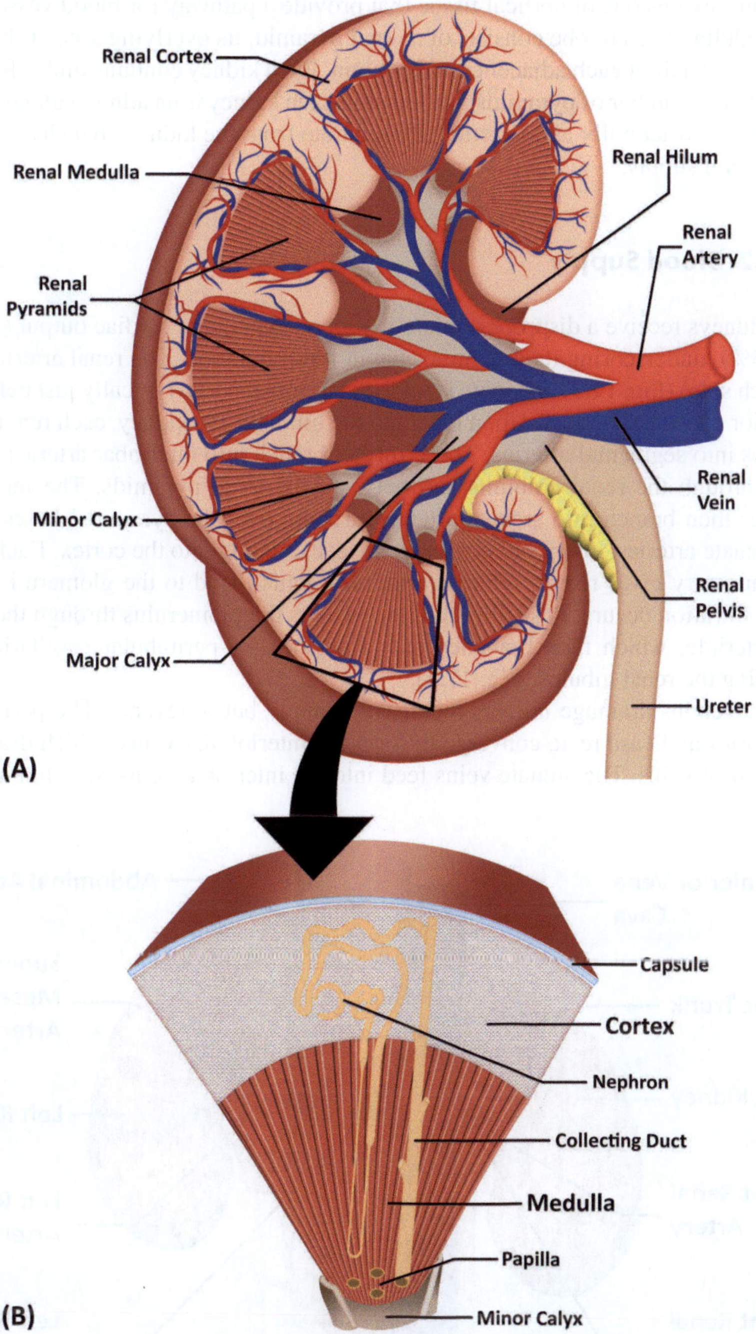

Fig. 17.2 The structure of the kidney. (**a**) Coronal section of the right kidney. (**b**) A renal pyramid

columns, extensions of cortical tissue that provide a pathway for blood vessels into the medulla. A renal lobe consists of a renal pyramid, its overlying area of the renal cortex, and half of each adjacent renal column. Each kidney contains multiple lobes, equal to the number of pyramids. Sitting atop each kidney is an adrenal gland. These glands are structurally and functionally separate from the kidneys but share a common blood supply.

17.1.2 Blood Supply

The kidneys receive a disproportionately high percentage of cardiac output (around 20–25%), underscoring their significance in filtering blood. The renal arteries (one on each side) (Fig. 17.3) originate from the abdominal aorta, typically just below the superior mesenteric artery. Upon reaching the hilum of the kidney, each renal artery divides into segmental arteries. These further branch into interlobar arteries, which pass through the renal columns and between the renal pyramids. The interlobar arteries then branch into arcuate arteries, arching over the pyramids' bases. From the arcuate arteries, interlobular arteries arise, extending into the cortex. Each interlobular artery gives rise to afferent arterioles, which lead to the glomeruli, where blood filtration occurs. Post-filtration, blood exits the glomerulus through the efferent arteriole, which then branches into a network of peritubular capillaries surrounding the renal tubules (Fig. 17.4).

The venous drainage mirrors the arterial supply but in reverse. The peritubular capillaries and vasa recta converge to form the interlobular veins, which drain into the arcuate veins. The arcuate veins feed into the interlobar veins, which drain into

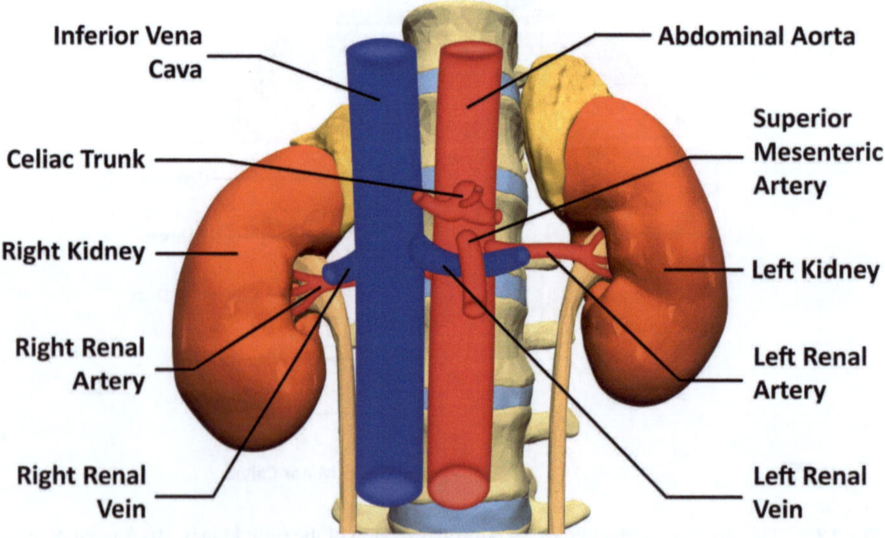

Fig. 17.3 Renal arteries and veins, anterior view

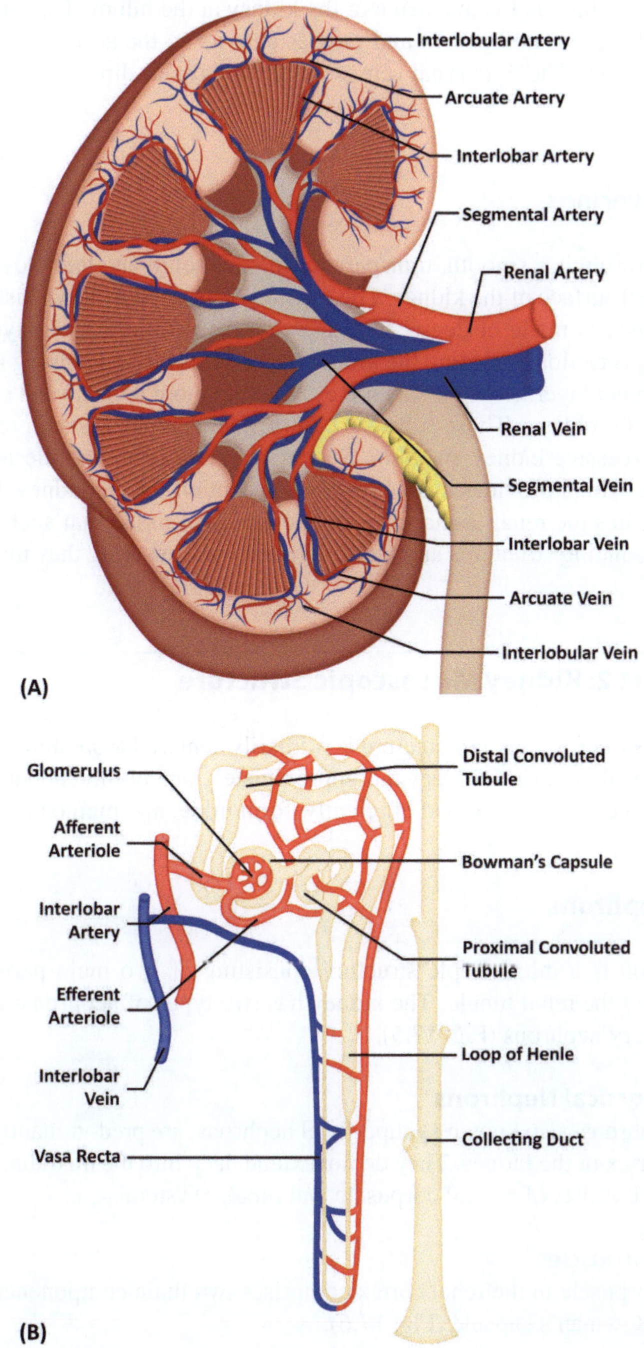

Fig. 17.4 Blood vessels of the kidney. (**a**) The frontal section illustrating major blood vessels. (**b**) Nephron blood vessels

the renal vein. The renal veins then exit the kidney at the hilum. The left renal vein is typically longer than the right and crosses anterior to the aorta to drain into the inferior vena cava. The right renal vein is shorter and drains directly into the inferior vena cava.

17.1.3 Coverings

The renal capsule is a smooth, transparent, fibrous membrane that closely adheres to the external surface of the kidney. The capsule is composed of two distinct layers. The outer layer is made of dense, irregular connective tissue. This layer provides mechanical protection, guarding the kidney against trauma and helping maintain its shape. The inner layer, closer to the kidney tissue, is more delicate and consists of a layer of myofibroblasts. These cells can contract, which may assist in resisting distention or excessive kidney swelling. Surrounding the renal capsule is a layer of perirenal fat, which provides cushioning and insulation for the kidney. Beyond the perirenal fat lies the renal fascia, a layer of connective tissue that anchors the kidneys to surrounding structures and the abdominal wall, ensuring they maintain their position.

17.2 Part 2: Kidney Microscopic Structure

The kidney's microscopic architecture is primarily centered around the nephron, its basic functional unit. Each kidney has approximately one million nephrons meticulously designed to process blood efficiently, form urine, and maintain homeostasis.

17.2.1 Nephrons

Each nephron is a microscopic structure consisting of two main parts: the renal corpuscle and the renal tubule. The kidney has two types of nephrons: cortical and juxtamedullary nephrons (Fig. 17.5).

17.2.1.1 Cortical Nephrons

Cortical nephrons, also known as superficial nephrons, are predominantly located in the outer cortex of the kidney. They do not extend deep into the medulla. Each cortical nephron consists of a renal corpuscle and tubular system.

Nephron Corpuscles

The renal corpuscle in the renal cortex comprises two main components: the glomerulus and Bowman's capsule (Fig. 17.6).

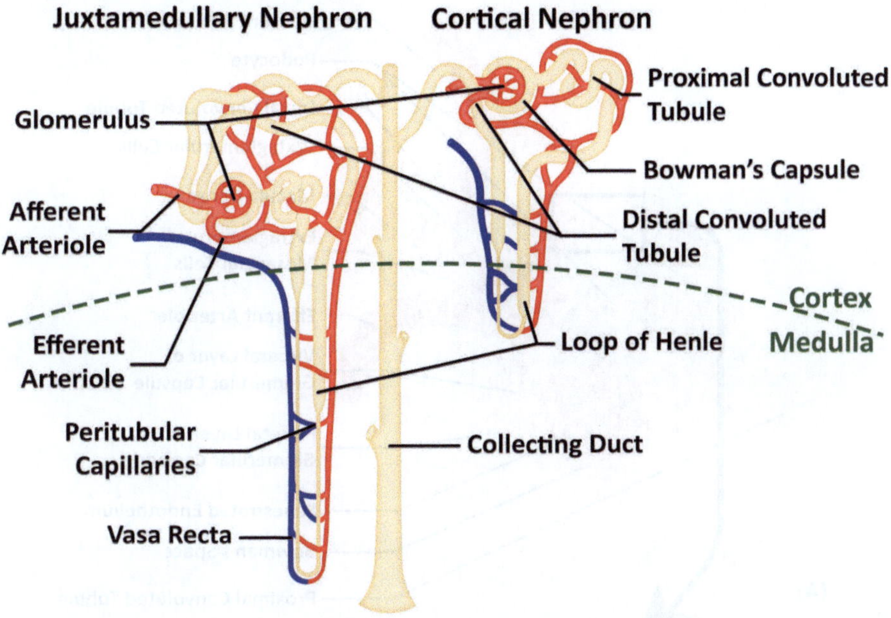

Fig. 17.5 Cortical and juxtamedullary nephrons and their blood vessels

Glomerulus

The renal glomerulus comprises a network of tiny blood vessels, or capillaries, and is intricately designed to facilitate efficient blood filtration, marking the first step in urine formation. The fenestrated endothelium lines the inside of the glomerular capillaries, directly interfacing with the blood that enters the glomerulus. It is one of the three layers of the glomerular filtration barrier, along with the glomerular basement membrane and the podocytes. The cells of this endothelium are flattened and form a continuous lining along the inner surface of the glomerular capillaries. They are connected by tight and adherens junctions (Page 52), which help maintain the barrier's integrity. Fenestrations (Fig. 17.6) are tiny pores within the endothelial cells, typically 70–100 nanometers in diameter. They allow the passage of water, ions, and small molecules from the blood into Bowman's space, initiating the formation of the glomerular filtrate. The fenestrated endothelium has a high density of fenestrations, covering up to 20% of the cell surface, significantly increasing its permeability.

Glomerular Basement Membrane

The glomerular basement membrane (GBM) is located between the endothelial cells of the glomerular capillaries and the podocytes. It acts as a semipermeable barrier, facilitating blood filtration while preventing the passage of proteins and cells. Type IV collagen is the most abundant component of GBM. It forms a mesh-like structure that provides strength and determines the pore size of the GBM. Laminins,

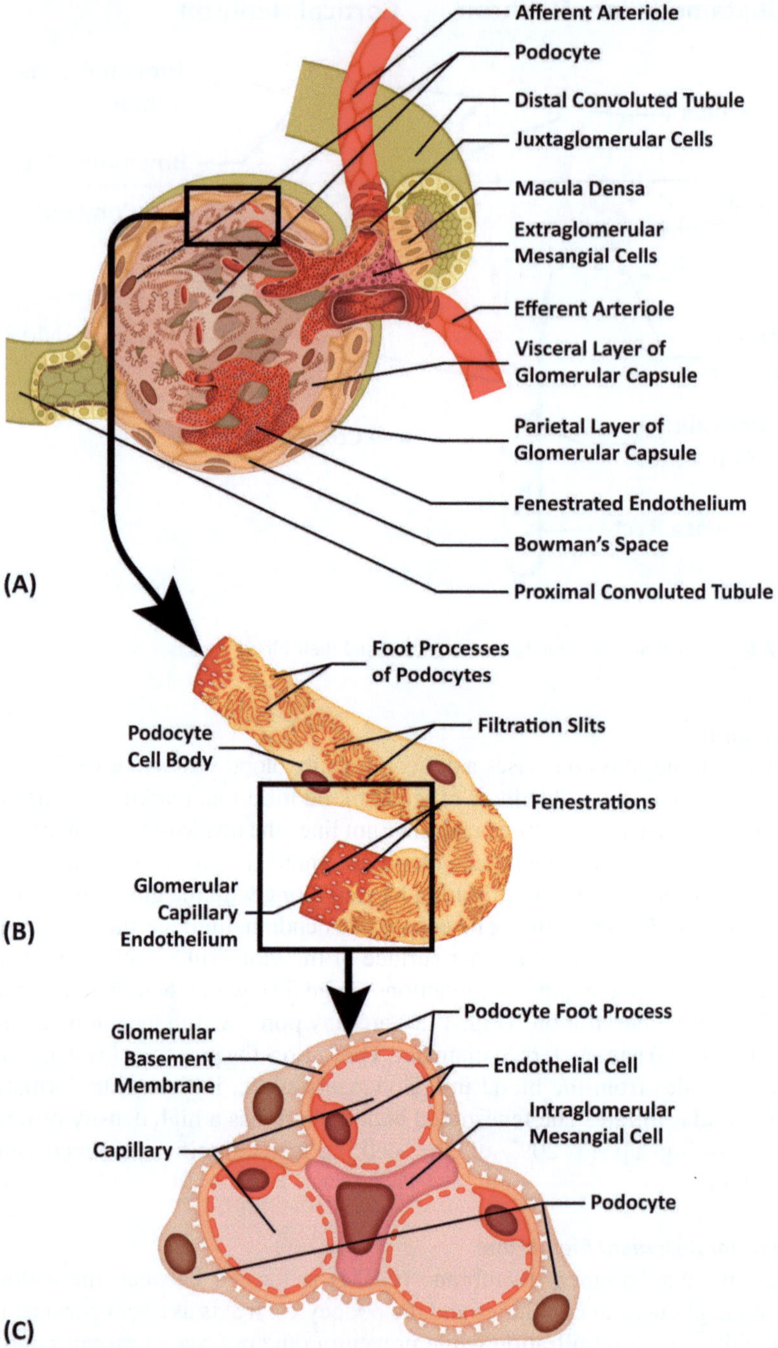

Fig. 17.6 (a) The glomerulus and Bowman's capsule structure. (b) Podocytes. (c) Microscopic section of the glomerular capillaries and podocytes

proteoglycans, and fibronectin are other constituents of GBM. Proteoglycans, such as heparan sulfate proteoglycans, carry negatively charged side chains that contribute to the GBM's charge-selective properties. Mesangial cells (Fig. 17.6) are found in the mesangium, the supportive tissue within the glomerulus. This area lies between the capillary loops and extends to the vascular pole of the glomerulus. They are near the glomerular capillaries and the GBM.

Blood enters the glomerulus through the afferent arteriole (Fig. 17.6). This arteriole originates from an interlobular artery that branches off the renal artery. Upon entering the glomerulus, the afferent arteriole branches into the glomerular capillaries. The extensive capillary network provides a large surface area for filtration. This is critical for maximizing the amount of blood that can be filtered at any given time. After passing through the glomerular capillaries, the blood is collected into the efferent arteriole. The efferent arteriole is narrower than the afferent arteriole, a feature that contributes to maintaining high pressure within the glomerular capillaries, essential for effective filtration. After leaving the efferent arteriole, the blood enters the peritubular capillaries (in cortical nephrons) or the vasa recta (in juxtamedullary nephrons). These capillary networks exchange substances with the renal tubules and are crucial for reabsorption and secretion. Glomerular capillaries are the only ones in the body between two arterioles instead of between an arteriole and a venule.

Glomerular Capsule

The glomerular capsule, also known as Bowman's capsule, is a cup-shaped structure that encases the glomerulus (Fig. 17.6). The glomerular capsule comprises two layers: an outer parietal layer and an inner visceral layer. The outer parietal layer of the capsule is made of simple squamous epithelium. Its primary function is structural, providing a sturdy yet flexible container for the filtrate produced in the glomerulus. It also plays a role in maintaining the shape of the capsule and protecting the delicate structures within. The inner visceral layer lies adjacent to the glomerulus and comprises specialized cells called podocytes (Fig. 17.6). They have a cell body from which several primary processes emanate.

These primary processes further branch into numerous secondary or foot processes (pedicels). They interlock with foot processes from adjacent podocytes, leaving narrow gaps known as filtration slits. Filtration slits are the gaps between the interdigitating foot processes. A thin, diaphragm-like structure bridges them called the slit diaphragm (Fig. 17.7). The slit diaphragm is composed of several proteins, including nephrin and podocin. The slit diaphragm functions as a size-selective filter, allowing the passage of water and small solutes while preventing the loss of larger molecules like albumin from the bloodstream into the urine. Bowman's space is located between the two layers of the glomerular capsule. This space surrounds the glomerulus. Blood is filtered through the glomerular capillaries into Bowman's space. The filtrate in Bowman's space is similar to plasma but lacks significant proteins. It includes water, electrolytes, glucose, amino acids, and waste products. The filtrate in Bowman's space then flows into the proximal convoluted tubule, the first segment of the nephron's tubular system.

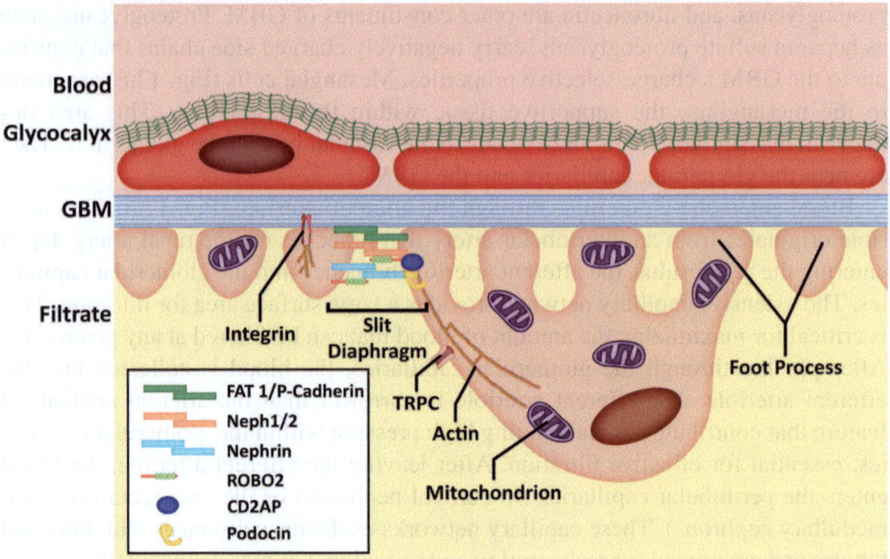

Fig. 17.7 Microscopic structure of the filtration membrane

Nephron Tubular System

The nephron tubular system (Fig. 17.5) is a highly organized structure that begins at Bowman's capsule and extends through several distinct segments, each with specific roles in urine formation.

Proximal Convoluted Tubule

The proximal convoluted tubule (PCT) is the first segment of the nephron's tubular system, immediately following Bowman's capsule. The PCT originates from Bowman's capsule and is in the renal cortex. It coils around the glomerulus, hence the term "convoluted." The PCT is lined with a single layer of cuboidal epithelial cells (Fig. 17.6). These cells are characterized by their prominent brush border—a dense array of microvilli on the apical surface facing the tubular lumen. The microvilli significantly increase the surface area for reabsorption. The basolateral membrane of these cells, facing the interstitium, contains numerous mitochondria, providing the ATP necessary for active transport processes. The PCT reabsorbs approximately 65–70% of the glomerular filtrate. This includes essential solutes like glucose, amino acids, sodium ions, bicarbonate, and water. The reabsorption of sodium ions is particularly crucial as it drives the reabsorption of other solutes and water. In addition to reabsorption, the PCT is involved in the selective secretion of substances into the tubular fluid.

Nephron Loop

The nephron loop (Fig. 17.5), also known as the loop of Henle, extends into the renal medulla and is divided into different segments: descending and ascending limbs. The first part of the descending limb is continuous with the proximal convoluted tubule and has the same cellular structure. The rest of the descending limb consists of a simple squamous epithelium, forming a thin segment. The early part of the ascending limb is thin with simple squamous epithelium. Most ascending limbs are thick and lined with cuboidal or low columnar epithelial cells.

Distal Convoluted Tubule

The distal convoluted tubule (DCT) (Fig. 17.6) is in the renal cortex and begins from the ascending limb of the loop of Henle, extending to the connecting tubule. The tubule is lined with cuboidal epithelial cells, distinct from the proximal convoluted tubule (PCT) cells in that they lack the brush border (microvilli). This gives the DCT a smoother luminal surface. These cells have extensive basolateral infoldings with mitochondria, supporting active transport processes.

Connecting Tubules

The connecting tubules are located in the renal cortex, bridging the gap between the DCT and the initial segment of the collecting duct. They follow the DCT and lead into either a cortical or a medullary collecting duct, depending on the nephron's location within the kidney. The tubule is lined with a simple cuboidal epithelium, which differs slightly in structure from the cells of the DCT. The cells in the connecting tubule exhibit characteristics intermediate between those of the DCT and the collecting duct, suggesting a transitional function.

Peritubular Capillaries

Peritubular capillaries arise (Fig. 17.5) from the efferent arterioles of the glomerulus. After blood is filtered in the glomerulus, it enters these capillaries. They form a dense network around the tubules of the nephron, particularly the proximal and distal convoluted tubules. In cortical nephrons, the peritubular capillaries are derived from efferent arterioles of glomeruli located in the renal cortex and primarily surround the cortical tubular segments. The walls of peritubular capillaries are highly permeable, allowing for the efficient exchange of substances between the blood and the surrounding tubular fluid. The extensive network and the close association with the renal tubules provide a large surface area for exchange, facilitating efficient reabsorption and secretion processes.

17.2.1.2 Juxtamedullary Nephrons

Juxtamedullary nephrons (Fig. 17.5) are located at the renal cortex and the renal medulla boundary. Their name is derived from their proximity to the medulla ("juxta-" meaning "near" and "medullary" referring to the renal medulla). The glomeruli of juxtamedullary nephrons are located deep in the cortex, close to the medulla. These glomeruli are generally larger than those of cortical nephrons. Like other nephrons, each juxtamedullary nephron has a Bowman's capsule where the initial blood filtration occurs. The PCT is similar in structure to cortical nephrons, lined with epithelial cells with a brush border to increase surface area for reabsorption. The defining feature of juxtamedullary nephrons is their long loop of Henle, which extends deep into the renal medulla. This structure is crucial for the nephron's ability to concentrate urine. The loop has descending and ascending limbs, with the descending limb permeable to water and the ascending limb impermeable to water but actively transporting salts. The DCT connects to the Henle loop and drains into the connecting duct.

The vasa recta arise from the efferent arterioles of the juxtamedullary glomeruli. Unlike the efferent arterioles of cortical nephrons that form peritubular capillaries, these arterioles descend into the medulla as the vasa recta. These vessels have a unique "hairpin" loop structure. They descend into the renal medulla, make a U-turn, and ascend back toward the cortex, paralleling the long loops of Henle of juxtamedullary nephrons. The vasa recta comprise capillaries and larger vessels, including arterioles and venules. The descending part of the vasa recta primarily consists of arterioles, while the ascending part is predominantly composed of venules.

17.2.2 Collecting Duct

The collecting ducts (Fig. 17.5) begin in the renal cortex and extend through the medulla to the renal papillae. They are formed by the convergence of multiple nephrons' connecting tubules. Collecting ducts are typically divided into two main segments: the cortical collecting duct (located in the cortex) and the medullary collecting duct (extending through the medulla). The ducts are lined with two primary types of epithelial cells: principal cells and intercalated cells.

Principal cells are responsible for sodium and water reabsorption and potassium secretion. They contain receptors for antidiuretic hormone (ADH) and aldosterone, which regulate water and sodium reabsorption. Intercalated cells exist in two forms, type A and type B, and are involved in acid-base balance. Type A intercalated cells secrete hydrogen ions and reabsorb bicarbonate, helping to manage acidosis, while type B intercalated cells do the opposite, playing a role in managing alkalosis.

17.2.3 Juxtaglomerular Complex (JGC)

The juxtaglomerular complex (JGC) (Fig. 17.8) is a specialized structure formed by interacting different cell types from the glomerulus and the renal tubule. The JGC is situated at the renal glomerulus's vascular pole, where the afferent arteriole enters and the efferent arteriole exits the glomerulus. Specifically, it is located where the DCT comes into direct contact with the arterioles feeding into and exiting from the glomerulus. Like the glomeruli they are associated with, juxtaglomerular complexes are located in the renal cortex. The JGC consists of three main types of cells: juxtaglomerular (JG) cells, macula densa cells, and extraglomerular mesangial cells (Fig. 17.8). JG cells are predominantly located in the walls of the afferent arterioles at the point where the arterioles enter the glomerulus. To a lesser extent, they can also be found in the efferent arterioles. These cells are situated near the region where

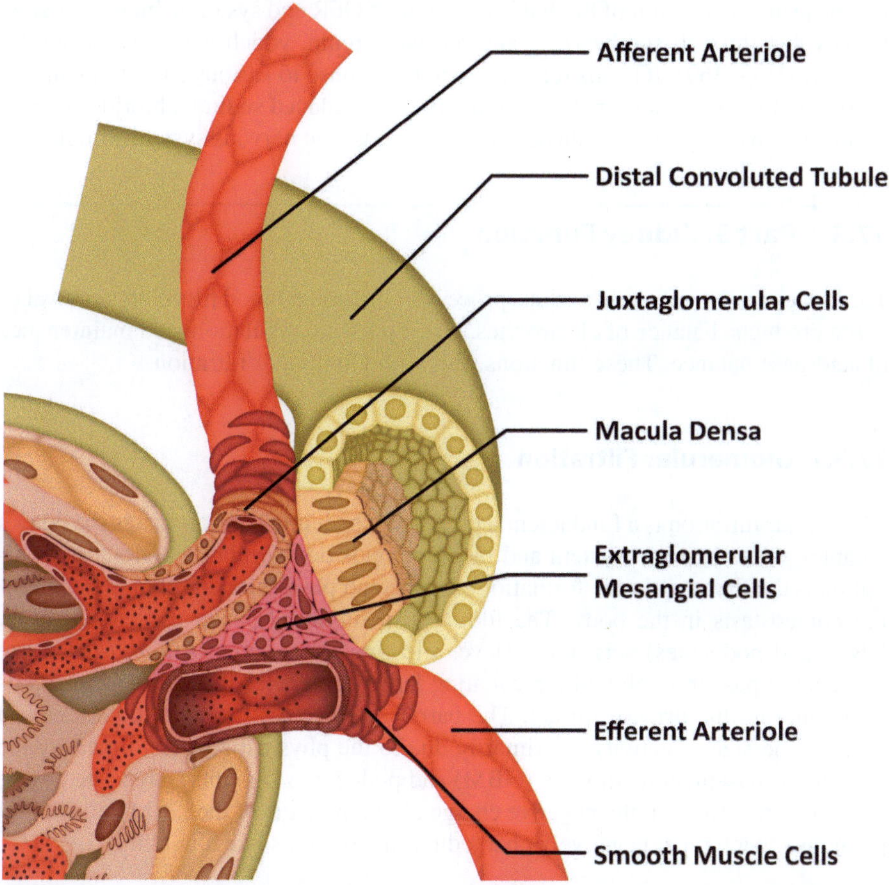

Fig. 17.8 Juxtaglomerular complex (JGC)

the afferent arteriole comes into contact with the distal convoluted tubule (DCT). JG cells are modified smooth muscle cells. They have a more granular cytoplasm than typical smooth muscle cells, owing to the presence of secretory granules. These granules predominantly contain renin, an enzyme that regulates blood pressure and fluid balance (Section 13.2.2.2). The macula densa is situated at the junction where the thick ascending limb (TAL) of the loop of Henle transitions into the distal convoluted tubule (DCT). It is located adjacent to the glomerulus of the same nephron, specifically near the vascular pole where the afferent and efferent arterioles are found. This structure is composed of a group of closely packed epithelial cells. These cells are taller and more densely nucleated than the surrounding tubular cells, giving the macula densa a distinctive appearance under a microscope. Extraglomerular mesangial cells are found in the space adjacent to the macula densa and are closely associated with the juxtaglomerular (JG) cells. These cells are distinguishable by their elongated, irregular shape and possess a central nucleus.

The primary function of the JGC is to regulate GFR and systemic blood pressure. It does this through synthesizing and releasing renin, which activates the RAAS pathway (Page 469). JG cells release renin in response to various stimuli, including decreased blood pressure in the afferent arteriole, reduced sodium chloride concentration as sensed by the macula densa, and sympathetic nervous system stimulation.

17.3 Part 3: Kidney Function

The kidneys are involved in various processes, including blood filtration, removal of waste products, balance of electrolytes, blood pressure regulation, and maintenance of acid-base balance. These functions start with glomerular filtration.

17.3.1 Glomerular Filtration

Glomerular filtration is a fundamental process in the kidneys that initiates the body's complex waste removal system and blood composition regulation. Glomerular filtration is the first step in the formation of urine and plays a crucial role in maintaining homeostasis in the body. The filtration membrane (glomerular endothelium, GBM, and podocytes) acts like a sieve, allowing water, small solutes, and waste products to pass through while preventing larger molecules like proteins and cells from entering the urinary filtrate. The barrier is selective based on both size and charge. The size selectivity is primarily due to the physical barrier formed by the glomerular basement membrane (GBM) and podocyte slit diaphragms. The charge selectivity results from the negative charge of components of the GBM and slit diaphragms, which repel negatively charged plasma proteins like albumin.

17.3.1.1 Pressures in Glomerulus

Glomerular filtration is heavily influenced by various pressures within the glomerulus (Fig. 17.9). These pressures determine the rate at which blood is filtered and are critical in maintaining the body's balance of fluid and electrolytes. Hydrostatic pressure in glomerular capillaries (blood pressure) is the primary force driving fluid out of the blood and into Bowman's capsule. A higher hydrostatic pressure in the glomerular capillaries increases the glomerular filtration rate (GFR), while a decrease in this pressure lowers the GFR. Oncotic pressure in glomerular capillaries (colloid osmotic pressure) is exerted by plasma proteins, primarily albumin, which are too large to pass through the filtration membrane. It opposes the filtration of fluid into Bowman's capsule. The oncotic pressure in the glomerular capillaries increases as blood passes through the glomerulus because fluid filtration concentrates plasma proteins. Hydrostatic pressure in Bowman's capsule is exerted by the fluid already present in Bowman's capsule and the proximal convoluted tubule. It opposes the outward movement of fluid from the glomerular capillaries. An increase in this pressure, which can occur due to urinary tract obstructions, can decrease the GFR. Oncotic

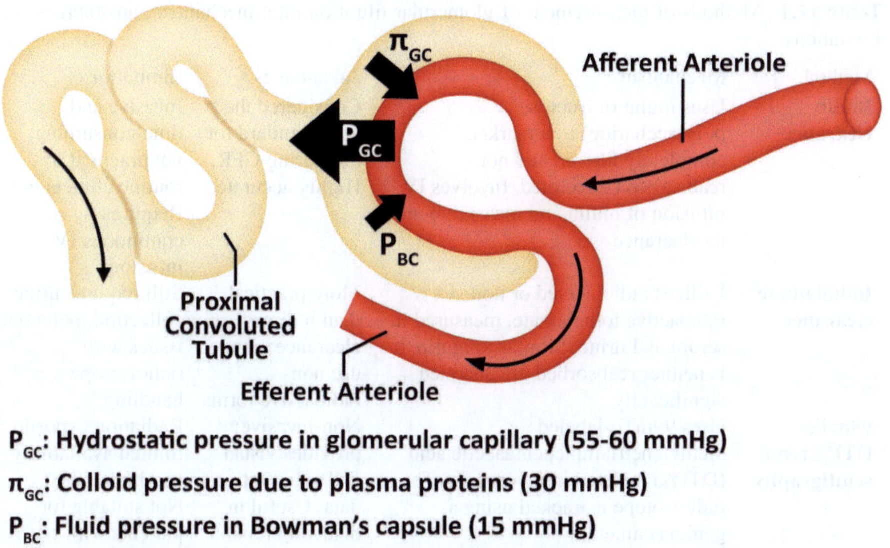

Fig. 17.9 The glomerulus's pressures, determining the filtration rate

pressure in Bowman's capsule is considered negligible, as the filtrate in Bowman's capsule is virtually free of proteins.

$$\text{Net Filtration Pressure}(\text{NFP}) = (\text{Hydrostatic pressure in glomerular capillaries}) - (\text{Oncotic pressure in glomerular capillaries} + \text{Hydrostatic pressure in Bowman's capsule})$$

GFR represents the rate at which blood is filtered in the kidneys' glomeruli. It's a key measurement for assessing kidney function and health (Table 17.1).

17.3.1.2 Regulation of Glomerular Filtration

Regulation of glomerular filtration involves various mechanisms to maintain a stable glomerular filtration rate (GFR). The kidney has the intrinsic ability to maintain a constant GFR despite fluctuations in systemic blood pressure. The myogenic mechanism (Fig. 17.10) involves the smooth muscle cells in the walls of the afferent arterioles. When blood pressure rises, the increased stretch of the arteriole walls triggers these cells to contract, reducing the lumen diameter (vasoconstriction), which decreases blood flow into the glomerulus and maintains a steady GFR. Conversely, a drop in blood pressure leads to vasodilation (widening of the arteriole), increasing blood flow and maintaining GFR. Tubuloglomerular feedback (Fig. 17.11) loop involves the macula densa. The macula densa senses changes in the filtrate's flow rate or ion concentration (especially NaCl). If GFR is high, the filtrate's flow rate and NaCl concentration increase. The macula densa responds by

Table 17.1 Methods of measurement of glomerular filtration rate: mechanism, advantages, and limitations

Method	Mechanism	Advantages	Limitations
Inulin clearance	Uses inulin (a fructose polysaccharide) as a marker, completely filtered and not reabsorbed or secreted. Involves IV infusion of inulin and measurement of its clearance	Considered the gold standard for measuring GFR. Highly accurate	Invasive and time-consuming; not practical for routine clinical use. Requires continuous IV infusion
Iothalamate clearance	Utilizes radiolabeled or non-radioactive iothalamate, measured in serum and urine. Similar to inulin, it is neither reabsorbed nor secreted significantly	More practical than inulin clearance; can use non-radioactive forms	Still requires urine collection; potential issues with radioisotope handling
99mTc-DTPA renal scintigraphy	Uses 99mTc-labeled diethylenetriaminepentaacetic acid (DTPA) injected intravenously. The radioisotope is tracked using a gamma camera	Non-invasive; provides visual and quantitative data. Useful in detecting renal scars and differential renal function	Radiation exposure; limited availability and higher cost. Not suitable for patients with severe renal impairment
Cystatin C clearance	Based on the serum levels of cystatin C, a protein produced by all nucleated cells, filtered, and catabolized by the kidneys	Less influenced by muscle mass compared to creatinine. Can be measured with a simple blood test	More expensive than creatinine testing; less widely available. Influenced by factors other than renal function, like inflammation
eGFR equations	Estimates GFR based on serum creatinine levels, age, sex, and body size, using formulas like CKD* and MDRD**	Non-invasive and easy to calculate; widely used in clinical practice	Less accurate in certain populations (e.g., very young, very old, extreme body weights). Assumes stable renal function

*The Cockcroft-Gault equation is expressed as follows:

$$\text{CrCl} = \frac{(140 - \text{Age}) \times \text{Weight}(\text{kg})}{72 \times \text{Serum Creatinine}(\text{mg/dL})}$$

**The simplified MDRD equation is expressed as:
eGFR (mL/ min /1.73m^2) = 175 × Serum Creatinine$^{-1.154}$ × Age$^{-0.203}$ × (0.742 if female) × (1.212 if Black)

signaling the afferent arteriole to constrict, reducing blood flow and GFR. If GFR is low, the opposite occurs. The sympathetic nervous system can also influence GFR, primarily under stress conditions. During intense sympathetic stimulation (as in stress or severe hemorrhage), norepinephrine is released, and the sympathetic nervous system is activated, leading to vasoconstriction of the afferent arterioles. This

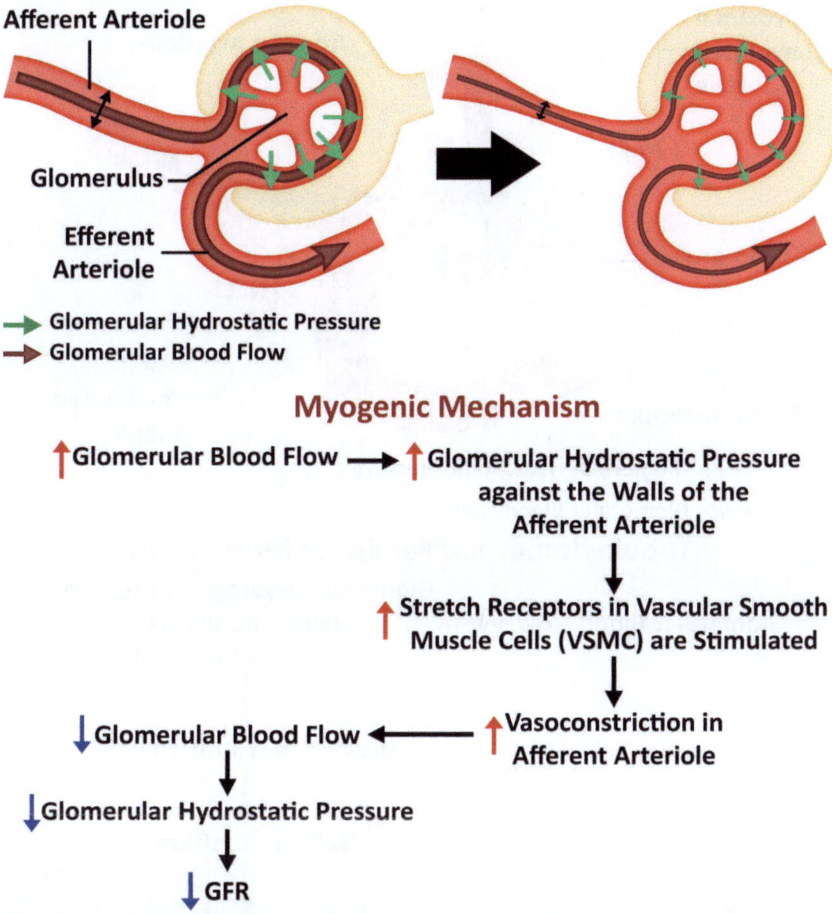

Fig. 17.10 Regulation of glomerular filtration: myogenic mechanism

response reduces renal blood flow and GFR, directing blood flow to more critical organs like the heart and brain. Additionally, sympathetic stimulation can enhance the release of renin, activating the renin-angiotensin-aldosterone system (RAAS). RAAS is initiated by the juxtaglomerular cells in response to reduced renal blood pressure, decreased sodium chloride concentration in the DCT, or sympathetic nervous system stimulation. This system ultimately produces angiotensin II, a potent vasoconstrictor that can constrict the efferent arterioles, increasing hydrostatic pressure in the glomerulus and raising GFR. Stimulated by angiotensin II, aldosterone enhances sodium reabsorption, indirectly affecting GFR. Atrial natriuretic peptide (ANP), released by the heart's atria in response to increased blood volume, relaxes the mesangial cells in the glomerulus, increasing the surface area available for filtration and thereby increasing GFR. ANP also antagonizes the effects of the RAAS, leading to vasodilation of the afferent arterioles and increased GFR.

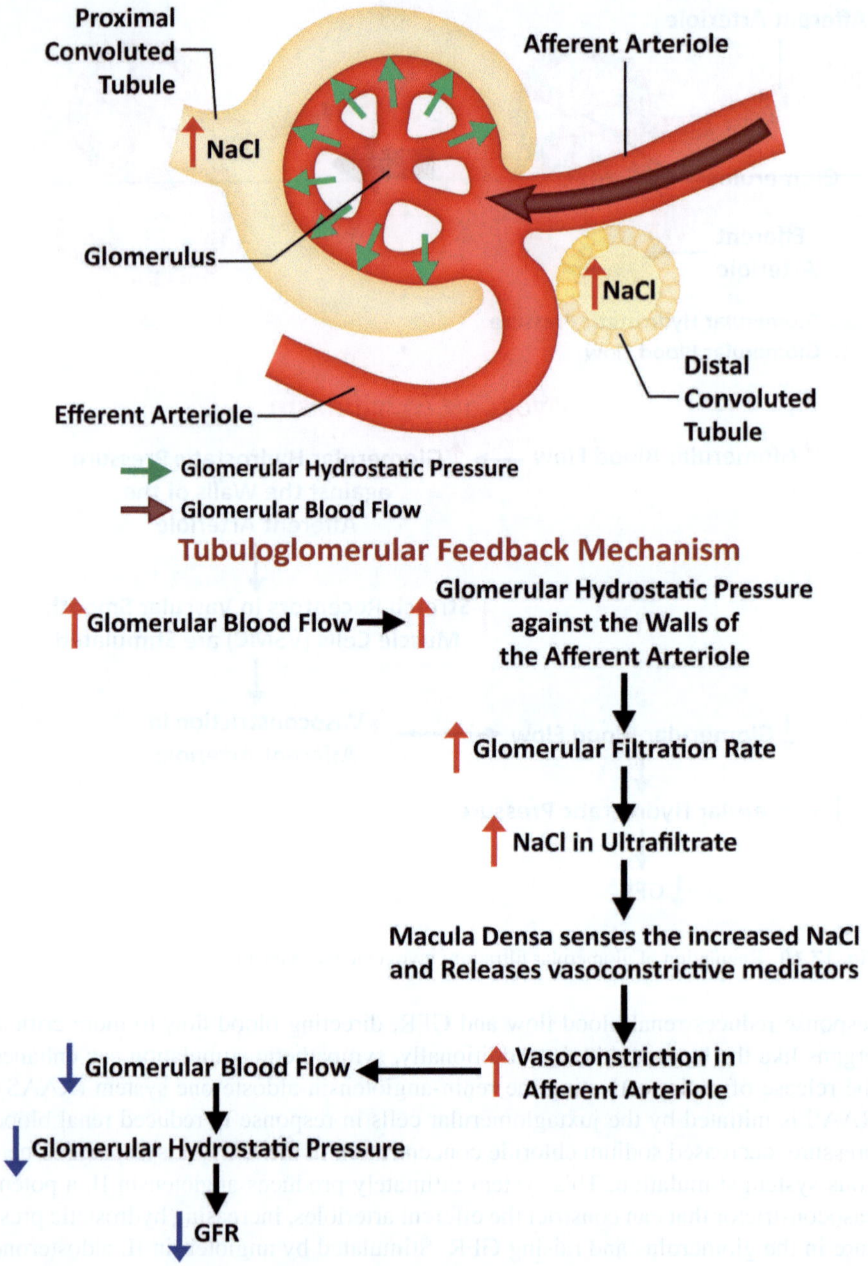

Fig. 17.11 Regulation of glomerular filtration: tubuloglomerular feedback

17.3.2 Tubular Reabsorption

Tubular reabsorption in the nephrons ensures the body retains essential substances while filtering out waste. The tubular system and collecting duct selectively reabsorb substances from the filtrate into the blood. The reabsorptive capacity is not infinite. The renal threshold is the blood concentration level of a substance at which its reabsorption in the kidneys is maximized, and any excess begins to be excreted in the urine. There is a maximum rate at which the renal tubules can reabsorb many substances. This is known as the transport maximum (Tm).

Reabsorption Mechanisms

- Active Transport: The energy-dependent process, primarily for Na^+ reabsorption, drives most of the other reabsorptive processes.
- Passive Transport: Driven by concentration gradients; necessary for water and some solute movement.
- Secondary Active Transport: It utilizes the electrochemical gradient established by primary active transport, essential in glucose and amino acid reabsorption.
- Aquaporins: Water channels that facilitate water reabsorption, regulated by ADH.

17.3.2.1 Proximal Convoluted Tubule

The proximal convoluted tubule (PCT) is responsible for reabsorbing a significant portion of the glomerular filtrate. Although hormones do not heavily influence the PCT's activity, it is sensitive to changes in blood pressure, volume status, and plasma solute concentrations.

Active transport of sodium (Na^+) is the primary driver for reabsorption in the PCT. Na^+/K^+-ATPase pump, located on the basolateral membrane, transports Na^+ out of the tubular cells into the interstitial fluid and creates a concentration gradient that facilitates Na^+ movement from the tubular lumen into the cell. Na^+ reabsorption is coupled with the reabsorption of glucose, amino acids, phosphate, and lactate via secondary active transport. Water follows the osmotic gradient created by solute reabsorption, especially Na^+. Approximately 65–70% of the filtrate volume is reabsorbed. Aquaporin channels are integral in facilitating water movement across cell membranes.

Glucose and amino acids are cotransported with Na^+ into the tubular cells and then move into the interstitial fluid by facilitated diffusion. Under normal physiological conditions, glucose and amino acids are almost completely reabsorbed in the PCT. Carbonic anhydrase facilitates the conversion of bicarbonate into carbon dioxide (CO_2) and water within the tubular cell. This mechanism is vital for maintaining the acid-base balance of the body. Approximately 50% of the urea in the filtrate is passively reabsorbed due to the concentration gradients established by water reabsorption.

17.3.2.2 Nephron Loop

The nephron loop is essential for the concentration and dilution of urine, significantly contributing to the kidney's ability to maintain fluid and electrolyte balance. The thin descending limb is highly permeable to water but relatively impermeable to solutes. Water is reabsorbed osmotically into the surrounding interstitial fluid, increasing the filtrate's concentration as it moves deeper into the medulla. There is minimal solute reabsorption in this segment. The thin ascending segment is permeable to ions, particularly Na^+ and Cl^-, but impermeable to water. This segment is more prominent in juxtamedullary nephrons. Thick ascending limb (TAL) actively reabsorbs Na^+, K^+, and Cl^- through the $Na^+/K^+/2Cl^-$ cotransporter on the luminal membrane. This segment is impermeable to water, so as solutes are reabsorbed, the filtrate becomes progressively more dilute. The reabsorption of ions, mainly Na^+, helps establish an electrochemical gradient that drives the passive reabsorption of other ions, such as magnesium (Mg^{2+}) and calcium (Ca^{2+}).

The unique arrangement of the loop of Henle and the vasa recta (the blood vessels that run parallel to the loop) creates a "countercurrent multiplication system," which is essential for the kidney's ability to concentrate urine. As the filtrate descends in the descending limb, water is reabsorbed osmotically into the surrounding interstitial fluid, concentrating the filtrate. In the ascending limb, ions (especially Na^+ and Cl^-) are actively pumped into the interstitial fluid, diluting the filtrate. The term "countercurrent" refers to the fact that the filtrate flows in opposite directions in the two limbs of the loop. The "multiplication" aspect pertains to the interaction between the flow of filtrate and the active transport of ions, which multiplies the concentration gradients established in the renal medulla. As the ascending limb actively transports ions into the interstitial fluid, the descending limb's permeability to water increases the filtrate concentration. This creates a progressively increasing osmotic gradient in the medulla. Urea, reabsorbed in the inner medullary collecting duct, also contributes to the osmolarity of the interstitial fluid in the medulla, enhancing the countercurrent mechanism. Vasa recta run parallel to the loop of Henle. As blood flows down the vasa recta, it picks up ions and loses water to the interstitium. On the way back up, it loses ions and gains water. This exchange helps maintain the high osmolarity in the medulla without washing out the gradient. The reabsorption of water in the descending limb and solutes in the ascending limb generates a significant osmotic gradient in the renal medulla, crucial for water reabsorption in the later parts of the nephron, particularly the collecting duct.

17.3.2.3 Distal Convoluted Tubule

The distal convoluted tubule (DCT) lies in the renal cortex and is comparatively shorter and less coiled than the PCT. The DCT actively reabsorbs sodium (Na^+) and chloride (Cl^-) ions. This is primarily mediated through the sodium-chloride symporter located on the luminal membrane. This process is regulated by hormones such as aldosterone, which increases sodium reabsorption. Calcium (Ca^{2+}) reabsorption occurs via passive and active mechanisms regulated by parathyroid hormone (PTH). PTH stimulates the insertion of calcium channels on the luminal membrane, facilitating increased calcium reabsorption. Unlike the PCT, the DCT is relatively impermeable to water. However, antidiuretic hormone (ADH) can increase its permeability to water. Magnesium (Mg^{2+}) is reabsorbed in the DCT, primarily by a passive paracellular mechanism influenced by the electrochemical gradient.

17.3.2.4 Collecting Duct

The primary function of collecting ducts is to reabsorb water from the urine. The ADH regulates this reabsorption. When the body needs to conserve water, ADH is released, making the collecting ducts more permeable to water. Water moves passively through aquaporin channels in the collecting duct cells in response to the concentration gradient created by ADH. This reabsorption helps concentrate urine and conserve water. The collecting ducts also play a role in reabsorbing sodium (Na^+) and secreting potassium (K^+). This helps maintain electrolyte balance in the body. Sodium reabsorption is regulated by hormones like aldosterone, which acts on the cortical collecting ducts. Aldosterone enhances sodium reabsorption and potassium secretion. The collecting ducts can secrete hydrogen ions (H^+) into the urine, which helps regulate blood pH. Hormones like aldosterone influence this process and can help the body maintain acid-base balance. In the inner medullary collecting ducts, urea recycling occurs. Urea, a waste product, is reabsorbed from the urine and returned to the interstitium of the renal medulla, contributing to the high osmolarity of the medullary interstitium. This is important for the concentration of urine. The collecting ducts can also reabsorb and secrete other ions, such as bicarbonate (HCO_3^-) and calcium ions (Ca^{2+}), depending on the body's needs and hormonal regulation.

17.3.3 Tubular Secretion

The secretion in the nephron complements filtration and reabsorption. It involves actively transferring substances from the blood into the renal tubular fluid. This mechanism is essential for excreting waste products, regulating acid-base balance, and eliminating excess ions and drugs from the body.

17.3.3.1 Proximal Convoluted Tubule

The PCT secretes various organic anions (such as bile acids, oxalate, and urate) and cations (including creatinine and certain drugs) through active transport mechanisms. Basolateral transporters initially take these substances from the blood into the tubular cells, and then apical transporters secrete them into the tubular lumen. The PCT secretes H^+ into the tubular fluid, helping to reclaim bicarbonate (HCO−) and excrete excess acid. This involves the Na^+/H^+ exchanger and H^+-ATPase pumps located on the apical membrane. The Na^+/H^+ exchanger allows for the reabsorption of Na^+ while secreting H^+ into the tubular lumen. In the tubular fluid, secreted H^+ combines with buffers, primarily bicarbonate, forming carbon dioxide (CO_2) and water (H_2O), which then re-enter the tubular cells. Ammonia is produced in the PCT cells from the deamination of glutamine. NH_3 diffuses readily across cell membranes and is secreted into the tubular fluid. In the tubular fluid, NH_3 combines with H^+ to form ammonium (NH_4^+), which is then excreted in the urine. This process contributes to acid elimination.

17.3.3.2 Nephron Loop

Compared to other parts of the nephron, such as the proximal convoluted tubule (PCT) and the distal convoluted tubule (DCT), the role of the nephron loop in secretion is relatively limited. The thick ascending limb of the nephron loop plays a minimal role in urea, potassium (K^+), and hydrogen (H^+) secretion.

17.3.3.3 Distal Convoluted Tubule

The secretory mechanisms in the DCT are crucial for maintaining the body's acid-base balance, potassium homeostasis, and overall fluid balance. The primary secretory activity in the DCT is the secretion of potassium (K^+). This process is primarily regulated by aldosterone. Aldosterone increases the synthesis and activity of Na^+/K^+-ATPase pumps and specific potassium channels (ROMK) on the apical membrane of the DCT cells. This results in increased reabsorption of sodium (Na^+) and the secretion of K^+ into the tubular fluid. The DCT also participates in acid-base homeostasis by secreting hydrogen ions (H^+). This is crucial for the excretion of acid and the maintenance of blood pH. H^+ secretion occurs via H^+-ATPase and H^+/K^+-ATPase pumps located on the apical membrane of the DCT cells. These pumps actively transport H^+ into the tubular fluid, often in exchange for K^+. The DCT secretes small amounts of calcium (Ca^{2+}) and magnesium (Mg^{2+}), but its reabsorptive capacity for these ions overshadows this.

17.3.3.4 Collecting Duct

Principal cells in the collecting ducts are responsible for potassium secretion. This process is influenced by aldosterone, which enhances potassium excretion. Intercalated cells are also responsible for the secretion of hydrogen ions (H^+) and the reabsorption of bicarbonate (HCO_3^-), which are crucial for maintaining acid-base balance. There are two types of intercalated cells—type A and type B. Type A cells secrete H^+ and reabsorb bicarbonate, aiding in acid elimination. In contrast, type B cells do the reverse, helping in alkaline conditions. The collecting ducts also function to secrete ammonia (NH_3) and urea.

17.3.4 Reabsorption Mechanisms

Reabsorption and secretion in the nephron complement filtration. It involves actively transferring substances from the renal tubular fluid into the blood and vice versa. Each segment of the nephron—the proximal convoluted tubule (PCT), the distal convoluted tubule (DCT), and the collecting duct—contributes to these processes in distinct ways.

17.3.4.1 Sodium Reabsorption

Sodium reabsorption in the proximal convoluted tubules (PCT) is essential for maintaining fluid and electrolyte balance in the body. The PCT reabsorbs most filtered sodium (Na^+), water, and other solutes.

Mechanisms of sodium reabsorption include primary active transport across the tubular cells and passive diffusion through paracellular spaces. The fundamental driver of sodium reabsorption is the Na^+-K^+-ATPase pump (primary active transport) located on the basolateral membrane of the PCT cells. This pump actively transports Na^+ from the cell into the interstitial fluid in exchange for K^+, creating a low intracellular Na^+ concentration. The low intracellular Na^+ concentration sets up a concentration gradient that drives the entry of Na^+ from the tubular fluid into the cells. Various cotransporters and antiporters on the apical membrane facilitate this entry. Sodium is reabsorbed with other solutes like glucose, amino acids, bicarbonate, phosphate, and lactate through cotransport mechanisms. Alongside transcellular pathways, there is also paracellular reabsorption of Na^+, driven by the electrochemical gradient established by the active transport processes. This is facilitated by tight junctions between cells, which are relatively leaky in the PCT, allowing passive movement of ions and water.

Cotransporters

Sodium-Glucose Cotransport (SGLT): The SGLTs on the apical membrane reabsorb Na^+ along with glucose. This is a crucial mechanism for glucose reabsorption, tightly linked to Na^+ reabsorption (Section 16.4.1).

Sodium-Amino Acid Cotransport: Like glucose, amino acids are reabsorbed with Na^+ through specific cotransporters (Section 16.4.2).

Sodium-Bicarbonate Reabsorption: A significant portion of bicarbonate reabsorption is linked to Na^+ transport. This involves the conversion of bicarbonate (HCO_3^-) to carbon dioxide (CO_2) within the cell, which then diffuses back into the tubular fluid and recombines with H^+ to form HCO_3^- again (Section 17.3.4.5).

Sodium-Phosphate and Sodium-Lactate Cotransport: Phosphate and lactate are also reabsorbed with Na^+ via specific cotransport mechanisms (Section 17.3.4.9).

17.3.4.2 Water Reabsorption

Water reabsorption in the renal tubules allows the kidneys to regulate body fluid volume and maintain water balance. Most water reabsorption occurs in the proximal tubules, loop of Henle, distal convoluted tubules, and collecting ducts of the nephrons.

Water movement in renal tubules is primarily driven by osmosis. In response to differences in solute concentrations, water moves passively across the tubular epithelial cells and through the paracellular pathway.

In the PCT, tight junctions are "leaky," allowing for significant paracellular transport of water and solutes. In the loop of Henle, DCT, and CD, tight junctions are more "tight," especially in the ascending limb of the loop of Henle, where they prevent the back-leak of ions into the tubular fluid.

Aquaporin water channels are present on tubular cells' apical and basolateral membranes. These channels facilitate the rapid movement of water molecules across the cell membranes. Aquaporins (Fig. 17.12) are a family of small integral membrane proteins forming channels to transport water across cell membranes. Aquaporins have a narrow pore diameter and a specific selectivity filter that allows

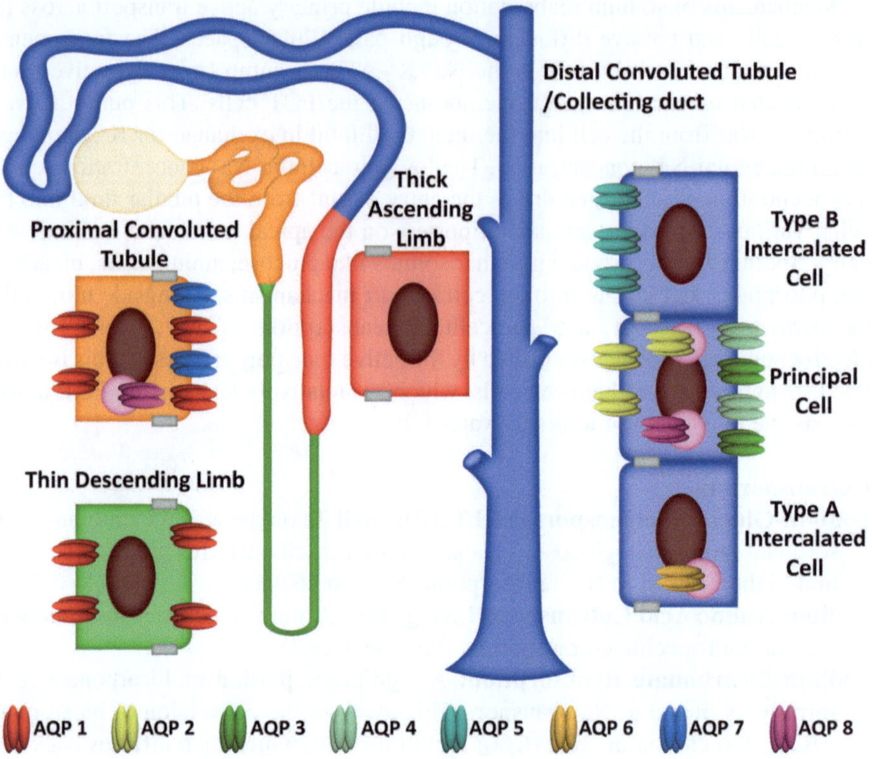

Fig. 17.12 Aquaporins (AQP) and their distribution in tubular cells

the rapid passage of water molecules while excluding ions and other solutes. The structure of aquaporins balances hydrophobic and hydrophilic regions to facilitate the selective passage of water molecules. There are several types of aquaporins:

- Aquaporin-1 (AQP1): Found in the proximal convoluted tubule and descending limb of the loop of Henle, AQP1 allows for passive water reabsorption in these segments, driven by the osmotic gradient.
- Aquaporin-2 (AQP2): Present in the apical membrane of the principal cells in the collecting ducts, AQP2 is crucial for water reabsorption in the collecting duct. It is regulated by antidiuretic hormone (ADH), which increases the number of AQP2 channels in the apical membrane, enhancing water reabsorption and urine concentration. ADH regulates AQP2 trafficking to the apical membrane via the cAMP signal transduction pathway.
- Aquaporin-3 and Aquaporin-4 (AQP3 and AQP4): Located in the basolateral membrane of the principal cells in the collecting duct, these aquaporins facilitate the exit of water from the cells into the interstitial fluid and eventually into the bloodstream.

Proximal Convoluted Tubule

In the proximal tubules, approximately 65–70% of filtered water is reabsorbed in an isosmotic manner, meaning that both water and solutes are reabsorbed in proportion, maintaining the osmolarity of the tubular fluid.

Nephron Loop

The loop of Henle creates a hypertonic interstitium in the medulla of the kidney, which is crucial for concentrating urine. In the descending limb of the loop of Henle, water is reabsorbed passively through aquaporins due to the increasing osmolarity of the interstitial fluid. The ascending limb is relatively impermeable to water but actively transports sodium ions (Na^+) out of the tubular fluid, contributing to the establishment of the hypertonic medullary interstitium.

Collecting Duct

The collecting ducts are the primary site of regulated water reabsorption. The extent of water reabsorption here is determined by the presence or absence of antidiuretic hormone (ADH). In the absence of ADH, aquaporin-2 (AQP2) channels are internalized from the apical membrane. Hence, the collecting ducts become relatively impermeable to water, resulting in the production of dilute urine. When ADH is present, it binds to receptors on the basolateral membrane of collecting duct cells, leading to the insertion of AQP2 channels into the apical membrane. This allows water to be reabsorbed from the tubular lumen into the interstitium, resulting in concentrated urine.

17.3.4.3 Nutrient Reabsorption

In the renal tubules, most nutrients, including glucose, amino acids, and vitamins, are reclaimed from the glomerular filtrate and returned to the bloodstream.

Glucose Reabsorption

Glucose is freely filtered through the glomerular filtration barrier. Once filtered, glucose in the tubular fluid enters the lumen of the PCT. In the PCT cells' luminal (apical) membrane, there are sodium-glucose cotransporters, specifically sodium-glucose cotransporters 1 and 2 (SGLT1 and SGLT2). SGLT2 is responsible for the initial reabsorption of about 90% of filtered glucose. It is a high-capacity, low-affinity transporter located in the early segment of the PCT. SGLT1 reabsorbs the remaining 10% of glucose. It is a low-capacity, high-affinity transporter found in the distal part of the PCT. The PCT is where virtually all the filtered glucose is reabsorbed, preventing its loss in urine. The glucose reabsorption process is coupled with the active transport of sodium (Na^+), utilizing the sodium gradient maintained by the Na^+/K^+-ATPase pump on the basolateral membrane. As Na^+ is actively transported out of the cell into the interstitial fluid, it creates a low intracellular Na^+ concentration, driving the entry of Na^+ and glucose from the tubular fluid into the cell.

Once inside the tubular cells, glucose is transported across the basolateral membrane (facing the interstitial fluid) into the peritubular capillaries. Glucose transport across the basolateral membrane is facilitated by glucose transporters (GLUTs), particularly GLUT2 and GLUT1. GLUT2 is responsible for transporting glucose from the PCT cells into the interstitial fluid. GLUT1 may play a role in the later segments of the renal tubules. After crossing the basolateral membrane, glucose enters the interstitial fluid and diffuses into the peritubular capillaries, ultimately returning to the systemic circulation.

A maximal renal threshold for glucose reabsorption is known as the transport maximum (Tm). When the blood glucose level exceeds this threshold, the capacity of SGLTs is saturated, and excess glucose is excreted in the urine (in conditions such as diabetes mellitus).

Amino Acid Reabsorption

The majority of amino acid reabsorption occurs in the PCT of the nephron. The reabsorption of amino acids involves specific transporters on the luminal and basolateral membranes of the renal tubular cells.

Amino acids are freely filtered through the glomerular filtration barrier. Once filtered, amino acids are delivered to the lumen of the PCT. Amino acid reabsorption in the PCT is primarily mediated by sodium-dependent amino acid transporters (SAATs) or sodium-amino acid cotransporters located on the PCT cells' luminal (apical) membrane. SAATs are the primary transporters in amino acid reabsorption in the PCT. They utilize the electrochemical gradient of sodium (Na^+), maintained by the Na^+/K^+-ATPase pump on the basolateral membrane. There are several types of SAATs:

- System B0+: Transports cysteine, neutral, and basic amino acids.
- System X-AG: Transports anionic (acidic) amino acids like aspartate and glutamate.
- System y + L: Transports basic amino acids.

Some sodium-independent amino acid transporters facilitate the facilitated diffusion of amino acids across the tubular epithelial cells. They are less common and typically function in the reabsorption of certain specific amino acids.

Once inside the tubular cells, amino acids exit across the basolateral membrane (facing the interstitial fluid) into the peritubular capillaries. Various amino acid transporters, such as systems L, A, and others, facilitate the transport of amino acids across the basolateral membrane. After crossing the basolateral membrane, amino acids enter the interstitial fluid and diffuse into the peritubular capillaries, ultimately returning to the systemic circulation.

Similar to glucose, there is a maximal renal threshold (transport maximum) for the reabsorption of amino acids (due to saturation of transporters). When plasma levels of amino acids are extremely high, exceeding this threshold, excess amino acids will appear in the urine (aminoaciduria).

Vitamin Reabsorption

The reabsorption of vitamins in the nephrons ensures the conservation of these essential nutrients.

Most B vitamins, including B1 (thiamine), B2 (riboflavin), B3 (niacin), B5 (pantothenic acid), B6 (pyridoxine), B7 (biotin), B9 (folate), and B12 (cobalamin), are reabsorbed in the PCT through sodium-dependent cotransporters or facilitated diffusion mechanisms. The reabsorption rates can be influenced by the body's vitamin status and nutritional needs.

Vitamin C (ascorbic acid) is also reabsorbed in the PCT using sodium-dependent transporters. Its reabsorption follows saturation kinetics, meaning that at higher plasma concentrations (as seen with supplementation), reabsorption becomes less efficient, and excess vitamin C is excreted in the urine.

Fat-soluble vitamins (A, D, E, K) are generally handled differently by the body compared to water-soluble vitamins. They are absorbed with dietary fats in the intestines and are less dependent on renal reabsorption as they are stored in body tissues. The kidneys play a minor role in directly handling these vitamins. However, the kidneys are crucial in activating vitamin D. The liver initially hydroxylates vitamin D to 25-hydroxyvitamin D, after which the kidneys convert it into its active form, 1,25-dihydroxyvitamin D (calcitriol), via 1-alpha hydroxylase enzyme.

17.3.4.4 Chloride Reabsorption

Chloride ions (Cl^-) are actively reabsorbed primarily in the renal tubules, particularly in the proximal convoluted tubule (PCT), the thick ascending limb of the loop of Henle, the distal convoluted tubule (DCT), and the collecting ducts.

Chloride reabsorption begins in the proximal convoluted tubule (PCT), where it is passively reabsorbed along with sodium ions (Na^+). Sodium ions are actively transported from the tubular lumen into the tubular cells by sodium-glucose cotransporters (SGLTs) and sodium-proton exchangers (NHEs) on the luminal membrane. Chloride ions follow sodium passively through chloride channels. Chloride reabsorption in the PCT helps maintain electrical neutrality and promotes water movement through aquaporin channels.

In the thick ascending limb of the loop of Henle, chloride reabsorption is a crucial step in establishing the osmotic gradient in the renal medulla, which is essential for concentrating urine. Chloride ions are actively transported out of the tubular lumen into the interstitial fluid by the apical membrane sodium-potassium-2 chloride cotransporter (NKCC2), also known as the bumetanide-sensitive cotransporter. The energy for this transport is derived from the Na^+/K^+-ATPase pump on the basolateral membrane, which pumps sodium ions out of the tubular cells into the interstitial fluid.

In the DCT, chloride reabsorption occurs primarily through chloride-bicarbonate exchangers on the apical membrane, such as the pendrin transporter. In the connecting tubule and collecting ducts, aldosterone, a hormone that regulates sodium and potassium balance, influences chloride reabsorption. Aldosterone increases the activity of sodium-chloride cotransporters (NCC) on the apical membrane of tubular cells.

Chloride ions exit the tubular cells across the basolateral membrane through chloride channels and transporters, such as the chloride channel Kb (ClC-Kb).

17.3.4.5 Bicarbonate Reabsorption

Most bicarbonate (HCO_3^-) reabsorption occurs in the proximal convoluted tubule (PCT) and the collecting ducts, with the PCT being the primary site. Bicarbonate reabsorption involves several complex mechanisms, including luminal and basolateral transport processes.

Bicarbonate ions are filtered at the glomerulus and enter the lumen of the PCT as a component of the glomerular filtrate. Carbonic anhydrase II (CA-II) is the most important isoform of carbonic anhydrase (CA) enzymes. CA-II is primarily found in the proximal convoluted tubules (PCT) and is involved in the reabsorption of HCO_3^- and the regulation of pH in the kidneys. To be reabsorbed, bicarbonate ions must be converted into a form that can be transported across the PCT cells' luminal (apical) membrane. Carbonic anhydrase II (CA-II) catalyzes the conversion of HCO_3^- and H^+ into carbonic acid (H_2CO_3). H_2CO_3 rapidly dissociates into carbon dioxide (CO_2) and water (H_2O). The generated CO_2 readily diffuses across the apical membrane into the tubular cell. Inside the tubular cell, CO_2 combines with H_2O under the influence of cytoplasmic carbonic anhydrase enzymes to regenerate HCO_3^- and H^+.

In the collecting ducts, bicarbonate ion reabsorption involves the luminal sodium-bicarbonate cotransporter (NBCe1-A) located on the apical membrane of collecting duct cells. NBCe1-A facilitates the reabsorption of bicarbonate ions in exchange for sodium ions (Na^+).

Bicarbonate ions are transported across the cells' basolateral membrane (facing the interstitial fluid) into the peritubular capillaries, facilitated by a sodium-bicarbonate cotransporter (NBCn1) and chloride-bicarbonate exchangers (AE1) on the basolateral membrane.

17.3.4.6 Calcium Reabsorption

Calcium ion (Ca^{2+}) reabsorption occurs at multiple sites along the renal tubules, including the proximal tubule, the thick ascending limb of the loop of Henle, the distal convoluted tubule, and the connecting tubule.

Calcium reabsorption in the proximal tubule is primarily passive and paracellular, which occurs through the tight junctions between tubular cells. As sodium ions (Na^+) are actively reabsorbed through sodium-glucose cotransporters (SGLTs) and sodium-phosphate cotransporters (NPT2), water follows passively, creating an osmotic gradient. This gradient drives the paracellular reabsorption of Ca^{2+} along with other cations. The thick ascending limb actively reabsorbs sodium ions (Na^+), potassium ions (K^+), and chloride ions (Cl^-) through the sodium-potassium-2 chloride cotransporter (NKCC2), also known as the bumetanide-sensitive cotransporter. Calcium reabsorption in this segment is also secondary to the transport of sodium and potassium ions. Calcium transport in the thick ascending limb is regulated by parathyroid hormone (PTH), which enhances calcium reabsorption in response to low blood calcium levels.

The DCT and the connecting tubule are responsible for fine-tuning calcium reabsorption and responding to hormonal signals. In the DCT, calcium reabsorption is regulated by the hormone aldosterone. Aldosterone stimulates the reabsorption of

Na^+, so Ca^{2+} is reabsorbed through the paracellular pathway. Parathyroid hormone (PTH) also acts on the DCT and enhances calcium reabsorption by increasing the activity of calcium channels (TRPV5) on the apical membrane, facilitating active calcium reabsorption. In the connecting tubule, calcium reabsorption is regulated by the sodium-calcium exchanger (NCX1), which exchanges Na^+ for Ca^{2+}.

In the collecting ducts, the reabsorption of calcium ions is minimal under normal circumstances. However, it can be influenced by PTH and other factors. PTH increases calcium reabsorption in these segments by enhancing the activity of calcium channels.

17.3.4.7 Magnesium Reabsorption

The majority of magnesium (Mg^{2+}) reabsorption occurs in the thick ascending limb of the loop of Henle. Still, some regulation and fine-tuning of magnesium levels also occur in the DCT and the collecting ducts. The thick ascending limb actively reabsorbs Na^+, K^+, and Cl^- through the sodium-potassium-2 chloride cotransporter (NKCC2), the bumetanide-sensitive cotransporter. The active reabsorption of these ions generates a positive electrical charge in the tubular lumen, which drives the paracellular reabsorption of cations, including Mg^{2+}.

In the DCT and connecting tubule, the reabsorption of magnesium ions is finely regulated. Parathyroid hormone (PTH) plays a significant role in magnesium reabsorption in these segments. PTH increases magnesium reabsorption by enhancing the activity of the transient receptor potential cation channel subfamily M member 6 (TRPM6) on the apical membrane of tubular cells. TRPM6 is a magnesium channel that allows the active reabsorption of magnesium ions from the tubular lumen into the tubular cells. PTH is released in response to low blood magnesium levels.

Under normal conditions, the collecting ducts' reabsorption of magnesium ions is minimal. However, in certain situations, such as severe magnesium deficiency, the collecting ducts can reabsorb more magnesium to help conserve this essential mineral.

17.3.4.8 Potassium Reabsorption

Most potassium (K^+) reabsorption occurs in the proximal tubule and the thick ascending limb of the loop of Henle. Approximately 65–70% of filtered potassium in the proximal tubule is passively reabsorbed with Na^+. It occurs via a paracellular mechanism and is directly proportional to water and Na^+ movement. The Na^+-K^+-ATPase causes Na^+ to move out of the proximal tubule cell and drives K^+ into the cell. The extrusion of Na^+ creates an osmotic gradient and an electrochemical gradient. Water moves from the PCT down the osmotic gradient and Cl^- down the electrochemical gradient. K^+ is reabsorbed and follows Cl^- into the bloodstream.

In the thick ascending limb of the loop of Henle, roughly 20% of K^+ is reabsorbed through the transcellular pathway.

Na^+-K^+-ATPase on the basolateral membrane pumps Na^+ out into the bloodstream and pumps K^+ into the thick ascending limb, keeping the sodium concentration in the cell low. This creates a gradient for the sodium-potassium-2 chloride

(NKCC2) cotransporter on the apical membrane. NKCC2 pumps Na^+, K^+, and 2 Cl^- into the cell from the lumen.

Intracellular K^+ can enter the bloodstream through the K^+-Cl^- symporter or the K^+ uniporter.

When the body attempts to preserve potassium, around 10% of filtered potassium is reabsorbed in the DCT via the transcellular pathway. Structurally, the initial collecting tubule and cortical collecting duct are both composed of 70% principal cells (secretion of K^+) and 30% intercalated cells (reabsorption of K^+). The apical H^+-K^+-ATPase drives H^+ into the lumen and K^+ into the intercalated cell. Once inside the cells, the basolateral K^+ channel allows the K^+ to leak out into the bloodstream.

17.3.4.9 Phosphate Reabsorption

The majority of phosphate (P_i) reabsorption occurs in the PCT. The mechanisms involved in phosphate reabsorption are highly regulated and involve specific transporters on the luminal and basolateral membranes of tubular cells. P_i is filtered at the glomerulus and enters the tubular fluid. It is present in both inorganic (HPO_4^{2-}) and organic forms. Phosphate reabsorption in the PCT is primarily mediated by sodium-phosphate cotransporters (NPTs) located on tubular cells' luminal (apical) membrane. Two types of NPTs are involved in phosphate reabsorption: NPT2a and NPT2c. These cotransporters actively transport phosphate ions from the tubular lumen into the tubular cells against their concentration gradient. The energy for this active transport is derived from the sodium gradient established by the Na^+/K^+-ATPase pump on the basolateral membrane. Once inside the tubular cells, phosphate ions exit through phosphate transporters (P_iT1 and P_iT2) on the basolateral membrane and enter the peritubular capillaries to return to the bloodstream. In the DCT and collecting ducts, phosphate reabsorption is minimal and occurs passively.

17.3.5 Secretion Mechanisms

17.3.5.1 Potassium Secretion

The mechanisms of potassium (K^+) secretion in the renal tubules involve intricate processes in both the distal convoluted tubules and the collecting ducts of the nephron. One of the primary mechanisms for potassium secretion occurs in the principal cells of the DCT and the cortical collecting ducts (CCD). Aldosterone plays a central role in this process. In response to low plasma potassium levels or high aldosterone levels, aldosterone binds to mineralocorticoid receptors on the principal cells. This binding activates the Na^+/K^+-ATPase pump on the basolateral membrane of the principal cells.

This pump actively transports Na^+ ions into the cell while pumping potassium ions out, creating a concentration gradient. Potassium channels, such as ROMK (renal outer medullary potassium), are present on the apical membrane of the principal cells. These channels facilitate the movement of potassium ions from the cell into the lumen of the tubule, contributing to potassium secretion. The connecting tubule (CNT), located between the DCT and the CCD, also participates in potassium secretion. This process is flow-dependent, meaning the filtrate's flow rate influences the secretion rate. High flow rates in the CNT increase potassium secretion. This is because an increased flow rate leads to a greater luminal positivity due to sodium reabsorption in the DCT. The increased luminal positivity enhances the driving force for potassium secretion through apical potassium channels. In addition to active secretion, passive paracellular potassium secretion can occur in the distal nephron segments. This involves the movement of potassium ions through tight junctions between tubular cells. Various factors, including changes in the luminal potassium concentration and the electrochemical gradient across the tubular epithelium, can influence the paracellular secretion of potassium.

17.3.5.2 Hydrogen Ion Secretion

The majority of hydrogen (H^+) ion secretion occurs in the PCT. Here, hydrogen ions are actively transported from the cytoplasm of tubular cells into the lumen of the nephron. The primary transporters responsible for this process are the sodium-hydrogen exchangers (NHEs), specifically NHE3. NHE3 exchanges Na^+ from the lumen of the tubule for hydrogen ions from the tubular cell cytoplasm. Hydrogen ion secretion also occurs in the DCT and the collecting ducts. In these segments, hydrogen ions are primarily secreted by the intercalated cells, specifically the type A intercalated cells. Proton secretion in these cells involves the activity of proton pumps, particularly the H^+-ATPase pump, located on the luminal membrane of intercalated cells. This pump actively transports hydrogen ions from the cytoplasm into the tubular lumen. A luminal potassium-hydrogen exchanger (kaliuretic proton pump) allows for the simultaneous secretion of hydrogen ions and reabsorption of potassium ions (K^+). Aldosterone can indirectly influence hydrogen ion secretion. It primarily acts on the principal cells in the collecting ducts to regulate sodium and potassium transport. Aldosterone increases sodium reabsorption and potassium secretion, affecting the electrochemical gradient and indirectly impacting hydrogen ion secretion.

In metabolic acidosis or increased acid load, the kidneys can utilize ammonia (NH_3) to help buffer and excrete excess H^+.

17.3.5.3 Ammonium Secretion

The renal tubules primarily secrete ammonium ions (NH_4^+) in response to metabolic acidosis or an increased acid load in the body. The proximal convoluted tubules (PCT) are central in producing ammonium ions. The tubular cells take up glutamine, an amino acid filtered through the glomerulus. Within the tubular cells, glutamine is metabolized by glutaminase enzyme to produce NH_4^+. The NH_4^+ are then actively secreted into the tubular lumen through the action of specific transporters, such as the ammonia transporters (RhCG and RhBG), located on the luminal membrane of tubular cells. In the medullary collecting duct, type A intercalated cells are primarily responsible for secreting ammonium ions into the urine. H^+-K^+-ATPase pumps on the luminal membrane of type A intercalated cells actively pump both H^+ and K^+ into the tubular lumen, increasing luminal positivity. The increased luminal positivity facilitates the secretion of ammonia (NH_3) through the luminal ammonia transporter (RhCG) in the apical membrane of type A intercalated cells, which in the lumen combines with H^+ and forms NH_4^+. The NH_4^+ combines with Cl^- in the tubular lumen to form ammonium chloride (NH_4Cl), which is then excreted in the urine.

The secretion of ammonium ions in the renal tubules is tightly regulated to maintain acid-base homeostasis in the body. When the body experiences metabolic acidosis or an increased acid load, the kidneys increase the production and secretion of ammonium ions to help excrete excess hydrogen ions. Conversely, in situations of metabolic alkalosis, the secretion of ammonium ions is decreased to conserve these ions and retain hydrogen ions in the tubules. Aldosterone can indirectly affect ammonium secretion by regulating Na^+ and K^+ transport in the distal nephron. The reabsorption of sodium and secretion of potassium can influence the luminal electrical potential, affecting ammonium ion secretion.

17.3.5.4 Organic Acid and Base Secretion

The kidneys are essential for eliminating endogenous and exogenous substances, including drugs, toxins, and metabolic waste products like uric acid.

The secretion of organic acids primarily occurs in the PCT. Organic acids, such as endogenous uric acid, and exogenous substances, like certain drugs and metabolites, are actively taken up from the peritubular capillaries into the renal tubular cells. Transporters responsible for this step include organic anion transporters (OATs), specifically OAT1 (SLC22A6) and OAT3 (SLC22A8), located on the basolateral membrane of the tubular cells. Once inside the tubular cells, the organic acids may undergo intracellular processing, including metabolism or binding to carrier proteins. These processes can alter the chemical properties of the organic acids, making them more amenable to secretion. The processed or unprocessed organic acids are actively secreted into the lumen of the proximal tubule. This secretion is mediated by transporters on the luminal membrane of tubular cells, such as multidrug resistance-associated proteins (MRPs) and breast cancer resistance protein (BCRP). Proton-coupled exchange mechanisms are often involved, with H^+ being exchanged for organic acids. Organic acids, now in the tubular lumen, are eliminated in the urine.

The secretion of organic bases follows a similar pattern but involves different transporters and mechanisms:

Organic cation transporters (OCTs), such as OCT2 (SLC22A2), located on the basolateral membrane of the tubular cells, are responsible for taking up these bases from the peritubular capillaries into the renal tubular cells. Transporters on the luminal membrane of tubular cells, such as multidrug and toxin extrusion proteins (MATEs), mediate the secretion of processed or unprocessed organic bases into the lumen of the proximal tubule.

17.3.5.5 Urea Secretion

Urea is a nitrogenous waste product formed in the liver during protein metabolism and is excreted primarily through the kidneys, maintaining nitrogen balance. The initial step in the excretion of urea is filtration at the glomerulus. Urea is a small, water-soluble molecule that is freely filtered from the blood into Bowman's space, along with other small molecules and ions. After filtration, some of the filtered urea is reabsorbed in the PCT. Approximately 50–70% of the filtered urea is reabsorbed in the PCT. Urea reabsorption in the PCT is primarily passive. It occurs through the paracellular pathway, driven by the concentration gradient created by the reabsorption of water and solutes (e.g., sodium and glucose) in this tubule segment.

Urea secretion mainly occurs in the medullary collecting duct, particularly in the inner medullary collecting duct (IMCD) and the papillary ducts. The secretion of urea is a crucial component of the renal countercurrent system, which helps in urine concentration and water preservation. The urea secretion in these segments is facilitated by urea transporters, particularly the UT-A1 and UT-A3 transporters, which are expressed on the apical membrane of the collecting duct cells. These transporters mediate the active secretion of urea from the tubular cells into the tubular lumen. ADH stimulates the secretion of urea.

17.3.5.6 Bicarbonate Secretion

While the primary role of the renal tubules is to reabsorb bicarbonate (HCO_3^-) to prevent its loss, there are specific mechanisms for HCO_3^- secretion in certain segments of the renal tubules when needed to excrete excess bicarbonate or regulate blood pH. The main segments involved in bicarbonate secretion are the PCT, the DCT, and the collecting ducts.

Bicarbonate secretion in the PCT is minimal under normal conditions. Cases where bicarbonate secretion occurs in the PCT typically involve the following steps: bicarbonate is transported from the peritubular capillaries into the tubular cells of the PCT through the action of sodium-bicarbonate cotransporters (NBCe1) on the basolateral membrane. Inside the tubular cells, bicarbonate combines with H^+ to form carbonic acid (H_2CO_3) under the influence of cytoplasmic carbonic anhydrase enzymes. H_2CO_3 dissociates into H_2O and carbon dioxide (CO_2). CO_2 can diffuse into the tubular lumen, where it combines with water to form H_2CO_3 again. H_2CO_3 dissociates into HCO_3^- and H^+ and excreted.

17.3.5.7 Peptide and Protein Secretion

While small peptides and proteins are typically filtered through the glomerulus, they are reabsorbed in the proximal tubules. However, specific larger peptides and proteins, such as hormones like insulin, glucagon, parathyroid hormone, and erythropoietin (EPO); some enzymes, such as angiotensin-converting enzyme (ACE); and specific cytokines and growth factors, are actively secreted into the renal tubules to be excreted in the urine. Active secretion of peptides and proteins in the renal tubules happens through the endocytosis-exocytosis mechanism, an energy-dependent process in PCT and DCT. Large proteins and peptides are endocytosed into the tubule cells. Inside the cells, they may or may not be processed. The last stage is exocytosis of the protein-containing vesicle through the luminal membrane of the tubule cells, releasing the proteins into the tubular lumen.

17.3.6 Concentrating Urine

The urine concentration is essential for maintaining fluid and electrolyte balance. This process involves the reabsorption of water from the filtrate produced by glomerular filtration, leading to the excretion of more concentrated urine than the original plasma.

17.3.6.1 Countercurrent Multiplier

Countercurrent multiplication (Fig. 17.13) takes place in the loop of Henle. This system's efficiency lies in its ability to create a high osmotic gradient in the renal medulla, which is essential for water reabsorption from the collecting ducts and thereby concentrates the urine.

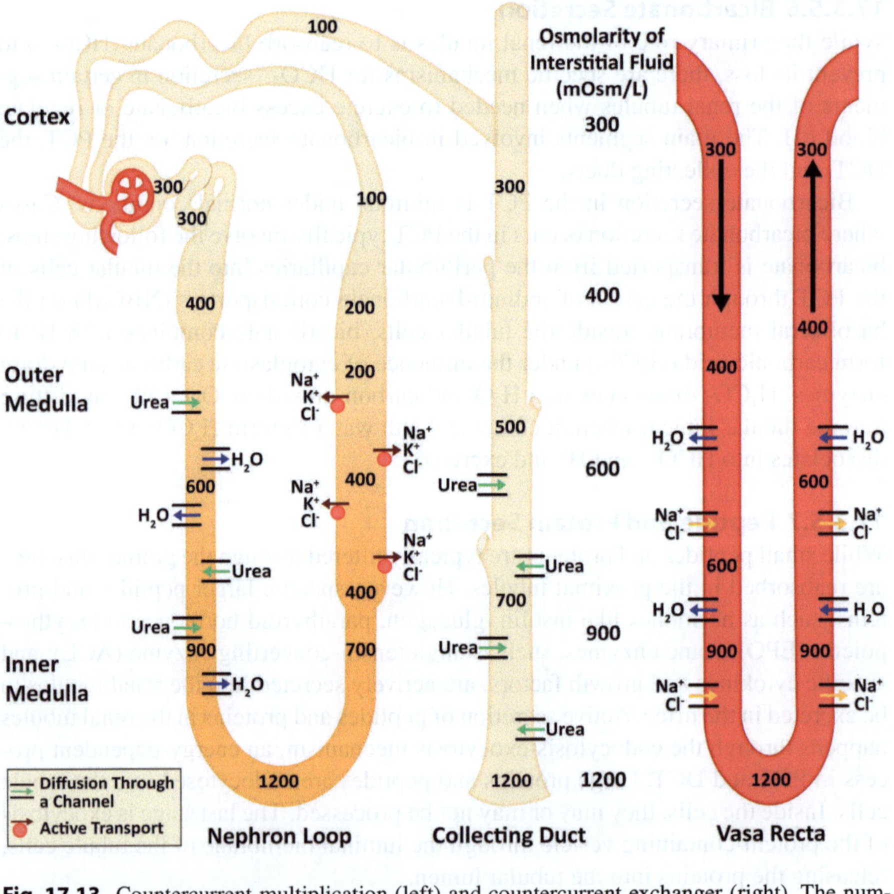

Fig. 17.13 Countercurrent multiplication (left) and countercurrent exchanger (right). The numbers in the illustration show the osmolality (mOsm/L)

Mechanism of Countercurrent Multiplication In the descending limb, water passively exits the tubular fluid due to the osmotic gradient created by the increased osmolarity in the medulla. This makes the fluid in the descending limb more concentrated. The thick ascending limb actively pumps Na^+ and Cl^- ions out into the surrounding interstitial fluid, creating a high osmolarity in the renal medulla. Due to its impermeability to water, this limb dilutes the tubular fluid by removing solutes without allowing water to follow.

The countercurrent flow of the fluid in the loop (descending fluid becoming more concentrated and ascending fluid becoming more diluted) multiplies the osmotic gradient in the medulla. This is why it's termed "countercurrent multiplication." Urea also contributes to the osmolarity of the renal medulla. It is partially reabsorbed in the inner medullary collecting ducts and secreted into the thin ascending limb, further enhancing the osmotic gradient.

The osmotic gradient created by countercurrent multiplication allows for water reabsorption from the collecting ducts, leading to the production of concentrated urine. This is crucial for water conservation, especially in states of dehydration. The extent of water reabsorption in the collecting ducts is regulated by ADH, which adjusts the permeability of the ducts to water in response to the body's hydration status.

17.3.6.2 Countercurrent Exchanger

The countercurrent exchanger (Fig. 17.13), working in tandem with the countercurrent multiplier system in the loop of Henle, is an essential component of the urinary concentration mechanism. This exchanger primarily involves the vasa recta, the capillary networks that supply blood to the kidneys' medulla.

Mechanism of Countercurrent Exchange As blood descends into the medulla in the descending limb of the vasa recta, it becomes more concentrated due to the loss of water and gain of solutes (mainly Na^+ and Cl^-), which move into the blood from the interstitial fluid. In the ascending limb, the blood loses solutes and gains water, diluting as it moves back toward the cortex. This movement of solutes and water occurs primarily due to the high osmolarity of the medullary interstitial fluid created by the countercurrent multiplier system in the loop of Henle. The countercurrent exchange system efficiently allows the blood to supply nutrients and remove waste from the medullary tissue without disrupting the osmotic gradient.

The countercurrent exchanger is critical for maintaining the renal medulla's high osmolarity. Without it, the blood flowing through the medulla would wash away the gradient, making it impossible for the kidneys to produce concentrated urine. The countercurrent exchanger sustains the hyperosmotic environment in the medulla, which is essential for reabsorbing water from the collecting ducts, which ADH regulates.

17.3.6.3 Role of Urea

Urea plays a significant role in the kidney's ability to concentrate urine. While urea is often considered a waste product of protein metabolism, its role in renal physiology is complex and essential for maintaining the osmotic balance necessary for effective urine concentration.

Urea is generated primarily in the liver as a byproduct of protein and amino acid metabolism. Approximately 40–50% of the urea filtered by the glomerulus is reabsorbed in the PCT. Further reabsorption occurs in the inner medullary collecting ducts under the influence of ADH. When ADH levels are high, as in conditions of dehydration, there is increased permeability to urea in the inner medullary collecting ducts. This accumulates urea in the renal medulla, significantly contributing to the medullary osmotic gradient. The accumulation of urea in the medulla complements the counter-current multiplication and exchange mechanisms operating in the loop of Henle and vasa recta. It enhances the osmotic gradient essential for water reabsorption from the collecting ducts. ADH upregulates the expression of urea transporters, particularly UT-A1 and UT-A3, in the inner medullary collecting ducts. This increases urea reabsorption and thus contributes to the high osmolality of the renal medulla.

The high osmolarity in the medulla, to which urea contributes, is crucial for water reabsorption from the tubular fluid as it passes through the collecting ducts. This process results in the production of concentrated urine. The kidney balances the excretion and recycling of urea to maintain osmotic balance. In states of dehydration, more urea is recycled to enhance water reabsorption, whereas in hydrated states, less urea is recycled, resulting in more dilute urine.

17.3.6.4 Formation of Dilute Urine

When hydration levels are high, ADH secretion is suppressed. This reduction in ADH decreases the permeability of the collecting ducts to water. The countercurrent multiplier system in the loop of Henle creates an osmotic gradient, but with low ADH, the collecting duct remains relatively impermeable to water. Without significant ADH effects, water is not effectively reabsorbed in the DCT and collecting duct. Consequently, the fluid remains dilute. The result is a larger volume of dilute urine with low osmolarity, which helps eliminate excess water and maintain electrolyte balance.

17.3.6.5 Formation of Concentrated Urine

In response to dehydration or high plasma osmolarity, ADH is secreted by the posterior pituitary gland. ADH increases the permeability of the collecting ducts to water by stimulating the insertion of aquaporin-2 channels. The high osmolarity in the medulla, maintained by the countercurrent multiplier mechanism in the loop of Henle and urea recycling, provides the osmotic gradient necessary for water reabsorption from the collecting ducts. The countercurrent multiplier system in the loop of Henle concentrates the tubular fluid by reabsorbing sodium and chloride in the thick ascending limb, which is impermeable to water. In the DCT and collecting duct, under the influence of ADH, water is reabsorbed into the surrounding hyper-osmotic medullary interstitial fluid, leading to an increased concentration of the remaining tubular fluid. Increased urea recycling in the medullary collecting ducts also contributes to the medullary osmolarity, aiding in water reabsorption. Aldosterone, another hormone, primarily affects sodium and potassium balance. It prompts sodium reabsorption (and water indirectly) in the DCT and collecting duct, which can also contribute to urine concentration. The result is a smaller volume of highly concentrated urine with high osmolarity, conserving water and efficiently eliminating waste.

17.3.7 Urine

Urine is a complex aqueous solution composed primarily of water and diverse dissolved substances. These substances are waste products of bodily metabolism, excess nutrients, electrolytes, and other compounds. The chemical composition of urine can vary depending on factors such as diet, hydration status, health conditions, and medication use. Urine has several physical characteristics (Table 17.2) that can provide insights into a person's health and hydration status. Several key constituents are commonly found in urine (Table 17.3).

Table 17.2 Urine characteristics: normal values and clinical significance

Characteristic	Description and normal values	Clinical significance
Color	Ranges from pale yellow to deep amber, depending on hydration	Darker urine can indicate dehydration, while very light urine often reflects adequate hydration. Some foods, medications, and diseases can also alter urine color
Transparency	Typically, clear	Cloudiness can suggest the presence of substances like phosphates, pyuria, or infection
Odor	Mild odor, often described as "nutty"	Strong or unusual odors may be due to dietary factors, dehydration, or metabolic disorders. Ammonia-like smell can indicate infection
pH	Ranges from 4.5 to 8.0, with an average around 6.0	A consistently high or low pH can indicate certain metabolic or renal disorders. Diet can also influence urine pH
Specific gravity	Normally ranges from 1.005 to 1.025	Low specific gravity can indicate overhydration or impaired renal concentrating ability. High specific gravity suggests dehydration or, rarely, glycosuria
Volume	Averages about 800 to 2000 mL/24 hours (varies based on hydration and kidney function)	Abnormally high volumes may indicate conditions like diabetes insipidus or chronic kidney disease, while low volumes can be due to dehydration or acute kidney injury

Table 17.3 Urine chemical components: normal values and clinical significance

Chemical component	Normal values	Clinical significance
Urea	12–20 g/24 hours	Elevated levels may indicate a high-protein diet, high-protein catabolism, or reduced renal function
Creatinine	Men: 14–26 mg/kg/24 hours Women: 11–20 mg/kg/24 hours	Altered levels can suggest changes in muscle mass or kidney function
Uric acid	250–750 mg/24 hours	Increased levels can be seen in gout, a high purine diet, or renal insufficiency. Decreased levels are less common
Sodium	40–220 mEq/24 hours	Abnormal excretion may indicate kidney disease, dietary intake, or fluid balance disorders
Potassium	25–125 mEq/24 hours	Variations may indicate changes in dietary intake, renal function, or certain metabolic disorders
Chloride	110–250 mEq/24 hours	Altered levels can reflect dietary intake, fluid balance, or kidney function
Calcium	100–300 mg/24 hours	High levels may indicate hypercalciuria, which can be associated with kidney stones, hyperparathyroidism, or other conditions. Low levels are less common
Phosphate	400–1300 mg/24 hours	Elevated levels can be associated with kidney disease or parathyroid disorders; low levels can result from malabsorption
Magnesium	75–150 mg/24 hours	Abnormal levels may be related to dietary intake, kidney function, or metabolic conditions
Protein	Less than 150 mg/24 hours	Proteinuria may indicate kidney damage, urinary tract infection, or other renal diseases
Glucose	Normally not present or minimal (<130 mg/24 hours)	Presence of glucose (glycosuria) can indicate diabetes mellitus or renal glycosuria
Ketones	Normally not present	Presence can indicate fat metabolism alteration due to diabetes, starvation, or a high-fat diet
Bilirubin	Normally not present	The presence of bilirubin can suggest liver disease or biliary obstruction
Hemoglobin	Normally not present	Presence can be due to hemolysis, bleeding in the urinary tract, or some renal diseases
Nitrites	Normally not present	Positive nitrites can indicate bacterial infection, commonly urinary tract infection
Leukocyte esterase	Normally not present	Presence suggests inflammation or infection in the urinary tract, often seen in urinary tract infections

17.4 Part 4: Collecting System

The collecting system in the kidney is responsible for transporting, storing, and eventually excreting urine. This system comprises several structures that work in a coordinated manner to ensure efficient urine handling, from its formation in the nephrons to its excretion from the body.

17.4.1 Renal Papillae

The renal papilla forms the renal pyramids' apex in the kidney's medulla (Fig. 17.14). Each kidney contains several renal papillae, corresponding to the number of renal pyramids present (usually 8 to 18 in humans). They point toward the center of the kidney, where they interface with the minor calyces. Each papilla is conical or pyramid-shaped, projecting into a minor calyx. The size of the papillae can vary, but they are typically a few millimeters in diameter. The surface of the renal papilla is covered by transitional epithelium, also known as urothelium, which is continuous with the lining of the calyces, pelvis, and ureters. The apex of each papilla contains the openings of several (usually 10 to 20) ducts of Bellini, or papillary ducts. These ducts are the terminal portions of the collecting ducts where urine exits into the minor calyx.

17.4.2 Minor Calyces

Each kidney typically contains several minor calyces (usually 8 to 18), corresponding to the number of renal pyramids. Minor calyces are cup-shaped or funnel-like structures. They vary in size but are generally small, with a diameter sufficient to snugly encapsulate the renal papilla (Fig. 17.15). The interior of each minor calyx is lined with transitional epithelium. The epithelium is specialized to be stretchable and impermeable, characteristics essential for accommodating fluctuations in urine volume while preventing absorption. Beneath the epithelium, there is a layer of smooth muscle. This layer facilitates the peristaltic movements that propel urine

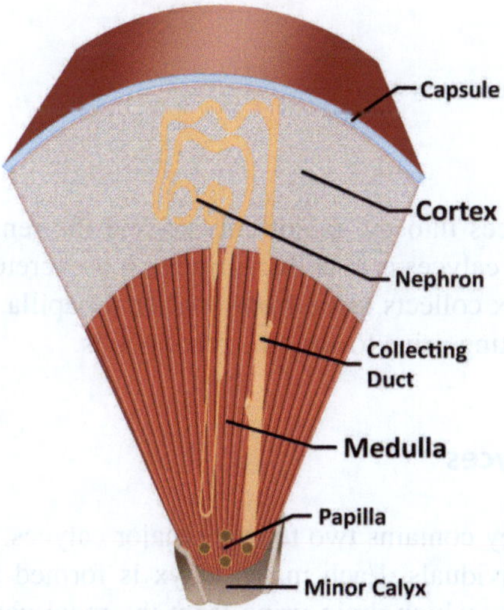

Fig. 17.14 Renal pyramid and associated minor calyx

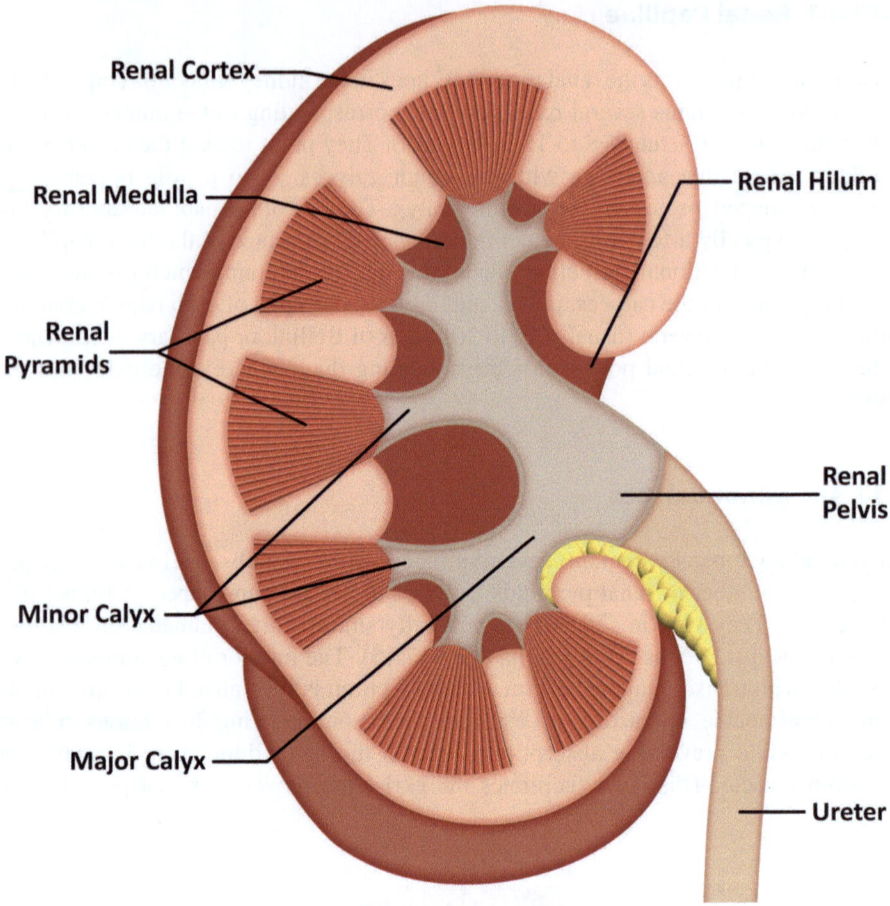

Fig. 17.15 Urinary collection system

from the minor calyces into the major calyces and the renal pelvis. The primary function of the minor calyces is to collect urine as it is excreted from the renal papillae. Each minor calyx collects urine from one renal papilla. They act as the initial conduits for transporting urine toward the renal pelvis.

17.4.3 Major Calyces

Typically, each kidney contains two to three major calyces. However, this number can vary among individuals. Each major calyx is formed by the convergence of several minor calyces, which drain urine from the renal papillae at the tips of the renal pyramids. The major calyces are located in the renal sinus, a cavity within the kidney that is filled with fat and connective tissue. They converge to form the renal pelvis, which narrows to become the ureter (Fig. 17.15). Their microscopic structure is similar to the minor calyces. The primary function of the major calyces is to collect urine from the minor calyces and transport it to the renal pelvis.

17.4.4 Renal Pelvis

The renal pelvis is a funnel-shaped reservoir for collecting urine from the major calyces and transporting it to the ureter (Fig. 17.15). Situated within the renal sinus, the renal pelvis is centrally located in the kidney where the major calyces converge. The size and shape of the renal pelvis can vary among individuals and change depending on the volume of urine it contains. The renal pelvis is lined with transitional epithelium. The wall of the renal pelvis contains layers of smooth muscle fibers. These muscles are not as well defined as the ureter but can produce peristaltic movements that help propel urine toward the bladder. The renal pelvis collects urine from the major calyces and narrows as it exits the kidney to become the ureter.

17.4.5 Ureters

The ureters are the conduits for urine transport from the kidneys to the bladder (Fig. 17.1). The ureters originate at the renal pelvis of each kidney, located in the retroperitoneal space. They descend along the posterior abdominal wall to enter the pelvis and connect to the urinary bladder. There are two ureters, one for each kidney. They run bilaterally, parallel to the vertebral column. Each ureter is approximately 25 to 30 cm long and 3 to 4 mm in diameter. The diameter can vary, especially at three physiological constrictions: the ureteropelvic junction (UPJ), where the ureter crosses the iliac vessels, and the ureterovesical junction (UVJ) (Fig. 17.16). The innermost layer of the ureter is lined with transitional epithelium. This type of epithelium is unique to the urinary tract and is highly specialized to accommodate stretching and changes in volume. The transitional epithelium cells can appear cuboidal when the ureter is empty and flattened when it is distended. This adaptability prevents damage to the cells during the passage of urine. Beneath the epithelium lies the lamina propria, a connective tissue layer providing support and elasticity. The muscularis layer consists of smooth muscle fibers arranged in two or three layers. Typically, there is an inner longitudinal layer and an outer circular layer of muscle fibers. In the lower part of the ureter, a third outer longitudinal layer may be present. The arrangement of smooth muscle fibers facilitates effective peristalsis. The contraction of these muscles in a coordinated wave-like manner propels urine downward. Adventitia is a layer of loose connective tissue that envelops the ureter. It provides structural support and anchors the ureter in place. This layer contains the blood vessels, lymphatics, and nerves that supply the ureter.

The ureters enter the bladder obliquely at the lower end, forming the ureterovesical junction (UVJ). There is one UVJ for each ureter situated at the posterior aspect of the bladder. The ureters penetrate the bladder wall obliquely and run within the

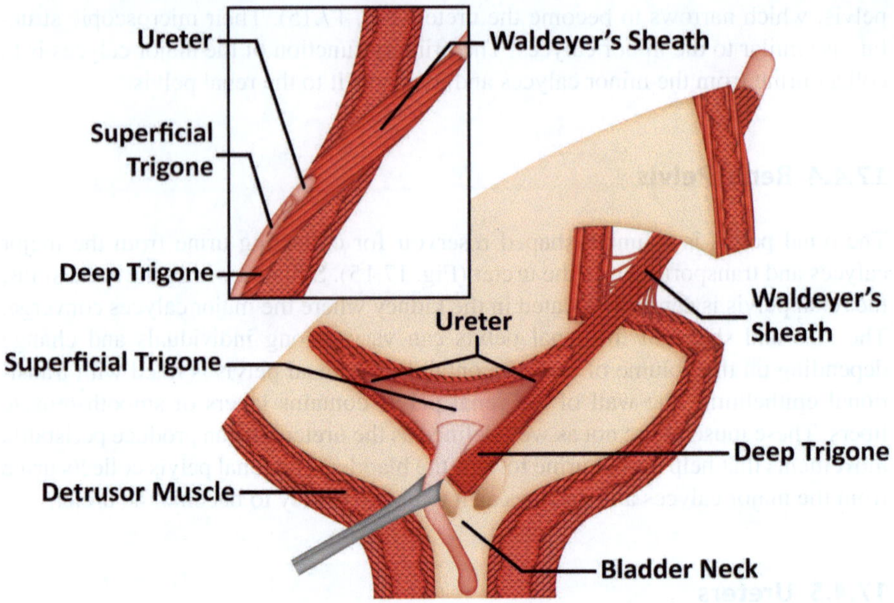

Fig. 17.16 Structure of the ureterovesical junction and trigone

bladder wall a short distance before opening into the bladder lumen. This oblique course, typically about 1.5 to 2 cm long, is crucial for the valve-like mechanism preventing urine reflux. A specialized fibromuscular tissue, Waldeyer's sheath, surrounds the distal ureter at the UVJ. This structure contributes to the competence of the junction to prevent reflux. The UVJ is an anti-reflux mechanism. When the bladder fills and the detrusor muscle expands, it compresses the intramural part of the ureter (the portion within the bladder wall), effectively closing the ureteral orifice and preventing backflow. During bladder contraction (urination), the increased pressure within the bladder does not lead to reflux due to the oblique course of the ureter and the valve-like action of the ureteral orifice.

17.4.6 Urinary Bladder

The urinary bladder is a hollow, muscular organ that stores and periodically eliminates urine.

17.4.6.1 Anatomy

The bladder is in the pelvic cavity, posterior to the pubic symphysis. In males, it lies in front of the rectum, and in females, it is anterior to the vagina and inferior to the uterus. When empty, the bladder resembles a flattened sphere and is located within the pelvis (Fig. 17.17). As it fills with urine, it expands and can extend into the lower abdominal cavity. Its size and shape vary depending on the urine volume and the

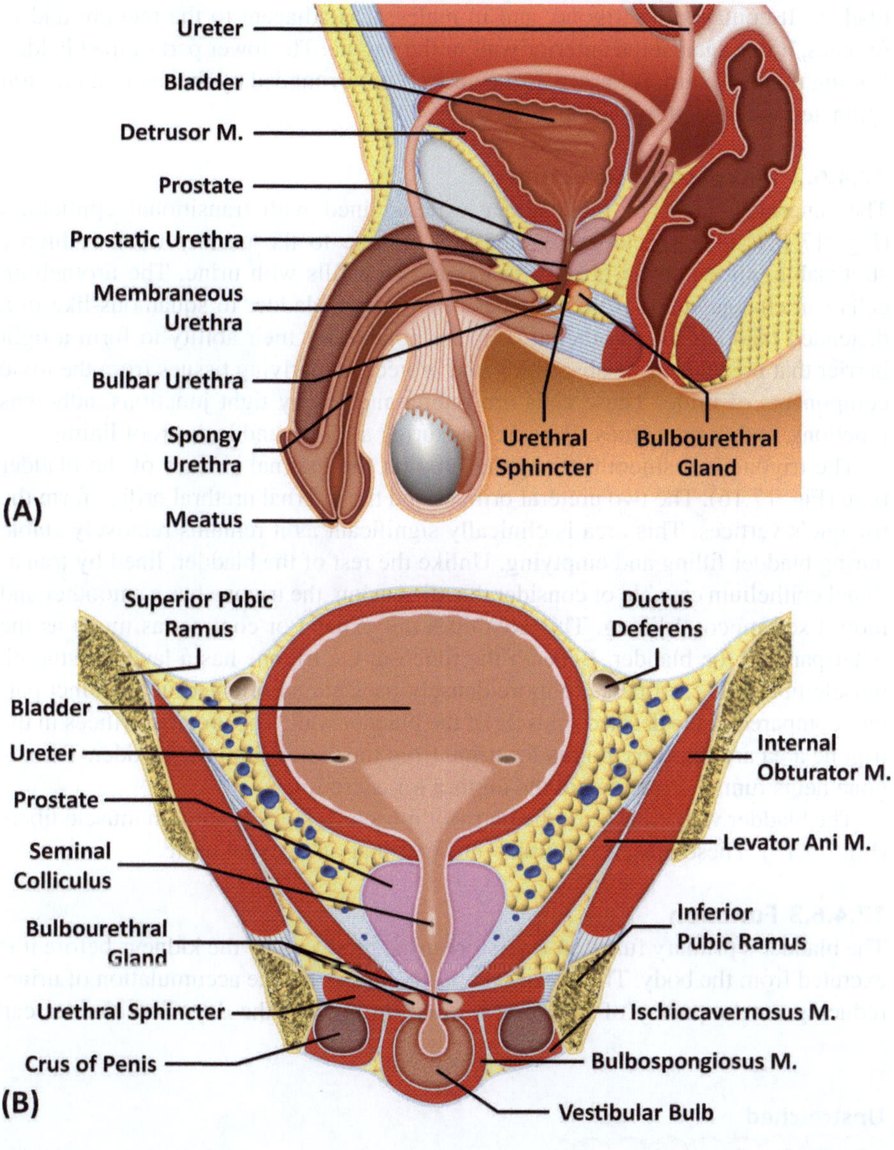

Fig. 17.17 Male bladder and urethra. (**a**) Midsagittal section. (**b**) Coronal section

individual's age, gender, and overall anatomy, typically holding about 400–600 mL of urine.

The anterior portion of the bladder, called the apex, points toward the upper front of the pelvic cavity and is connected to the umbilicus by the median umbilical ligament. The body is the central portion of the bladder, between the apex and the base, which expands as the bladder fills with urine. The base is the posterior surface of the bladder. It contains the trigone, and in males, it is adjacent to the rectum, and in females, it lies against the anterior wall of the vagina. The lower part of the bladder, leading to the urethra, is known as the neck. It is surrounded by the internal urethral sphincter, which controls urine release.

17.4.6.2 Microscopic Structure

The innermost layer of the bladder wall is lined with transitional epithelium (Fig. 17.18). This specialized epithelium, unique to the urinary tract, is highly stretchable, allowing the bladder to expand as it fills with urine. The urothelium cells can change shape from cuboidal in an empty bladder to squamous-like in a distended bladder. The urothelial cells are notable for their ability to form a tight barrier that prevents urine absorption and protects underlying tissues from the toxic components of urine. These cells are interconnected by tight junctions, adherens junctions, and desmosomes (Page 52), ensuring a robust and leak-proof lining.

The trigone is a smooth, triangular area on the internal surface of the bladder base (Fig. 17.16). The two ureteral orifices and the internal urethral orifice form the triangle's vertices. This area is clinically significant as it remains relatively stable during bladder filling and emptying. Unlike the rest of the bladder, lined by transitional epithelium capable of considerable distension, the trigone has a smoother and more fixed mucosal lining. This area does not expand or contract as much as the other parts of the bladder. Beneath the mucosa, the trigone has a layer of smooth muscle fibers. These fibers are more densely packed and oriented in a distinct pattern compared to the detrusor muscle of the bladder wall. The ureteric orifices in the trigone area are the entry points for urine from the ureters into the bladder. The trigone helps funnel urine toward the urethra for excretion.

The bladder wall comprises the detrusor muscle, a layer of smooth muscle fibers (Fig. 17.17). These muscles contract during urination to expel urine.

17.4.6.3 Function

The bladder's primary function is to store urine produced by the kidneys before it is excreted from the body. This storage capacity allows for the accumulation of urine, reducing the frequency of urination needed throughout the day. The bladder can

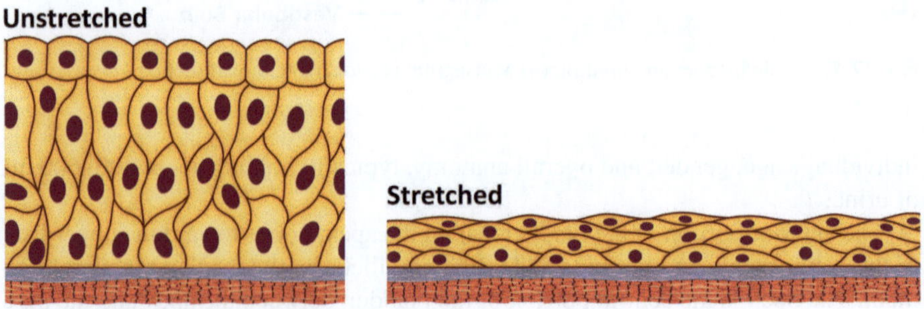

Fig. 17.18 Transitional epithelium

typically hold about 400–600 mL of urine, although this capacity can vary among individuals. The bladder facilitates the controlled release of urine through a process called micturition. This involves the coordination of the muscular bladder wall (detrusor muscle), the internal urethral sphincter, and the external urethral sphincter. The bladder's design, particularly the angle at which the ureters enter and the structure of the ureterovesical junction, helps prevent the backflow of urine into the ureters and kidneys. This mechanism protects the kidneys from potential damage due to increased pressure or infection. The bladder lining contains glycosaminoglycans, which form a protective barrier that helps prevent bacterial adherence and reduces the risk of urinary tract infections.

17.4.7 Urethra

The urethra is responsible for conducting urine from the urinary bladder to the outside of the body. Its structure varies significantly between males and females, mainly due to differences in length and function related to the reproductive system.

17.4.7.1 Male Urethra

The male urethra (Fig. 17.17) is significantly longer than its female counterpart, measuring approximately 19–20 cm long. The diameter of the urethra varies along its length, typically being the widest in the prostatic urethra and narrowest at the membranous urethra. The male urethra is divided into four main segments, each with unique anatomical features:

- Prostatic Urethra: The prostatic urethra is about 3–4 cm long. It begins at the base of the urinary bladder and passes through the prostate. The prostate gland sends several small prostatic ducts to the prostatic urethra.
- Membranous Urethra: It is approximately 1–2 cm long and passes through the urogenital diaphragm (Section 9.3.4.10), a muscular structure below the prostate. This is the narrowest and least dilatable part of the urethra. It is surrounded by the external urethral sphincter, which provides voluntary control over urination.
- Bulbourethral (or Bulbar) Urethra: About 3–4 cm long, it extends from the urogenital diaphragm to the beginning of the penile urethra. Encased within the bulb of the penis, this segment is surrounded by the bulbospongiosus muscle. It includes the openings of the bulbourethral (Cowper's) glands.
- Spongy (or Penile) Urethra: The length of the spongy urethra is approximately 15 cm, making it the longest segment, running along the length of the penis in the corpus spongiosum. It opens at the external urethral orifice at the tip of the penis (glans). Along its course, it receives the openings of the urethral glands.

17.4.7.2 Female Urethra

The female urethra (Fig. 17.19) measures approximately 4–6 cm long. It has a variable diameter, typically around 6 mm, but can expand to allow the passage of urine. The urethra is embedded in the anterior wall of the vagina and extends from the bladder's internal urethral orifice to the external urethral orifice in the vulva. It runs

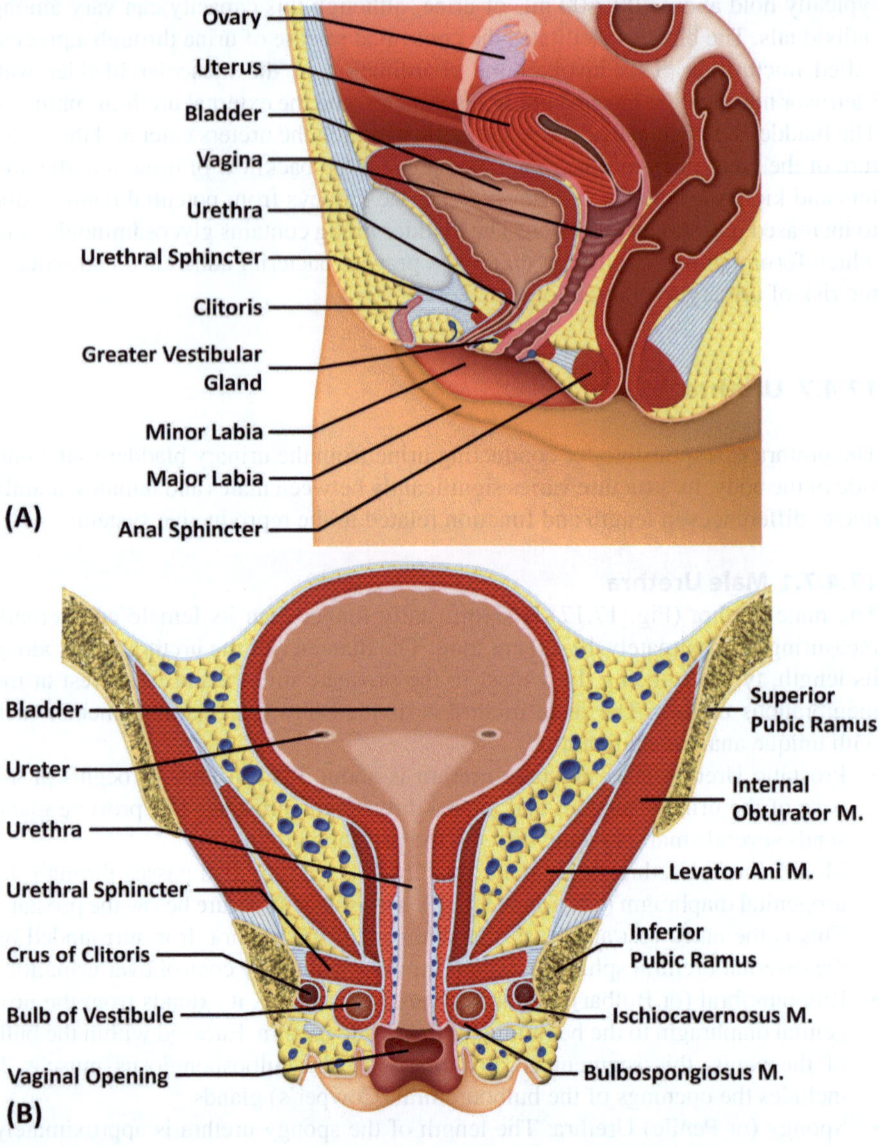

Fig. 17.19 Female bladder and urethra. (**a**) Midsagittal section. (**b**) Coronal section

a relatively straight course, slightly angling toward the front of the body. The external urethral orifice, or meatus, is located between the clitoris and the vaginal opening in the vulvar vestibule. Small periurethral (Skene's) glands near the external urethral orifice secrete mucus. They are homologous to the male prostate gland.

The innermost layer of the urethra is lined with a type of epithelium that varies along its length. In males and females, the part closest to the bladder is lined with transitional epithelium, which becomes stratified columnar and finally stratified squamous near the external urethral orifice.

17.4.7.3 Urethral Sphincters

There are two main sphincters: the internal urethral sphincter and the external urethral sphincter (Figs. 17.16 and 17.19). The internal urethral sphincter is situated at the junction of the bladder and the urethra, known as the bladder neck. It is composed of smooth muscle fibers that are continuous with the detrusor muscle of the bladder. This sphincter is involuntary, meaning the autonomic nervous system regulates it. The primary function of the internal sphincter is to maintain continence, especially when there is increased abdominal pressure, such as coughing or sneezing. It also prevents the backflow of semen into the bladder during ejaculation in males (retrograde ejaculation). The sympathetic nervous system predominantly controls the internal sphincter. Sympathetic activation causes it to contract, thereby closing the urethral opening and retaining urine in the bladder. The external urethral sphincter is located in the urogenital diaphragm, inferior to the internal sphincter. In males, it surrounds the membranous urethra, while in females, it is positioned near the mid-urethra. This sphincter is composed of striated muscle fibers, and unlike the internal sphincter, it is under voluntary control. It forms a circular band around the urethra and can be consciously contracted to inhibit urination. The pudendal nerve (S2–S4) innervates the external sphincter, which provides the voluntary control necessary for its function.

17.4.8 Urination

Urination, or micturition, is the process of expelling urine from the urinary bladder to the outside of the body. It can be divided into two phases: the storage phase and the voiding phase.

Storage Phase: During this phase, urine is collected in the bladder, gradually stretching to accommodate the increasing volume. The internal urethral sphincter (involuntary smooth muscle) and the external urethral sphincter (voluntary skeletal muscle) remain contracted to retain urine in the bladder.

Voiding Phase: This phase involves urinating, where urine is expelled from the bladder.

Urination is controlled by the micturition reflex, commonly known as the urination reflex. It is primarily a spinal reflex mediated by the parasympathetic nervous system and modulated by higher brain centers in the cerebral cortex and the pontine micturition center (PMC). This reflex coordinates the filling and emptying of the bladder, balancing the need to expel urine regularly with storing it conveniently between voiding.

The urination reflex consists of several steps:

Filling Phase: As the bladder fills with urine, its wall stretches. Mechanoreceptors (stretch receptors) in the bladder wall detect this stretching.

Reflex Initiation: When the bladder reaches a certain level of distension (usually 200–400 mL of urine), the stretch receptors send signals to the sacral spinal cord, triggering the micturition reflex.

Afferent Signals: The reflex's afferent pathways travel from the bladder to the spinal cord and then to the brain, particularly the pontine micturition center and the cerebral cortex.

Efferent Responses: In response to these signals, efferent impulses are sent back to the bladder via the parasympathetic pelvic nerves. These impulses cause the detrusor muscle (the muscular layer of the bladder wall) to contract and the internal urethral sphincter to relax, facilitating urine flow toward the urethra.

Voluntary Control: The external urethral sphincter, controlled by somatic nerves (pudendal nerve), can be voluntarily contracted to delay urination. The brain exerts conscious control over micturition, allowing individuals to choose an appropriate time and place for urination.

Voiding Phase: When it is suitable to void, the brain sends signals to relax the external urethral sphincter. The detrusor muscle contracts, the internal urethral sphincter relaxes, and urine is expelled from the bladder.

Suggested Readings

Advanced Physiology and Pathophysiology: Essentials for Clinical Practice, 2nd Ed. Nancy Tkacs, Linda Herrmann, Springer Publishing Company, 2024

Anatomy & Physiology, S. Meloni, et al. Independent Publisher, 2023

Anatomy & Physiology: The Unity of Form and Function, 10th Ed. Kenneth S. Saladin, McGraw Hill, 2023

Gray's Anatomy for Students, 5th Ed. Richard L. Drake, et al. Elsevier, 2020

Human Anatomy & Physiology, 11th Ed. Elaine N. Marieb and Katja Hoehn, Pearson, 2022

Moore's Clinically Oriented Anatomy, 9th Ed. Keith L. Moore, et al. Lippincott Williams & Wilkins, 2020

Netter Atlas of Human Anatomy: Classic Regional Approach, 8th Ed. Frank H. Netter, Elsevier, 2022

Principles of Anatomy and Physiology, 16th Ed. Gerard J. Tortora, Bryan H. Derrickson, Wiley, 2020

Fluid, Electrolyte, and Acid-Base Balance 18

In this chapter, you will learn:

- Body fluid compartments: distinction between intracellular fluid (ICF) and extracellular fluid (ECF).
- Fluid balance: mechanisms maintaining fluid balance for homeostasis, including input and output regulation.
- Sensing mechanisms for water regulation: role of osmoreceptors and baroreceptors in monitoring blood osmolarity and pressure.
- Hormonal regulation of water: functions of antidiuretic hormone (ADH), aldosterone, and atrial natriuretic peptide (ANP).
- Electrolyte functions and homeostasis: functions and regulation of electrolytes like sodium, potassium, calcium, and magnesium.
- Acid-base homeostasis: processes that maintain blood pH within a narrow range.
- Buffer systems: chemical systems (bicarbonate, phosphate, and protein buffers) that minimize pH changes.
- Respiratory regulation: control of CO_2 concentration in blood through respiratory rate and depth adjustments.
- Renal regulation: handling of bicarbonate, hydrogen ions, phosphate, and ammonium ions to maintain acid-base balance.
- Phosphate functions and regulation: renal handling and hormonal influences affecting phosphate levels.
- Anions in physiological processes: roles of bicarbonate, chloride, and other anions in acid-base balance and metabolic processes.
- Trace elements: roles of elements like iron, zinc, copper, selenium, and iodine.
- Vitamins: importance of water-soluble and fat-soluble vitamins in physiological functions.

The intricate and harmonious balance of the body's fluids, electrolytes, and acid-base levels is essential for maintaining life and health. This balance is critical because it influences and regulates various bodily functions, including but not

H. Hosseinzadeh, *Fundamentals of Medicine for Biomedical Engineering*,
https://doi.org/10.1007/978-3-031-74085-5_18

limited to cell function, nerve impulses, muscle contractions, and the overall maintenance of the internal environment.

18.1 Part 1: Body Fluids

Body fluids, primarily water, serve as the medium through which various essential substances are transported throughout the body. They play a crucial role in numerous physiological processes, including nutrient delivery, waste removal, temperature regulation, and the facilitation of chemical reactions.

18.1.1 Compartments

The human body's fluid can be broadly categorized into two main compartments: intracellular fluid and extracellular fluid compartments.

18.1.1.1 Intracellular Fluid

Intracellular fluid (ICF) is the fluid within the body's cells. It constitutes about 60% of the total body water or approximately 40% of an adult's body weight. This fluid is rich in potassium, magnesium, and phosphate ions (Table 18.1). It also contains proteins and soluble organic compounds, which contribute to the osmotic balance and provide essential building blocks for cellular activities. ICF is the medium where various intracellular processes occur, including enzymatic reactions, cellular respiration, and protein synthesis. It also plays a vital role in maintaining cell structure and function.

18.1.1.2 Extracellular Fluid

Extracellular fluid (ECF) refers to all body fluid outside the cells. It accounts for about 40% of the total body water or approximately 20% of an adult's body weight.

Table 18.1 Intracellular and extracellular fluid components: normal values

Component	Intracellular fluid	Extracellular fluid	
		Interstitial fluid	Plasma
Sodium (Na^+)	10–12 mEq/L	135–145 mEq/L	135–145 mEq/L
Potassium (K^+)	140–150 mEq/L	4–5 mEq/L	3.5–5 mEq/L
Calcium (Ca^{2+})	<0.0001 mEq/L	1.2–1.3 mEq/L	2.2–2.6 mEq/L
Magnesium (Mg^{2+})	18–20 mEq/L	1.5–2 mEq/L	1.5–2.5 mEq/L
Chloride (cl^-)	3–4 mEq/L	110–117 mEq/L	98–107 mEq/L
Bicarbonate (HCO_3^-)	7–10 mEq/L	25–30 mEq/L	22–28 mEq/L
Phosphate (PO_4^{3-})	40–120 mmol/L	1–2 mmol/L	0.8–1.5 mmol/L
Protein	High (varies)	Low (1–2 g/dL)	6–8 g/dL
Glucose	Low (varies)	80–110 mg/dL	70–110 mg/dL
pH	6.0–7.2	7.35–7.45	7.35–7.45
Sulfate (SO_4^{2-})	Low concentration	Low concentration	Low concentration
Anions (general)	Proteins, phosphates	Cl^-, HCO_3^-	Cl^-, HCO_3^-

ECF is rich in sodium, chloride, and bicarbonate ions (Table 18.1). Unlike ICF, it has a lower concentration of potassium. The composition of ECF is critical for maintaining the osmotic balance and pH of the body. The ECF facilitates the transport of nutrients, oxygen, and waste products to and from the cells. It plays a key role in maintaining the internal environment, allowing cells to function effectively.

The ECF can be further subdivided into several compartments:

- Interstitial Fluid: This fluid surrounds the cells, filling the spaces between them. It forms the immediate microenvironment, allowing ions, proteins, and nutrients to move across the cell membrane.
- Plasma: The liquid component of blood, plasma comprises about 20–25% of the ECF. It is the transport medium for cells and various substances, such as glucose, hormones, and blood proteins.
- Transcellular Fluid: These are specialized fluids, including cerebrospinal fluid (in the brain and spinal cord cavities), synovial fluid (in joints), aqueous humor and vitreous body (in the eyes), and pleural, pericardial, and peritoneal fluids (in the respective cavities).

18.1.2 Fluid Movement Among Compartments

The movement of fluid among different compartments in the human body is a dynamic and highly regulated process critical for maintaining homeostasis. This fluid movement occurs between the two primary fluid compartments: intracellular fluid (ICF) and extracellular fluid (ECF).

Mechanisms of fluid movement include:

Osmosis: The primary mechanism for fluid movement among compartments is osmosis, which is the movement of water across a semipermeable membrane from a region of lower solute concentration (higher water concentration) to one of higher solute concentration (lower water concentration).

Diffusion: Diffusion is the movement of solutes between compartments, moving from areas of higher concentration to lower concentration. This includes the movement of electrolytes, gases, and other soluble substances.

Filtration: This occurs primarily in the capillaries, where fluid is forced out of the blood plasma into the interstitial space due to blood pressure and then reabsorbed.

Active Transport: Some substances are moved against their concentration gradient through active transport, which requires energy. This is important for maintaining different concentrations of electrolytes inside and outside the cells.

18.2 Part 2: Water Regulation

The regulation of water content in the human body is essential for maintaining homeostasis. Water is the most abundant molecule in the body, crucial for numerous physiological functions, and a complex interplay of physiological mechanisms and systems tightly controls its balance.

18.2.1 Sensing Mechanisms

Osmoreceptors: Located primarily in the hypothalamus, osmoreceptors monitor the osmolarity of the blood. When osmolarity rises (indicating dehydration), these receptors trigger thirst and release antidiuretic hormone (ADH) to conserve water.

Baroreceptors: These receptors in the blood vessels sense changes in blood pressure (Section 13.2.2.2). A drop in blood pressure, which can indicate reduced blood volume and hence water loss, also stimulates ADH release.

18.2.2 Regulation Mechanisms

Hormonal Regulation:

- Antidiuretic Hormone (ADH): Produced by the hypothalamus and released by the posterior pituitary gland, ADH increases water reabsorption in the kidneys, reducing urine output and conserving water in the body.
- Aldosterone: This hormone, secreted by the adrenal cortex, promotes sodium (and consequently water) reabsorption in the kidneys, helping to maintain blood volume and pressure.
- Atrial Natriuretic Peptide (ANP): Secreted by heart cells in response to increased blood volume and pressure, ANP decreases sodium reabsorption, increases urine production, and decreases water retention.

Renal Regulation:

- Kidney Function: The kidneys regulate water, filter blood, and produce urine. They adjust urine volume and concentration based on hydration levels and hormonal signals.

Behavioral Responses:

- Thirst Mechanism: This is the primary driver for water intake. When the body needs water, the hypothalamus triggers thirst, prompting an individual to drink fluids.
- Dietary Intake: Water intake through drinking and food consumption directly influences the body's water content.

Integration with Other Systems:

- Cardiovascular System: Maintains blood pressure and volume closely related to the body's water status.
- Respiratory System: Exhales water vapor, contributing to water loss, especially in dry or cold conditions.
- Digestive System: Absorbs water from ingested food and drink; also loses water through feces.

Environmental and Lifestyle Factors:

- Temperature and Humidity: Influence sweating, which is a significant route of water loss.
- Physical Activity: Increases water loss through sweat and respiration.
- Diet: Certain foods and beverages (like those high in sodium) can affect water retention and requirements.

18.3 Part 3: Electrolytes

18.3.1 Sodium

Sodium distribution within the body is a finely tuned aspect of homeostasis. An average adult body contains about 100 grams of sodium, which can vary based on diet, health, and other factors. Most of the body's sodium is found in the ECF, including blood plasma and interstitial fluid. Sodium is the principal cation (positively charged ion) in the extracellular space, contributing significantly to the osmotic balance and volume of the ECF. A small proportion of the body's sodium is located within cells. Compared to the extracellular space, ICF has a much lower concentration of sodium. Although present in smaller amounts, intracellular sodium is crucial for certain cellular functions, including regulating cell volume and various metabolic processes. Bones act as a reservoir of sodium in the body. Some sodium is exchangeable and can be mobilized to maintain electrolyte balance under certain conditions. Sodium is a major component of blood plasma, where its concentration is closely regulated, typically between 135 and 145 mEq/L. In the fluid surrounding cells, sodium helps to maintain the osmotic pressure that governs the movement of water and other substances into and out of cells.

18.3.1.1 Functions of Sodium

Fluid Balance and Osmoregulation: Sodium is the primary cation in ECF and plays a crucial role in regulating the total amount of water in the body. It helps maintain osmotic pressure and fluid balance inside and outside cells.

Nerve Impulse Transmission: Sodium ions are essential for generating and propagating action potentials in nerve cells. The influx of sodium ions into neurons (depolarization) and their subsequent removal (repolarization) is fundamental to nerve impulse transmission (Section 10.1.2.4). Sodium is also involved in neurotransmitter release at synapses, facilitating communication between neurons (Section 10.1.3).

Muscle Contraction: Sodium plays a role in muscle function, including initiating muscle contractions. The flow of sodium into muscle cells triggers a series of events that result in muscle fiber contraction (Section 9.1.3). In muscle cells, sodium works with other electrolytes, such as potassium and calcium, to regulate contraction and relaxation.

Acid-Base Balance: Sodium participates in some of the body's buffer systems that help maintain pH balance. Sodium bicarbonate, for instance, is a crucial buffer in the blood.

$$NaHCO_3 + H_2O \rightarrow H_2CO_3 + NaOH$$

$$H_2CO_3 \leftrightarrow HCO_3^- + H^+ \leftrightarrow H_2CO_3 \leftrightarrow H_2O + CO_2\,(gas)$$

Nutrient Absorption and Transport: Sodium helps transport certain nutrients across cell membranes. For example, glucose and amino acid transport in the intestines (Sections 16.4.1 & 16.4.2) and kidneys (Section 17.3.4.3) often occur via sodium-dependent cotransport mechanisms.

Blood Pressure and Volume Regulation: Sodium's ability to retain water directly impacts blood volume and blood pressure. This relationship is central to hypertension (high blood pressure) pathophysiology. Sodium levels influence the renin-angiotensin-aldosterone system (RAAS), a hormone system that regulates blood pressure and fluid balance.

Cellular Function and Metabolic Reactions: Sodium ions regulate cell volume and are involved in various metabolic reactions, including the electron transport chain (Section 4.4.2.3) and enzymatic processes.

Impact on Bone Health: Research suggests a complex relationship between sodium and bone health, with high sodium intake potentially affecting calcium balance and bone density.

18.3.1.2 Regulation

Renal Regulation: The kidneys are the primary regulators of sodium balance in the body. They filter the blood, reabsorbing needed sodium and excreting excess through urine. The renin-angiotensin-aldosterone system (RAAS) is activated in response to low blood pressure or sodium levels. Renin, released by the kidneys, converts angiotensinogen to angiotensin I, which is then converted to angiotensin II. Angiotensin II stimulates aldosterone secretion from the adrenal cortex, promoting sodium reabsorption and potassium excretion in the kidneys, thereby increasing blood pressure and volume.

Hormonal Control: Aldosterone, secreted by the adrenal cortex, increases sodium reabsorption in the distal convoluted tubules and collecting ducts of the kidneys, leading to an increase in blood volume and pressure. Antidiuretic hormone (ADH), while primarily regulating water balance, also influences sodium levels indirectly by controlling the amount of water reabsorbed in the kidneys. Atrial natriuretic peptide (ANP), produced by the heart in response to an increase in blood pressure and volume, opposes the action of aldosterone, promoting sodium excretion to decrease blood volume and pressure.

Dietary Intake: The amount of sodium consumed in the diet can significantly affect its regulation. Excessive sodium intake can lead to increased blood pressure and edema, while inadequate intake can lead to hyponatremia. Sodium is absorbed from dietary sources through the gastrointestinal tract, with the rate of absorption influenced by nutritional needs and the body's homeostatic mechanisms.

18.3.2 Potassium

Potassium distribution is finely regulated and balanced between various compartments, ensuring proper cellular function and overall health. The human body contains approximately 140 grams of potassium, most of which (about 98%) is intracellular, making it the most abundant cation within cells. The high intracellular potassium concentration (about 140–150 mEq/L) is essential for maintaining cell volume, membrane potential, and cellular homeostasis. ECF, including blood plasma, has a much lower potassium concentration, typically between 3.5 and 5.0 mEq/L. Despite its relatively low concentration in the ECF, potassium is crucial for various functions, including nerve impulse conduction, muscle contraction, and cardiovascular function.

Skeletal muscles are major reservoirs of potassium. Other tissues, such as the liver and red blood cells, also contain significant amounts. The distribution of potassium in these tissues is essential for their specific physiological roles, such as muscle excitability and metabolic functions. The sodium-potassium pump (Na^+/K^+-ATPase), an active transport mechanism, maintains the high intracellular concentration of potassium (and low sodium concentration) against concentration gradients, using ATP for energy. Various potassium channels in cell membranes facilitate the passive movement of potassium ions, playing a role in activities like resting and action potential generation in muscle fibers and neurons (Sections 9.1.3.2 & 10.1.2.3).

18.3.2.1 Functions of Potassium

- Maintenance of Cell Membrane Potential: Potassium is vital in establishing and maintaining cell membrane potential. The high intracellular potassium concentration, relative to its extracellular concentration, is crucial for the electrochemical gradient across cell membranes. The rapid influx and efflux of potassium ions are integral to the action potential mechanism in neurons.
- Muscle Function: Potassium is pivotal in muscle contractions, including cardiac muscle. The movement of potassium ions influences muscle cells' excitation and

subsequent contraction. In the heart, potassium's role in regulating electrical impulses is critical for maintaining a healthy heart rhythm and function.

- Regulation of Blood Pressure: Potassium can help regulate blood pressure by causing vasodilation, or the widening of blood vessels, which reduces vascular resistance. The balance between sodium and potassium is essential in controlling blood pressure, with higher potassium intake often helping to mitigate the hypertensive effects of high sodium consumption.
- Metabolic Functions: Potassium is a cofactor for several enzymes and plays a role in metabolism, including synthesizing proteins and glycogen. It regulates insulin and glucose uptake in cells, impacting glucose metabolism.
- Bone Mineral Density: Potassium may influence bone health by helping preserve calcium, essential for bone strength and density.
- Neurological Function: Proper potassium levels are essential for brain health, influencing cognitive functions and nerve transmission.

18.3.2.2 Regulation

- Dietary Intake: Potassium is obtained primarily through nutritional sources, including fruits, vegetables, meat, and dairy products. It is efficiently absorbed in the gastrointestinal tract, predominantly in the small intestine, and this absorption is not tightly regulated.
- Renal Regulation: The kidneys are the primary organs maintaining potassium balance by adjusting the amount excreted in the urine. Potassium is freely filtered at the glomerulus and reabsorbed in the proximal tubule and loop of Henle. The fine-tuning of potassium excretion occurs in the distal tubules and collecting ducts. Aldosterone stimulates the excretion of potassium in the distal tubules and collecting ducts. It is released in response to elevated blood potassium levels or low blood sodium levels and by activating the renin-angiotensin system.
- Cellular Distribution: The Na^+/K^+-ATPase pump maintains high intracellular potassium levels. The pump moves three sodium ions out of the cell and two potassium ions into the cell for each ATP molecule hydrolyzed. Changes in the pH balance can cause shifts in potassium between the ICF and ECF. For example, acidosis can lead to hyperkalemia as potassium moves out of cells, while alkalosis can cause hypokalemia as potassium moves into cells.
- Hormonal Influences: Insulin promotes potassium uptake into cells, particularly in muscle and liver tissues, lowering blood potassium levels. As beta-adrenergic agonists, catecholamines stimulate potassium uptake into cells by activating the Na^+/K^+ pump.
- Osmolality Changes: Changes in osmolality can influence potassium distribution. For example, hyperosmolarity can lead to the movement of potassium out of cells.
- Exercise: Physical activity can cause transient increases in plasma potassium levels due to the release from working muscles.

18.3.3 Calcium

Calcium is a vital mineral in the human body, renowned for its crucial roles in skeletal structure, cellular processes, and various physiological functions. The human body contains approximately 1–1.5 kg of calcium, making it the most abundant mineral. About 99% of the body's calcium is stored in the bones and teeth, with the remaining 1% distributed in blood, extracellular fluid, and soft tissues. In bones and teeth, calcium is primarily found as calcium phosphate salts (like hydroxyapatite), providing structural rigidity and strength. Bones also serve as a reservoir for calcium, releasing it into the bloodstream under regulated conditions to maintain constant concentrations of calcium in other body parts.

Calcium concentrations are tightly regulated in blood and extracellular fluid, generally between 8.5 and 10.5 mg/dL. Approximately half of the calcium in the blood is ionized (free), and this form is physiologically active. The remainder is bound to proteins (mainly albumin) or complexed with other anions such as bicarbonate. Intracellularly, the concentration of free calcium ions is much lower than in the extracellular fluid, typically around 100 nM, which is crucial for cell signaling functions. Within cells, calcium is stored in organelles like the endoplasmic reticulum and mitochondria, and it is released into the cytosol in response to various signals.

18.3.3.1 Functions of Calcium

- Bone Structure and Health: Most of the body's calcium (about 99%) is stored in the bones and teeth, providing structural strength and rigidity. Calcium phosphate salts (like hydroxyapatite) are crucial for forming and maintaining the skeletal structure.
- Muscle Function: Calcium is vital in muscle contraction, including skeletal, cardiac, and smooth muscle tissues. The influx of calcium ions into muscle cells triggers the interaction between actin and myosin, the proteins responsible for muscle contraction (Page 227). In cardiac muscle, calcium is essential for regulating the force and duration of heart contractions, impacting heart rate and rhythm.
- Nervous System Function: Calcium ions are vital in transmitting signals from one nerve cell to another. They facilitate the release of neurotransmitters at synapses. Calcium entry into nerve endings triggers the exocytosis of neurotransmitter-containing vesicles, enabling nerve signal transmission (Section 10.1.3.1).
- Blood Clotting (Coagulation): Calcium plays a role in various stages of the coagulation cascade, acting as a cofactor for several blood clotting factors. Calcium also aids in the function and aggregation of platelets, which are essential for forming blood clots (Section 13.3.2).
- Cellular Functions: Calcium is a second messenger in many intracellular signaling pathways. It regulates various cellular processes, including gene expression, cell division, and apoptosis (programmed cell death). Calcium ions activate certain enzymes and modulate their activity, influencing metabolic pathways.
- Hormonal Secretion: Calcium regulates the secretion of various hormones. For example, it plays a role in insulin release from the pancreas.

- Vascular Function: Calcium ions are involved in the contraction and relaxation of blood vessels (vasoconstriction and vasodilation), affecting blood flow and blood pressure.
- Stabilization of Cell Membranes: Calcium helps stabilize cell membranes and influences membrane permeability and electrical properties.

18.3.3.2 Regulation

Calcium levels are tightly controlled through a complex interplay of hormonal signals, intestinal absorption, renal excretion, and bone resorption and formation.

Hormonal Regulation:

- Parathyroid Hormone (PTH): Secreted by the parathyroid glands, PTH is the primary regulator of calcium levels. It increases blood calcium levels by stimulating bone resorption (release of calcium from bones), enhancing renal reabsorption of calcium, and activating vitamin D synthesis in the kidneys.
- Vitamin D (Calcitriol): The active form of vitamin D, calcitriol, is synthesized in the kidneys under the influence of PTH. It facilitates the intestinal absorption of calcium and, to a lesser extent, phosphate. It also aids in bone remodeling and reabsorption.
- Calcitonin: Produced by the thyroid gland's parafollicular cells (C cells), calcitonin lowers blood calcium levels by inhibiting bone resorption and enhancing kidney calcium excretion. Its role in calcium homeostasis is less prominent than PTH and calcitriol.

Intestinal Absorption: Calcium is absorbed from the diet in the small intestine, with efficiency influenced by factors like age, dietary needs, and vitamin D levels. Vitamin D enhances the intestinal absorption of calcium by upregulating the expression of calcium-binding proteins in the intestinal lining (Section 16.4.6.4).

Renal Regulation: The kidneys filter calcium, most of which is reabsorbed into the bloodstream. PTH and calcitriol regulate this reabsorption. The amount of calcium excreted in the urine is a balance between what is filtered and what is reabsorbed. Factors like excess dietary calcium, phosphate, and certain diuretics increase excretion.

18.3.4 Phosphate

Phosphate is critical in various physiological processes, including energy metabolism, bone formation, and cellular function. The human body contains about 500–800 grams of phosphate, of which approximately 85% is found in bones and teeth, with the rest distributed in soft tissues and extracellular fluid. In bones and teeth, phosphate is primarily present as calcium phosphate salts, such as hydroxyapatite, which provide rigidity and strength. A significant amount of phosphate within the body resides intracellularly. The phosphate concentration in plasma is usually between 2.5 and 4.5 mg/dL. Only a small fraction of body phosphate is in the extracellular fluid, but this fraction is crucial for metabolic functions.

18.3.4.1 Functions of Phosphate

- Bone and Teeth Mineralization: Phosphate, in combination with calcium, forms calcium phosphate crystals such as hydroxyapatite, which are essential for the hardness and strength of bones and teeth. Adequate phosphate levels are crucial for maintaining bone density and overall skeletal health.
- Energy Storage and Metabolism: Phosphate is a crucial component of adenosine triphosphate (ATP), the primary energy carrier in cells. The energy stored in the phosphate bonds of ATP is used for various cellular activities. In muscle cells, phosphate is stored as creatine phosphate (Section 9.1.3.9), a rapid energy source during high-intensity muscular activities.
- Nucleic Acids and Genetic Material: Phosphate is essential for synthesizing nucleotides, the building blocks of nucleic acids. Thus, phosphate forms the backbone of DNA and RNA molecules and plays a critical role in storing and expressing genetic information.
- Cell Membrane Structure: Phosphate is a component of phospholipids, which are fundamental to the structure and function of cell membranes (Section 4.1.2.1). Phospholipids contribute to membrane fluidity and facilitate cell signaling.
- Cell Signaling and Regulation: Phosphate-containing molecules like cyclic AMP (cAMP) act as second messengers in various cell signaling pathways, translating external signals into cellular responses. Phosphorylation, the addition of a phosphate group to a protein, is a key mechanism for activating or deactivating proteins, thereby regulating their function.
- Acid-Base Balance: Phosphate is a buffer in intracellular and extracellular fluids, helping maintain the body's acid-base balance (Section 18.4.1.2).
- Enzymatic Reactions: Phosphate groups are involved in the activation of many enzymes, influencing diverse biochemical pathways.

18.3.4.2 Regulation

- Renal Regulation: The kidneys are the primary organs regulating phosphate levels in the body. They filter phosphate from the blood and can either reabsorb it or excrete it in the urine. Phosphate reabsorption in the kidneys is primarily controlled by the action of hormones such as PTH and fibroblast growth factor 23 (FGF23), with PTH generally increasing renal excretion of phosphate and FGF23 reducing its reabsorption.
- Hormonal Influences: PTH is secreted in response to low serum calcium levels, indirectly decreasing phosphate reabsorption in the kidneys to maintain the calcium-phosphate product within a normal range. Fibroblast growth factor 23 (FGF23), produced by bone cells, decreases phosphate reabsorption in the kidneys and reduces vitamin D synthesis, thus lowering serum phosphate levels. Active vitamin D (calcitriol) increases intestinal absorption of phosphate. However, it also stimulates the production of FGF23, which counters this effect by promoting phosphate excretion in the kidneys.

- Intestinal Absorption: Phosphate is absorbed from the diet in the small intestine. Various factors, including active vitamin D, dietary habits, and the gastrointestinal tract's health, influence the absorption efficiency. Vitamin D enhances the intestinal absorption of phosphate, thus increasing its availability to the body.

18.3.5 Anions

Anions are negatively charged ions that play crucial roles in various physiological processes in the human body. These include bicarbonate (HCO_3^-), chloride (Cl^-), phosphate ($HPO_4^{2-}/H_2PO_4^-$), sulfate (SO_4^{2-}), and organic acids.

18.3.5.1 Distribution in Body

- Bicarbonate (HCO_3^-): Bicarbonate is the most significant extracellular anion and acts as the primary buffer in the blood. The kidneys regulate bicarbonate concentration by reabsorbing or excreting it. The lungs also play a role by regulating carbon dioxide (CO_2) levels, which are in equilibrium with bicarbonate.
- Chloride (Cl^-): Chloride is the most abundant anion in the ECF.
- Sulfate (SO_4^{2-}): Sulfate ions are present in smaller concentrations and are primarily the product of protein metabolism.
- Organic Acids: Various organic acids (like lactate, acetate, and citrate) occur in the body, often as intermediates in metabolic pathways. These anions are vital in energy production cycles, such as glycolysis (Section 4.4.1.1) and the citric acid cycle (Section 4.4.2.2).
- Proteins and Amino Acids: Certain proteins and amino acids can act as anions, especially when they lose a proton and gain a negative charge. These are predominantly found intracellularly and contribute to the anionic environment of the cell.

18.3.5.2 Function

Anions are essential for numerous physiological functions in the human body. They play critical roles in maintaining acid-base balance, osmolarity, and membrane potential and are involved in various metabolic processes. Anions contribute to establishing and preserving membrane potential, which is essential for nerve impulse transmission and muscle contraction. The movement of anions, along with cations like sodium and potassium, is necessary for the electrical activity of neurons and muscle cells (Sections 9.1.3.2 & 10.1.2.3). Anions are part of various buffer systems in the body, which are crucial for maintaining pH homeostasis in blood, intracellular fluid, and urine. Anions interact with cations (such as sodium, potassium, calcium, and magnesium) to maintain electrolyte balance and cellular functions. The distribution of anions and cations is crucial for maintaining fluid balance, osmolarity, and blood pressure.

Bicarbonate (HCO_3^-)

Bicarbonate is a major component of the body's buffer system, especially in blood. It helps to maintain the pH balance by neutralizing acids produced in metabolism.

Bicarbonate is closely linked to the respiratory system (Section 15.2.9); carbon dioxide (CO_2) and water interact to form carbonic acid (H_2CO_3), which dissociates into H^+ and HCO_3^-, a reaction catalyzed by the enzyme carbonic anhydrase.

Chloride (Cl^-)
As the most abundant extracellular anion, chloride is vital in maintaining fluid balance and osmolarity in the body. It is a crucial component of hydrochloric acid (HCl) in the stomach, which is necessary for digestion (Section 16.3.4.3). Cl^- participates in maintaining electrical neutrality across cell membranes and is involved in cell volume regulation.

Sulfate (SO_4^{2-})
Sulfate is involved in the synthesis of sulfur-containing amino acids and proteins. In the liver, sulfate conjugates with various substances, aiding in detoxification.

Organic Acids (e.g., Lactate, Acetate)
Organic acids like lactate are crucial intermediates in metabolic pathways, such as glycolysis. Lactate, produced during anaerobic metabolism, can be converted back to glucose in the liver, a process called gluconeogenesis.

18.4 Part 4: Acid-Base Balance

The intricate balance between acids and bases, known as acid-base homeostasis, is essential for properly functioning various physiological processes. In the context of the human body, acids are substances that can donate hydrogen ions (H^+) or protons. The most notable acid in the body is carbonic acid (H_2CO_3), a byproduct of metabolic processes involving carbon dioxide (CO_2). On the other hand, bases are compounds that can accept hydrogen ions. Bicarbonate (HCO_3^-) is a primary base in the body, playing a vital role in buffering acids. One of the primary roles of acids and bases is to regulate the pH of bodily fluids, especially blood, which typically maintains a slightly alkaline pH of about 7.35 to 7.45. This pH range is critical for various enzymatic and biochemical processes. Many enzymes function optimally within a specific pH range, and deviations from this range can impair their activity and, subsequently, numerous metabolic processes.

18.4.1 Chemical Buffer Systems

Chemical buffer systems maintain the pH of body fluids, especially blood, within a narrow, optimal range. These systems consist of a weak acid and its conjugate base, which can absorb excess H^+ ions (when the solution becomes too acidic) or OH^- ions (when it becomes too alkaline), thus minimizing changes in pH. The primary buffer systems in the body are the bicarbonate buffer system, the phosphate buffer system, and the protein buffer system.

18.4.1.1 Bicarbonate Buffer System

The bicarbonate buffer system is the primary buffer against pH changes in the extracellular fluid, including blood. Keeping the pH within the narrow range necessary for physiological processes is essential. Components of the bicarbonate buffer system include:

Carbonic Acid (H_2CO_3): This weak acid forms when carbon dioxide (CO_2) dissolves in water.

$$CO_2 + H_2O \rightleftharpoons H_2CO_3$$

Bicarbonate Ion (HCO_3^-): Bicarbonate acts as the conjugate base in this system.

$$H_2CO_3 \rightleftharpoons H^+ + HCO_3^-$$

When the blood's pH decreases (becomes more acidic), there is an increase in H^+. HCO_3^- combines with these excess H^+ to form H_2CO_3, reducing the acidity and raising the pH back toward normal. Conversely, if the pH increases (becomes more alkaline), H_2CO_3 dissociates into HCO_3^- and H^+, thereby increasing the H^+ concentration and lowering the pH.

Carbonic acid is in equilibrium with CO_2 and water. The respiratory system regulates the concentration of CO_2 in the blood by either increasing or decreasing the rate and depth of breathing (Section 15.2.10.2). The enzyme carbonic anhydrase in red blood cells catalyzes the reversible reaction between CO_2 and water to form carbonic acid, which then dissociates into bicarbonate and hydrogen ions. The kidneys regulate the bicarbonate buffer system by reabsorbing bicarbonate from the urine or excreting it when necessary. They can also generate new bicarbonate ions for blood plasma (Section 15.2.9).

18.4.1.2 Phosphate Buffer System

The phosphate buffer system plays a significant role in buffering pH changes in body fluids and is especially important in the renal handling of hydrogen ions. The system consists of a mixture of dihydrogen phosphate ions ($H_2PO_4^-$, a weak acid) and hydrogen phosphate ions (HPO_4^{2-}, its conjugate base).

$$H_2PO_4^- \rightleftharpoons H^+ + HPO_4^{2-}$$

In an acidic environment (excess H^+), HPO_4^{2-} combines with H^+ to form $H_2PO_4^-$, thereby reducing acidity. Conversely, in an alkaline environment (deficiency of H^+), $H_2PO_4^-$ dissociates to release hydrogen ions and form HPO_4^{2-} ions, thus increasing the acidity.

The phosphate buffer system is particularly effective in the intracellular environment with high phosphate concentrations. It helps buffer pH changes that result from cellular metabolic processes, such as the production of lactic acid and other metabolic acids. In the kidneys, this buffer system helps neutralize urine acidity.

The renal tubules can secrete H^+ into the urine while reabsorbing bicarbonate ions. The $H_2PO_4^-$ formed in the tubular fluid is excreted in the urine, effectively removing H^+ from the body. Bones, which store large amounts of phosphates, can also contribute to this buffer system.

Bone resorption can release phosphate ions into the bloodstream, affecting the buffer capacity.

18.4.1.3 Protein Buffer System

Proteins are the most abundant buffers in the body, particularly in the intracellular fluid and plasma. Protein amino acids contain carboxyl (–COOH) and amino (–NH_2) groups, which can act as acids and bases, respectively. Proteins can bind to H^+ (acting as bases) or release H^+ (acting as acids) depending on the pH. Hemoglobin in red blood cells is a significant buffer, helping to control pH changes during CO_2 transport.

18.4.2 Respiratory Regulation of Acid-Base Equilibrium

The respiratory regulation of H^+ is a component of the body's acid-base homeostasis, working closely with chemical buffer systems and renal mechanisms. This respiratory regulation primarily involves the control of CO_2 concentration in the blood, which directly influences the levels of H^+. CO_2 is a byproduct of cellular metabolism, and its concentration in the blood is directly linked to H^+ levels through the following reaction:

$$CO_2 + H_2O \rightleftharpoons H_2CO_3\,(\text{carbonic acid}) \rightleftharpoons H^+ + HCO_3^-\,(\text{bicarbonate})$$

An increase in CO_2 leads to an increase in H^+, lowering pH (more acidic), while a decrease in CO_2 leads to a reduction in H^+, raising pH (more alkaline).

The primary method of respiratory regulation is through changes in the rate and depth of breathing (ventilation). Increased ventilation expels more CO_2, reducing its concentration and lowering H^+ concentration. Conversely, decreased ventilation leads to CO_2 retention, increasing H^+ concentration. Chemoreceptors in the medulla (central chemoreceptors) and the carotid and aortic bodies (peripheral chemoreceptors) sense changes in CO_2 and pH. They send signals to the respiratory centers in the brain to adjust ventilation (Section 15.2.10.2). Respiratory compensation is a rapid response mechanism, occurring within minutes to hours. The respiratory system interacts closely with the bicarbonate buffer system. Changes in CO_2 levels shift the equilibrium of the carbonic acid-bicarbonate buffer system, thereby influencing blood pH.

18.4.3 Renal Regulation of Acid-Base Equilibrium

Renal regulation of acid-base equilibrium involves multiple mechanisms related to the handling of bicarbonate (HCO_3^-), hydrogen ions (H^+), phosphate (HPO_4^{2-}), and ammonium (NH_4^+) ions. These mechanisms help maintain the body's acid-base balance and blood pH within a narrow and tightly regulated range. The renal compensation is a delayed response mechanism.

18.4.3.1 Bicarbonate (HCO_3^-)

Bicarbonate is filtered at the glomerulus and is mainly reabsorbed in the proximal convoluted tubule (PCT). Bicarbonate reabsorption occurs through sodium-dependent bicarbonate cotransporters (NPTs) on PCT cells' apical and basolateral membranes (Section 17.3.4.4). In cases of metabolic acidosis (low blood pH), excess bicarbonate can be generated by the kidneys through the secretion of H^+ in the distal convoluted tubule (DCT) and collecting ducts (Section 17.3.4.4). The secreted H^+ ions combine with filtered bicarbonate ions to form carbonic acid (H_2CO_3), which can be converted into carbon dioxide (CO_2) and water (H_2O). This process generates new bicarbonate ions, which are reabsorbed into the bloodstream to help raise blood pH.

18.4.3.2 Hydrogen Ion (H^+)

In the DCT and collecting ducts, excess hydrogen ions are actively secreted into the tubular lumen by hydrogen ion pumps (H^+-ATPases) and hydrogen-potassium ion exchangers (H^+/K^+ exchangers). The hormone aldosterone stimulates the secretion of hydrogen ions and is a crucial mechanism for lowering blood pH when it becomes too alkaline (high pH).

18.4.3.3 Phosphate (HPO_4^{2-})

Phosphate ions are filtered at the glomerulus and are primarily reabsorbed in the proximal tubule. Phosphate reabsorption occurs through sodium-phosphate cotransporters (NPTs) on PCT cells' apical and basolateral membranes. In cases of metabolic acidosis, phosphate reabsorption can be reduced, leading to increased phosphate excretion.

18.4.3.4 Ammonium (NH_4^+)

When blood becomes too acidic, the kidneys can produce ammonia (NH_3) from glutamine metabolism in the renal tubular cells. Ammonia combines with excess H^+ to form ammonium ions (NH_4^+), which are less acidic. Ammonium ions can be excreted in the urine, effectively eliminating excess hydrogen ions and helping to raise blood pH.

18.5 Part 5: Trace Elements

Although present in minute amounts in the human body, trace elements play indispensable roles in a myriad of physiological processes. They are involved in enzyme functions, hormonal activity, immune system maintenance, and structural stability of proteins and nucleic acids (Table 18.2).

Table 18.2 Trace elements in the human body: distribution in the body, function(s), and regulation

Trace element	Distribution in body	Function	Regulation of levels
Iron (Fe)	Hemoglobin, myoglobin, various enzymes	Oxygen transport, electron transfer, enzyme activity	Absorption regulated by hepcidin in the intestine; stored in ferritin and hemosiderin
Zinc (Zn)	Enzymes, transcription factors	Enzyme catalysis, protein synthesis, immune function	Regulated by dietary intake, absorption, and protein binding
Copper (Cu)	Enzymes (cytochrome c oxidase, others)	Enzyme activity, iron metabolism, connective tissue formation	Regulated by absorption in the gut and binding to proteins like ceruloplasmin
Selenium (Se)	Selenoproteins, enzymes	Antioxidant defense, thyroid hormone metabolism	Regulated by dietary intake and incorporation into selenoproteins
Manganese (Mn)	Enzymes (superoxide dismutase, others)	Enzyme function, bone development, amino acid metabolism	Regulated by absorption in the gut and biliary excretion
Fluoride (F)	Bones, teeth	Bone and tooth mineralization, resistance to dental caries	Regulated by intake from water, food, or supplements
Chromium (Cr)	Insulin receptors, enzymes	Enhances insulin action, carbohydrate, and lipid metabolism	Regulated by dietary intake and excretion
Iodine (I)	Thyroid hormones	Synthesis of thyroid hormones, metabolic regulation	Regulated by dietary intake and thyroid hormone synthesis
Molybdenum (Mo)	Enzymes (xanthine oxidase, others)	Enzyme activity, purine metabolism, detoxification	Regulated by dietary intake and excretion in urine
Cobalt (Co)	Vitamin B12 (cobalamin)	Red blood cell production, nerve function, DNA synthesis	Primarily regulated by vitamin B12 intake
Nickel (Ni)	Enzymes, hormone activation	Enzyme function, gene expression	Regulated by dietary intake and excretion
Silicon (Si)	Connective tissues, bones, skin	Bone formation, connective tissue strength	Regulated by dietary intake and excretion
Vanadium (V)	Enzymes	Enzyme function, possibly insulin-like effects	Regulated by dietary intake and excretion
Boron (B)	Cell membranes, bone	Bone health, cognitive function, hormone regulation	Regulated by dietary intake and excretion

18.6 Part 6: Vitamins

Vitamins are essential organic compounds required in small quantities for various physiological functions. They are crucial for growth, development, and maintaining overall health. The human body cannot synthesize most vitamins in sufficient amounts, so they must be obtained from the diet or, in some cases, synthesized by gut microbiota (Table 18.3).

Table 18.3 Vitamins: type, distribution in the body, necessary daily dose, function, and regulation

Vitamin	Type	Distribution in body	Daily dose needed	Function	Regulation in body
Vitamin B1 (Thiamine)	Water-soluble	Nervous system, muscles, heart	1.1–1.2 mg (adults)	Carbohydrate metabolism, nerve function	Dietary intake; not stored extensively in the body
Vitamin B2 (Riboflavin)	Water-soluble	Eyes, skin, nerves	1.1–1.3 mg (adults)	Energy production, cell function, fat metabolism	Dietary intake; not stored extensively in the body
Vitamin B3 (Niacin)	Water-soluble	All body tissues	14–16 mg NE (adults)	Digestive function, skin health, nerve function	Dietary intake; can be synthesized from tryptophan
Vitamin B5 (Pantothenic acid)	Water-soluble	Everywhere in the body	5 mg (adults)	Synthesis of coenzyme A, fatty acid synthesis	Dietary intake; widespread in foods
Vitamin B6 (Pyridoxine)	Water-soluble	Muscle, liver, brain	1.3–1.7 mg (adults)	Amino acid metabolism, red blood cell production	Dietary intake; stored in muscle tissue
Vitamin B7 (Biotin)	Water-soluble	Liver, hair, eyes	30 μg (adults)	Fat and carbohydrate metabolism, amino acid synthesis	Dietary intake; synthesized by gut bacteria
Vitamin B9 (Folate)	Water-soluble	Liver, leafy green vegetables	400 μg (adults)	DNA synthesis, red and white blood cell production	Dietary intake; stored in the liver
Vitamin B12 (Cobalamin)	Water-soluble	Liver, brain, blood	2.4 μg (adults)	Nerve function, red blood cell production, DNA synthesis	Dietary intake from animal products; stored in the liver
Vitamin C (Ascorbic acid)	Water-soluble	Various tissues	75–90 mg (adults)	Antioxidant, collagen synthesis, immune function, iron absorption	Dietary intake; water-soluble, not stored in the body
Vitamin A	Fat-soluble	Liver, retina, skin	700–900 μg (adults)	Vision, immune function, cell growth, reproduction	Stored in the liver; dietary intake from animal and plant sources
Vitamin D	Fat-soluble	Bones, teeth, muscles	15–20 μg (adults)	Calcium absorption, bone health, muscle function, immune response	Synthesized in skin from sunlight; dietary intake
Vitamin E (tocopherol)	Fat-soluble	Immune system, skin, eyes	15 mg (adults)	Antioxidant, cell protection, immune function	Dietary intake; fat-soluble, stored in the body's fatty tissues
Vitamin K	Fat-soluble	Liver, bones	90–120 μg (adults)	Blood clotting, bone metabolism	Dietary intake; partly synthesized by gut bacteria

Suggested Reading

Advanced Physiology and Pathophysiology: Essentials for Clinical Practice, 2nd Ed. Nancy Tkacs, Linda Herrmann, Springer Publishing Company, 2024

Anatomy & Physiology, S. Meloni, et al. Independent Publisher, 2023

Anatomy & Physiology: The Unity of Form and Function, 10th Ed. Kenneth S. Saladin, McGraw Hill, 2023

Clinical Nutrition, 2nd Ed. Rebecca J. Stratton, et al. Wiley-Blackwell, 2013

Gray's Anatomy for Students, 5th Ed. Richard L. Drake, et al. Elsevier, 2020

Human Anatomy & Physiology, 11th Ed. Elaine N. Marieb and Katja Hoehn, Pearson, 2022

Introduction to Human Nutrition, 3rd Ed. Hester H. Vorster, et al. Wiley-Blackwell, 2019

Moore's Clinically Oriented Anatomy, 9th Ed. Keith L. Moore, et al. Lippincott Williams & Wilkins, 2020

Netter Atlas of Human Anatomy: Classic Regional Approach, 8th Ed. Frank H. Netter, Elsevier, 2022

Principles of Anatomy and Physiology, 16th Ed. Gerard J. Tortora, Bryan H. Derrickson, Wiley, 2020

Reproductive System

19

In this chapter, you will learn:

- General structure of the female reproductive system
- Development of ovarian follicles
- The process of ovulation
- The concept and importance of the menstrual cycle
- Phases of the endometrial cycle
- Hormonal regulation in females
- Female sexual response cycle
- Structure and function of the female breast
- Structure of the male reproductive system
- The process of spermatogenesis
- Structure of mature sperm
- Hormonal regulation in males
- Male sexual response cycle
- Methods and effectiveness of various contraception options
- Assisted reproductive technologies (ART)

The human reproductive system (Fig. 19.1) is a complex and sophisticated biological system designed to ensure the continuation of the species through sexual reproduction. This system encompasses a range of organs, hormones, and processes, each intricately linked and coordinated to facilitate the creation of new life. It is bifurcated into two primary components based on biological sex—the male and female reproductive systems—each with distinct structures and functions yet interdependent for reproduction.

H. Hosseinzadeh, *Fundamentals of Medicine for Biomedical Engineering*,
https://doi.org/10.1007/978-3-031-74085-5_19

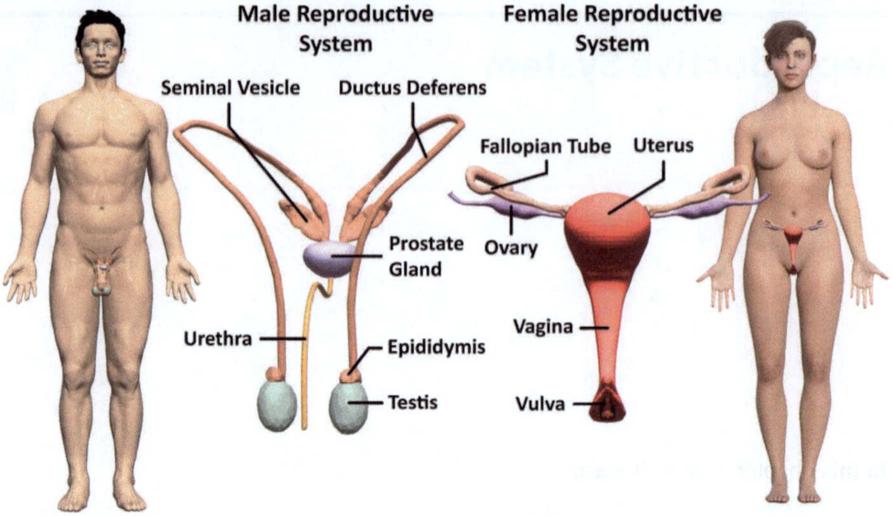

Fig. 19.1 The reproductive system

19.1 Part 1: Female Reproductive System Structure

In females, the reproductive system includes the ovaries, which produce eggs (ova) and various hormones; the fallopian tubes, where fertilization of the egg by a sperm typically occurs; the uterus, a muscular organ where the fertilized egg implants and develops into a fetus; the cervix, the lower part of the uterus that opens into the vagina; the vagina itself, which serves as the canal for menstrual flow, sexual intercourse, and childbirth; and the external genitalia (Fig. 19.2).

19.1.1 Ovary

The ovaries (Fig. 19.3) are in the pelvic cavity on either side of the uterus.

19.1.1.1 Anatomy

Each ovary, roughly almond-shaped, measures approximately 3–5 cm in length, 1.5–3 cm in width, and 0.5–1.5 cm in thickness. However, these dimensions can vary with age, hormonal status, and during the menstrual cycle.

Each ovary is covered by the germinal epithelium, a remnant of the peritoneum, which is closely adherent to an underlying dense fibrous layer known as the tunica albuginea. Beneath the tunica albuginea lies the ovarian cortex, which contains the ovarian follicles at various stages of development. The ovaries are held in place by several structures: ovarian ligament, attaching the ovary to the uterus; suspensory ligament of the ovary, containing the ovarian artery and vein and attaching the ovary to the pelvic wall; mesovarium, part of the broad ligament of the uterus, connecting

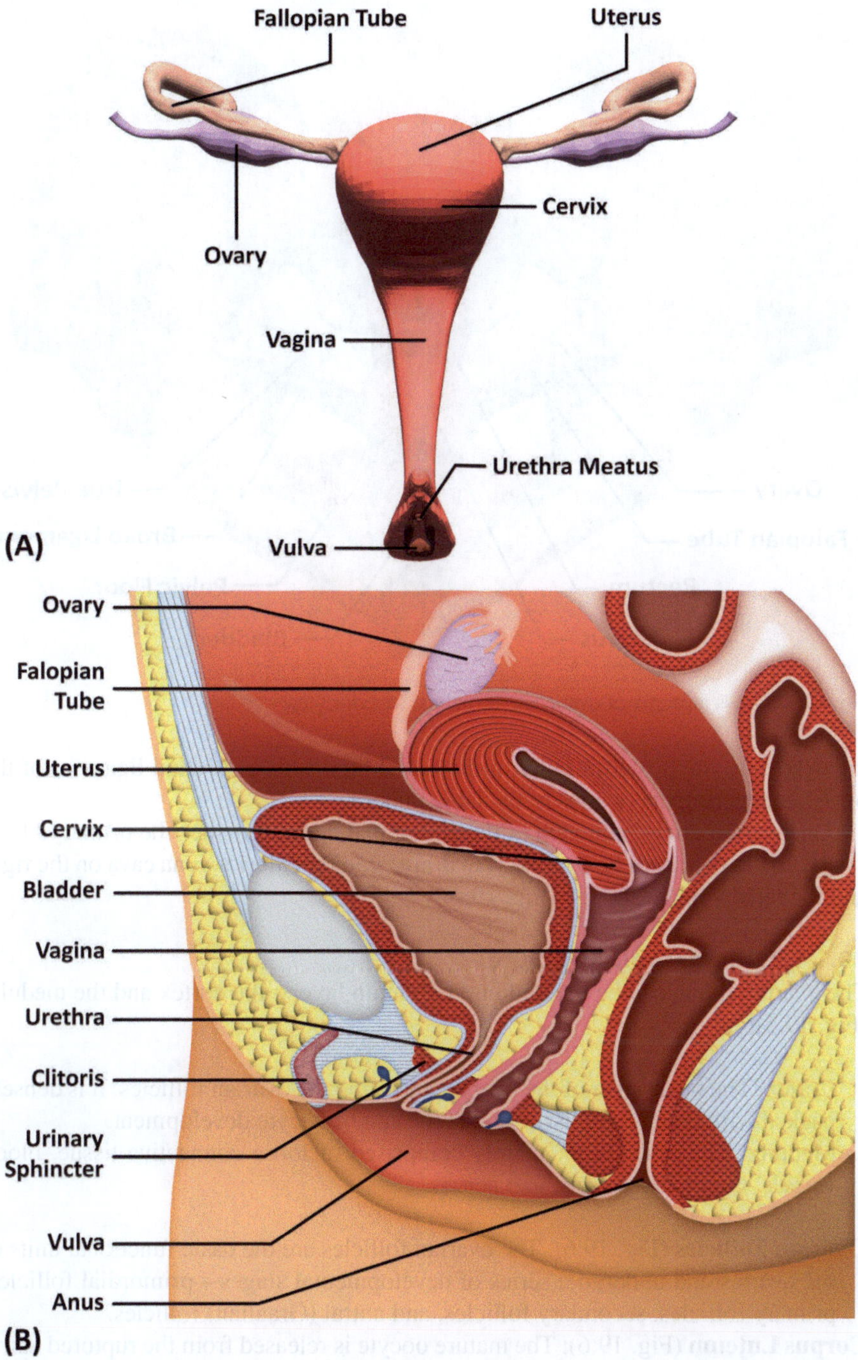

Fig. 19.2 The female reproductive system structure. (**a**) Anterior view. (**b**) Sagittal section

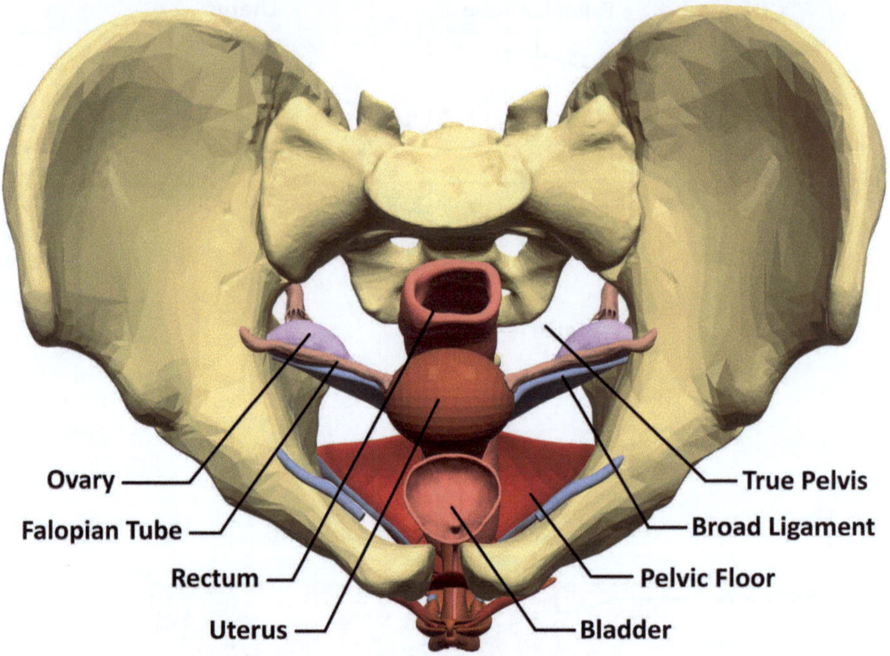

Fig. 19.3 Gross anatomy of ovaries in the pelvis, superior view

the ovaries to the posterior layer of the broad ligament; and round ligament of the ovary, connecting the ovaries to the labia majora.

The ovarian artery, a branch of the abdominal aorta, provides the primary blood supply to the ovaries. The ovarian vein drains into the inferior vena cava on the right and the renal vein on the left (Fig. 19.4).

19.1.1.2 Microscopic Structure

The ovarian tissue can be divided into two main layers: the cortex and the medulla (Fig. 19.5).

- Cortex: The outer layer of the ovary, containing the ovarian follicles. It is densely packed with connective tissue and is the site of oocyte development.
- Medulla: The inner layer, mainly comprised of loose connective tissue, blood vessels, lymphatic vessels, and nerves.

Ovarian Follicles (Fig. 19.6): The ovarian follicles are the basic functional units of the ovaries and undergo a series of developmental stages—primordial follicles, primary follicles, secondary follicles, and antral (Graafian) follicles.

Corpus Luteum (Fig. 19.6): The mature oocyte is released from the ruptured antral follicle during ovulation. The remaining granulosa and theca cells undergo

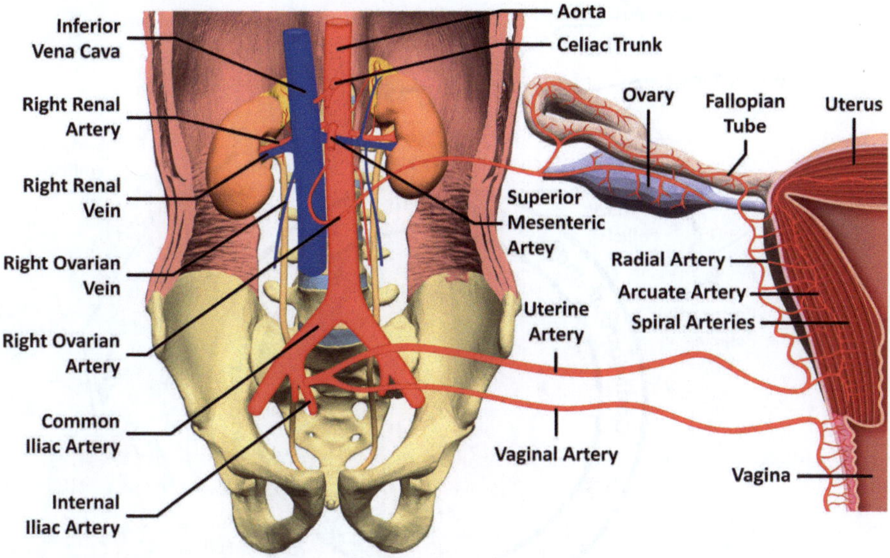

Fig. 19.4 Blood vessels of the female reproductive system

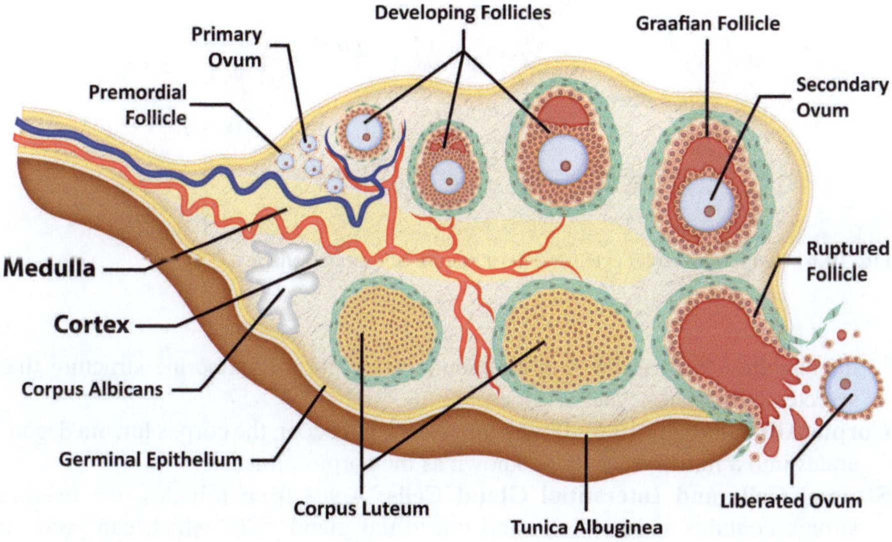

Fig. 19.5 Internal structure of the ovary

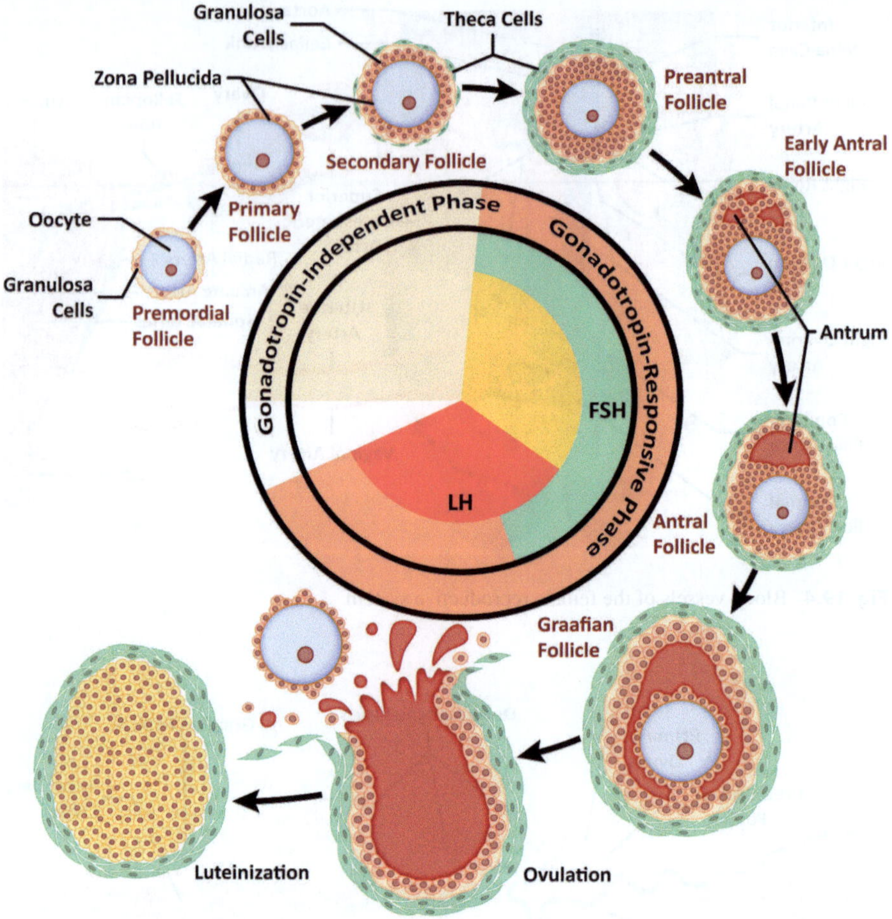

Fig. 19.6 Ovarian follicular cycle and the structure of ovarian follicles

luteinization to form the corpus luteum, a temporary endocrine structure that secretes progesterone and estrogen.

Corpus Albicans (Fig. 19.6): If pregnancy does not occur, the corpus luteum degenerates into a fibrous scar tissue known as the corpus albicans.

Stromal Cells and Interstitial Gland Cells: Apart from follicles, the ovarian stroma contains stromal cells and interstitial gland cells, which can produce androgens and estrogens.

The dense vascular network within the medulla is crucial for delivering hormones and nutrients. The ovarian innervation plays a role in follicular development and ovulation.

19.1.1.3 Age-Related Changes

Throughout a woman's life, the structure of the ovaries changes:

Before Puberty: The ovaries are relatively inactive, containing primarily primordial follicles.

Reproductive Years: Cyclic changes occur in the ovaries, characterized by follicular development, ovulation, and corpus luteum formation. A mix of follicles at various stages of development is present.

Postmenopause: Ovarian function declines, leading to a decrease in size and vascularity. A decreased number of follicles, with more fibrous tissue, is seen.

19.1.2 Uterine Tubes

19.1.2.1 Anatomy

The uterine tubes (Fig. 19.7) are a pair of narrow, tubular structures extending from the lateral edges of the uterus toward the ovaries. They measure approximately 10–12 cm in length and vary in diameter along their length, being wider at the ovarian end and narrower as they approach the uterus. The uterine tubes are divided into four distinct segments, each with unique anatomical features:

Infundibulum: The funnel-shaped distal end of the tube, closest to the ovary, has finger-like projections called fimbriae that help capture the ovulated egg.

Ampulla: This is the longest and widest section, where fertilization usually occurs. It has a thick, muscular wall and a highly folded mucosal lining to increase its surface area.

Isthmus: This is a narrower, more muscular segment adjacent to the uterus. It has a thicker wall and a smaller lumen than the ampulla.

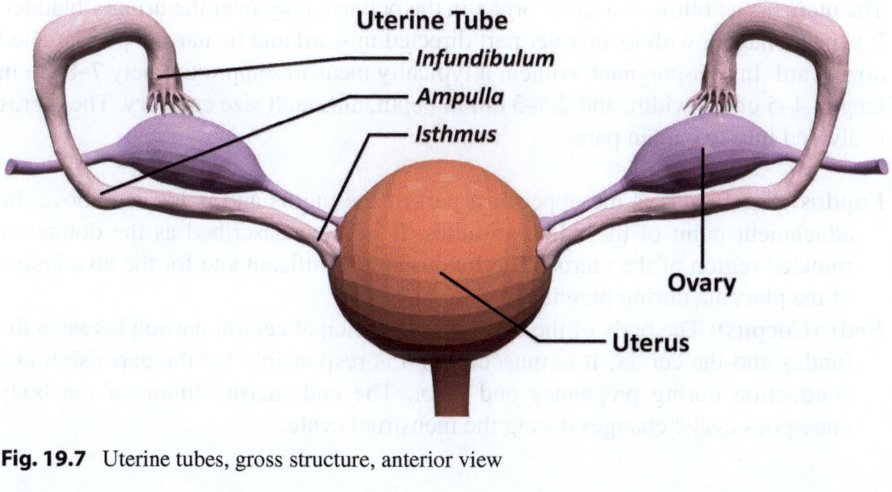

Fig. 19.7 Uterine tubes, gross structure, anterior view

Intramural or Interstitial Part: This segment traverses the uterine wall and opens into the uterine cavity. It is the shortest and narrowest part of the tube.

19.1.2.2 Microscopic Structure

Each uterine tube consists of several layers:

Mucosa: The innermost layer, lined with ciliated columnar epithelial and secretory cells. The cilia beat toward the uterus, aiding the movement of the egg.
Muscularis: A middle layer of smooth muscle fibers arranged in circular and longitudinal layers, facilitating peristaltic movements.
Serosa: The outermost layer, a part of the visceral peritoneum that covers the tubes.

The blood supply to the uterine tubes is primarily from the ovarian and uterine arteries (Fig. 19.4). Venous drainage parallels the arterial supply.

19.1.2.3 Function

The primary functions of the uterine tubes include:

- Transport of the ovum from the ovary to the uterus
- Providing the optimal environment for fertilization
- Early transport of the fertilized egg to the uterine cavity for implantation

19.1.3 Uterus

The uterus (Fig. 19.8) is a central organ in the female reproductive system. It regulates the menstrual cycle, implants fertilized ovum, and develops fetuses during pregnancy.

19.1.3.1 Anatomy

The uterus is a hollow, muscular organ in the pelvic cavity over the urinary bladder. It is pear-shaped, with its broader part directed upward and its narrow part situated downward. In non-pregnant women, it typically measures approximately 7–8 cm in length, 4–5 cm in width, and 2.5–3 cm in depth, although size can vary. The uterus is divided into two main parts:

Fundus: The fundus is the uppermost part of the uterus and is located above the attachment point of the fallopian tubes. It is often described as the domed or rounded region of the uterus. The fundus is a significant site for the attachment of the placenta during pregnancy.
Body (Corpus): The body of the uterus is the principal central portion between the fundus and the cervix. It is muscular and is responsible for the expansion and contraction during pregnancy and labor. The endometrial lining of the body undergoes cyclic changes during the menstrual cycle.

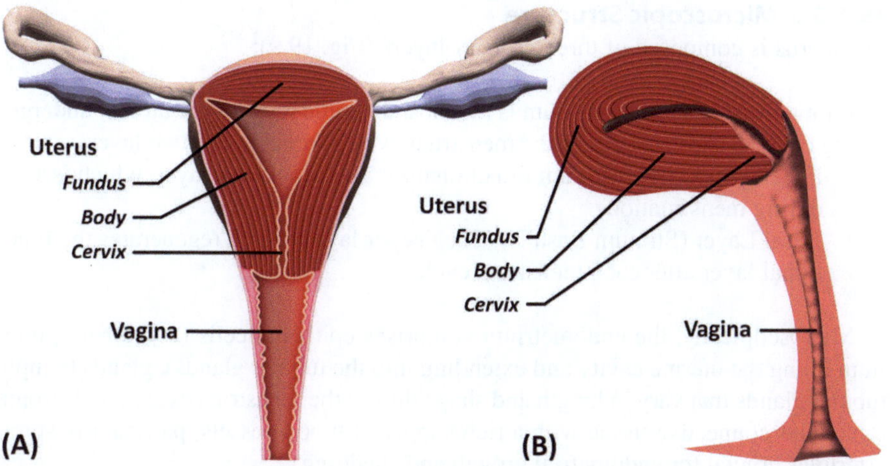

Fig. 19.8 Lower female reproductive organs. (**a**) Coronal section, anterior view. (**b**) Sagittal section

Cervix: The cervix is the lower, cylindrical part of the uterus that extends into the vagina. It measures approximately 2.5–3 cm in length and around 2.5 cm in diameter. The cervix is divided into two main regions: ectocervix, which is the portion of the cervix that protrudes into the vagina, and endocervix, which is the canal that connects the ectocervix to the uterine cavity. External os is the opening of the cervix into the vagina. It appears as a small, circular hole in women who have never given birth (nulliparous). In women who have given birth (parous), it may appear as a slit. Internal os is the opening into the uterus, separating the cervical canal from the uterine cavity. The cervical canal connects the external os to the internal os. It contains cervical glands that produce mucus.

19.1.3.2 Ligaments

The uterus is held in position by several ligaments within the pelvic cavity, which include:

Broad Ligament: A double layer of peritoneum extending from the sides of the uterus to the pelvic walls and floor

Round Ligaments: Extend from the uterine fundus through the inguinal canal to the labia majora

Uterosacral Ligaments: Extend from the cervix to the sacrum, providing posterior support

Cardinal Ligaments: Attach the cervix to the lateral pelvic walls

The primary blood supply is the uterine artery branch of the internal iliac artery (Fig. 19.4). Venous drainage is through the uterine veins into the internal iliac vein.

19.1.3.3 Microscopic Structure

The uterus is composed of three primary layers (Fig. 19.9):

Endometrium: The endometrium is the most dynamic layer of the uterus, undergoing cyclical changes during the menstrual cycle. It consists of two layers:

- Functional Layer (Stratum Functionalis): The superficial layer, which is shed during menstruation
- Basal Layer (Stratum Basalis): The deeper layer, which regenerates the functional layer after each menstrual cycle

Microscopically, the endometrium comprises epithelial cells (columnar epithelium lining the uterine cavity and extending into the tubular glands), glands (simple tubular glands that vary in length and shape during the menstrual cycle), and stroma (a cellular connective tissue with a rich supply of blood vessels, particularly spiral arterioles crucial for endometrial growth and shedding).

Myometrium: The myometrium is primarily composed of smooth muscle cells. These cells are arranged in interlacing bundles and layers with varying orientations, allowing the uterus to contract efficiently.

Perimetrium: The perimetrium is a serous layer, continuous with the peritoneal lining of the pelvic cavity. It comprises a thin mesothelium layer supported by a connective tissue base.

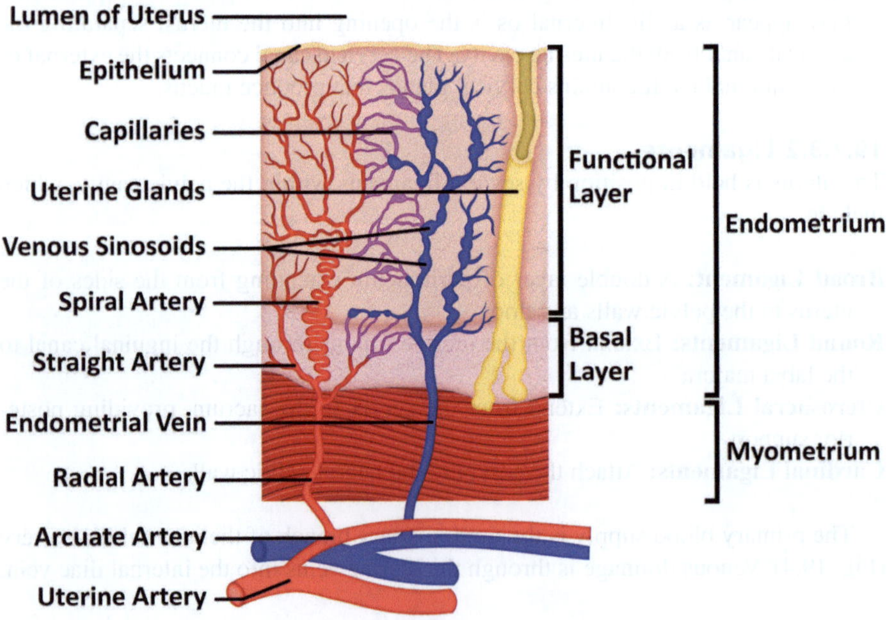

Fig. 19.9 Microscopic structure of the uterus

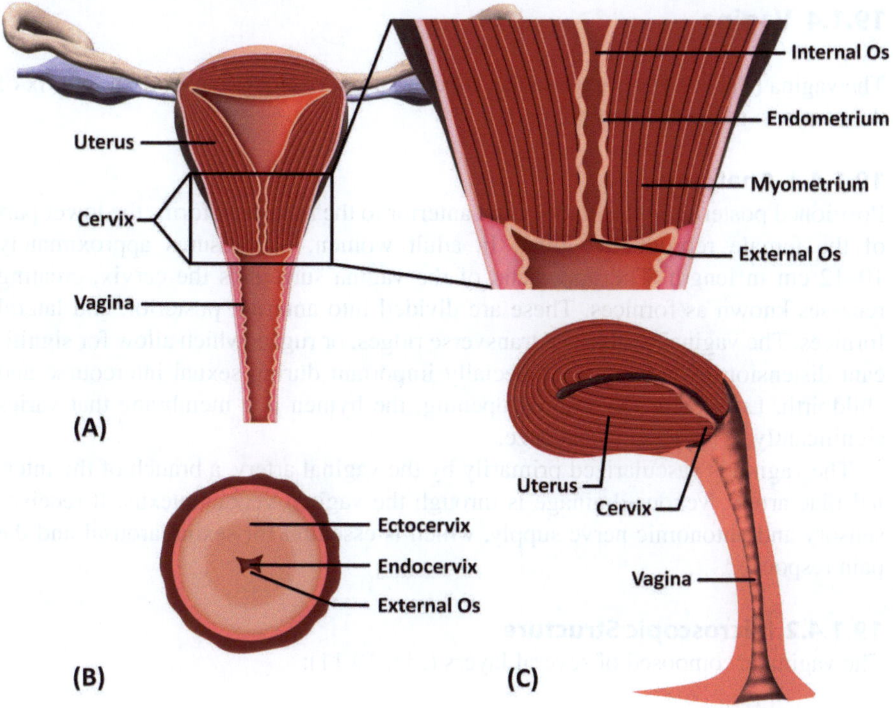

Fig. 19.10 The cervix. (**a**) Frontal section, anterior view. (**b**) Intravaginal view. (**c**) Sagittal view

The cervix comprises two primary types of epithelial cells, reflecting its anatomical division into the ectocervix and endocervix (Fig. 19.10). Ectocervix is covered by stratified squamous epithelium, similar to the vaginal epithelium, protecting against the acidic vaginal environment. The endocervix is lined by a columnar epithelium, which contains mucus-secreting glands. The area where these two epithelia meet is known as the transformation zone. This region is of particular clinical importance as it is the most common site for cervical neoplasia, including cervical cancer. The endocervical canal contains numerous mucin-secreting glands. These glands produce cervical mucus, which plays a vital role in fertility by facilitating or impeding sperm passage depending on the menstrual cycle phase. Beneath the epithelium lies the stroma, composed of dense fibrous connective tissue intermixed with smooth muscle fibers. This layer provides structural support and flexibility, essential for cervical dilation during childbirth.

19.1.3.4 Function

The uterus plays a critical role in:

- Menstrual cycle regulation
- Providing a site for implantation of the fertilized ovum
- Supporting fetal development during pregnancy
- Facilitating childbirth through myometrial contractions

19.1.4 Vagina

The vagina (Fig. 19.10) is a muscular, tubular structure extending from the cervix of the uterus to the exterior of the body.

19.1.4.1 Anatomy

Positioned posterior to the bladder and anterior to the rectum, it forms the lower part of the female reproductive tract. In adult women, it measures approximately 10–12 cm in length. The upper end of the vagina surrounds the cervix, creating recesses known as fornices. These are divided into anterior, posterior, and lateral fornices. The vaginal walls have transverse ridges, or rugae, which allow for significant distension of the vagina, especially important during sexual intercourse and childbirth. Located at the vaginal opening, the hymen is a membrane that varies significantly in appearance and size.

The vagina is vascularized primarily by the vaginal artery, a branch of the internal iliac artery. Venous drainage is through the vaginal venous plexus. It receives sensory and autonomic nerve supply, which is essential for sexual arousal and the pain response.

19.1.4.2 Microscopic Structure

The vagina is composed of several layers (Fig. 19.11):

Mucosal Layer: It is the innermost lining of the vagina. The vaginal mucosa is lined with non-keratinized, stratified squamous epithelium. This epithelium undergoes cyclic changes under hormonal influence and is thicker during the reproductive years. The epithelial cells of the vagina contain glycogen, which is broken down into lactic acid by lactobacilli, maintaining the vagina's acidic environment. Lamina propria is a connective tissue layer beneath the epithelium, containing blood vessels, lymphatics, and nerve fibers. It supports the epithelium and responds to hormonal changes.

Muscularis Layer: It is the middle layer of the vaginal wall. This layer consists of smooth muscle fibers arranged in longitudinal and circular bundles. These muscles provide the structural integrity and elasticity necessary for vaginal functions, including accommodating penile penetration during sexual intercourse and stretching during childbirth.

Adventitial Layer: Adventitia is the outermost connective tissue layer. It is a layer of loose connective tissue that connects the vagina to surrounding pelvic structures. It contains blood vessels, lymphatics, and nerves.

Estrogen significantly influences the vaginal microscopic anatomy, particularly affecting the thickness of the epithelium and the richness of the vascular supply.

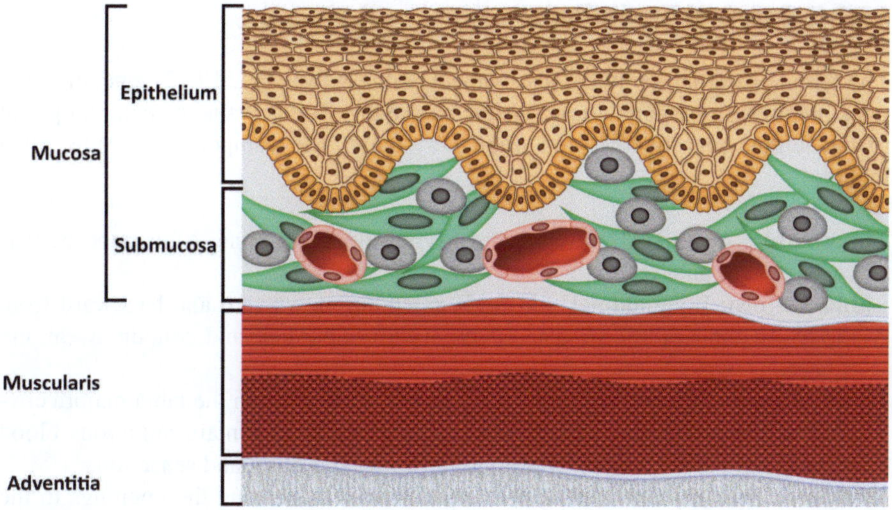

Fig. 19.11 Microscopic structure of the vaginal wall

19.1.4.3 Microbiota

The vaginal microbiota (a group of microbes normally living inside the vagina) is a complex ecosystem that plays a crucial role in maintaining vaginal health and preventing infections. The vaginal microbiota consists of a diverse community of microorganisms: *Lactobacillus* species are the predominant bacteria in the vagina of healthy, reproductive-age women. They play a crucial role in maintaining the vaginal pH at a low level (around pH 3.5–4.5), which is necessary for inhibiting the growth of pathogenic organisms. Other bacterial genera present include *Gardnerella*, *Streptococcus*, *Ureaplasma*, *Mycoplasma*, *Bifidobacterium*, and various anaerobes. Fungi, predominantly *Candida* species, and an array of viruses, including human papillomavirus (HPV), also inhabit the vaginal ecosystem.

Factors influencing vaginal microbiota include:

- Hormonal Changes: Estrogen and progesterone levels significantly influence the composition of the vaginal microbiota.
- Sexual Activity: The introduction of new microbes through sexual contact can alter the vaginal ecosystem.
- Antibiotics and Douching: These can disrupt the natural balance of vaginal microorganisms, leading to overgrowth of harmful bacteria.

19.1.4.4 Function

The vagina plays several vital roles in the female reproductive system:

- Menstrual Flow: Acts as the outlet for the menstrual flow from the uterus.
- Sexual Intercourse: Accommodates the penis during sexual intercourse and is an integral part of sexual arousal and pleasure.
- Childbirth: The birth canal through which the fetus is delivered.

19.1.5 External Genitalia

The female external genitalia, also known as the vulva (Fig. 19.12), encompasses the structures of the genital tract visible externally. The vulva is the external portal to the female reproductive and urinary systems. The components of the vulva include:

Mons Pubis: A rounded, fatty area overlying the pubic bone. After puberty, it is covered with pubic hair.

Labia Majora: Longitudinal skin folds extending downward and backward from the mons pubis. They are also covered with pubic hair and contain sweat and sebaceous glands.

Labia Minora: Smaller, hairless folds of skin located within the labia majora covered with non-keratinized stratified epithelium. They contain numerous blood vessels and nerve endings, contributing to sexual arousal and sensation.

Vestibule: The area enclosed by the labia minora, containing the openings to the urethra (urethral meatus) and vagina (vaginal orifice).

Bartholin's Glands: Located on either side of the vaginal orifice, these glands secrete fluid that lubricates the vulva.

Clitoris (Fig. 19.13): A small, erectile structure located at the anterior junction of the labia minora above the urethral opening. It is a highly sensitive organ, rich in nerve endings and blood vessels, primarily designed for sexual stimulation and arousal. Components of the clitoris include:

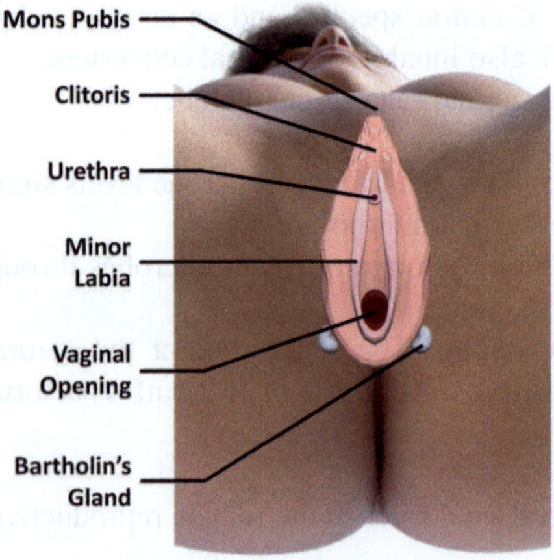

Fig. 19.12 Female external genitalia

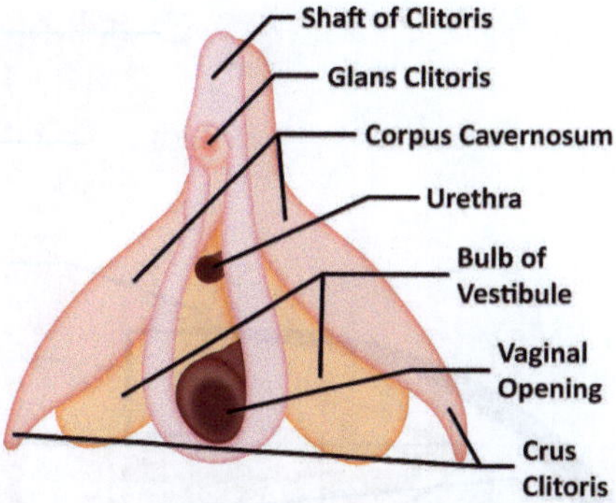

Fig. 19.13 The erectile tissues of the female external genitalia

- Glans Clitoris: The visible portion of the clitoris, located at the top of the vulva. The glans is a highly sensitive structure covered by a hood formed by the labia minora.
- Clitoral Body (Shaft): Extends internally from the glans and is made of erectile tissue. It can become engorged with blood and firm during sexual arousal.
- Crura of the Clitoris: Two elongated structures originating at the lower end of the pubic bone and extending around the vaginal opening. These are also composed of erectile tissue.
- Bulbs of the Vestibule: Located on either side of the vaginal opening, these bulbs are erectile and engorge during sexual arousal, contributing to the tightening of the vaginal orifice.

19.1.6 Breast

The female breast (Fig. 19.14) is a key anatomical structure, playing a vital role in lactation, aesthetics, and identity.

19.1.6.1 Anatomy

The breasts are paired structures on the anterior chest wall, primarily composed of glandular tissue and fat. Breast size and shape vary greatly among individuals and are influenced by genetics, age, hormonal status, and body weight. The skin over the breast is similar to elsewhere on the body but can stretch significantly during pregnancy and lactation. The nipple is a cylindrical projection located roughly at the center of the breast, containing multiple openings for the milk ducts. The pigmented area surrounding the nipple is called the areola, containing sebaceous glands (Montgomery glands) that provide lubrication.

The breasts receive blood from the internal thoracic (mammary), lateral thoracic, and intercostal arteries. Lymphatic drainage from the breasts primarily goes to the axillary lymph nodes, with some drainage to the parasternal and clavicular nodes. This is significant in the context of breast cancer.

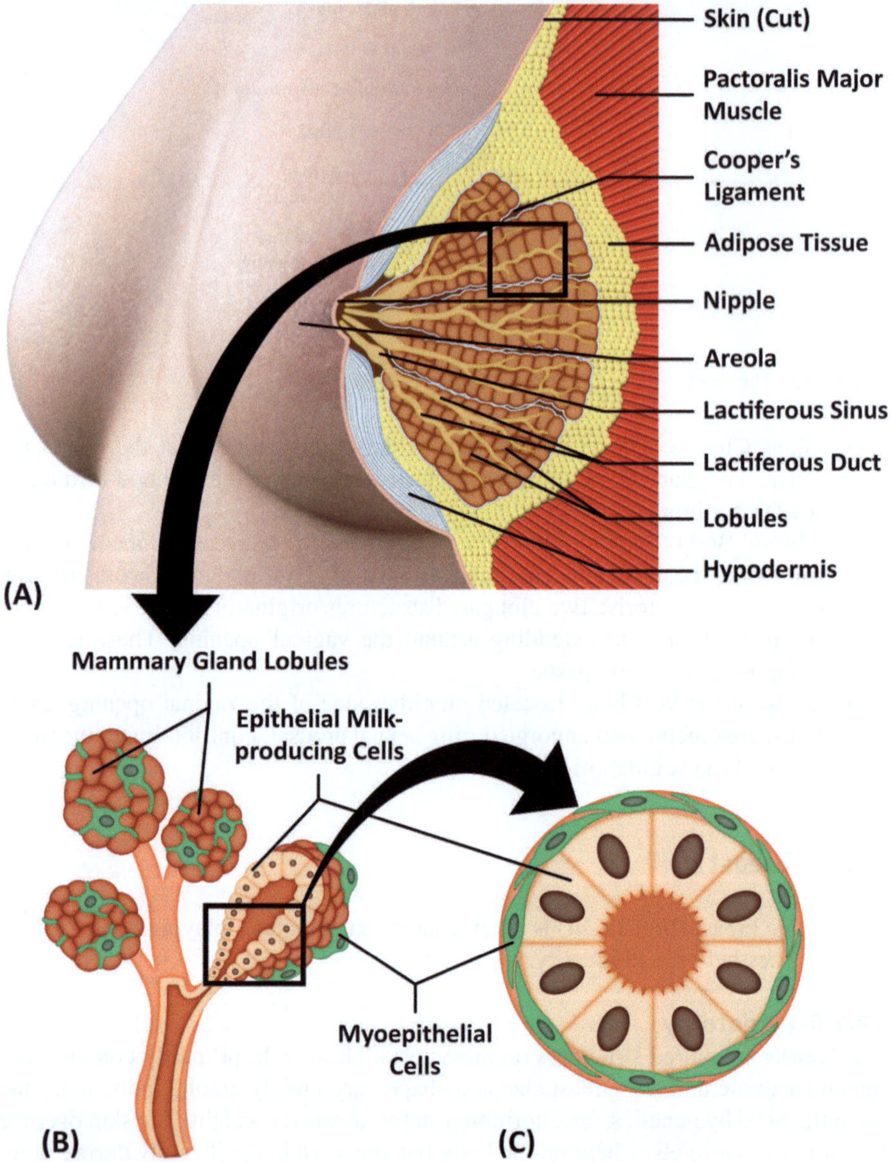

Fig. 19.14 The breast (mammary gland). (**a**) Gross structure, sagittal section. (**b**) Structure of lactating lobular glands (**c**) Microscopic structure of the lactating ducts

19.1.6.2 Microscopic Structure

The breast comprises glandular tissue and adipose (fat) tissue supported by a framework of connective tissue (Fig. 19.14).

Glandular Tissue: Lobules are the functional units of the breast, containing the alveoli, which are milk-producing glands. During lactation, the alveoli are active and engorged with milk. Milk ducts transport milk from the lobules to the nipple. They are lined with a layer of epithelial cells and myoepithelial cells.

Myoepithelial Cells: These cells surround the alveoli and ducts. They contract during lactation to aid in the expulsion of milk and function similarly to smooth muscle cells.

Adipose Tissue: Adipose tissue makes up most of the breast volume, providing padding and contributing to its size and shape.

Connective Tissue: Cooper's ligaments are fibrous bands that provide structural support to the breast, maintaining its shape and integrity. The connective tissue stroma, which contains blood vessels, lymphatics, and nerves, interweaves with the glandular and adipose tissues.

Breast tissue is highly responsive to hormonal changes, particularly estrogen and progesterone, which influence the development and function of the glandular structures.

19.1.6.3 Developmental Changes

The female breast undergoes developmental changes throughout a woman's life, which can be divided into several key stages:

Infancy: At birth, female infants have rudimentary breast tissue, including a small number of mammary glands and ducts. In the first few weeks of life, female infants may experience a transient period of breast enlargement due to the influence of maternal hormones passed through breastfeeding.

Childhood and Prepuberty: During childhood, breast development is minimal, and breasts are typically flat and undeveloped. As girls approach puberty, typically between the ages of 8 and 13, the onset of breast development, known as "thelarche," begins. Thelarche is characterized by the development of breast buds, small lumps beneath the nipple area resulting from the proliferation of mammary ducts, and the growth of glandular tissue. Hormonal changes, particularly the increase in estrogen, trigger this development.

Puberty: Puberty marks the most significant stage of breast development. The breast buds continue to grow and become more prominent. The areolas often enlarge and darken in color. The mammary glands and ducts continue to develop and branch out. Fat tissue accumulates in the breasts, contributing to their size and shape.

Adolescence and Young Adulthood: Breast development typically continues through adolescence and into the late teens and early 20 s. During this period, breasts often reach their mature size and shape. Hormonal fluctuations throughout the menstrual cycle can lead to temporary changes in breast size and tenderness.

Pregnancy and Lactation: During pregnancy, the breasts undergo further changes in preparation for breastfeeding. Hormones, primarily prolactin and oxytocin, stimulate the enlargement of mammary glands and ducts. The breasts become fuller, firmer, and more sensitive. After childbirth, breast milk production is initiated, and the breasts continue to change in response to lactation.

Menopause and Aging: With the onset of menopause (usually around the age of 45–55), hormonal activity decreases, primarily estrogen and progesterone. The breasts may undergo involution, a process where glandular tissue is gradually replaced by fatty tissue, leading to a reduction in breast size and firmness. The skin of the breasts may lose elasticity, leading to sagging.

19.2 Part 2: Female Reproductive System Physiology

The physiology of the female reproductive system encompasses a complex interplay of anatomical structures, hormonal cycles, and biological processes essential for human reproduction. This system facilitates the creation, fertilization, and development of a new human life and plays a significant role in the broader aspects of women's health and well-being.

19.2.1 Oogenesis

Oogenesis (Fig. 19.6), the process of egg formation in females, begins before a female is born and continues throughout her reproductive years. Oogenesis involves forming and maturing oocytes (egg cells) within the ovaries. Unlike spermatogenesis in males, which occurs continuously, oogenesis is a discontinuous process closely linked to the menstrual cycle. Oogenesis begins in fetal life. Primordial germ cells migrate to the developing ovaries and differentiate into oogonia. Oogonia undergo rapid mitotic division, increasing their numbers. Around mid-gestation, oogonia enter the first meiotic division, becoming primary oocytes. This process is halted at the prophase stage (Section 4.1.3.2). Each primary oocyte is surrounded by a layer of flat granulosa cells, forming a primordial follicle. These remain dormant until puberty. At the onset of puberty, a small number of primordial follicles are recruited during each menstrual cycle. These follicles undergo a series of developmental stages, becoming primary, secondary, and tertiary (antral) follicles. The secondary oocyte completes the first meiotic division just before ovulation (releasing a secondary oocyte from the dominant follicle in the ovary). It begins the second meiotic division but stops at metaphase II (Section 4.1.3.2), completed only if fertilization occurs. Post-ovulation, the ruptured follicle transforms into the corpus luteum, which secretes progesterone and estrogen to support a potential pregnancy. Not all the ovaries' oocytes and primordial follicles develop into secondary and tertiary oocytes. Most oocytes undergo atresia, a process of degeneration and absorption, at various stages of development. With advancing age, the reserve of follicles diminishes, leading to menopause, the end of a woman's reproductive years.

19.2.2 Ovarian Follicles

Ovarian follicles (Fig. 19.6) are dynamic structures within the ovaries. From birth, the ovaries contain a finite number of follicles, each with the potential to mature and release an oocyte during the menstrual cycle. Each follicle contains a single oocyte (egg) surrounded by supportive cells.

19.2.2.1 Types of Ovarian Follicles

Primordial Follicles: The most basic form, consisting of a primary oocyte surrounded by a single layer of flat granulosa cells. These follicles are present from birth and can remain dormant for years.

Primary Follicles: Characterized by an enlarged oocyte and a multilayered granulosa cell layer.

Secondary Follicles: These are distinguished by the theca folliculi, a layer of connective tissue, and blood vessels surrounding the granulosa cells.

Tertiary (Antral) Follicles: These are mature follicles with a fluid-filled cavity called the antrum. The oocyte is located in a cumulus oophorus, surrounded by granulosa cells.

19.2.2.2 Folliculogenesis

Usually, several primordial follicles are activated to become primary follicles. These primary follicles grow and differentiate into secondary and then tertiary follicles. Usually, one follicle becomes dominant and matures, while others undergo atresia (degeneration).

19.2.2.3 Hormonal Regulation

Follicle-stimulating hormone (FSH) promotes the growth and maturation of follicles, and luteinizing hormone (LH) surge leads to final maturation and ovulation of the dominant follicle (Fig. 19.6). Granulosa cells of the growing follicles produce estrogen, providing nutrients to the growing oocyte. Theca cells interact with granulosa cells to facilitate steroidogenesis (hormone production).

19.2.2.4 Ovulation

The mature tertiary follicle releases the secondary oocyte into the fallopian tube. Post-ovulation, the ruptured follicle transforms into the corpus luteum, secreting progesterone and estrogen to support a potential pregnancy.

19.2.3 Ovarian Cycles

The ovarian cycle (Fig. 19.15) is a series of recurring physiological changes in the ovaries, integral to the reproductive system, occurring concurrently with the menstrual cycle. It typically spans about 28 days and can be divided into three distinct phases: the follicular phase, ovulation, and the luteal phase.

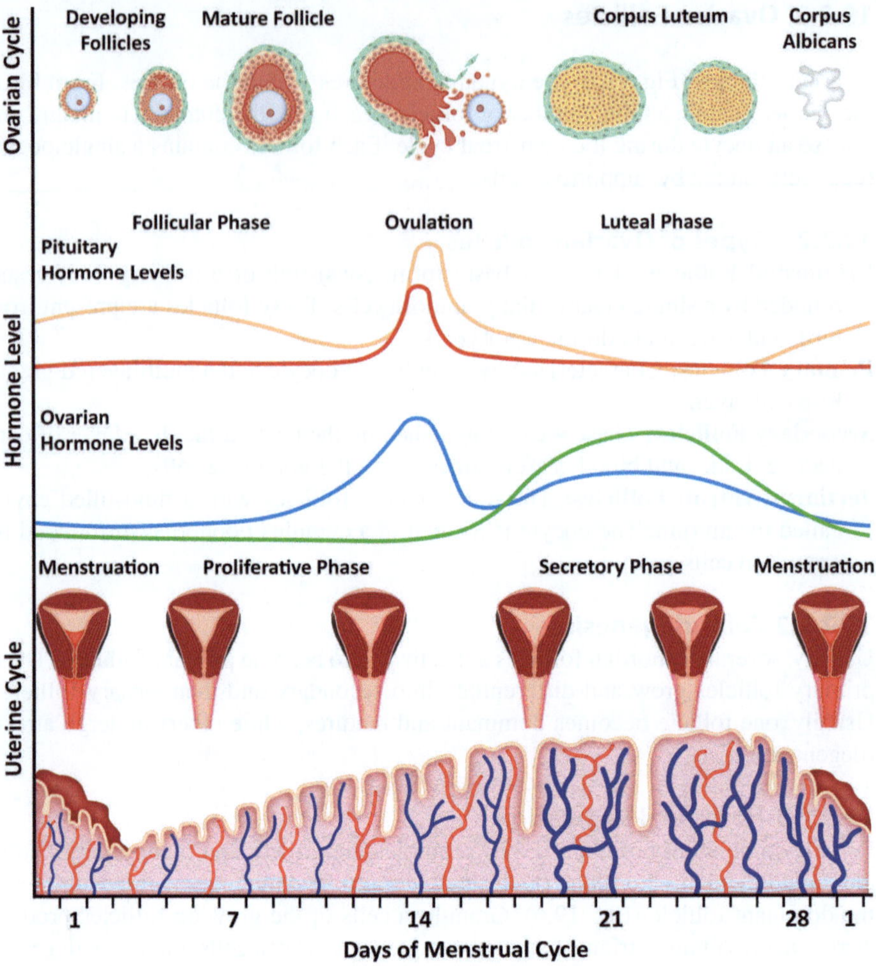

Fig. 19.15 Ovarian and uterine cycle. Correlation of anterior pituitary and ovarian hormones with the ovary and uterus structural changes

19.2.3.1 Follicular Phase

The follicular phase begins on the first day of the menstrual cycle and is characterized by the development of ovarian follicles. Under the influence of FSH, multiple follicles start to grow, but usually only one reaches full maturity.

Developing follicles produce estrogen, which stimulates the lining of the uterus (endometrium) to thicken.

19.2.3.2 Ovulation

As the levels of estrogen rise, they eventually reach a threshold that triggers a surge in LH from the anterior pituitary gland. This LH surge typically occurs around the middle of the menstrual cycle, approximately 24–36 h before ovulation. The LH surge triggers the release of enzymes that weaken the follicular wall. Within 24–36 h after the LH surge, the mature follicle ruptures, expelling the mature egg (ovum) from the ovary. This event is ovulation. The released egg is then captured by the fimbriae (finger-like projections) of the fallopian tube and is transported into the fallopian tube.

19.2.3.3 Luteal Phase

After ovulation, the remaining part of the follicle within the ovary transforms into a temporary endocrine structure called the corpus luteum. The corpus luteum secretes progesterone and some estrogen, which help maintain the thickened endometrial lining in preparation for a potential pregnancy. If fertilization and implantation do not occur, the corpus luteum degenerates after 10–16 days. The decline in progesterone and estrogen levels triggers menstruation, and the cycle begins anew.

The ovarian and menstrual cycles (Section 19.2.3) are closely linked, with changes in the ovaries causing endometrial changes in the menstrual cycle. While the average length of the ovarian cycle is 28 days, it can vary significantly among individuals and can be influenced by factors like stress, health, and lifestyle.

19.2.4 Ovarian Hormones

The ovaries produce several hormones, the most prominent of which are estrogen and progesterone. They also produce small amounts of testosterone and other androgens.

Estrogen: Estrogens, including estradiol (Fig. 19.16), estrone, and estriol, are synthesized mainly in the ovaries. Estradiol is the most potent and prevalent form. Estrogen regulates the menstrual cycle, promotes the development of female secondary sexual characteristics, and maintains the reproductive tissues. It also significantly affects bone health, cardiovascular function, and skin and hair health. Estrogen levels are regulated by the hypothalamic-pituitary-ovarian axis, peaking during the follicular phase of the menstrual cycle.

Progesterone: Progesterone (Fig. 19.16) is produced in the ovaries by the corpus luteum post-ovulation, and LH stimulates its production. Progesterone prepares

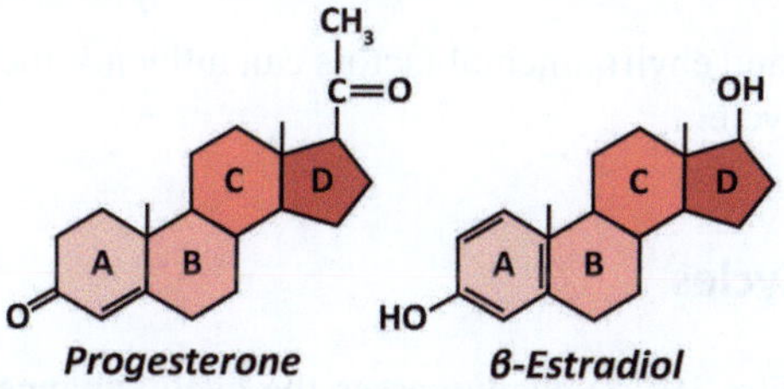

Fig. 19.16 Structure of the primary ovarian hormones

the endometrium for implantation and supports early pregnancy. It also regulates the menstrual cycle and affects libido, bone health, and mood. Progesterone levels peak during the luteal phase of the menstrual cycle and drop if fertilization does not occur, leading to menstruation.

Androgens: The ovaries produce androgens in smaller quantities, including testosterone and androstenedione. While typically associated with male characteristics, ovarian androgens play roles in female libido, bone health, and the development of muscle mass. Gonadotropins influence androgen production.

The onset of puberty sees an increase in estrogen and progesterone, leading to the development of secondary sexual characteristics and the commencement of menstrual cycles. During reproductive years, hormone levels fluctuate cyclically, regulating ovulation and menstruation. By the beginning of menopause, ovarian function declines, leading to decreased hormone levels and marking the end of reproductive capability.

19.2.5 Regulation of Ovarian Cycles

The ovarian cycle regulation is closely tied to the hypothalamic-pituitary-ovarian axis. The hypothalamus produces gonadotropin-releasing hormone (GnRH), which controls the release of hormones from the pituitary gland. The pituitary gland responds to GnRH by secreting follicle-stimulating hormone (FSH) and luteinizing hormone (LH). In response to FSH and LH, ovaries produce estrogen and progesterone, which regulate the hypothalamus and pituitary through feedback mechanisms.

During the follicular phase, FSH promotes the growth and maturation of ovarian follicles. Rising estrogen levels (Fig. 19.15) from the maturing follicles initially inhibit GnRH, FSH, and LH production (negative feedback) but eventually stimulate their release (positive feedback) as the levels peak. The LH surge, triggered by the peak in estrogen levels, leads to the release of the mature egg from the dominant follicle (ovulation). Post-ovulation, the ruptured follicle transforms into the corpus luteum, which secretes progesterone and some estrogen. High progesterone levels inhibit GnRH, FSH, and LH production, preventing further follicular development during this phase. If fertilization does not occur, the corpus luteum degenerates, causing a drop in estrogen and progesterone levels. The reduction in hormone levels releases the inhibition on the hypothalamus and pituitary, leading to the start of a new cycle.

Stress, nutrition, and environmental factors can influence the regularity and function of the ovarian cycle.

19.2.6 Uterine Cycles

The uterine cycle (Fig. 19.15) encompasses the cyclic changes in the uterus, particularly in the endometrium, in response to ovarian hormones. This cycle, commonly known as the menstrual cycle, typically spans approximately 28 days, though it can vary among individuals. It is divided into menstrual, proliferative, and secretory phases.

Menstrual Phase: This phase typically lasts 3–7 days and is characterized by the shedding of the endometrial lining, which results in menstrual bleeding. It is triggered by a decrease in progesterone and estrogen levels due to the degeneration of the corpus luteum if fertilization does not occur.

Proliferative (Follicular) Phase: This phase begins after menstruation and lasts until ovulation. Under the influence of estrogen secreted by developing ovarian follicles, the endometrium undergoes regeneration, thickening, and vascularization. Estrogen also influences the cervix, causing cervical mucus to become thin and permeable to sperm.

Ovulation: It marks the transition from the proliferative to the secretory phase.

Secretory (Luteal) Phase: This phase starts immediately following the ovulation and ends with the onset of the next menstruation. During the secretory phase, under the influence of progesterone from the corpus luteum, the endometrium becomes more glandular and secretes nutrients, preparing for the potential implantation of a fertilized egg. Cervical mucus becomes thick and impenetrable to sperm, a protective mechanism post-ovulation.

19.2.7 Female Sexual Response

Female sexual function and response encompass a complex interplay of physiological, emotional, and psychological factors. The female sexual response cycle is typically described in four stages: excitement, plateau, orgasm, and resolution. This response is a coordinated interaction between the nervous, vascular, and endocrine systems, as well as psychological factors.

19.2.7.1 Phases

Excitement Phase: This phase is initiated by physical or psychological erotic stimuli. It is associated with increased heart rate, blood pressure, and breathing rate. Vasocongestion leads to swelling of the clitoris, labia, and vaginal walls, along with vaginal lubrication. The person feels sexual desire and pleasure.

Plateau Phase: It is a continuation and intensification of the excitement phase. During this phase, genital engorgement and sensitivity continue to increase. The inner two-thirds of the vagina expands, the clitoris retracts slightly, and the uterus engorges and elevates, often associated with heightened sexual arousal and anticipation of orgasm.

Orgasm Phase: Orgasm is the peak of sexual pleasure, characterized by involuntary muscle contractions and release of sexual tension. It is associated with rhythmic contractions of the pelvic floor muscles, uterus, and anal sphincter, with increased heart rate, blood pressure, and respiratory rate. Orgasms vary significantly in intensity and duration among women and may or may not be accompanied by ejaculation (release of fluid from Skene's glands).

Resolution Phase: During this phase, the body returns to its pre-arousal state, with decreased genital engorgement, relaxation of muscles, decrease in heart rate and blood pressure, and dissipation of sexual tension. The woman feels relaxation, contentment, and emotional intimacy, although some women may experience a quick return to the non- aroused state without these feelings.

19.2.7.2 Regulation

Role of Hormones: Estrogen and progesterone influence libido, vaginal lubrication, and overall sexual responsiveness. Testosterone, although at lower levels than males, plays a role in sexual desire and arousal.

Neural Control: The neural control of the female sexual response involves an interplay between the central and peripheral nervous systems. This neural control governs the physiological changes during the sexual response cycle. The brain plays a central role in the initiation and coordination of the female sexual response. Key brain regions involved in this process include:

- The limbic system, which is associated with emotions and motivation, can influence sexual desire and arousal.
- The hypothalamus regulates hormone production and can influence sexual motivation and response.
- The cerebral cortex, responsible for conscious thought and decision-making, can process sensory input and influence sexual desire.

Sensory stimulation, such as erotic imagery, touch, or sexual fantasies, can activate the cerebral cortex and lead to sexual arousal. Arousal also involves increased blood flow to the genital region, which both the CNS and the PNS mediate.

The autonomic nervous system, a division of the PNS, plays a crucial role in sexual response. The sympathetic nervous system is responsible for the initial phase of sexual arousal. It stimulates the release of nitric oxide (NO) and relaxes the smooth muscles in the blood vessels of the genitalia, leading to increased blood flow. The parasympathetic nervous system becomes dominant during sexual arousal, leading to further relaxation of smooth muscles and increased engorgement of the genital tissues.

During sexual stimulation, sensory signals from the genitalia are transmitted to the brain, leading to the experience of sexual pleasure. The CNS coordinates the release of oxytocin and endorphins during orgasm, which contribute to feelings of bonding and pleasure.

19.3 Part 3: Male Reproductive System Structure

In males, the reproductive system (Fig. 19.17) primarily consists of the testes, where spermatozoa (sperm cells) are produced; the epididymis, where sperm is stored and matures; the vas deferens, through which sperm travels during ejaculation; the prostate and seminal vesicles, which produce seminal fluid to nourish and transport sperm; and the penis, through which sperm is delivered to the female reproductive system.

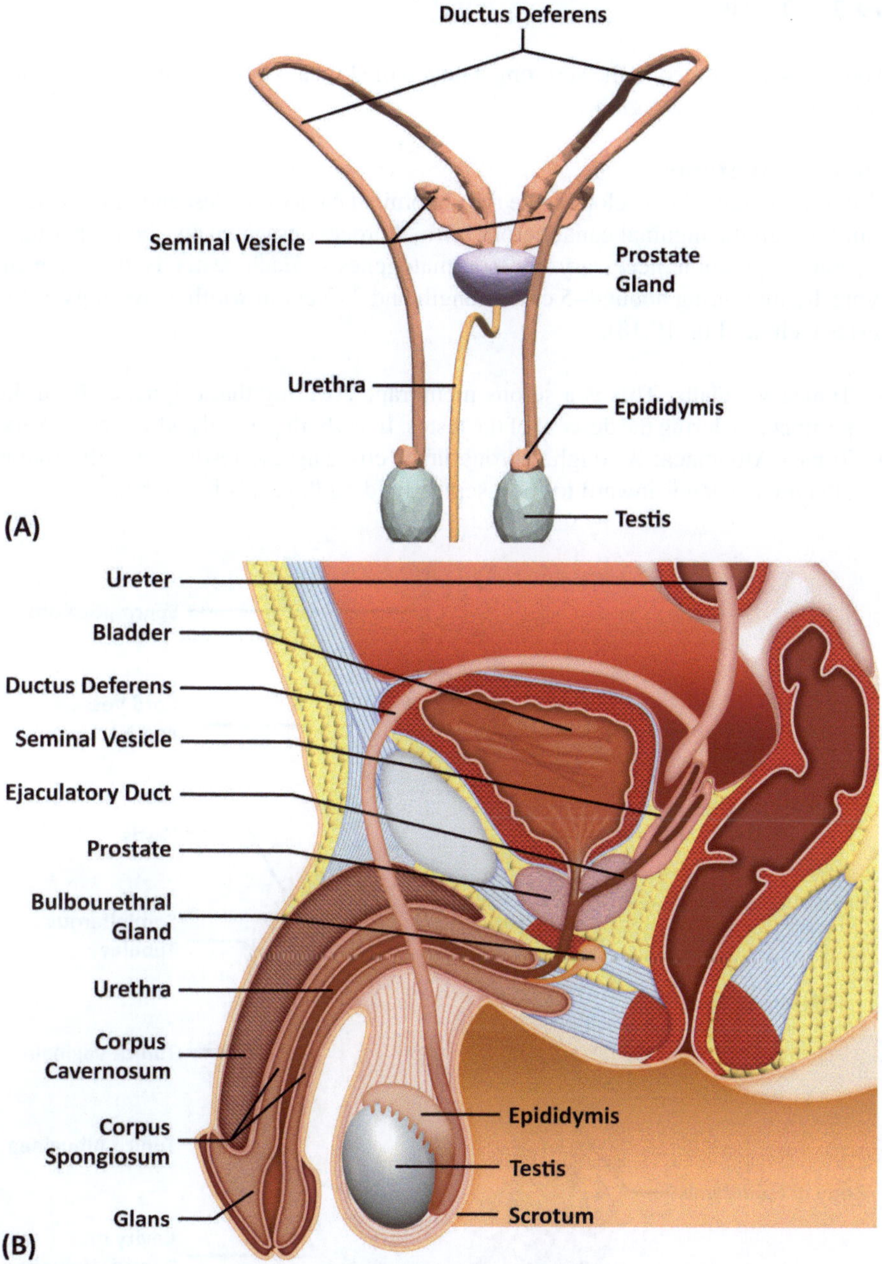

Fig. 19.17 The male reproductive system structure. (**a**) Anterior view. (**b**) Sagittal section

19.3.1 Testis

The testes are located in the scrotum, a pouch of skin and muscle situated below the penis in the inguinal region.

19.3.1.1 Anatomy

The testes originally develop inside the abdominal cavity and descend into the scrotum through the inguinal canal before birth, a process essential for the proper temperature regulation necessary for spermatogenesis. Each testis is oval-shaped, typically measuring about 4–5 cm in length and 2–3 cm in width. Coverings of the testes include (Fig. 19.18):

- Tunica Vaginalis: This is a serous membrane covering that originates from the peritoneum during the descent of the testes. It has both parietal and visceral layers.
- Tunica Albuginea: A tough, fibrous layer covering the testis itself, the tunica albuginea extends inward to form septa, dividing the testis into lobules.

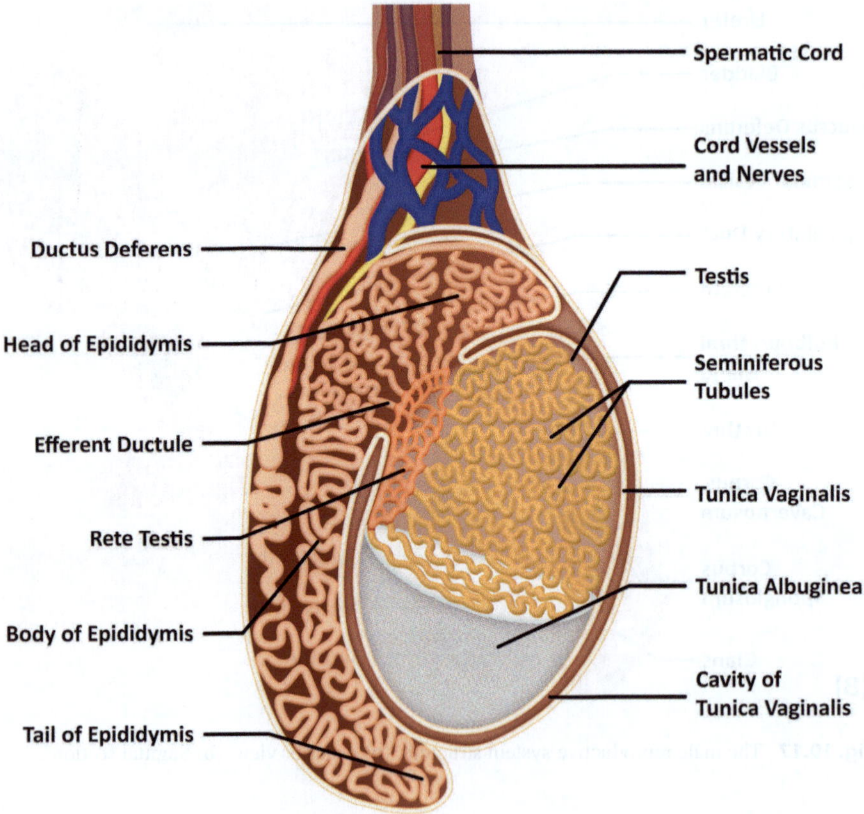

Fig. 19.18 Testis and associated structures

Internal Structure: Each testis comprises 200–300 wedge-shaped lobules containing 1–3 tightly coiled seminiferous tubules where sperm is produced. The mediastinum testis is a thickened area of the tunica albuginea that extends into the testis from its posterior border and provides a pathway for blood vessels, lymphatics, and nerves.

The testicular arteries (Fig. 19.19) are branches of the abdominal aorta that supply arterial blood to the testes. The pampiniform plexus of veins assists in cooling arterial blood entering the testes.

19.3.1.2 Microscopic Structure

Seminiferous Tubules (Fig. 19.20): Seminiferous tubules are coiled structures composed of a stratified epithelium surrounding a central lumen. The epithelium is composed of two main types of cells: germ cells (spermatogenic cells) are the actual sperm-producing cells, which undergo a series of mitotic and meiotic

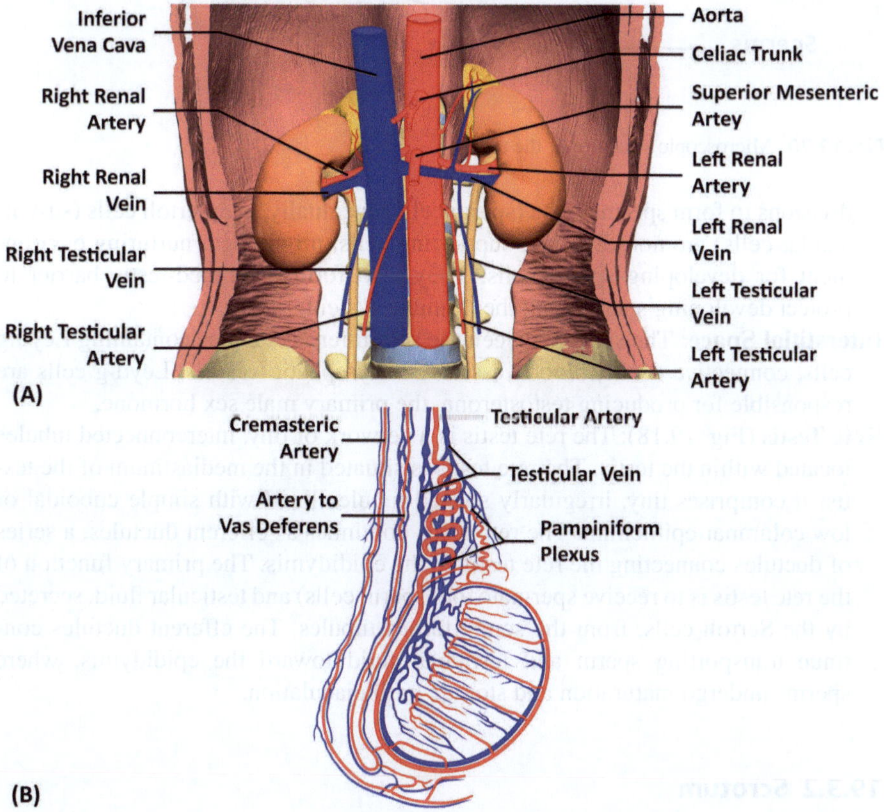

Fig. 19.19 Blood vessels to male reproductive system. (**a**) Testicular arteries and veins. (**b**) Vessels in the spermatic cord and testis

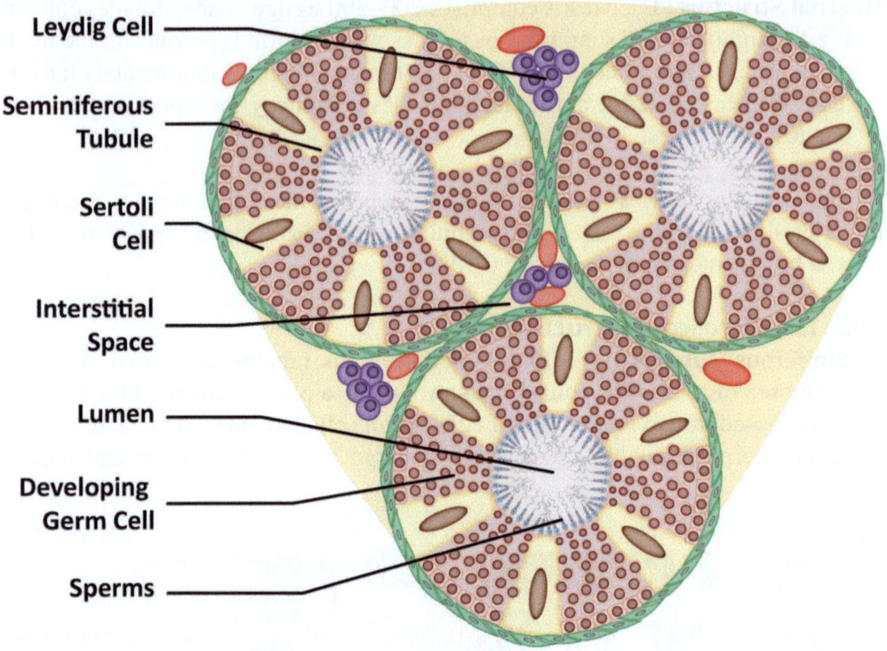

Fig. 19.20 Microscopic structure of the testis

divisions to form spermatozoa (sperm cells) eventually, and Sertoli cells (sustentacular cells) are non-dividing, supporting cells, providing a nurturing environment for developing sperm cells. They also form the blood-testis barrier to protect developing sperm from the immune system.

Interstitial Space: The space between the seminiferous tubules containing Leydig cells, connective tissue, blood vessels, and lymphatic vessels. Leydig cells are responsible for producing testosterone, the primary male sex hormone.

Rete Testis (Fig. 19.18): The rete testis is a network of tiny, interconnected tubules located within the testis. The rete testis is situated in the mediastinum of the testis. It comprises tiny, irregularly shaped tubules lined with simple cuboidal or low columnar epithelium. The rete testis continues as efferent ductules, a series of ductules connecting the rete testis to the epididymis. The primary function of the rete testis is to receive spermatozoa (sperm cells) and testicular fluid, secreted by the Sertoli cells, from the seminiferous tubules. The efferent ductules continue transporting sperm and testicular fluid toward the epididymis, where sperms undergo maturation and storage until ejaculation.

19.3.2 Scrotum

The scrotum (Fig. 19.21) is an external pouch that hangs below the penis. It is often darker than the rest of the skin and has a wrinkled appearance. The size of the

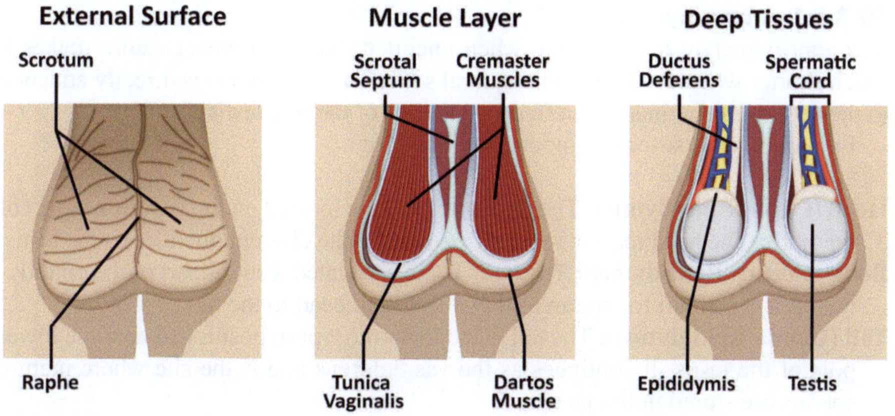

Fig. 19.21 The scrotum

scrotum varies among individuals and changes in response to temperature (contracts in cold and relaxes in warm temperatures).

Layers of Scrotum The outermost layer of the scrotum is formed by skin, thin yet durable, with hair follicles, sweat glands, and sebaceous glands. Beneath the skin, the dartos fascia and muscles are located. This layer contains smooth muscle fibers. The dartos muscle contracts in response to cold temperatures, wrinkling the scrotal skin and pulling the testes closer to the body for warmth. The next layer is the external spermatic fascia, derived from the external oblique muscle's aponeurosis. Deep to the external spermatic fascia, the cremaster muscle and fascia are located, a continuation of the internal oblique muscle. This layer plays a role in temperature regulation by raising or lowering the testes. Deep to this layer is internal spermatic fascia, derived from the transversalis fascia. The deepest layer is tunica vaginalis, a serous membrane with parietal and visceral layers enveloping the testes and a portion of the spermatic cord. The scrotum is divided into two compartments by a midline septum, each housing a testis, epididymis, and lower part of the spermatic cord.

The scrotum regulates the temperature of the testes, essential for optimal sperm production, which requires a temperature slightly lower than the body's core temperature.

19.3.3 Epididymis

The epididymis (Fig. 19.18) is a tightly coiled, elongated structure located at the posterior margin of each testis.

19.3.3.1 Anatomy

It is approximately 6–7 cm long when uncoiled, but its compact nature makes it much shorter when observed in its natural state. The epididymis is directly attached to the testis and continues the efferent ductules of the rete testis.

Epididymis has several components, including:

Head (Caput Epididymis): The uppermost part is located on the superior aspect of the testis. The head receives spermatozoa from the efferent ductules of the testis.

Body (Corpus Epididymis): The body is the elongated, central part of the epididymis. It is a conduit for sperm to travel from the head to the tail.

Tail (Cauda Epididymis): The tail is the lowermost part, positioned near the lower pole of the testis. It continues as the vas deferens and is the site where mature sperms are stored until ejaculation.

19.3.3.2 Microscopic Structure

The microscopic structure of epididymis is formed by:

Epithelial Lining: The epididymal duct is lined with pseudostratified columnar epithelium, characterized by principal and basal cells. Principal cells are tall and columnar, with stereocilia (long, non-motile microvilli) on their apical surface, which increase the surface area for absorption and secretion. Basal cells are smaller and located near the basement membrane, and they are thought to be involved in the regeneration of principal cells.

Lamina Propria: Beneath the epithelium is a thin layer of connective tissue, the lamina propria, which contains blood vessels and lymphatics.

Smooth Muscle Layer: Surrounding the lamina propria is a layer of smooth muscle. This layer is relatively thin in the head but becomes thicker in the body and thickest in the tail, aiding in sperm transport.

19.3.3.3 Function

The epididymis is essential for the maturation of spermatozoa, a process that takes about 12–14 days. During this time, sperm acquires motility and the ability to fertilize an ovum. The tail of the epididymis serves as a storage reservoir for mature sperm.

19.3.4 Ductus Deferens

The ductus deferens (Fig. 19.17), commonly known as the vas deferens, is a muscular tube that begins at the tail of the epididymis, ascends through the spermatic cord, traverses the inguinal canal, enters the pelvic cavity, and ends at the ejaculatory ducts.

19.3.4.1 Anatomy

The ductus deferens is approximately 30–40 cm long and 2–3 mm in diameter. The ductus deferens components include:

Head (Ampulla): The ampulla, located near the terminal part of the epididymis, is an enlargement of the duct to store a small quantity of sperm.

Body: The body is the longest part of the ductus deferens. It ascends along the posterior border of the testis and through the inguinal canal.

Tail: It is the terminal portion, which joins with the duct of the seminal vesicle to form the ejaculatory duct.

The ductus deferens is enveloped in the spermatic cord as it ascends through the inguinal canal. The fascial layers that cover it include the internal spermatic fascia, the cremasteric fascia, and the external spermatic fascia.

19.3.4.2 Microscopic Structure

Its wall consists of mucosa (the innermost layer, composed of pseudostratified columnar epithelium and underlying lamina propria), muscularis (a prominent layer, composed of three layers of smooth muscle [inner longitudinal, middle circular, and outer longitudinal], responsible for the peristaltic movement that propels sperm), and adventitia (the outermost layer, composed of loose connective tissue).

19.3.4.3 Function

The primary function of the ductus deferens is to transport sperm from the epididymis to the urethra in preparation for ejaculation. It also serves as a secondary site for sperm storage.

19.3.5 Spermatic Cords

The spermatic cord (Fig. 19.18) connects the testis to the abdominal cavity. It contains a combination of blood vessels, nerves, and other structures necessary for the functioning of the testis and the transport of sperm: ductus deferens, testicular artery, pampiniform venous plexus (network of veins surrounding the testicular artery, helping regulate the temperature of the testis, and draining deoxygenated blood away from the testis), artery to the ductus deferens, sympathetic nerves, lymphatic vessels, and connective tissue and fascia. The cremaster muscle (a continuation of the abdominal internal oblique muscle) surrounds the spermatic cord. Contraction of the cremaster muscle helps regulate the position of the testis within the scrotum in response to changes in temperature and other stimuli.

19.3.6 Seminal Vesicles

The seminal vesicles (Fig. 19.17) are located posterior to the bladder and anterior to the rectum. They lie above the prostate gland and are closely associated with the ductus deferens.

19.3.6.1 Anatomy

The seminal vesicles are paired, tubular structures that appear convoluted and sac-like. Each seminal vesicle is approximately 5–10 cm long, though the size can vary among individuals. The outer surface of the seminal vesicles is irregular due to their lobulated composition. They are generally yellowish-brown in color. Internally, each seminal vesicle consists of a coiled tube that, when unraveled, can measure up to several centimeters in length. The tube forms multiple diverticula, giving the seminal vesicles a spongy texture. Each seminal vesicle joins with the vas deferens to form the ejaculatory duct on its respective side.

19.3.6.2 Microscopic Structure

Microscopically, the seminal vesicles are tubuloalveolar glands, meaning they have a structure composed of both tubular and alveolar elements. They are divided into multiple lobules, each containing its secretory unit. The inner lining of the seminal vesicles is composed of pseudostratified columnar epithelium. These epithelial cells are secretory, producing seminal fluid. Interspersed among them are ciliated cells that help move the fluid through the glandular structure. The stroma consists of connective tissue that supports the glandular tissue. The stroma also contains a rich supply of blood vessels and nerves that regulate the function of the seminal vesicles. Surrounding the epithelium is a layer of smooth muscle fibers. The contraction of these muscles aids in expelling the seminal fluid during ejaculation.

19.3.6.3 Function

The seminal vesicles produce a significant portion (about 60–70%) of the ejaculate volume. This fluid is alkaline, which helps to neutralize the acidity of the vaginal tract. The secretion contains fructose (providing energy to sperm), prostaglandins (facilitating motility and viability of sperm), and fibrinogen (contributing to the coagulation of semen post-ejaculation).

19.3.7 Ejaculatory Ducts

Each ejaculatory duct (Fig. 19.17) is formed by the convergence of the duct from a seminal vesicle and the vas deferens. These ducts travel through the prostate gland and open into the urethra. Typically, each duct is about 1–2 cm long and relatively narrow in diameter. They pass through the prostate gland to terminate at the verumontanum on the urethra. The ejaculatory ducts are lined by a pseudostratified columnar epithelium, similar to that found in the vas deferens and seminal vesicles. The duct is surrounded by a thin layer of smooth muscle, which aids in the expulsion of semen during ejaculation.

The primary role of the ejaculatory ducts is to transport semen, a mixture of sperm and seminal fluid, into the urethra during ejaculation. The muscular activity of the ducts is coordinated with the contraction of the urethral sphincter and other muscles involved in ejaculation.

19.3.8 Prostate

The prostate (Fig. 19.17) is below the bladder and in front of the rectum. It encircles the urethra, which passes through it.

19.3.8.1 Anatomy

The gland is typically the size of a walnut, measuring approximately 3 cm in length, 4 cm in width, and 2 cm in depth. Its shape resembles that of an inverted cone or a chestnut. The prostate is enclosed in a fibrous capsule rich in smooth muscle fibers. The prostate has two lateral lobes, an anterior lobe, a posterior lobe, and a median or middle lobe.

19.3.8.2 Microscopic Structure

Microscopically, the prostate contains numerous glandular acini responsible for secreting prostatic fluid. A layer of epithelial cells lines these glands, typically pseudostratified or simple columnar. The epithelial lining is mainly comprised of secretory luminal cells and basal cells. The luminal cells are responsible for secreting prostatic fluid, while the basal cells are thought to act as progenitor cells for the luminal epithelium. The stroma of the prostate is composed of smooth muscle fibers and fibroelastic connective tissue. This stromal component not only provides structural support but also plays a role in the contractility of the gland during ejaculation. The secretions of the prostatic glands are slightly alkaline and contain enzymes, including prostate-specific antigen (PSA), acid phosphatase, and proteolytic enzymes.

19.3.8.3 Function

The secretion of the prostate constitutes about 20–30% of the volume of semen. This fluid enhances the viability and motility of sperm. During ejaculation, the smooth muscle in the prostate contracts, helping to expel semen.

19.3.9 Bulbourethral Glands

The bulbourethral glands, also known as Cowper's glands (Fig. 19.17), are located bilaterally at the base of the penis, below the prostate gland, and within the deep perineal pouch. These glands are small, about the size of a pea, and round or oval. They are palpable and firm, surrounded by a thin fibrous capsule. The ducts of the bulbourethral glands, around 2.5 cm long, extend through the perineal membrane and open into the spongy part of the urethra.

The bulbourethral glands secrete a clear, mucus-like fluid known as pre-ejaculate. This fluid is released into the urethra upon sexual arousal. The pre-ejaculate fluid lubricates the urethra for the passage of semen and neutralizes any residual acidity, which could harm the sperm.

19.3.10 Urethra

The male urethra (Fig. 19.17) extends from the urinary bladder to the external urethral orifice at the tip of the penis.

19.3.10.1 Anatomy

It is approximately 18–20 cm long, significantly longer than the female urethra. The male urethra is conventionally divided into four distinct segments:

Preprostatic Urethra: About 0.5–1.5 cm long, surrounded by the internal urethral sphincter and begins at the bladder neck.

Prostatic Urethra: Approximately 3–4 cm long, it passes through the prostate gland. The prostatic urethra is the widest part of the urethra. It includes the seminal colliculus (verumontanum), a landmark for the openings of the ejaculatory ducts and the prostatic utricle.

Membranous Urethra: About 1–2 cm long, it passes through the urogenital diaphragm. It is the shortest and narrowest part of the urethra and is surrounded by the external urethral sphincter, providing voluntary control over urination.

Spongy (Penile) Urethra: Approximately 15 cm in length, it runs along the length of the penis. The corpus spongiosum encloses the spongy urethra that ends at the external urethral orifice. The bulbourethral glands open into this section.

19.3.10.2 Microscopic Structure

The urethra comprises three primary layers: the mucosa, submucosa, and muscular. The type of epithelium varies along the length of the urethra: the prostatic urethra has transitional epithelium, similar to the bladder, and the membranous and penile urethra have stratified and pseudostratified columnar epithelium. The distal spongy urethra is lined with stratified squamous epithelium, particularly near the external urethral orifice. Mucus-secreting urethral glands (Littre's glands) are present, particularly in the spongy urethra. The submucosal layer consists of connective tissue with a rich vascular network and some elastic fibers. It provides support to the mucosal layer and houses the urethral glands. The muscular layer varies between segments: in the preprostatic urethra, where the internal urethral sphincter is located, it is smooth muscle, and in the membranous urethra, where the external urethral sphincter is located, it consists of skeletal muscle, providing voluntary control over urination.

19.3.10.3 Function

The epithelial lining acts as a barrier against pathogens and mechanical damage. The urethral glands lubricate the urethra, protecting the epithelium and facilitating the passage of urine and semen. The primary function of the urethra is to transport urine from the bladder to the outside of the body. During ejaculation, it serves as a passage for semen, which enters the urethra from the ejaculatory ducts.

19.3.11 Penis

The penis (Fig. 19.22), as the male external genitalia, extends from the lower abdomen and is divided into the root, body (shaft), and glans.

Root: Anchored to the pubic arch, the root consists of the bulb of the penis and the paired crura. The bulb is attached to the urogenital diaphragm, while the crura attach to the pubic bones.

Body (Shaft): The elongated, cylindrical part extending from the root. It is composed of three columns of erectile tissue—two corpora cavernosa (paired dorsal erectile bodies, each surrounded by a tough fibrous sheath called the tunica albuginea) and one corpus spongiosum (a single ventral erectile body that contains the urethra, expanding at the distal end to form the glans penis).

Glans Penis: The conical-shaped distal end of the corpus spongiosum, it covers the ends of the corpora cavernosa and is often more sensitive.

The skin of the penis is relatively mobile and devoid of adipose tissue. It extends over the shaft and partially or entirely covers the glans. Prepuce (foreskin) is a double-layered fold of smooth skin and mucosa that covers and protects the glans penis. In circumcised males, the prepuce is surgically removed.

The penis serves as a conduit for urine to exit the body via the urethra. It is the critical organ for sexual intercourse. The corpora cavernosa and corpus spongiosum fill with blood to produce an erection, allowing for vaginal penetration and ejaculation.

19.4 Part 4: Male Reproductive System Physiology

The male reproductive system is not merely a conglomeration of organs and tissues but a sophisticated network harmoniously integrated to achieve the ultimate goal of reproduction. It encompasses not only the production and delivery of sperm but also the synthesis of male sex hormones, which are pivotal in maintaining both reproductive capabilities and secondary sexual characteristics.

19.4.1 Sperm Production

Sperm production, also known as spermatogenesis, is the process by which male germ cells (spermatozoa or sperm) are produced in the testes.

19.4.1.1 Spermatogenesis

Spermatogenesis is the complex and finely tuned process of producing spermatozoa, the male reproductive cells. It occurs within the seminiferous tubules of the testes and has several stages (Fig. 19.23).

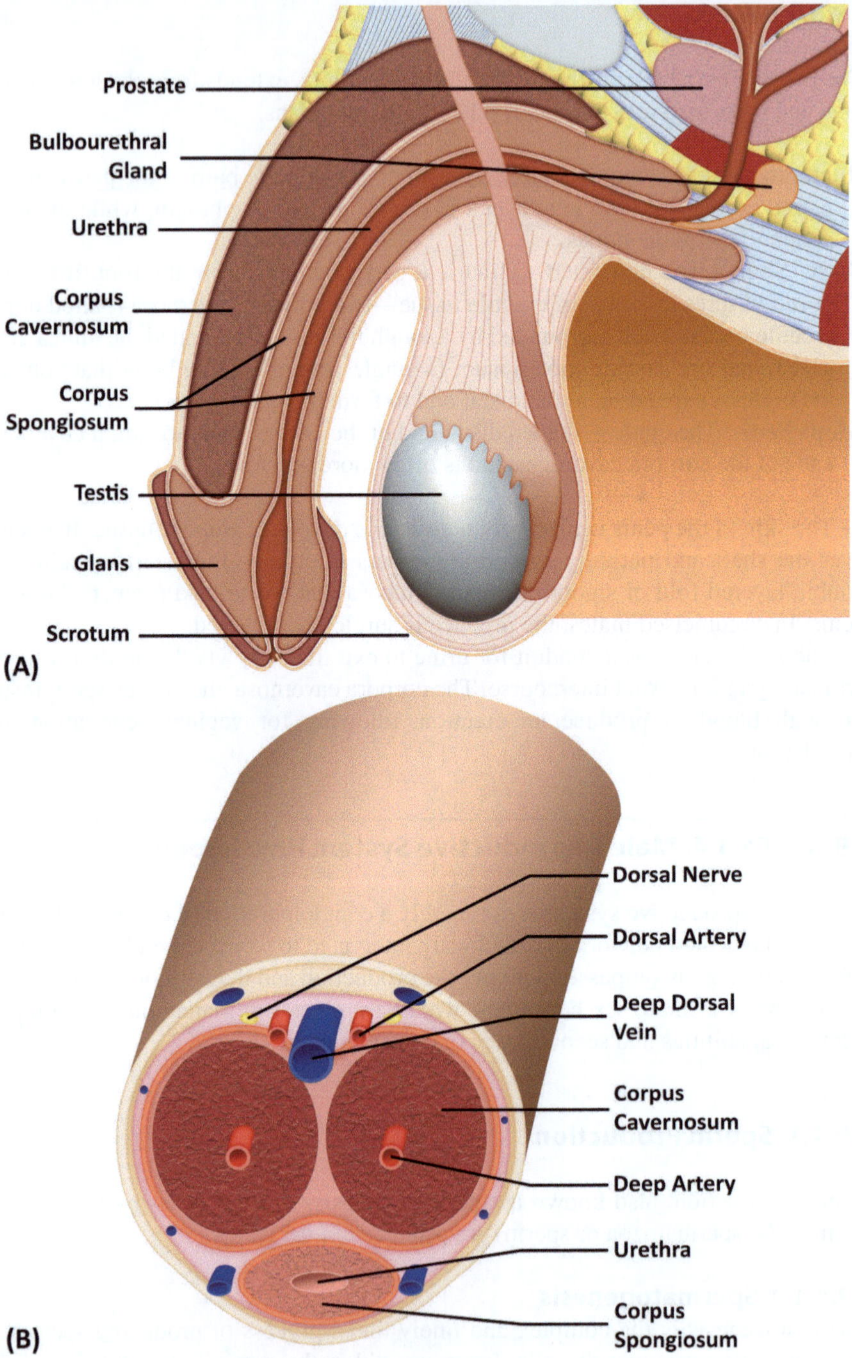

Fig. 19.22 The penis. (**a**) Sagittal section. (**b**) Coronal section

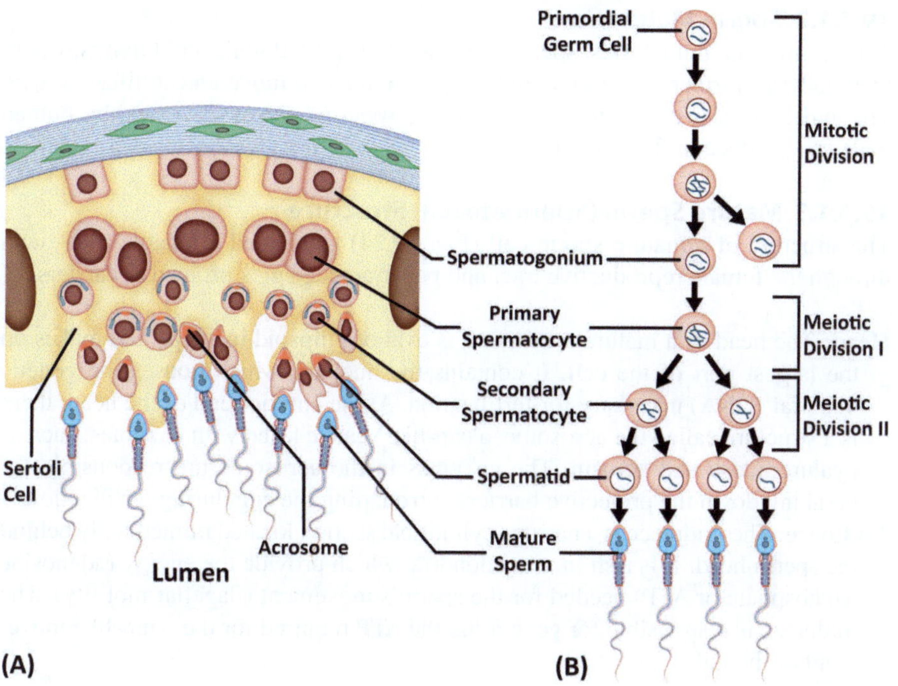

Fig. 19.23 Spermatogenesis. (**a**) A seminiferous tubule section showing the spermatogenesis process. (**b**) Matching the spermatogonia division (mitosis and meiosis) with the spermatogenesis process

Spermatocytogenesis: Spermatocyte originates from spermatogonial stem cells located along the basal membrane of the seminiferous tubules. Spermatogonial stem cells undergo mitotic division, resulting in primary spermatocytes.

Meiosis I: Diploid primary spermatocytes undergo the first meiotic division to form haploid secondary spermatocytes. **Meiosis II:** Secondary spermatocytes quickly enter the second meiotic division to produce spermatids.

Spermiogenesis: Spermatids, initially round cells, undergo morphological changes to transform into mature spermatozoa. Then, they undergo developmental changes, including the formation of the acrosome, condensation of nuclear material, development of the flagellum for motility, and shedding of excess cytoplasm.

The hypothalamic-pituitary-gonadal axis regulates spermatogenesis. Gonadotropin-releasing hormone (GnRH), secreted by the hypothalamus, stimulates the pituitary gland to release follicle-stimulating hormone (FSH) and luteinizing hormone (LH). FSH acts on Sertoli cells to promote the nourishment of developing sperm cells. LH stimulates Leydig cells in the testes to produce testosterone, which is crucial for spermatogenesis.

19.4.1.2 Sperm Maturation

After formation in the testes, spermatozoa are transported to the epididymis, where they undergo further maturation and gain the ability to move and fertilize an egg. The entire process of spermatogenesis takes about 64–72 days. A healthy human male can produce millions of sperm daily.

19.4.1.3 Mature Sperm (Spermatozoa) Structure

The structure of a mature sperm cell (Fig. 19.24) is adapted to enable it to swim through the female reproductive tract and penetrate the egg's protective barriers.

Head: The head of a mature sperm cell is oval or ellipsoid in shape and makes up the largest part of the cell. It contains the nucleus, which houses the genetic material (DNA) necessary for fertilization. At the anterior end of the head, there is a structure called the acrosome, a cap-like vesicle filled with enzymes, such as hyaluronidase and acrosin. The enzymes in the acrosome are responsible for breaking down the protective barriers surrounding the egg during fertilization.

Midpiece: The midpiece is a narrow, cylindrical section located immediately behind the sperm head. It is rich in mitochondria, which provide the energy (adenosine triphosphate or ATP) needed for the sperm's movement (flagellar motility). The midpiece is responsible for generating the ATP required for the whip-like movement of the tail.

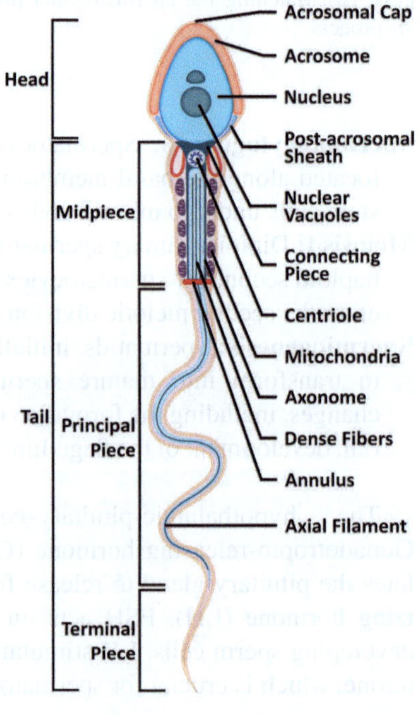

Fig. 19.24 Mature sperm structure

Tail (Flagellum): The tail, or flagellum, is a long, whip-like structure extending from the midpiece. It comprises microtubules and is responsible for the sperm's motility or swimming capability. The tail propels the sperm forward in a corkscrew-like motion, allowing it to navigate the female reproductive tract in search of an egg.

The plasma membrane encases the entire sperm cell, providing structural integrity. It contains specific receptors that allow the sperm to recognize and bind to the zona pellucida, a glycoprotein layer surrounding the egg. The mature sperm cell contains a centriole in the midpiece, which is essential for the formation of the sperm's tail and its motility. The nucleus of the sperm head contains a compacted and highly condensed single set of chromosomes (haploid) that carries the genetic information to be contributed to the embryo upon fertilization.

19.4.2 Testicular Hormones

The primary hormones secreted by the testes are testosterone and inhibin. In addition to testosterone, the testes produce smaller amounts of other androgens, such as dihydrotestosterone (DHT) and androstenedione.

Testosterone is synthesized in the Leydig cells in the interstitial space between the seminiferous tubules. Cholesterol is the precursor for testosterone synthesis. The process involves several enzymatic steps catalyzed by enzymes like 17α-hydroxylase and 17,20-lyase. Testosterone is essential for male sexual differentiation during fetal development. It also promotes the development of male secondary sexual characteristics during puberty, such as increased muscle and bone mass, deepening of the voice, and growth of body hair. Testosterone stimulates spermatogenesis within the seminiferous tubules. It is the main stimulator of the libido, influencing sexual drive and behavior. This hormone has significant anabolic effects on muscle and bone tissue.

Inhibin is produced by the Sertoli cells within the seminiferous tubules. It primarily inhibits the secretion of follicle-stimulating hormone (FSH) from the anterior pituitary gland, forming part of a feedback mechanism to regulate the rate of spermatogenesis.

The hypothalamic-pituitary-gonadal (HPG) axis regulates testicular hormone secretion. GnRH, secreted by the hypothalamus, stimulates the pituitary to release LH and FSH. LH stimulates testosterone production in Leydig cells, and FSH acts in conjunction with testosterone to stimulate spermatogenesis and inhibin production in Sertoli cells.

19.4.3 Penile Erection

A penile erection involves an intricate interplay of vascular, neurological, hormonal, and psychological factors, enabling the penis to become rigid and erect, facilitating sexual intercourse. The penis contains three cylindrical bodies of erectile tissue: two dorsally located corpora cavernosa and one ventral corpus spongiosum (Fig. 19.22). The blood necessary for erection is supplied to the penis via the internal pudendal arteries, which branch into the dorsal and cavernosal arteries.

19.4.3.1 Mechanism of Erection

Sexual arousal is processed in the brain, which sends signals down the spinal cord. The parasympathetic nervous system, as the primary mediator of erection, activates the release of NO from endothelial cells and nerve terminals. NO stimulates the production of cyclic guanosine monophosphate (cGMP) in smooth muscle cells, leading to relaxation and increased blood flow.

Sexual arousal, by triggering the release of NO, causes dilation of the cavernosal arteries and trabecular smooth muscle relaxation. Arterial dilation leads to increased blood flow into the sinuses of the corpora cavernosa. The relaxed smooth muscles in the corpora cavernosa allow it to expand. The expanding corpora cavernosa compress the subtunical venular plexuses against the tunica albuginea, reducing venous outflow and trapping blood within the erectile bodies.

Testosterone, while not directly involved in the erection mechanism, is essential for normal libido and erectile function. It modulates the CNS response to sexual stimuli.

19.4.3.2 Phases of Penile Erection

Flaccid Phase: The penis is in a non-erect state. Arterioles and cavernous sinusoids are constricted, limiting blood flow. Sympathetic nervous system dominance maintains the penis in a flaccid state.

Tumescence (Engorgement) Phase: Triggered by sexual arousal, which can be psychological (central arousal) or reflexogenic (peripheral arousal), the parasympathetic nervous system is activated, leading to the release of NO. NO causes increased blood flow into the cavernous sinusoids, leading to swelling and elongation of the penis.

Full Erection Phase: As the sinusoids fill with blood, they expand against the tunica albuginea, compressing the venous plexuses and reducing venous outflow. The intracavernosal pressure rises, and the erection is maintained by a balance of continuous arterial inflow and restricted venous outflow. The corpus spongiosum also becomes engorged, but to a lesser extent, ensuring the urethra remains open for ejaculation.

Detumescence (Flaccidity) Phase: Following ejaculation or cessation of sexual arousal, there is a decrease in parasympathetic activity and an increase in sympathetic activity, resulting in the constriction of penile arteries and reducing blood inflow. Meanwhile, the smooth muscles contract, decreasing pressure on the tunica albuginea and allowing the venous plexuses to drain blood from the erectile tissues.

19.4.4 Ejaculation

Ejaculation is the expulsion of semen from the male reproductive tract. Ejaculation occurs in two primary phases: emission and expulsion.

Emission Phase: This phase is triggered by heightened sexual arousal and stimulation. Seminal vesicles and prostate gland secrete fluids that move into the prostatic urethra, mixing with sperm transported from the epididymis. Under sympathetic control, the internal urethral sphincter contracts to prevent retrograde ejaculation into the bladder.

Expulsion Phase: The bulbospongiosus and ischiocavernosus muscles (Table 9.13), under somatic control, contract rhythmically, forcing semen through the urethra. Sensory nerves in the penis relay information to the spinal cord, intensifying rhythmic contractions and leading to the expulsion of semen. Relaxation of the external urethral sphincter allows semen to be ejaculated out of the body.

Ejaculation is primarily a spinal reflex mediated by the spinal ejaculation center located in the lumbar and sacral spinal cord. Afferent impulses from genital stimulation travel to the spinal cord, and efferent pathways coordinate the emission and expulsion processes. Testosterone, essential for normal libido and sexual function, influences the sensitivity of the genitalia and the response of the ejaculatory mechanism.

19.4.5 Semen

Semen, a complex fluid ejaculated during the male orgasm, consists of spermatozoa and seminal plasma. Semen provides a medium for sperm to travel from the male reproductive tract to the female reproductive tract. Seminal plasma is a combination of secretions from the seminal vesicles, prostate gland, and bulbourethral glands. Seminal vesicles contribute approximately 60–70% of the seminal fluid. This secretion is rich in fructose (providing energy to sperm), prostaglandins (contributing to sperm motility), and fibrinogen (for semen coagulation). The prostate gland contributes around 20–30% of the semen volume. Its alkaline secretion is essential for sperm viability and motility. Bulbourethral glands secrete a mucus-like fluid that lubricates and neutralizes acidity in the urethra.

Semen typically has a slightly alkaline pH range of 7.2–8.0. The average ejaculate volume is between 1.5 and 5 milliliters. Normally, semen contains 15 to more than 200 million sperm per milliliter. Semen initially coagulates upon ejaculation due to enzymes from the prostate and then liquefies within 20–30 min, facilitating sperm motility.

19.4.6 Male Sexual Response

The male sexual response encompasses the stages of sexual arousal, erection, orgasm, ejaculation, and resolution.

19.4.6.1 Stages

Arousal: This stage is triggered by psychological (thoughts, feelings) and physical (sensory stimulation) factors. It involves increased heart rate, blood pressure, and blood flow to the genital area, leading to penile erection. Nitric oxide (NO) is critical in initiating the vascular changes necessary for erection.

Erection: Arterial dilation and venous constriction in the penis result in engorgement of erectile tissues (corpora cavernosa and corpus spongiosum) with blood, controlled by a balance between parasympathetic (erection-inducing) and sympathetic (erection-inhibiting) nervous systems.

Orgasm: Orgasm is primarily a reflex mediated by the spinal cord, triggered by sexual stimulation. Involuntary rhythmic contractions of muscles in the pelvic floor, specifically the bulbospongiosus and ischiocavernosus muscles, occur during orgasm. Orgasm involves heightened activity in several brain areas, including those responsible for pleasure and reward. Dopamine, serotonin, and oxytocin play significant roles in the experience of pleasure and the orgasmic process. Orgasm leads to a surge in oxytocin and prolactin. Although orgasm and ejaculation are often simultaneous, ejaculation (the release of semen) is a separate physiological process.

Ejaculation: First, the emission phase, which is the movement of sperm and seminal fluid into the urethra, happens. The sympathetic nervous system regulates it. Then, the expulsion phase, which is the forceful expulsion of semen by rhythmic contractions of the pelvic floor muscles and urethral bulb, happens.

Resolution and Refractory Period: Post-ejaculation, a reduction in sexual arousal leads to detumescence (penile flaccidity). Following ejaculation, a refractory period appears, during which further erection and orgasm are not possible. The duration varies among individuals.

19.4.6.2 Regulation

Sexual arousal is processed in the brain, influencing the initiation and maintenance of erection and ejaculation. Spinal reflexes are involved in the control of erection and ejaculation, especially in response to direct genital stimulation.

Testosterone is the key to sexual desire, erectile function, and spermatogenesis. LH and FSH regulate testicular function, indirectly influencing sexual response. Perceptions, memories, and fantasies are integral to sexual arousal and desire.

19.5 Part 5: Contraception

Contraception, the deliberate control of fertility through the utilization of various contraceptive methods, has enabled individuals and societies to exert agency over the timing and number of offspring, profoundly influencing demographic patterns, gender dynamics, and socio-economic structures. Different contraceptive methods are available, from hormonal contraceptives and barrier devices to intrauterine devices and permanent sterilization techniques.

19.5.1 Hormonal Contraceptives: Pills, Patches, and Injections

Hormonal contraceptives represent a cornerstone of modern contraception, providing individuals with highly effective options to control fertility.

19.5.1.1 Mechanisms of Action

Hormonal contraceptives exert their contraceptive effects through the manipulation of the endocrine system. The primary goal is to prevent ovulation, thereby inhibiting the release of an egg from the ovaries for fertilization.

Combined Oral Contraceptives (COCs): Combined oral contraceptives, often referred to as "the pill," contain synthetic versions of both estrogen and progestin. These hormones work together to suppress ovulation by preventing the LH surge, thickening cervical mucus to impede sperm penetration, and altering the uterine lining, making it less receptive to implantation.

Progestin-Only Contraceptives: Progestin-only contraceptives, including progestin-only pills (POP), injectables, implants, and intrauterine devices (IUDs), primarily act by thickening cervical mucus and inhibiting sperm mobility. Some progestin-only methods may also affect the endometrial lining, making it less hospitable for implantation.

19.5.1.2 Types of Hormonal Contraceptives

Hormonal contraceptives come in various forms to accommodate individual preferences and needs. These include:

Combined Oral Contraceptives (COCs)

- Monophasic COCs: Maintain a constant dose of hormones throughout the menstrual cycle
- Multiphasic COCs: Vary hormone levels to mimic natural hormonal fluctuations
- Extended-Cycle COCs: Allow fewer menstrual periods per year by extending the hormone-active phase

Progestin-Only Contraceptives

- Progestin-Only Pills (POP): Must be taken consistently at the same time daily for maximum efficacy
- Injectable Contraceptives: Administered by a healthcare provider every few months
- Implantable Contraceptives: Small rods placed under the skin, protecting for several years
- Progestin-Releasing Intrauterine Devices (IUDs): Long-acting contraception inserted into the uterus

Other Hormonal Contraceptives

- Contraceptive Patches: Deliver hormones through the skin
- Vaginal Rings: A flexible ring inserted into the vagina, releasing hormones over 3 weeks

19.5.2 Barrier Methods

Barrier methods of contraception constitute a diverse group of contraceptive techniques designed to prevent pregnancy by physically blocking the entry of sperm into the cervix and uterus. Unlike hormonal methods that rely on synthetic hormones to regulate fertility, barrier methods provide a non-hormonal and highly effective alternative for individuals seeking contraception.

19.5.2.1 Male Condoms

Male condoms are perhaps the most recognizable and widely used barrier method of contraception. These latex or polyurethane sheaths are worn over the erect penis and act as a physical barrier, preventing sperm from reaching the cervix. Condoms provide a dual benefit by reducing the transmission of sexually transmitted infections (STIs).

19.5.2.2 Female Condoms

Female condoms, made of polyurethane or nitrile, are inserted into the vagina before intercourse. These pouch-like devices offer an internal barrier, preventing sperm from reaching the cervix. Like male condoms, female condoms offer partial protection against STIs.

19.5.2.3 Diaphragms and Cervical Caps

Diaphragms and cervical caps are dome-shaped, silicone or latex devices that are inserted into the vagina and placed over the cervix before sexual activity. They act as barriers to sperm and are used with spermicide for added effectiveness.

Spermicide is applied to the device before insertion to immobilize sperm. They may increase the risk of urinary tract infections (UTIs).

19.5.3 Intrauterine Devices

Intrauterine devices (IUDs) represent a highly effective and long-acting form of contraception that has gained widespread popularity due to their reliability and convenience. IUDs are small, T-shaped devices that are inserted into the uterus by a healthcare provider.

19.5.3.1 Mechanisms of Action

IUDs prevent pregnancy through two primary mechanisms:

Physical Barrier: IUDs create a physical barrier within the uterus that impedes the movement of sperm, preventing them from reaching the egg.

Hormonal Regulation (For Hormonal IUDs): Hormonal IUDs release a low, continuous dose of progestin locally within the uterus. This hormone thickens cervical mucus, inhibits sperm mobility, and sometimes suppresses ovulation.

IUDs can be removed at any time, and fertility typically returns promptly, allowing individuals to plan for future pregnancies.

19.5.3.2 Types of IUDs

IUDs come in two main types:

Copper IUD (Non-hormonal): This type of IUD contains copper with spermicidal properties. It remains effective for up to 10 years, does not contain hormones, and can be used during breastfeeding.

Hormonal IUD: This type of IUD releases progestin into the uterus. It offers protection for 3–7 years, depending on the specific type. The hormone content may result in lighter or no periods for some individuals.

19.5.4 Spermicides

Spermicides are chemical agents designed to immobilize and kill sperm, offering a non-hormonal and barrier-independent method of contraception. While they are less effective on their own compared to other contraceptives, spermicides can be used alone or in combination with barrier methods for added protection.

Spermicides work by disrupting sperm cell membranes and enzymes, rendering sperm immobile and unable to fertilize an egg. Spermicides create a chemical barrier within the vagina or on the cervix. They contain active ingredients such as nonoxynol-9, octoxynol-9, or benzalkonium chloride, which immobilize sperm upon contact.

Spermicides are available in various forms, catering to individual preferences and needs. Common types of spermicides include:

Foam: Spermicidal foam is dispensed from an aerosol can into the vagina, where it expands to create a barrier. It requires insertion shortly before each act of intercourse.

Gel: Spermicidal gel is applied directly into the vagina using a pre-filled applicator. It must be inserted shortly before each sexual encounter.

Film or Suppositories: Spermicidal film or suppositories are thin sheets or solid forms inserted into the vagina. They dissolve, releasing the active ingredient. It must be applied shortly before each act of intercourse but sooner than other forms of spermicides, allowing time for dissolution.

19.5.5 Emergency Contraception

Emergency contraception, often referred to as the "morning-after methods," is a critical component of modern reproductive healthcare. It provides a time-sensitive option for preventing unintended pregnancies following unprotected intercourse or contraceptive failure. Emergency contraception is available in two main forms:

Emergency Contraceptive Pills (ECPs): These are oral medications designed to be taken after unprotected intercourse. The most common ECPs contain levonorgestrel or ulipristal acetate as active ingredients. Levonorgestrel-based ECPs are typically effective within 72 h, while ulipristal acetate is effective up to 120 h (5 days) after intercourse.

Copper Intrauterine Device (Copper IUD): The copper IUD can also be used as emergency contraception when inserted by a healthcare provider within a specific timeframe after unprotected intercourse. The copper IUD is highly effective as emergency contraception when inserted within 5 days of unprotected intercourse and can also provide long-term contraception.

Emergency contraception prevents pregnancy through the following mechanisms:

Delayed or Inhibited Ovulation: ECPs may delay or inhibit the release of an egg from the ovaries, thereby preventing fertilization.

Impaired Sperm Function: ECPs can alter cervical mucus, making it difficult for sperm to reach the egg.

Disruption of Implantation: In rare cases, emergency contraception may alter the uterine lining, making it less receptive to implantation if fertilization has occurred.

19.5.6 Surgical Methods

Surgical methods of contraception, often referred to as permanent sterilization, represent a long-lasting and highly effective option for individuals or couples who have completed their family planning and wish to prevent future pregnancies. These methods involve surgical procedures that either block or sever the reproductive pathways, rendering individuals permanently infertile. Surgical contraception methods are highly effective, and the failure rate is extremely low. They are permanent, eliminating the need for ongoing contraception. Reversing these procedures is often complex, expensive, and not always successful. Surgical contraception encompasses two main types:

19.5.6.1 Tubal Ligation (Female Sterilization)

Tubal ligation involves blocking or sealing the fallopian tubes, preventing the passage of eggs from the ovaries to the uterus. Standard methods include cutting, tying, or placing clips on the fallopian tubes (Fig. 19.25). Tubal ligation can be performed

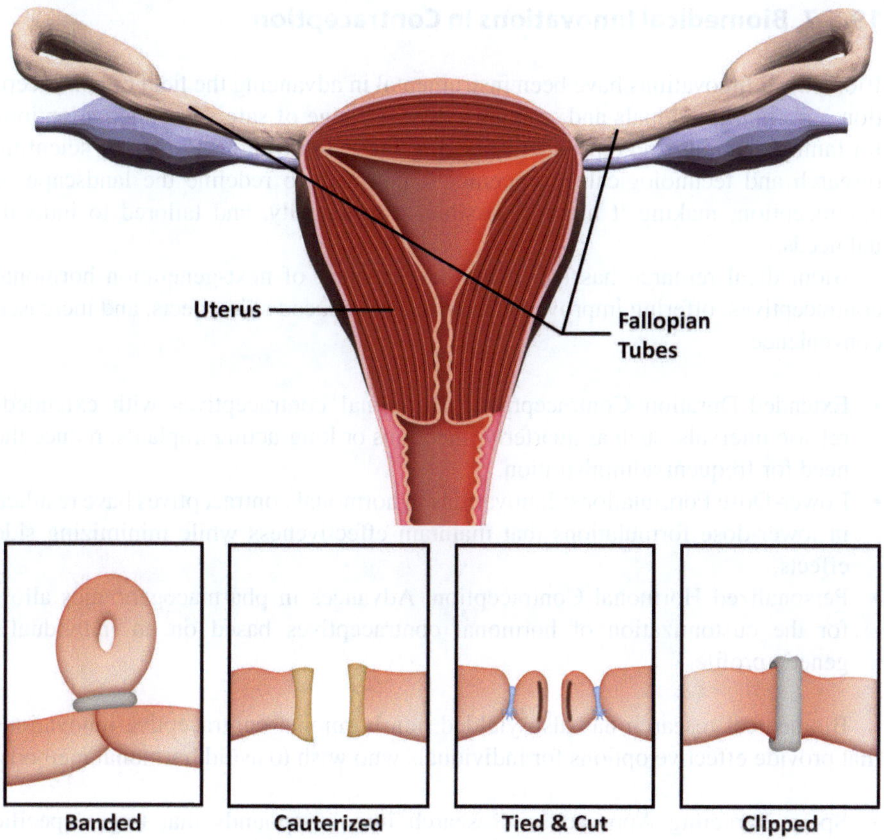

Fig. 19.25 Methods of tubal ligation

laparoscopically, through a small incision in the abdomen, or at the time of another abdominal procedure, especially a cesarean section.

19.5.6.2 Vasectomy (Male Sterilization)

Vasectomy involves the cutting or sealing of the vas deferens, the tubes that transport sperm from the testes to the urethra. This procedure prevents sperm from being ejaculated during sexual intercourse. Vasectomy is a relatively simple outpatient procedure performed under local anesthesia. The vas deferens are located and either cut or sealed to prevent the passage of sperm. It takes a few months and several ejaculations to clear any remaining sperm from the reproductive tract after a vasectomy.

19.5.7 Biomedical Innovations in Contraception

Biomedical innovations have been instrumental in advancing the field of contraception, offering individuals and couples a diverse range of safe and effective options for family planning and reproductive health. These innovations, driven by scientific research and technological advancements, continue to redefine the landscape of contraception, making it more accessible, user-friendly, and tailored to individual needs.

Biomedical research has led to the development of next-generation hormonal contraceptives, offering improved effectiveness, reduced side effects, and increased convenience:

- Extended-Duration Contraceptives: Hormonal contraceptives with extended-release intervals, such as quarterly injections or long-acting implants, reduce the need for frequent administration.
- Lower-Dose Formulations: Innovations in hormonal contraceptives have resulted in lower-dose formulations that maintain effectiveness while minimizing side effects.
- Personalized Hormonal Contraception: Advances in pharmacogenomics allow for the customization of hormonal contraceptives based on an individual's genetic profile.

Biomedical research has also yielded non-hormonal contraceptive innovations that provide effective options for individuals who wish to avoid hormonal methods:

- Sperm-Targeting Approaches: Research into compounds that target specific sperm proteins may lead to highly effective non-hormonal contraceptives.
- Immunocontraceptives: Biomedical advancements in immunology have paved the way for developing vaccines or antibodies that target reproductive proteins, providing non-hormonal contraceptive options.

Innovations in male contraceptives have been a focal point of biomedical research:

- Non-hormonal Male Contraceptives: Researchers are exploring non-hormonal methods, such as ultrasound-based approaches, which temporarily disrupt sperm production.
- Reversible Vasectomy: Innovative techniques are being developed to make vasectomy reversible, providing men with the option to regain fertility.

Biomedical innovations have led to more user-friendly and efficient delivery systems for contraceptives:

- Intravaginal Rings: Intravaginal rings are being developed for both hormonal and non-hormonal contraceptions, providing long-acting, user-controlled options.

- Biodegradable Implants: Biodegradable implants offer contraception for a specific duration and then safely degrade within the body, eliminating the need for removal.

The integration of digital health technologies has revolutionized contraception:

Smartphone Apps and Wearables: Mobile apps and wearable devices can help individuals track their menstrual cycles, monitor fertility, and receive contraceptive reminders.

Telemedicine and Telecontraception: Telemedicine platforms enable individuals to consult with healthcare providers remotely for contraceptive prescriptions and guidance.

19.6 Part 6: Reproductive Technologies and Assisted Reproduction

The realm of assisted reproduction has witnessed a remarkable transformation in recent decades. Advances in science and technology have brought forth a wealth of innovative techniques and treatments that offer new avenues for individuals and couples facing fertility challenges. Assisted reproductive technology (ART) encompasses a wide range of treatments, from in vitro fertilization (IVF) to gamete donation and surrogacy, each tailored to address specific causes of infertility or genetic concerns.

19.6.1 In Vitro Fertilization (IVF)

In vitro fertilization (IVF) is one of reproductive medicine's most groundbreaking and transformative developments. IVF involves the manipulation of reproductive cells, with eggs (ova) and sperm being combined in a laboratory setting.

Fertilized embryos are cultured and monitored for viability before being transferred into the uterus. The IVF procedure includes:

Ovulation Stimulation: IVF begins with controlled ovarian hyperstimulation to induce the development of multiple eggs.

Egg Retrieval: A minor surgical procedure retrieves mature eggs from the ovaries.

Sperm Collection and Fertilization: Sperm from a partner or donor is collected and used to fertilize the retrieved eggs.

Embryo Culture and Selection: Fertilized embryos are cultured, with the most viable ones selected for transfer. Preimplantation genetic testing (PGT) may be employed to screen for genetic abnormalities.

Embryo Transfer: One or more healthy embryos are transferred into the uterus, with any surplus embryos often cryopreserved for future use.

Future Horizons of IVF: The future of IVF holds great promise, with ongoing research into technologies such as artificial gametes, gene editing, and mitochondrial replacement therapy. These advancements can potentially enhance the success rates and accessibility of IVF.

19.6.2 Intracytoplasmic Sperm Injection (ICSI)

Intracytoplasmic sperm injection (ICSI) is a groundbreaking technique within the realm of assisted reproductive technology (ART) that has redefined the treatment of male infertility. Since its introduction in the 1990s, ICSI has provided new hope and possibilities for individuals and couples facing challenges related to sperm quality or quantity. The development of ICSI emerged from the broader advancements in ART, driven by the need to address severe male infertility.

19.6.2.1 Procedure

The ICSI procedure includes:

Oocyte Retrieval: ICSI typically begins with retrieving mature eggs (ova) from the ovaries, similar to in vitro fertilization (IVF).

Sperm Collection: Sperm from a partner or a donor is collected and prepared for injection.

Microscopic Injection: A single sperm is selected and injected directly into the cytoplasm of a mature egg, bypassing any natural barriers.

Embryo Culture and Transfer: Fertilized embryos are cultured and monitored for viability before being transferred into the uterus, following a similar process to IVF.

ICSI is primarily used in cases of severe male infertility, including low sperm count, poor sperm motility, or abnormal sperm morphology. ICSI may also be employed in situations such as previous IVF failure, frozen sperm, or genetic concerns.

19.6.2.2 Advancements

IMSI (Intracytoplasmic Morphologically Selected Sperm Injection): IMSI utilizes high-magnification microscopy to select the best-quality sperm for injection, potentially improving success rates.

PICSI (Physiological Intracytoplasmic Sperm Injection): PICSI mimics the natural selection of sperm by assessing their ability to bind to a hyaluronan-coated dish, enhancing the likelihood of successful fertilization.

Future Prospects of ICSI: The future of ICSI holds potential for further advancements in sperm selection techniques, genetic screening, and the treatment of previously untreatable forms of male infertility. These developments can potentially increase success rates and broaden the scope of individuals who can benefit from ICSI.

19.6.3 Sperm and Ovum Banking

Sperm and ovum banking, gamete cryopreservation or banking, is a revolutionary facet of modern reproductive medicine. It allows individuals and couples to safeguard their reproductive potential for medical, social, or personal reasons. Gamete cryopreservation involves rapidly cooling sperm or eggs to extremely low temperatures, preserving their biological integrity. Cryoprotectants prevent ice crystal formation and cellular damage during freezing and thawing.

19.6.3.1 Sperm Banking

Sperm banking allows individuals to donate or preserve sperm for future use. Donor sperm is used in cases of male infertility, single parenthood, or same-sex couples wishing to conceive. Sperm banking is also essential for individuals undergoing medical treatments that may impair fertility, such as chemotherapy or radiation therapy.

Sperm is collected, processed, and mixed with a cryoprotectant before being divided into small vials and frozen in liquid nitrogen.

19.6.3.2 Ovum (Egg) Banking

Egg banking permits women to donate or preserve their eggs for various reasons, including fertility preservation before cancer treatment or delayed family planning. Donor eggs are also used in assisted reproduction for women with infertility or genetic concerns. In vitro maturation (IVM) is an emerging technique that allows the retrieval and cryopreservation of immature eggs, providing an alternative to traditional egg freezing.

Egg freezing involves stimulating the ovaries, retrieving mature eggs, and freezing them using vitrification.

19.6.4 Biomedical Engineering in Assisted Reproduction

Biomedical engineering has emerged as a vital and transformative force in assisted reproduction, redefining the landscape of reproductive medicine.

19.6.4.1 Microfluidics and Sperm Sorting

Microfluidic devices enable precise sorting and selection of sperm based on parameters like motility, morphology, and DNA integrity. These devices offer a higher degree of sperm enrichment, enhancing the chances of successful fertilization.

19.6.4.2 Lab-on-a-Chip Systems for Embryo Culture

Lab-on-a-chip systems provide controlled and miniaturized culture environments for embryos. These systems allow for real-time monitoring of embryo development and provide more stable conditions for optimal growth. They can also offer non-invasive assessment of embryo quality, reducing the need for invasive techniques such as embryo biopsy.

19.6.4.3 Artificial Intelligence and Machine Learning

Predictive Algorithms: Artificial intelligence and machine learning algorithms are being employed to analyze vast patient and embryo information datasets and predict the likelihood of successful IVF outcomes.

Personalized Treatment Plans: Biomedical engineers are developing AI-driven systems that create personalized treatment plans, optimizing hormone dosages and timing for each patient.

19.6.4.4 Cryopreservation and Vitrification

Biomedical engineering has improved cryopreservation and vitrification techniques for eggs, sperm, and embryos, resulting in higher post-thaw survival rates.

19.6.4.5 Future Prospects and Advancements

The future of biomedical engineering in assisted reproduction holds promise for further advancements, including 3D printing for creating artificial reproductive organs, stem cell-based gametes, and novel diagnostic tools for assessing embryo viability.

Suggested Reading

Advanced Physiology and Pathophysiology: Essentials for Clinical Practice, 2nd Ed. Nancy Tkacs, Linda Herrmann, Springer Publishing Company, 2024

Anatomy & Physiology, S. Meloni, et al. Independent Publisher, 2023

Anatomy & Physiology: The Unity of Form and Function, 10th Ed. Kenneth S. Saladin, McGraw Hill, 2023

Gray's Anatomy for Students, 5th Ed. Richard L. Drake, et al. Elsevier, 2020

Human Anatomy & Physiology, 11th Ed. Elaine N. Marieb and Katja Hoehn, Pearson, 2022

Moore's Clinically Oriented Anatomy, 9th Ed. Keith L. Moore, et al. Lippincott Williams & Wilkins, 2020

Netter Atlas of Human Anatomy: Classic Regional Approach, 8th Ed. Frank H. Netter, Elsevier, 2022

Principles of Anatomy and Physiology, 16th Ed. Gerard J. Tortora, Bryan H. Derrickson, Wiley, 2020

Index

H. Hosseinzadeh, *Fundamentals of Medicine for Biomedical Engineering*,
https://doi.org/10.1007/978-3-031-74085-5

E

H

I

P

V

MIX
Papier aus verantwortungsvollen Quellen
Paper from responsible sources
FSC® C105338

If you have any concerns about our products,
you can contact us on
ProductSafety@springernature.com

In case Publisher is established outside the EU,
the EU authorized representative is:
Springer Nature Customer Service Center GmbH
Europaplatz 3, 69115 Heidelberg, Germany

Printed by Libri Plureos GmbH
in Hamburg, Germany